SATISH BHAGWAT
July 13, 1942

Standard Handbook
of Hazardous Waste
Treatment and Disposal

OTHER McGRAW-HILL HANDBOOKS OF INTEREST

For more information about McGraw-Hill materials, call 1-800-2-MCGRAW in the United States. In other countries, call your nearest McGraw-Hill office.

Standard Handbook
of Hazardous Waste
Treatment and Disposal

HARRY M. FREEMAN, Editor
U.S. Environmental Protection Agency
Hazardous Waste Engineering Research Laboratory

McGraw-Hill Book Company

New York   St. Louis   San Francisco   Auckland
Bogotá   Hamburg   London   Madrid   Milan   Mexico
Montreal   New Delhi   Panama   Paris   São Paulo
Singapore   Sydney   Tokyo   Toronto

Library of Congress Cataloging-in-Publication Data

Standard handbook of hazardous waste treatment and disposal / Harry M.
  Freeman, editor.
    p. cm.
    ISBN 0-07-022042-5
    1. Hazardous wastes—Handbooks, manuals, etc. 2. Hazardous waste
treatment facilities—Handbooks, manuals, etc. I. Freeman, Harry,
date
TD1032.S73 1988   88-14079
628.4′4—dc19   CIP

    4567890   DOCDOC   9654321

ISBN 0-07-022042-5

Cover Photo: Courtesy of Chemical Waste Management, Riverdale, Ill.

The editors for this book were Betty Sun, Susan Thomas, and Ruth Mendelsohn; the
designer was Mark E. Safran; and the production supervisor was Dianne L. Walber. It
was set in Times Roman and was composed by the McGraw-Hill Book Company,
Professional & Reference Division Composition unit.

For more information about other McGraw-Hill materials, call 1-800-2-MCGRAW in the
United States. In other countries, call your nearest McGraw-Hill office.

Printed and bound by R. R. Donnelley & Sons Company.

CONTENTS

CONTRIBUTORS

Donald G. Ackerman, Jr.* *Scientific Director, SITEX Environmental, Inc., Research Triangle Park, N.C.*
SECTION 8.8: Oceanic Incineration

Aloysius A. Aguwa *Technical Center, General Motors Corporation, Warren, Mich.*
SECTION 6.3: Electrolytic Recovery Techniques

Anthony A. Albert *Manager, Project Sales, Minerals Systems Division, Boliden Allis, Inc., Milwaukee, Wis.*
SECTION 8.2: Rotary Kilns

Clark C. Allen, Ph.D. *Research Triangle Institute, Research Triangle Park, N.C.*
SECTION 6.9: Thin-Film Evaporation

Douglas C. Ammon *Clean Sites, Inc., Alexandria, Va.*
SECTION 1.4: Underground Storage Tanks

David C. Anderson *Vice President, K. W. Brown and Associates, Inc., College Station, Tex.*
SECTION 10.4: Surface Impoundments; SECTION 10.5: Soil Liners; SECTION 10.7: Aboveground Disposal

Richard N. L. Andrews *Professor and Director, Institute for Environmental Studies, University of North Carolina, Chapel Hill, N.C.*
SECTION 3.1: Siting of Hazardous-Waste Facilities

Robert W. Benedict *Manager, Engineering and Development, Waste Tech Services, Inc., Golden, Colo.*
SECTION 8.3: Fluidized-Bed Thermal Oxidation

Joan B. Berkowitz *President, Risk Science International, Washington, D.C.*
SECTION 6.6: Solvent Extraction

Joan V. Boegel *Polaroid, Waltham, Mass.*
SECTION 6.8: Air Stripping and Steam Stripping

George Brant *Associated Technologies, Inc., Charlotte, N.C.*
SECTION 6.2: Distillation

Craig J. Brown, P.E. *Vice President and General Manager, Eco-Tec Limited, Pickering, Ontario, Canada*
SECTION 6.5: Ion Exchange

K. W. Brown *Professor, Soil and Crop Sciences Department, Texas A&M University, College Station, Tex.*
SECTION 10.7: Aboveground Disposal

James L. Buelt *Pacific Northwest Laboratory, Richland, Wash.*
Section 8.9: Molten-Glass Processes

Michael E. Burns *U.S. Enviromental Protection Agency, Office of Solid Waste, Washington, D.C.*
SECTION 2.1: Quantities and Sources of Hazardous Waste

John F. Chadbourne, Ph.D. *Director of Environmental Services, General Portland, Inc., Dallas, Tex.*
SECTION 8.5: Cement Kilns

Neville K. Chung *Metcalf & Eddy, Inc., Wakefield, Mass.*
SECTION 7.2: Chemical Precipitation

S. Robert Cochran *Interface, Inc., Alexandria, Va.*
SECTION 1.4: Underground Storage Tanks

William M. Copa *Vice President, Technical Services, Zimpro, Inc., Rothschild, Wis.*
SECTION 8.6: Wet Oxidation

Frank L. Cross, Jr., P.E. *President, Cross/Tessitore Associates, P.A., Orlando, Fla.*
SECTION 4.4: Infectious Wastes; SECTION 8.16: Infectious-Waste Incineration

V. R. Daiga *Surface Combustion, Inc., Midland-Ross Corporation, Toledo, Ohio*
SECTION 8.7: Pyrolysis Processes

Peter Daley *Director, Research and Development, Chemical Waste Management, Inc., Riverdale, Ill.*
CHAPTER 11: Comprehensive Hazardous-Waste Treatment Facilities

Steve G. DeCicco *IT Enviroscience, Knoxville, Tenn.*
SECTION 8.13: Mobile Thermal Treatment Systems

B. Tod Delaney *Ground/Water Technology, Inc., Rockaway, N.J.*
SECTION 7.7: Evaporation

Paul R. de Percin *U.S. Environmental Protection Agency, Hazardous Waste Engineering Research Laboratory, Cincinnati, Ohio*
SECTION 10.8: Air Pollution from Land Disposal Facilities

Luis F. Diaz *President Cal Recovery Systems, Inc., Richmond, Calif.*
SECTION 9.3: Composting of Industrial Wastes

Stephen K. Dietz *Westat, Inc., Rockville, Md.*
SECTION 2.1: Quantities and Sources of Hazardous Waste

Gerry Dorian *U.S. Environmental Protection Agency, Office of Solid Waste, Washington, D.C*
SECTION 4.6: Household Hazardous Wastes

M. Pat Esposito *Bruck, Hartman & Esposito, Inc., Cincinnati, Ohio*
SECTION 4.3: Dioxin Wastes

Gordon M. Evans *Economist, U.S. Environmental Protection Agency, Hazardous Waste Engineering Research Laboratory, Cincinnati, Ohio*
SECTION 14.2: Cost Perspectives for Hazardous-Waste Management

Edward G. Fochtman *Chemical Waste Management, Inc., Riverdale, Ill.*
SECTION 7.4: Chemical Oxidation and Reduction

Richard C. Fortuna *Hazardous Waste Treatment Council, Washington, D.C.*
SECTION 1.1: Hazardous-Waste Treatment Comes of Age

Wayne B. Gitchel *Rothschild, Wis.*
SECTION 8.6: Wet Oxidation

Douglas E. Gladstone, Ph.D. *Vice President, Environmental Strategies Corporation, Vienna, Va.*
SECTION 3.3: Environmental Risk Assessment and Audit Principles

John A. Glaser, Ph.D. *U.S. Environmental Protection Agency, Hazardous Waste Engineering Research Laboratory, Cincinnati, Ohio*
SECTION 9.6: Enzymatic Systems

Clarence G. Golueke *Director of Research and Development, Cal Recovery Systems, Inc., Richmond, Calif.*
SECTION 9.3: Composting of Industrial Wastes

Paul G. Gorman *Principal Chemical Engineer, Midwest Research Institute, Kansas City, Mo.*
SECTION 8.17: Guide for Incinerator Trial Burns

Charles N. Haas *Pritzker Department of Environmental Engineering, Illinois Institute of Technology, Chicago, Ill.*
SECTION 6.3: Electrolytic Recovery Techniques

Fred D. Hall, P.E. *PEI Associates, Inc., Cincinnati, Ohio*
SECTION 8.15: Hazardous Waste as Fuel in Industrial Processes

Clark W. Heath, Jr., M.D. *Bureau of Preventive Health Services, South Carolina Department of Health and Environmental Control, Columbia, S.C.*
SECTION 3.2: Health Effects of Hazardous Wastes

James A. Heist, P.E. *Vice President, Research and Development, Heist Engineering Corporation, Raleigh, N.C.*
SECTION 6.10: Freeze-Crystallization

Louis R. Hovater *Hovater Engineers, Laguna Hills, Calif.*
SECTION 10.2: Synthetic Linings

Robert L. Hoye *Project Manager, PEI Associates, Inc., Cincinnati, Ohio*
SECTION 4.5: Mining Wastes

S. Jackson Hubbard *Project Officer, U.S. Environmental Protection Agency, Office of Research and Development, Water Engineering Research Laboratory, Cincinnati, Ohio*
SECTION 4.5: Mining Wastes

Gary E. Hunt *Head, Technical Assistance, Pollution Prevention Pays Program, North Carolina Division of Environmental Management, Raleigh, N.C.*
SECTION 5.1: Minimization of Hazardous-Waste Generation

Robert L. Irvine *Professor, Department of Civil Engineering, University of Notre Dame, Notre Dame, Ind.*
SECTION 9.1: Aerobic Processes

Ruby H. James, *Southern Research Institute, Birmingham, Ala.*
CHAPTER 13: Sampling and Analysis of Hazardous Wastes

Stephen C. James *U.S. Environmental Protection Agency, Hazardous Waste Engineering Research Laboratory, Cincinnati, Ohio*
SECTION 12.2: Treatment and Containment Technologies

Larry D. Johnson *U.S. Environmental Protection Agency, Environmental Monitoring Systems Laboratory, Research Triangle Park, N.C.*
CHAPTER 13: Sampling and Analysis Techniques for Hazardous Wastes

Philip C. Kearney *Pesticide Degradation Laboratory, Agricultural Research Service, USDA, Beltsville, Md.*
SECTION 7.3: Photolysis

Yen-Hsung Kiang, Ph.D. *Four Nines, Inc., Conshohocken, Pa.*
SECTION 8.14: Catalytic Incineration

Alfred Kornel, Ph.D. *U.S. Environmental Protection Agency, Hazardous Waste Engineering Research Laboratory, Cincinnati, Ohio*
SECTION 7.5: Dehalogenation Processes

Eugene E. Kupchanko   *Project Engineer, West Vancouver, B.C., Canada*
SECTION 3.4: Case Study of a Successful Facility Siting

C. C. Lee, Ph.D.   *U.S. Environmental Protection Agency, Hazardous Waste Engineering Research Laboratory, Cincinnati, Ohio*
SECTION 8.12: Plasma Systems

Li-Shiang Liang   *Director of Process and Engineering Development, Memtek Corporation, Billerica, Mass.*
SECTION 7.1: Filtration and Separation

Frances M. Lynn   *Research Assistant Professor, Instiute for Environmental Studies, University of North Carolina, Chapel Hill, N.C.*
SECTION 3.1: Siting of Hazardous-Waste Facilities

Janet M. MacNeil   *Staff Scientist, Environmental Strategies Corporation, San Jose, Calif.*
SECTION 6.7: Membrane Separation Technologies

Edward Martin   *Peer Consultants, Rockville, Md.*
SECTION 14.1: Control and Disposal Indices for Comparing Waste-Management Alternatives

Howard B. Mason   *Manager, Energy Engineering Department, Acurex Corporation, Mountain View, Calif.*
SECTION 8.4: Hazardous Waste as Fuel for Boilers

Paul H. Mazzocchi   *Department of Chemistry and Biochemistry, University of Maryland, College Park, Md.*
SECTION 7.3: Photolysis

Mark M. McCabe   *Senior Scientist, Remediation Technologies, Inc., Concord, Mass.*
SECTION 4.1: Waste Oil

Drew E. McCoy   *McCoy and Associates, Denver, Colo.*
SECTION 4.2: PCB Wastes

Thomas P. McVeigh   *Quality Assurance Manager, SPER Division, Roy F. Weston, Inc., Westchester, Pa.*
SECTION 12.1: Sampling and Monitoring of Remedial-Action Sites

Lynne M. Miller   *President, Environmental Strategies Corporation, Vienna, Va.*
SECTION 3.3: Environmental Risk Assessment and Audit Principles

Michael Modell   *Modell Development Corporation, Cambridge, Mass.*
SECTION 8.11: Supercritical-Water Oxidation

Robert E. Mourninghan   *U.S. Environmental Protection Agency, Hazardous Waste Engineering Research Laboratory, Cincinnati, Ohio*
SECTION 8.15: Hazardous Waste as Fuel in Industrial Processes

Frederick C. Novak   *Director of Marketing, H2M Group, Melville, N.Y.*
SECTION 7.6: Ozonation

Robert A. Olexsey   *U.S. Environmental Protection Agency, Hazardous Waste Engineering Research Laboratory, Cincinnati, Ohio*
SECTION 8.4: Hazardous Waste as Fuel for Boilers

E. T. Oppelt   *U.S. Environmental Protection Agency, Hazardous Waste Engineering Research Laboratory, Cincinnati, Ohio*
SECTION 14.1: Control and Disposal Indices for Comparing Waste-Management Alternatives

Tan Phung, Ph.D.   *Woodward Clyde, Santa Ana, Calif.*
SECTION 9.4: Land Treatment of Hazardous Wastes

Bill Quan   *Project Manager, Solid Waste Management Program, Office of The Chief Administrative Officer, City and County of San Francisco, Calif.*
SECTION 5.2: Waste Exchanges

George P. Rasmussen   *Vice President of Engineering, Waste Tech Services, Inc., Golden, Colo.*
SECTION 8.3: Fluidized-Bed Thermal Oxidation

John R. Reinhardt   *Tufts University, Center for Environmental Management, Medford, Mass.*
SECTION 1.2: Summary of Resource Conservation and Recovery Act Legislation and Regulations

Rosemary Robinson, P.E.   *Cross/Tessitore and Associates, P.A., Orlando, Fla.*
SECTION 4.4: Infectious Wastes; SECTION 8.16: Infectious-Waste Incinerators

Tony N. Rogers,   *Research Triangle Institute, Research Triangle Park, N.C.*
SECTION 6.2: Distillation

Paul J. Rogoshewski   *Chemical Engineer, Environmental Consulting Services, Jamaica Plain, N.Y.*
SECTION 12.2: Treatment and Containment Technologies

David E. Ross, P.E.   *SCS Engineers, Long Beach, Calif.*
SECTION 10.1: Hazardous-Waste Landfill Construction: The State of the Art

Joseph J. Santoleri   *Principal Consultant, Four Nines, Inc., Plymouth Meeting, Pa.*
SECTION 8.1: Liquid-Injection Incinerators

George M. Savage   *Cal Recovery Systems, Inc., Richmond, Calif.*
SECTION 9.3: Composting of Industrial Wastes

Charles F. Schaefer   *Manager, Pyro Processing Development, Minerals System Division, Boliden Allis, Inc., Milwaukee, Wis.*
SECTION 8.2: Rotary Kilns

Roger N. Schecter   *Director, Pollution Prevention Pays Program, North Carolina Division of Environmental Management, Raleigh, N.C.*
SECTION 5.1: Minimization of Hazardous-Waste Generation

T. J. Schultz   *Surface Combustion Inc., Toledo, Ohio*
SECTION 8.7: Pyrolysis Systems

P. R. Sferra, Ph.D.   *Biologist, U.S. Environmental Protection Agency, Hazardous Waste Engineering Research Laboratory, Cincinnati, Ohio*
SECTION 9.5: Biodegradation of Environmental Pollutants

J. K. Shah   *Surface Combustion Inc., Toledo, Ohio*
SECTION 8.7: Pyrolysis Systems

Steve I. Simpson   *Associated Technologies, Inc., Charlotte, N.C.*
SECTION 6.9: Thin-Film Evaporation

John M. Smith   *J. M. Smith and Associates, Cincinnati, Ohio*
SECTION 8.10: Deep-Shaft Wet-Air Oxidation

Rick Stanford   *Roy F. Weston & Co., Washington, D.C.*
SECTION 1.3: Summary of CERCLA Legislation and Regulations

Ronald B. Stone   *GEOMIN, Inc., Tulsa, Okla.*
SECTION 10.6: Disposal in Mines and Salt Domes

Lori Tagawa   *Hydrogeologist, SCS Engineers, Long Beach, Calif.*
SECTION 10.1: Hazardous-Waste Landfill Construction: The State of the Art

Michael F. Torpy *General Manager, Laboratory Services, ReTec, Kent, Washington*
SECTION 9.2: Anaerobic Processes

William L. Troxler *IT Enviroscience, Knoxville, Tenn.*
SECTION 8.13: Mobile Thermal Treatment System

Ronald J. Turner *U.S. Environmental Protection Agency, Hazardous Waste Engineering Research Laboratory, Cincinnati, Ohio*
SECTION 7.7: Evaporation

Ronald A. Venezia *President, SITEX Environmental, Inc., Research Triangle Park, N.C.*
SECTION 8.8: Oceanic Incineration

Gregory A. Vogel *The MITRE Corporation, McLean, Va.*
SECTION 2.2: Hazardous-Waste Management Capacity

Thomas C. Voice *Department of Civil and Environmental Engineering, Michigan State University, East Lansing, Mich.*
SECTION 6.1: Activated-Carbon Adsorption

Don L. Warner *School of Mines and Metallurgy, University of Missouri-Rolla, Rolla, Mo.*
SECTION 10.3: Subsurface Injection of Liquid Hazardous Wastes

Roger S. Wetzel *Manager, Waste Site Engineering, Science Applications International Corporation, McLean, Va.*
SECTION 12.3: Site Remediation

Peter A. Wilderer *Professor, University of Hamburg-Harburg, Hamburg, West Germany*
SECTION 9.1: Aerobic Processes

Alan Wilds *President, BDT, Inc., Clarence, N.Y.*
SECTION 6.4: Hydrolysis

Carlton C. Wiles *U.S. Environmental Protection Agency, Hazardous Waste Engineering Research Laboratory, Cincinnati, Ohio*
SECTION 7.8: Solidification and Stabilization Technology

Thomas D. Wright, P.E. *GSA/Waste Management, Santa Ana, Calif.*
SECTION 10.1: Hazardous-Waste Landfill Construction: The State of the Art

Edward C. Yang *Planning Research Corporation, McLean, Va.*
SECTION 1.3: Summary of CERCLA Legislations and Regulations

Craig M. Young *Manager, Quality Control and Analysis, Waste Tech Services, Inc., Golden, Colo.*
SECTION 8.3: Fluidized-Bed Thermal Oxidation

---

*Deceased

PREFACE

Undoubtedly, *the* environmental issue in industrial countries throughout the world in the 70s and 80s has been and will continue to be how to best manage hazardous waste. The issue has manifested itself in the form of two questions: how to prevent environmental deterioration caused by the generation of hazardous wastes, and how to effectively clean up the problems caused by past examples of improper disposal. Both these questions are complex, depending for resolution on many social and technical factors. This seemingly simple statement can be appreciated very quickly by anyone charged with resolving even small parts of the hazardous-waste issue. Just a brief investigation of the issues brings the realization that not only are there many technical processes for properly managing, treating, and disposing of hazardous wastes, but also the technical options are enmeshed in a formidable legislative and regulatory framework.

I felt that there was clearly a need for an extensive handbook to

1. Provide a summary of relevant laws and regulations

2. Provide an overview of the extent of the "hazardous-waste problem" and explore various issues integral to the problem

3. Explore institutional options such as waste minimization and waste exchanges

4. Describe in detail the state of the art for the many physical, chemical, thermal, and biological processes for treating hazardous wastes

5. Outline the acceptable disposal options

6. Review the unique hazardous-waste issues and processes associated with remedial action

I contacted a hundred or so recognized experts in the various fields associated with the issues and asked them to join with me in the project. This book is the result of our efforts.

The contributing authors have written extremely well, and provided you, the reader, with chapters not only summarizing what is the state of the art in the various areas, but also sharing with you the results of their having worked directly in the subject areas for many years.

As one who has struggled with the seemingly infinite amount of information related to hazardous-waste treatment and disposal, I hope you will find this compilation useful in carrying out your work. While the Handbook is designed to be used as a reference for resolving specific problems and not intended to be read from cover to cover, I can tell you that there is a world of interesting information included between these covers. For those inclined to browse, you will find such activity rewarding.

Hazardous-waste problems will certainly not just go away. However, I believe their resolutions are certainly within our abilities. I hope this handbook will in some small way hasten the day when hazardous waste is no longer *the* environmental problem it currently is.

ACKNOWLEDGMENTS

Of course, major acknowledgment is due to the contributing authors, a group of unusual individuals who not only know a lot about their subjects but who can write as well. Others without whom this book would not have happened are Vicki Hahn and Rick Prairie. Thank you.

Harry M. Freeman
Editor

ABOUT THE EDITOR

HARRY M. FREEMAN is a Research Program Manager with the U.S. Environmental Protection Agency's Hazardous Waste Engineering Research Laboratory in Cincinnati, Ohio, responsible for various hazardous waste minimization and alternative technology research programs being supported by the USEPA. He is a Registered Professional Engineer, a member of the American Institute of Chemical Engineers, A Diplomate of the American Academy of Environmental Engineers, and a member of the International Juggling Association.

GUIDE FOR USING
THE HANDBOOK

This Handbook is intended to be a reference book for engineers, industrial managers, government officials, researchers, teachers, and others involved with the management, regulation, treatment, and disposal of hazardous chemical wastes. In general, the subjects covered include management, technical, and regulatory considerations for hazardous-waste generation, treatment, and disposal and for similar concerns related to cleaning up abandoned hazardous-waste disposal sites.

The book can be used both to familiarize oneself with broad areas of hazardous-waste regulation and/or technology, and to review in depth the state of the art for specific alternative treatment and disposal processes.

Included in the book are discussions of virtually every existing and emerging treatment or disposal technology currently being pursued as an alternative to land disposal. Many of the processes are discussed in more than one chapter, reflecting their inclusion in more than one emphasis area. The reader is directed to the extensive index at the end of the book for direction to the various subjects.

Chapter 1: Laws and Regulations. In the United States, federal regulations pertaining to hazardous wastes are developed under the authority of two major pieces of legislation: the 1984 Amendments to the Resource Conservation and Recovery Act (RCRA) and the 1987 Superfund Reauthorization. These two bills for cradle to grave management of hazardous wastes and cleanup actions respectively, are covered in Chap. 1. Other sections of this chapter provide insight into the deliberation leading to these bills, and discuss an emerging significant environmental issue, underground storage tanks.

Chapter 2: Hazardous-Waste Characteristics, Quantities, and Treatment Capacities. Any discussion of hazardous-waste problems involves large numbers and many statistics. The U.S. Environmental Protection Agencies (EPA) and others have carried out rather extensive surveys to gather information about the extent of these problems and current capabilities for mitigating them. Much of the information from these surveys is analyzed in the two sections of Chap. 2.

Chapter 3: Hazardous Waste Topics and Issues. The "hazardous-waste problem" involves much more than treatment and disposal technology. In fact, these broader issues provide the context in which the technical solutions are developed and applied. Chapter 3 explores risk assessments, liability insurance, and health effects, as well as siting, the issue which is arguably the most important consideration related to hazardous waste. Included here is a case study of the only successful siting of a major new treatment facility in North America in the past 10 years.

Chapter 4: Special Hazardous Wastes.   Several categories of hazardous wastes have for various reasons received special attention from the public and writers of environmental legislation. These wastes streams (waste oil, PCB wastes, dioxins, infectious wastes, mining wastes and household hazardous wastes) are each the subject of a section of Chap. 4. The individual discussions include the factors unique to the waste streams and possible treatment technology options. These technology options are also discussed further in later chapters of the book.

Chapter 5: Waste Minimization and Recycling.   As increasingly stringent new environmental regulations are introduced, hazardous-waste management strategies that incorporate waste reduction through process changes and increased recycling will become more important. The two sections of Chap. 5 explore two of the institutional approaches being pursued by individuals and organizations throughout the world: i.e., waste minimization and waste exchanges. Minimization, now in its formative stages as a preferred strategy for hazardous-waste management, will certainly become more of an integral part of national waste-management plans throughout the world, and waste exchanges should assume importance as recycling becomes more prevalent.

Chapter 6: Hazardous-Waste Recovery Processes.   The 10 processes included in Chap. 6 are generally recognized as processes useful for recovering materials of value from waste streams so that the materials can be re-used. These are chemical engineering processes designed for hazardous-waste applications. A matrix relating the processes to suitable hazardous-waste streams is shown in Table 1.

TABLE 1   Applicability of Recovery Processes (Chap. 6) to Various Waste Streams

Treatment processes (Handbook Sections)	Corrosives	Cyanides	Halogenated solvents	Nonhalogenated solvents	Chlorinated organics	Other organics	Oily wastes	PCBs	Aqueous with metals	Aqueous with organics	Reactives	Contaminated soils	Liquids	Solids/sludges	Gases
Carbon adsorption (6.1)									X	X	X		X		X
Distillation (6.2)			X	X	X	X							X		
Electrolytic Recovery (6.3)									X				X		
Hydrolysis (6.4)											X		X		X
Ion exchange (6.5)	X								X	X			X		
Solvent Extraction (6.6)			X	X	X	X					X	X			
Membranes (6.7)									X	X			X		
Air and steam stripping (6.8)			X	X	X	X				X			X		
Thin-film evaporation (6.9)			X	X	X								X		
Freeze-crystallization (6.10)			X	X	X	X			X	X			X		

Chapter 7: Physical and Chemical Treatment. The eight processes detailed in Chap. 7 are used in various schemes for treating hazardous waste streams for either volume reduction, detoxification, or stabilization. They are not significantly different from the processes in Chap. 6. Their inclusion in Chap. 7 rather than Chap. 6 is a reflection of the end point of the processes, i.e., treatment rather than recovery and re-use. Applicability of these physical and chemical treatment processes to various hazardous-waste streams is summarized in matrix form in Table 2.

Chapter 8: Thermal Processes. Thermal processes are designed to break down hazardous waste through either combustion or pyrolysis by exposing the material to high temperature in a controlled environment. Thermal processes can, in a matter of seconds, destroy waste materials that might otherwise take many years to deteriorate in a landfill. These processes are very important as elements in a hazardous-waste management plan.

The types of thermal processes are legion, as illustrated by the large number of sections in Chap. 8. In general, each of the various thermal processes is appropriate for a particular type of waste stream. A matrix outlining applicability of thermal processes to organic hazardous waste streams is shown in Table 3. The last part of Chap. 8 is a discussion of the requirements for trial burns, a necessary part of the permitting process for any incinerator.

Chapter 9: Biological Processes. Biological processes, or processes using living microorganisms, have been used for many years to treat municipal and industrial wastewater. Many of the advantages of biological processes translate equally well to selected hazardous waste streams. Consequently, biological treatment can be an integral part of an alternative technology system.

The first four sections of Chap. 9 discuss in detail four somewhat traditional biological treatment processes: aerobic processes, anaerobic processes, compost-

TABLE 2 Applicability of Physical and Chemical Processes (Chap. 7) to Various Waste Streams

Treatment processes (Handbook Sections)	Corrosives	Cyanides	Halogenated solvents	Nonhalogenated organics	Chlorinated organics	Other organics	Oily wastes	PCBs	Aqueous with metals	Aqueous with organics	Reactives	Contaminated solids	Liquids	Solids/sludges	Gases
						Hazardous waste streams							Form of waste		
Separation/filtration (7.1)		X	X	X	X	X			X	X			X		X
Chemical Precipitation (7.2)	X								X				X		
Photolysis (7.3)													X		
Chemical oxidation/reduction (7.4)		X							X				X		
Dehalogenation (7.5)			X		X			X			X	X	X		
Ozonation (7.6)		X	X		X				X				X		X
Evaporation (7.7)			X	X	X	X	X						X	X	
Solidification (7.8)	X	X									X	X	X		

TABLE 3 Applicability of Thermal Systems (Chap. 8) to Various Waste Streams

Thermal systems (Handbook Sections)	Corrosives	Cyanides	Halogenated solvents	Nonhalogenated organics	Chlorinated organics	Other organics	Oily wastes	PCBs	Aqueous with metals	Aqueous with organics	Reactives	Contaminated soils	Infectious wastes	Liquids	Solids/sludges	Gases
Liquid injection (8.1)			X	X	X	X	X							X		X
Rotary kilns (8.2)			X	X	X	X	X	X				X	X	X	X	X
Fluid beds (8.3)			X	X	X	X	X	X				X		X	X	X
Boilers (8.4)			X	X	X	X								X		X
Cement kilns (8.5)			X	X	X	X	X	X						X	X	
Wet Oxidation (8.6)	X		X							X				X		
Pyrolysis (8.7)			X	X	X	X						X		X	X	
Oceanic Incineration (8.8)			X	X	X	X	X							X		
Molten glass (8.9)			X	X	X	X	X			X				X	X	X
Deep shaft (8-10)			X	X	X					X				X		
Supercritical water (8-11)			X	X	X	X				X				X		
Plasma (8-12)			X	X	X	X	X							X	X	
Mobile units (8-13)			X	X	X	X	X	X				X		X	X	X
Catalytic incineration (8-14)				X												X
Industrial processes (8-15)	X	X		X	X									X	X	
Infectious-waste incineration (8.16)													X	X	X	

ing, and land treatment. The last two sections of the chapter address some of the newer developments.

Chapter 10: Land Storage and Disposal. Regardless of the effectiveness of the type of alternative processes covered in Chap. 6 through 9, there will always be residuals that must eventually be deposited in, or on, the land. The eight parts of Chap. 10 address the many aspects of and requirements for operation of land facilities of various types. The information contained in these chapters reflects current state of the art design for the various facilities. These designs reflect all the requirements of current environmental regulations.

Chapter 11: Comprehensive Hazardous-Waste Treatment Facilities. This chapter describes in detail the possibilities for combining the many alternative treatment processes into an effective, integrated system for treating the thousands of different hazardous waste streams that typically must be managed by commercial treatment operations. Also addressed are political and institutional questions related to comprehensive facilities.

Chapter 12: Remedial-Action Techniques and Technology. Currently a significant part of "the hazardous-waste problem" is cleaning up abandoned waste-treatment facilities and disposal sites. The requirements of these cleanups have created a field of technology that spans the fields of construction techniques and hazardous-waste treatment processes. This field of technology, remedial-action technology, is the subject of the three sections of Chap. 12. These sections address the three major aspects of site cleanup.

Chapter 13: Sampling and Analysis Techniques for Hazardous Waste. A highly technical field such as hazardous-waste treatment and disposal depends upon a body of generally recognized and accepted techniques for identifying and characterizing waste chemicals and environmental emissions. Such a body of techniques has evolved out of work of the EPA and others. These techniques are covered in Chap. 13.

Chapter 14: Engineering Considerations. The authors of the two sections of this chapter discuss approaches to deciding between various alternative processes. While those presented here are certainly not the only accepted ways of making such decisions, they should certainly be reviewed by decision makers charged with actually selecting a waste-treatment or disposal process.

Harry M. Freeman
U.S. Environmental Protection Agency
Hazardous Waste Engineering Research Laboratory

SYMBOLS AND ABBREVIATIONS USED IN THIS BOOK

Å	angstrom(s)
AA	atomic adsorption
AACE	American Association of Coast Engineers
AAS	atomic-absorption spectroscopy
ac	alternating current
AC	asphaltic concrete
ACS	American Chemical Society
ADP	automated data processing
AFVC	absorption freeze vapor compression
ANSI	American National Standards Institute
APEGs	alkaline polyethylene glycolates
APEGM	alkaline polyethylene glycolate monomethyl ether
API	American Petroleum Institute
APU	acid-purification unit
ASTM	American Society for Testing and Materials Standards
ATA	anaerobic toxicity assay
ATC	Arc T
atm	atmosphere(s)
B[a]P	benzo[a]pyrene
BH	baghouse
BMP	biochemical methane potential
BOD	biological oxygen demand
BOD$_5$	five-day biological oxygen demand
b.p.	boiling point
BS&W	bottom sediments and water
Btu	British thermal unit(s)
°C	degree(s) Celsius
ca	circa
CAA	Clean Air Act
CB	chlorobenzenes
CBD	Commerce Business Daily
CDB	Consolidated Data Bureau
CE	combustion efficiency
CEC	cation-exchange capacity
CEI	Committee of Environmental Improvement

CERCLA	Comprehensive Environmental Response Compensation and Liability Act
CFR	*Code of Federal Regulations*
CFS	continuous-flow system
CGI	combustible-gas indicator
CGL	comprehensive general liability (insurance)
CIS	Chemical Information System
CKD	cement-kiln dust
CLP	Contract Laboratory Program
CMA	Chemical Manufacturers Association
CMB	completely mixed batch
CMF	completely mixed flow
CMP	corrugated-metal pipe
COD	chemical oxygen demand
COG	coke-oven gas
cP	centipoise
CP	concentration profile; critical point
CPE	chlorinated polyethylene
CQA	construction quality-assurance (programs)
CRF	capital recovery factor
CSPE	chlorosulfonated polyethylene
CWA	Clean Water Act
CYI	Control Years Index
d	day(s)
dc	direct current
DDE	1,1-dichloro-2,2-bis(4-chlorophenyl)ethene
DDT	dichlorodiphenyltrichloroethane
DE	destruction efficiency
DEHPA	di-2-ethylhexyl phosphate
DF	decontamination factor
DGME	diethylene glycolate monomethyl ether
DHS	Department of Health Services (California State)
DHT	down-hole temperature
DI	disposal index, Department of Interior
DIPE	diisopropyl ether
DMF	N,N-dimethylformamide
DNA	deoxyribonucleic acid
DOC	dissolved organic carbon
DOD	Department of Defense
DOE	Department of Energy
DOFASCO	Dominion Foundaries and Steel Company
DOT	Department of Transportation
DRE	destruction and removal efficiency
dscf	dry standard cubic feet
dscm	dry standard cubic meters
dscfm	dry standard cubic feet per minute
dscmm	dry standard cubic meters per minute
DT	drying tube
DVB	divinylbenzene
ECA	Environmental Council of Alberta
ED	electrodialysis (p. 7.6.14—another meaning—au queried)
EDB	ethylene dibromide

EDTA	ethylenediaminetetraacetic acid
e.g.,	exemplia gratis (for example)
EI	electron-impact (mode)
EOS	Environmental Oceanic Services
EP	extraction procedure (toxicity test)
EPA	Environmental Protection Agency
EPDM	ethylene propylene diene monomer
EPRI	Electric Power Research Institute
eq	equivalent(s)
ESB	Employee Supplemental Benefits
ESE	extended-surface electrolysis
ESP	electrostatic precipitator
°F	degrees(s) Fahrenheit
FID	flame-ionization detector
FIFRA	Federal Insecticide, Fungicide, and Rodenticide Act
FIT	Field Investigation Team
FML	flexible-membrane liner
FPD	freezing-point depression
FR	*Federal Register*
FRG	Federal Republic of Germany
FRP	fiber-reinforced plastic
ft	feet
FTMS	Federal Test Method Standard
g	gram(s)
GAC	granular activated carbon
gal	gallon(s)
GC	gas chromatograph
Gcal	gigacalorie(s)
g-cal	gram-calorie(s)
GC/MS	gas chromatograph mass spectrometry
G&A	general and administration (expenses)
gfd	gal/(ft·d)
gpm	gallon(s) per minute
gpd	gallon(s) per day
gr	gram(s)
GW	groundwater; gigawatt(s)
h	hour(s)
ha	hectare(s)
HDO	hydrogen deuterium oxide
HDPE	high-density polyethylene
HETP	height equivalent of a theoretical plate
HF	hyperfiltration
HHS	(Department of) Health and Human Services
HHV	higher heating value
HHW	household hazardous waste
HMX	cyclotetramethylene tetranitramine
hp	horsepower
HPLC	high-performance liquid chromatography
HPVC	high-density polyvinyl chloride
HS	hazardous substance(s)
HSA	high-surface-area (reactor)
HSL	Hazardous Substances List

HSWA	Hazardous and Solid Waste Amendments
HTU	height of a transfer unit
HW	human wastes
HWDMS	Hazardous Waste Data Management System
HWERL	Hazardous-Waste Engineering Research Laboratory
HWF	hazardous-waste fuel
HxCDDs	hexachlorodibenzodioxins
HxCDFs	hexachlorodibenzofurans
ICAP	inductively coupled argon-plasma (emission spectroscopy)
ID	induced-draft (fan); inside diameter
i.e.,	id est (that is)
IMO	International Maritime Organization
IOA	International Ozone Association
IFC	isolation flux chamber
ISV	in situ vitrification
J	joule(s)
K	kelvin(s)
K	mass-transfer coefficient
KD	Kuderna-Danish (concentration gas chromatography)
kg	kilogram(s)
kg-cal	kilogram-calorie(s)
K_m	Michaelis constant
kN	kilonewton(s)
kmol	kilomole(s)
KPEG	polyethylene glycol potassium hydroxide
kW	kilowatt(s)
kWh	kilowatt-hour(s)
L	liter(s)
lb	pound(s)
LC	liquid chromatography
LEL	lower explosion limit
LF	landfill
LFG	landfill gas
LHV	lower heating value
LI	liquid injection
m	meter(s)
M	molar
m-	$meta$-
MBTH	3-methyl-2-benzothiazolinone hydrazone
MCL	maximum (permissible) concentration limits
MDQ	minimum detectable quantity
MEK	methyl ethyl ketone
MEP	multiple extraction procedure
meq	milliequivalents
μ	microns (change to μm)
μg	microgram(s)
μR	microroentgen(s)
μm	micrometer(s)
mg	milligram(s)
mgd	million gallon(s) per day
MIBK	methyl isobutyl ketone
min	minute(s)

mL	milliter(s)
MM5	Modified Method Five train
mm Hg	millimeter(s) of mercury
MN	meganewton(s)
mo	month(s)
mol wt	molecular weight
MPa	megapascal(s)
mrem	millirem(s)
MS	mass spectrometry
ms	millisecond(s)
MW	molecular weight
n-	normal (isomer)
N	normal (solution)
NaPEG	sodium polyethylene glycolate
NCP	National (Oil and Hazardous Substances) Contingency Plan
NEDS	National Emmission Inventory Data Base
NFPA	National Fire Protection Association
NIOSH	National Institute for Occupational Safety and Health
NPC	Northern Petrochemical Company
NPDES	National Pollutant Discharge Elimination System (permit)
NPL	National Priorities List
NTIS	National Technical Information Service
NTU	number of a transfer unit
NYSDEC	New York State Department of Environmental Conservation
o-	*ortho-*
OCC	Occidental Chemical Corporation
OD	outside diameter, oxygen demand
OERR	Office of Emergency and Remedial Response
O&M	operation and maintenance
OHS	Occupational Health Services
OLR	organic loading rates
OMB	Office of Management and Budget
ONWI	Office of Nuclear Waste Isolation
OSC	on-scene coordinator
OSHA	Occupational Safety and Health Administration
OSWER	Office of Solid Waste and Emergency Response
p-	*para-*
Pa	pascal(s)
PAC	powdered activated carbon
PACT	powdered activated-carbon treatment
PAH	polynuclear aromatic hydrocarbons
PAT	purge-and-trap analysis
PCBs	polychlorinated biphenyls
PCDDs	polychlorinated dibenzodioxins
PCE	perchloroethylene
PCP	pentachlorophenol
PeCDDs	pentachlorodibenzodioxins
PeCDFs	pentachlorodibenzofurans
PEGM	polyethylene glycol monoalkyl ethers
% (wt/wt)	percent (weight for weight)
PF	plug flow (reactors)
PFLT	Paint Filter Liquids Test

pH	hydrogen-ion activity
PICs	products of incomplete combustion
PID	photoionization detector
PITs	products of incomplete treatment
PLIA	Pollution Liability Insurance Association
PNAs	polynuclear aromatic (hydrocarbons)
PNL	Pacific Northwest Lab
POHCs	principal organic hazardous constituents
POTW	publicly owned (wastewater) treatment works
PP	priority pollutants
ppb	part(s) per billion
ppm	part(s) per million
psi, psia, psig	pound(s) per square inch, pounds per square inch (absolute), pounds per square inch (gage)
PTFE	polytetrafluoroethylene
PVC	polyvinyl chloride
QA/QC	quality assurance and quality control
R	roentgen(s)
RCC	Resource Conservation Company (Washington State)
RCRA	Resource Conservation and Recovery Act
R&D	research and development
RD&D	research, development, and demonstration
RDF	refuse-derived fuel
RD/RA	remedial design and remedial action
RDX	cyclotrimethylene trinitramine
rem	roentgen equivalent in man
RF	radio frequency
RFP	request for proposal
r/min	revolution(s) per minute
RIA	regulatory impact analysis
RI/FS	remedial investigation and feasibility study
RO	reverse osmosis
s	second(s)
SARA	Superfund Amendments and Reauthorization Act
SASS	Source Assessment Sampling System
SBR	sequencing batch reactor
SCBA	self-contained breathing apparatus
scfm	standard cubic feet per minute
SCW	supercritical water
SCWO	supercritical-water oxidation
SDWA	Safe Drinking Water Act
SI	surface impoundment
SIC	Standard Industrial Code
SIM	selected ion monitoring
SITE	Superfund Innovative Technology Evaluation
SPDES	State Pollutant Discharge Elimination System (permit)
SPL	spent pot lining
SS	stainless steel
S/S	solidification and stabilization
SSU	Saybolt Seconds Universal
STEL	short-term exposure limit
SVPOHC	semivolatile principal organic hazardous constituents

SW	solid wastes
t	metric tonne(s) (the English ton is never abbreviated)
TAT	Technical Assistance Team
TC	temperature control
TCDDs	tetrachlorodibenzodioxins
TCDFs	tetrachlorodibenzofurans
TCE	trichloroethylene
TCLP	toxicity characteristics leaching procedure
TDI	toluene diisocyanate
TDS	total dissolved solids
THC	total hydrocarbons
THM	trihalomethane
TIC	total ion current
TLV	threshold limit value
TNO	The Netherlands Organization (for Applied Scientific Research)
TNT	trinitrotoluene
TOC	total organic carbon
TOD	total oxygen demand
TOX	total organic halides
TSCA	Toxic Substances Control Act
TSDF	treatment, storage, and disposal facility
TSS	total suspended solids
TUHC	total unburned hydrocarbon
TWA	time-weighted average
UASB	upflow anaerobic-sludge blanket
UF	ultrafiltration; urea-formaldehyde resin
UIC	underground injection control
UNC	University of North Carolina
U.S.C.	*United States Code*
U.S. DA	U.S. Department of Agriculture
U.S. EPA	U.S. Environmental Protection Agency
U.S. GS	U.S. Geological Survey
U.S. OWRT	U.S. Office of Water Research and Technology
UST	underground storage tank
UV	ultraviolet
%v/v	percent (volume per volume)
VA	volatile acids
VCM	vinyl chloride monomer
VOA	volatile organic analysis
VOC	volatile organic carbon
VOST	volatile-organic sampling train
VPOHC	volatile principal organic hazardous constituents
VSS	volatile suspended solids
VTR	vertical-tube reactor
W	watt
w.c.	water column
wk	week(s)
wt %	weight percent
WWTP	wastewater treatment process
yr	year(s)

CHAPTER 1

LAWS AND REGULATIONS

SECTION 1.1

HAZARDOUS-WASTE TREATMENT COMES OF AGE

Richard C. Fortuna*

Executive Director
Hazardous Waste Treatment Council

Winston Churchill once said that "success is going from failure to failure without loss of enthusiasm." Most would agree that this epigram of Churchill's aptly summarizes the history of the hazardous-waste management program in the United States since the enactment of the Resource Conservation and Recovery Act (RCRA) in 1976, and the subsequent passage of the "Superfund" cleanup law in 1980. Despite the buoyant expectations created by these statutes, it is now clear that the difficulty in establishing a comprehensive system and creating incentives for preventive management and proper cleanup has exceeded all expectations. After 10 years of trial, error, and unfulfilled promises, we are now only beginning the transition from previous land disposal practices to the treatment of hazardous wastes. While the shortcomings of the RCRA and the Superfund programs have been a recent discovery to the nation, they have been a long-standing impediment to those who developed and sought to use alternative forms of treatment technology that protect human health and the environment.

The enactment of the Superfund Amendments and Reauthorization Act (SARA) in 1986 and the Hazardous and Solid Waste Amendments (HSWA), which amend RCRA, in 1984, represents both the culmination of our 10-year flirtation with failure and a commitment to prevent the propagation of future problem sites and uncontrolled management practices. Collectively, these statutes are a watershed in environmental legislation in terms of their stringency, specificity, and scope.

As a result of these new laws, the United States is in the midst of the most profound and sweeping transition in hazardous-waste management practices that this nation or any nation has witnessed. While various nations of Western Europe have taken significant strides in control of hazardous-waste management practices, these efforts have been sporadic, not coordinated and concerted. The United States has set a course toward a day in the not too distant future when it

---

*Richard C. Fortuna was a principal architect of the 1984 RCRA Amendments, while serving as Staff Toxicologist to the House Energy and Commerce Committee from 1981 to 1983. He is also the author of *Hazardous Waste Regulation: The New Era*, an analysis and guide to RCRA and the 1984 Amendments (McGraw-Hill, 1987).

can look on its hazardous-waste management system and truly declare it to be second to none. Moreover, for the first time since the original enactment of our basic hazardous-waste statute, the RCRA, in 1976, there is a sense of certainty in this field; certainty that public health will be protected and certainty of a mandate that allows for investment in more costly but permanent treatment technology. But the burgeoning of activity and interest in hazardous-waste management and the confidence in the future growth of the field is a recent development.

Until recently the firms farthest out on the limb were those that made the greatest investments in technology. Treatment firms that were seeking to do the best possible treatment, albeit at high cost, were finding themselves in the worst possible market position. Why was this the case? It certainly was not because industry did not have technologies. It was not because it did not have the methods, the means, or the know-how to properly treat hazardous waste. It was not because it was trying and failing to render wastes nonhazardous or to destroy or reclaim them.

Reports by state and congressional commissions reinforced these findings. Then why weren't these technologies taking off years ago? Why weren't people beating a path to the doorway of these firms to limit these liabilities? A brief review of recent history points to a few clear principles in this regard.

In 1980, I was working on the original Superfund bill on the staff of Congressman John D. Dingell. It was clear that we needed to begin the cleanup of abandoned sites, but we did not know exactly what "cleanup" meant: was it 50 totally clean sites, or 500 imminent hazards contained? There was considerable uncertainty about how to distribute responsibility, manage these projects, set priorities among them, or determine what took precedence: costs or effectiveness. Also in 1980, the first set of regulations to implement the 1976 RCRA law were issued. They will long be remembered as being lengthy (three volumes), landmark (given that nothing else had preceded them)—and full of loopholes.

In 1982, when the Congress began reexamining the Superfund and RCRA laws, several unexpected discoveries were made. The laws that were supposed to prevent creation of future Superfund sites were so riddled with deficiencies that the holes were bigger than the regulatory doughnut. More wastes, generators, and facilities were exempt from regulation than were subject to it. The nation discovered to its chagrin that "legal" dumping was a far more prevalent threat to public health than illegal dumping. If left unchecked, the RCRA regulatory system itself would be the leading cause of the generation of future Superfund sites.

The process of implementing the Superfund and RCRA laws throughout the early eighties reinforced many of the fundamental lessons that were emerging for those responsible for congressional oversight. In the area of hazardous-waste cleanup we learned the hard way that simply attempting to cap, pump, and treat or to redispose wastes was more of a shell game than a solution. In the RCRA area we continued to rely on unrestricted land disposal as the primary method of hazardous-waste management. In fact the hazardous-waste laws and regulations were written as if the two universal axioms—"There is no free lunch" and "You get what you pay for"—somehow did not apply to the management of hazardous waste.

We learned the hard way that hazardous waste, unless restrained by regulation, is like water running downhill—it follows the path of least regulatory control and least cost. Moreover, it became clear that we could not expect firms to invest in technologies and to compete against unrestricted land disposal practices where cost alone is allowed to dictate the choice of management method and where a lack of proper regulation indirectly subsidizes the status quo.

Finally, we came to recognize that discretion had become its own worst enemy: too much was just as bad as too little. The Environmental Protection Agency (EPA) could not be saddled with numerous tasks and tight deadlines, and at the same time be forced to choose from an unlimited number of regulatory paths without clear and specific direction from the legislation itself. No administration, no matter how well intentioned, could institute solutions without changes in the statute and without the unambiguous support of the Congress. The 1986 SARA and 1984 HSWA Amendments Act with specificity to provide this program with the direction and backing it needs to institute and maintain the transition toward treatment technology.

In the 1986 SARA Amendments Congress dealt with many of the ambiguities and uncertainties that riddled the remedial action program throughout its early years. In general, Congress established a much more proscriptive program with a definite set of expectations. While some may criticize the approach taken as overly proscriptive and one that severly limits EPA's discretion, it was the absence of congressional guidance in 1980 that contributed to the lack of progress we have witnessed to date. It is also true that the regulated community would rather face the unpleasant than the uncertain. While nobody wants to spend more than they have to, the least desirable situation is one in which total uncertainty and ambiguity govern the process. It was this fundamental fact, coupled with the lack of any visible progress under a discretionary program, that brought the Congress to the current approach, summarized below:

- Requiring the use of permanent remedies, rather than capping and containment, whenever possible.
- Shifting the relationship of costs and effectiveness by ensuring that we first establish the necessary level of effectiveness and then determine the least-cost method of achieving it, not the converse.
- Requiring EPA to select on-site remedial actions that meet legally applicable or relevant and appropriate standards, requirements, and criteria of federal environmental law and any more stringent state environmental or facility-siting laws. There is a presumption in favor of using cleanup standards based upon existing environmental statutes such as the Safe Drinking Water Act standards and the new RCRA requirements (i.e., land disposal bans).
- With the goal of expediting remedial and removal actions the new statute also provides for limited indemnification of a response-action contractor against any liability for negligence in carrying out response action. The statute exempts from RCRA permitting procedures those *remedial actions* conducted on site, provided that RCRA standards for hazardous-waste treatment, storage, and disposal are observed.
- With the goal of stimulating the siting and expansion of new and existing treatment capacity, the statute contains a provision to cut off Superfund support to those states that fail to provide necessary treatment, storage, and disposal capacity for wastes generated within their borders.

Congress's desire for certainty in the prevention of future waste-management problems was first expressed in the 1984 RCRA Amendments—the forerunner of SARA. They established very specific restrictions on land disposal and surface impoundments, and closed many regulatory loopholes. However, the most enduring and fundamental reform issuing from this legislation was a restructuring of the regulatory decision-making process itself. The legislation: (1) establishes a national policy which states that land disposal is to be the method of last resort in the management of hazardous wastes, (2) institutes presumptive prohibitions against the land disposal of all hazardous wastes, and (3) supports the presumptions with a statutory set of minimum controls and self-implementing total prohibitions (or "hammers," as they have come to be known) in the event that the EPA does not act to override or modify them by a specified date. This decision-making structure was intended not to punish the EPA, but rather to assist it in playing catch-up on the backlog of overdue decisions that lay before it. Rather than forcing the EPA to specifically justify a restriction for each waste under a generally permissive scheme, this decision-making structure allows it to focus on the specific exemptions to a general prohibition. In many ways land disposal is now presumptively guilty until proven innocent. In addition, the regulated community now knows that unless it provides the EPA with good faith assistance in developing a regulation, it will have to live with a general rule that has no exceptions, legitimate or otherwise.

The 1986 SARA and 1984 HSWA amendments to RCRA accomplished their primary goal, namely to establish a sense of certainty in the market for alternative technologies in hazardous-waste management, and thus to make a transition to the use of alternative technologies possible. The impacts of these amendments are unmistakable:

- Thousands of land disposal units have closed rather than certify compliance with groundwater and financial responsibility requirements; many of these units were already slated to close or would have closed rather than seek final RCRA permits.
- Hundreds of unlined and leaking surface-impoundment units are being taken out of service.
- The long-term demand for thermal-destruction, impending-reclamation, and quality stabilization services is rapidly increasing due to the land disposal ban.
- A significant acceleration in the development of and demand for both mobile and stationary technology has emerged in response to the pretreatment, waste-minimization, and corrective-action requirements of RCRA.
- There has been a proliferation of interest in commercial hazardous-waste management on the part of firms that traditionally have managed only the wastes they generate.

The new demand and the changes it brings with it pose their own set of challenges, primarily in the permitting of RCRA facilities and timely expansion of capacity. Unless we can "permit" this transition to permanent treatment, all the laws and regulations in the world will never make it off the paper. We are now being challenged to ensure that new capacity is permitted on a timely basis, and that there is sufficient flexibility to modify facilities in order to meet the demands of the new laws. This includes a program for the permitting of mobile treatment units, which must now be repeatedly permitted at each site they visit. Moreover, many states are becoming more interested in stopping the flow of wastes into their territory than in being partners in a national scheme that stems the flow of wastes into land disposal facilities.

While this struggle and tension as we attempt to apply the statute is inevitable and significant, firms now know that it is worthwhile to endure these difficulties, for the market now exists and will continue to expand. The 1984 Amendments to RCRA and the demand they have created for permanent treatment did not occur by chance. The statute and the changes took more than three years to put into place, an effort that, beginning in 1982, spanned two Congresses, and enjoyed overwhelming bipartisan support. Similarly, Superfund took more than three years to develop, and shares a broad bipartisan consensus. Together these landmark statutes mark the end of the beginning for the national program by establishing the beginning of the end of unrestricted land disposal. They represent an enduring commitment to creation of a hazardous-waste management system in the United States that we can at last look on with pride and a sense of certainty rather than look back on with chagrin.

SECTION 1.2

SUMMARY OF RESOURCE CONSERVATION AND RECOVERY ACT LEGISLATION AND REGULATION

John R. Reinhardt

Tufts University
Center for Environmental Management

This chapter provides an overview of existing legislation and regulation, and a context in which to evaluate current and future regulatory activity. Furthermore, it provides the regulatory and legislative context for the waste-treatment technologies discussed throughout this handbook. Section 1.2.1 provides a summary description of the regulatory framework, followed by a summary of the 1984 Amendments to RCRA in Section 1.2.2. The chapter concludes with Section 1.2.3, a summary of the important upcoming regulatory developments.

The Resource Conservation and Recovery Act (RCRA) of 1976 is the primary legislation controlling hazardous-waste management. RCRA originated as a law with a comprehensive scope; it instituted "cradle to grave" control of hazardous waste requiring massive and complex regulation due to the diversity of industrial and commercial operations under its purview. To best understand the present framework of regulations, one must first examine the evolution of the RCRA legislation and its amendments.

The first comprehensive hazardous-waste legislation was passed by Congress in 1976 as a set of amendments to the Solid Waste Act of 1965, and it became commonly known as RCRA. Two years later, the U.S. Environmental Protection Agency (EPA) responded with proposed regulations which were partially rejected as too complex. After revisions, the regulations were finalized in May, 1980. Still further major amendments to RCRA were signed into law in 1980 and

1984. The amendments of November 1984 (Public Law 98-616) marked a strong departure from providing general guidance for regulatory implementation.

These latter amendments blurred the distinction between legislative and regulatory activities by including many "hammer" provisions which require specific regulatory actions according to a rigid schedule. Furthermore, these hammer provisions specify minimum requirements in the legislation itself that will supersede existing regulations in the event that the EPA does not adhere to the schedule.

The next section, 1.2.1, discusses the present regulatory framework and Section 1.2.2 discusses the 1984 RCRA amendments.

1.2.1 THE FRAMEWORK OF REGULATIONS UNDER RCRA

Regulations under RCRA are generic in the sense that they are written without reference to specific industries (i.e., they apply to all industries generating hazardous waste except for specific exclusions). The responsibility of hazard identification lies with industry. The regulations describe responsibilities of each of the primary hazardous-waste managers, i.e., generators, transporters, treaters, storers, and disposers. This chapter is organized in the same manner.

It is important to note that this chapter discusses only the federal regulations. Under RCRA, states have the option of developing their own hazardous-waste programs. State programs must meet or exceed the stringency of the federally mandated program. Most states have programs authorized by the EPA. Frequently these state programs are more stringent than RCRA because they control smaller quantities and wider variety of waste types than specified under federal law. While the federal regulations provide the minimum requirements, full compliance with the law requires conformity with applicable state regulations as well.

Generator Responsibilities and Waste Identification

The first task for any waste generator or handler is to determine whether the waste is hazardous. Waste classified as hazardous must meet two criteria. The first criterion applied is whether the material is a solid waste, and the second step is determining whether the solid waste is hazardous. Solid wastes include any liquid, solid, semisolid, or contained gas that is discarded or stored prior to discarding (Part 40 of the *Code of Federal Regulations* Subpart 261.2, i.e., 40 *CFR* 261.2). Frequently this involves chemical sampling and analysis to distinguish specific waste components. Sampling and analysis techniques are discussed in Chap. 13. The degree of regulatory control applying to a particular facility is determined by the precise amounts of waste generated and the waste-management methods used by that particular facility. Once material is determined to be solid waste, there are two general methods for determining whether the solid waste is hazardous. First, does the waste exhibit one of the four hazard characteristics?

- Ignitability
- Reactivity
- Corrosivity
- Extraction-procedure toxicity

These four characteristics are defined in 40 *CFR* 261. The second method of identification is to locate the waste in the lists of hazardous wastes in the following regulations:

- Nonspecific-source hazardous wastes—in 40 *CFR* 261.31
- Specific-source hazardous wastes—in 40 *CFR* 261.32
- Acutely hazardous wastes—in 40 *CFR* 261.33(e)
- Generally hazardous wastes—in 40 *CFR* 261.33(f)

The regulations also incorporate a gross degree of hazard into the stringency of control applied to hazardous wastes. For example, acutely hazardous wastes are regulated in quantities of 1 Kg (2.2 lb) per month or more. Other hazardous wastes, i.e., generally hazardous wastes, are fully regulated in quantities of more than 1,000 Kg (2,200 lb) per month. The 1984 amendments regulate most hazardous-waste generators producing between 100 Kg (220 lb) and 1,000 Kg (2,200 lb) per month. (Generator responsibilities are discussed later in this section under the heading Additional Generator Responsibilities.) Since many states have regulated smaller quantities and wider varieties of waste for several years, and authorized state programs supersede federal regulations, generators should consult individual state regulations to resolve any questions.

Generators must comply with guidelines for storage of their hazardous waste; otherwise they will be subject to storage-facility permit requirements (see *Treatment, Storage, and Disposal Facility Owner and Operator Requirements* on p. 1.13). These guidelines for allowable generator storage times vary depending on the amount of hazardous waste generated on a monthly basis. Generators of more than 1,000 Kg (2,200 lb) have 90 days before they are classified as storage facilities under RCRA regulations. Generators of between 100 and 1,000 Kg (220 and 2,200 lb) can store up to 180 days before they are required to have a storage permit. This class of smaller generators has an additional 90 days if they have to transport the waste off site farther than 322 Km (200 mi). On-site management involves one facility without transfers. Off-site management involves shipment to commercial management facilities.

Hazardous-waste regulations also address the issue of mixing hazardous wastes with nonhazardous or solid wastes. The amount of hazardous waste is defined in the regulations to include the total amount in any mixture of hazardous waste with solid waste, regardless of the resulting concentrations of hazardous waste. For example, 3.8 L (1 gal) of spent hazardous solvent mixed with a holding tank of 3,880 L (1,000 gal) of nonhazardous water results in 3,884 L (1,001 gal) of hazardous waste. The "mixture" rule is contained in 40 *CFR* 261.3(b)(2).

A number of special wastes and waste generators are excluded from RCRA requirements. They may be covered under other environmental laws and regulations. The regulation that lists the exclusions, 40 *CFR* 261.4, has four major exclusion categories with about 18 specific listings under them. The four major exclusion categories are (1) materials which are not solid wastes, (2) solid wastes that are not hazardous wastes, (3) hazardous wastes that are exempted from certain regulations, and (4) laboratory samples. The primary exclusion subcategories include:

- Domestic sewage
- Irrigation return flows
- Special nuclear or nuclear by-product material as defined by the Atomic Energy Act

- Mining overburden returned to the site
- Cement-kiln dust waste
- Fly ash, bottom ash waste, slag waste, and flue-gas emission-control waste from fossil fuel combustion
- Drilling wastes in oil, gas, and geothermal exploration

However, some exclusion categories are under review by mandate of the November 1984 amendments to RCRA. The trend is toward granting fewer exclusions.

A number of special wastes also deserve mention and are covered in other sections of this handbook. The first is waste oil. Waste oils are covered under the Used Oil Recycling Act of 1980 and the 1984 RCRA amendments. The EPA's proposed rule of March 10, 1986, regulates waste oils except those that are recycled. Waste oils are discussed more fully in Sec. 4.1. A discussion of infectious wastes, not regulated under RCRA, is found in Sec. 4.4. Polychlorinated biphenyls (PCBs) are regulated under the Toxic Substances Control Act in 40 *CFR* 760 and therefore preempted from regulation under RCRA. PCBs are the subject of Sec. 4.2. Dioxins are large groups of chemicals that are contaminants and by-products of certain combustion processes. The most publicized sources of dioxins are pesticide production and combustion processes. Dioxins are not specifically listed as RCRA hazardous waste. However, the 1984 RCRA amendments require a study of dioxin emissions from waste-to-energy facilities. Dioxins are covered in Sec. 4.3.

The final special waste entry is asbestos. While it is regulated under the Clean Air Act, Clean Water Act, Toxic Substances Control Act, Occupational Safety and Health Act, and Superfund Act, it is not an RCRA hazardous waste.

Recycled waste is either partially regulated or not regulated. The EPA asserts that some recycled materials are solid waste and that those come under the regulations. Other recycled materials are not solid wastes and, subsequently, are exempt from RCRA. In addition each state with an authorized hazardous-waste program may add its own interpretation of what constitutes recycling. Generally, the EPA deems recycling a preferred management method and places fewer regulatory controls on recycling than other methods. For a complete discussion of the reasoning of the federal recycling regulations for hazardous waste see the following *Federal Register* citations: 50 *FR* 614, January 4, 1985, and 50 *FR* 49164, November 29, 1985. The codified regulations can be found in 40 *CFR* 266.

Additional Generator Responsibilities

Once an establishment determines that its waste is regulated, a number of responsibilities follow. The first is to obtain an EPA identification number. This unique number is used systematically to track waste from generation to disposal through the manifest and biennial report. Each generator must fill out a multicopy form, called a manifest, which accompanies waste upon leaving the facility and continues with it to its ultimate disposal. Manifests are essential to RCRA's control strategy. The tracking system for manifests includes the EPA identification number; name and address of transporter, treater, storer, or disposer; and a description of the wastes, including type, quantities, and official classifications. States require from four to eight copies of the manifest for each shipment. Copies of the form are sent back to the generator from the transporter upon completion of delivery. If a generator has not received a copy of the manifest within 35 days of initial dispatch, then the generator is required to investigate. After 45 days without confirmation of delivery, the generator must file an "exception report" with the appropriate regional EPA office. Finally, a provision mandated by a 1984 amendment to the RCRA requires that manifests also contain certification that the generator is minimizing hazardous waste as much as economically practical and that the chosen method of disposal minimizes risk to human health and the environment.

Furthermore, generators must keep all records of hazardous-waste activity, including manifests, for three years and submit biennial reports that summarize on-site and off-site hazardous-waste activity during the previous calendar year.

Transporter Responsibilities

Transporters of hazardous waste by highway, rail, water, or air are required to obtain EPA identification numbers. Hazardous-waste transporters are not allowed to accept shipments of hazardous waste without completed manifests. Transporters must use only facilities with RCRA permits (see *Treatment, Storage, and Disposal Facility, Owner and Operator Requirements* on p. 1.13). The transporter cannot store waste for more than 10 days without becoming subject to storage facility regulations. Also, transporters that mix hazardous waste may be subject to generator regulations [see 40 *CFR* 263.10(c)(2)].

Transporters are subject to joint EPA and Department of Transportation regulations concerning spilling and reporting of spills. They are required to notify the National Response Center within 24 hours when the spill exceeds reportable quantities (see 40 *CFR* 302). Reportable quantities range from 1 lb (0.45 Kg) to 5,000 lb (2,250 Kg). If a spilled substance is not listed, then the default reportable quantity equals 1 lb (0.45 Kg). Lastly, transporters must retain their records for a minimum of three years.

Treatment, Storage, and Disposal Facility Owner and Operator Responsibilities

Treatment, storage, and disposal facilities (TSDF) are defined in 40 *CFR* 260.10 as follows:

- *Treatment:* "Any method, technique, or process, including neutralization, designed to change the physical, chemical, or biological character composition of any hazardous waste so as to neutralize it, or render it nonhazardous or less hazardous or to recover it, make it safer to transport, store, or dispose of, or amenable for recovery, storage, or volume reduction"
- *Storage:* "The holding of hazardous waste for a temporary period, at the end of which the hazardous waste is treated, disposed, or stored elsewhere"
- *Disposal:* "The discharge, deposit, injection, dumping, spilling, leaking, or placing of any solid waste or hazardous waste into or on any land or water so that any constituent thereof may enter the environment or be emitted into the air or discharged into any waters, including groundwaters" (40 *CFR* 265.1)

Because of the variety of facility types and the number of potential environmental contamination routes, the regulations that concern these facilities are complex, with many different conditions. In general, all TSDFs are regulated except for the following:

- Farmers disposing of pesticides from their own use
- Owners and operators of totally enclosed treatment facilities
- Owners and operators of a neutralization unit or a wastewater-treatment plant
- Persons responding to a hazardous-waste spill or discharge
- Facilities that reuse, recycle, or reclaim hazardous waste
- Generators accumulating waste within the allowed time periods
- Transporters storing manifested wastes less than 10 days[1]

The regulations for owners and operators of TSDFs can be divided into two categories: standards of performance and permit requirements. Performance standards are contained in 40 *CFR* 264 and 265. The performance standards in 40 *CFR* 265 are interim standards for facilities in existence before November 19, 1980. These interim standards are intended to allow for gradual compliance of older facilities. The regulations in 40 *CFR* 265 are generally classified as "good housekeeping" procedures. These older TSDFs will eventually be expected to comply with the full performance standards in 40 *CFR* 264. The performance standards in 40 *CFR* 264, for TSDFs built after November 19, 1980, contain design and operating criteria. The permit requirements contained in 40 *CFR* 270 are generally based on the performance standards in 40 *CFR* 264.

Both interim and full performance standards begin with general provisions that apply to all TSDFs (40 *CFR* 264 and 265 Subparts A through H). The remainder of each of the performance standard regulations, 40 *CFR* 264 and 265, is devoted to provisions for specific facility types. Both interim and full performance standards cover containers, tanks, surface impoundments, waste piles, land treatment, landfills, and incinerators. Underground-injection, thermal treatment, and chemical, physical, and biological treatment facilities are covered only in 40 *CFR* 265. The general provisions address the following issues:

- *Waste analysis* to ensure accurate information concerning properties of the waste for adequate treatment, storage, or disposal
- *Security measures* to prevent unauthorized entry with resultant injury or environmental damage
- *Inspections* to assess the facility and its potential problems
- *Training* of facility personnel to reduce mistakes that could threaten human health, safety, and the environment
- *Compliance with location standards* that include floodplain, earthquake, and hydrological considerations

For a complete discussion of these requirements see 40 *CFR* 264.10 through 264.18. Other general provisions cover the following areas:

- Preparedness and prevention, 40 *CFR* 264.30–264.37
- Contingency plan and emergency procedures, 40 *CFR* 264.50–264.56
- Manifest system, record keeping, and reporting, 40 *CFR* 264.70–264.77
- Groundwater monitoring, 40 *CFR* 264.90–264.94
- Closure and postclosure, 40 *CFR* 264.110–264.120

• Financial requirements, including liability insurance requirements, 40 *CFR* 264.140–264.178

The remaining sections in 40 *CFR* 264 and 265 pertain to specific TSDF types. All the facility-specific regulations for land disposal facilities emphasize groundwater protection. This emphasis is clearly seen in the extensive design criteria and location standards that apply to all types of TSDFs. For example, new landfills are required to install double liners and leakage-collection systems. They must also carefully monitor groundwater quality to ensure that these protective systems are effective, and they must take corrective action should the protective systems fail.

TSDFs have a number of significant responsibilities under the regulations covering closure and postclosure and financial requirements. Closure occurs when wastes are no longer accepted for treatment, storage, or disposal. Postclosure regulations apply only to disposal facilities during the period of time after closure; they require owners and operators to conduct maintenance and monitoring to ensure the integrity of the facility. The financial requirements are established to ensure that the TSDF owners will be able to pay for the stringent closure and postclosure requirements.

Enforcement of these detailed performance standards is based on the Subtitle C permit requirement. Subtitle C is the part of RCRA dealing with hazardous waste. There are three exceptions to this rule. One is for TSDFs that have approved permits under the Safe Drinking Water Act (Underground Injection Control permit), the Clean Water Act (National Pollution Discharge Elimination System permit), or the Marine Protection, Research and Sanctuaries Act (Ocean Dumping permit). They can obtain an RCRA permit-by-rule. Permits-by-rule are based on the similarity of the permitting requirements of these three regulations. Another permit exception is made for emergencies with "imminent and substantial endangerment of human health or the environment." The third exception covers test periods for new incinerators or land treatment facilities.

The permit process is divided into two stages: acquisition of Part A and Part B permits. Part A permits are interim permits for existing facilities. Facilities with interim permits will eventually be required to apply for full permits (i.e., Part B permits). New facilities must simultaneously apply for Part A and Part B permits. Part A permits contain general information and a description of hazardous-waste activities at the site. Part B permits contain extensive technical requirements, described in 40 *CFR* 264 and 270. Part A permits serve as interim permits for many existing facilities until the EPA requires submission of applications for full Part B permits. The 1984 RCRA amendments have set a schedule for submission of Part B by TSDF type.

RCRA's Relation to CERCLA

The Comprehensive Environmental Response Compensation and Liability Act (CERCLA, also referred to as Superfund) provides the EPA with authority to clean up groundwater contamination due to active or inactive dump sites. This act also places strict and several liabilities on responsible parties associated with the contamination. The potential liabilities from Superfund actions are a major motivation for present generators to comply with all hazardous-waste regulations. However, full compliance with all RCRA regulations is no defense against liability under actions under CERCLA.[2]

RCRA is a preventative regulatory policy designed to cover hazardous-waste management. CERCLA's mission is to remedy problems caused by past mismanagement of hazardous waste, so it is a remedial policy. There are three general areas of interaction between provisions of these two hazardous-waste laws. The three areas of overlap are the following: (1) disposal of CERCLA site wastes, (2) reactions to imminent hazards, and (3) corrective actions for hazard-substance releases from TSDFs. These are briefly discussed below.

For the most part, hazardous waste from cleanup operations at CERCLA sites must be managed in accordance with applicable RCRA requirements. CERCLA wastes taken off site must be handled in accordance with the generator and transporter regulations. CERCLA wastes treated, stored, or disposed on site or off site must comply with TSDF regulatory provisions. CERCLA waste transported off site needs a manifest.

Circumstances where there is an imminent hazard from past and present TSDFs are regulated under Sections 7003 of RCRA and 106 of the Superfund act. They both require responsible persons to take necessary cleanup actions for situations posing imminent hazards to groundwater. RCRA regulations are more comprehensive than those of CERCLA because they cover both hazardous-waste and solid-waste imminent hazards. On the other hand, CERCLA has the authority to place responsibility on past users of the TSDF, as well as on the past owners. In general, Superfund's more comprehensive authority usually makes it the prime authority in imminent-hazard cases. RCRA's authority in these cases is usually used to bolster the government's action.

The final direct connection between RCRA and CERCLA is in the area of corrective actions. The corrective-action requirements cover releases from hazardous-waste and solid-waste facilities. The EPA has stated that it will require corrective actions when they are necessary to protect human health and the environment. Furthermore, corrective actions can range beyond facility boundaries where needed. Until the 1984 RCRA amendments, CERCLA provided the only authority for EPA to require corrective action in response to a release of hazardous waste to the environment. The 1984 RCRA amendments allow the EPA to require TSDFs, either through the permit process or administrative order, to take steps ranging from remedial site investigations to corrective actions. This applies to TSDFs that received a permit after November 8, 1984. However, the regulations issued under this amendment also cover waste piles, landfills, surface impoundments or land treatment facilities that were active after July 26, 1982, or closed after January 26, 1983. In contrast, prior to the amendments, TSDFs were required only to give financial assurance of meeting proper closure and postclosure requirements. This is one of the major provisions of the 1984 RCRA amendments which are discussed in Sec. 1.2.2.

1.2.2   1984 RCRA AMENDMENTS

This section begins with an overview of the November 1984 RCRA amendments (Public Law 48-616), sometimes called the Hazardous and Solid Waste Amendments (HSWA). The following section summarizes them by discussing their general nature and categorizing them into eight subject areas. These eight categories are then discussed individually.

Overview

The 54 amendments to RCRA in the November 1984 act cover a wide range of topics concerning hazardous and solid waste. As the government tightens controls over hazardous waste, the distinction between solid and hazardous wastes becomes less definite. These amendments sharpen the regulatory thrust of RCRA, and contain many specific provisions. As previously mentioned, there are several provisions that contain detailed regulatory requirements and schedules that go into effect upon EPA's failure to meet the legislative deadlines. These are called "hammer" provisions. More specifically, the 1984 RCRA amendments contain 89 specific deadlines. Of the 89 deadlines, 40 are for industry and non-EPA agencies and 49 pertain to the EPA. As of 1986 about 42 of the deadlines had passed. Several of these deadlines have been met, while others have not. Some of the slipped deadlines involve interagency disputes with the Office of Management and Budget over the costs of proposed regulations. For specific deadline references see the original text of 1984 RCRA amendments.

In the preamble to some regulations of July 15, 1985, the EPA estimated that the first-year costs of the requirements in the 1984 RCRA amendments would range between $1.9 and $14.5 billion. Another EPA estimate states that the regulated community will increase from about 60,000 to 2,000,000 with the 1984 RCRA amendments' addition of small-quantity generators and underground storage tank operators.[4] These speculative estimates are based on many assumptions concerning the resolution of controversial issues current in 1987. A discussion of the future direction of hazardous-waste regulation is provided later in Sec. 1.2.

The sheer number of amendments is overwhelming at first glance. Table 1.2.1 presents a list of the amendments by number and a shortened version of each one's title. For example, Amendment 201 covers Land Disposal of Hazardous Waste while number 102 discusses Dioxins from Resource Recovery Facilities. Table 1.2.1 also serves as a reference guide for Table 1.2.2, which loosely categorizes the amendments for the purpose of this discussion.

There are seven major categories in Table 1.2.2, including one, labeled "Hazardous Waste," which has eight subcategories. The category name is provided in the left-hand column, the amendment number is provided in the middle column, and the right column references the section of RCRA to be amended (Public Law 94-580, which replaced the previous language in the Solid Waste Disposal Act). The categories in Table 1.2.2 are not intended to be exhaustive nor are they for use outside the context of this discussion. However, these categories provide a convenient format for discussion of the 1984 RCRA amendments. The eight subcategories of hazardous-waste amendments will be covered under *Hazardous Waste Regulation Categories* on p. 1.21. The six other major categories are reviewed here.

Administrative Amendments. This category includes titles, budget authorizations, findings and objectives, several state authorization amendments, an effective-date provision, and technical and clerical amendments. The most significant amendment in this group is Section 226, which requires states to make information concerning TSDFs of hazardous waste available to the public to have an authorized program. Prior to the 1984 amendments states were required to divulge only a permit applicant's name and address.

Financial Amendments. This category contains two financial responsibility amendments. Section 205, Direct Action, delineates the financial responsibilities of a guarantor, i.e., a person, other than the owner or operator, who provides

TABLE 1.2.1   1984 RCRA Amendments

Amendment number	Amendment short title
1	Short Title and Table of Contents
2	Authorizations for Fiscal Years 1985–1991
101	Findings and Objectives
102	Dioxins from Resource Recovery Facilities
103	Ombudsman
201	Land Disposal of Hazardous Waste
202	Minimum Technological Requirements
203	Ground Water Monitoring
204	Burning and Blending of Hazardous Wastes
205	Direct Action
206	Continuing Releases at Permitted Facilities
207	Corrective Action beyond Facility Boundaries Underground Tanks
208	Financial Responsibility for Corrective Action
209	Mining Waste and Other Special Wastes
211	Authority for Permit to Construct Hazardous Waste Treatment, Storage, or Disposal Facilities
212	Permit Life
213	Interim Status Facilities
214	New and Innovative Treatment Technologies
215	Existing Surface Impoundments
221	Small Quantity Generator Waste
222	Listing and Delisting of Hazardous Waste
223	Clarification of Household Waste Exclusion
224	Waste Minimization
225	Basis of Authorization
226	Availability of Information
227	Interim Authorization of State Programs
228	Application of Amendments to Authorized States
229	Federal Facilities
230	State-Operated Facilities
231	Mandatory Inspections
232	Federal Enforcement
233	Interim Status Corrective Action Orders
234	Effective Date of Regulations
241	Management of Used Oil
242	Recovery and Recycling of Used Oil
243	Expansion During Interim Status
244	Inventory of Federal Agency Hazardous Waste Facilities
245	Export of Hazardous Waste
246	Domestic Sewage
247	Exposure Information and Health Assessments
301	Size of Waste-to-Energy Facilities
302	Subtitle D Improvements

TABLE 1.2.1 1984 RCRA Amendments (*Continued*)

Amendment number	Amendment short title
401	Citizen Suits
402	Imminent Hazard
403	Enforcement
404	Public Participation in Settlements
405	Interim Control of Hazardous Waste Injection
501	Use of Recovered Materials by Federal Agencies
502	Technical and Clerical Amendments
601	Underground Storage Tank Regulation
701	Report to Congress on Injection of Hazardous Waste
702	Extending the Useful Life of Sanitary Landfills
703	Uranium Mill Tailings
704	National Ground Water Commission

Source: Beveridge and Diamond, "Summary and Analysis of the Hazardous and Solid Waste Amendments of 1984," *Hazardous Waste and Hazardous Materials Journal,* Mary Ann Liebert, vol. 2, no. 1, spring 1985, pp. 114 and 115.

TABLE 1.2.2 Categorization of the 1984 RCRA Amendments

Amendment category	Amendment number	Codified RCRA section number
Administrative	1, 2, 101, 103, 225, 226, 227, 228, 234, 502	2007, 2007, 1002, 2008, 3006, 3006, 3006, 3006, 3010, 3007
Financial	205, 208	3004, 3004
Enforcement	229, 230, 231, 232, 233, 245, 401, 402, 403, 404	3007, 3007, 3007, 3008, 3008, 3017, 7002, 7003, 7003, 7003
Solid waste	301, 302	4001, 4010
Reports	102, 244, 246, 701, 702, 704	1006, 3016, 3018, Title VII, 8002, Title VII
Miscellaneous	501	6002
Hazardous waste		
Land disposal	201, 202, 203, 215	3004, 3004, 3004, 3005
Permits	206, 211, 212, 213, 243, 247	3004, 3005, 3005, 3005, 3015, 3019
Underground storage tanks	207, 601	3004, 9003
Technology	214, 224	3005, 3005
Mining	209, 703	3004, Title VII
Small-quantity generators	221, 223	3001, 3001
Waste oil	241, 242	3014, 3014
Miscellaneous	204, 222, 405	3004, 3001, 7010

evidence of financial responsibility for an owner or operator of a hazardous-waste facility. The most important aspect of this amendment is to provide claimants with direct access to guarantors, such as insurance companies. The other financial amendment, Section 208, requires the EPA to set standards for evidence of financial responsibility by hazardous-waste TSDFs for corrective actions required under Section 207 (discussed in under *Hazardous-Waste Regulation Categories* on p. 1.21).

Enforcement Amendments.   The first three amendments in the enforcement category, Sections 229, 230, and 231, essentially cover inspection-activity schedules of federal, state, and private hazardous-waste facilities respectively. The remaining sections under this category pertain to a variety of enforcement issues. The short-term EPA enforcement priorities include facilities that:

- Present an immediate threat to public health or the environment
- Lack or have inadequate groundwater monitoring
- Require corrective actions
- Fail to comply with closure and postclosure requirements
- Qualify as new facilities
- Qualify as federal facilities[5]

Section 232 clarifies prohibited actions; this section covers remedial actions, penalties, and the statutory definition of knowing endangerment. Section 233 authorizes the EPA to require corrective actions of environmental releases of hazardous waste from interim-status TSDFs.

The next section listed, 245, regulates the export of hazardous waste. Provisions under Section 245 include notification and consent of the receiving country.

Sections 401 through 404 address enforcement aspects of imminent hazards. The wording on citizen suits, under Section 401, gives private citizens standing to sue generation transporters and TSDF owners and operators in cases of imminent and substantial endangerment. Section 402 clarifies the imminent hazard from releases of hazardous and solid wastes (as discussed under RCRA's Relation to CERCLA), covering topics from the definition of "imminent hazard" to extension of liability to past generators. Sections 403 and 404 contain procedures for notification of the EPA in imminent-hazard cases.

Solid-Waste Amendments.   The only amendments solely related to solid waste are in Sections 301 and 302. Section 301 requires states to incorporate consideration of recycling in their comprehensive state solid-waste plans that include waste-to-energy facilities. This provision was written foreseeing the potential conflict created by the need of waste-to-energy facilities for recyclable paper, which has good combustion properties. The other solid-waste amendment, 302, requires the EPA to study the adequacy of current criteria for solid-waste management facilities. Furthermore, it requires the agency to revise the facility criteria for solid-waste facilities, hazardous waste from small-quantity generators, and households by March 1988.

Reports Amendments.   The next six amendments listed in Table 1.2.2 require the EPA to submit reports, along with some mandatory issuance of regulations to follow. They pertain to a number of disparate areas. The first amendment, Section

102, requires a report describing the emission of dioxins from municipal solid-waste-burning waste-to-energy facilities. Section 244 requires an inventory of all past and present federally owned hazardous-waste TSDFs. Section 701 also requires the EPA and the state to develop an inventory of injection wells used for hazardous-waste disposal.

Section 246 of the amendments requires EPA to report after investigating the issue of hazardous wastes in publicly owned wastewater-treatment works (POTWs). Congress wants the EPA to determine whether the POTW exemption under present RCRA regulations is warranted. Section 702 requires the EPA to study and produce a series of reports on methods of extending the useful life of solid-waste landfills. The first report was submitted to Congress in October 1986. The others followed by early 1987. The final report amendment, Section 704, establishes the National Groundwater Commission. This amendment delineates membership on the commission as well as its responsibilities.

Miscellaneous Amendment. The single amendment falling into the miscellaneous category is Section 501, pertaining to federal procurement policies. This amendment applies to all federal agencies that procure materials that are recyclable. It requires these agencies to examine present policies and change them to maximize use of recovered materials within the guidelines produced by the EPA.

Hazardous-Waste Regulation Categories

The 23 amendments that deal primarily with hazardous waste are divided for the purposes of our discussion here into the eight subcategories that follow.

Land Disposal. One of the most important intentions of the 1984 RCRA amendments is to tighten restrictions on land disposal of hazardous wastes and promote other, more environmentally sound, waste-handling methods. Many of the 23 amendments in the major category of hazardous waste are relevant to various aspects of land disposal. However, only four deal primarily with land disposal. Section 201, a very detailed amendment, defines land disposal to include landfills, surface impoundments, waste piles, injection wells, land treatment facilities, salt-dome formations, salt-bed formations, and underground mines or caves. It contains many schedules for prohibiting certain waste types from specified facilities. One of the most dramatic provisions of Section 201 is the schedule set for the EPA to review all listed RCRA hazardous wastes for possible land disposal prohibition. The schedule requires EPA review of the first third of listed wastes that are high-volume and or high-hazard by August 1988. The second third of listed wastes are to be reviewed by June 1989, and the final third, all listed wastes, by May 1990. Other specific issues specified in Section 201 include:

- Placement of nonhazardous liquid wastes in hazardous landfills
- Placement of hazardous waste in salt-dome and salt-bed formations, underground mines, and caves (prohibited before standards are finalized)
- Disposal of free liquids in landfills
- Monitoring and controlling of air emissions from hazardous TSDFs
- Environmental soundness of deep-well injection disposal
- Barring the use of dioxin-contaminated wastes as dust suppressant

Section 202, entitled Minimum Technological Requirements, primarily concerns land disposal. This amendment specifies double liners and leachate-collection systems for newly constructed landfills or expansions of existing landfills. It also specifies liner requirements for other land disposal facilities. Furthermore, groundwater monitoring is specified for land disposal facilities. This section sets a schedule for the EPA to incorporate these provisions into the permitting process. Section 202 also contains a provision requiring hazardous-waste incinerator permits to specify destruction levels and removal efficiencies. Finally, this amendment addresses location standards for new and existing TSDFs.

Section 203 specifies that groundwater monitoring requirements apply to all landfills, surface impoundments, waste piles, and land treatment facilities, regardless of location and groundwater protection measures. However, Congress did allow for a few exceptional circumstances as grounds for a variance.

The last land disposal amendment pertains to existing surface impoundments. It generally mandates that facilities meet minimum technological requirements for new facilities, specified under Section 202. There are three grounds for exceptions to this requirement; possible exceptions are for landfills:

1. Having one liner and no evidence of leakage
2. Located farther than 0.4 km (0.25 mi) from an underground drinking water source
3. Being in full compliance with groundwater-monitoring requirements under its permit

Any one of the above may be grounds for an exemption. These exempted surface impoundments must meet closure standards. Furthermore, this amendment modifies conditions for surface impoundments used for prohibited wastes under Amendment 201.

Permits.   Section 206 specifies that TSDFs permitted after November 1984 are required to take corrective action for any releases, regardless of whether the facility is active or inactive. This amendment does allow TSDFs to obtain a permit for a continued release, if they schedule corrective actions and provide financial assurance for the corrective actions. Section 211 clarifies the overlapping jurisdiction of RCRA and the Toxic Substance Control Act (TSCA) concerning incineration of polychlorinated biphenyls (PCBs). It allows incineration of PCB without an RCRA permit if the facility has appropriate approvals under the Toxic Substances Control Act (TSCA). The next permit amendment, 212, fixes permit life for TSDFs at 10 years. It also contains a requirement for a five-year review process for land disposal facilities. These reviews must include consideration of changes in technologies relevant to these facilities. The next two permit amendments address interim-status TSDFs. Section 213 requires termination of interim status for facilities that have not submitted full permit applications or are not in compliance with groundwater-monitoring requirements. This amendment is designed to speed up the permitting process that Congress judged to be too slow. It also requires the EPA to process permits in order of facility priority. The priorities are shown in the required schedule, which lists land disposal facilities first, incinerators next, and all other TSDFs third. The other amendment covering interim-status facilities, Amendment 243, requires minimum technological requirements for expansions of these facilities or for that part of the facility receiving waste after May 1985.

The last amendment pertaining to permits, Amendment 247, requires landfill and surface-impoundment permit applications after August 1985 to contain information concerning the potential public exposure to hazardous waste from the facility. Previously submitted permits will also have to be amended to comply with this rule. In a July 15, 1985, *Federal Register* Notice, the EPA did not make this information requirement mandatory for issuance of the permit. This provision is very significant because of the added effort required for compliance and its potential for raising public awareness.

Underground Storage Tanks. Section 601 adds a new program to RCRA covering underground storage tanks. The targets of this program are tanks that leak petroleum products (except heating oil and substances that are regulated under Superfund). Hazardous substances covered under RCRA are specifically exempt from these underground storage tank provisions because they are already regulated. This amendment also specifies rigid schedules for notification, for leak detection and correction or control, and for standard setting for new tanks. Furthermore, it specifies underground tanks as tanks that are physically a minimum of 10% underground. It also mentions requiring the EPA to specify financial responsibility standards, but does not set a deadline.

This amendment involves a very large number of small establishments, including federal underground storage facilities. Along with the small-quantity generator provisions, it substantially enlarges the RCRA regulated community. There are, however, a number of significant exemptions including farm and residential tanks of less than 4,260 L (1,100 gal), septic tanks, pipeline facilities, and others. There are also a number of studies required of the EPA under this provision. The other underground storage tank provision relates to the extent of corrective actions taken by owners and operators. Section 207 requires corrective actions to go beyond property boundaries to protect human health and the environment. These corrective actions require consent of the property owners. It also requires the EPA to develop and finalize regulations for underground tanks that cannot be entered for inspection.

Technology. Section 214 authorizes the EPA to issue one-year permits for experimental facilities for a maximum total duration with renewals of four years. The congressional intent was that this provision apply to non-land-related storage and treatment methods of pilot scale. The regulations generally define pilot-scale facilities as managing less than 15,000 kg per month.[7]

Section 224 concerns use of waste-minimization techniques. This amendment is not strictly related to technological methods, but is strongly related to selection of technology for waste-management purposes. It requires all hazardous-waste generators to certify on all manifests that they are reducing their wastes as much as economically practical and that their chosen method of treatment, storage, and disposal minimizes human health and environmental impacts. Certification will be confirmed by the presence of a signature in the appropriate space on the manifest. TSDFs must also certify to the above provisions to qualify for a permit. Both of the certification requirements took effect September 1, 1985. Furthermore, the provision requires biennial reports to identify source-reduction efforts. This identification includes a comparison of past generation patterns.

Lastly this amendment requires the EPA to report on the desirability and feasibility of waste reduction by all generators.

Mining. Section 209 authorizes the EPA to modify certain requirements for land disposal facilities to accommodate certain mining wastes, fly-ash waste, bottom-ash waste, slag waste, flue-gas emission-control waste, or cement-kiln dust waste, if they come under RCRA requirements. The other mining amendment merely states that the 1984 RCRA amendments do not alter any provisions of the Uranium Mill Tailings Radiation Control Act of 1978.

Small-Quantity Generators.   Section 221, Small Quantity Generators Waste, requires the EPA to regulate generators of between 100 and 1,000 kg (220 and 2,200 lb) of generally hazardous waste per month. Prior to the 1984 RCRA amendments only generators of amounts greater than 1,000 kg (2,200 lb) per month came under regulation. The EPA issued regulations on March 24, 1986, under Section 221 of the RCRA amendments that will bring this new group of hazardous-waste generators under almost full regulation. There are an estimated 175,000 small-quantity generators that have come under regulation. Many of these generators are in the service sector, e.g., dry cleaners. The regulated minimum of 100 kg (220 lb) per month is equivalent to about one-half of a 55-gal drum, or about 213 L. Consequently, the EPA is conducting a large educational program to raise private industry awareness of these new regulations. The only regulatory exceptions made for small-quantity generators are for longer allowable storage periods before they are subject to TSDF permitting requirements and some reduced reporting requirements, such as not filing exception reports.

The other provision covering small-quantity generators pertains to household hazardous waste. Section 223 clarifies the exemption of waste-to-energy facilities from hazardous-waste regulations. Operators of these facilities are exempt if they establish procedures to prevent combustion of household and small-quantity generator hazardous waste in their facilities.

Waste Oil.   The EPA has issued proposed regulations for regulating waste oil as an RCRA hazardous waste as prescribed by Section 241. Used motor oil that will be recycled is exempt. This amendment potentially covers a large number of small businesses. The provision directs special attention to assessing the impact of regulatory controls on this segment. Amendment 242 clarifies congressional intent concerning control of waste oil as a hazardous waste. Specifically, it states that recycling regulations should not discourage recovery and recycling but should protect human health and the environment.

Miscellaneous.   There are three amendments that are lumped into the miscellaneous category. Section 222 has very broad implications, while the other two amendments are specific in nature. Section 222, Listing and Delisting of Hazardous Waste, specifically directs the EPA to list additional wastes containing chlorinated dioxins and chlorinated dibenzofurans and other halogenated dioxins and dibenzofurans where appropriate. Other wastes that must be explicitly considered for listing include:

- Chlorinated aliphatics
- Dioxin
- Dimethyl hydrazine
- Toluene diisocyanate (TDI)
- Carbamates
- Bromacil
- Linuron
- Organobromines

- Solvents
- Refining wastes
- Chlorinated aromatics
- Dyes and pigments
- Inorganic chemical industry wastes
- Lithium batteries
- Coke by-products
- Paint production wastes
- Coal-slurry pipeline effluent

This amendment also directs development of additional hazardous-waste characteristics, including measures of toxicity. These characteristics would be added to the four characteristics contained in 40 *CFR* 261. The EPA is also required to correct the deficiencies in the Extraction Procedure Toxicity test as a predictor of the leachate potential of wastes. Finally, this amendment contains scheduled requirements for the EPA processing of delisting applications and spells out public commenting procedures. Section 204 addresses the producers, burners, distributors, and marketers who blend hazardous wastes as a fuel. It requires these persons to notify the EPA and states that they will eventually be subject to technical standards. Single- and double-family residences are exempt from notification and record-keeping requirements. This amendment goes further by requiring labeling of fuels blended with hazardous waste. Other areas covered by this section include petroleum refineries, de minimis quantities of hazardous waste in fuels, and permit requirements for cement kilns located in urban areas burning hazardous wastes. De minimis quantities are quantities that are considered "insignificant." The last amendment, Section 405, bans the underground injection of hazardous waste into a geologic formation within 0.25 mi (0.4 km) of an underground drinking source after May 1985.

The 1984 RCRA amendments refine the regulatory controls over hazardous-waste management. In total, they address many different aspects because many different problems have arisen during the first eight years of the administration of the original law. This section clearly shows the pertinence of these amendments to persons involved in any way with the handling of hazardous waste.

1.2.3 FUTURE DIRECTIONS OF HAZARDOUS-WASTE REGULATION

The 1984 RCRA amendments attempt to close the regulatory loopholes in the "cradle to grave" network of controls covering hazardous-waste management. Their breadth and specificity underscore the dramatic changes the RCRA amendments will have on hazardous-waste management practices. This is especially true for the eight-year period from 1984 to 1992. There are numerous deadlines during these years for the hazardous-waste management industry and its regulators.

A series of "hammer" provisions in the 1984 RCRA amendments delineates presumptive prohibitions with minimum prescriptive requirements. The EPA's record to date for meeting the hammer provision deadlines has been mixed. Its ability to meet the remaining deadlines depends on the availability of resources to the EPA. Furthermore, the EPA's ability to meet the deadlines depends on the number and intensity of the conflicts over regulatory strategies it has with the Office of Management and Budget (OMB). The OMB has authority through Executive Order 12291 to review regulations by executive agencies.

With so many different provisions, it is difficult to identify the most important amendments. However, one of the most prevalent themes throughout the 1984 Amendments is the promotion of management options that minimize land disposal. This is true for both hazardous-waste and solid-waste management.

The policy of shunning land disposal in favor of incineration and other methods gives rise to a complex process subject to many influences. One factor is the complexity of the permitting process for land disposal facilities, with closure and postclosure costs, that encourages more incineration and other treatment facilities. Additionally, the high cost and uncertainty of environmental liability insurance encourages incineration because the waste is "destroyed." Destroyed in this context means visibly destroyed. Liability with land disposal extends for many years after initial deposition in the facility. Consequently, some industry segments favor incineration. There are a number of important deadlines and regulations to monitor in the coming years. Their promulgation and implementation will have significant impacts on hazardous-waste management. Five of the most far-reaching amendment areas are:

- *Land disposal ban for listed substances:*   Congress is requiring the EPA to review the top third of listed wastes that are high-volume and high-hazard for possible prohibition from land disposal by August 1988. It further requires the EPA to review the second third of listed wastes by June 1989 and the last third of listed wastes by May 1990.

- *Permit schedule:*   Congress has mandated the full permitting of TSDFs and started with land disposal facilities. As of the November 1985 deadline there were many missing permit applications from the initial list of establishments that applied for interim-status permits. This is now under study to ascertain the extent and nature of the problems.

- *Adding a toxicity characteristic:*   Adding another characteristic to the present list of four will make it possible to form a more comprehensive regulatory network to identify hazardous wastes.

- *Corrective action:*   Congress requires permits granted after enactment of the 1984 RCRA amendments to contain provisions for waste-pile, landfill, surface-impoundment, and land treatment facilities to take corrective actions for environmental releases. This provision covers past and present releases. It closes a loophole in liability left unregulated between RCRA and Superfund.

- *Small-Quantity Generators and Underground Storage Tanks:*   This newly regulated community could range as high as 2,000,000 establishments. The EPA will require a large degree of voluntary compliance with these provisions because of the exorbitant amount of agency resources required to force compliance. Mobile treatment units for hazardous waste, now an issue under Superfund, many have significant advantages for these smaller establishments.

Depending on how these provisions are promulgated and enforced, hazardous-waste management in the United States could change dramatically over the next 10 years.

This chapter provides a framework that allows professionals in the field to place the methods and techniques described in the rest of this handbook in the context of RCRA regulations. This chapter's discussion clearly indicates the complexity of even the present federal regulatory coverage and its dynamic nature. A description of the many state programs would take several volumes—and hazardous-waste practitioners will have to comply with applicable state hazardous-waste regulations. This chapter can do no more than provide a basis for understanding individual state provisions.

The implementation of the 1984 RCRA amendments marks the next phase of federal hazardous-waste controls. One certainty is continued change as the program evolves to try to equitably and efficiently protect human health and the environment.

1.2.4   REFERENCES

1. U.S. Environmental Protection Agency, *RCRA Orientation Manual,* EPA/530-SW-86-001, Washington, D.C., January 1986.

2. J. Quarles, *Federal Regulation of Hazardous Wastes: A Guide to RCRA,* Environmental Law Institute, Washington, D.C., 1982, p. 183.

3. Ibid, p. 167.

4. *Environment Reporter,* Bureau of National Affairs, Washington, D.C., March 7, 1986, p. 2038.

5. *Environment Reporter,* Bureau of National Affairs, Washington, D.C., November 1, 1985, p. 1163.

6. McCoy Associates, *1985 Index to Hazardous Waste Regulations,* McCoy and Associates, Lakewood, Colorado, 1985, p. 147.

7. Ibid.

1.2.5   FURTHER READING*

Beveridge and Diamond, "Summary and Analysis of the Hazardous and Solid Waste Ammendments of 1984," *Hazardous Waste and Hazardous Materials Journal,* vol. 2, no. 1, Spring 1985.

Brown, M., and W. Truitt, "EPA Development of New RCRA Regulations, Guidance Brings Understanding of Far-Reaching Effects of 1984 Amendments," *Environment Reporter,* Bureau of National Affairs, Washington, D.C., pp. 1154–1163.

Code of Federal Regulations, Part 40, Subparts 260–270 (40 *CFR* 260–270). Updated annually.

Greenberg, M., and R. Anderson, *Hazardous Waste: The Credibility Gap,* Center for Urban Policy Research, New Brunswick, New Jersey, 1984.

Resource Conservation and Recovery Act of 1976, Public Law 98-616, November 8, 1984.

Note: Legislation is found in the *U.S. Code Annotated* or bound separately. Each version of the legislation has a different numbering system for each section. Regulations derived from legislation are first bound in the *Federal Register (FR).* This contains valuable information concerning reasoning and background information for the regulations. The *Federal Register* will contain proposed versions of regulations for public comment. Once the finalized version is published in the *Federal Register,* the regulation is published in the *Code of Federal Regulations (CFR).* The *CFR* is published annually in July.

SECTION 1.3

SUMMARY OF CERCLA LEGISLATION AND REGULATIONS AND THE EPA SUPERFUND PROGRAM

Richard Stanford
Roy F. Weston, Inc.
Washington, D.C.

Edward C. Yang
Planning Research Corporation
McLean, Virginia

1.3.1 INTRODUCTION

Over the past 15 years the Congress and the EPA have developed a comprehensive program to protect human health, welfare, and the environment. Initial statutory and regulatory activities emphasized reducing current and future emissions of pollutants into the environment. These statutes, and regulations promulgated under them, focused on specific environmental media, such as air (the Clean Air Act), surface waters (the Federal Water Pollution Control Act, later the Clean Water Act), and the land surface [the Federal Insecticide, Fungicide, and Rodenticide Act (FIFRA), and the Solid Waste Disposal Act]. Other initial statutes, such as the Toxic Substances Control Act (TSCA), sought to reduce the amounts of hazardous chemicals generated in the United States.

In the late 1970s the national environmental program began to address two new areas. These were (1) providing remedies for environmental problems that were the result of past practices, and (2) increasing emphasis on immediate response to environmental emergencies caused by the release of hazardous materials. These two areas of environmental protection are addressed in the Comprehensive Environmental Response, Compensation, and Liability Act of 1980

(CERCLA). The former initiative is generally known as the remedial-response program under CERCLA, while the latter initiative is generally known as the removal- (or emergency-) response program under CERCLA.

This new program, known as the Superfund program because of early estimates of the size of the Hazardous Substances Response Trust Fund established by Title II (now by Part I of Title V) of CERCLA, differs fundamentally from other EPA environmental programs. First, it addresses environmental problems with a multimedia approach. CERCLA provides the EPA with authority to remedy releases of hazardous substances into all environmental media. Second, it addresses environmental problems that are the result of past practices as well as current emergencies. Third, rather than being a regulatory program, the Superfund program provides the EPA with the authority to take direct action to abate releases into the environment.* Under CERCLA the government may respond to any release into the environment of any hazardous substance, or pollutant or contaminant. In addition, the EPA is empowered to compel private parties to respond to releases of hazardous substances, or to recover the costs of government-sponsored response to such releases.

During the initial Superfund program several major areas of hazardous-waste management were identified that required increased national attention. It was quickly learned that our knowledge concerning the applicability and performance of technologies to remedy releases of hazardous substances was inadequate to meet the challenge posed by the number of uncontrolled hazardous-waste sites and the magnitude of the problem. Efforts to determine appropriate response actions were frustrated by the sometimes inconsistent and uncoordinated requirements of media-specific environmental regulations. Procedures available to evaluate the nature and extent of the problem caused by releases of hazardous substances, and to determine the effect of releases upon human health, welfare, and the environment, were inadequate to identify the appropriate extent of remedy.

Recognizing these problems, Congress sought to address them in the recent passage of the Superfund Amendments and Reauthorization Act of 1986 (SARA, Public Law 99-499). SARA established significant new requirements under CERCLA and added new focus to the Superfund program. Among the major new provisions were:

- Preference for remedies that incorporate permanent treatment or destruction of wastes
- Codifying existing cleanup standards and expansion to include the application of state standards
- Increased opportunities for state and citizen involvement
- Expansion of removal authorities
- Focus on public health
- Increased focus on research and training
- Special attention to the roles and requirements of other federal agencies
- New enforcement authorities and procedures
- Requirements for community and state preparedness

---

*A predecessor to the Superfund was found in Section 311 of the Water Pollution Control Act of 1972 which established a $35 million revolving trust fund to address problems resulting from oil spills. However, no provision was made to deal with damage to land resources resulting from contamination by hazardous substances.

- Requirements for implementation of corrective actions for underground storage tanks

This chapter presents an overview and summary of CERCLA, as amended by SARA, and the Superfund program that has been established by the EPA to implement its provisions. This chapter emphasizes the aspects of CERCLA and the Superfund program that affect, and are affected by, hazardous-waste technologies. Thus, many aspects of the legislation and the Superfund program are not discussed. Aspects of the legislation and the program that are not discussed include, for example, the preliminary activities performed in order to initially identify sites and evaluate their priority for response, enforcement considerations, community and state involvement, and the basis for the revenues that form the Hazardous Substances Response Trust Fund.

1.3.2 SCOPE OF CERCLA AUTHORITIES

CERCLA gives the federal government broad authority to respond to releases or to threats of releases of hazardous substances, and to pollutants or contaminants that may present an imminent and substantial danger to the public health or welfare or to the environment. A "release" is defined to include not only the spilling or leaking of a substance into the environment, but also the abandonment or discarding of barrels, containers, or other closed receptacles that contain regulated substances. Section 101(14) of CERCLA defines the substances that are considered as hazardous. These are:

- Any substance designated pursuant to Section 311(b)(2)(a) of the Federal Water Pollution Control Act (Clean Water Act)
- Any element, compound, mixture, solution, or substance designated pursuant to Section 102 of CERCLA (reportable quantities)
- Any hazardous waste having the characteristics identified under or listed pursuant to Section 3001 of the Solid Waste Disposal Act (except those wastes for which regulation has been suspended by Act of Congress)
- Any toxic pollutant listed under Section 307(a) of the Federal Water Pollution Control Act
- Any hazardous air pollutant listed under Section 112 of the Clean Air Act
- Any imminently hazardous chemical substance or mixture with respect to which the Administrator of EPA has taken action pursuant to Section 7 of the Toxic Substances Control Act (TSCA)

Petroleum, including crude oil or any fractions that are not otherwise specified in the above list of substances, natural gas, natural gas liquids, liquefied natural gas, or synthetic gas usable for fuel, are not hazardous substances regulated by CERCLA.

CERCLA distinguishes two broad categories of response that may be undertaken to abate the threat posed by the release of hazardous substances, pollutants, or contaminants. The Act provides for long-term actions that are consistent with permanent remedy, known as remedial-response actions, and shorter-term actions in response to immediate danger, known as removal actions. Both of these response actions include not only the activities associated with the cleanup of the released substances and contaminated materials, but also a wide range of other activities including legal, fiscal, economic, engineering, architectural, and other studies, and other planning as necessary to direct response actions.

The basic operating document which sets forth procedures to be followed in providing response under the Superfund program is the National Oil and Hazardous Substances Pollution Contingency Plan (NCP; 40 *CFR* 300; 50 *FR* 47912, November 20, 1985). Section 105 of CERCLA requires that the NCP be revised by June 1988 to incorporate the provisions of SARA, and from time to time thereafter. Among other things, the NCP must specify procedures, techniques, materials, equipment, and methods to be employed in identifying, removing, or remedying releases of hazardous substances. Subpart F of the NCP (Sections 300.61 through 300.71) describes these technical procedures to be followed in determining and implementing appropriate response actions. Figure 1.3.1 shows the current NCP process for identifying appropriate response actions. Removal activities are described at Section 300.65 of the NCP, remedial actions are described at Section 300.68. Another important section is Section 300.70, which provides a list of demonstrated technologies appropriate for remedial actions at CERCLA sites.

1.3.3   KEY PROVISIONS OF CERCLA AND THE NCP

Several key provisions of CERCLA define the policies, goals, and requirements of the removal and remedial response programs, and establish a program of research, development and demonstration of alternative or innovative treatment technologies. The following paragraphs describe these key provisions of CERCLA, and the sections of the NCP that provide further guidance on the implementation of the Superfund removal and remedial programs.

Emergency Response

The emergency-response (or removal) program is established by Sections 101(23), 104(a)(1) through (a)(4), 104(b)(1), and 104(c)(1) of CERCLA. Guidance for the implementation of the removal program is set forth as Section 300.65 of Subpart F of the NCP. Together, these documents form the overall structure and scope of the Superfund hazardous-substances emergency-response program.*

Section 101(23) of CERCLA broadly defines a removal action to include the following:

- Cleanup or removal of released hazardous substances from the environment
- Actions as may be necessary in the event of the threat of release of hazardous substances into the environment
- Actions required to monitor, assess, and evaluate the release or threat of release of hazardous substances
- Disposal of removed material
- Other actions as may be necessary to prevent, minimize, or mitigate damage to the public health or welfare or to the environment which may otherwise result from a release or threat of release

---

*Subpart E of the NCP addresses response to oil spills pursuant to Section 311 of the Clean Water Act.

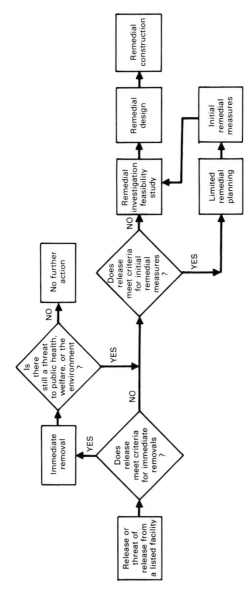

FIG. 1.3.1 NCP process for identifying appropriate response actions.

Removal actions may be performed to respond to any release that threatens the public health or welfare or the environment. Generally, the factors that EPA considers in determining the need for removal actions are given at Section 300.65(b)(2) of the NCP and include:

- Actual or potential exposure to hazardous substances or pollutants or contaminants by nearby populations, animals, or food chain
- Actual or potential contamination of drinking water supplies or sensitive ecosystems
- Hazardous substances or pollutants or contaminants in drums, barrels, tanks, or other bulk storage containers, that may pose a threat of release
- High levels of hazardous substances or pollutants or contaminants in soils largely at or near the surface, that may migrate
- Weather conditions that may cause hazardous substances or pollutants or contaminants to migrate or be released
- Threat of fire or explosion
- Availability of other appropriate federal or state response mechanisms
- Other situations or factors which may pose threats to public health or welfare or the environment

Releases of hazardous substances do not have to present an imminent and substantial endangerment, and the site does not have to be listed on the National Priorities List (NPL) in order for a removal action to be initiated. Removal actions are limited, however, to 12 months' duration and $2 million unless (1) continued response actions are immediately required to prevent an emergency, there is an immediate risk to public health or welfare or the environment, and assistance will not otherwise be provided on a timely basis; (2) a remedial response can be undertaken; or (3) continued response action is otherwise appropriate and consistent with the remedial action to be taken.

Remedial Response

The remedial response program is established by Sections 101(24), 104(a)(1), 104(b)(1), 104(c)(2) through (c)(9), 104(e), 117, and 121 of CERCLA. Guidance for the implementation of the remedial program is set forth as Section 300.68 of Subpart F of the NCP.

Section 101(24) of CERCLA defines remedial actions to be those actions consistent with permanent remedy taken instead of or in addition to removal actions in the event of a release or threatened release of a hazardous substance into the environment. Remedial actions are taken to prevent or minimize the release of hazardous substances so that they do not migrate to cause substantial danger to present or future public health or welfare or the environment.

Section 104 of CERCLA establishes broad remedial-response authorities. In general, the EPA is empowered to take any response measure, consistent with the NCP, that is seen to be necessary to protect the public health or welfare or the environment. However, the EPA does not have the authority to implement remedial measures unilaterally. Both Section 104 and Section 121 of CERCLA require the EPA to coordinate all remedial-response activities with the affected state (or states). The EPA cannot implement remedial measures unless the affected state has entered into a contract or a cooperative agreement with the

EPA,and has (1) assured that it will provide all future maintenance of the actions, (2) assured the availability of an acceptable hazardous-waste disposal facility (if required), and (3) assured payment of 10% of the costs of the remedial action at a privately owned site, or at least 50% of the costs of the remedial action at a site that is wholly or partially publicly owned (sites on Indian lands are exempt from cost sharing).

In addition to requiring that response actions be coordinated with the affected state or states, CERCLA requires extensive public participation in the development and selection of remedial actions. Section 117 requires that a notice and brief analysis of proposed remedial-action plans be published, and that the plan be made available to the public for review and comment. Groups of individuals that may be affected by the release of the hazardous substances can receive grants up to $50,000 (or greater if approved by the EPA) to retain expert technical assistance to review the remedial-action plans.

Perhaps the most far-reaching new statutory requirements affecting the remedial-response program are those of Section 121 regarding cleanup standards, or the question that is popularly phrased as "How clean is clean?" This section requires the EPA to give preference to remedies that permanently address the release of hazardous materials. In addition, the section specifies the environmental standards that cleanup activities must meet, and seeks to establish order out of the often contradictory collection of environmental standards, criteria, and advisory documents that may be applicable to response activities. Section 300.68(i) of the NCP requires that remedial actions attain or exceed applicable or relevant and appropriate federal public health and environmental requirements that have been identified for the specific site. However, no specific guidance is given on how to identify such applicable or relevant and appropriate requirements. The revised NCP, now being prepared to address the provisions of SARA, will provide this specific guidance.

Research, Development, and Demonstration

A major new provision of CERCLA is Section 311 which establishes programs of research, development, training, and demonstrations. Several new programs are established within the Department of Health and Human Services (HHS) as well as in the EPA.

The statute establishes within the Department of Health and Human Services a basic research program that focuses on:

- Advanced techniques for the detection, assessment, and evaluation of the effects on human health of hazardous substances
- Methods to assess the risks to human health presented by hazardous substances
- Methods and technologies to detect hazardous substances in the environment and basic biological, chemical, and physical methods to reduce the amount and toxicity of hazardous substances

There is also established within HHS a program, to be funded through grants, cooperative agreements, and contracts, that provides for training in several areas of the management of hazardous substances. The major areas of training are:

- Short courses and continuing education for state and local health- and environmental-agency personnel and other personnel engaged in the handling of hazardous substances, in the management of facilities at which hazardous substances are located, and in the evaluation of the hazards to human health presented by such facilities

- Graduate or advanced training in environmental and occupational health and safety and in the public-health and engineering aspects of hazardous-waste control

- Graduate training in the geosciences, including hydrogeology, geological engineering, geophysics, geochemistry, and related fields, necessary to meet professional personnel needs in the public and private sectors

CERCLA also establishes a program of research within the EPA which is coordinated with the HHS program through an advisory council. Activities sponsored by the EPA involve conducting research with respect to the detection, assessment, and evaluation of the effects on and risks to human health of hazardous substances and detection of hazardous substances in the environment. These grants, cooperative agreements, and contracts will be made to at least five institutions of higher learning in the United States. The Congress has given the EPA a goal of having 10 such hazardous-substance research centers, or "Centers of Excellence."

This new section of CERCLA also establishes a program of research, development, and demonstration of alternative or innovative treatment technologies for use in remedial actions under the Superfund program. This program is generally known as the Superfund Innovative Technology Evaluation (SITE) program. The major aspects of this new program are discussed in detail on p. 1.42.

1.3.4 THE EPA SUPERFUND PROGRAM

The program that has been established within the EPA to administer and implement the provisions of CERCLA is popularly known as the Superfund program. Officially, the program is administered by the Office of Emergency and Remedial Response (OERR), within EPA's Office of Solid Waste and Emergency Response (OSWER). The Superfund program is divided into two distinct programs, each having responsibility for implementing a different area of response. The two areas of response are (1) removal (or emergency) response, which is administered by the Emergency Response Division of OERR, and (2) remedial response, which is administered by the Hazardous Site Control Division of OERR. Occasionally, these two types of response are required at a single site. When this happens the emergency response is coordinated by the remedial project officer, who maintains authority for nonemergency activities during the removal and who continues to direct response activities after the removal actions are completed. These two response programs are discussed in detail below.

Removal Program

The Superfund removal program provides response to immediate threats to public health, welfare, and the environment. The Superfund removal program has developed out of the original EPA oil spill response program, established under the original NCP and Section 311 of the Clean Water Act, to include providing response to the release of hazardous substances.

The removal program may respond to any type of incident, is not limited to situations that present an imminent and substantial endangerment, and is not limited to response at sites on the NPL. Removal-program personnel perform cleanup activities or stabilize site conditions based on three general categories of site situation. These categories are:

- Classic emergency removal actions—where cleanup or stabilization of the site must be initiated within hours or days
- Time-critical removal actions—where cleanup or stabilization of the site must be initiated within six months
- Removal actions that are not time-critical—where cleanup or stabilization of the site may be delayed for six months or more

Sites where a removal action may be required undergo a preliminary site assessment. The on-scene coordinator (OSC) is responsible for conducting this assessment as quickly as possible, often in a matter of hours, but occasionally it may take several weeks. The assessment is based on observations and readily obtained information regarding the substances involved and other site characteristics. Special consideration is paid to situations involving any of the following: contamination of drinking water; floodplains and wetlands; radioactive wastes; contamination of Native American lands; and situations involving evacuation and relocation of nearby populations.

The technologies employed in removal actions are generally straightforward and do not require lengthy, site-specific engineering studies or design. Section 300.65(c) of the NCP identifies types of removal activities that may be appropriate for specific situations. Although the list is not exhaustive, and does not limit the response manager, the OSC, from taking other actions as appropriate, it does provide an indication of the level of technologies generally employed in response actions. The NCP suggests the use of the following technologies in certain response situations:

- Fences, warning signs, or other security or site-control precautions—to be taken where humans or animals have access to the release
- Drainage controls (e.g., runoff or run-on diversion)—where precipitation or runoff from other sources may enter the release area from other areas
- Stabilization of berms, dikes, or impoundments—where needed to maintain the integrity of the structure
- Capping of contaminated soils or sludges—where needed to reduce migration of hazardous substances or pollutants or contaminants into soil, groundwater, or air
- Use of chemicals and other materials to retard the spread of the release or to mitigate its effects—where the use of such chemicals will reduce the spread of the release
- Removal of highly contaminated soils from drainage or other areas—where removal will reduce the spread of contamination
- Removal of drums, barrels, tanks, or other bulk containers that contain or may contain hazardous substances or pollutants or contaminants—where it will reduce the likelihood of spillage; leakage; exposure to humans, animals, or food chain; or fire or explosion
- Provision of alternative water supply—where it will reduce the likelihood of exposure of humans or animals to contaminated water

Response actions are limited in duration to one year, and in expense to $2 million, unless extensions are authorized by the EPA. Otherwise, a removal response is terminated when the site has been stabilized so that remedial actions can be taken or continued, or when there is no longer the threat of release of hazardous substances, pollutants, or contaminants that prompted the action.

Remedial Program

Perhaps the most well known area of the Superfund program is the remedial-response program. This program addresses those releases of hazardous materials, pollutants, or contaminants that require complex, long-term response. In general, the NCP established a four-step process by which the nature and extent of the problem is determined, potential remedial actions and technologies are identified and evaluated, and the appropriate remedy is selected. The four steps are:

• Site evaluation and scoping of response actions
• Initial evaluation and screening of technologies
• Detailed evaluation of technologies
• Selection of remedy

The selection and approval of the remedy are generally performed during the phase of the Superfund process known as the remedial investigation and feasibility study (RI/FS). The RI/FS phase is followed by the phase of the Superfund process known as the remedial design and remedial action phase, or RD/RA. Each of these phases of remedial response is described in detail in the following paragraphs.

Remedial Investigation and Feasibility Study.   As a result of SARA, Section 121(b)(1) of CERCLA requires an assessment of permanent solutions and alternative treatment technologies or resource-recovery technologies that, in whole or in part, will result in a permanent and significant decrease in the toxicity, mobility, or volume of the hazardous substance, pollutant, or contaminant. The most technically rigorous phase of this assessment is known as the remedial investigation and feasibility study (RI/FS). Figure 1.3.2 shows the stages of this process and illustrates the interdependency of the two efforts. The remedial investigation and the feasibility study are generally performed concurrently. The nature and extent of the release, as well as the physical conditions of the site, are determined during the remedial investigation. The information obtained during the remedial investigation is used to evaluate the appropriateness of the remedial alternatives. The evaluation of technologies is performed during the feasibility study. Information needed to evaluate the competing technologies that are identified as potentially applicable serves to focus the effort of the remedial investigation.

EPA has adopted a phased approach to conducting the RI and FS. Both the RI and the FS are performed in three phases. The following paragraphs briefly describe the content and relationship of these phases.

Phase I Remedial Investigation—Project Scoping.   The first phase of the remedial investigation involves developing a description of the current site situation. The primary objective of this phase is to summarize existing information on hazardous-waste sources, pathways, and receptors, and to evaluate potential im-

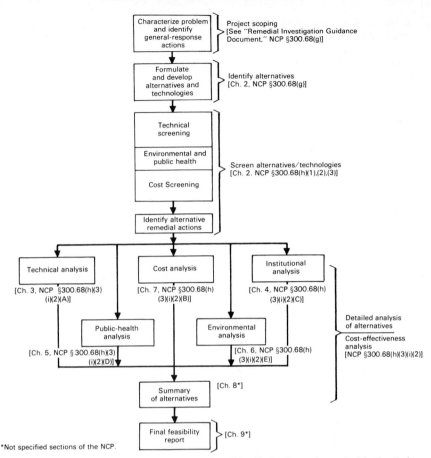

FIG. 1.3.2 Feasibility-study process. NCP sections identified refer to those cited in the *Federal Register*, July 16, 1982. (Guidance on Feasibility Studies under CERCLA, *U.S. EPA, EPA/540/G-85/003, June 1985*

pacts on public health, welfare, and the environment. This information is used to identify general-response actions during the first phase of the FS.

Phase I Feasibility Study—Development of Alternatives. The first phase of the feasibility study involves the development of general-response actions and the screening of alternative remedial technologies. Obviously inappropriate response actions are eliminated based on existing site information obtained during the Phase I remedial investigation. Also identified at this stage are the applicable, relevant, and appropriate standards that will determine the overall response objectives. The potential remedial alternatives focus the effort to collect data during the Phase II remedial investigation.

Phase II Remedial Investigation—Field Investigations and Site Characterization. This phase of the remedial investigation consists of developing the various management and site-activities plans that will govern the conduct of the field investigations, as well as the actual performance of field sampling and analysis.

Before any actual on-site activities are performed, the following plans must be prepared:

- Site-management plan
- Sampling plan
- Quality-assurance and quality-control plan
- Health and safety plan
- Community relations plan

The information obtained during the field investigation is used to determine the following:

- The need for any remedial actions
- The extent of any remedial actions
- The feasibility of any potential remedial actions

In order to make these determinations, sufficient quantitative data must be collected and analyzed to characterize the site in terms of the nature and extent of any contamination, and any adverse effects on public health, and any adverse effects on the environment. While the latter two assessments are actually made during the feasibility study, all of these assessments rely upon the data collected during the site characterization.

A site characterization is performed to determine the nature of any contaminants on the site and to identify the nature of any pathways for the migration of the contaminant to the public and the environment. Sufficient physical data are also collected during the site characterization to allow the engineering evaluation of remedial technologies.

Phase II Feasibility Study—Screening of Alternatives. Following the full field investigation, possible remedial alternatives are evaluated to eliminate those that are obviously inappropriate, i.e., those that cannot be implemented given site conditions, their effectiveness, or their cost. The general screening of alternatives focuses further remedial-investigation efforts to obtain information necessary to prepare conceptual designs of the remedy. Such information is obtained through bench-scale and pilot-scale treatment studies. Occasionally, such further studies are not required to fully evaluate alternative remedial actions and the RI/FS effort proceeds directly to the third phase of the FS.

Phase III Remedial Investigation—Bench- and Pilot-scale Testing. Once specific remedial technologies are identified and combined to form comprehensive remedial-response scenarios, bench-scale and pilot-scale tests are often required to obtain performance and cost information necessary for complete evaluation of the alternatives. The scope of bench- and pilot-scale testing is generally limited to treatability and materials-testing activities. Bench- and pilot-scale testing also determines any scaleup factors in the remedial design that are necessary for the detailed evaluation of alternatives performed in the third phase of the FS.

Phase III Feasibility Study—Detailed Evaluation of Alternatives. The third phase of the feasibility study involves the detailed evaluation of the remaining remedial alternatives. The alternatives are evaluated with respect to institutional constraints, their effectiveness in protecting public health and the environment, the feasibility of implementation considering site conditions, and their relative costs. This phase of the RI/FS results in the preliminary identification of the most appropriate remedial-response alternative.

The engineering analysis of alternative remedial technologies involves an assessment of their relative performance, reliability, implementability, and safety. The performance of a technology is determined by its effectiveness, or the ability to perform intended functions, and its useful life. The reliability of a technology is determined by the amount and complexity of operation and maintenance activities that are required to maintain its effectiveness. New statutory language promotes the evaluation of new and innovative technologies by making it clear that a remedial action may be selected whether or not acceptable performance has been achieved in practice at any other facility or site that has similar characteristics. The implementability of a technology is determined by the ability to actually build, construct, or implement the technology under the specific site conditions, and the time required to implement the remedy and obtain desired results. Finally, the safety of each technology is assessed in terms of any short- or long-term threats to nearby communities and the environment, and to workers during construction.

The effects of each prospective remedial action are evaluated in terms of the level of protection that will be afforded by its implementation. The public-health analysis generally consists of (1) a baseline evaluation that establishes the nature of the threat before remedial actions are undertaken, (2) an exposure assessment to determine the degree to which surrounding populations are exposed to hazardous materials after implementing the remedy, and (3) a risk assessment to determine the actual threat to health and welfare posed by the residual exposure. The public-health analysis is generally performed using various fate and transport models that are modified to simulate the effects of remedial measures. Results are usually given in terms of the incremental risk of cancer incurred by exposure to residual contaminants, or the relationship of the concentration of remaining contaminant to federal health and environmental standards and criteria.

Each alternative is evaluated in order to determine which federal public health and environmental requirements are applicable or relevant and appropriate. The requirements for community-relations activities, coordination with other federal agencies, and any permitting requirements are also determined.

Each alternative is evaluated in terms of the environmental benefits expected, and any adverse environmental effects that may result from its implementation. These environmental effects are quantified to the extent possible, and measures to mitigate adverse effects are identified.

The capital, as well as operation and maintenance costs of each alternative are estimated. These estimates generally are within -30 to $+50\%$ of the final-design estimated costs. The cost of the alternative is presented in terms of its net present value, as well as in terms of the distribution of costs over time.

Selection of the Appropriate Remedy. The initial CERCLA statute required only that the appropriate remedial action be cost-effective and, to the extent practicable, be in accordance with the NCP. Section 121 of CERCLA, added by SARA, now requires the EPA to favor certain categories of response actions and response technologies. In addition to being cost-effective and consistent with the NCP, the EPA is now required to select remedial actions that utilize permanent solutions and alternative treatment technologies or resource-recovery technologies to the maximum extent practicable. Remedial actions in which treatment permanently and significantly reduces the volume, toxicity, or mobility of the hazardous substances are to be preferred over remedial actions not involving such treatment. The off-site transport and disposal of hazardous substances or contaminated materials without treatment is to be the least-favored alternative remedial action when treatment technologies are available.

Remedial Design and Remedial Action.   Following the selection of the appropriate remedial action, the project is passed to the U.S. Army Corps of Engineers for design and construction. The Corps of Engineers solicits consulting architectural and engineering firms to perform these activities. Generally, the design of a project proceeds in three phases, involving a preliminary design; an intermediate design; and a prefinal–final design that is converted at the end of the third phase to a final design. The preliminary design generally addresses not less than 30% of the total project design. The intermediate design phase is optional, and is generally necessary only for complex projects. The intermediate design review is performed at the 60% completion stage of the design. The prefinal design includes a 95% complete set of design documents, and after review and approval of these, the final design is prepared with a 100% complete set of documents. The technical review of these documents ensures:

- The biddability and constructability of the design
- The accuracy of the construction cost estimate
- Utilization of currently accepted construction practices and techniques
- The ability of a construction contractor to submit a fair and reasonable bid based upon the bid schedule included in the specifications
- The accuracy of any estimated quantities of materials specified in the design
- That the responsibilities and liabilities of the construction contractor and the government are clearly defined and detailed in the design documents

   After the design of the remedy is approved the Corps of Engineers assists the EPA in the implementation of the remedy. In general, the Corps of Engineers is responsible for conducting procurement activities for the remedial action and managing the contractor. The Corps of Engineers monitors and provides oversight of construction activities to ensure compliance with all contract requirements. Finally, the Corps of Engineers conducts final inspections of the remedy and certifies that it was completed in accordance with the contract requirements.

1.3.5 SUPERFUND INNOVATIVBE TECHNOLOGY EVALUATION (SITE) PROGRAM

Section 311(b) of CERCLA is a new provision, added by SARA, that establishes an Alternative or Innovative Treatment Technology Research and Development Program. The demonstration-assistance program consists of several parts, including:

- The publication of a solicitation and the evaluation of applications for demonstration projects utilizing alternative or innovative technologies
- The selection of sites that are suitable for the testing and evaluation of innovative technologies
- The development of detailed plans for innovative technology demonstration projects
- The supervision of such demonstration projects and the providing of quality assurance for data obtained
- The evaluation of the results of alternative innovative technology demonstration projects and the determination of whether or not the technologies used are effective and feasible

The following sections discuss the background behind the development of the SITE program and the structure of the technology-demonstration portion of the program.

Background. Experience with the early Superfund program raised concern in the Congress and in the EPA that existing technologies were inadequate to deal with the problems caused by the uncontrolled releases of hazardous substances. Early site-remediation efforts relied heavily on a combination of removal of contaminated materials to other disposal facilities and construction of physical barriers to prevent the migration of remaining hazardous wastes into the groundwater and the surrounding environment. Congress found that while such methods are likely to reduce the risk of release of hazardous substances in the short term, serious questions remain about the safety, efficacy, and cost of such measures over the long term. In addition, the original Superfund could not be used for research other than research directly relevant to short-term emergency and remedial-response activities such as site assessment and sampling. The Superfund program primarily looked to other related programs, such as the Resource Conservation and Recovery (RCRA) program for research into hazardous-waste technologies.

As a result of congressional hearings, and findings by the Office of Technology Assessment and EPA's Science Advisory Board, Congress included provisions in SARA that establish an innovative technology research and demonstration program in EPA. This program, called the Superfund Innovative Technology Evaluation (SITE) program, is administered jointly by EPA's Office of Research and Development and Office of Emergency and Remedial Response. The primary purpose of the SITE program is to enhance the development and demonstration, and thereby establish the commercial availability, of innovative technologies at Superfund sites as alternatives to the containment systems that are presently being used. The SITE program, as developed by the EPA, consists of four activities; these are:

- To identify and, where possible, remove impediments to the development and commercial use of alternative technologies
- To conduct a demonstration program of the more promising innovative technologies to establish reliable performance and cost information for site characterization and cleanup decision making
- To develop procedures and policies that encourage selection of available alternative treatment remedies at Superfund sites
- To structure a development program that nurtures emerging technologies

The following section discusses the demonstration aspects of the SITE program; further information about the entire program is available in OSWER Directive 9380.2-3, *Superfund Innovative Technology Evaluation (SITE) Strategy and Program Plan* (EPA/540/G-86/001, December 1986).

Demonstration Program. Early deliberations over the reauthorization of CERCLA quickly focused on the need for a research, development, and demonstration (RD&D) program. A group of experts composed of individuals from large and small companies, academia, state governments, environmental groups, and consulting engineering firms, known as the SITE Strategy Review Group, recommended that the EPA establish a demonstration program to evaluate new technologies at Superfund sites. The demonstration program established pursuant to this recommendation, and under Section 311 of CERCLA, focuses on innovative and alternative treatment technologies. Section 311(b)(10) of CERCLA defines the term "alternative or innovative treatment technologies" to mean those technologies, including proprietary or patented methods, which permanently alter the composition of hazardous waste through chemical, biological, or physical means so as to significantly reduce the toxicity, mobility, or volume (or any combination thereof) of the hazardous waste or contaminated materials being treated. The term also includes technologies that characterize or assess either the extent of contaminants or the stresses imposed by the contaminants on complex ecosystems at sites.

The EPA has established several categories of alternative technologies according to the status of their development. Figure 1.3.3 illustrates the relationship between the status of development technologies and their classification in the SITE program. A distinction is made between an alternative technology that is currently available, one that is innovative, and one that is emerging. These distinctions are:

- *Available Alternative Technology:* One that is fully proven and is currently in routine commercial or private use.

- *Innovative Alternative Technology:* Any fully developed technology for which cost or performance information is incomplete, thus hindering routine use at CERCLA hazardous-waste sites. An innovative alternative technology requires full-scale field testing before it is considered proven and available for routine use.

- *Emerging Alternative Technology:* An alternative technology in an earlier stage of development; the research has not yet successfully passed laboratory- or pilot-scale testing.

The SITE demonstration program focuses on the evaluation of innovative alternative technologies. SARA requires that at least 10 demonstrations be performed annually.

The SITE demonstration program consists of two parts: (1) searching for and identifying prospective innovative alternative technologies, and (2) evaluating the performance of those technologies through demonstration. The search for prospective innovative alternative technologies is accomplished primarily by soliciting developers of technologies through advertisements in the *Commerce Business Daily (CBD)*. Any person and any public or private nonprofit entity may submit an application to EPA in response to the solicitation. Prospective alternative innovative technologies are evaluated by EPA based on several considerations, including:

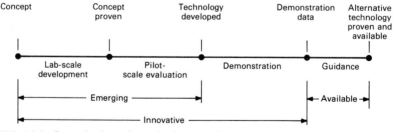

FIG. 1.3.3 Stages in alternative-technology development.

- The ability to provide a permanent solution
- The ability to be used on site rather than requiring off-site transport of hazardous materials
- Having lower costs than current technologies
- Having the potential for significantly better performance than current technologies
- Producing emissions, effluents, or residues that are easy to manage
- Ease and safety of operations

Once technologies are selected for a demonstration, EPA will determine the scale and location of the demonstration project. The demonstration project may involve bench- or pilot-scale evaluations before a full-scale demonstration is performed. Technologies will usually be conducted at uncontrolled hazardous-waste sites; however, they may be evaluated at the developer's site, or at an EPA test and evaluation facility. The objectives of the demonstration will be determined on a case by case basis, but will generally include evaluating:

- Performance and design characteristics
- Characteristics of the wastes handled and treated
- Destruction and removal efficiencies
- Process residues and wastes generated
- Operations and maintenance requirements
- Operational safety
- Mass flow–energy balances for feed and types and quantities of residues
- Setup, startup, decontamination, and takedown procedures
- Instrumentation and control characteristics
- Quality-assurance data necessary to ensure compliance with EPA test, measurement, and analytical guidelines

Upon completion of the demonstration project, EPA will prepare a technology-evaluation report. The report will describe the technology and its purpose in terms of its operating range, its costs, and a general assessment of its usefulness with regard to addressing the threats posed by Superfund hazardous-waste sites.

SECTION 1.4
UNDERGROUND STORAGE TANKS

Douglas C. Ammon
Clean Sites, Inc.
Alexandria, Virginia

S. Robert Cochran
Interface, Inc.
Alexandria, Virginia

1.4.1 INTRODUCTION

An estimated 3 to 5 million underground tanks in the United States are used to store liquid petroleum and chemical substances. Government and industry sources estimate that in 1986 between 100,000 and 400,000 of these tanks and associated piping systems may leak. Tank leaks can result in the contamination of subsurface soils, migration of toxic or explosive vapors, and contamination of groundwater and surface water. Environmental and tank-monitoring, tank-replacement, product-recovery, and corrective-action technologies will reduce the potential of future releases or remedy prior releases from underground storage tanks.

Industry sources estimate that the average cost of tank cleanup is approximately $70,000; however, should tank removal and treatment of surrounding soils be required, costs may approach or exceed $1 million. The corrective-action costs may exceed $1 million, if underground tank releases have significant effects on groundwater, surface water, drinking-water supplies, reservoirs, sewers, or utility trenches. The primary factors affecting corrective-action costs are the magnitude of release, the toxicity of the substance, the hydrogeological setting, and the environmental criteria, standards, levels, or objectives applicable to the cleanup.

Concern for the status of underground storage tanks has resulted in the development of legislation regarding the registration and monitoring of existing tanks, design and installation of new tanks, and corrective actions for releases. At the federal level, the 1984 Hazardous and Solid Waste Amendments added to the Resource Conservation and Recovery Act (RCRA) a new Subtitle I, Regulation of Underground Storage Tanks. Numerous states have passed laws requiring similar, often more stringent, regulations.

1.4.2 UNDERGROUND STORAGE TANK PROFILE

An underground storage tank under RCRA Subtitle I is a tank that stores "regulated substances" and that has at least 10% of its volume below the surface of the ground ("volume" here includes volume of piping connected to the tank). Regulated substances include hazardous chemical products regulated under CERCLA, and also petroleum products, including crude oil and refined products that are liquid at standard conditions of temperature and pressure. Underground tanks containing hazardous waste, as defined under RCRA, are regulated under Subtitle C.

Millions of tanks have been installed underground for use in storing many materials. The configurations of these installations vary to suit several constraints including geography of the site, material stored, insurance-underwriter requirements, government regulations, and the owner's operation. The installation will likely have most of these basic components:

• Tank or tanks
• Antiflotation anchorage
• Piping system
• Pumps
• Means for gauging the level of liquid in the tank

Tanks are usually placed underground to store materials that are hazardous because of their flammability or combustibility. Many tanks were installed underground on the premise that they would never develop leaks; thus, little consideration was given to the consequences of any leakage that might develop.

Two broad categories of underground tank applications are gasoline stations and industrial or commercial installations. Underground tanks in gasoline stations are used for storage of gasoline (leaded, unleaded, and premium grades), diesel fuel, and waste oil (may contain some gasoline). Industrial or commercial underground tank installations store a wide variety of materials including solvents and various hydrocarbons. They may provide storage for new materials or for waste products. These installations usually will have individual tanks for each stored material and not multiple tanks connected by common suction piping (as commonly found in gas stations). System piping in industrial installations is not always located below grade but is run on overhead racks if they are available. In commercial installations, such as dry-cleaning establishments, system piping will generally be found below grade into the building.

Underground tanks can be found in various sizes, shapes, and construction materials. Metal tanks are usually of welded construction and have some type of exterior coating or cathodic protection against corrosion. Tanks fabricated from fiberglass, epoxy, or other nonmetallic material may be found in newer installations; these generally do not require any coating for corrosion resistance. Newer installations might also have tanks with double-wall construction. The annular space between the walls is used in various ways to detect product leaks from the inner tank wall or groundwater leaks from the outer tank wall. Secondary containment systems, generally a flexible-membrane liner surrounding the tank or tanks, are also found in newer installations.

Depth of tank burial varies, but the tank top is generally not more than one tank diameter below finished grade. In some installations, tanks will be found in mounds, partially buried below grade. Underground tanks are usually paved over if traffic is to travel above them. Manholes, caps, or other hardware are provided to cover and protect tank appurtenances such as fill connections and gauge pipes. Some large tanks are equipped with a manhole to allow access into the tank. Unless the water table is well below the tank excavation, an anchor system is present to prevent the tank from floating out of the ground.

A method for tracking inventory, e.g., a metering stick inserted in the fill tube, is generally part of any underground storage tank installation. More elaborate systems might employ bubbler-type pneumatic level sensors or use a differential-pressure instrument.

1.4.3 FAILURE MODES

Leaks can occur from storage tanks and associated piping and pumping via leaks from corrosion; from system rupture due to overloading, external stresses, or puncture; and from faulty construction or installation. Overfilling is another source of a tank-system release.

The most common failure mode for underground tank systems is corrosion of the tank or piping. Corrosion may be traced to failure of the corrosion-protection system due to such problems as "holidays" in the coating or taping system, depletion of the sacrificial anodes, inadvertent switchoff of impressed current, or corrosion from the inside by the stored product. Many leaks resulting from corrosion will be encountered in older tanks that have no corrosion protection at all.

Fiberglass-reinforced plastic and other nonmetallic tanks and piping are inherently corrosion-resistant, but these materials have lower structural strength than steel and require greater care during installation. Reinforced ribs are usually incorporated to withstand both the internal stresses from the stored liquid and the external loads. Some resins used for tank construction may lose structural strength when exposed to certain chemicals, while others may dissolve, soften, or become brittle in acidic or saline soil environments. Failure to observe the design limitations and to follow the handling and installation requirements for tanks can lead to tank failure.

A substantial percentage of underground storage system leaks occur in piping systems. This is probably because the majority of piping systems utilize threaded joints which, without adequate corrosion protection, are vulnerable to corrosion from outside. Piping joints may also leak because of improper sealing during installation or loosening due to vibrations, traffic loadings, temperature cycling, frost heave, and settlement of backfill.

Overfilling is another source of leaks or spills from tanks. Spills of this type can be caused by human error, failure of shutoff valves of the delivery source, and failure of the tank-level indicator. These spills are generally small (less than a few gallons) but can potentially be much larger if the equipment is left unattended during tank filling or there is a failure of shutoff valves.

1.4.4 LEAK DETECTION

Leaks should be identified early and initial assessments should be performed promptly to minimize the adverse effects of releases from underground storage tanks. Leaks are often discovered through an odor in groundwater, in a confined space, or near a utility conduit. Unfortunately, discovery from odors or vapors may occur well after the initiation of the leak. A successful tank-monitoring program should provide early detection, occur on a regular basis, and employ accurate leak-detection systems. Once a release is discovered, accurate characterization of the extent of release and the associated migration pathways are important.

The four classes of methods for detecting leaks in underground storage tanks are: (1) volumetric (quantitative) leak testing and leak-rate measurement, (2) nonvolumetric (qualitative) leak testing, (3) inventory controls, and (4) monitoring of environmental effects (outside detection). A state of the art review for leak detection is given by the EPA in a 1986 publication.[1]

Volumetric leak testing is based on detection of a change in tank volume by measuring parameters such as liquid level, temperature, pressure, and density. This includes methods which use an air-bubbling system to monitor pressure changes resulting from changes in product level in the tank. Other methods detect level changes by using either a J-tube manometer, a laser beam and its reflection, or a dipstick-type device. Another approach is to measure any volume change by maintaining a constant level.

Nonvolumetric leak testing determines the presence of a leak by qualitative methods, usually by using a second material other than the product (a tracer material). The tracer material, usually helium, is generally used to pressurize a tank and a leak is indicated by either loss of pressure or the detection of the tracer gas outside the tank by a sniffer mass spectrometer. Because of the rapid diffusivity of helium, it can escape through a leak in the tank as small as 0.013 cm (0.005 in).[2] This type of testing can be used to determine whether there is a leak, while volumetric testing may be used to determine the leak rate.

Inventory control is perhaps the simplest and most economical detection method. Leaks are detected by keeping records of tank inventories and noting any unexplained change in liquid levels or amounts. Inventory monitoring can be performed by gauge-stick or electronic-level measurement or by weight monitoring using pressure and density measurements. Although the problem of keeping records of tank inventory is complicated by the fact that petroleum products and other chemicals are volatile and losses due to evaporation are possible, the inventory method should be used as a first and convenient method for gross leak monitoring.

Environmental-effects monitoring typically identifies leaks by monitoring wells outside the storage tanks. Other environmental-monitoring techniques could be employed such as volatile organic carbon vapor analyzers, suction lysimeters, and other similar techniques. Environmental-effects monitoring may not be able to distinguish which tank is leaking if multiple sources are present and also may detect other sources and interferences. These methods do, however, provide a direct measure of the environment and once installed a method to frequently check for leaks.

1.4.5 CORRECTIVE ACTIONS

Corrective-action responses for leaking underground storage tanks can be divided into two categories. Initial responses are directed at controlling the immediate impacts resulting from newly discovered or sudden releases. If an emergency situation exists, immediate responses may include notification of fire departments, evacuation, removal of ignition sources, and venting of flammable vapors. Permanent corrective measures are directed at cleanup of the releases to some acceptable level for protection of human health and the environment. A flowchart of corrective-action responses is presented in Fig. 1.4.1. The degree of detail and the amount of resources applied in these steps is a function of site-specific conditions. Some of the more critical site-specific conditions include the volume of release, the time over which the release transpired, the site hydrogeologic conditions, and the proximity of environmentally sensitive communities and human receptors.

During initial-response actions, focus is on alleviating the immediate threat to public health and safety. Figure 1.4.2 illustrates the typical initial-response process. The critical element of initial-response actions is the timing associated with implementation. Effective implementation commonly dictates that field deployment occur within hours of the discovered release. Options that may be applicable in initial-response actions are presented in Table 1.4.1.

The selection of more permanent corrective actions requires a higher level of analysis and involves a broader range of potential technologies. Table 1.4.2 presents a listing of technologies that may be used in permanent corrective actions for four classes of release profile.

During permanent corrective action the focus is on site characterization and assessment to determine the need for further action beyond source control, on calculating transport rates, on assessing the hazard to the environment and from potential human exposure, on collecting data for selection and design of corrective actions, and on determining the effectiveness of corrective actions. These activities may involve hydrogeologic, environmental, and atmospheric monitoring, chemical analysis, fate and transport modeling, and analysis of corrective-action effectiveness.

When selecting corrective actions, the physiographic location of the release should be considered. Often leaks associated with gasoline stations are located within residential communities or light commercial areas. As a result, corrective-action planning must take into consideration the following logistic and engineering factors:

- Limitations on the amount of land available for implementing corrective actions
- Restricted property boundaries requiring easements to conduct off-site activities
- Close proximity to densely populated zones
- Complexities associated with subsurface cultural features (sewer, water, and electrical utilities, basements, wells, subways, and tunnels)

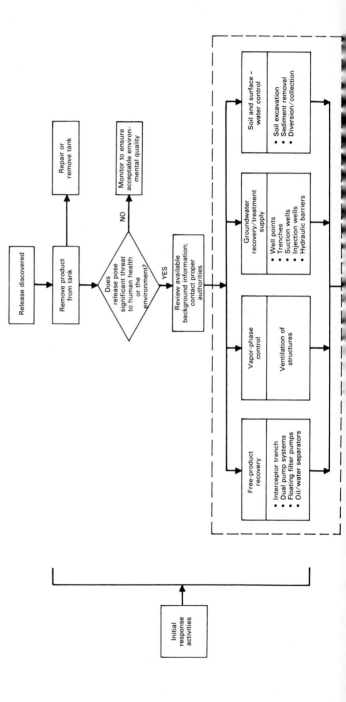

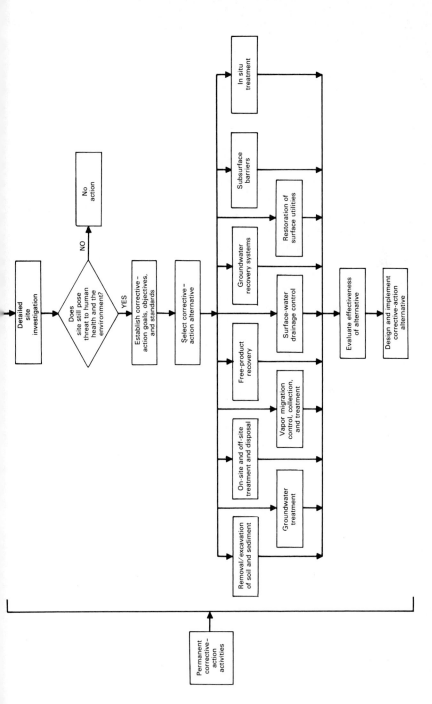

FIG. 1.4.1 Underground storage tank corrective-action flowchart.

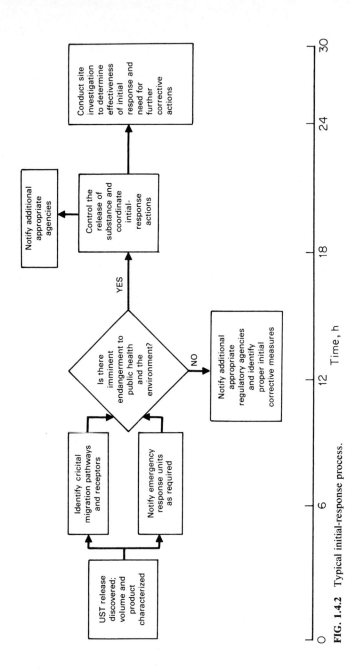

FIG. 1.4.2 Typical initial-response process.

TABLE 1.4.1 Potential initial-response situations and associated response options.

Situation	Tank repair or removal	Free-product recovery	Groundwater recovery and treatment	Subsurface barriers	Soil excavation	Vapor migration control and collection	Sediment removal	Surface-water diversion drainage	Treatment of central water supply or location of alternative supply	Treatment of point-of-use water supply or location of alternative supply	Restoration of utility, water, and sewer lines	Evacuation of nearby residents	Restricted egress and ingress
Groundwater contamination													
Existing public or private wells	•	•	•	•					•	•			
Potential future source of water supply	•	•	•										
Hydrologic connection to surface water	•	•	•					•					
Soil contamination													
Potential for direct human contact: nuisance or health hazard	•			•	•								
Agricultural use	•				•								
Potential source of future releases to groundwater		•	•	•	•								
Surface-water contamination													
Drinking-water supply	•	•	•				•	•	•	•			
Source or irrigation water	•	•	•				•	•	•	•			
Water-contact recreation	•						•						
Commercial or sport fishing	•												
Ecological habitat	•												
Other hazards													
Danger of fire or explosion	•					•					•	•	•
Property damage to nearby dwellings	•										•	•	•
Vapors in dwellings	•					•					•	•	•

TABLE 1.4.2 Potentially applicable corrective-action technologies relative to release volume and chemical characteristics

Technology	Small- to moderate-volume recent gasoline/petroleum release (gas station/tank farm)	Large-volume or long-term chronic gasoline/petroleum release (gas station/tank farm)	Release from tanks containing hazardous substances (organic)	Release from tanks containing hazardous substances (inorganic)
Removal/excavation of tank, soil, and sediments				
Tank removal	•	•	•	•
Soil excavation	•	•	•	•
Sediment removal		•	•	•
On-site and off-site treatment and disposal				
Solidification/stabilization		•	•	•
Landfilling		•	•	•
Landfarming	•	•	•	
Soil washing			•	
Thermal destruction		•	•	
Aqueous-waste treatment	•	•	•	
Deep-well injection				•
Free-product recovery				
Dual pump systems	•	•	•	
Floating Filter pumps	•	•	•	•
Surface oil/water separators	•	•	•	
Groundwater recovery systems				
Groundwater pumping	•	•	•	•
Subsurface drains	•	•	•	•
Subsurface barriers				
Slurry walls		•	•	•
Grouting		•	•	•
Sheet piles				
Hydraulic barriers	•	•	•	•
In situ treatment				
Soil flushing	•	•	•	•
Biostimulation	•	•	•	•
Chemical treatment			•	•
Physical treatment				
Groundwater treatment				
Air stripping	•	•	•	
Carbon adsorption	•	•	•	
Biological treatment	•	•	•	
Precipitation/flocculation/sedimentation				•
Dissolved-air flotation				•
Granular-media filtration				•
Ion exchange/resin adsorption				•
Oxidation/reduction				•
Neutralization				•
Steam stripping			•	
Reverse osmosis				•
Sludge dewatering				•
Vapor migration control, collection, and treatment				
Passive collection systems		•	•	
Active collection systems		•	•	
Ventilation of structures	•	•	•	
Adsorption			•	
Flaring				
Surface-water/drainage controls				
Diversion/collection systems	•	•	•	•
Grading	•	•	•	•
Capping		•	•	•
Revegetation	•	•	•	•
Restoration of contaminated water supplies and sewer lines				
Alternative central water supplies		•	•	•
Alternative point-of-use water supplies	•	•	•	•
Treatment of central water supplies		•	•	•
Treatment of point-of-use water supplies		•	•	•
Replacement of water and sewer lines		•	•	•
Cleaning/restoration of water and sewer lines	•	•	•	•

Typical site problems are presented in Table 1.4.3 along with associated categories of corrective-action technology. Once appropriate corrective-action categories have been identified, the following technical, institutional, environmental, and public-health factors may need to be evaluated:

- Performance, including effectiveness, time to achieve given level of responses, and useful life
- Reliability, including operation and maintenance requirements
- Implementability relative to site conditions
- Safety and health of nearby communities and workers
- Public-health and environmental factors, including the ability to mitigate exposure concerns, effects on environmentally sensitive areas, and exceedance of environmental standards
- Costs, including capital and operations and maintenance

TABLE 1.4.3 Matrix of general corrective-action categories for specific site problems

Site problem	Removal/excavation of soil and sediments	On-site and off-site treatment and disposal of contaminants	Free-product recovery	Groundwater recovery systems	Subsurface barriers	In situ treatment	Groundwater treatment	Vapor migration control, collection, and treatment	Surface-water drainage controls	Restoration of contaminated water supplies and sewer lines
Volatilization of chemicals into air								•		
Hazardous particulates released to atmosphere	•							•		
Dust generation by heavy construction or other site activities										
Contaminated site runoff	•					•			•	
Erosion of surface by water									•	
Surface seepage of released substance	•		•						•	
Flood hazard or contact of surface water body with released substance	•								•	
Released substance migrating vertically or horizontally			•	•	•					
High water table, which may result in ground-water contamination or interfere with other corrective action				•	•					
Precipitation infiltrating site and accelerating released substance migration							•		•	
Explosive or toxic vapors migrating laterally underground							•	•		
Contaminated surface water, groundwater, or other aqueous or liquid waste	•	•		•	•	•	•			
Contaminated soils	•	•				•				
Toxic and/or explosive vapors that have been collected		•						•		
Contaminated stream banks and sediments	•	•								
Contaminated drinking-water distribution system		•		•	•		•			•
Contaminated utilities								•		•
Free product in groundwater and soils	•	•	•	•	•	•	•			

• Institutional factors, including regulatory compliance (local, state, and federal regulations), potential responsible parties, and financial considerations

1.4.6   REFERENCES

1. U.S. Environmental Protection Agency, *Tank Detection Methods: A State-of-the-Art Review,* EPA/600/2-86/011, Hazardous Waste Engineering Research Laboratory, Cincinnati, Ohio, 1986.
2. Ibid.

1.4.7   FURTHER READING

American Petroleum Institute, *Installation of Underground Petroleum Storage Systems,* API Publication 1615, Washington, D.C., 1979.

American Petroleum Institute, *Underground Spill Cleanup Manual,* API Publication 1628, Washington, D.C., 1980.

American Petroleum Institute, *Recommended Practice for Abandonment or Removal of Used Underground Service Station Tanks,* API Publication 1604, Washington, D.C., 1981.

American Petroleum Institute, *Recommended Practice for the Interior Lining of Existing Steel Underground Storage Tanks,* API Publication 1631, Washington, D.C., 1983.

American Petroleum Institute, *Cathodic Protection of Underground Petroleum Storage Tanks and Piping Systems,* API Publication 1632, Washington, D.C., 1983.

American Petroleum Institute, *Groundwater Monitoring and Sample Bias,* API Publication 4367, Washington, D.C., 1983.

American Petroleum Institute, *Treatment Technology for Removal of Dissolved Gasoline Components from Groundwater,* API Publication 4369, Washington, D.C., 1983.

American Petroleum Institute, *Feasibility Studies on the Use of Hydrogen Peroxide to Enhance Microbial Degradation of Gasoline,* API Publication 4389, Washington, D.C., 1985.

American Petroleum Institute, *Test Results of Surfactant Enhanced Gasoline Recovery in a Large-Scale Aquifer,* API Publication 4390, Washington, D.C., 1985.

American Petroleum Institute, *Detection of Hydrocarbons in Groundwater by Analysis of Shallow Soil Gas/Vapor,* API Publication 4394, Washington, D.C., 1985.

American Petroleum Institute, *Laboratory Study on Solubilities of Petroleum Hydrocarbons in Groundwater,* API Publication 4395, Washington, D.C., 1985.

American Petroleum Institute, *Field Evaluation of Well Flushing Procedures,* API Publication 4405, Washington, D.C., 1985.

American Petroleum Institute, *Subsurface Venting of Hydrocarbon Vapors from an Underground Aquifer,* API Publication 4410, Washington, D.C., 1985.

American Petroleum Institute, *Literature Survey: Hydrocarbon Solubilities and Attenuation Mechanisms,* API Publication 4414, Washington, D.C., 1985.

American Petroleum Institute, *Literature Survey: Unassisted Natural Mechanisms to Reduce Concentrations of Soluble Gasoline Components,* API Publication 4415, Washington, D.C., 1985.

American Petroleum Institute, *Cost Model for Selected Technologies for Removal of Gasoline Components from Groundwater,* API Publication 4422, Washington, D.C., 1986.

Atlas, R. M. (ed.), *Petroleum Microbiology,* MacMillan, New York, 1984.

de Pastrovich, T. L., Y. Baradat, R. Barthel, A. Chiarelli, and D. R. Fussell, *Protection of Groundwater from Oil Pollution,* CONCAWE Report #3/79, CONCAWE Secretariat,

Babylon-Kantoren A, Koningin Julianaplein 30-9, 2595 AA The Hague, Netherlands, 1979.

National Fire Protection Association, *Recommended Practice for Control of Flammable and Combustible Liquids and Gases in Manholes, Sewers, and Similar Underground Structures*, NFPA 328-1982, Quincy, Massachusetts.

National Water Well Association, *Petroleum Hydrocarbons and Organic Chemicals in Groundwater*, Proceedings 1984 NWWA/API Conference, Worthington, Ohio, 1984.

National Water Well Association, *Petroleum Hydrocarbons and Organic Chemicals in Groundwater Prevention, Detection, and Restoration*, Proceedings 1985 NWWA/API Conference, Worthington, Ohio, 1985.

U.S. Environmental Protection Agency, *State-of-the-Art of Aquifer Restoration*, EPA 600/2-84/182a and b, R. S. Kerr Environmental Research Laboratory, Ada, Oklahoma, 1984.

U.S. Environmental Protection Agency, *Handbook: Remedial Actions at Waste Disposal Sites (Revised)*, EPA 625/6-85/006, Hazardous Waste Engineering Research Laboratory, Cincinnati, Ohio, 1985.

U.S. Environmental Protection Agency, *Leachate Plume Managements*, EPA 540/2-85/004, Hazardous Waste Engineering Research Laboratory, Cincinnati, Ohio, 1985.

U.S. Environmental Protection Agency, *Underground Storage Tank Corrective Action Technologies*, EPA 625/6-87/015, Hazardous Waste Engineering Research Laboratory, Cincinnati, Ohio, 1987.

CHAPTER 2

HAZARDOUS-WASTE CHARACTERISTICS, QUANTITIES, AND TREATMENT CAPACITIES

SECTION 2.1
QUANTITIES AND SOURCES OF HAZARDOUS WASTES

Stephen K. Dietz
Vice President
Westat, Inc.

Michael E. Burns
Senior Program Analyst
Office of Solid Waste
U.S. Environmental Protection Agency

2.1.1 INTRODUCTION

This section presents the best available information pertaining to the numbers of active hazardous-waste generators and treatment, storage, and disposal facilities in the United States, and to the national quantities of hazardous wastes generated and managed (i.e., treated, stored, and disposed of) annually. It is important to highlight, however, that substantial degrees of uncertainty permeate all the figures presented in this section of the Handbook. Accordingly, readers are encouraged to use this information cautiously and to pay particular attention to qualifications attached to the various estimates presented.

A continuing anomaly of the hazardous-waste system in the United States is the lack of consolidated information describing its nature, scope, and extent. While certain states have assembled detailed inventories of all hazardous-waste activities occurring within their boundaries, hazardous-waste management information at the national level is highly incohesive and characterized by wide margins of uncertainty. The lack of detailed national data results primarily from the structure of the hazardous-waste regulatory program developed by the U.S. Environmental Protection Agency (EPA) under Subtitle C of the Resource Conservation and Recovery Act. As noted in Sec. 1.2, Congress directed the EPA to delegate authority for administering the hazardous-waste regulatory program to the states. Furthermore, states are encouraged to expand the scope and extent of the hazardous-waste regulatory program beyond the minimum national requirements promulgated by the EPA. Accordingly, most information on hazardous-waste management activities flows directly to the states, as opposed to EPA, and efforts to compile state data into national profiles are complicated by variations in the nature and scope of the state programs to which they pertain.

While a consolidated national hazardous-waste information system remains a goal for the future (and is, in fact, actually under development in the EPA's Office of Solid Waste and Emergency Response), a variety of information on hazardous waste and hazardous-waste management practices has been developed through a wide range of diverse and largely independent information-collection activities conducted since the RCRA program first took effect in 1980. The major information sources available to practitioners in the field are described in Subsec. 2.1.2. These information sources range from simple inventories of regulated or potentially regulated hazardous-waste generators and facilities and closed or abandoned waste sites to detailed waste-, site-, process-, and industry-specific studies of hazardous wastes and waste-management practices. The information presented in this chapter is drawn from a variety of these sources and is intended to provide a general overview of the quantities and sources of hazardous wastes. To the extent permitted under EPA and other federal regulations governing the management of confidential business information, practitioners in need of specific hazardous-waste information should consult these sources directly.

Data presented in this chapter pertain primarily to hazardous wastes and hazardous-waste handlers subject to regulation under Subtitle C of RCRA, and are drawn primarily from two sources: The EPA's Hazardous Waste Data Management System (HWDMS) and its associated Consolidated Data Base (CDB), maintained by the Office of Solid Waste; and the *National Survey of Hazardous Waste Generators and Treatment, Storage, and Disposal Facilities Regulated Under RCRA in 1981,* a detailed statistical survey covering hazardous-waste management practices as of 1981. The survey was conducted for the Office of Solid Waste by Westat, Inc., of Rockville, Maryland, and completed in April 1984. The HWDMS and CDB sources contain the most current data on the number of hazardous-waste treatment, storage, and disposal facilities regulated under RCRA, and contain lists of all entities that have notified EPA that they are or may be hazardous-waste generators or transporters. Data on quantities of hazardous wastes generated and managed are generally not available from the HWDMS and CDB sources. Westat's 1981 national survey represents the only source of comprehensive national data on hazardous-waste quantities. While its 1981 estimates are increasingly dated, the general consensus at the EPA is that they provide a reasonable indication of the magnitude and nature of RCRA-regulated hazardous wastes.

2.1.2 SOURCES OF STATISTICAL ESTIMATES

There are a variety of sources of data on hazardous-waste generation and management in the United States. These information sources or data bases differ in terms of their purpose (e.g., regulatory versus research and planning), scope (e.g., what industry sectors are included), content (i.e., what specific variables or measures are recorded), and time frame. Unfortunately no single data base exists that is comprehensive, extensive, and current! For this reason we have provided brief descriptive summaries of five different information sources so that practitioners and researchers may be aware of these various sources, their attributes, and their limitations.

EPA's Official Data from the Hazardous Waste Data Management System (HWDMS)

The purpose of the Hazardous Waste Data Management System is regulatory rather than planning or research. It contains a record for every facility in the United States that has applied for an EPA permit to treat, store, or dispose of hazardous wastes regulated under RCRA. It also includes information on the type of process that "will be" used (i.e., container, tank, waste pile, surface impoundment, injection well, landfill, land application, ocean disposal, or incinerator). While the HWDMS seems to serve its regulatory purposes well, it has some definite drawbacks for certain research and planning purposes, namely:

1. There is no current information on quantity of waste managed.
2. The process types (tank, landfill, injection well, etc.) are not reliable representations of the processes currently in use; of the technologies listed in the part A permit in 1980, many were *not* confirmed as currently in use in Westat's 1981 survey (confirmation rate ranged from only 30% for waste piles up to 74% for storage containers).*
3. Many of the facilities (20%) were not managing hazardous waste at all in 1981.
4. In recent years, two-thirds of the facilities with interim permits have officially withdrawn their permit applications.

EPA National Survey of Hazardous Waste in 1981, Conducted by Westat, Inc.

The first national probability survey on hazardous-waste generation and management was conducted for EPA by Westat, Inc. in 1982. It provided the first statistically reliable estimate of (1) the numbers of facilities generating, treating, storing, and disposing of hazardous waste, (2) the types of processes used, and (3) the characteristics of such facilities. The results of this survey were published in the report, *National Survey of Hazardous Waste Generators and Treatment, Storage and Disposal Facilities Regulated Under RCRA in 1981,* by Westat, Inc., April 1984, available from the U.S. National Technical Information Service. The data is also available on computer file at EPA. The disadvantage of this information is that it is somewhat dated; however, it was the most reliable data available in 1986 and therefore has been used heavily in the statistical estimates of this chapter.

Report on the Telephone Verification Survey of Hazardous Waste Treatment, Storage, and Disposal Facilities Regulated Under RCRA in 1981, Westat, Inc., Rockville, Maryland, November 11, 1982.

National Update Survey on Hazardous-Waste Facilities under RCRA

At the EPA's request, Westat designed a sample for a follow-up survey to update the 1981 estimates. The data are scheduled to be collected by Research Triangle Institute in 1986, and revised figures should be available in 1987.

EPA Biennial Report

Every two years, starting in 1981, EPA has conducted a census of hazardous-waste treatment, storage, and disposal facilities. A limited number of key information items are included in the survey, including: facility identifiers, waste description (DOT code and EPA waste number), and amount of each waste type for each facility to which (or from which) waste was shipped.

Despite the limited number of information items, the attractiveness of this data source is that it is kept current (it is updated every two years) and that it is a "census" rather than a sample survey, thus eliminating sampling error. However, there are some problems with the Biennial Report data. Approximately two-thirds of the states have been authorized to conduct their own collection, processing, and tabulation of the data. These "authorized" states are then supposed to provide the EPA with a standard set of tabulations so that the EPA can combine these results into "national statistics" by combining authorized states' data with EPA-tabulated data for the remaining states. The problems with the current process appear to grow out of the substantial differences in the data-collection requirements of the various authorized states and the difficulties involved in getting authorized states to report a truly standardized set of figures to the EPA. In addition, improvements in the data-editing, verification, and data-gap-filling procedures are needed to produce reliable figures. EPA has initiated a study to validate information from the Biennial Report and develop improved data-preparation procedures and quality controls. Therefore, an improved source of hazardous-waste data is anticipated from the 1987 Biennial Report.

2.1.3 HAZARDOUS-WASTE GENERATION IN THE UNITED STATES

This section presents best available estimates of the numbers of hazardous-waste generators regulated under Subtitle C of RCRA and the quantities of RCRA-regulated hazardous wastes annually generated. As noted in Subsec. 2.1.1, large degrees of uncertainty permeate the estimates presented in this subsection and the next, which addresses the management (treatment, storage, and disposal) side of the hazardous-waste system. Furthermore, different factors account for the uncertainties inherent in each set of estimates. Hazardous-waste generators are regulated differently from treatment, storage, and disposal facilities, accounting for greater difficulty in estimating the number of regulated generators than in estimating the number of regulated facilities. Estimating regulated quantities of hazardous waste is considerably more difficult than estimating the numbers of regulated handlers. Discussions of the specific factors affecting certainty (or uncertainty) that accompany each of the estimates are presented in these two subsections.

Numbers of Hazardous-Waste Generators

Total Number of Generators. At the end of April 1986, EPA's Hazardous Waste Data Management System listed 68,265 entities as RCRA-regulated hazardous-waste generators. However, this number at once overestimates and underestimates the true population of hazardous-waste generators, depending upon how the population of hazardous-waste generators regulated under Subtitle C of RCRA is defined. As noted above, hazardous-waste generators are regulated differently under RCRA than are facilities. Once a hazardous-waste treatment, storage, or disposal facility applies for and is granted interim status or becomes fully permitted, it remains regulated under Subtitle C of RCRA until it undertakes and completes formal closure proceedings approved of by the administering government (its state, if authorized, or EPA). The regulatory status of hazardous-waste generators, however, can change from year to year and even from month to month. Hazardous-waste generators are regulated by EPA under Subtitle C of RCRA only when the hazardous wastes they generate exceed quantities specified in EPA's regulations. Accordingly, a generator may be regulated under RCRA in one year but not in the next, or vice versa, a fact which severely complicates the process of estimating the actual number of hazardous-waste generators in any one year.

All generators of federally regulated types and quantities of hazardous waste were required to submit notification forms to the EPA in 1980 or at the time they commenced generating hazardous wastes. The 68,265 figure represents the sum total of all entities that have submitted such forms to EPA since 1980. It is highly unlikely, however, that all these entities generated hazardous waste during 1986. In fact, when Westat, Inc., conducted an extensive survey of the nearly 56,000 generators listed in HWDMS on August 1, 1982 (questionnaires were mailed to a sample of approximately 11,000 of these generators, and more than 90% were completed and returned), it found that in 1981 only 25%, or roughly 14,100, had actually generated quantities of hazardous wastes regulated by EPA under RCRA (see Fig. 2.1.1). Of the 75 percent that indicated they were not regulated under RCRA during 1981, nearly half stated that they had never generated hazardous wastes and did not expect to at any point during the next five years. Smaller percentages indicated that they were small-quantity generators or potential future generators. Many firms submitted notification forms as "protective" filings either to obtain an EPA hazardous-waste identification number in case they should ever need one for transportation purposes or to guard against charges that they failed to comply with EPA requirements.

If the 25% confirmation rate found by Westat in 1981 were to have held true in 1986, roughly 17,000 of the 68,265 entities listed in HWDMS could have been expected to have generated federally regulated quantities of hazardous wastes in 1986. However, Westat's estimate may have understated the size of the federally regulated generator population. When the EPA compiled state summaries of 1983 Biennial Report submissions, it counted roughly 43,000 active 1983 generators. Of these, however, roughly 13,000 were identified as small-quantity generators regulated solely by individual states and not by EPA, leaving up to 30,000 federally regulated hazardous-waste generators. Due to a number of anomalies associated with the 1983 Biennial Report compilation process, however, the 30,000 count may still contain large numbers of state-regulated generators not subject to federal regulation. Accordingly, the EPA has never formally issued the 30,000 figure as an estimate of the number of federally regulated hazardous-waste gen-

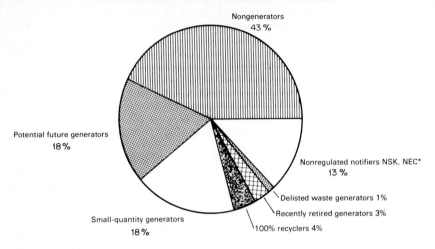

FIG. 2.1.1   Portion of 68,265 notifiers that generated RCRA-regulated quantities of hazardous wastes in 1981. NSK = not specified by kind. NEC = not elsewhere classified.   (*HWDMS, 6-1-86*)

erators. Nonetheless, it is likely that more than 25% of the 68,265 HWDMS-listed entities generated federally regulated quantities of hazardous waste in 1986.

Except for the 43,000 figure developed from the 1983 Biennial Report, which is clouded by considerable uncertainties, no estimate is available for the number of active hazardous-waste generators regulated *either* by EPA or individual states. As noted in previous sections, states are given the latitude under RCRA to regulate hazardous-waste handlers and activities not covered under federal law. Accordingly, the 68,265 entities listed in HWDMS may not include certain classes of state-regulated generators (such as state-regulated small-quantity generators or generators of large volumes of low-hazard wastes that are regulated in some states but not by EPA).

Finally, the April 1986 HWDMS count of 68,265 surely understates the size of the population of hazardous-waste generators that will be regulated by EPA under RCRA in 1987 and beyond. Commencing November 8, 1986, the federally regulated population of hazardous-waste generators included an estimated 125,000 entities generating between 100 and 1,000 Kg (220 and 2,200 lb) of hazardous waste per month. These previously exempt small-quantity generators are now subject to the same handling requirements (with a few minor exceptions) imposed upon "large-quantity" generators regulated since 1980. One can expect that any problems encountered in estimating the size of the original population of federally regulated hazardous-waste generators will be magnified considerably in the future.

Figure 2.1.2 compares all the estimates presented in this section, indicating the extreme variation across efforts to quantify the size of the population of RCRA-regulated hazardous-waste generators.

Number of Generators by EPA Region.   While there is considerable disagreement between the number of hazardous-waste generators listed in HWDMS and the number estimated by Westat to be active hazardous-waste generators, the distribution of generators across EPA's 10 administrative regions is characterized sim-

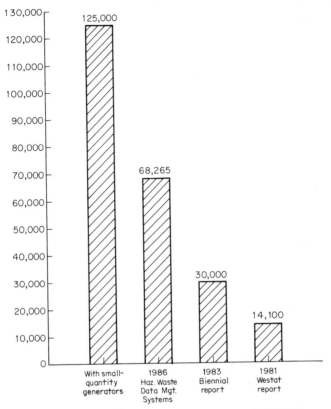

FIG. 2.1.2   Number of hazardous-waste generators.   (*HWDMS*)

ilarly by both HWDMS and the Westat survey. Estimates drawn from both sources of the portions of the federally regulated population of hazardous-waste generators located in each of EPA's regions are presented in Table 2.1.1, and Westat's estimates of the actual number of active 1981 generators in each region are shown in Fig. 2.1.3.

As indicated in Table 2.1.1 and Fig. 2.1.3, EPA's Region 5, representing the industrial states of Illinois, Indiana, Michigan, Minnesota, Ohio, and Wisconsin, contains the largest single portion of the population of federally regulated hazardous-waste generators. Six of EPA's regions, covering the remaining areas of heavy industrial activity in the United States, are closely grouped in their respective percentages of the federal population (and are virtually indistinguishable within the statistical confidence intervals achieved for these estimates in Westat's survey). The three remaining EPA regions, stretching from the midwestern farm belt through the northern plains and Rockies to the northwestern sections of the country, account for relatively small proportions of the total population of hazardous-waste generators in the United States.

Number of Generators by Industry Type.   As indicated in the introduction to the chapter, hazardous wastes are generated in nearly every sector of the U.S. econ-

TABLE 2.1.1 Comparison of HWDMS and Westat Survey Distributions of Hazardous-Waste Generators across EPA Regions

Rank	Region	Percent of Westat estimated active generators in 1981	Percent of HWDMS listed generators as of:	
			8/1/82	5/15/86
1	5	23.0	22.4	18.5
2	9	14.1	10.2	18.1
3	4	13.0	13.1	10.3
4	2	11.8	12.4	21.0
5	1	10.2	8.0	6.1
6	3	9.9	9.7	8.6
7	6	8.8	14.3	7.7
8	7	5.1	4.6	4.8
9	10	2.4	3.1	3.5
10	8	1.7	2.2	1.4
Total numbers		14,100	55,739	68,265

omy, ranging from traditional heavy industries to food production, service industries (including hospitals), and federal, state, and local governmental agencies. Despite its broad scope, however, the population of federally regulated hazardous-waste generators (prior to regulation of small-quantity generators) was concentrated in the manufacturing sector (Standard Industrial Codes 2000–3999). Figure 2.1.4 indicates the proportions of generators in major industrial sectors.

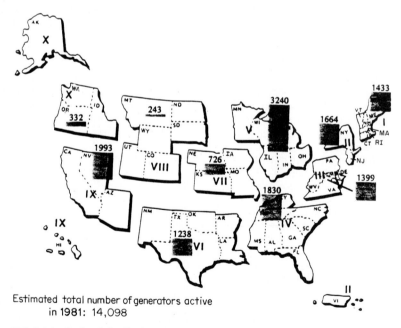

Estimated total number of generators active in 1981: 14,098

FIG. 2.1.3 Regional distribution of hazardous-waste generators in 1981.

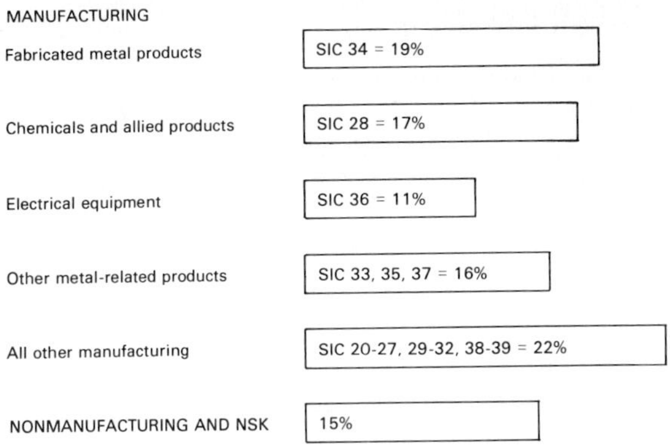

MANUFACTURING

Fabricated metal products

| SIC 34 = 19% |

Chemicals and allied products

| SIC 28 = 17% |

Electrical equipment

| SIC 36 = 11% |

Other metal-related products

| SIC 33, 35, 37 = 16% |

All other manufacturing

| SIC 20-27, 29-32, 38-39 = 22% |

NONMANUFACTURING AND NSK

| 15% |

FIG. 2.1.4 Number of hazardous-waste generators by industry type. SIC = Standard Industrial Code. NSK = not specified by kind.

Figures are presented as percentages only because of the uncertainty about the actual number of generators and the number in each group.

As illustrated in Fig. 2.1.4, manufacturing industries account for approximately 85% of the total population of federally regulated hazardous-waste generators. Metal-related industries (fabricated metals and other metal-related products) account for a substantial proportion of this group. The chemicals and allied products group (SIC 28) is second only to the fabricated metal products group in terms of its share of the generator population, and if expanded to include petroleum-related industries (SIC 29) would represent the single largest generator group. The petrochemical industries dominate the hazardous-waste picture when quantities generated are examined (see under *Quantity of Hazardous Waste Generated* on p. 2.13).

Nonmanufacturing generators operate in many service industries, ranging from transportation to maintenance and health care. The population of federally regulated generators will be dominated by these sectors once small-quantity generators are regulated, although the influx of the large numbers of these generators is not expected to significantly increase the quantities of hazardous wastes generated (see under *Quantity of Hazardous Waste Generated* on p. 2.13).

Number of Generators Shipping Waste Off Site. As observed in Westat's 1981 survey, most generators (62%) ship all of the hazardous wastes they generate to off-site treatment, storage, and disposal facilities within 90 days of generation (thereby avoiding regulation under the permitting requirements established for treatment, storage, and disposal facilities). Slightly less than one-quarter of the generators (22%) manage at least some of their wastes on site either prior to or in addition to shipping hazardous wastes to off-site facilities for final disposition. Finally, less than one in five generators (only 16%) manage all of their hazardous wastes completely on site, with no shipments of hazardous wastes to off-site facilities. Figure 2.1.5 illustrates the breakout of the generator population into these three categories.

Readers are encouraged to contrast these figures with the information on the distribution of hazardous-waste quantities managed on site versus off site presented under *Quantity of Hazardous Waste Generated* (p. 2.13). While most gen-

Number of generators

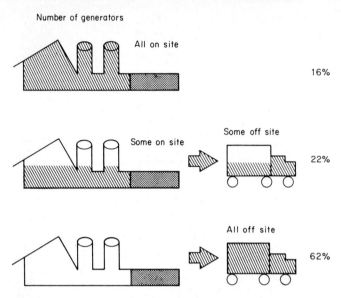

FIG. 2.1.5 Number of generators shipping hazardous waste off site in 1981.

erators ship some or all of their wastes off site for final disposition, the overwhelming percentage of the quantity of hazardous wastes generated annually is managed on site (95% of the total quantity of hazardous wastes generated annually are managed entirely on site; only 5% are shipped off site). This apparent anomaly is explained by the fact that a few very large generators account for most of the hazardous wastes generated annually, generally as large wastewater streams, and are able to take advantage of economies of scale in constructing on-site treatment and disposal facilities (see under *Quantity of Hazardous Waste Generated* on p. 2.13). Most generators, however, find it much more economical to ship the relatively small quantities of hazardous wastes that they generate to off-site commercial hazardous-waste management facilities.

Many of the generators falling into the "ship some wastes off site and manage some wastes on site" category are simply storing their hazardous wastes for more than 90 days prior to off-site shipments, and are not engaged in actually treating or disposing of hazardous wastes on site. As explained in Subsec. 2.1.2, generators who store hazardous waste on site in tanks or containers for less than 90 days are exempt from regulation as hazardous-waste storage facilities. Since compliance costs for facilities (even storage-only facilities) are generally significantly greater than compliance costs for generators, generators have economic incentives to take advantage of the 90-day storage exemption, and most generators plan their hazardous-waste shipments accordingly. In many cases, however, generators do not feel compelled to take advantage of this exemption, and allow their wastes to accumulate on site for more than 90 days before shipping them off site for treatment or disposal. Other generators in the "some on-site, some off-site" category actually treat or dispose of some wastes in on-site processes, while shipping other types of hazardous wastes off site, and some treat their hazardous wastes on site and then ship treatment residuals (which are often still hazardous wastes) to off-site facilities for final disposition.

Quantity of Hazardous Waste Generated

While some of the estimates presented in the previous section were drawn from EPA's Hazardous Waste Data Management System, HWDMS does not provide useful estimates of hazardous-waste quantities. Accordingly, all the estimates presented in this section are drawn from Westat's 1981 survey, which at the time of this writing remains the only authoritative source of national data on quantities of hazardous waste.

Prior to presenting the quantity estimates for generation of hazardous wastes, it is important for readers to understand more specifically what the estimates include and exclude. Specifically excluded from the generation estimates are:

- Quantities of wastes regulated as hazardous solely by states and by EPA
- Quantities of RCRA-regulated hazardous wastes that are managed exclusively in processes that are exempt from regulation under RCRA (e.g., RCRA-regulated hazardous wastewaters that are managed in tank treatment systems exempt from RCRA due to regulation under the National Pollutant Discharge Elimination System, provided such hazardous wastewaters do not pass through surface impoundments)
- Quantities of waste that are exempt from regulation as hazardous wastes under RCRA (e.g., wastes discharged to sewers or POTWs, wastes generated by installations engaged in extraction or benefication of minerals, radioactive wastes
- Quantities of wastes generated during emergency- or remedial-response actions taken under the Comprehensive Environmental Response Liability and Compensation Act (Superfund cleanups)

Specifically included in the generation estimates are:

- Large volumes of wastewaters that have been mixed with hazardous wastes
- Residues from the treatment of listed hazardous wastes

Readers requiring precise descriptions of these inclusions and exclusions are encouraged to consult Westat's April 1984 summary report, as well as the EPA hazardous-waste regulations that they attempt to track (see Subsec. 2.1.1).

Quantity Generated. The best national estimate of the quantity of RCRA-regulated hazardous waste generated in 1981 is 264 million t. This includes solid hazardous waste and associated wastewater since both are hazardous waste under RCRA. This estimate, based on combined results from national EPA-sponsored surveys conducted by Westat, Inc., in 1982 and 1983 is subject to substantial sampling error ($\pm$ 50% at the 95% confidence level). The reason for this unusually wide confidence interval is the inherently skewed distribution of generator sizes (i.e., relatively few very large generators produce the vast majority of the total quantity of hazardous waste generated in the nation) and the fact that generators were sampled with approximately equal probability regardless of their size.

Despite this large degree of uncertainty, the 264 million t estimate produced by

Westat's survey served to reshape the EPA's understanding of the dimensions of the hazardous-waste management system in the United States. Prior to the survey, the quantity of hazardous waste generated annually had been estimated to be in the neighborhood of 40 million t. This earlier estimate was, however, based upon studies of the quantities of hazardous waste received at commercial hazardous-waste management facilities and did not, accordingly, adequately address the huge quantities of RCRA-regulated hazardous wastes managed entirely on site.

Size Distribution.   The size distribution of the population of hazardous-waste generators is highly skewed. A few extremely large generators account for most of the 264 million t of hazardous waste generated annually. In fact, as indicated in Fig. 2.1.6, more than 95% of the total quantity generated is accounted for by fewer than 10% of the generators.

The hazardous wastes generated by these extremely large generators are generally comprised mostly of wastewaters that have either been mixed with hazardous wastes or that result from large wastewater-treatment systems utilizing large volumes of water during the treatment process. The solids contents of these streams is often in the parts per million range. Dry-weight estimates of hazardous-waste quantities (excluding water content) are not available, since the

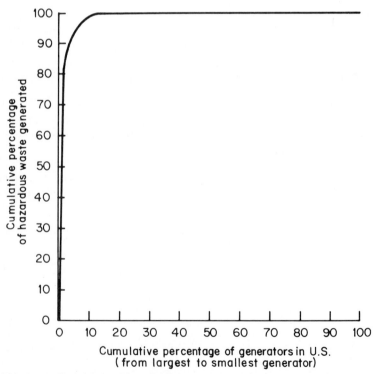

FIG. 2.1.6 Comparison of cumulative distributions of quantity of hazardous waste generated in 1981 and the number of generators.

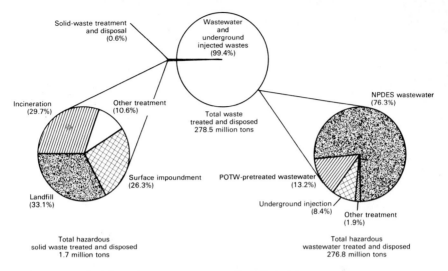

FIG. 2.1.7   1984 CMA hazardous-waste survey results (725 plants).

entire quantity of these dilute streams are specifically regulated under RCRA. The Chemical Manufacturer's Association has, however, developed partial estimates of quantities of hazardous waste generated by its members, breaking out "solid" hazardous wastes from hazardous wastewaters. In 1984, the most recent survey year available before this handbook went to press, members reported to CMA that they generated 1.7 million tons of "solid" hazardous wastes and 276.8 million t of hazardous wastewaters (see Fig. 2.1.7). Readers are reminded that CMA members are not the only manufacturers regulated under RCRA, and should note that the CMA survey is voluntary and not statistical in nature (i.e., the results cannot be used to extrapolate population estimates even for the chemical industry). CMA believes, however, that its estimates account for most of the quantities of hazardous waste generated by its membership, since most of the larger companies responded. While CMA's total estimates appear compatible with EPA's 1981 (and 1985) quantity estimates, they may not provide a very good indication of the "solids" portion of the 264 million tons of hazardous waste estimated to be generated annually across all industries (solids contents are likely to be greater in other industries).

On Site vs. Off Site.   Nearly all of this waste (96%) was managed on site (i.e., at the generator facility). Most of the waste shipped off site was shipped to some other (usually commercial waste-management) firm while only 18% (of the 4% shipped off site) went to a separate location owned by the generating firm.

Industry Sectors.   According to Fig. 2.1.8, the chemical and petroleum industries account for most (71%) of the hazardous waste generated. In fact, the chemical and petroleum industries may account for as much as 85% of the total quantity generated. Under either estimate, however, the lopsided distribution of hazardous-waste quantities across industry types differs substantially from the more even distribution of the number of hazardous-waste generators.

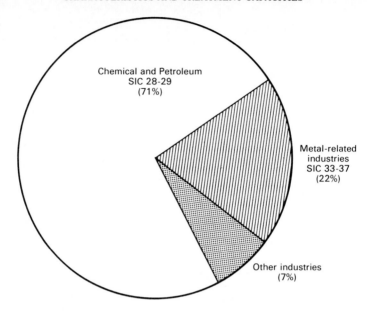

FIG. 2.1.8 Quantities of hazardous waste generated in 1981 by industry type.

Quantities Recycled. Very little hazardous waste was recycled in 1981 (only 4% of the total generated) even though 40% of generators say they recycle some hazardous waste. Not counted in these figures, however, are currently unknown quantities of otherwise hazardous waste that are recycled and, accordingly, exempt from regulation under RCRA. Most (81%) of the hazardous waste that was recycled in 1981 was recycled on site and the rest was shipped off site to be recycled. There are reasons to expect that the percentage of hazardous waste recycled has increased since 1981 as a result of increasing concerns about long-term liabilities associated with hazardous-waste disposal, restrictions on land-based management of hazardous waste, congressional proposals for taxes on hazardous waste, and increasing recognition of the resource value of constituents contained in hazardous wastes.

Types of Waste. Certain types of hazardous waste are far more prevalent than others among the types of waste generated, as shown in Table 2.1.2. (When reviewing Table 2.1.2, it is important to remember that a waste may fall in more than one "EPA-listed category" and more than one "characteristics category," and therefore percentages can add up to more than 100%.)

2.1.4 IMPACTS OF FUTURE CHANGES IN THE RCRA HAZARDOUS-WASTE MANAGEMENT SYSTEM

Although quantitative data relating to anticipated future rates of hazardous-waste generation have not been developed to date, it is appropriate nonetheless to con-

TABLE 2.1.2   Generated Waste in Order of Prevalence of Type

Waste code	Type of waste	Waste, % by weight
"F" or "K" wastes	Spent solvents, process wastewaters and sludges, and listed industry wastes	167
D003	Reactive wastes	52
D002	Corrosive wastes	35
D004–D017	Toxic wastes	10
D001	Ignitable wastes	1
"U" or "P" wastes		1
Unspecified		5

clude this chapter with a discussion of the factors that are likely to contribute to increases or decreases in the numbers of RCRA-regulated hazardous-waste generators and TSDFs and the annual quantities of regulated hazardous waste generated and managed in future years.

The quantity of RCRA-regulated hazardous wastes generated at any point in time is essentially affected by two factors:

- The specific nature and scope of the RCRA hazardous-waste regulatory program at that time
- The nature of the industrial and other activities that actually result in hazardous by-products

In its 1984 amendments to RCRA, Congress enacted a number of major new provisions in the Subtitle C hazardous-waste management system that will have profound effects on the nature and scope of RCRA-regulated waste-generation and management practices over the next 5 to 10 years. The impact of the change in the small-quantity generator cutoff—reduced from 1,000 kg/mo to 100 kg/mo (2,200 lb/mo to 220 lb/mo)—on the number of regulated hazardous-waste generators has already been noted (see under *Total Number of Generators* on p. 2.7). EPA expects that as many as 125,000 new entities will join the universe of RCRA-regulated generators as a result of this change. While the agency expects the level of regulatory attention applied to this new class of generators to be small relative to the attention focused on the larger generators, all of these previously exempt generators will be required to obtain RCRA identification numbers from EPA. As noted previously, however, the influx of these large numbers of new generators is not expected to have a significant impact on the quantities of hazardous waste generated each year. The impact of these additional quantities may, however, be more significant in relation to commercial waste-management capacity, since few of these generators manage their wastes on site.

Another major provision enacted by Congress in the HSWA has already had its effect: the requirement for certification of compliance with groundwater-monitoring and financial responsibility regulations by November 8, 1986, as a condition for maintaining interim status for land disposal operations. Approximately two-thirds of the population of land disposal facilities closed or are in the process of closing as a direct result of this provision. A second HSWA provision that is expected to reduce the number of land disposal facilities is the requirement for all surface impoundments to retrofit to comply with liner and other permitting standards originally intended only for newly constructed units. The retrofit requirements take effect in 1988, by which time many existing surface impoundments are expected to cease operations. Many of these were already beginning to close in 1986 [in fact, many of the facilities that closed under the (LOIS) provisions were surface impoundments that owners and operators did not intend to retrofit].

Another major provision of the HSWA that will have a significant impact on waste-management practices is the restriction on land disposal of hazardous wastes. Congress dictated that land disposal of untreated hazardous waste should cease according to a specific schedule between 1986 and 1990. The EPA is in the process of developing treatment standards that wastes must comply with prior to placement in land disposal units. These requirements are expected to reduce the volumes of wastes land-disposed and increase volumes going to treatments, particularly incineration.

The EPA is also considering the regulation of mining wastes and other large-volume wastes that are currently exempt from regulation under RCRA. To the extent that these wastes are in fact brought under regulation, the quantities of wastes included in the RCRA hazardous-waste management system will increase greatly.

The EPA has also taken steps to bring into the RCRA regulatory system the burning and blending of hazardous wastes in boilers and furnaces, and a variety of activities relating to the management of used oil. The EPA estimates that as many as 1,000 large on-site burners of hazardous-wastes fuel supplements and approximately 50,000 generators, intermediaries, and burners of off-specification used oils will be added to the RCRA system once these provisions have been fully implemented.

The EPA is also beginning to rethink its hazardous-waste identification system. Changes to the basic structure of this system, which currently has three tiers (broad characteristic tests for ignitability, corrosivity, reactivity, and extract toxicity; specific listings of individual industrial-waste streams; and delistings of specific waste streams), could substantially increase the quantities of wastes regulated as hazardous under RCRA or, conversely, could release many hazardous wastes from control under the RCRA Subtitle C regulatory system.

Implementation of industrial pretreatment standards by EPA and various states may also affect future hazardous-waste generation. Currently, an unestimable quantity of otherwise hazardous wastes are exempted from regulation under RCRA because they pass legally through public sewer systems into Publicly Owned Treatment Works (POTWs). While the quantities of such wastes cannot be estimated by this survey, information from other sources suggests that such quantities are indeed large. The Massachusetts Bureau of Solid Waste Disposal, for example, estimates that the quantity of hazardous waste generated annually in Massachusetts could increase by more than 50 percent with the implementation of pretreatment standards and the installation of industrial pretreatment processes by currently exempt generators. Similar experiences in other states would add considerably to the quantities of hazardous waste regulated under RCRA in future years.

Increased industrial output, prompted by improved economic conditions, could also increase hazardous-waste generation in future years. In 1981, the national economy was in a slump, resulting in reduced industrial output. The survey was not able to determine the relationship between changes in the levels of industrial production and changes in the quantities of hazardous waste generated. However, it is generally believed that there exists a direct relationship between these two factors. Thus, hazardous-waste generation would be expected to increase during periods of improved economic conditions.

Finally, increased cleanup of abandoned or closed hazardous-waste sites will generate additional quantities of hazardous waste that require proper disposal under RCRA. The survey observed some such quantities being generated during the 1981 calendar year. Implementation of the Superfund program by the EPA and the states has advanced since 1981 and is expected to continue to advance in coming years, with resulting increases in the quantities of hazardous waste to be managed.

While the factors stated above are among the factors that are expected to result in increases in hazardous-waste generation in future years, a number of other factors are expected to contribute to decreases in future hazardous-waste generation. Prominent among these are proposals in Congress and some states to adopt specific taxes on the generation and/or disposal of hazardous waste. These taxes are often intended, among other things, to provide economic disincentives toward the actual generation of hazardous waste. These and other existing or future economic factors are expected to encourage industries to engage in greater "source reduction" and "source separation" to eliminate or reduce their generation of hazardous waste.

Further regulatory actions by the Congress and the EPA that may encourage or require reductions in hazardous-waste generation include proposed bans on the land disposal and underground injection of certain hazardous wastes. As disposal options for hazardous waste become more limited and more costly, generators will be encouraged to reduce their hazardous-waste generation.

Finally, one of the purposes behind the enactment of the Resource Conservation and Recovery Act was to encourage the conservation of natural resources through the use, reuse, recycling, and reclamation of materials contained in industrial and other waste streams. As the value of natural resources increases, economic incentives for such recycling activities are expected to increase, resulting in increased efforts by generators to turn their waste streams into useful, valuable commodities.

SECTION 2.2
HAZARDOUS-WASTE
MANAGEMENT CAPACITY

Gregory A. Vogel
The MITRE Corporation
McLean, Virginia

This section focuses on the capacity of hazardous-waste management units that are operating within the Resource Conservation and Recovery Act (RCRA) regulatory program under interim status on a final permit. RCRA hazardous wastes are defined in 40 *CFR* 261 to include wastes that exhibit the characteristics of ignitability, corrosivity, reactivity, or EP toxicity; those wastes from listed specific or nonspecific sources; or specified discarded commercial chemical products. Both hazardous and nonhazardous wastes could be managed in any type of unit. For the purposes of this study, it will be assumed that the entire capacity of RCRA-regulated units can be dedicated to hazardous-waste management. Under the Hazardous and Solid Waste Amendments of 1984 (HSWA), some solid-waste management units at RCRA-regulated facilities may be subject to RCRA corrective-action authority if specific conditions exist. The capacity of such solid-waste management units is not addressed in this study.

The design capacity of hazardous-waste management units is used throughout this report. The design, or nameplate, capacity is the throughput or volume for which a unit is initially engineered. Actual throughputs may be higher than the design capacity because of process modifications or they may be lower because of improper design or lack of hazardous-waste input.

2.2.1 CAPACITY DATA SOURCES

EPA Sources

A primary source of capacity information is the RCRA Part A and Part B permit applications received by the U.S. Environmental Protection Agency (EPA) from hazardous-waste management facilities. The Part A application is a brief, coded notification form requiring the specification of unit capacities among other items. However, the responses do not always conform to the intent of the form and the Part A data are not always reliable. For example, an incinerator capacity might be incorrectly entered on a Part A if a waste is sent off site to be incinerated. A reviewer would not know whether an incinerator existed at the facility or not. The MITRE Corporation found from a 1981 telephone survey that approximately half of the facilities indicating an incinerator capacity on the Part A applications actually had such units.[1] The due dates of the Part A applications have passed except for new facilities to be constructed or units brought into the RCRA program through recent amendments.

The Part B permit applications require narrative descriptions of waste-management practices and they are less likely to contain erroneous information. Part B applications from existing land disposal facilities were due by November 8, 1985, those from incinerators were due by November 8, 1986, and those for other facilities were due by November 8, 1988. Several projects to summarize technical data from Part B applications are in progress and the results are not available at present.

The progress of RCRA Part A and Part B applications through the permitting process is tracked by the Hazardous Waste Data Management System (HWDMS), a computerized data base maintained at the EPA's National Computer Center at Research Triangle Park, North Carolina. All data from the Part A applications are entered into this data base at the EPA regional offices, so it does contain a limited amount of capacity information. Most other data entries are a record keeping of schedules and compliance activities for internal EPA use. Another data base that will contain technical information from Part B applications is being developed by the Oak Ridge National Laboratories at Oak Ridge, Tennessee, for the EPA Hazardous Waste Engineering and Research Laboratory at Cincinnati, Ohio. Initially, the Hazardous Waste Control Technology Data Base at Oak Ridge will contain information from hazardous-waste incinerators and may be expanded to include other waste-management technologies.

A widely quoted source of capacity information is the National Survey of RCRA Facilities conducted by Westat, Inc., of Rockville, Maryland in 1981.[2] Questionnaires were mailed to approximately one-third or one-half of the hazardous-waste management facilities known to EPA at that time. The responses were tabulated in a computerized data base and published in the referenced document.

Commercial Sources

Directories.   Several directories of commercial hazardous-waste management facilities have been published. Although these directories generally do not provide capacity information, they are useful for indicating the technologies available at specific commercial facilities. Some directories are listed to assist readers seeking to dispose of wastes off site:

- *Industrial and Hazardous Waste Management Firms:* Published annually by Environmental Information, Ltd., of Minneapolis, Minnesota.
- *Hazardous Waste Services Directory:* J. J. Keller and Associates, Inc., of Neenah, Wisconsin.
- *Hazardous Wastes Management Reference Directory:* Published annually by Rimbach Publishing of Pittsburgh, Pennsylvania.

- *Pollution Equipment News Catalog and Buyers Guide:* Published annually by Rimbach Publishing of Pittsburgh, Pennsylvania.
- *Pollution Engineering Consultants and Services Telephone Directory,* and the *Environmental Control Telephone Directory:* Published separately and annually by Pudvan Publishing Co. of Northbrook, Illinois.

Services. Several services for locating hazardous-waste transporters and management firms are also available.

- The National Solid Waste Management Association of Washington, D.C., is available at (212) 659-4613.
- The Institute of Chemical Waste Management, a member institute of the National Solid Waste Management Association, offers a *Directory of Hazardous Waste Treatment and Disposal Facilities.*
- The Government Refuse Collection and Disposal Association of Silver Spring, Maryland, at (301) 585-2898, also offers assistance.
- The Northeast Industrial Waste Exchange of Syracuse, New York, at (315) 422-6572, is a nonprofit government-supported information clearinghouse for waste exchanges. It maintains a computerized service that lists (1) potential buyers, sellers, and brokers for hazardous-waste transactions and (2) the wastes that are available.
- The Southern Waste Information Exchange of Tallahassee, Florida, at (904) 644-5576, offers services similar to those of the Northeast Industrial Waste Exchange.
- The Industrial Materials Exchange Service of Springfield, Illinois, at (217) 782-0450, also offers clearinghouse services similar to those of the Northeast Industrial Waste Exchange.

Additional information about commercial waste-management capacity is found in Subsec. 2.2.3.

2.2.2 WASTE-MANAGEMENT CAPACITY ESTIMATES ITEMIZED BY TECHNOLOGY

Most of the information on hazardous-waste management capacity given in this subsection is obtained from Part A permit application information entered into the HWDMS, except as noted. Table 2.2.1, developed by the EPA from the HWDMS data, presents an overview of the number of facilities subject to RCRA regulation by EPA region as of July 31, 1985, prior to the effective date of HSWA.[3] Large-quantity generators produce more than 1,000 kg (2,200 lb) of hazardous waste per month. The EPA estimates that between 100,000 and 130,000 small-quantity generators producing from 100 to 1,000 kg (220 to 2,200 lb) of hazardous waste per month were brought into the RCRA program through HSWA. The EPA also estimates that approximately 1,500 treatment, storage, and disposal facilities will withdraw from the RCRA program because of stricter technical requirements under HSWA and inability to comply with financial requirements. Because the impacts of these factors are still uncertain, the hazardous-

TABLE 2.2.1 Facilities Subject to RCRA Regulation, July 31, 1985

EPA region	Large-quantity generators	Transporters	Treatment, storage, and disposal facilities
1	4,358	557	379
2	11,453	1,559	379
3	4,737	1,177	418
4	6,280	1,542	563
5	10,937	2,909	1,312
6	4,536	1,258	698
7	1,836	481	279
8	659	313	116
9	5,931	1,848	445
10	2,137	699	128
	52,864	12,343	4,961

Source: U.S. EPA, *The New RCRA—A Fact Book*, Office of Solid Waste and Emergency Response, Washington, D.C., October 1985.

waste storage, disposal, and treatment capacities prior to HSWA (July 1985) are discussed in this chapter.

Hazardous-Waste Storage Capacity

Hazardous wastes may be stored in tanks, surface impoundments, containers, or piles. It is estimated that there are 5,455 storage tanks in the RCRA regulatory program, with a total design capacity of 757 million L (200 million gal).[5] The annual waste-management capacity may be up to 25 times greater than the design capacity. Approximately 58% of the tank capacity is in aboveground closed carbon-steel units with a median capacity of 20,819 L (5,500 gal). Approximately 16% of the capacity is in ongrade, closed carbon-steel units with a median capacity of 795,000 L (210,000 gal). Approximately 15% of the capacity is in underground carbon-steel tanks with a median capacity of 15,140 L (4,000 gal).[5] Other tank designs are underground fiberglass and stainless steel tanks, and in-ground, open concrete and carbon-steel units. In addition to these RCRA-regulated tanks, there are approximately 6,400 accumulation tanks with a total capacity of 890 million L (235 million gal) that store wastes for less than 90 days and therefore do not require an RCRA permit.[5] Waste-management facilities with storage tanks have an average of four such units.

There are approximately 631 storage surface impoundments with a total design capacity of 71.5 billion L (18.9 billion gal), based on information obtained from the HWDMS in June 1986. The actual waste-management capacity may be up to 25 times greater than the design capacity. Typically, storage surface impoundments are larger than treatment surface impoundments and they are smaller than disposal impoundments. A storage surface impoundment of median size is estimated to have a capacity of 113 million L (29.9 million gal). Waste-management facilities with surface impoundments are estimated to have an average of three units.

It is estimated that there are approximately 299 hazardous-waste storage piles with a total capacity of 7.15 million m^3 (9.36 million yd^3), based on information obtained from the HWDMS in June 1986. The median capacity of a waste pile is approximately 24,000 m³ (31,000 yd³). If the maximum height of a waste pile is 3 m (10 ft), corresponding to the maximum operating height of a front-end loader, the base area of a median-size pile would be about 0.8 ha (2 acres).

The volume of waste estimated to be stored in containers is 757 million L (200 million gal).[2] The most prevalent storage container is a 55-gal steel drum, meaning that the annual inventory of hazardous-waste storage containers is equivalent to 3.6 million drums.

Hazardous-Waste Treatment Capacity

Hazardous wastes are treated in tanks, surface impoundments, incinerators, and other thermal devices such as cement kilns and boilers. It is estimated that there are 3,569 treatment tanks in the RCRA regulatory program with a total design capacity of 90 million L (23.7 million gal).[5] The annual waste-treatment capacity of these tanks may be up to 365 times greater than the design capacity.[2] Approximately 36% of the tank treatment capacity is in open, aboveground carbon-steel tanks with a median design capacity of 8,700 L (2,300 gal). Closed tanks of similar design account for 24% of the total design capacity. Approximately 20% of the total design capacity is in open, concrete in-ground tanks with a median capacity of 14,000 L (3,700 gal). The remaining capacity is distributed among open aboveground tanks with a median capacity of 227,000 L (60,000 gal) and in-ground open carbon or stainless steel tanks with a median capacity of 14,000 L (3,700 gal).[5] Waste-management facilities with treatment tanks have an average of five such units.

There are 441 treatment surface impoundments in the RCRA program with a total design capacity of 195 billion L (51.5 billion gal), based on data obtained from the HWDMS in June 1986. The annual waste-management capacity may be up to 180 times greater than the design capacity. The median design capacity of the treatment impoundments is approximately 44.3 million L (11.7 million gal).

There are approximately 221 hazardous-waste incinerators in the RCRA program with a total annual design capacity of 3 million t, or 3 billion L (800 million gal) of hazardous waste.[6] Approximately two-thirds of the design capacity is in use and the remaining one-third of the capacity would be available if all the incinerators operated continuously. Rotary-kiln incinerators account for 37% of the total design capacity, liquid injection units have 38%, fume incinerators have 11%, hearth incinerators have 10%, and other types of incinerators, such as fluidized bed and infrared units, account for the remaining 4%. Approximately 90% of the rotary kilns and 40% of other incinerator types are equipped with some air-pollution control equipment. The average design thermal rating of an incinerator is 8.3 million kg-cal/h (33 million Btu/h). The averaged characteristics of the incinerated hazardous wastes are a heating value of 4767 cal/g (8582 Btu/lb, a solids content of 8% and a water content of 50%. Approximately 46% of the wastes are halogenated and the halogenated wastes have an average halogen content of 33%.[6]

Hazardous wastes can be thermally treated in the approximately 250 cement kilns that are operating in the United States. Based on current waste-destruction practices, cement kilns have a potential annual hazardous-waste treatment capacity ranging from 2 to 6 million t, or 2 to 6 billion L (530 to 1,600 million gal).[6] Less than 2% of the potential capacity is presently in use. The characteristics of hazardous wastes that are destroyed in cement kilns are typically limited to heating values ranging from 4444 to 10,000 cal/g (8000 to 18,000 Btu/lb), water and chlorine contents less than 10%, solids content less than 12%, and a pH ranging from 4 to 11.

Hazardous wastes can also be thermally treated in industrial boilers. The waste must have a heating value greater than 8300 cal/g (15,000 Btu/lb) to be used effectively as a fuel supplement. Such wastes are typically fired at 5% of the thermal load to the boiler. There are approximately 195,000 industrial boilers rated less than 2.52 kg-cal/h (10 million Btu/h), 38,025 rated between 2.52 and 25.2 kg-cal/h (10 and 100 million Btu/h), and 5,045 rated greater than 25.2 kg-cal/h (100 million Btu/h).[7] If all these units burn hazardous wastes for 7,200 h/yr, the potential deduction capacity of industrial boilers would be 44.1 million t/yr.

Hazardous-Waste Disposal Capacity

Hazardous wastes can be disposed in landfills, land treatment facilities, injection wells, and surface impoundments. A survey of the HWDMS in March 1983 indicated that there were 535 landfills in the RCRA program at that time. In June 1986, the HWDMS indicated that there are 297 landfills with a total capacity of approximately 14.2 billion m^3 (18.6 billion yd^3). The active area of the landfills is approximately 10% of the total capacity. A median-size landfill has a volume of 46.6 million m^3 (61 million yd^3), a depth of 7.62 m (25 ft), and a surface area of 630 ha (1,556 acres) based on the HWDMS data.

There are 138 land treatment facilities in the RCRA program based on the June 1986 HWDMS survey. The author estimates that there are two or three land treatment units per facility, indicating that there may be 275 to 400 units. A median-size land treatment unit has an area of 275 ha (678 acres). Wastes are typically applied at a rate of 7,663 L/ha (5,000 gal/acre) up to a frequency of every second day.[8] Using these statistics, a median-size land treatment unit can dispose of 2,343 million L (619 million gal) of waste per year and the total annual waste-management capacity of all land treatment areas would be approximately 323 billion L (85.4 billion gal). The actual utilization of this potential capacity was not specified in the available literature.

In 1983, there were 114 deep-well injection facilities that disposed of 13.2 billion L (3.5 billion gal) of hazardous waste per year.[9] The June 1986 HWDMS survey indicated that there were 109 injection wells with a total design capacity of 3.86 trillion L (1.02 trillion gal). Most of these facilities are located in Texas and Louisiana and the total number does not include 65,000 other wells in Texas that are used for oil-field brine injection.[9] The average well depth is approximately 1,676 m (5,500 ft) and some wells are 3,048 m (10,000 ft) deep. Well depth is governed by finding a geologic formation that will totally contain the injected waste. The median capacity of an injection well is 35.5 billion L (9.39 billion gal).

There are 181 disposal surface impoundments with a total design capacity of 29.7 billion L (7.85 billion gal) of hazardous waste according to the June 1986 HWDMS survey. The median design capacity of a disposal impoundment is 164 million L (43.4 million gal). Wastes are left in the impoundments to either volatilize or seep into the ground.

The quantities of hazardous waste managed in 1980 for each technology are listed in Table 2.2.2, based on data from the National Survey.[2] The geographic distribution of the waste-management technologies is presented in Table 2.2.3; landfills, incinerators, cement kilns, storage and treatment tanks, surface impoundments, waste piles, and land treatment facilities are included. Table 2.2.3

TABLE 2.2.2 Quantities of Hazardous Waste Managed in 1980 Listed by Technology

Technology	Millions of metric tons
Treatment	
Treatment tanks	33.0
Treatment surface impoundments	62.9
Incineration	1.9
Other technologies	17.4
Disposal	
Injection well	32.6
Landfill	3.0
Land application	0.4
Disposal surface impoundments	19.3
Other technologies	0.8
Storage	
Storage surface impoundments	53.5
Storage tanks	19.3
Containers	0.8
Piles	1.5
Other storage	1.1
Total	247.5

Source: Ref. 2.

includes the number of waste-management units in each of the 10 EPA regions and the median design capacity for each type of unit. Most of the information was obtained from the HWDMS in June 1986. Several geographic trends may be noted in the data. For example, Regions 5 and 6 generally have the largest population of management units, the regions with the smallest median size for their units are generally located in the northeast, and the largest units are located in the west.

2.2.3 COMMERCIAL WASTE-MANAGEMENT CAPACITY

It was estimated in 1984 that there were approximately 130 commercial waste-management facilities in the United States under RCRA.[10] The numbers of commercial facilities in 1980 and 1984 are itemized by technology in Table 2.2.4. The number of commercial hazardous-waste management facilities did not change significantly over the four-year period, and the volume of wastes managed probably did not change greatly.

The numbers of facilities presented in Ref. 10, 11, and 12 correlate well. For example, 50 landfills and land treatment facilities are identified in Ref. 10, 51 facilities are identified in Ref. 11, and 49 facilities are identified in Ref. 12. MITRE identified 26 commercial incinerators in the RCRA program with a total annual capacity of 781,000 t.[6] However, 34% of this capacity has not yet been constructed and only 13 facilities were operating in mid-1986.

TABLE 2.2.3 Number and Median Design Capacity of Hazardous-Waste Management Units Located in Each EPA Region

	Injection wells		Landfills		Land treatments		Piles		Incinerators	
EPA region	No. of units	Capacity, 10^9 L (10^9 gal)	No. of units	Capacity, hectares (acres)	No. of units	Capacity, hectares (acres)	No. of units	Capacity, 10^3 m^3 (10^3 yd^3)	No. of units	Capacity, kg/h (lb/h)
1	1	0.09 (0.025)	20	36 (88)	1	0.8 (2)	24	57 (75)	4	735 (1,620)
2	3	0.02 (0.005)	24	34 (83)	8	19 (46)	20	80 (104)	30	1,107 (2,440)
3	0	0	38	580 (1,433)	6	40 (99)	31	32 (42)	21	1,152 (2,540)
4	8	7,300 (1,930)	24	638 (1,576)	9	23 (57)	35	7(9)	38	1,470 (3,240)
5	18	18,900 (5,000)	49	538 (1,330)	20	32 (80)	71	11 (14)	44	2,454 (5,410)
6	63	25,000 (6,600)	89	802 (1,982)	44	21 (51)	59	24 (31)	64	1,960 (4,320)
7	4	9,500 (2,500)	10	190 (470)	8	456 (1,128)	8	11 (15)	5	
8	2	9,800 (2,600)	13	369 (913)	12	35 (87)	11	22 (29)	1	
9	6	306,600 (81,000)	22	1,983 (4,900)	17	1,441 (3,561)	25	7 (9)	13	1,750 (3,860)
10	4	314,200 (83,000)	8	8 (19)	8	883 (2,181)	15	19 (25)	1	
Total no. of units	109		297		138		299		221	
Median design capacity		35,500 (9,390)		630 (1,556)		274 (678)		24 (31.3)		1,764 (3,890)
Total design capacity		$3,876 \times 10^9$ L ($1,024 \times 10^9$ gal)		187,000 ha (462,127 acres)		17,656 h (43,629 acres)		7.1×10^6 m^3 (9.36×10^6 yd^3)		10.6×10^3 kg/h 4.83 tons/h

*SI = Surface impoundments

Disposal SI*		Storage SI*		Treatment SI*		Storage tanks		Treatment tanks	
No. of Units	Capacity, 10^3 L (10^3 gal)	No. of Units	Capacity 10^6 L (10^6 gal)	No. of units	Capacity, 10^6 L (10^6 gal)	No. of units	Capacity, 10^6 L (10^6 gal)	No. of units	Capacity, 10^3 L (10^3 gal)
13	1,230 (325)	41	106 (28)	33	38 (10)	356	49 (13)	352	8,860 (2,340)
8	26 (17)	47	23 (6)	38	11 (3)	629	87 (23)	528	13,290 (3,510)
14	485 (128)	44	68 (18)	39	8 (2)	455	64 (17)	343	8,630 (2,280)
16	140 (37)	91	34 (9)	63	11 (3)	1,045	144 (38)	321	8,060 (2,130)
17	208 (55)	106	42 (11)	57	8 (2)	1,356	189 (50)	910	22,860 (6,040)
71	114 (30)	187	261 (69)	143	114 (30)	866	121 (32)	564	14,190 (3,750)
6	15 (4)	23	8 (2)	8	4 (1)	164	23 (6)	135	3,410 (900)
6	644 (170)	21	68 (18)	12	4 (1)	91	11 (3)	54	1,360 (390)
22	227 (60)	59	95 (25)	43	4 (1)	422	57 (15)	303	7,610 (2,010)
8	8(2)	12	4 (1)	5	8 (2)	71	11 (3)	54	1,480 (390)
181		6631		441		5,455		3,564	
	164 (43.4)		113 (29.9)		44 (11.7)		0.139 (0.0367)		25 (6.64)
	29.7×10^9 L (7.85×10^9 gal)		71.5×10^9 L (18.9×10^9 gal)		19.5×10^9 L (5.15×10^9 gal)		757×10^6 L (200×10^6 gal)		$89.7 \times 10^6 6$ L (23.7×10^6 gal)

TABLE 2.2.4 Commercial Waste-Management Units

Waste-management technology	1980 waste-management volume, 1,000s of metric tons	1980 population	1984 population
Landfill	2571	40	41
Land treatment	537	11	8
Incineration	396	23	11
Deep-well injection	788	9	10
Chemical treatment	2346	33	21*
Resource recovery	424	33	43†
			140‡

Sources: Refs. 6, 10, 11, 12.
*Surface impoundments.
†Waste oil.
‡Solvent recovery.

Known commercial facilities, other than those listed in Table 2.2.4, include 140 solvent-recovery facilities,[10] 15 thermal treatment facilities other than incinerators (such as cement kilns and sulfuric acid roasters),[12] and 43 waste-oil reclaimers.[12] The EPA has identified 12,343 waste transporters in the RCRA program[4] and 533 commercial transporters are listed in Ref. 11.

2.2.4 FUTURE TRENDS

At the beginning of 1985, after two years of full RCRA program operation, approximately 7% of the 4,900 hazardous-waste storage, incineration, and land disposal facilities determined to require RCRA permits by EPA had been issued permits. Another 66% of the potential RCRA facilities had been issued draft permits. Approximately 950 facilities closed and 30 were denied permits. The average processing time for permit issuance was 420 days.[4] The EPA has established deadlines for Part B submittals; all land disposal applications were required by October 1985, incinerator applications were required by October 1986, and applications from other facilities are required by October 1988.

Because of the complexity of the RCRA permitting process and limited EPA staff resources, the number of RCRA permits issued is not expected to increase rapidly, despite the regulatory deadlines. In addition, new technical standards are being promulgated that impose more requirements on facilities seeking permits. For example, approximately 1,000 land disposal facilities should have closed by November 8, 1985, because of the HSWA provisions. Additional wastes are being considered for inclusion in the RCRA program, e.g., mining wastes and used crankcase oil. RCRA enforcement authorities have been expanded by HSWA. Such actions further burden RCRA staff and impair the permitting process.

Despite these limitations, predictions for rapid growth in hazardous-waste management are still being made. Arthur D. Little, Inc., of Cambridge, Massachusetts, predicts an eightfold increase in the 1990 commercial hazardous-waste management industry over the 1983 level, accelerating to a 13-fold increase by 1995.[13] For comparison, Booz-Allen and Hamilton, Inc., predicted that incin-

eration capacity would nearly double from 1980 to 1982.[11] Instead, nearly 100 incinerators have withdrawn from the RCRA program between 1980 and 1985.[6]

This author is not as optimistic as several EPA consultants. Small-quantity
generators and traditional land disposal users will be seeking waste-management
capacity that either may not exist or is not being planned.The federal government
may have to either phase regulatory activities, construct and operate waste-
management facilities, or provide insurance for private facilities if a shortfall of
waste-management capacity develops.

2.2.5 REFERENCES

1. E. Keitz et al., *Hazardous Waste Control Technology Data Base—A Profile of Incineration Facilities,* MTR-82W170, prepared for the U.S. EPA Office of Solid Waste by
   The MITRE Corporation, McLean, Virginia, November 1982.

2. *National Survey of Hazardous Waste Generators and Treatment, Storage and Disposal Facilities Regulated Under RCRA in 1981,* prepared for the U.S. EPA Office of
   Solid Waste by Westat, Inc., Rockville, Maryland, April 20, 1984.

3. U.S. Environmental Protection Agency, *The New RCRA—A Fact Book,* Office of Solid
   Waste and Emergency Response, Washington, D.C., October 1985.

4. P. Guerrero, "Hazardous Waste Facility Permits Under RCRA," *proceedings of the
   National Conference on Hazardous Wastes and Environmental Emergencies, May
   1985, Cincinnati, Ohio,* Hazardous Materials Control Research Institute, Silver Spring,
   Maryland, June 1985.

5. *Hazardous Waste Tanks Risk Analysis,* draft report prepared for the EPA Office of
   Solid Waste by ICF Inc., and Pope-Reid Associates, Inc., March 1986.

6. G. Vogel et al., *Incinerator and Cement Kiln Capacity for Hazardous Waste Treatment,* WP-85W397, prepared for the U.S. EPA Hazardous Waste Engineering Laboratory, Cincinnati, Ohio, by The MITRE Corporation, McLean, Virginia, June 1986.

7. C. Castaldini et al., *A Technical Overview of the Concept of Disposing of Hazardous
   Wastes in Industrial Boilers,* prepared for the U.S. EPA Hazardous Waste Engineering
   Laboratory, Cincinnati, Ohio, by The Acurex Corporation, Mountain View, California,
   January 1981.

8. G. Vogel and D. O'Sullivan, *Air Emission Control Practices at Hazardous Waste Management Facilities,* MTR-83W89, prepared for the U.S. EPA Office of Solid Waste by
   The MITRE Corporation, McLean, Virginia, June 1983.

9. Statement of John Mobley, President, Mobley Industries, Austin, Texas, on behalf of
   Institute of Chemical Waste Management, National Solid Waste Management Associ-
   ation, before the U.S. House of Representatives, Committee on Merchant and Marine
   Fisheries, Subcommittee on Fisheries and Wildlife Conservation and the Environment,
   May 12, 1983.

10. C. Baty and C. Perket, "1984 Summary Report: Commercial Industrial and Hazardous
    Waste Facilities," *Pollution Engineering,* **6**(7):24 (1984).

11. *Off-site Hazardous Waste Management Capacity,* draft report prepared for the U.S.
    EPA Office of Planning and Evaluation by Booz-Allen & Hamilton, Inc., Bethesda,
    Maryland, August 1980.

12. *Industrial and Hazardous Waste Management Firms, 1985 Directory,* Environmental
    Engineering and Management, Ltd., Minneapolis, Minnesota, 1985.

13. *New Hazardous Waste Technologies for the United States,* prepared for the EPA Of-
    fice of Solid Waste by Arthur D. Little, Inc., Cambridge, Massachusetts, 1985.

CHAPTER 3

HAZARDOUS-WASTE TOPICS AND ISSUES

SECTION 3.1
SITING OF HAZARDOUS-WASTE FACILITIES

Richard N. L. Andrews
Professor and Director
Institute for Environmental Studies
University of North Carolina at Chapel Hill

Frances M. Lynn
Research Assistant Professor
Institute for Environmental Studies
University of North Carolina at Chapel Hill

3.1.1 PRINCIPLES OF SITING

The siting of hazardous-waste facilities differs from most other aspects of hazardous-waste management in that it usually cannot be accomplished by technical and economic choices alone. It requires a synthesis of two distinct selection procedures: a technical screening process based upon engineering, environmental, and economic suitability, and a public-approval process based upon legality and political acceptability.

Traditionally, the siting of most industrial facilities has been considered a prerogative of business enterprises, based upon their own technical and economic criteria. Such decisions were subject only to general government regulatory constraints whose terms were specified in advance, such as land-use zoning and air- and water-pollution discharge permits. Public involvement was limited and often favorable, due to anticipated benefits of economic development.

Under recent federal and state laws, however, hazardous-waste management facilities now face far more rigorous requirements for public review, permitting, and site approval, even though other industrial facilities may handle equally hazardous materials. The reasons for this special treatment include the special hazards of the materials they handle, public concern about their potential impacts, and a widely publicized record of problems resulting from past management practices.

The result is that hazardous-waste facility siting may involve not only traditional technical site selection and zoning and permit approval, but an open-ended process of negotiation with both public officials and citizens of the host community. This process typically includes a full assessment of the facility's potential impacts, an open process of public discussion, and sometimes enforceable agreements concerning prevention, mitigation, or compensation of adverse effects that may go well beyond previously specified zoning and permit requirements. Such agreements are sometimes required by law, but even where they are not they are a prudent practice for obtaining public acceptance of proposed facilities.

3.1.2 SITE SUITABILITY

A technical screening process is used to identify sites that are potentially suitable for a hazardous-waste facility. There are three sets of criteria: characteristics of the facility, characteristics of the site, and economic and legal feasibility. This process is described first for convenience, but should where possible be integrated with the public-approval process from the start rather than sequenced before it (see, e.g., Refs. 1, 2, and 3).

Facility Characteristics

Characteristics of the proposed facility weigh heavily in both the suitability and the acceptability of possible sites. Chief among these characteristics are purpose, effectiveness, resource requirements and impacts, and scale.

Site screening depends first on the *purpose* of a proposed facility: what is the market or need for its services, what types of wastes is it to handle, and is it intended for on-site or proprietary wastes only, for commercial off-site shipments, or for limited single-purpose disposal of wastes from a public cleanup operation? Not all hazardous wastes are highly toxic; many are defined as hazardous because of other properties, such as corrosive or flammable materials. A proprietary or limited-purpose facility, especially one for wastes of relatively low toxicity, may be less difficult to site than commercial facilities for highly toxic wastes.

Effectiveness is closely related to purpose. Figure 3.1.1 shows the primary technological options ordered by their effectiveness in reducing hazardous wastes. Most preferred are those that reduce waste generation at the source, followed by those that recover and recycle such materials into other economic processes. Less preferred are technologies that treat or destroy hazardous wastes, and least preferred those that merely store or landfill them. This hierarchy has been adopted by some states as a basis for policy goals.[4]

Resource requirements and impacts include both specific constraints affecting site suitability, and additional considerations affecting acceptability. Resource requirements include adequate land area, transport access, and energy and water supply; some facilities have additional special requirements, such as appropriate

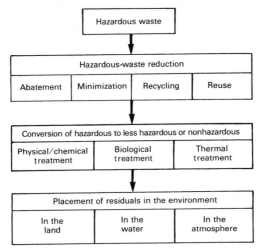

FIG. 3.1.1 Waste-management options.

soils and geology for landfills or other underground storage. Expected impacts may include the facility's own waste streams, odors, noise, traffic, and others; these may limit the sites for some types of facilities to existing industrial areas or others well-buffered from residential or recreational areas.

The magnitudes of resource requirements and impacts depend both on the technologies selected and on the *scale* of the facility proposed. Large integrated or comprehensive treatment facilities may have economies of scale compared to some smaller or limited-purpose facilities, but they also have larger resource requirements and may impose greater impacts on their host community.

Site Characteristics

Given the characteristics of a possible facility, a site-screening process is used to identify candidate sites for consideration. Figure 3.1.2 shows one system of criteria for this purpose.

Level I criteria are used to exclude clearly inappropriate areas from consideration, such as flood-hazard areas, public water supply watersheds, seismic-risk zones, designated natural preservation areas, and those too small for the proposed facility.

Level II criteria are then used to identify sites having favorable characteristics, such as sites of existing facilities, sites designated for industrial use, sites in counties where substantial quantities of the wastes are generated, sites close to major transportation routes, and publicly owned lands available for this purpose. Note that while many recent proposals for commercial facilities have favored rural sites, persuasive arguments can also be made for facilities in major generator counties, where they could provide significant benefits for local waste disposal rather than merely for regional or interstate markets.

For sites satisfying both level I and level II criteria, Level III criteria are used to compare additional site-specific characteristics in greater detail. Examples include natural factors such as geology, soils, surface-water and groundwater qual-

Level I: Exclusion Areas

Flood-hazard areas and wetlands
Public water supply watersheds
Groundwater aquifers, wellfields, recharge areas
Designated state and national parks, forests, natural and historic areas, wildlife
  refuges
Seismic-risk zones
Sites too small for proposed facility

Level II: Favorable Areas

Sites of existing facilities
Sites designated for industrial use
Sites in major generator counties
Sites close to major transportation routes
Public lands available for this use

Level III: Site-Specific Trade-offs

Geology (depth to bedrock, solid vs. fractured)
Soils (permeability, pH, etc.).
Groundwater (water-table depth, quality)
Surface water (flow, quality)
Air (nonattainment areas)
Ecological habitats
Transportation access, utilities
Surrounding and downstream land uses
Population density
Special site considerations (ownership, archaeology, history)

FIG. 3.1.2 Examples of site-screening criteria (*After L. B. Cahill and W. R. Holman, "Siting Waste Management Facilities," chap. 7 in J. J. Peirce and P. A. Vesilind (eds.),* Hazardous Waste Management, *Ann Arbor Science, Ann Arbor, Michigan, 1981, pp. 87–97*)

ity, ecological habitats, and cultural and infrastructure factors such as transportation access, surrounding land uses, population density, and special considerations such as archaeological sites. A particularly important consideration is minimizing the potential for groundwater contamination, since groundwater is an important resource whose purification, once contaminated, requires great expense over long time periods.

Economic and Legal Feasibility

Physically suitable sites must also be evaluated for economic and legal feasibility: they must provide satisfactory benefits to the developer for the costs involved, and they must satisfy permit and site-approval requirements.

The feasibility of on-site options is relatively straightforward to estimate, based on conventional calculations of capital and operating costs. Note however that waste-reduction and recycling alternatives should also be considered, since they may often prove more cost-effective than treatment or disposal facilities.[5]

The feasibility of a commercial facility, however, depends heavily on the demand for commercial disposal services and the relationship between its prices and those of its competitors. For hazardous wastes, these factors in turn are primarily influenced not by ordinary market forces, but by how vigorously government regulates waste generators and alternative disposal facilities. This circumstance imposes a major uncertainty on feasibility estimates for such facilities. Facilities for disposal of wastes from government cleanup operations face similar uncertainties as to "how safe is safe?"—i.e., how much treatment government sponsors will require, and at what cost.

3.1.3 REGULATORY SETTING

Within the range of technically feasible sites, hazardous-waste facility siting requires at least two forms of regulatory approval: facility approval by a federal or federally approved state agency, and site approval by local or state government. Other permit requirements may also apply; such requirements should be identified by competent legal counsel as early as possible in the siting process.

Some variations in requirements may occur in cases where a siting proposal is initiated by government itself (such as a state-owned or military facility), or where a temporary site is needed immediately to cope with an emergency situation (for instance, a major accident or spill).

Federal and State Permit Requirements

RCRA permit. Federal law requires that every facility that treats, stores, or disposes of hazardous wastes obtain a permit to operate, either from the federal Environmental Protection Agency (EPA) or from an EPA-approved state program (Resource Conservation and Recovery Act of 1976, 42 U.S.C. 6901-87). The permit application must describe in detail all aspects of design, operation, and maintenance of the facility, and must show compliance with applicable federal and state requirements (40 *CFR* Parts 264, 265, 270). The draft permit is subject to public notice and comment, and hearings if requested; the permit itself, if issued, is good for a stated time period which may not exceed 10 years.

Federal permit requirements include both administrative and technical standards. Administrative requirements include general facility standards, record-keeping and reporting requirements, preparedness and prevention practices, and contingency plans and emergency procedures. Technical requirements include general standards for groundwater monitoring, closure and postclosure care procedures, and financial responsibility, and specific standards for design and operation of particular types of facilities (surface impoundments, incinerators, landfills, etc.). States are free to adopt additional or more stringent requirements, and many have done so.

Other permits. Each facility must also obtain any permits required by other applicable federal and state laws. Examples include permits for water usage, air- and water-pollutant discharges, underground injection, highway rights-of-way, leasing of state lands, coastal location, and handling of low-level radioactive wastes if such wastes are to be received.

Local Powers

In addition to federal and state powers, local governments have relatively broad authority to regulate facility siting. Three local powers most directly affect siting proposals: the power to control proposed uses of land; to regulate health hazards, nuisances, and construction (often including specific references to solid wastes and hazardous substances); and to assess taxes and fees on facilities operating in their jurisdiction.

Most local governments exercise primary jurisdiction in land-use controls, such as planning and zoning ordinances, site-plan review requirements, and control of easements and rights-of-way. Many now use a "special use permit" procedure for major proposals such as hazardous-waste facilities, since it allows more flexibility in incorporating special safeguards and in triggering subsequent compliance review and reconsideration. Most local governments also require building permits; and some local health departments can also require facility permits, or at least impose specific restrictions on siting or operating procedures.

Many local governments may also assess not only general ad valorem property taxes, but privilege license taxes, tipping fees, or gross-receipts taxes to recover special costs anticipated from a hazardous-waste facility. In addition, increasing numbers of local governments charge substantial application fees to recover the costs of technical and legal review of proposed hazardous-waste facilities.[6,7] It is in the interest of site applicants to encourage this practice, since competent review by independent professionals accountable to the community is often essential to public acceptance of a proposed facility.

State and Local Siting Approaches

In addition to permits as described above, over half the states in the United States have enacted special site-approval processes for hazardous-waste facilities.[8,9,10] These processes are of four types.

Local veto reaffirms the traditional preeminence of local governments in land-use controls, though state or federal agencies can disapprove facility permits as well. As of 1983, 8 states had explicitly adopted this approach; 31 others are presumed to use it by default.

Override creates authority for a state board or agency to supersede local disapproval of siting proposals, either as a general practice or on appeal. Michigan, for instance, exemplifies the former; North Carolina the latter. Some such boards include ad hoc representatives of the local jurisdiction, while others do not. As of 1983, 19 states had established one or another form of override authority.

Preemption gives the state itself full power to initiate and approve siting proposals. This practice is often followed for state disposal of cleanup residues, but as of 1983 had been attempted only in seven states for general-purpose commercial facilities (e.g., Arizona, New York, Minnesota). Some additional states have since authorized this approach as a default option if no privately initiated facilities are sited by specified dates (e.g., North Carolina).

Finally, *procedural requirements* have been adopted by several states to try to assure fair treatment and reasonable decisions for all affected parties while leaving primary siting controls at the local level. Massachusetts and Wisconsin, for example, both leave clear authority to local governments but establish explicit negotiation procedures by which that authority must be exercised; they also restrict both the grounds on which local governments may deny approval, and their freedom to make exclusionary changes in local ordinances once a proposal is initiated.

Few facilities have yet been sited under any of these approaches.[10] Traditional overtures to local governments have often been blocked by community opposition, although some new facilities have been sited under this approach. Approaches relying on state override and preemption laws—advocated to overcome local opposition—have succeeded in some cases, but in others have simply exacerbated local opposition to outside threats. Procedural requirements are less disruptive of traditional local governance, but have in some cases imposed heavy burdens of cost and effort on the community as well as the applicant.

The likely model for most future site-approval processes is continued development of forms of dual jurisdiction by local and state governments. Primary local jurisdiction in land-use controls, taxation, and some public-health regulations should be expected to continue, but with provisions for state override in some states and with continued refinement of information requirements, negotiation and public-involvement procedures, and safeguards for parties involved. Override and preemption provisions will continue to be used to site government facilities, and in some states to site commercial facilities or as incentives against exclusionary local decisions; but they should not be relied upon merely to defeat public opposition.

3.1.4 SITE ACCEPTABILITY

Site approval for hazardous-waste facilities requires not only technical suitability and permit approval, but acceptability of the facility to the host community. Such acceptance is frequently denied, sometimes out of parochial self-interest but more often out of legitimate public distrust towards sponsors' claims. While sponsors and even government regulatory agencies are often confident that serious contamination is unlikely, past facilities are in fact many of today's hazards, belying the reassurances of their sponsors. These hazards are also widely perceived as lethal, involuntarily imposed, beyond individual control, persistent, memorable, and unfair—and accordingly unwelcome.[13] Public concern therefore focuses on the possible consequences themselves—rather than their assertedly low probability—and on how siting decisions are made; and demands enforceable safeguards rather than merely reassurances.[12,13,14,15]

At a minimum, public acceptance normally requires two steps of the sponsors beyond technical analysis and state permit approval: substantive information concerning potential impacts, and willingness to engage in an open process of communication and negotiation with the host community.

Thorough information about a proposed facility is not by itself a panacea for public opposition, but it is a necessary prerequisite to public acceptance. Information on possible impacts must be gathered carefully and competently for a specific set of alternative proposals, since much of it will be specific to particular sites and to the concerns of the communities in which they are located. Impacts will also vary substantially with the wastes handled and the type and scale of operation.

Environmental Impacts

Environmental impacts include potential effects on ecological processes, on human health, and on the esthetics of the natural and cultural landscape. For any given facility, these impacts will depend on the volumes and types of wastes to be managed; on the design and operation of the facility, including safeguards and

Groundwater contamination
Surface-water contamination
Air pollution
Leaks, spills, accidents
Destruction of wildlife habitat, natural areas, wetlands
Loss of any unique site features (e.g. archaeological)
Permanent contamination of site
Contamination of nearby crops, fisheries
Traffic congestion
Odors
Noise
Visual ugliness
Effects on character of community: magnet for other
    heavy industry, image as dumping ground

FIG. 3.1.3   Examples of possible environmental impacts.

mitigation measures; and on the location and characteristics of the site, especially the pathways by which humans or other species might be exposed to either anticipated or accidental releases of hazardous materials.

Many states require applicants to submit an environmental impact statement as part of the application process, describing in detail the anticipated need for a facility, the alternatives considered, and the likely impacts of each. Even where not required, preparation of such a statement is a prudent practice: it provides early warning to the sponsors of potentially controversial impacts, demonstrates to the public that the sponsors have been thorough in evaluating site suitability, and provides information that will usually be requested during the site-review process in any case.

Figure 3.1.3 lists examples of potential environmental impacts that should be addressed, taking into account both expected and possible accidental effects from each element of the facility's operation (e.g., transport, storage, treatment, incineration). Environmental impacts in the absence of a facility should also be addressed: if the facility will in fact help to remedy disposal practices currently inuse that are *more* hazardous, that point should be made and factually supported. Safeguards to prevent or mitigate such impacts should also be discussed, but frankness about all possibilities will generally be trusted more readily than mere reassurances.

In practice this list should be used only as a starting point, to be refined on a site-specific basis through discussions with a citizen advisory committee that is representative of the host community.

Socioeconomic Impacts

A hazardous-waste management facility may also have significant social and economic impacts on a community. Many of these are similar to those of other industrial facilities; others are specific to hazardous-waste facilities. While some of these impacts may be negative unless prevented or compensated, others will be beneficial.[6,7] Figure 3.1.4 lists the major categories of economic impact that may result from a hazardous-waste facility.

Costs of proposal review (technical and legal)
Employment and income (plus secondary services)
Purchases of goods and services (capital and annual)
Public-service demands (water, sewer, energy, etc.)
Payments to local governments (property taxes; gross-receipts taxes, tipping fees, etc.)
Transportation-related costs (capital costs, maintenance, accidents)
Effects on property values
Monitoring and inspection
Emergency response
Closure and postclosure care (monitoring and management, loss of taxes from site; cleanup?)

FIG. 3.1.4   Examples of possible economic impacts.

Even if a proposed facility is not constructed, local governments may face significant costs for legal and technical review and related public-information activities. Typical costs for review without additional data collection may range from $10,000 to $25,000; total costs have been as high as $180,000 in one case, and might run far higher if an action were subsequently challenged in court. Many states allow local governments to recapture such expenses through application fees, technical-assistance grants, or other sources; communities that cannot do so may face a significant economic impact for proposal review alone.

In addition to impacts, moreover, facility sponsors should be prepared to explain and justify the economic *need* for the proposed facility, with specific reference to their assumptions about types, volumes, and sources of materials, market area, and effectiveness and economic feasibility of the proposed facility compared to source-reduction and competing treatment and disposal options.[14]

Impact Prevention, Mitigation, and Compensation

Many possible adverse impacts of a proposed facility can be prevented, mitigated, or compensated. *Preventive* measures, such as monitoring and ongoing local oversight, help to keep an adverse impact from occurring in the first place. *Mitigative* measures, such as funds for road maintenance, serve to reduce or reverse an adverse impact. *Compensatory* measures, such as property-value guarantees or donation of parkland, provide one type of benefit in exchange for a benefit lost or potentially lost. Six states now require the developer to negotiate the issue of compensation and incentives with the proposed host community; seventeen have policies endorsing such practices.[16] Figure 3.1.5 lists examples of measures agreed to by some existing facilities.

Of these measures, preventive steps are usually valued most highly by host communities. Compensation payments, and additional goodwill gestures such as free waste disposal for the town's residents, typically are effective only when issues of prevention and mitigation are first resolved.[17,18]

One measure useful in establishing good working relationships with the host community is to create an ongoing mechanism for public oversight of the facility's operation and management. An upstate New York firm, for instance,

Preventive Measures

Natural barriers (site selection)
Engineered safeguards (facility design)
Safeguards in operating procedures
Financial assistance for community's technical review of proposal
Monitoring (on site and off site)
Training of inspectors and monitoring professionals
Ongoing local oversight of plant operations

Mitigative Measures

Emergency-response capability
Road construction and maintenance
Payments to community for water supply, wastewater treatment, energy use, extra
  security services
Siting agreements allowing renegotiation of operating practices if adverse impacts
  occur

Compensatory Measures

Liability insurance
Payments to finance postclosure perpetual care of site, and in lieu of future taxes
  foregone
Property value guarantees, other guaranteed compensation
Provision of parks and open space
Discounted or free local waste disposal

FIG. 3.1.5 Illustrative measures to prevent, mitigate, or compensate for impacts.

has agreed to the creation of a citizen advisory committee which will meet monthly to review facility operations; similar mechanisms may increase public confidence in other cases.[6,13]

3.1.5 ELEMENTS OF A SITING PROCESS

The cornerstone of a successful siting process is early, open, and ongoing communication among all affected parties. The main building blocks of this process include joint fact finding, a community review committee, independent technical assessment, and a fair and open process of negotiation with the host community.

Joint Fact Finding

It is not wise for a developer to approach a community with a preselected site, elaborate plans, and a permit application ready to be filed with federal or state regulatory authorities. This often leads to community intransigence and to die-hard opposition in the requisite public hearing. This so-called "decide-announce-defend" procedure has frequently failed in the past.[3]

An alternative approach, based on research on citizen attitudes about siting, is a process of joint fact finding, in which a facility developer contacts a representative cross section of the community—not just the chamber of commerce and government officials—and begins a process of information sharing in order to gradually develop understanding and agreement about the facts and possible impacts of proposed facilities.[2,11,19,20]

This process must begin as early as possible in the sponsor's own technical screening process, in order to incorporate into site-screening criteria the concerns of community residents as well as those of the sponsors themselves. It is essential to this process that site screening be treated as provisional, molded both by technical constraints and community concerns, not simply as a decision to be ratified by public approval at its conclusion.[20]

Community Review Committee

To accomplish joint fact finding it is necessary to work with a local review committee that is broadly representative of the citizens of the host community. This committee may serve not only to review and advise on the proposal, but also as an educational and sometimes a decision-making group.

The ideal situation is to find a representative environmental advisory body already established. In one of the few cases of a successful siting of a new facility (Greensboro, North Carolina), a Hazardous Waste Task Force had been in operation for over 5 years and an Environmental Advisory Board had existed for over 15 years prior to the siting of the facility. Citizens in Greensboro were familiar with the issues of hazardous waste and were prepared in relatively short order to work with the developer (who had been a regular attendee at Task Force meetings) on reviewing proposals and suggesting modifications.

If the local government does not volunteer to establish such a committee itself, sponsors may request that it do so and indicate their willingness to work openly with it. The first step in establishing such a body is to make sure that a genuine cross section of community spokespersons is included. These could include local government officials (especially representatives from the health, planning, fire, and police departments); citizens with scientific background; representatives from environmental, civic, labor, and industry groups; and as soon as possible, those who might be immediately affected by a facility's location.

Establishment of a formal review committee is not a signal for communication with the community at large to stop. It is in the interest of a developer to be available and, indeed, to solicit opportunities to meet informally with individuals and groups of citizens and to be willing to answer any questions. Members of a citizen review committee must also keep in regular contact with the larger community, and be sure that nonmember citizen concerns can be voiced as the review proceeds. Educational activities might include sponsoring a variety of public forums where citizens are able to air their concerns, and to learn more about the developer's proposals and about the progress that the formally constituted committee is making.

Independent Technical Assessment

Community access to independent technical review is an important factor not only in achieving public acceptance, but also in assuring good outcomes. At the least, it assures the community that all essential questions have been addressed, that the assertions made are reasonable, and that uncertainties are either resolved or at least accurately acknowledged. In some cases, it also surfaces additional ideas that will improve the design or operation of the facility; and in a few, it may show that the site is not appropriate after all, thus saving the developer the cost of later remedies.

To succeed, independent technical assessment requires the cooperation of the facility developer, both in supplying needed information and sometimes in providing financial assistance to the community to hire their own consultants. In the case of the successful siting in Greensboro, "reading/assessment groups" of citizens, including scientists from local colleges and industries, evaluated the hazardous-waste firm's proposals. In three other North Carolina communities, the state provided funds with which the community could hire independent consultants to conduct independent assessments of the developers' proposals and help them consider other technical alternatives. Similar processes are provided by law in some other states.

The developer should offer whatever assistance may be welcome in the community's review process. The citizen committee, however, must have full authority to frame the questions and to hire, and monitor the work of, consultants.[19]

Negotiation and Mediation

Independent technical review is especially useful as part of a larger process of formal or informal negotiation between a community and a developer. Three states, Massachusetts, Wisconsin, and Rhode Island, require negotiation between a developer and a host community. In other states, such as Texas and North Carolina, community-review mechanisms which encourage developer-community interactions are encouraged although not required. In Texas, participants from environmental groups, citizen organizations, public interest groups, industry, local, state and federal governments, and academia developed the so-called "Keystone Siting Process," the goal of which is to help parties understand their counterpart's positions in a manner which is not destructive to either the applicant or the community. It is important to this process that the citizen review committee and negotiation process be "interest based" rather than "position based": that is, that their focus be on raising and resolving all concerns, rather than merely on compromising adversaries' demands.[2,11]

The training manual which emerged from the original meetings at the Keystone Center contains an outline of the types of issues that are likely to be important in any siting situation, detailed instructions for establishing local review committees, and a chapter on the role of the developer in the review and negotiation process.[2]

In some cases, *mediation* may be a useful addition to the negotiation process. Mediation is defined as "the assistance of a neutral third party in a negotiation process." Such assistance may not be essential, but it can often help to increase the likelihood that a negotiation process will be successful.[11] The use of this procedure is currently increasing in the United States.

The rules for negotiation between a community and a developer are not mysterious, nor are they set in stone. Negotiation is simply a process by which developers can work with communities to see whether their respective needs can be incorporated into an overall agreement for the management and monitoring of a hazardous-waste facility. It is not a panacea for sometimes irreconcilable differences, but it may offer a more viable process than past approaches to siting publicly acceptable hazardous-waste facilities.

3.1.6 REFERENCES

1. The Keystone Center, *Siting Waste Management Facilities in the Galveston Bay Area: A New Approach,* Keystone, Colorado, 1982.

2. The Keystone Center, *Keystone Siting Process Training Manual,* Keystone, Colorado, 1985.

3. L. E. Susskind, "The Siting Puzzle: Balancing Economic and Environmental Gains and Losses," *Environmental Impact Assessment Review,* **5**:157–163 (1985).

4. National Academy of Sciences, *Reducing Hazardous Waste Generation,* National Academy Press, Washington, D.C., 1985.

5. R. N. L. Andrews and A. G. Turner, "Control Options," chap. 6 in L. B. Lave and A. C. Upton (eds.), *Toxic Chemicals, Health, and the Environment,* Praeger, New York, 1986, plus references cited therein.

6. M. A. Smith, F. M. Lynn, R. N. L. Andrews, R. Olin, and C. Maurer, *Costs and Benefits to Local Government Due to Presence of a Hazardous Waste Management Facility and Related Compensation Issues,* Governor's Waste Management Board, Raleigh, North Carolina, 1985, pp. 195–204.

7. M. A. Smith, F. M. Lynn, and R. N. L. Andrews, "Economic Impacts of Hazardous Waste Facilities," *Hazardous Waste,* **3**(2) (1986).

8. S. G. Hadden, J. Veillette, and T. Brandt, "State Roles in Siting Hazardous Waste Disposal Facilities: From State Preemption to Local Veto," chap. 10 in J. P. Lester and A. O'M. Bowman (eds.), *The Politics of Hazardous Waste Management,* Duke University Press, Durham, North Carolina, 1983, pp. 196–211.

9. R. N. L. Andrews and T. K. Pierson, "Hazardous Waste Facility Siting Processes: Experience from Seven States, *Hazardous Waste,* **1**:(3)377–386 (1984).

10. ———, "Local Control or State Override: Experiences and Lessons to Date," *Policy Studies Journal,* **14**(1):90–99 (1985).

11. G. Bingham and D. S. Miller, "Prospects for Resolving Hazardous Waste Siting Disputes Through Negotiation," *Natural Resources Lawyer,* **17**(3):473–489 (1984).

12. F. J. Popper, "LP/HC and LULUs: The Political Uses of Risk Analysis in Land-Use Planning," *Risk Analysis,* **3**(4):255–263 (1983).

13. M. L. P. Elliot, "Improving Community Acceptance of Hazardous Waste Facilities Through Alternative Systems for Mitigating and Managing Risk," *Hazardous Waste,* **1**(3):397–410 (1984).

14. J. S. Hirschhorn, "Siting Hazardous Waste Facilities," *Hazardous Waste,* **1**(3):423–429 (1984).

15. R. C. Mitchell and R. T. Carson, "Property Rights, Protest, and the Siting of Hazardous Waste Facilities," *Papers and Proceedings of the American Economics Association,* Nashville, Tenn., **76**(2):285–290 (1986).

16. K. Hawthorne, *The Use of Compensation and Incentives in Siting Low Level Radioactive Waste Management Facilities and Hazardous Waste Management Facilities,* Governor's Waste Management Board, Raleigh, North Carolina, 1986, 14 pp.

17. U.S. Environmental Protection Agency, *Siting of Hazardous Waste Facilities and Public Opposition,* Report No. SW-809, Washington, D.C., 1979.

18. K. E. Portney, "Allaying the NIMBY Syndrome: The Potential for Compensation in Hazardous Waste Treatment Facility Siting," *Hazardous Waste,* **1**(3):411–421 (1984).

19. C. P. Ozawa and L. Susskind, "Mediating Science-Intensive Policy Disputes," *Journal of Policy Analysis and Management,* **5**(1):23–29 (1985).

20. P. M. Sandman, *Getting to Maybe: Some Communication Aspects of Siting Hazardous Waste Facilities,* Hazardous Waste Advisory Council, New Jersey Hazardous Waste Facilities Siting Commission, Trenton, New Jersey, 1985.

3.1.7   FURTHER READING

Anderson, R., and M. Greenberg, "Hazardous Waste Facility Siting," *J. Amer. Planning Assoc.*, **48**:204–218 (1982).

Andrews, R. N. L., R. J. Burby, and A. G. Turner, *Hazardous Materials in North Carolina: A Guide for Decisionmakers in Local Government*, UNC Institute for Environmental Studies, Chapel Hill, North Carolina, 1986.

Bacow, L. S., and J. F. Milkey, "Overcoming Local Opposition to Hazardous Waste Facilities: The Massachusetts Approach," *Harvard Environmental Law Review* **6**:265–305 (1982).

Bacow, L. S., and D. R. Sanderson, *Facility: Siting and Compensation: A Handbook for Communities and Developers*, MIT Center for Energy Policy Research, Cambridge, Massachusetts, 1980.

Bingham, G., *Resolving Environmental Disputes: A Decade of Experience*, The Conservation Foundation, Washington, D.C., 1985.

Bingham, G., and T. Mealey, *Negotiating Hazardous Waste Facility Siting and Permitting Agreements*, The Conservation Foundation, Washington, D.C., 1988.

Bishop, A. B., "An Approach to Evaluating Environmental, Social, and Economic Factors in Water Resources Planning," *Water Resources Bulletin*, **8**(4):43–53 (1972).

Boyle, S., *Siting New Hazardous Waste Management Facilities Through a Compensation and Incentives Approach: A Bibliography*, CPL Bibliographies, Chicago, 1983.

Clark-McGlennon Associates, *A Decision Guide for Siting Acceptable Hazardous Waste Management Facilities in New England*, New England Regional Commission, Boston, Massachusetts, 1980.

————, *Criteria for Evaluating Sites for Hazardous Waste Facilities*, New England Regional Commission, Boston, Massachusetts, 1980.

Duerksen, C., *Environmental Regulation of Industrial Plant Siting*, The Conservation Foundation, Washington, D.C., 1983.

Greenberg, M. R., R. F. Anderson, and K. Rosenberger, "Social and Economic Effects of Hazardous Waste Management Sites, *Hazardous Waste*, **1**(3):387–396 (1984).

Kasperson, R. E., "Hazardous Waste Facility Siting: Community, Firm, and Government Perspectives," in National Academy of Engineering, *Hazards: Technology and Fairness*, National Academy Press, Washington, D.C., 1986, pp. 118–144.

Morell, D. L., and C. Magorian, *Siting Hazardous Waste Facilities: Local Opposition and the Myth of Preemption*, Ballinger, Cambridge, Massachusetts, 1982.

National Audubon Society, *Siting Hazardous Waste Facilities: A Handbook*, National Audubon Society, Tavernier, Florida, 1983.

O'Hare, M., L. S. Bacow, and D. R. Sanderson, *Facility Siting and Public Opposition*, Van Nostrand Reinhold, New York, 1983.

Seley, J. E., *The Politics of Public Facility Siting*, Lexington Books, Lexington, Massachusetts, 1983.

Susskind, L. E., and J. Cruikshank, *Dealing With Differences: Practical Approaches to Resolving Disputes*, Basic Books, New York, 1987.

Tarlock, A. D., "Siting New or Expanded Treatment, Storage, or Disposal Facilities: The Pigs in the Parlors of the 1980s," *Natural Resources Lawyer*, **17**:429–461 (1984).

Taylor, G. C., *Socioeconomic Analysis of Hazardous Waste Management Alternatives: Methodology and Demonstration*, EPA-600/5-81-001, U.S. EPA, Cincinnati, Ohio, 1981.

U.S. Environmental Protection Agency, *Using Compensation and Incentives When Siting Hazardous Waste Management Facilities*, Report No. SW-942, Washington, D.C., 1982.

U.S. Environmental Protection Agency, *Using Mediation When Siting Hazardous Waste Management Facilities*, Report No. SW-944, Washington, D.C., 1982.

U.S. Environmental Protection Agency, *RCRA Orientation Manual*, EPA-530-SW-86-001, Washington, D.C., 1986.

HEALTH EFFECTS OF HAZARDOUS WASTE

Clark W. Heath, Jr., M.D.

Director
Bureau of Preventive Health Services
South Carolina Department of Health
and Environmental Control

Environmental health risks, natural or of human origin, are an ever-present feature of human life. As modern industrial society has developed over the past century, the portion of these risks arising from human activity has grown and changed. Initially, infectious diseases were the major concern. With urbanization came residential crowding and the need to deal with human sanitary waste and heightened potentials for food and waterborne epidemics. As microbiological knowledge advanced, however, such infectious disease problems have come under control, and the health risks associated with chemical or radiation exposures have commanded increasing attention. At first, these chemical and radiation problems were principally recognized in occupational settings where exposures were particularly intense and sustained. At present, however, public-health concerns extend to the potential for such toxic exposures in all phases of human activity, occupational and nonoccupational.

This shift in public-health focus away from microbiologic problems toward toxic workplace exposures, and now more and more toward nonoccupational exposures, has been accompanied by a shift in concern away from acute illness arising promptly after relatively high doses of environmental exposure toward concern for delayed health effects resulting from lower doses of toxic exposure. Unfortunately, risk assessment with respect to delayed illness and low-dose exposure is a much more ambiguous process than risk assessment for acute illness in high-dose settings. For many aspects of this process, in fact, methods for toxic measurement and for epidemiologic evaluation are often quite inadequate to provide complete answers to questions raised by society regarding potential health effects. Faced with these inadequacies, we must often rely upon risk extrapolations instead of direct measurements of health effects to provide a basis for functional risk assessment.

This section focuses on hazardous chemicals and addresses the ways in which we identify and measure the health risks posed by exposure to waste chemicals in particular. The reader should be aware, however, that the same kinds of risk and the same methods for risk assessment are involved in all phases of human contact with manufactured toxic chemicals: their role in manufacturing processes, their transportation to the marketplace, their use as consumer goods, as well as their eventual disposal as waste material. It is also important to realize that much of what we know concerning toxic-chemical risk assessment has grown in parallel with comparable knowledge regarding similar risks associated with exposure to ionizing radiation. In many instances, in fact, the health risks are identical (cancer, genetic abnormalities), and the methods for risk assessment, first developed for radiation exposures, have found direct application with regard to toxic-chemical exposures.

Whatever the setting, risk assessment is critically dependent on three types of data: (1) data concerning the nature of potentially toxic exposure, (2) data concerning the existence and nature of disease or biologic abnormalities, and (3) data concerning the degree to which exposure and biologic abnormalities may be etiologically linked in any particular setting. To understand the process of assessing environmental health risks, it is useful to evaluate available methodologies in the context of these three categories.

3.2.1 EXPOSURE

Without concrete and often extensive information regarding chemicals present in a particular waste-site situation it is extremely hard and often impossible to achieve any quantitative assessment of the health risks involved.

Document the Toxins. The first step in the risk-assessment process is therefore to document the toxins present and to characterize the exposure setting (Fig. 3.2.1). If written records of the dump site's contents exist, these should be reviewed. Identification or verification of contents may also require some systematic sampling for purposes of laboratory testing and chemical identification of dump-site material. Either approach has its limitations, since it is impossible to test all of a site's contents or to be certain that chemicals recorded as dumped still exist in the site in their original form.

Evaluate Properties of Chemicals. Once an inventory of site contents is complete to the extent possible, the physical and toxic properties of identified chemicals must be evaluated. For an experienced toxicologist, most chemicals identified may be familiar in terms of physical nature, animal or human toxicity, and manner of absorption, metabolism, storage, and excretion in biologic systems. However, since many chemicals in a waste-disposal setting are discarded by-products by definition, a substantial proportion of them may well be chemicals for which full physical and toxicologic evaluations do not exist. It is also possible that interactions within mixtures of disposed chemicals may produce progeny chemicals

1. Inventory of chemicals present at the site.
   a. Review of available records regarding materials deposited at the site (amounts, dates, names of chemicals, physical condition, types of storage containers, etc.).
   b. Testing of site contents (especially if records are deficient or absent). Sampling of drums and containers, detection of potential chemical mixtures.
2. Evaluation of toxicologic data for chemicals detected at the site. Review of scientific literature, consultation with experienced toxicologists regarding physical and chemical properties; known animal and human toxicity; modes of absorption, metabolism, storage, and excretion by living tissues.
3. Measurement of environmental spread of chemical contamination.
   a. Assessment of principal environmental vehicles likely to carry chemicals from site to potentially exposed populations.
   b. Appropriate sampling of water (surface, ground), air, soil, and food.
4. Measurement of chemicals or chemical products in tissue of exposed persons (serum, fat, liver, skin, hair, etc.) (applicable only to chemicals which persist in tissues, unless exposures are very recent or are ongoing).

FIG. 3.2.1 Assessment of exposure at a chemical waste site.

for which no formal toxicologic information may exist. These various possibilities, together with the unknown toxicities of chemical mixtures themselves, make it often difficult to achieve as comprehensive an inventory of disposal-site materials as might ideally be required.

Measure Spread of Contamination. The third step in exposure assessment is to characterize the extent to which waste-site chemicals have migrated from the site, or the potential for such migration, and the pathways by which migration would be most likely. This requires variable degrees of environmental sampling and testing, the nature of which is determined by the features of each particular site. In one situation, the nature of chemicals present, the physical arrangement of the site, and local climatologic factors may focus concern on potential spread of toxic materials by airborne scatter of dust. In another setting, concern may be more with soil contamination, seepage into groundwater, and migration through connecting aquifers. Or potential pathways may be multiple, with greater emphasis on one environmental modality than another.

Sampling and testing should be designed and conducted on the basis of these kinds of environmental-engineering considerations, together with the all-important consideration of potential populations at risk. A waste site quite remote from any human residences or community activities will obviously not require the degree of risk assessment (if any is required at all) that will be needed for a site located in the midst of a community or in proximity to surface or groundwater from which a community may obtain drinking water.

Test Exposed Persons. A fourth phase of exposure assessment is to test for the presence of toxic chemicals or chemical metabolites in the tissues of exposed persons. Such measurements are clearly the best, most direct way in which to demonstrate actual exposure, provided the chemicals detected can be conclusively attributed to the waste source under suspicion and not to more widespread and ubiquitous exposures. For example, polychlorinated biphenyls (PCBs) detected in tissue might arise from exposure to a particular waste site but at the same time could reasonably be considered the result of diverse dietary exposures which have recently been commonplace in developed countries. Testing of tissues for the presence of chemicals, of course, is useful only for those chemicals which persist and are stored in tissue (PCB in fat, for instance) or which leave stored traces of chemical metabolites. Chemicals which are rapidly cleared or metabolized (and most waste-chemical exposures fit this category) may be detectable in tissues only if exposure is very recent or ongoing.

Throughout these several phases of exposure assessment, the limitations and technical requirements of environmental sampling, specimen collection, and laboratory test procedures must be kept clearly in mind. Such work is usually expensive and hence must be carefully planned before being conducted. Quality-control procedures must be built into the specimen collection and testing process. Desired limits of testing sensitivity must be defined, with consideration given both to factors of cost and technical difficulty and to the degree of toxicity expected for humans. For chemicals for which low doses of exposure can be expected but which are known to persist in human tissue, and to be highly toxic, exposure assessment must obviously involve greater sensitivity of testing than for chemicals known to be less toxic or less biologically persistent. Given the wide variety of chemicals, toxicity behaviors, pathways, and modes of absorption, each waste-site situation can be expected to present its own special problems and hence must be evaluated in an individualized manner.

3.2.2 BIOLOGIC ABNORMALITIES

The range of biologic abnormalities potentially related to exposures to toxic-waste chemicals is as wide and diverse as the range of such chemicals themselves. Whether biologic effects will occur and be discernible, if exposures do take place, is a matter dependent on the nature and degree of actual exposure as well as on our technical capacity for linking exposure to biologic effect. The means available for making such links, and the limitations of those means, are discussed later in this chapter. First, however, it is necessary to consider the scope of potential biologic abnormalities and how they can be reliably identified, steps required prior to efforts to link abnormalities to exposures. Figure 3.2.2 summarizes the process.

Since different toxic chemicals can have quite different biologic effects, both as a result of differing chemical properties for individual target tissues and in relation to their doses and pathways of exposure, many different kinds of disease and biologic change need to be considered.

Acute effects, in which exposure to a toxin is followed relatively quickly by biologic effect (a latency of days or weeks), are usually related to high doses of toxin exposures. Skin rashes, acute neurologic effects, acute hepatic or renal disease, and acute gastrointestinal or cardiovascular illness constitute the usual range of such acute clinical effects.

Subclinical acute effects can also be seen: abnormalities in liver-enzyme function, nerve-conduction slowing, and chromosomal damage, for example. Given sufficient information concerning the nature of waste chemicals involved in any particular exposure situation, as well as some concept of degree or dose of exposure, one can choose which particular kinds of effects—clinical, subclinical, or both—and which particular organ systems should receive attention. Hence, for

1. Determination of the range of abnormalities that deserve investigation in a given situation, as suggested by:
   a. Severity or dose of exposure. Examine both acute and delayed or chronic effects if dose is high; focus principally on delayed or chronic effects if dose is low.
   b. Organ specificity and toxicity of chemicals involved. Nervous system, liver, bone marrow, kidney, gastrointestinal system, cardiovascular system, etc.
   c. Feasibility or appropriateness of studying subclinical effects. Nerve-conduction velocity, liver-function disorders, cytogenetic abnormalities, etc.
   d. Feasibility or appropriateness of studying effects in exposed animal populations, domestic or wild.
2. Accurate and valid identification of clinical illness or of biological abnormality.
   a. Diagnostic confirmation of clinical cases by review of medical or hospital records, either as the primary source of case ascertainment or as a check on other sources of case ascertainment (questionnaires, etc.).
   b. Use of adequate technical laboratory procedures with suitable means for quality control when testing for subclinical abnormalities.
3. Complete ascertainment of abnormalities in the populations chosen for study (whether entire populations or samples of populations).
   a. Assure thoroughness of case-finding efforts by using multiple, overlapping sources for case identification (medical records, death certificates, etc.).
   b. Assure comparability of case-finding efforts in exposed and unexposed populations by minimizing opportunities for selective case reporting or interview biases.

FIG. 3.2.2 Assessment of biological abnormalities in a population.

exposure to neurotoxic pesticides, neurologic abnormalities should obviously be a major focus for investigation. For toxins such as benzene and its derivatives, hematologic and subclinical cytogenetic abnormalities might be sought. Not uncommonly, of course, many different chemicals are involved in any given exposure situation. This requires that a wide range of different biologic abnormalities receive consideration. Whatever the setting, it remains a basic principle that, in deciding which abnormalities to seek and study, one first should consider the particulars of the exposure scenario.

Whether exposure is high-dose or low-dose, biologic effects that are chronic or delayed require attention, in addition to ones that are acute and appear to occur after short latency. Here the process of linkage to possible toxic exposure etiologies is usually difficult and complex. The health effects of concern are principally those of cancer, genetic abnormalities, and complications of reproduction and parturition such as sterility, congenital malformations, and low birth weight. Subclinical effects such as abnormal subcellular genetic markers and cytogenetic changes also require attention. In addition to human observations, investigations may consider biological abnormalities in potentially exposed nonhuman populations such as domestic pets or wild animals. If this approach is chosen, the rationale will be that illnesses in animals may conceivably serve as markers for potential effects in human populations.

Whatever parameters of biologic abnormality are used in any particular toxic-waste investigation, it is essential that instances of abnormality be accurately identified and that ascertainment of such cases either be complete or be based on sound sampling techniques. Accurate recording of clinical illness requires review of medical records in hospitals, clinics, or physicians' offices to confirm diagnoses. Where other nonmedical sources of case finding are used, such as death or birth certificates or community survey information, all diagnoses should be checked against medical records before being accepted. When it is deemed essential to use unconfirmed health data, such as the results of household surveys, in an investigation, the investigators should recognize the great potential for both false-positive and false-negative observations, that potential being skewed according to individual household perceptions. Accuracy of subclinical information, in turn, depends on the accuracy of technical procedures used. Inadequate methods for collecting, processing, and analyzing specimens will obviously compromise validity of results. If subclinical outcomes are to be used in any particular study, investigators must take special care that appropriate techniques are used in all its phases.

Complete case ascertainment is as crucial as accurate case diagnosis. One without the other will seriously flaw any effort to draw conclusions about health risks when illness patterns in exposed and unexposed groups are ultimately compared. If the object in a particular study is to ascertain all illness of a particular sort in a population, it is usually desirable to use several different means of case ascertainment so that complete case finding can be assured. Hence a search for cases of cancer might make use of death certificates as well as medical records in all clinics or hospitals (primary or referral) serving a particular area. On the other hand, the study might rely on sampling techniques for estimating disease frequencies. Questionnaires regarding particular illnesses might be administered to a sample of households in exposed and unexposed areas, for instance. In such a setting, not only will it be important to confirm both positive and negative reports through medical-record reviews, but the framework by which samples are drawn will need to have been carefully designed. Any technical flaw which may permit oversampling of one particular exposure group will detract from the study's validity, as will any major differences that occur in the manner of questionnaire administration and response. In the latter regard, the inevitable subjectivity of questionnaire data is usually an unavoidable problem. Although one may try hard to have questionnaires administered blindly (by phone or in person), thus perhaps controlling subjectivity of response introduced by the questioner, bias on the part of the responder, depending on his or her perception of exposure, is virtually impossible to neutralize. The use of questionnaire surveys may be the least desirable and most subjective approach for measuring potential illness patterns in hazardous-waste settings, given the potential for public hysteria and misperceptions when issues regarding waste disposal come under scrutiny in a community.

3.2.3 LINKAGE OF EXPOSURE WITH BIOLOGIC ABNORMALITIES

The process of establishing etiology, and specifically of linking hazardous-waste exposure to human biologic abnormalities, is essentially a matter for epidemiologic analysis. Epidemiologic investigations describe frequencies of illness in populations and relate those frequencies to various characteristics of those populations (characteristics which may include degrees of exposure to hazardous waste). Such studies constitute our principal means for establishing etiol-

1. Limitations on information regarding exposure doses and biologic abnormalities (see Figs. 3.2.1 and 3.2.2).

2. Long and variable latency between exposure and biologic outcome.

3. Nonspecificity of biologic outcomes with respect to particular risk factors and exposures. Need to adjust analyses for the potential etiologic contributions of competing risk variables.

4. Low frequency (relative rarity) of diseases under study. Low expected incidence requires relatively large sample sizes, especially with need to satisfy power requirements and to stratify data while adjusting for the possible effects of competing variables.

FIG. 3.2.3   Linkage of chemical exposures with biologic abnormalities: epidemiologic limitations.

ogies of human illness since direct experimentation is not an ethical option and since questions of etiology must often be addressed some time after exposure has ceased.

The success of epidemiologic analysis in either suggesting or discounting etiologic relationships rests ultimately (as discussed in the two prior subsections) with the validity and accuracy of information obtained regarding chemical exposures and biologic abnormalities. Given accurate and valid information to work with, however, the epidemiologist must still be aware of three additional features which can profoundly affect the capacity of any study to link exposures to abnormalities. These features are (see Fig. 3.2.3): (1) disease latency, (2) biologic nonspecificity (or multiple causation), and (3) relative frequency of disease occurrence. Simple and quickly convincing epidemiologic analyses, akin to direct observation of the cause-effect process, are usually attainable only when the latent period between exposure and disease onset is short (a situation which is rarely encountered in the absence of a high exposure dose). This analytic advantage conferred by short latency may be further enhanced, of course, if at the same time the incidence of disease in question is greatly increased above its usual level of frequency, if the disease is clinically unusual or even unique, or if the disease is particularly associated with the specific clinical exposure (that is, is relatively "specific" as a toxicologic consequence of exposure to the chemical). Unfortunately, most questions of health effects from hazardous-waste exposure arise in settings where dose is likely to have been low, latency long, and illness outcomes multifactorial in nature (not identifiably specific for a particular chemical cause). Under such circumstances epidemiologic analysis requires great care and patience and even so may not often yield conclusive results.

Latency

It almost goes without saying that cause-effect relationships which involve only a short time interval between cause and effect are much easier to define than long-interval relationships. When latency is measured in months or years (as is the case with pregnancy outcomes or cancers), it becomes increasingly difficult to distinguish the exposure of interest from intervening exposures of other sorts. This is especially hard when the illness in question is not particularly specific for the exposure being studied. In high-dose–short-latency settings, firm etiologic conclusions can often be reached on the basis of merely one or two disease case reports. When latency is long, however, conclusions become more a matter of comparing illness frequencies in exposed and unexposed groups, statistical comparisons which may require substantial numbers of cases to provide stable rates. Long latencies are also associated with greater variability in individual latency lengths (a latency of 20 years may reasonably encompass individual latencies ranging from 15 to 25 years, for instance). The time framework within which observations are made must therefore be adjusted accordingly.

Biologic Nonspecificity (Multiple Causation)

There would be little need for complex epidemiologic studies if each individual case of illness or biologic abnormality were identifiable as arising from a particular cause or blend of causes, regardless of latency intervals. Obviously, quite the opposite is true for most abnormalities. A particular case of cancer or of congenital malformation may have any of a number of causes (chemical exposures, radiation exposures, genetic influences, etc.), and at present we have no specific markers by which cases which are clinically indistinguishable can be identified according to their particular etiologic origins (unlike cases of sore throat that may be distinguished by bacterial culture techniques, for instance).

Until the day arrives when methods are at hand for case-specific "etiology testing," our only recourse is to apply epidemiologic techniques for separating out competing risk factors. To do such analysis, it is first essential to obtain information in each case of illness regarding whatever risk factors may be considered pertinent to that form of illness. If a study is focusing on low birth weight as a possible outcome of hazardous-waste exposure, exposures to risk factors other than hazardous waste will need to be taken into account: factors such as maternal age, parity, and smoking patterns which are known to influence frequency of low birth-weight occurrence. Then, during later epidemiologic analysis, comparisons between population groups exposed and not exposed to waste can be adjusted for any differences with respect to these competing risk factors, differences which might otherwise tend to obscure or exaggerate the true local relationship between low birth weight and hazardous-waste exposure.

An unfortunate methodologic side effect of this need to adjust epidemiologic data for competing risk variables is that total sample-size requirements are accordingly increased. The more variables that need to be entered into the analysis, the greater the need to subdivide the population during analysis into groupings according to multiple-risk-factor patterns. The numerical size of each subgrouping is therefore a critical consideration for determining overall sample size. Where a total "exposed" population is of limited size, there are obvious limitations imposed on effective epidemiologic analysis.

Frequency of Disease

Since epidemiologic investigations are ultimately concerned with measuring rates of disease occurrence in populations, it follows that the nature of particular investigations will be very much influenced by the underlying "expected" frequencies of the diseases being studied. If the disease under study is common, a significant difference in disease rates between exposed and unexposed populations may quite possibly be demonstrable in small population groups or in limited samples of larger populations. If the disease is uncommon, however, effective analysis may not be possible, particularly if rather small differences in disease incidence are being sought (two to threefold differences, for instance). In designing an epidemiologic study, therefore, the investigator must at the start address three issues regarding the study's ultimate size and feasibility: (1) the size of the population available for study, (2) the underlying frequency or incidence of the disease to be studied, and (3) the degree of risk increase which the study will be expected to be able to show (i.e., the statistical power of the study). Together with considerations of exposure and dose, these practical aspects regarding disease frequency and population size impose very real, and often insurmountable, restrictions on what one may reasonably expect from epidemiologic analyses in any given hazardous-waste situation.

3.2.4 EVALUATION OF CASE "CLUSTERS"

Situations requiring assessment of possible hazardous-waste health effects can take either of two forms. As implied in the preceding sections, one entails assessing the toxic potential of a known waste-disposal site and then seeking patterns of human illness which might result from site exposure. The other, however, constitutes the opposite scenario: the epidemiologist is called upon to assess an apparently unusual pattern of disease in a community and, in the process, to seek a possible environmental cause. The latter situation is often referred to as a disease case "cluster." Reports of alleged case clusters not infrequently focus on cancer or on reproductive abnormalities such as spontaneous abortions or congenital malformations.

Techniques for assessing such reports of case clusters (and they can take many different forms) do not differ from what has already been described. First, one must define the scope of disease to be studied as well as the time and place boundaries of the alleged disease-cluster problem. This can be quite an arbitrary process and often calls for quite pragmatic decisions, dictated heavily by the nature of the concern as originally voiced. One proceeds by ascertaining case-incidence patterns from reliable medical sources, and then by calculating incidence rates to be compared statistically with an appropriate source of expected incidence. If a clear excess of disease in the population is confirmed, if case features seem unusual (striking clinical similarities, peculiar age or sex distribution, etc.), and if community concern is particularly increased, the investigator may proceed to obtain more detailed information for each case and for the community with respect to possible risk factors. This often includes a search for possible hazardous-waste exposures in the local area.

Data may be collected only for patients with disease and their households, or from matched control households as well, especially if specific local risk hypotheses exist. The object will be, in either case, to identify patterns of community risk exposure shared strongly enough by the cases or the case households so that a plausible explanation for the particular disease cluster can be offered. Although each cluster report warrants some degree of investigation, it should be recognized that in practice, such cluster investigations rarely if ever unearth any convincing explanation for the cluster's origin. This is not unexpected, if one considers the open-endedness of the assessment approach, the multiplicity of available hypotheses, and the many methodologic difficulties involved in retrospective investigations of this sort.

One pertinent explanation available for any case cluster is that it may represent a chance aggregation of cases, occurring close together in time and place, which may arise merely in the course of random case occurrence. Although such an explanation can not be tested except through exclusion of other more biologically meaningful hypotheses, it remains a distinct possibility and one for which statistical probabilities can be calculated (i.e., the statistical likelihood that a case cluster of a given size will occur by chance, given a particular population size and underlying disease frequency). The statistical "significance" of the individual cluster itself, of course, calculated in comparison with expected disease frequencies, is of limited value in assessing a case cluster's statistical meaning, let alone its biologic meaning. Since any case cluster can come to attention purely on the basis of its "clusteredness," the calculation of a statistical significance level will merely confirm the existence of an increased disease frequency (perhaps on a chance basis) when that increased frequency was already intuitively obvious. The critical test will lie in the search for possibly meaningful biologic explanations for the cluster. The extent to which that search is pursued will be determined by the nature of the local problem and, to some extent, by the degree of concern present in the local community. Factors which may influence one to conduct detailed investigations may include the nature of the disease in question (is it clinically unusual, diagnostically homogeneous, very rare? does it affect an unusual age group?) and the existence of any communitywide hypothesis which might plausibly bear some etiologic relationship to the case cluster under consideration.

3.2.5   EXTRAPOLATION OF RISK ESTIMATES

It is common toxicologic practice to estimate risks in particular settings by extrapolating from risks measured in other settings. Thus, low-dose toxic-chemical risks in humans are often estimated by extrapolation from high-dose experimental observations in laboratory animals. Whenever such an extrapolation is calculated, safety factors are introduced into the equation which allow not only for probable interspecies differences but also for differences in dose levels. Dose extrapolations often involve selecting one of several mathematical models by which unmeasurable low-dose effects are predicted on the basis of measured high-dose effects. The most common, and often most conservative, mathematical model is a straight-line or linear projection (effect varies directly with dose). Alternatively, various curvilinear models can be used which make allowances for differences in such things as toxin absorption or tissue repair at different dose levels. Most extrapolation procedures of this sort will project a conservative risk estimate. Since that estimate is commonly used as part of the process of assessing health risks in particular hazardous-waste settings, investigators should clearly understand the rather complex and often arbitrary means by which such risk estimates are derived as well as the degree to which the estimate is conservative.

Risk extrapolation of a different kind is implicit when an investigator chooses to measure a subclinical biologic effect such as chromosome aberrations or liver-function abnormalities and then seeks to translate those observations in terms of risk of eventual clinical illness. This sort of biologic extrapolation is a most uncertain process, since the relationships between observed subclinical changes and future disease are often not well known. In choosing to incorporate subclinical endpoints into an epidemiologic investigation or into a risk-assessment procedure, one must first have given close attention to the manner in which the results of such analysis will be interpreted.

3.2.6 SUMMARY

Assessment of the potential human biologic effects of exposure to hazardous chemical wastes requires (1) careful assessment of local exposure conditions, pathways, and doses, (2) thorough characterization of biologic endpoints, whether these be clinical diseases or subclinical effects, and (3) appropriate epidemiologic techniques for testing the strength of associations between exposure and biologic patterns. Epidemiologic analysis can be severely hampered by long latency, clinical nonspecificity (multifactorial etiology), and limited population sizes or low frequencies of disease occurrence. Risk assessments are further compromised by the frequent necessity to extrapolate high-dose animal toxicologic data to low-dose human exposures and to use subclinical observations as a theoretical basis for predicting future human disease.

3.2.7 FURTHER READING

Bloom, A. D. (ed), *Guidelines for Studies of Human Populations Exposed to Mutagenic and Reproductive Hazards,* March of Dimes Birth Defects Foundation, New York, 1981, 163 pp.

Caldwell, G. G., and C. W. Heath, Jr., "Case Clustering in Cancer," *Southern Med. J.,* **69**:1598–1602 (1976).

Clark, C. S., C. R. Meyer, P. S. Gartside, et al., "An Environmental Health Survey of Drinking Water Contamination by Leachate from a Pesticide Waste Dump in Hardeman County, Tennessee," *Arch. Environ. Health*, 37:9–18 (1982).

Halperin, W., R. Altman, A. Stemhagen, et al., "Epidemiologic Investigation of Clusters of Leukemia and Hodgkin's Disease in Rutherford, New Jersey," *J. Med. Soc. New Jersey,* **77**:267–273 (1980).

Heath, C. W., Jr., "Field Epidemiologic Studies of Populations Exposed to Waste Dumps," *Environ. Health Perspect.,* 48:3–7 (1983).

Heath, C. W., Jr., M. R. Nadel, M. M. Zack, Jr., et al., "Cytogenetic Findings in Persons Living Near the Love Canal," *J. Amer. Med. Assoc.,* 251:1437–1440 (1984).

Hoffman, R. E., P. A. Stehr-Green, K. B. Webb, et al., "Health Effects of Long-term Exposure to 2,3,7,8-Tetrachlorodibenzo-*p*-Dioxin," *J. Amer. Med. Assoc.,* **255**:2031–2038 (1986).

Janerich, D. T., W. S. Burnett, G. Feck, et al., "Cancer Incidence in the Love Canal Area," *Science*, **212**:1404–1407 (1981).

Rowley, M. H., J. J. Christian, D. K. Basu, et al., "Use of Small Mammals (Voles) to Assess a Hazardous Waste Site at Love Canal, Niagara Falls, New York," *Arch. Environ. Contam. Toxicol.,* 12:383–397 (1983).

Vianna, N. J., and A. K. Polan, "Incidence of Low Birth Weight Among Love Canal Residents," *Science,* 226:1217–1219 (1984).

World Health Organization, *Guidelines on Studies in Environmental Epidemiology,* Environmental Health Criteria Document No. 27, WHO, Geneva, 1983, 351 pp.

SECTION 3.3
ENVIRONMENTAL RISK-ASSESSMENT AND AUDIT PRINCIPLES

Lynne M. Miller
President
Environmental Strategies Corporation
Vienna, Virginia

Douglas E. Gladstone, Ph.D.
Vice President
Environmental Strategies Corporation
Vienna, Virginia

3.3.1 INTRODUCTION

Effective management of environmental liabilities will ultimately protect corporate assets while protecting environmental resources. Adequate management of environmental liabilities is possible only after a thorough process of risk identification, evaluation, and reduction. This process, often referred to as environmental-risk assessment, evaluates the potential for liabilities, impairment, or claims resulting from environmental pollution arising from a facility. An important premise of a risk assessment is that regulatory compliance alone is an insufficient indicator of the potential for pollution liabilities; this is because impairment is just as likely to result from areas that fall outside environmental regulations. Regulatory compliance is a minimum standard or baseline for a risk assessment. A useful procedure to evaluate regulatory compliance is an environmental audit, a procedure that determines whether a facility meets appropriate regulatory and nonregulatory standards. Methodology for assessments and audits is discussed in this chapter. While both are important corporate tools to manage environmental exposures, other uses are also discussed, particularly relating to pollution insurance.

3.3.2 ENVIRONMENTAL RISK ASSESSMENT

An environmental risk assessment is a snapshot of a facility's potential to cause environmental impairment. For a company with multiple locations, the first step in the risk-assessment process involves selection of sites to be assessed, taking into account corporate resources and the objective of the risk-assessment program. Some companies will target only those locations with known environmental problems, while others will ultimately assess all locations, but may rank facilities in priority order to select those to be assessed initially. Factors used to select locations may include type of operations, nature of past operations, existence of factors that increase environmental exposures (such as untested underground tanks, unlined lagoons, on-site waste disposal, private wells in area), prior complaints or environmental permit problems, and setting of facility. It is often helpful to send questionnaires to all facilities to gather the information needed to make site selections.

For the selected locations, plant personnel should compile data that will be used in the risk-assessment process. Such data would include:

- Description of waste-disposal practices (past and present)
- Past site operation and ownership
- Inspection reports relating to environmental and engineering practices
- Site plans and location maps
- Descriptions of site operations and processes
- Environmental permits (such as air, sewer, hazardous waste, etc.)
- Information on underground tanks
- Inventory of chemical storage

The next step in the risk-assessment process involves an on-site inspection of the facility to review the compiled data, meet with operating personnel, observe site operations, and gather information needed for the risk-assessment process. Information is gathered to evaluate four critical factors: the environmental pathways, the receptor populations, management and practices, and the properties of the materials. The information is used to evaluate the potential for each factor to contribute to pollution liabilities caused by sudden events or through gradual emissions, leaks, or discharges. The integration and interaction of these factors determines the overall potential for pollution impairment, liabilities, or claims from a facility.

Evaluation of the first of the factors, the *environmental pathways,* involves analysis of the pathways materials could take to escape from the facility. All potential pathways are represented, including escape through surface water, through the air, and through the soil and groundwater. Both direct and indirect discharges are evaluated. The potential for emissions, discharges, leaks, or spills to move off site is evaluated. Groundwater is considered to be an off-site pathway, whether it is beyond the facility boundary or not. The effectiveness of pollution-control devices is assessed under this factor since they act as impediments to movement through the pathways.

The *receptor populations* are those that can be affected by the materials that move along the environmental pathways. The receptor populations evaluated are broader than the human population that may reside nearby. The existence of potentially affected populations of domestic animals or wildlife is also determined. The nature and quality of inorganic natural resources, such as surface water and groundwater, are also assessed, as are impacts on publicly owned treatment works. In evaluating the surrounding human populations, the existence of any political or community opposition to the facility is factored into the analysis. Also, the potential for other industry in the area to affect the assessed facility is evaluated.

Given that for most industrial facilities environmental pathways and potentially affected target populations will exist, it is important to determine how the *management and practices* at the facility reduce or eliminate the potential for off-site impacts. The existence and effectiveness of environmental programs is evaluated. The facility's dealings with the regulatory agencies are reviewed as is the compliance with environmental permits. The evaluation of management and controls occurs not only for current operations but also for past practices. At some facilities, the current management can be environmentally astute; however, past owners or managers may have had a more cavalier attitude, leaving a legacy of high environmental risk.

It is obvious that the seriousness of the risks due to the above factors will hinge on the *properties of the materials* at the facility. Materials evaluated include raw materials, intermediates, products, and wastes, depending on the potential for a release of the substance. The effects of short and long-term exposures are evaluated as mitigated or exacerbated by the behavior of the material in the environment. The potential to cause health effects and other types of damage is evaluated.

The overall risk from the facility depends to a large extent on the integration and interaction of the four factors. Stringent environmental controls and the lack of pathways can mitigate the risks from extremely hazardous materials, while the impact from a facility may be minor despite the freer use of materials if those materials are more innocuous.

A risk assessment is designed to be a tool for facilities to use to control their exposures. Therefore, one of the most important parts of the risk-assessment report is a set of recommendations to reduce risks, listed in priority order. One of the most common shortcomings found by evaluators is that a facility's managers have an insufficient amount of information about the facility to thoroughly understand and evaluate the exposures. Accordingly, the need for additional sampling or monitoring is a frequent recommendation.

A periodic assessment update is recommended so that a company can evaluate the current environmental risk-management concerns. The update would include evaluating the effectiveness of risk-reduction programs and would incorporate any new environmental monitoring data. A corporation's risk assessment should be updated when operations or environmental personnel change substantially. Updating is especially important when an acquisition is being contemplated.

3.3.3　ENVIRONMENTAL AUDIT

While the primary focus of an environmental risk assessment is the identification of environmental liabilities, an environmental audit focuses both on regulatory compliance and environmental liabilities. The use of environmental audits has been encouraged by the U.S. Environmental Protection Agency (EPA) in a policy statement issued in July 1986.[1]

An environmental audit is not inspectors wearing green eyeshades using long checklists. It is an analytical tool that, when used by facility personnel, will allow them to determine not only if their facility meets the appropriate regulatory and nonregulatory standards, but also why the facility may be lacking in some areas. Facility personnel will be in the best position to institute the appropriate remedies. Making facility personnel the auditors also helps to ensure that the aims and goals of the audit are not forgotten once an inspector leaves the facility, since the "inspector" is always there.

The purpose of an environmental audit is to provide corporate management with the assurance that their facilities will not be the source of liabilities for the corporation or its officers and directors. Through reporting requirements, corporate inspections, and other mechanisms of corporate oversight, the audit can be tailored so that it is under the amount of corporate control appropriate to the structure of a particular company.

The audit will run more smoothly if the audit procedures are designed to match the way the company is run. Facilities that are accustomed to operating fairly autonomously will not necessarily respond well to intense scrutiny from corporate headquarters. Centralizing audit procedures, however, may be desirable. Centralization of the audit can have advantages beyond allowing corporate officials to keep close track of the potential for liabilities at a particular division. Similar operations will frequently have similar environmental requirements and problems. Having the audit centrally coordinated prevents a situation in which people are inventing solutions anew each time a problem arises.

Once the initial meeting to structure the scope of the audit are completed, the facilities that will be part of the audit program are visited and the audit itself is initiated.

There are numerous approaches to environmental auditing. The one described below is that developed and used by Environmental Strategies Corporation.

The Risk Review

The first step in developing an audit for a specific facility is a site visit. The visit, which usually lasts one day, allows the audit team to become familiar with the operations of the facility and to determine the extent and substance of any environmental practices and procedures that may already be in place. The site visit consists of considerable discussion with plant personnel and a tour of the facility. It concludes with an exit briefing in which facility personnel are informed of the preliminary findings, any problem areas requiring immediate attention, and the scope of the subsequent audit steps.

The site visit provides the baseline of information on which the audit is constructed. The first product of the audit is the risk-review report. The risk-review report is very similar to the environmental risk-assessment report.

The risk review serves a number of functions in the audit process. It will highlight imminent problems that probably should be addressed before completion of the audit. It not only describes areas of obvious regulatory noncompliance, but also addresses those unregulated areas that could lead to environmental liabilities. The risk review indicates how these practices, while unregulated, can nonetheless lead to third-party liabilities, and recommends ways in which these shortcomings can be corrected. By indicating those areas in which the potential is greatest for creating environmental liabilities at a facility, the risk review focuses the development of the audit manual. The risk review serves as a base which allows the audit manual to be organized by functional areas, such as wastewater, air, and wastes, rather than into the structure imposed by the regulations.

Recent legislation and court decisions make it clear that corporations can be held accountable for the environmental consequences of their past actions. The risk review examines potential hazards related to earlier activities as closely as possible. The fact that the industrial practices may have been in full accord with the standards and norms of an earlier period may not serve as an effective defense. The research into prior practices will include a review of the status of offsite waste disposal that may have been contracted for in prior years.

The Environmental Audit Manual

The other major product generated by the ESC environmental audit process is the environmental audit manual. The manual provides facility personnel with a tool that can be used to become familiar with the appropriate regulations and with procedures that can be followed to ensure compliance. The manual also contains the procedures necessary to ensure that good environmental practices are used where there are no regulatory guidelines. The manual serves as the focal point for all environmental documents at the facility, including permits, contingency plans, and monitoring data.

The audit manual consists of a number of packaged units, each of which deals with the regulations and procedures for a specific functional area. Examples of the areas covered in these packages are hazardous and nonhazardous wastes, wastewater, air emissions, and underground tanks. The regulations covered in each package are those that specifically apply to the facility. The procedures as outlined in the packages are extracted from all appropriate sections of the regulations, and contain all relevant sections of federal, state, and local regulations.

Some of the most common environmental liabilities faced by corporations, and some of the most expensive, are not incurred from regulatory noncompliance. The audit process goes beyond the relatively simple question of technical compliance with mandated requirements. In many areas of environmental and health risks, there are no effective standards that serve as protective safety demarcation lines, inside of which it can be argued that no damage can be caused and hence there can be no liability exposure. Compliance with regulatory standards is a minimum requirement rather than a guarantee of zero liability. The audit manual contains packages that outline procedures that will ensure a facility is using sound environmental practices, whether mandated by regulation or not.

Follow-up

An audit manual that contains all the appropriate material can be a large and formidable document. Its value will be lost, however, if it is used only to decorate the plant manager's bookshelf. Corporate personnel must also be briefed about the manuals; this will enable them to coordinate the effort and respond appropriately when faced with facility managers' questions.

Both corporate and facility personnel should be imbued with the concept that the audit is a dynamic process. As operations change, on the one hand, and regulations change, on the other, the manual will have to be supplemented or portions deleted, and the procedures altered. Various packages in the manual indicate the kinds of changes in status which would necessitate the incorporation of additional, appropriate packages.

3.3.4 USES OF ASSESSMENTS AND AUDITS

Environmental assessments and audits can be powerful tools to unite corporate and facility personnel in the struggle against the accrual of environmental liabilities. The risk assessment identifies environmental exposures, while the audit focuses the facility manager's attention on the specific duties that must be performed to ensure compliance with the law. Both together can serve as a basic tool for facility managers to use to communicate on environmental matters with corporate headquarters. Assessments and audits must be updated to reflect changes in plant operations, and also new requirements and changes in legal and scientific interpretations. They demonstrate tangible evidence of a good faith attempt by the company to comply with its legal and social obligations.

Not only are assessments and audits useful in evaluating ongoing operations, they are particularly useful in the acquisition process. Too often corporations evaluating proposed facilities for acquisition do not adequately consider environmental liabilities. Some corporations are not aware that they can be held directly responsible for environmental problems caused by the prior owner. Such liability can be controlled, but not eliminated, by contract. The seller may agree to indemnify the purchaser for environmental problems that resulted from operations for a certain period of time before the sale of the property. Only a thorough environmental risk assessment with sampling can ascertain the state of the facility at the time of sale. It is essential to have this baseline of information for future legal action. Environmental assessments may identify liabilities so enormous that the acquisition is not carried out. Other times, an assessment by the purchaser is used as the basis for requiring the seller to make certain changes before the sale, such as removing underground tanks or performing environmental sampling. Environmental assessments may also identify areas of risk that serve to reduce the purchase price of the property. Environmental assessments are being used by some financial institutions before foreclosure or loan commitments.[2]

Environmental assessments are also used in support of pollution-liability insurance, which is described in the next section. Underwriters use the assessment as an important tool in determining whether a facility will be offered insurance, and if it is, what the terms of coverage and pricing will be.[3] A company seeking coverage should anticipate that risk assessments will be required before coverage can be offered. Depending on the type of operations, the number of locations, and the insurance company involved, assessments may be required for a minimum of a quarter of all locations to be covered or may be required for all locations. Usually the insurance company will provide a list of independent consulting firms that are acceptable to perform the assessments, although sometimes the insurance company may use their own in-house professionals.

3.3.5 RISK ASSESSMENTS AND POLLUTION INSURANCE

Financial Responsibility Requirements

The Resource Conservation and Recovery Act (RCRA) requires certain hazardous-waste facilities to demonstrate financial responsibility for bodily injury and property damage to third parties.[4] Owners or operators of hazardous-waste treatment, storage, or disposal facilities must show evidence of liability coverage for sudden and accidental occurrences at minimum limits of $2 million. Owners or operators of hazardous-waste surface impoundments, landfills, or land treatment facilities must demonstrate liability coverage for non-sudden and accidental occurrences at levels of $3 million per occurrence and $6 million annual aggregate. Until recently, either commercial insurance or a financial test was used to meet the liability requirements. Primarily because of the constrained insurance market, a corporate guarantee was added as a mechanism to meet liability requirements, effective September 9, 1986.[5] Some state requirements vary, so it is advisable to check with state regulators.

Comprehensive General Liability Insurance

Until recently, most facilities used their comprehensive general-liability (CGL) insurance policy to demonstrate coverage for sudden and accidental occurrences. The CGL policy covers numerous liability risks, and until about 1985, policies included coverage for sudden and accidental pollution (while excluding gradual pollution). The CGL policies issued today will almost certainly exclude all pollution claims. There are several reasons for the removal of pollution coverage from the CGL policy, most notably the large number of pollution claims being paid under these policies. Most insurance companies had not anticipated and had not priced the policy to reflect pollution claims under the CGL policy. The present judicial trend of holding insurers responsible under old CGL policies for pollution claims has heightened insurers aversion to covering pollution exposures. Typically, CGL policies have been written on an "occurrence-based" form, meaning that the insured loss must have happened during the policy period regardless of when the claim is filed. This has been the basis of many successful pollution claims. Compare this to a "claims-made" approach, which the industry has recently introduced for the CGL policy: the insured loss must have happened after the effective date of the policy and be reported during the policy term.

Pollution Liability Insurance

A separate type of coverage for pollution has been available since the mid-1970s. Environmental-impairment liability (or pollution legal liability) insurance provides claims-made coverage for third-party bodily injury and property damage resulting from pollution. Most of the pollution coverage written in the past was for gradual pollution incidents, but the removal of sudden coverage from the CGL has prompted most insured parties to request coverage for both sudden and gradual pollution.

Until 1980, there was only one (London-based) provider of pollution insurance, but the increased demand for coverage with the imposition of RCRA financial responsibility requirements as well as industry's heightened awareness of pollution liability caused several U.S. companies to offer pollution coverage in 1980. The number of providers of coverage peaked at 11 in 1983.

Presently, there are only two major providers of pollution insurance: American International Group (National Union Fire Insurance Co.) and a reinsurance pool of about 20 insurance companies, the Pollution Liability Insurance Association (PLIA).[6] American International Group will accept submissions from all potential insured parties, while PLIA only accepts companies who are existing clients of a PLIA member. Presently, the maximum limits written per policy are $12.5 million per occurrence and $12.5 million annual aggregate for American International Group and $2 million per occurrence and $4 million annual aggregate for PLIA. The decline in the number of companies writing pollution insurance is attributable to several factors, including the lack of reinsurance support for the payment of pollution losses under several programs, the existence of a "hard insurance" cycle for the insurance industry in which premiums are higher and capacity lower for most coverages, and large net underwriting losses in the property and casualty markets.

Because the availability of pollution insurance is diminished and the CGL no longer provides sudden-pollution coverage, companies are finding it difficult to transfer pollution-liability exposures. Particularly affected are facilities that must demonstrate financial responsibility under RCRA. The 1984 amendments to the statute required all land disposal facilities to have shown evidence of financial responsibility (as well as groundwater monitoring) by November 8, 1985, or automatically lose their interim-permit status.

Coverage

Policies are offered on an individual named-site basis, and only for site owners and operators. Pollution policies are intended to provide coverage for claims that the insured is legally obligated to pay for third-party bodily injury or property damage resulting from the unexpected, gradual (sometimes also sudden) discharge of materials from specifically listed locations. The policies provide a rather broad definition of pollution conditions: the discharge, dispersal, release, or escape of smoke, vapors, soot, fumes, acids, alkalis, toxic chemicals, liquids or gases, waste materials, or other irritants, contaminants, or pollutants into or upon land, the atmosphere, or any watercourse or body of water.

Pollution policies are written for a 12-month period. Each policy is written with a limit for each pollution incident and a total limit for the policy year. Policies typically will not include coverage for damages for any of the following: liability incurred by noninsured parties that is assumed under contract; injury to employees; willful or knowing violation of a law; damage to property owned, occupied, rented, used or in the control, care, or custody of the insured; fines and penalties; and preexisting pollution conditions. The wording of the policies varies. It is important to read individual policies to determine specific coverages and exclusions.

Commercial Alternatives to Insurance

As the availability of pollution insurance has decreased, corporations are exploring alternatives to commercial insurance coverage. These alternatives have included self-insurance programs, the formation of "captives" or pooling arrangements, and an increased emphasis on environmental risk management.

Self-Insurance. *Self-insurance* takes a wide variety of forms. The simplest self-insurance mechanism is the retention of risk by a company without funding for it. An obvious disadvantage of this approach is that losses are paid out of a company's working capital. Another way to self-insure is to create an internal reserve of funds that will be available to pay losses. Using either mechanism, it is essential that a corporation fully understand potential environmental liabilities so that the exposure can be managed adequately. This need has resulted in an increased reliance on principles of environmental risk assessment and audit.

Captives. Another mechanism that can be considered is the formation of a "captive" insurance company, a subsidiary formed by a parent company for the purpose of writing all or some of the parent's loss exposures. *Association captives* are owned jointly by several parent companies, while a *pure captive* is owned by one company. Several groups are actively exploring the formation of *pollution captives*. Major hurdles these groups must deal with are the lack of reinsurance support and funding commitments from the members.

3.3.6 CONCLUSION

As environmental liabilities become more onerous and the transfer of such liabilities through insurance becomes more difficult, corporations are recognizing that it is more important than ever to understand their potential pollution exposures. Because companies are self-assuming a large portion of their pollution exposures today, the incentive to identify pollution liability has increased greatly. Corporations are turning to the relatively new fields of environmental risk assessment and auditing. A risk assessment focuses on potential pollution liabilities while an audit concentrates on regulatory concerns. Used together, they are powerful tools to assist a corporation in identifying and reducing environmental exposures.

3.3.7 REFERENCES

1. Environmental Auditing Policy Statement, *Federal Register,* vol. 51, no. 131, pp. 25,004–25,010, July 9, 1986.
2. M. J. Murphy, "Environmental Risks Pose Hidden Liabilities," *American Bankers Association Journal,* April 1986, p. 96.
3. D. Forbes, "Environmental Impairment Liability Risk Assessments Open Corporate Eyes to Festering Pollution Problems," *Risk Management,* April 1984, p. 116.
4. 40 *CFR* 264.147 and 265.147.
5. Standards Applicable to Owners and Operators of Hazardous Waste Treatment, Storage and Disposal Facilities; Liability Coverage, *Federal Register,* vol. 51, no. 133, pp. 25,350–25,356, July 11, 1986.
6. R. A. Finlayson, "General Liability Policy May Be Best Chance for Pollution Cover," *Business Insurance,* March 17, 1986, p. 25.
7. ———"Backers of Three New EIL Captives Raring To Go," *Business Insurance,* July 14, 1986, p. 3.

3.3.8 FURTHER READING

For general readings on environmental risk assessment see:

All-Industry Research Advisory Council, *Risk Assessment for Pollution Liability: A Survey of Insurers and Environmental Consultants,* Oak Brook, Illinois, December 1985.

Finlayson, R. A., "Environmental Studies Useful, But Risky," *Business Insurance,* March 17, 1986, pp. 22–24.

Miller, L. M., "Environmental Risk Assessment," chap. 31 in H. Bhatt, R. Sykes and T. Sweeney (eds). *Management of Toxic and Hazardous Wastes,* Lewis Publishers, Chelsea, Michigan, 1985, pp. 367–371.

For general readings on environmental auditing see:

Cahill, L., and F. Kane, *Environmental Audits,* Government Institutes, Inc., Rockville, Maryland, December 1985.

Trilling, B. J., "Implementing Environmental Audit Findings Through Performance Appraisal," *Environmental Advisor,* July 1984, p. 2.

For a general discussion of the pollution exclusion see:

Faron, R. S., and R. E. Freer, "Assessing the Future in Environmental Liability Coverage," *Toxics Law Reporter,* June 6, 1986, pp. 52–57.

Gladstone, D., "Pollution Exclusion: Exclusion Stronger Than Critics Say," *Business Insurance,* October 1984.

Rich, B. W., "Environmental Litigation and the Insurance Dilemma," *Risk Management,* December 1985, pp. 34–41.

For a general discussion of pollution insurance see:

All-Industry Research Advisory Council, *Pollution Liability: The Evaluation of a Difficult Insurance Market,* Oak Brook, Illinois, 1985.

Sterling, D. C., *Environmental Impairment Liability: An Insurance Perspective,* Chartered Property and Casualty Underwriters Monograph 9, 1984, Malvern, Pennsylvania.

CASE STUDY OF A SUCCESSFUL FACILITY SITING

Eugene E. Kupchanko, P. Eng.

North Vancouver, British Columbia,
Canada

3.4.1 BACKGROUND

The selection of hazardous-waste sites has generally met with severe public opposition throughout the western world. The construction of treatment plants is essential. However, the perception that they create and the publicized mismanagement of hazardous wastes over the past decade necessitates the orchestration of a site-location program to minimize adverse reactions.

The influence of public opinion in site selection has certainly become a predominant issue. Lack of communication and cooperation has created adversary situations and direct public opposition to the siting of hazardous-waste facilities. An inappropriate site-selection strategy has caused problems for governments and private developers alike.

It was not until the spring of 1979 that hazardous wastes became an issue in Alberta, Canada. A private firm proposed to build a disposal plant capable of handling the wastes in western Canada. The proposal developed into a highly volatile issue, and the company's action served as the primary catalyst for the public's awareness of the hazardous-waste problem. So strong was the reaction to the proposal that by fall of that year the Minister of the Environment placed a moratorium with respect to the construction of a hazardous-waste facility. He also announced the formation of a committee to report on the management of hazardous wastes, with public hearings to be held across the province upon receipt of the report.

No one knew it at the time, but this was the commencement of an intensive five-year hazardous-waste investigation and implementation program within the province. As with any complex issue of this nature, the activities were surrounded with a great deal of uncertainty, and no one really envisioned the magnitude of such an undertaking.

3.4.2 DEFINING THE PROBLEM

Over the next 16 months, the hazardous-waste issue was defined through two bodies: the Hazardous Waste Committee and the Environment Council of Alberta. During this period consultants were engaged to prepare background reports on hazardous-waste issues and the hazardous-waste strategy was being formulated for the province.

The Hazardous-Waste Management Committee

In response to public pressure regarding the proposed hazardous-waste facility, a six-member group, the Hazardous Waste Committee, was established to examine the problems with this waste issue. The committee consisted of three private citizens and three public servants.

The three key recommendations from the list of 20 were:

- "The implementation of a plan to treat industrial and other special wastes should begin immediately."
- "An integrated waste treatment facility should be established in Alberta to manage industrial and other special wastes."
- "The Alberta Government should be the leader in managing industrial wastes by developing comprehensive legislation, as well as establishing and, if necessary, operating the waste management facilities. By seeking the participation of industry, other levels of government and citizens, the provincial government will ensure that industrial waste, some of which can be hazardous, will be managed for the protection of the environment and public health."[1]

On the basis of this report, the Environment Council of Alberta conducted public hearings later that year.

Environment Council of Alberta

The Environment Council of Alberta (ECA) is a provincial crown corporation reporting directly to the Minister of the Environment. The traditional role of the organization is to conduct a continuing review of the government's environmental policies, and because of citizen concern about siting a facility, The Environment Minister requested that public hearings be held on the subject of hazardous wastes.

Information bulletins and other material were distributed to provide the public with background material on the subject of hazardous waste. Also, public-information meetings were held in 25 locations throughout the province.

Following the meetings, hearings were held in 16 centers throughout the province. The centers were chosen to reflect the distribution of the population and the interest expressed in the problem.

Approximately 1,000 people attended the public hearings, over half of whom registered as private citizens.

The final report[2] contained 51 recommendations, broken down into 13 areas of concern including legislative control, organization and function, system design and site selection, treatment licensing, manifest system, fees, regulation, enforcement, intergovernmental agreements, environmental trust, and target date.

One of the most significant recommendations concerned the need for umbrella legislation and the location of the plant (or plants), and its ownership. It was suggested that the government form an Alberta Crown Corporation and proceed with the site-selection process.

With the receipt of the draft report from the Environment Council of Alberta, the Minister of the Environment proceeded to the implementation phase.

3.4.3 IMPLEMENTATION

Both the Hazardous Waste Management Committee and the Environment Council made recommendations for the implementation of the hazardous-waste program. The Environment Council recommended that in the implementation phase a site-selection committee be established comprised of technical experts and representatives from the public.[2]

The implementation structure consisted of two bodies, the Hazardous Waste Team and the Task Force, which were to work closely together. The Hazardous Waste Team consisted of private citizens, some of whom had previously served on the Hazardous Waste Management Committee. The Task Force, comprised of specialized government staff, assisted in the public participation, communications, and mapping portions of the project.

Hazardous Waste Team

The Hazardous Waste Team was established to develop and implement the site-selection process which would lead to the eventual siting of the hazardous-waste plant. They were seen as being at arm's length from government, and once formed, quickly established a hazardous-waste resources office to respond to current concerns. It was at this time that a strong community-involvement program was commenced.

The Task Force

A number of Alberta Environment staff members were assigned to assist the Hazardous Waste Team with the implementation phase of the program. The objective of these technical specialists, was to develop and provide the resources for the public participation program and to assist in the technical evaluation of candidate sites. Consultants were also engaged to assist the Task Force in this respect.

After reviewing previous studies the Task Force refined the criteria to be used for site evaluation. Their most prominent achievement was the development of the system by which environmentally unsuitable areas could be eliminated from further investigation, a technique known as constraint mapping. The dedicated Task Force members also devoted the many long days of personal time required to make the community-involvement program a success.

The Invitational Process

The criteria for site selection were based on environmental and social acceptability. The Task Force developed a step by step process, known as the invitational process. Entrance into the community was on invitation only, and linked closely with a constraint-mapping process.

First, invitations were sought from regional planning commissions, municipalities, and interest groups for informational meetings and further discussions on the siting process. Each municipality or agency had to request that an overview and a constraint map be done. Once the constraint maps were finalized, they were presented to that municipality or agency. The municipality or agency then needed to decide whether to go further with the specific site assessment or not. If they agreed to proceed, further meetings were held to outline prospective sites and to form citizen committees. The agency or municipality was again contacted for agreement to go ahead with specific site assessment. If they were in agreement, then further meetings and discussions were held including the contacting of site owners and neighbors to explain the program and the exploratory drilling work to be done. Exploratory drilling was then finalized and the data evaluated. These reports were presented and discussed with the municipality or agency. The municipality or agency would then decide if more detailed (site-specific) assessment was warranted. The detail would include more drilling, environmental and socioeconomic evaluation, and archaeological and historical evaluations. At any time they so wished, a municipality or agency could withdraw from the site-selection process. There was absolutely no obligation on their part.

Mapping the Constraints

The technical approach to locating a site was based upon criteria by which environmentally unsuitable areas in the province could be eliminated from further investigation, a process known as *constraint mapping*. The most desirable land parcel would exhibit low capabilities for all land-use possibilities. A system of overlay mapping was used to assess land-use capability, or areas of potential conflict. The criteria for eliminating sites were defined under four main headings: physical, biological, land use, and human. Using these criteria and combining the existing data on provincial conditions, maps outlining each constraint were prepared. By overlaying these maps a provincial composite map was prepared.

Three levels of study were incorporated into the siting component of the program. The provincewide assessment provided a preliminary overview in graphic form, showing the constraint-free areas on a large scale. The regional analysis provided more detailed information to define areas of potential interest or constraint. After further clarification of issues such as landownership, a detailed level of study was undertaken including a site-specific drilling program to verify the geophysical suitability for a hazardous-waste facility.

To site a facility for a broad spectrum of hazardous wastes, a number of environmental parameters must be assessed. This assessment provides an excellent overview of land-use possibilities and defines four general categories of constraints. The sequence of constraints detailed here follows the natural progression from the stable environment through to the human activities reliant on biophysical limitations.[3.8]

Physical constraints: The physical constraints consisted of the category of environmental resources that were essentially nonrenewable. They included surficial and bedrock geology, surface water, groundwater, topography, and climate. Stability and permeability of the surface were of prime importance. Surface water, groundwater flows, and climatic factors were also significant as it is necessary to locate facilities where they are least likely to contaminate surface water or groundwater.

Biological constraints: A high priority was assigned to the preservation of natural systems. Areas exhibiting consistently high productivity of vegetation (forests and crops) and important wildlife habitats were delineated. The mapping of biological constraints was prepared to ensure that valuable lands were not taken out of production and wildlife species not significantly affected.

Land-use constraints: Land-use constraints consisted of readily identifiable activities or land-tenure dispositions and included agriculture, resource-extraction transportation corridors, and parks and reservations. The land-use constraints represent major difficulties not only in terms of human factors, but also with regard to jurisdiction leading to severe land-use conflicts or incompatibilities.

Human constraints: The cultural, or human, constraints are the ones that relate to the public participation program. Interrelationships between people and their environment, and the preservation of culturally important lands, were taken into consideration. These included sites of archaeological or historical interest and prime recreational land.

To provide the required data base for this type of study, it was necessary to maintain contact with experts on a continuing basis, to maintain liaison with numerous government agencies, and to interview various special interest groups so that all available expertise could be fully utilized.

Throughout the project, information on site suitability was prepared and presented graphically. This technique served to provide an immediate impression of suitable areas for the project-management team. It also provided the basis for visual displays for use in the public-information and participation programs.

As each constraint map was overlaid onto the provincial base map, all the areas which were unsuitable appeared as a dark color, while moderately suitable areas were shaded in a lighter tone. Constraint-free areas showed up as white (see Fig. 3.4.1). More attention was paid to those areas in which civic officials indicated an interest in the potential siting of the waste facility. Those regions that invited detailed studies were involved in both a technical-assessment and an intense public participation program.

The major thrust of the constraint-mapping process was to utilize the natural factors of the land to guard against human failure. This was seen as particularly important with regard to the landfilling of residues and the possibility of groundwater contamination.

While the constraints were specified, no weights were assigned, and how they would be applied in a particular circumstance was left open. It was always up to the local municipality to make the final choice of areas to be eliminated. This was seen as a positive move, a sensible approach when dealing with public issues.

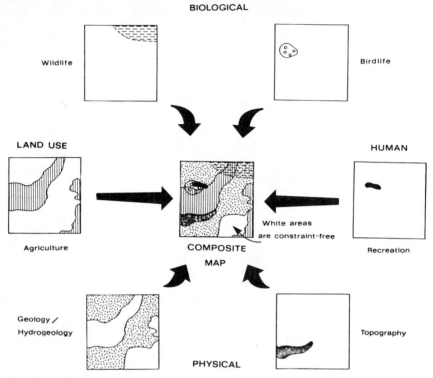

FIG. 3.4.1   The constraint-mapping process.

Some communities however, suggested potential sites that were not constraint-free. Similarly, some of the constraints were considered to be mutually exclusive, for example, access to water for the facility and protection of the water source (surface water or aquifer). Constraint mapping is defensible because it rules out areas where the environment or public health would be seriously jeopardized.

Raising Awareness

The experience of the Rural Education and Development Association was utilized for the field program. This organization was deeply involved in public education in rural Alberta. Six part-time field staff were hired with 60 centers tentatively selected as community workshop sites. Field staff worked with communities, approaching formal (elected representatives) and informal leaders to gain support for the process and to further obtain community contacts. For seven months the field staff contacted community leaders and held informal workshops. Field staff also attended and in some cases facilitated meetings where municipal council or special interest groups extended an invitation. This resulted in 59 organizational meetings and 65 community informational workshops.

The workshops were advertised in several ways including personal contact or word of mouth, newspaper advertisements, posters, and public-service announcements. The purpose of these workshops was to provide a forum for sharing and discussing concerns, facilitating selection of a local action committee.

The agenda consisted of an introduction to the history of the waste-management program, along with an explanation of the constraint-mapping process. A variety of concerns arose including risk and safety-system design, criteria for facility location, classification of wastes, recycling, legislation and enforcement, transportation, ownership and operation, socioeconomic impact, public involvement, potential for disposal of radioactive wastes, and acceptance of wastes from outside the province.

As a result of this border to border information program, invitations to assess suitability for a plant site were received from 52 of a possible 70 jurisdictions.

Field Testing

Hydrogeologic criteria require detailed field verification. Upon invitation, more detailed assessments of candidate areas within a region were prepared. If hydrogeologic office studies on potentially constraint-free land were favorable, test drilling was undertaken to verify suitability or eliminate sites from further consideration.

Sites for field testing were identified by the municipality and after being evaluated by an outside consultant. From the section of land provided for field testing, the consultant selected the best quarter-section, based on available data, field inspection, and preliminary drilling. An evaluation of the accuracy of the hydrologic constraint map was made and major anomalies identified. These were discussed with a committee of local citizens who monitored the drilling operations on behalf of the community.

Permission from the landowner or leaseholder for access was a condition for drilling to start. No drilling was performed if a leaseholder refused access or if a landowner did not indicate willingness to sell the land if the land was chosen for a site. Test holes were drilled, observation wells installed, and pumping tests conducted.

3.4.4 PUBLIC-INVOLVEMENT PROCESS

The principal policy in the public-involvement program was that environmental acceptability would be the main criterion for selection and that the siting criteria would be open to public scrutiny. The government would reserve the overall right of final selection. Given the potentially volatile nature of the siting issue, the main thrust of the program was to develop an environmentally sound procedure for site elimination (as opposed to site selection) and to link it to public involvement. The entire province was included in the site-elimination process (border to border) and thus all options were kept open. The overlay mapping technique revealed that almost every municipality within the province in fact had some constraint-free land.

From the site-elimination strategy, it was anticipated that alternate sites would become available and that communities would compete for the facility because of the economic benefits the plant may bring. Initially, a prevailing lack of interest throughout the province was perceived as a general lack of awareness about hazardous waste, and to generate interest, field staff tried to heighten awareness by presenting the issue in terms of the total waste perspective, with the hazardous-waste issue being considered as one small subset. The focus was on areas that the public would appreciate such as the individual responsibility for waste generation given our current lifestyle. Two concepts were conveyed: First, field staff emphasized that hazardous waste is only a small part of the overall waste problem which is familiar to every one. Second, by showing the public their role in waste generation and asking them to accept some responsibility, it was hoped that a realization might occur that they were part of the problem and they would volunteer to be part of the solution. The other objective of the field staff was to inform the public that the risks associated with a well-designed waste-management system were considerably less than those associated with waste mismanagement and that the risks associated with the facility would be no greater, and probably less, than those associated with any other industry.

Care was taken to ensure that any information was accurately transmitted to the participating communities. Results were presented and discussed in open forums and any suggestions by the local citizens were quickly incorporated. Assessment reports and any other information were made freely available. These assessments were useful in achieving a total awareness of the situation within a community. The assessment document was also useful to the community for future planning and zoning. The process was explicit, it occurred in a public forum, and guidelines required initiation of each phase by the potentially affected party. As the process progressed, many municipalities which were not interested in proceeding further withdrew. The meticulous, step by step public-involvement approach helped to prevent a premature escalation of conflict over contentious issues.

As actual alternate sites were identified within the municipalities, the awareness of affected citizens rose. A heightening of public concern was evidenced by the formation of citizen groups within the affected municipalities, necessitating frequent meetings. The meeting formats were an effective mechanism for the expression of negative and positive opinions. The process was lengthy and time-consuming, but while public acceptance could not be totally guaranteed in any one community, the probability of total rejection or failure was diminished. Both technical and nontechnical issues arising at these meetings could be addressed. Had the public-involvement process focused entirely on technical issues relating to the hazardous-waste plant, it is doubtful whether any sites would have surfaced.

Proponent and government agencies have concentrated on the technical, economic, and administrative aspects of hazardous-waste management for almost a decade with mixed success. Increased public opposition to siting of these facilities has placed a new focus on social issues. Social issues can create far more problems than technical ones. Only a few of these siting cases can be considered successful in terms of public acceptance of the facility. In the successful cases there was a strong commitment to fostering good community relations on the part of the sponsoring agency, a strong commitment to access to information, and the cooperative involvement of government, industry, and citizen groups.[4]

Based on experience in Alberta and elsewhere[5,6] in the siting of a hazardous-waste plant the following conclusions can be drawn:

- Opposition rises when the public perceives that the project does not solve a local problem.
- The credibility of the agency sponsoring the project is a critical factor in acceptance by the public. If the public-involvement and hearing process appears too closely allied with the permitting process, then credibility is undermined.
- The economic viability of the facility is not an issue. If the public perceives that

there is need for the protection of human health and the environment from hazardous wastes, then the benefits far outweigh the costs.

• The importance of the media cannot be overemphasized.

The strategy of keeping siting options open is significant, for it allows a flexibility of choice. This is of prime importance in rapidly changing circumstances. The integration of environmental and social criteria through constraint mapping and public involvement provided a legitimate means of entry into local communities. Finally, the plan to identify a site on the basis of elimination rather than selection as well as combining the hazardous-waste issue with general waste issues has a great deal of merit.

Usually in such an undertaking, there are strong pressures to proclaim a site, and when public acceptance cannot be guaranteed in any community, rejection of the selected site may occur as objections mount. From an outsider's perspective, one of the prime difficulties in understanding and dealing with public involvement is that it takes so much longer than anticipated and when hurried leads to unexpected results. An example of this is is the siting failures in North America over the past decade.

Another factor to consider is credibility. Expertise and perceived trustworthiness are closely related to credibility. A proponent or agency's trustworthiness can be undermined by previous performance or by perceived secrecy. No studies were kept confidential and all test results were immediately available for public scrutiny.

Perhaps the most important aspect to the successful siting is the view of the public. A strong policy on the right of the public to participate or to intervene is the foundation of any public involvement program in a democratic society.

3.4.5 COMMUNICATIONS

The communications program played an important and vital role in the successful siting program. The technique required to communicate a volatile issue requires a well-defined plan with clearly stated objectives. The communications program was based on two distinct strategies: one concerned media relationships, and the other involved the type of material distributed to the public.

The objective of any communications plan should be to inform a particular audience of all key factors surrounding the system so they can participate from a basis of knowledge in the final development of the system. This was accomplished through:

1. Provision of communication services to public participation personnel in their work with public groups
2. Provision of communication services to key govenment officials
3. Liaison with the media
4. Ensuring that the key personnel are accessible to the media

The target audience identified in this program included community leaders, municipal authorities, environmental groups, government departments, transporters, school and academic personnel, professional groups, local health boards, local social services, planners, and site-specific groups.

The general components for the communications plan included:

- Materials to assist at public meetings
- Displays and participation at local conferences and conventions
- A print campaign of brochures and handout materials
- A literature review on process made in other jurisdictions
- A publication and newspaper-clipping inventory
- Preparation of key officials for on-the-spot radio, television, and newspaper interviews
- An education program for schools
- Development of the mapping program for presentation at public meetings
- A newsletter outlining the issues, activities, and progress

A number of audiovisual presentations were also developed including a slide library and slide-tape shows.

The role of the field personnel was viewed as being essential to the communications program. From their knowledge of local circumstances, they became an integral two-way conduit of information and advocacy,—from the government to the local public and vice versa. Their input normally influenced the kind of communication tools that were used and in general the communications strategy responded to their needs. The complexity and magnitude of the siting problem was never fully realized until the issue continued to unfold. Consequently, the communication plan for the project became an essential part of the hazardous-waste strategy.

Media relations were emphasized in the communication process as it was recognized that this group could influence a broad audience. One should strive to maintain cooperation with the media, making information about hazardous-waste management fully available to them.[6] This "open door" policy was well received by the media as were the periodic individual briefing sessions with various environmental and media reporters. This strengthened rapport with the media and generally resulted in supportive articles and editorials. Regardless of the controversy, there was value in a continued media liaison while trying to communicate on a controversial and complex issue such as hazardous waste. This helped to sustain public interest over a long period of time. Direct contact with specific reporters on negative coverage is also advisable.

Reporters unfortunately seldom have the opportunity to do a thorough analysis of a subject, and for ease and clarity of presentation, controversial issues are often presented in a simplistic format. It takes knowledgeable and committed individuals to report issues as controversial as hazardous wastes accurately and fairly. In this case the media in general acted fairly, exposing both sides on the hazardous-waste issue. Without the media strategy, success in siting would probably not have been achieved.

3.4.6 SELECTING THE SITE

The siting of controversial plants does not just happen. It takes informed citizens to make an educated decision. Over 50 municipalities requested regional assessments. Although almost every region in the province contained environmentally acceptable areas, some municipalities offered environmentally unacceptable sites while other municipalities which offered potentially suitable sites had to withdraw because of public opposition.

On the basis of the combined efforts of citizen action groups and local governments, and with support from elected representatives, six potentially suitable areas agreed to go to plebiscite (a requirement for site selection). One later withdrew because of public opposition. In the five candidate areas, detailed drilling was performed to verify the suitability of the local geology.

A series of three seminars and numerous workshop discussion groups were held at the five candidate areas to help the public fully understand the background of the program and to address any concerns.

While the results were generally positive, results in two of the five locations where plebiscites were held were overwhelmingly in favor of hosting the hazardous-waste facility. In one location 79% of the voters were in favor of the plant, while in the other 77% of the voters supported the facility.

The ultimate siting decision rested with the provincial government and careful orchestration of the site-selection announcement was very important because of the heightened competition and public interest in hosting the facility. Prior to the announcement, a prebriefing of the key participants was held to clarify various positions on controversial issues. The evening before the announcement the news of the selection was given to the candidate communities informing them of the decision and why. A press conference including the key participants was held early the next morning.

On March 12, 1984, the Minister of the Environment announced that an area 20 k northeast of Swan Hills (125 km northwest of Edmonton) was to be the site for the Alberta hazardous-waste facility, making it the only community in North America that had agreed to host a facility in their backyards in the preceding six years. An unparalleled success.

3.4.7 REFERENCES

1. Alberta Environment, *Hazardous Waste Management in Alberta, A Report of the Hazardous Waste Management Committee,* Edmonton, Alberta, January 1980.
2. Environmental Council of Alberta, *Hazardous Waste Management in Alberta, Report and Recommendations,* Edmonton, Alberta, December 1980.
3. Alberta Environment, *Interim Report on Site Selection Program and Public Participation for the Integrated Hazardous Waste Management System,* Edmonton, Alberta, October 1981.
4. Clark-McGlennon Associates, *A Handbook on Siting Acceptable Hazardous Waste Facilities in New England,* New England Regional Commission, Boston, Massachusetts, November 1980.
5. Ontario Hydro, *Social Impacts of P.C.B. Waste Management,* Preliminary Report TG-07224, Social and Community Studies Section, Site Planning Department, Toronto, Ontario, February 1979.
6. N. M. Krawetz, *Hazardous Waste Management, A Review of Social Concerns and Aspects of Public Involvement,* Alberta Environment, Staff Report No. 4, Edmonton, Alberta, November 1979.
7. Alberta Environment, *Hazardous Waste Management, The Communications Program,* Edmonton, Alberta, Canada, January 28, 1981.
8. J. McQuaid-Cooke and K. J. Simpson, "Siting a Special Waste Treatment Facility in Alberta," APCA, 79th Annual Meeting, Minneapolis, Minnesota, June 22–27, 1986.

CHAPTER 4

SPECIAL WASTES

SECTION 4.1
WASTE OIL

Mark M. McCabe

Senior Scientist
Remediation Technologies, Inc.
Concord, Massachusetts

4.1.1 INTRODUCTION

Waste oil refers to lubricating oils that have gone through their intended use cycle and must be either disposed of or treated and re-used. Generally, the term embraces spent automotive lubricating oils and spent industrial oils including those used for lubrication, refrigeration, and process applications.

The fate of waste or used oils is currently a matter of great regulatory concern in the United States. The fact that the majority of waste oil originates from numerous small generators, through the disposal of spent automotive lubricating oils, makes it impossible to regulate the waste stream at its source.

In a report to Congress in 1981, the U.S. Environmental Protection Agency (EPA) estimated that 2.9 billion gal of oil are sold and that 1.2 billion gal of waste oil are generated in the United States on an annual basis. The study reported that 50% of the used oil is burned as fuel, while 30% is discharged to the land or sewers. The remainder (8%) is rerefined to produce a lube-oil base stock or used for oiling road surfaces.[1] It should be noted that the latter practice has been banned due to adverse environmental impacts.

The primary environmental concern regarding used oil, aside from the sheer volume of wastes generated, is that the oil has the potential to be contaminated either during use or from external sources. The contamination can result from physical or chemical changes of constituents or from blending with hazardous wastes during transport or storage. The presence of these hazardous contaminants presents a problem in that used oil has a significant potential for treatment and re-use as a fuel or lubrication base stock. Regulators are faced with the problem of proposing and enforcing standards that are appropriate to protect the environment but also sufficiently balanced that they do not promote illicit dumping or incineration of used oils, and waste a valuable resource.

This section provides a general characterization of waste oils and an overview of current disposal and recycling options.

4.1.2 CHARACTERIZATION OF WASTE OILS

Automotive and industrial oils are composed of an organic base stock and additive packages which have been developed for specific lubricating applications in order to significantly increase the performance and life of the oil. The lubrication base stock can be comprised of hundreds of thousands of organic constituents, the majority of which are polynuclear aromatics (PNAs). The additives, which may comprise up to 15% of the oil by volume, typically contain inorganic constituents such as sulfur, nitrogen, and trace metals.[2]

In recent years, studies have documented the presence of additional compounds such as chlorinated solvents in samples of waste oil. Some of these compounds are present in refined petroleum products, while others are introduced during use or are illegally blended with the used oil during storage. Studies have documented the fact that the concentration of chlorinated solvents increases dramatically, sometimes by two orders of magnitude, as waste oil moves from generators to processing facilities.[3] The presence of these compounds, which are suspected carcinogens and mutagens, is the basis of concern about waste oils in the environment. Table 4.1.1 lists some compounds of potential concern and potential concentration ranges of these constituents.

Polynuclear Aromatics

Polynuclear aromatic hydrocarbons are present in the petroleum base stock and can be produced during the use of the oil. Several of these compounds, primarily four-, five-, and six-ring structures, are known carcinogens and mutagens. Benzo[a]pyrene, B[a]P, is a prime example of a PNA which exhibits carcinogenic effects. A review of available data indicates that residual oils (No. 6 fuel oil) often contain higher levels of PNAs (20 to 100 mg/kg) than waste oils (<50 mg/kg) and, as a result, the substitution of waste oils for these virgin stocks would not have a significant effect on the environment.[3]

Halogenated Organics

Halogenated hydrocarbons may be produced in oil during normal use cycles by the reaction of base-stock hydrocarbons and halogenated compounds, typically inorganic chlorides, from the additive packages. Virgin lubricating oils typically contain less than 100 mg/kg of these compounds.[3] The presence of higher concentrations of these compounds (<100 mg/kg), however, is more likely the result of blending with chlorinated wastes during storage. Constituent analyses indicate that these contaminants are generally associated with degreasing solvents such as trichloroethane, trichloroethylene, and perchloroethylene.

TABLE 4.1.1 Contaminants of Potential Concern in Waste Oils

Organic contaminants	Probable source	Approximate concentration range, μg/L*
Aromatic hydrocarbons	Petroleum base stock	
Polynuclear (PNA)		
benzo[a]pyrene		360–62,000
benz[a]anthracene		870–30,000
pyrene		1,670–33,000
Monoaromatic	Petroleum base stock	
alkyl benzenes		900,000
Diaromatic	Petroleum base stock	
napthalenes		440,000
Chlorinated hydrocarbons		
trichloroethanes	May be formed chemically	18–1,800
trichloroethylenes	during use of contaminated oil	18–2,600
perchloroethylene		3–1,300
Metals†		
barium	Additive package	60–690
zinc		630–2,500
aluminum	Engine or metal wear	4–40
chromium		5–24
lead‡	Contamination from leaded gasoline	3,700–14,000

* Concentration data from Ref. 11–15.
† All concentrations are in mg/kg (ppm).
‡ Higher levels are no longer anticipated because of federal regulation of lead in gasoline.
Source: Ref. 1.

Trace Metals

Trace metals are introduced into the oil as part of the lubricating package or from external sources such as the wear of metal parts. Metals such as zinc, chromium, aluminum, and barium are examples of contaminants that are introduced in this manner, and that may have a detrimental effect when released to the environment. Lead, which is present in the most significant concentrations, up to 2,000 mg/kg, is primarily introduced through the use of leaded gasoline in automotive engines. Concentrations of lead in virgin oils are generally less than 10 mg/kg.[3]

The regulation of lead concentrations in automotive fuel will eventually remove this element as a primary concern. In 1985, the EPA promulgated standards requiring that the average concentration of lead be reduced from 0.29 g/L (1.1 g/gal) of gasoline to 0.1 g/L (0.5 g/gal) by mid-1985, and to 0.03 g/L (0.1 g/gal) by the start of 1986. A similar reduction in the concentration of lead in waste automotive oil should follow.[4]

4.1.3 RECYCLING AND DISPOSAL PRACTICES

Used oil is unique as a waste material in that there is a significant market demand for its recycling and re-use. Current activities in this area primarily relate to physical treatment to produce a specification fuel and rerefining to produce a lubrication base stock.

It should be noted, however, that a substantial quantity of waste oil, approximately 1.29 billion L (342 million gal) annually, is released to the environment through land and sewer disposal.[2] These disposal practices are generally carried out by those segments of the population that change their own automotive oil.

Recycling to produce fuel oils is the primary route of re-use of waste oils. These fuels provide a low-cost alternative to more expensive fossil fuels. Savings of up to 75% can be realized by using fuels recycled from waste oils.[5]

Rerefining to a lubrication base stock is also a viable option for re-use but the strict quality specifications required of the resulting product make it a far more costly and thus a less competitive alternative.

Recycling to Produce a Specification Fuel

The processing of waste oil to produce a specification fuel is generally accomplished using cursory treatment steps such as settling, filtration, and dehydration to produce a product that can be readily fired in a boiler. The specification criteria generally relate to physical considerations such as heat content, viscosity, flashpoint, and bottom sediments and water (BS&W).

Treatment steps for the processing of these fuels are minimal in order to maintain the significant cost advantage over virgin fuels. The physical treatment does not address the presence of chemical contaminants in the oil. And because of their extremely small particle size, filtration techniques do not effectively remove trace metals. More intensive treatment processes such as centrifugation, clay contacting, and distillation are currently available to address the problems of contaminants in used oils; however, the additional costs and marginal benefits associated with these treatment steps generally place fuel processors at a competitive disadvantage. In general, once used oils have been treated to this degree, it is considered to be more profitable to continue processing to produce a lube-oil base stock.[3]

Without regulation there is little or no incentive for recyclers to address the problem of hazardous contaminants. The prices for their products are largely dependent upon meeting the physical specifications associated with virgin fuel oils.

Regulation of Used-Oil Fuels

Current trends in the regulation of recycled oils to be used as fuels indicate that compositional specifications will be applied to identify and regulate recycled fuels that have been contaminated with hazardous constituents. Regulatory agencies are concerned about the potential exposure to emissions of toxic chemicals and trace metals, particularly since the majority of boilers firing these recycled fuels are inefficient, uncontrolled systems located in areas of high population density.[3] Federal specifications for used-oil fuels, as promulgated in November 1985, are presented in Table 4.1.2.[3]

The EPA estimates that 80% of the waste automotive oil generated in this country will be able to meet these specifications. Waste oils meeting these criteria may be burned in boilers of any size without regulation since these specifications have been determined to be protective under virtually all circumstances and fuels meeting them should not produce adverse environmental impacts. These

TABLE 4.1.2   Used-Oil Fuel Specification

Constituent or property	Allowable level
Arsenic	5 mg/kg maximum
Cadmium	2 mg/kg maximum
Chromium	10 mg/kg maximum
Lead	100 mg/kg maximum
Total halogens	4,000 mg/kg maximum
Flashpoint	37.7°C (100°F) minimum

Source: Federal Register, vol. 50, no. 23, 49,164–49,249.

specifications are generally equivalent to "worst case" levels found in virgin fuels or levels which would not pose a significant health risk upon burning.

The specification for total halogens includes a "rebuttable" presumption that oils containing more than 1,000 mg/kg of total halogens—even though concentrations are less than the 4,000 mg/kg specification—have been adulterated by blending with hazardous wastes. Suppliers of oils in this category must demonstrate by analysis that the oil does not contain more than 100 mg/kg of any one hazardous constituent. Oils that do not meet this requirement are regulated as a hazardous-waste fuel.

The burning of recycled oils that do not meet the referenced specifications would be limited to high-efficiency industrial boilers, industrial-process furnaces, or boilers demonstrating compliance with the performance standards for hazardous-waste incinerators.

Combustion

Waste oils are generally used as primary fuel in small commercial boilers and space heaters and as a supplementary fuel in larger industrial boilers and furnaces. Boilers firing waste-oil fuels have the potential to emit toxic constituents to the atmosphere as well as generate a solid waste, fly ash, that may contain trace metals and products of incomplete combustion. The concentration of these contaminants may be significant enough to require that the residual ash be regulated as a hazardous waste.[6]

Commercial Boilers.   As stated previously, the use of these fuels in smaller commercial boilers (rated heat capacities of less than 10×10^6 Btu/h) is a subject of regulatory concern because of the number and location of these systems. In addition, these units typically have very low stack heights—less than 12 m (40 ft)—which put any hazardous emissions in greater proximity to the general population.

Boilers of this type are generally of cast iron or firetube design and are used for generating heat in apartment buildings, greenhouses, and small institutional facilities such as schools and hospitals. These systems typically have limited combustion controls, cycle in an on-off mode, and operate at variable loads. Studies conducted on small commercial boilers firing traditional fossil fuels indicate that the majority of flue-gas emissions, particulate matter, smoke, and hydrocarbons, occur at the beginning and end of such cycles.[7] Load variation, generally for high and low output, is achieved through modulation of fuel flow rate to the boiler.

Compressed-air atomization, mechanical atomization, and rotary-cup atomization are considered to be the most common methods of firing fuel for these systems. Compressed-air and mechanical atomization are the most efficient means of delivering oil to the combustion chamber in that they present a fine spray of droplets for mixing with combustion air. The suspended particulate matter found in waste oils may present serious operating problems for these systems by constricting the orifices used to atomize fuel. For this reason, rotary-cup burners, which have been phased out for traditional combustion applications because of their inefficiency, are commonly used for the combustion of waste oils. These systems use a rotary cup and centrifugal force to deliver the fuel to the combustion zone. Though less efficient than other methods, rotary-cup burners are able to handle oils of higher viscosity and ash content without extensive pretreatment.

Boilers in this size range do not have integral pollution-control devices. This is of particular concern when firing waste oils in that these fuels generally contain high levels of ash and metal contaminants that can be emitted to the atmosphere as particulate matter.

Studies conducted by the EPA indicate that 30 to 50% of the lead fired in these systems is emitted to the atmosphere, and that these levels could, in areas of high population density, exceed the National Ambient Air Quality Standard for lead (1.5 $\mu g/m^3$ per calender quarter). Of additional concern is the fact that 80 to 90% of the lead emitted is submicron in nature and is considered respirable.[6]

Commercial boilers have been able to demonstrate destruction efficiencies of greater than 99.9% for organic constituents typically found in waste oils, indicating that less than 1% of the organic species fired are emitted to the atmosphere.[6]

Although regulations requiring that only specification fuels can be fired in these systems will ensure that there are no adverse impacts to the environment, the economic impact of additional pretreatment costs and potentially shrinking supplies of fuels that meet the specifications have not been fully determined. A regulatory impact analysis (RIA) has indicated a potential annual decrease of 15% 110 million L (29 million gal) in the supply of used-oil fuel for these systems.[8]

Space Heaters. Used-oil space heaters are typically sold to service stations and maintenance shops of firms with large fleets of vehicles. These heaters are rated at less than 0.5×10^6 Btu/h, fire 3.8 to 7.61 L/h (1 to 2 gal/h) of used-oil fuel, and are considered to be a reasonably efficient means of disposing of waste automotive oils that are generated on site. The two principal designs for these systems are air-atomization and vaporizing-pot heaters.

The atomization heater is similar to traditional combustion devices in that it uses a nozzle and orifice to deliver the fuel into the combustion chamber in the form of a fine spray. Units of this design appear to promote relatively efficient combustion resulting in higher destruction efficiencies for organic contaminants. However, they have been demonstrated to emit 75 to 95% of the metals introduced.[2,3] The metals from the fuel are entrained in the combustion airstream and, as a result of the small size of the combustion chamber, do not have an opportunity to settle out prior to venting to the atmosphere.

Vaporization heaters utilize a very simple design in which the oil is vaporized from a pan in the base of the combustion chamber. Low-boiling organics are volatilized and combusted in the area directly above the pot, while metals and less volatile organic species are deposited as a slag in the base of the heater. The metals emission rate from these units, 5 to 15% of the metals fired, is considered to be comparable to or lower than those from larger boilers.[2,3] It is estimated that 90% of the space heaters currently in use are of the vaporization-pot design.[3]

The EPA has provided a conditioned exemption from the prohibition of burning of off-specification fuels in used-oil space heaters. The referenced conditions are:[3]

1. That the heater burn only used oil that the owner or operator generates or used oil received from do-it-yourself oil changers who generate used oil as a household waste.
2. The heater be designed to have a maximum capacity of not more than 0.5×10^6 Btu/h.
3. The combustion gases from the heater be vented to the ambient air.

Industrial Boilers.   Industrial boilers are defined as utility or power boilers that supply heated or cooled air or steam for manufacturing processes. These boilers are generally of watertube construction and are rated at greater than 25×10^6 Btu/h. These systems typically fire a blend of recycled used oils and virgin fuels to maintain a consistent, trouble-free fuel supply that offers some economic advantage over virgin fuels.

Industrial boilers are equipped with more sophisticated combustion controls to better deal with variations in composition of used-oil fuels and have greater stack heights to enhance the ambient dilution of any emissions. These larger systems are also capable of achieving higher combustion temperatures and combustion-gas residence times appropriate for the efficient destruction of organic contaminants in the fuel.

Some boilers in this size range are equipped with pollution-control devices. The existing equipment is generally designed for the control of particulate matter (fly ash) but is probably not sufficient to adequately control submicron metal emissions. The effective removal of these materials from the flue-gas stream would require the use of high-efficiency (99%) control devices such as fabric filters and electrostatic precipitators. Data collected from the EPA National Emission Inventory Data Base (NEDS) indicates that only 10 to 20% of the boilers in this category are sufficiently controlled to remove particulate matter to a level of even 50%.[9,10]

Although more stringent regulations are pending, current federal regulations will allow used-oil fuels that do not meet the specifications for use in commercial-size systems to be fired in industrial boilers. This will enable recyclers and suppliers of used oil to divert off-specification product to the industrial sector for re-use as opposed to disposal. The EPA has estimated that under the current regulatory scheme the annual amount of waste-oil fuels fired in industrial boilers will increase by 44%.[8]

Rerefining

Rerefining represents the most esthetically appealing option for dealing with used oils. Technologies are currently available that have the capability to produce a product with characteristics similar to virgin lube oils. Rerefiners are currently not operating at system capacity due to the demand for waste oils as fuel. The lack of sufficient feedstock for rerefining makes it extremely difficult for commercial operations to produce a competitively priced product. However, as with fuel for the industrial-boiler market, the regulation of used-oil fuels may direct off-specification streams to the rerefining sector.

Virtually all rerefining processes start with a distillation step to remove water

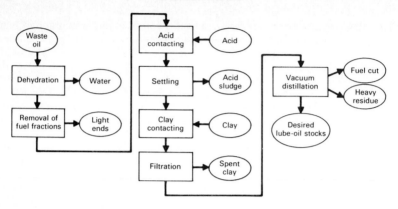

FIG. 4.1.1   Acid-clay rerefining process.

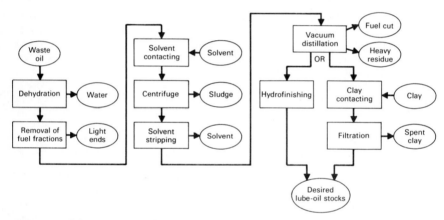

FIG. 4.1.2   Solvent treatment-distillation-finishing rerefining process.

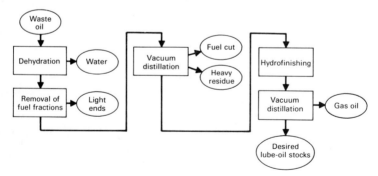

FIG. 4.1.3   Distillation-hydrofinishing rerefining process.

and light ends (the fuel fraction) from the feedstock. The most common rerefining processes are discussed below.

The acid-clay rerefining process has been the most commonly used in the United States for a number of years. The primary process involves mixing the feedstock with sulfuric acid in order to remove most of the contaminants contributed by the additive packages. These inorganic species, such as trace metals, form insoluble sulfates that settle out of the feedstock. The product is then subjected to a clay-contacting step for neutralization and improvement of the color and odor. There can be significant problems in disposing of the waste products (acid sludge and spent clay) from this process since these products often exhibit the characteristics of a hazardous waste. A schematic of this process is presented in Fig. 4.1.1.[2]

The other rerefining processes generally use a vacuum-distillation step in conjunction with various pretreatment and polishing steps to yield a lube-oil product.

Pretreatment using a combination of solvents such as butyl alcohol, isopropyl alcohol, and methyl ethyl ketone has proven to be effective in dissolving the lubricating oil and removing contaminants responsible for fouling in subsequent distillation steps.

Vacuum distillation at specific temperatures and pressures then provides a separation of lubricating base stock from the lighter fuel cut and heavy residues contained in the feedstock.

If required, the distilled lubrication cut is polished using either a clay-contacting or hydrofinishing (catalyst-assisted) step to remove remaining contaminants and improve the esthetic qualities of the product. Schematics of two additional rerefining processes are provided in Figs. 4.1.2 and 4.1.3.[2]

4.1.4 SUMMARY

The vast quantities of used oil generated on an annual basis in this country represent a great resource if managed in an environmentally sound manner. The fate and future of technologies associated with the re-use of these materials are, however, directly tied to the costs associated with environmental regulation and outside economic factors such as the price of crude oil. With appropriate regulation and environmental concern, the use of recycling and disposal practices—i.e., recycling to produce fuel, rerefining and combustion to produce energy in a wide variety of boilers and process furnaces—can continue to play an important, and necessary, role in the management of used oils.

4.1.5 REFERENCES

1. U.S. Environmental Protection Agency, *Listing of Waste Oil as a Hazardous Waste Pursuant to Section (8)(2), Public Law 96-463,* Washington, D.C., 1981.

2. N. F. Suprenant and W. H. Battye, *The Fate of Hazardous and Nonhazardous Wastes in Used Oil Disposal and Recycling,* DOE/BC/10375-6, U.S. DOE, GCA/Technology Division, Bedford, Massachusetts, October 1983.

3. *Hazardous Waste Management System; Burning of Waste Fuel and Used Oil Fuel in Boilers and Industrial Furnaces, Final Rule, Federal Register,* vol. 50, no. 23, pp. 49,164–49,249, November 1983.

4. *Federal Register,* vol. 50, pp. 9,386 and 9,400, March 7, 1985.

5. R. E. Hall, W. M. Cooke, and R. L. Barbour, "Comparison of Air Pollution Emissions from Vaporizing and Air Atomizing Waste Oil Heaters," *JAPCA,* **33**:(7)683 (July 1983).

6. P. F. Fennelly and M. M. McCabe, *Environmental Characterization of Disposal of Waste Oils in Small Combustors,* GCA/Technology Division, Bedford, Massachusetts, under EPA Contract No. 68-02-3168, January 1984.

7. G. R. Offen et al., *Control of Particulate Matter From Oil Burners and Boilers,* EPA-450/3-76-005, prepared for the U.S. EPA by Acurex Corp., Mountain View, Calif., 1976.

8. *Hazardous Waste Management System; Recycled Used Oil Standards, Proposed Rule, Federal Register,* vol. 50, no. 23, pp. 49,245–49,249, November 1985.

9. N. F. Suprenant et al., *Emissions Assessment of Conventional Combustion Systems,* vol. V, *Industrial Combustion Sources,* EPA-600/7-79-029b, U.S. EPA, GCA/Technology Division, Bedford, Mass., April 1978.

10. C. Shih et al., *Emissions Assessment of Conventional Stationary Combustion Systems,* vol. III, *Electric Utility Sources,* EPA-600/7-79-029d, U.S. EPA, IERL, Research Triangle Park, N.C., 1980.

11. F. O. Cotton et al., "Analysis of 30 Used Motor Oils," *Hydrocarbon Processing,* **56**(9):131–140 (September 1977).

12. D. W. Brinkman et al., *Environmental Resource Conservation, and Economic Aspects of Used Oil Recycling,* DOE/BETC/RI-80/11, U.S. DOE, Bartlesville, Oklahoma, April 1981.

13. K. D. Weinstein et al., *Enhanced Utilization of Used Lubricating Oil Recycling Process By Products,* DOE/BETC/10054-19, U.S. DOE, Bartlesville, Oklahoma, March 1982.

14. E. Peake and H. Parker, "Polynuclear Aromatic Hydrocarbons and the Mutagenicity of Used Crankcase Oils," *The Fourth International Symposium on Polynuclear Aromatic Hydrocarbons,* Battelle Columbus Labs, Columbus, Ohio, October 1979.

15. M. Hermann, et al., "Correlation of Mutagenic Activity With Polynuclear Aromatic Hydrocarbon Content of Various Mineral Oils," *The Fourth International Symposium on Polynuclear Aromatic Hydrocarbons,* Battelle Columbus Labs, Columbus, Ohio, October 1979.

SECTION 4.2
PCB WASTES

Drew E. McCoy
McCoy and Associates
Denver, Colorado

Polychlorinated biphenyls (PCBs) possess many useful characteristics. They are very stable, have a low vapor pressure, low flammability, high heat capacity, low electrical conductivity, and a high dielectric constant. Unfortunately, these chemicals can enter the human body through the skin, lungs, and gastrointestinal tract, and are accumulated in fatty tissue. The U.S. Environmental Protection Agency (EPA) has determined that PCBs may cause adverse reproductive effects, developmental toxicity, and tumor development in humans. They also tend to bioaccumulate in the food chain. Because of these health and environmental hazards, the use and disposal of PCBs is stringently regulated.

4.2.1 HISTORICAL USES OF PCBs

Figure 4.2.1 shows the molecular structure of the biphenyl molecule. Chlorine substitution can occur at any one or more of the ten numbered locations. As a result, there are 209 individual chemical species which fall under the generic classification of "PCBs."

Before the EPA implemented controls on the production and use of PCBs, the chemicals were used for the applications shown in Table 4.2.1. Because of their excellent electrical insulating properties, the largest application of PCBs is in electrical equipment. As shown in Table 4.2.2 PCB dielectric fluids have been marketed under a number of different trade names in different countries.

4.2.2 PCB REGULATIONS UNDER THE TOXIC SUBSTANCES CONTROL ACT

In 1976, Congress enacted the Toxic Substances Control Act (TSCA), which directed the EPA to control the manufacture, processing, distribution, use, dis-

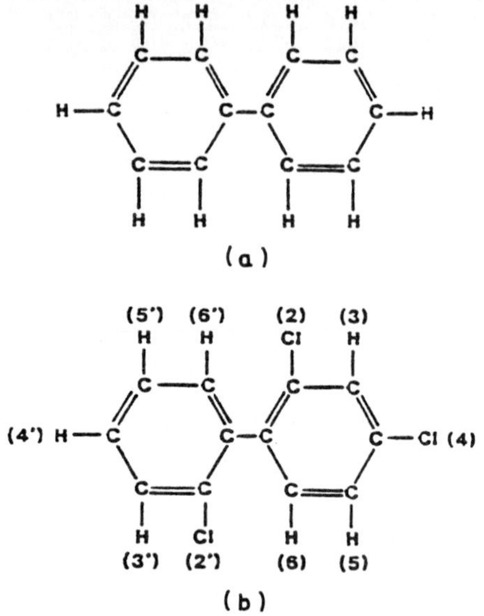

FIG. 4.2.1 PCB Molecular Structure. (*a*) The biphenyl molecule. (*b*) 2,2',4-trichlorobiphenyl. H = hydrogen. C = carbon. Cl = chlorine.

posal, and labeling of PCBs. The TSCA generally prohibits the use of PCBs after January 1, 1978. However, two exceptions were allowed: (1) PCBs can be used in a "totally enclosed manner," and (2) they can be used in a manner that the EPA finds "will not present an unreasonable risk of injury to health or the environment."

On May 31, 1979, the EPA began to issue regulations to implement the requirements of the TSCA. These regulations, located in Title 40 of the *Code of Federal Regulations,* Part 761, have been continuously amended and updated over the years and are likely to undergo further changes in the future.

The PCB regulations deal with four major topics:

TABLE 4.2.1 Applications of PCBs by Amount

Uses	Amount	
	t	tons
Transformers	150,000	165,000
Capacitors	290,000	320,000
Plasticizers	51,700	57,000
Hydraulic fluids and lubricants	36,280	40,000
Carbonless copy paper	20,000	22,000
Miscellaneous industrial uses	11,800	13,000
Heat-transfer fluids	9,070	10,000
Petroleum additives	900	1,000
	570,000	628,000

TABLE 4.2.2   Common Trade Names for Dielectric Fluids Containing PCBs

Apirolio (Italy)	Inerteen (Canada & U.S.)
Aroclor (U.S. & Great Britain)	Kennechlor
Aroclor B	Kanechlor (Japan)
Asbestol	MCS 1489
Askarel	Montar (U.S.)
Adkarel	Nepolin
Chlorextol	No-Flamol
Clophen (Germany)	Phenoclor (France)
Clorphen	Pydraul (U.S.)
Clorinol	Pyralene (France)
Diaclor	Pyranol (Canada & U.S.)
DK (Italy)	Pyroclor (Great Britain)
Dykanol	Saf-T-Kuhl
EEC-18	Santotherm FR (Japan)
Elemex	Solvol
Eucarel	Sorol (U.S.S.R.)
Fenclor (Italy)	Therminol FR (U.S.)
Hyvol	

1. Prohibitions on the manufacture, processing, distribution in commerce, and use of PCBs and PCB-contaminated items (Part 761, Subpart B)

2. Requirements for marking, labeling, and placarding PCB containers, equipment, storage areas, transport vehicles, etc. (Part 761, Subpart C)

3. Regulations on the storage and disposal of PCBs and PCB-contaminated items (Part 761, Subpart D)

4. Requirements for record keeping and reporting (Part 761, Subpart J)

The regulatory requirements that are applicable to the disposal of PCBs are described under *PCB Concentrations of Concern,* (p. 4.15) and *Acceptable Methods of PCB Disposal* (p. 4.16).

PCB Concentrations of Concern

Regulations under TSCA govern all forms and combinations of chemicals which contain the biphenyl molecule with one or more chlorine atom substitutions. They also apply not only to the PCB chemicals themselves, but to items which have been in contact with PCBs, such as containers, pipes, and electrical equipment.

Concentration ranges are the primary basis for determining how PCB-containing wastes must be handled; three concentration ranges are of concern (all expressed on a dry-weight basis):

1. *<50 ppm:*   Disposal of wastes containing less than 50 ppm PCB is not regulated, with one exception. Waste oil (including fuel, motor, gear, cutting, transmission, hydraulic, and dielectric oil) is strictly regulated if it contains any detectable concentration of PCBs. These waste oils cannot be used as a sealant, coating, dust-control agent, pesticide or herbicide carrier, or rust preventative.

2. *50 to 500 ppm:*   Contamination at this intermediate level is strictly regulated. However, several disposal options exist for these materials.

3. *>500 ppm:*   Contamination at concentrations above 500 ppm requires adherence to the most stringent disposal requirements.

Acceptable Methods of PCB Disposal

One of the primary components of the PCB regulations deals with allowable methods of PCB disposal. Table 4.2.3 shows a master listing of disposal methods that can be used for different types of PCB materials. Each of the approved PCB-disposal methods is described below.

PCB Incinerators.   Only high-efficiency incinerators that have been approved by the EPA can be used to destroy PCBs at concentrations greater than 50 ppm. Incinerators burning PCB liquids must meet the following criteria: 2-s dwell time at $1200 \pm 100°C$ ($2192 \pm 180°F$) and 3% excess oxygen; or 1.5-s dwell time at $1600 \pm 100°C$ ($2912 \pm 180°F$) and 2% excess oxygen in the stack gas. In either case, the combustion efficiency of these incinerators must be at least 99.9%. (This is not to be confused with destruction and removal efficiency, which is discussed below.)

Incinerators used to burn nonliquid PCBs are subject only to air-emission standards specifying that emissions cannot exceed 1 mg/kg (ppm) of PCBs introduced into the incinerator. Note that the destruction and removal efficiency (DRE) for the nonliquid PCBs is equivalent to 99.9999%. Although a DRE isn't specified for liquid PCBs, the EPA believes that the required combustion conditions will result in DREs in excess of 99.9999%.

Incinerators are discussed in Handbook Chap. 8.

PCB High-Efficiency Boilers.   TSCA regulations allow liquids containing PCBs in concentrations between 50 and 500 ppm to be disposed of in a high-efficiency boiler. The boiler must meet the following criteria:

- It must be rated at a minimum of 14.6 MW (50 million Btu/h).
- For gas- or oil-fired boilers, the carbon monoxide (CO) concentration in the flue gas must be less than 50 ppm.
- For coal-fired boilers, the CO concentration in the stack cannot exceed 100 ppm.
- Excess oxygen must be at least 3%.
- The waste cannot exceed 10% by volume of the total fuel fed to the boiler.
- Waste can only be fed into the boiler when it is at operating temperature (feed during start-up or shutdown is prohibited).
- Specific process-monitoring and operating procedures must be followed.
- EPA must be notified 30 days before such a burn takes place.

A number of utilities have obtained permits for incinerating PCB-contaminated fluids in high-efficiency boilers. These include Northeast Utilities, Potomac Electric and Power, Duke Power, Union Electric, and Washington Water and Power. In many cases, PCB burns are limited to the individual facility's own wastes.

Use of boilers for waste destruction is discussed in Handbook Sec. 8.4.

TABLE 4.2.3 Allowable Methods of PCB Disposal*

Type of PCB material	PCB incinerator	PCB high-efficiency boiler	PCB landfill	Approved alternative methods
Liquid PCBs, spills				
>500 ppm	x			Chemical dechlorination
50–500 ppm	x	x	x†	Chemical dechlorination
Nonliquid PCBs, spill residues, >50 ppm (contaminated soil, rags, or other debris)	x		x	
All dredged materials and municipal sewage-treatment sludge >50 ppm.	x		x	
PCB transformers‡ (>⅝?? PPM(	X		X	Transformer disposal
PCB large capacitors >50 ppm, 3 lb dielectric fluid	x			Capacitor disposal
PCB small capacitors <3 lb dielectric fluid or <100 in³	x		x	Municipal solid waste
PCB hydraulic machines containing >50 ppm				Municipal solid waste‡
Other PCB articles, >50 ppm	x		x‡	Municipal solid waste‡
PCB containers				
>500 ppm	x		x‡	
>50, <500 ppm				Municipal solid waste‡
Decontaminated				Re-use or municipal solid waste

*In general, disposal of waste containing <50 ppm PCBs is not regulated.
†Liquid cannot be ignitable; i.e., flashpoint cannot be below 140°F.
‡Drained of free-flowing liquid; hydraulic machines must be flushed with suitable solvent if PCB concentration >1,000 ppm.

4.17

PCB Landfills.   Landfilling of wastes with PCB levels of less than 50 ppm is not currently regulated—these wastes can be disposed in landfills permitted under the Resource Conservation and Recovery Act (RCRA) or even solid-waste landfills. However, because of potential future liability, this latter option is not recommended. In order to accept wastes having PCB concentrations between 50 and 500 ppm, special technical requirements must be met:

- The facility must have a liner or underlying soil with a permeability of less than 10^{-7} cm/s.
- The bottom of the landfill must be at least 15.2 m (50 ft) above the historical high groundwater table.
- The site must have monitoring wells and a leachate-collection system.
- The site must either be above the 100-year floodplain, or be diked to a level 0.6 m (2 ft) above the 100-year floodwater level.
- Specific operating and record-keeping requirements must be met.

Note that even though current regulations for new RCRA landfills are more stringent than the requirements for PCB landfills, not all RCRA landfills can accept PCB wastes at concentrations between 50 and 500 ppm. Special approval is required before an RCRA landfill can accept PCBs in this concentration range.

Alternative Disposal Methods.   The EPA can approve alternative treatment and disposal methods for managing PCB wastes. Physical, chemical, and biological treatment alternatives have been approved, with chemical detoxification being the most common.

A number of companies have developed processes to chemically destroy PCBs in dielectric fluids. In general, these companies provide a mobile treatment system which can travel to a client's site for detoxification of PCB liquids. In some cases, the processes can be used while transformers are still energized. Not only do these processes destroy the PCBs, but they also allow the dielectric fluids to be re-used. Depending on the specific process being used, PCB concentrations of 1,000 to 10,000 ppm can be accommodated.

The chemical detoxification processes currently approved by the EPA attack the PCB molecules at the chlorine-carbon bonds. Under the mild reaction conditions involved, chlorine is stripped from the biphenyl molecules, and the biphenyls then polymerize to form insoluble sludge. These sludges are subsequently filtered from the dielectric fluid.

The dechlorination reagents used by different companies are proprietary, but generally utilize metallic sodium and solvents to form an organosodium complex. This reagent reacts with the PCBs to form sodium chloride and polyphenylene polymer.

Other components of the dielectric fluid, such as inhibitors, acids, thiols, and chlorides, also react with the sodium reagent to form insoluble sludges. After the reaction step has been completed, a small amount of water may be added to destroy excess sodium reagent through the formation of sodium hydroxide. Solids, water, and dissolved gases are typically removed from the detoxified dielectric fluid by filtration and vacuum degassing.

The amount of waste generated during the chemical dechlorination reaction is typically 2 to 3% of the original fluid volume. Disposal of these wastes is normally handled by the company providing the detoxification service. Handbook Sec. 7.5 discusses dehalogenation processes.

4.2.3 PCB REGULATIONS UNDER THE RESOURCE CONSERVATION AND RECOVERY ACT

The Resource Conservation and Recovery Act (RCRA) does not directly control how PCB wastes must be managed. However, if a waste meets the definition of a hazardous waste under RCRA and it also contains PCBs, it could be subject to RCRA's land disposal prohibitions. For example, if a liquid waste is hazardous because it is ignitable (one of the characteristics of a hazardous waste) and it also contains more than 50 ppm of PCBs, it will be subject to the land disposal restrictions. However, if the PCBs are contained in a similar waste that is not ignitable or otherwise hazardous, it is not subject to the land disposal restrictions.

The land disposal restrictions under RCRA are codified in 40 *CFR* Part 268. For PCBs, these regulations require the initial generator of a waste to determine if (1) the waste is a hazardous waste according to RCRA definitions, (2) the waste is a liquid using the Paint Filter Liquids Test (Method 9095 of *Test Methods for Evaluation of Solid Wastes, Physical/Chemical Methods*, EPA Publication No. SW-846), and (3) the waste contains more than 50 ppm PCBs. If all of these conditions are met, and if the generator is not conditionally exempted from RCRA regulations (i.e., it generates more than 100 kg/month of hazardous wastes), the land disposal restrictions apply.

The scope of the land disposal restrictions for PCB-containing hazardous wastes is relatively limited. Basically, the land disposal option is foreclosed for liquid PCB wastes containing at least 50 but less than 500 ppm PCBs. (Recall that under the TSCA, materials containing 50 to 500 ppm PCBs can be disposed in an approved landfill.) The land disposal restrictions specify that these wastes must be destroyed in an incinerator or a high-efficiency boiler meeting the TSCA standards (40 *CFR* Parts 761.70 and 761.60, respectively). For PCB concentrations greater than or equal to 500 ppm PCBs, the wastes must be incinerated. Although an exemption from the land disposal prohibitions is available for wastes containing 50 to 500 ppm PCBs, it must be shown that the wastes will not migrate from the land disposal unit for as long as the waste remains hazardous. "No-migration exemptions" are exceedingly difficult to justify and are unlikely to be of any practical significance.

Finally, the land disposal restrictions specify that if there is a conflict between the PCB disposal restrictions of RCRA and the TSCA, the more stringent requirements apply.

4.2.4 EPA-APPROVED PCB-DISPOSAL FACILITIES

To safely dispose of PCB wastes, two criteria must be met. First, an EPA-approved facility must be identified that can handle the specific type of waste involved. Table 4.2.4 identifies commercial facilities that have EPA approval to manage PCB wastes (as of March 1988). Note that some companies offering mobile PCB-destruction technologies are approved to operate in all 10 regions.

Figure 4.2.2 shows the location of incinerators and landfills having EPA approval under TSCA to handle PCB wastes. (Recall that incinerators and landfills having RCRA permits may also have approval to manage wastes containing less

TABLE 4.2.4 Listing of EPA-Approved PCB-Disposal Facilities

Map key no.	Company	Address and phone
	TSCA incinerators	
1.	ENSCO	P.O. Box 1957 El Dorado, Ark. 71730 (501) 223-4160
2.	Rollins	P.O. Box 609 Deer Park, Tex. 77536 (713) 479-6001
3.	EPA Mobile Incinerator	Woodbridge Ave. Raritan Depot, Bldg. 10 Edison, N.J. 08837 (201) 321-6635
4.	Pyrotech Systems	1st Tennessee Bank Bldg. Third Floor Franklin, Tenn. 37064 (501) 863-7173
5.	General Electric	100 Woodlawn Ave. Pittsfield, Mass. 01201 (413) 494-3729
6.	SCA Chemical Services	1000 E. 111th St. 10th Floor Chicago, Ill. 60628 (312) 646-5700
7.	Pyrochem/Aptus	P.O. Box 907 Coffeyville, Kansas 67337 (316) 251-4782
	TSCA landfills	
8.	Casmalia Resources	559 San Ysidro Road P.O Box 5275 Santa Barbara, Calif. 93150 (805) 937-8449
9.	CECOS International	56th St. & Niagara Falls Blvd. Niagara Falls, N.Y. 14302 (716) 282-2676
10.	CECOS International	5092 Aber Rd. Williamsburg, Ohio 45176 (513) 720-6114
11.	Chem-Security Systems, Inc.	Star Route Arlington, Ore. 98712 (503) 454-2777
12.	Chemical Waste Management of Alabama, Inc.	P.O. Box 55 Emelle, Ala. 35459 (205) 652-9721
13.	Chemical Waste Management	Box 471 Kettleman City, Calif. 93239 (209) 386-9711

TABLE 4.2.4   Listing of EPA-Approved PCB-Disposal Facilities (*Continued*)

Map Key no. Company	Address and phone
14. Envirosafe Services, Inc., of Idaho	P.O. Box 417 Boise, Idaho 83701 (208) 384-1500
15. SCA Chemical Services	Box 200 Model City, N.Y. 14107 (716) 754-8231
16. U.S. Ecology, Inc.	Box 578 Beatty, Nev. 89003 (702) 553-2203
17. U.S. Pollution Control Inc.	Grayback Mountain Knolls, Utah 84074 (405) 528-8371

Company	Address and phone
Alternate thermal, physical, chemical, or biological treatment	
American Mobil Oil Purification Co.*	233 Broadway, 17th Floor New York, N.Y. 10279 (212) 267-7073
Chemical Decontamination Corporation*	5 Riga Lane Baltic Mews Industrial Park Birdsboro, Pa. 19508 (215) 582-2766
Chemical Waste Management, Inc.*	Model City Facility P.O. Box 200 Model City, N.Y. 14107 (716) 754-8231
Detox, Inc.	4800 Sugar Grove Blvd. Suite 210 Stafford, Tex. 77477 (713) 240-0892
ENSCO*	1015 Louisiana Street Little Rock, Ark. 72202 (501) 223-4100
Exceltech, Inc.	41638 Christy Street Fremont, Calif. 94539 (415) 659-0404
GA Technologies, Inc.*	P.O. Box 85608 San Diego, Calif. 92138 (619) 445-2517
General Electric	1 River Rd. Schenectady, N.Y. 12345 (518) 385-3134
J. M. Huber Corporation	P.O. Box 2831 Borger, Tex. 79007 (806) 274-6331

TABLE 4.2.4 Listing of EPS-Approved PCB-Disposal Facilities (*Continued*)

Company	Address and phone
National Oil Processing/Aptus	P.O. Box 1062 Coffeyville, Kansas 67337 (800) 345-6573
Niagara Mohawk Power Corp.	300 Erie Boulevard West Syracuse, N.Y. 13202 (315) 474-1511
PPM, Inc.*	1875 Forge St. Tucker, Ga. 30084 (404) 934-0902
Quadrex HPS, Inc.*	1940 N.W. 67th Pl. Gainesville, Fla. 32606 (904) 373-6066
Sun Environmental, Inc.*	1700 Gateway Blvd., S.E. Canton, Ohio 44707 (216) 452-0837
T & R Electric Supply Company, Inc.	Box 180 Colman, S.Dak. 57017 (800) 843-7994
Transformer Consultants*	P.O. Box 4724 Akron, Ohio 44310 (800) 321-9580
Trinity Chemical Co., Inc.	6405 Metcalf, Cloverleaf #3 Suite 313 Shawnee Mission, Kan. 66202 (913) 831-2290
Unison Transformer Services, Inc.	P.O. Box 1076 Henderson, Ky. 42420 (800) 544-0030

*Approved to operate in all U.S. EPA regions.
Source : U.S. EPA, March 1988.

than 50 ppm PCBs. However, because of company policy or more restrictive state regulations, not all such facilities handle PCBs.)

The second criterion for safely disposing of PCB wastes is that the waste must, in fact, be disposed at the selected facility. Although this second criterion may seem redundant, several very expensive damage suits have been filed in cases where PCB wastes were diverted from their intended disposal site to some type of illegal waste disposal.

Keep in mind that many firms simply act as waste brokers or transfer agents. Although they may claim that they are licensed to handle PCB wastes, they simply transfer the wastes to an EPA-approved facility. Where problems have occurred in the past, unscrupulous transfer agents or intermediate waste handlers have failed to deliver the wastes to an EPA-approved facility. Instead, they dumped the wastes illegally or simply accumulated the wastes until the company went out of business.

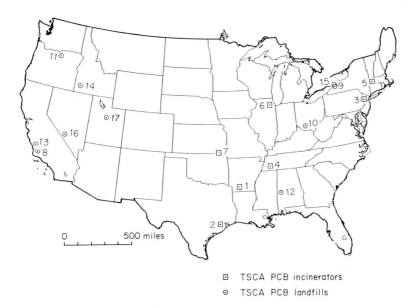

FIG. 4.2.2 Locations of TSCA-approved PCB incinerators and landfills. Numbers correspond to facilities located in Table 4.2.4.

Because elaborate schemes have been devised to deceive generators into believing that their wastes have been safely disposed, the generator must be very cautious in using the services of any intermediaries. At a minimum, the standing of the intermediary with state and federal regulatory agencies should be checked before relinquishing control of a waste shipment. Better yet, deal directly with the EPA-approved facility.

4.2.5 FURTHER READING

Additional information on the technology and regulations which apply to PCBs are listed here. The information in this section was derived from these sources as well as from numerous industry and EPA contacts.

Environment Canada, *Destruction Technologies for Polychlorinated Biphenyls (PCBs)*, Report EPS 3-EC-83-1, Environmental Protection Service, Environment Canada, Ottawa, Ontario, Canada KIA IC8, reprinted July 1983.

Environment Canada, *Handbook on PCBs in Electrical Equipment*, rev. ed., Environmental Protection Service, Environment Canada, Ottawa, Ontario, Canada KIA IC8, December 1982.

National Rural Electric Cooperative Association, *PCB Equipment, Operations and Management Reference Manual*, National Rural Electric Cooperative Association, 1800 Massachusetts Ave., N.W., Washington, D.C. 20036, March 1983.

U.S. Environmental Protection Agency, *The PCB Regulations Under TSCA: Over 100 Questions and Answers to Help You Meet These Requirements*, rev. ed. no. 3, U.S. EPA Office of Toxic Substances, 401 M Street S.W., Washington, D.C. 20460, August 1983.

SECTION 4.3

DIOXIN WASTES

M. Pat Esposito

Bruck, Hartman, & Esposito, Inc.
Cincinnati, Ohio

4.3.1 BACKGROUND

Dioxins are a family of aromatic compounds known chemically as dibenzo-*p*-dioxins. Each dioxin compound has as a nucleus a triple-ring structure consisting of two benzene rings interconnected to each other through a pair of oxygen atoms. Figure 4.3.1 shows the structural formula for the dioxin nucleus.

Many types of dioxin compounds are possible. In its unsubstituted form, all eight of the outer ring positions are occupied by hydrogen ions (H). Alternatively, partially or fully substituted forms are known to occur, with one or more of the hydrogens replaced by such common constituents as chlorine, bromine, amine groups, hydroxyl groups, methyl groups, or any combination of these. With the exception of the chlorinated forms, little is known about the physical, chemical, and toxicologic properties of most dioxins.

Most environmental interest in dioxins has centered on chlorinated dioxins, in which a chlorine atom occupies one or more of the eight positions on the benzene rings of the nucleus. Theoretically, there are 75 chlorinated dioxins, each with different physical, chemical, and toxicologic properties, differing only in the number and arrangement of chlorine atoms in each molecule.

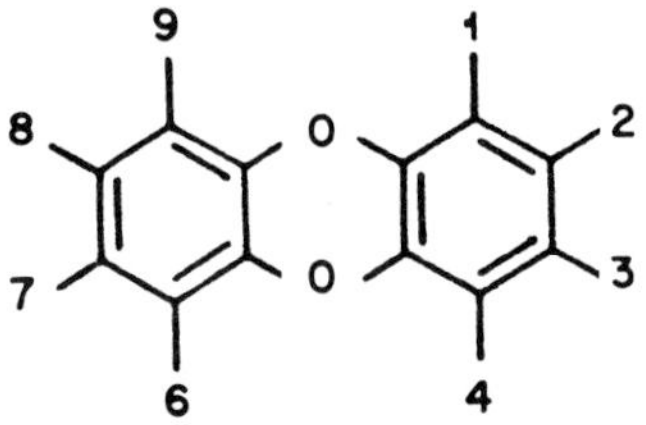

FIG. 4.3.1 The dioxin molecule.

Dioxin compounds do not appear to be constituents of any normal growing environment; rather, they are by-products of certain chemical manufacturing or combustion processes involving precursor compounds and heat. Their formation yield is very low—in the parts per trillion to parts per million range—but once formed, they are quite stable. For example, chlorinated dioxins were widely generated in the 1960s and 1970s as by-products of the mass production of chlorophenoxy herbicide compounds such as 2,4,5-T, Silvex, Erbon, and others for domestic and defensive use. Dioxin residuals from this period are still contaminating soils and process wastes associated with the production and use of these chemicals.

Concern about chlorinated dioxins has been largely directed toward the highly toxic tetra-, penta-, and hexachlorinated forms. Of the tetrachlorinated isomers, 2,3,7,8-tetrachlorodibenzo-p-dioxin (2,3,7,8-TCDD) has been of particular concern worldwide because it is one of the most toxic manufactured compounds known to exist. The pure compound is a colorless crystalline solid at room temperature and is only sparingly soluble. Its chemical and physical properties are summarized in Table 4.3.1. The biological degradation rate of the compound appears to be very slow compared with that of almost any other organic compound. Because it has a very low vapor pressure, it does not readily evaporate or volatilize to the atmosphere. Also, although it can be photolytically dechlorinated by ultraviolet radiation and thermally destroyed at high temperatures in combustion devices such as incinerators, neither mechanism is important as a natural decomposition pathway. The compound adheres tightly to soil particles and does not migrate or leach into ground or surface waters over time unless the contaminated soil particles themselves migrate via erosion processes. Thus, wastes and soils

TABLE 4.3.1   Chemical and Physical Properties of TCDD

CAS registry no. 1746-01-6	
Empirical formula	$C_{12}H_4Cl_4O_2$
Percent by weight	
Carbon	44.7
Oxygen	9.95
Hydrogen	1.25
Chlorine	44.1
Molecular weight	322
Vapor pressure, mmHg at 25°C	1.7×10^{-6}
Melting point, °C	305
Decomposition temperature, °C	>700
Solubilities, g/L	
o-Dichlorobenzene	1.4
Chlorobenzene	0.72
Benzene	0.57
Chloroform	0.37
n-Octanol	0.05
Methanol	0.01
Acetone	0.11
Water	2×10^{-7}

Source: Adapted from Refs. 1 and 2.

contaminated with 2,3,7,8-TCDD and other dioxins tend to be long-lived, persistent environmental problems for which no simple remedies have been found.

4.3.2 TECHNOLOGIES FOR TREATING DIOXIN WASTES

Considerable research has been conducted in recent years in search of treatment methods for destroying or detoxifying dioxins and other chemically similar compounds, such as PCBs. Both thermal and nonthermal technologies have been studied, and several appear feasible. Table 4.3.2 lists the most promising technologies together with cross references to the chapters in this handbook where discussions of engineering and design principles underlying each technology can be found. A summary of the research status and testing results for each treatment process is presented in Table 4.3.3.

Thermal Technologies

Because studies have shown that dioxins decompose by heating or oxidation at high temperatures (greater than 1000°C), thermal methods for treating these wastes have received a large amount of attention.[3-16] The U.S. Environmental Protection Agency (EPA) has indicated that incineration is the only treatment technology for dioxin-containing waste that has been demonstrated sufficiently.[17] Treatment standards for incineration and other thermal processes specified in Title 40 of the *Code of Federal Regulations* require the demonstration of 99.9999% destruction and removal efficiency (DRE) of the dioxin. Several of the thermal

TABLE 4.3.2 Treatment Technologies for Dioxins

Technology	Handbook section no.
Thermal technologies	
Stationary rotary-kiln incineration	8.2
Mobile rotary-kiln incineration	8.13
Liquid-injection incineration	8.1
Fluidized-bed incineration (circulating-bed combustor)	8.3
High-temperature fluid-wall destruction	8.13
Infrared incineration	8.13
Supercritical-water oxidation	8.11
Plasma-arc pyrolysis	8.12
In situ vitrification	8.9
Nonthermal technologies	
Solvent extraction	6.6
Stabilization/fixation	7.8
Ultraviolet (UV) photolysis	7.9
Chemical dechlorination	7.5
Biodegradation	9.5
Carbon adsorption	6.1

TABLE 4.3.3 Summary of Research on Treatment Processes for Dioxin Wastes

Process name	Applicable waste streams	Stage of development
Stationary rotary-kiln incineration	Solids, liquids, sludges.	Several approved and commercially available units for PCBs; not yet used for dioxins.
Mobile rotary-kiln incineration	Solids, liquids, sludges.	EPA mobile unit is permitted to treat dioxin wastes; ENSCO unit has been demonstrated on PCB wastes.
Liquid-injection incineration	Liquids or sludges with viscosity less than 10,000 ssu (i.e., pumpable).	Full-scale land-based units permitted for PCBs; only ocean incinerators have handled dioxin wastes.
Fluidized-bed incineration (circulating-bed combustor)	Solids, sludges.	GA Technologies mobile circulating-bed combustor has a TSCA permit to burn PCBs anywhere in the nation; not tested yet on dioxins.
High-temperature fluid-wall destruction	Primarily for granular contaminated soils, but may also handle liquids.	Huber stationary unit is permitted to do research on dioxin wastes; pilot-scale mobile reactor has been tested at several locations on dioxin-contaminated soils.
Infrared incinerator	Contaminated soils/sludges.	Pilot-scale, portable unit tested on waste containing dioxin; full-scale units have been used in other applications; not yet permitted for TCDD.
Molten salt	Solids, liquids, sludges; high-ash-content wastes may be troublesome.	Pilot-scale unit was tested on various wastes—further development is not anticipated.
Supercritical-water oxidation	Aqueous solutions or slurries with less than 20% organics can be handled.	Pilot-scale unit tested on dioxin-containing wastes—results not yet published.

Performance/destruction achieved	Cost	Residuals generated
Greater than six nines DRE for PCBs; greater than five nines DRE demonstrated on dioxin at combustion-research facility.	$0.25–$0.70/lb for PCB solids	Treated waste material (ash), scrubber wastewater, particulate from air filters, gaseous products of combustion.
Greater than six nines DRE for dioxin by EPA unit; process residuals delisted.	NA	Same as above.
Greater than six nines DRE on PCB wastes; ocean incinerators only demonstrated three nines on dioxin containing herbicide orange.	$200–$500/ton	Same as above, but ash is usually minor because solid feeds are not treated.
Greater than six nines DRE demonstrated by GA unit on PCBs.	$60–$320/ton for GA unit	Treated waste (ash), particulates from air filters.
Pilot-scale mobile unit demonstrated greater than five nines DRE on TCDD-contaminated soils at Times Beach (79 ppb reduced to below detection).	$300–$600/ton	Treated waste solids (converted to glass beads), particulates from baghouse, gaseous effluent (primarily nitrogen).
Greater than six nines DRE on TCDD-contaminated soil.	Capital cost of incinerator is $2–3 million; treatment costs are $200–$1,200/ton	Treated material (ash); particulates captured by scrubber (separated from scrubber water).
Up to eleven nines DRE on hexachlorobenzene; greater than six nines DRE on PCB using bench-scale reactor.	NA	Spent molten salt containing ash; particulates from baghouse.
Six nines DRE on dioxin-containing waste reported by developer, but not presented in literature; lab testing showed greater than 99.99% conversion of organic chloride for wastes containing PCB.	$0.32–$2.00/gal $77–$480/ton	High purity water, inorganic salts, carbon dioxide, nitrogen.

TABLE 4.3.3 Summary of Research on Treatment Processes for Dioxin Wastes (*Cont*)

Process name	Applicable waste streams	Stage of development
Plasma-arc pryolysis	Liquid waste streams (possibly low-viscosity sludges).	Prototype unit (same as full-scale) currently being field-tested.
In situ vitrification	Contaminated soil—soil type is not expected to affect the process.	Full-scale on radioactive waste; pilot-scale on organic-contaminated wastes.
Solvent extraction	Soils, still bottoms.	Full-scale still-bottoms extraction has been tested—pilot-scale soils washer needs further investigation.
Stabilization/fixation	Contaminated soil.	Laboratory-scale using cement and emulsified asphalt; lab tests also using K-20.
UV photolysis	Liquids, still bottoms, and soils can be treated if dioxin is first extracted or desorbed into liquid.	Full-scale solvent extraction/UV process was used to treat 4,300 gal of still bottoms in 1980; thermal desorption/UV process currently undergoing second field test.
Chemical dechlorination— APEG processes	Contaminated soil (other variations of the process used to treat PCB-contaminated oils).	Slurry process currently being field-tested at pilot scale; in situ process has been tested in the field.
Biological degradation— primarily in situ addition of microbes	Research has been directed toward in situ treatment of contaminated soils—liquids are also possible.	Currently laboratory scale—field testing in next year or two.

Source: Technical Resource Document, *Treatment Technologies for Dioxin Containing Wastes,* U.S. EPA, Cincinnati, Ohio, 1986.
NA = not available.

Performance destruction achieved	Cost	Residuals generated
Greater than six nines destruction of PCBs and CCl_4.	$300–$1,400/ton	Exhaust gases (H_2 and CO) which are flared; scrubber water containing particulates.
Greater than 99.9% destruction efficiency (DE) (not off-gas treatment system) on contaminated soil.	$120–$250/m³	Stable/immobile molten glass; volatile-organic combustion products (collected and treated).
Still-bottom extraction: 340 ppm TCDD reduced to 0.2 ppm; 60–90% removal from soils, but reduction to below 1 ppb not achieved	NA	Treated waste material (soil organic, liquid); solvent extract with concentrated TCDD.
Tests using cement showed decreased leaching of TCDD, but up to 27% loss of stabilized material due to weathering followed by leaching.	NA	Stabilized matrix (soil plus cement, asphalt, or other stabilization material); matrix will still contain TCDD.
Greater than 98.7% reduction of TCDD using solvent extraction/UV process—residuals contained ppm concentrations of TCDD; thermal desorption/UV process demonstrated reduction of TCDD in soil to below 1 ppb.	Cost of treating the 4,300 gal of still bottoms using solvent extraction/UV was $1 million; thermal desorption/UV estimated to cost $250–$1,250/ton	Solvent extraction/UV process generated treated still bottoms, a solvent extract stream, and an aqueous salt stream; thermal desorption/UV generates a treated soil stream and a solvent extract stream.
Laboratory research has demonstrated reduction of 2,000 ppb TCDD to below 1 ppb for slurry (batch process); laboratory and field testing of in situ process not as promising.	$296/ton in situ APEG process; $91/ton for slurry (batch) process	Treated soil containing chloride salts (reagent is recovered in the slurry process)
50–60% metabolism of 2,3,7,8-TCDD in a weeklong period under lab conditions using white rot fungus—reduction to below 1 ppb not achieved.	NA	Treated waste medium such as soil or water with TCDD metabolites depending on microorganisms

treatment technologies have demonstrated this performance on chlorinated compounds of one type or another (e.g., PCBs), but only four have demonstrated this level of performance on dioxin. These technologies are the EPA mobile rotary-kiln incinerator, the high-temperature fluid-wall reactor, the infrared incinerator, and possibly the supercritical-water oxidation process. Modar Inc. claims to have achieved "six nines" dioxin destruction with the latter technology, but has not yet released conclusive data showing this performance.[14]

Nonthermal Technologies

Several nonthermal technologies for treating dioxin wastes have also been investigated.[18–31] Many of the nonthermal technologies chemically alter the structure of the dioxin molecule in some way to render it less toxic (e.g., dechlorination, ultraviolet radiation, and biodegradation). Other treatment methodologies mobilize or immobilize the dioxin rather than degrade it (e.g., solvent extraction, and stabilization and fixation).

Nonthermal technologies are not evaluated in the same manner as the thermal technologies, in which stack-gas emissions are of primary concern. The primary residuals of nonthermal technologies are the waste matrices themselves, and stack gases are of little relative importance compared with thermal processes. The Centers for Disease Control have found that soils and other solid-waste residues containing less than 1 ppb of dioxins pose no significant health risk, and based on this, the EPA has recently proposed that only residues of dioxin waste treatment containing 1 ppb or less of dioxin may be landfilled. Thus, the ability of a nonthermal method to reduce dioxin levels in a waste to 1 ppb or less has become the main criterion by which these technologies are evaluated. Only two technologies—thermal desorption followed by UV photolysis, and chemical dechlorination—have been demonstrated capable of achieving this level of effectiveness.

4.3.3 REFERENCES

1. W. B. Crummett and R. H. Stehl, "Determination of Chlorinated Dibenzo-*p*-Dioxins and Dibenzofurans in Various Materials," *Environmental Health Perspectives,* 5:17–25, 1973.

2. National Research Council of Canada, *Polychlorinated Dibenzo-p-Dioxins: Criteria for Their Effects on Man and His Environment,* NRCC 18574, NRCC/CNRC Assoc. Comm. on Scientific Criteria for Environmental Quality, Ottawa, Canada, 1981.

3. R. W. Ross II, T. H. Backhouse, R. N. Vogue, J. W. Lee, and L. R. Waterland, *Pilot-Scale Incineration Test Burn of TCDD-Contaminated Toluene Stillbottoms from Trichlorophenol Production from the Vertac Chemical Company,* prepared for the U.S. EPA Hazardous Waste Engineering Research Laboratory, under EPA Contract No. 68-03-3267, Work Assignment 0-2, by the Acurex Corporation, Energy and Environmental Division, Acurex Technical Report TR-86-100/EE, Mountain View, Calif., January 1986.

4. R. A. Carnes, "U.S. EPA Combustion Research Facility Permit Compliance Test Burn," *Proceedings of the Eleventh Annual Research Symposium on Incineration and Treatment of Hazardous Waste,* sponsored by the U.S. EPA Hazardous Waste Engineering Research Laboratory, Cincinnati, Ohio, April 29–May 1, 1985, EPA 600/9-85-028, September 1985.

5. *Dioxin Trial Burn Data Package, EPA Mobile Incineration System at the James Denney Farm Site, McDowell, Missouri,* prepared for the U.S. EPA Hazardous Waste

Engineering Research Laboratory, under EPA Contract No. 68-03-3069, by IT Corporation, Torrance, Calif., June 21, 1985.

6. *Interim Summary Report on Evaluation of Soils Washing and Incineration as On-Site Treatment Systems for Dioxin-Contaminated Materials,* prepared for the U.S. EPA Hazardous Waste Engineering Research Laboratory, under EPA Contract No. 68-03-3069, by IT Corporation, June 7, 1985.

7. U.S. Environmental Protection Agency, *Assessment of Incineration as a Treatment Method for Liquid Organic Hazardous Waste,* Background Report 1, *Description of Incineration Technology,* U.S. EPA Office of Planning and Evaluation, Washington, D.C., March 1985.

8. P. L. Daily, "Performance Assessment of Portable Infrared Incinerator, Shirco Infrared Systems, Inc.," preprinted extended abstract of paper presented before the Division of Environmental Chemistry, American Chemical Society, 191st National Meeting. New York, April 13–18, 1986, **2b**:(1).

9. H. M. Freeman, "Innovative Thermal Processes for the Destruction of Hazardous Waste," in *Separation of Heavy Metals and Other Trace Contaminants, AIChE Symposium Series,* **81**:(243) (1985).

10. D. P. Y. Chang and N. W. Sorbo, "Evaluation of a Pilot Scale Circulating Bed Combustor with a Surrogate Hazardous Waste Mixture," *Proceedings of the Eleventh Annual Research Symposium on Incineration and Treatment of Hazardous Waste,* sponsored by the U.S. EPA Hazardous Waste Engineering Research Laboratory, Cincinnati, Ohio, April 29–May 1, 1985, EPA 600/9-85-028, September 1985.

11. W. S. Rickman, N. D. Holder, and D. T. Young, "Circulating Bed Incineration of Hazardous Waste," *Chemical Engineering Progress,* March 1985.

12. J. Boyd, H. D. Williams, and T. L. Stoddard, "Destruction of Dioxin Contamination by Advanced Electric Reactor," preprinted extended abstract of paper presented before the Division of Environmental Chemistry, American Chemical Society, 191st National Meeting, New York, April 13–18, 1986, **26**:(1).

13. IR. R. Bellobono, E. Selli, and L. Veronese, "Destruction of Dichloro- and Trichlorophenoxyacetic Acid Esters Containing 2,3,7,8-Tetrachlorodibenzo-*p*-dioxin by Molten Salt Combustion Technique," *Acqua-Aria,* January 1982.

14. W. Killiley, Modar, Inc., telephone conversation with Lisa Farrell, GCA Technology Division, Inc., February 25, 1986.

15. *Technologies Applicable to Hazardous Waste,* briefing prepared for the U.S. EPA Hazardous Waste Engineering Research Laboratory by Metcalf & Eddy, Inc., Washington, D.C., May 1985.

16. U.S. Environmental Protection Agency, *Status of Dioxin Research in the U.S. Environmental Protection Agency,* U.S. EPA Office of Research and Development, August 1985(b).

17. *Federal Register,* vol. 51, p. 1,733, January 14, 1986.

18. M. P. Esposito, T. O. Tiernon, and F. E. Dryden, *Dioxins,* EPA 600/2-80-197, U.S. EPA, Cincinnati, Ohio, November 1980, pp. 263–264.

19. C. J. Sawyer, "Environmental Health and Safety Considerations for a Dioxin Detoxification Process" in *Detoxification of Hazardous Waste,* chap. 18, 1982.

20. R. Helsel, et al., "Technology Demonstration of a Thermal Desorption/UV Photolysis Process for Decontaminating Soils Containing Herbicide Orange," reprint extended abstract, presented before the Division of Environmental Chemistry, American Chemical Society, New York, April 1986.

21. C. T. Ward and F. Matsumura, "Fate of 2,3,7,8-Tetrachlorodibenzo-*p*-dioxin (TCDD) in a Model Aquatic Environment," *Archives of Environmental Contamination and Toxicology,* **7**:(3)349–357 (1978).

22. G. F. Vanness, et al., "Tetrachlorodibenzo-*p*-dioxins in Chemical Wastes, Aqueous Effluents and Soils," *Chemosphere,* **9**:(9)553–63 (1980).

23. A. L. Young, "Long-Term Studies on the Persistence and Movement of TCDD in a Natural Ecosystem," in R. E. Tucker et al. (eds.), *Human and Environmental Risks of Chlorinated Dioxins and Related Compounds,* Plenum Press, New York, 1983, pp. 173–190.

24. A. DiDominico et al., "Accidental Release of 2,3,7,8-Tetrachlorodibenzo-*p*-dioxin (TCDD) at Seveso, Italy," *Ecotoxicology and Environmental Safety,* **4**:(3)282–356 (1980).

25. J. A. Bumpus, M. Tien, D. A. Wright, and S. D. Aust, "Biodegradation of Environmental Pollutants by the White Rot Fungus *Phanerocheate Chrysosporium,*" presented at U.S. EPA Hazardous Waste Engineering Research Laboratory Eleventh Annual Research Symposium, Cincinnati, Ohio, April 29–May 1, 1985.

26. J. F. Quensen and F. Matsumura, "Oxidative Degradation of 2,3,7,8-tetrachlorodibenzo-*p*-dioxin by Microorganisms," *Environmental Toxicology,* **2**:(3)261–268 (1983).

27. W. H. Vick, S. Denzer, W. Ellis, J. Lambauch, and N. Rottunda, *Evaluation of Physical Stabilization Techniques for Mitigation of Environmental Pollution From Dioxin-Contaminated Soils,* interim report, summary of progress-to-date, EPA Contract No. 68-02-3113, Work Assignment no. 36, submitted to the U.S. EPA Hazardous Waste Engineering Research Laboratory by SAIC/JRB Associates, June 1985.

28. J. R. Ryan and J. Smith, "Land Treatment of Wood Preserving Wastes," *Proceedings of the National Conference on Hazardous Wastes and Hazardous Materials,* March 4–6, 1986, Atlanta, Georgia, pp. 80–86.

29. D. C. Ayres, D. P. Levy, and C. S. Creaser, "Destruction of Chlorinated Dioxins and Related Compounds by Ruthenium Tetroxide," in *Chlorinated Dioxins and Dibenzofurans in the Total Environment II,* Butterworth Publishers, Stoneham, Massachusetts, 1985.

30. J. A. Bumpus, M. Tien, D. A. Wright, and S. D. Aust, "Oxidation of Persistent Environmental Pollutants by a White Rot Fungus," *Science,* **228**:1,434–1,436 (1985).

31. H. R. Buser and H. J. Zehnder, "Decomposition of Toxic and Environmentally Hazardous 2,3,7,8-Tetrachlorodibenzo-*p*-dioxin by Gamma Irradiation," *Experientia 41* (1985), pp. 1,082–1,084.

SECTION 4.4
INFECTIOUS WASTE

Frank L. Cross, Jr., P.E.
President
Cross/Tessitore & Associates, P.A.
Orlando, Florida

Rosemary Robinson
Project Engineer
Cross/Tessitore & Associates, P.A.
Orlando, Florida

An infectious waste is one that presents the hazard of causing disease. Although infectious wastes are generated at many facilities, the largest generators are[1]

- The health care industry
- Academic and industrial research laboratories
- The pharmaceutical industry
- Veterinary facilities
- The food, drug, and cosmetics industries

Our discussion of infectious waste will be primarily directed at that generated by hospitals and will cover the entire hospital waste-management program. In addition, toxic and hazardous wastes generated by hospitals and appropriate regulations for these wastes will be discussed.

4.4.1 QUANTITIES AND TYPES OF INFECTIOUS WASTE

In order to evaluate infectious-waste quantities, it is necessary to categorize waste streams according to important characteristics. Solid-waste classification

TABLE 4.4.1 Classification of Wastes

Principal components	Sources	Approx. composition, wt %	Moisture content, %	Incombustible solids, %	Btu value/lb of refuse as fired
		Class 0: Trash			
Highly combustile waste. Paper, wood, cardboard cartons, and up to 10% treated papers, plastic, or rubber scraps.	Commercial, industrial	100 trash	10	5	8500
		Class 1: Rubbish			
Combustile waste, paper, cartons, rags, wood scraps, combustile floor sweepings.	Domestic, commercial, industrial	20 garbage	25	10	6500
		Class 2: Refuse			
Rubbish and garbage.	Residential	50 rubbish 50 garbage	50	7	4800
		Class 3: Garbage			
Animal and vegetable wastes.	Restaurants, hotels, markets, institutional, commercial, clubs	65 garbage 35 rubbish	70	5	2500
		Class 4: Animal solids and organics			
Carcasses, organs, solid organic wastes.	Hospitals, laboratories, abattoirs, animal pounds, etc.	100 animal and human tissue	85	5	1000

Source: Courtesy of C. E. Raymond.[2]

systems categorize wastes according to uniform properties such as moisture content, ash content, and overall composition. Based on these parameters, the heat content of the waste can be determined for incineration purposes. Hospital wastes should be classified further to include infectious-waste properties. These waste-classification systems will be discussed further in this subsection.

Table 4.4.1 outlines typical municipal-waste categories and corresponding parameters. These data are helpful in analyzing hospital waste streams.

Estimates of hospital waste generation range from 4.54 to 9.08 kg/d per bed (10 to 20 lb/d per bed). Based on the classification system in Table 4.4.1, typical waste composition for hospitals categorized according to incineration characteristics is shown in Table 4.4.2.

TABLE 4.4.2 Hospital Waste Characteristics

		Heat of combustion	
Waste type	Weight %†	J/kg	(Btu/lb)
Class 0*	70	(8,500)	19.7×10^6
Plastic	15	(19,500)	45.24×10^6
Class 3*	10	(4,500)	10.44×10^6
Class 4*	5	(1,000)	2.32×10^6

*See Table 4.4.1.
†Courtesy of Simonds Manufacturing Corporation.

For the categories shown in Table 4.4.2, wastes are identified only according to uniform incineration characteristics. Infectious wastes are present in each category of that classification system. In terms of infectious properties, the categories summarized in Table 4.4.3 are representative for hospital wastes.

TABLE 4.4.3 Hospital Waste Composition

		Generation rate based on (9 kg/bed/day) 20 lb/bed/day	
Waste type	Weight %	kg/bed/day	(lb/bed/day)
Pathological	0.5	(0.10)	0.045
Infectious	10.0	(2.00)	0.900
General/administrative (noninfectious)	50.0	(10.00)	4.500
Food	30.0	(6.00)	2.700
Cardboard	9.5	(1.90)	0.850
Total	100.0	(20.00)	9.000

Source: Ref. 3.

According to the U.S. Environmental Protection Agency (EPA) in its *Draft Manual For Infectious Waste Management*,[1] the following types of wastes are considered infectious:

- Isolation wastes
- Cultures and stocks of etiologic agents
- Blood and blood products
- Pathological wastes
- Other wastes from surgery and autopsy
- Contaminated laboratory wastes

- Sharps (hypodermic needles, etc.)
- Dialysis-unit wastes
- Animal carcasses and body parts
- Animal bedding and other wastes from animal rooms
- Discarded biologicals
- Contaminated food and other products

More recently (1986), these categories were amended into the following infectious-waste categories:

- Isolation wastes
- Cultures and stocks of infectious agents and associated biologicals
- Human blood and blood products
- Pathological wastes
- Contaminated sharps
- Contaminated animal carcasses, body parts, and bedding
- Miscellaneous contaminated wastes

This more recent list appears in the updated EPA manual, *EPA Guide for Infectious Waste Management,* which had not yet been released at the time of this writing.

In summary, infectious wastes are generated from a variety of sources, many of which may seem somewhat innocuous. Also, the generation rates of infectious (including pathological) wastes may seem less than catastrophic, E[0.95 kg/d per bed (2.1 lb/d per bed) from Table 4.4.3]; however, with respect to a 500-bed hospital, the generation rate of infectious wastes totals over 454 kg/d (1,000 lb/d). That's over 454 kg (1,000 lb) that requires treatment prior to disposal. In light of this, alternative methods of treatment, transport, and disposal deserve special consideration.

4.4.2 ALTERNATIVE METHODS OF TREATMENT, TRANSPORT, AND DISPOSAL

There are many methods available to ensure efficient handling and disinfection of infectious wastes. Table 4.4.4 outlines several series of alternatives for the collection, transport, and disposal of hospital wastes. Waste is generally collected in color-coded plastic bags (infectious waste in red bags—hence "red-bag waste"—Fig. 4.4.1) The waste is then transported for internal or external processing.

Included in Table 4.4.4 are several methods of treating infectious wastes including sterilization, incineration, irradiation, and chemical disinfection. Each hospital must evaluate the best solid-waste management plan for its individual situation. However, the treatment method most preferred by hospitals for infectious-waste destruction is incineration (see Sec. 8.16). Figure 4.4.2 traces possible treatment and disposal steps with incineration as the primary treatment method.

One of the areas of waste disposal of increasing concern at hospitals is dis-

TABLE 4.4.4 Alternative Methods of Treatment, Transport, and Disposal of Hospital Wastes

Type of waste	Storage: Internal – Red bag	Storage: Internal – White bag	Storage: Internal – Sharps container	Storage: External – Loose in container	Storage: External – Compactor	Transport: Internal – Carts	Transport: Internal – Hydraulically	Transport: Internal – Pneumatically	Transport: Internal – Chutes	Transport: External – Trucks	Volume reduction (on-site) – Incineration	Volume reduction (on-site) – Compaction	Volume reduction (on-site) – Wet pulping	Disinfection On-Site – Sterilization	Disinfection On-Site – Incineration	Disinfection On-Site – Irradiation	Disinfection On-Site – Chemical disinfection	Disinfection Off-site – Incineration	Disinfection Off-site – Sterilization	Disposal – Sanitary landfill	Disposal – Incineration	Disposal – Recycle
Pathological wastes*	X					X			X		X			X	X	X	X					
Infectious wastes	X					X	X	X	X	X	X		X	X	X	X		X	X		X	
Administrative wastes		X		X	X	X	X	X		X	X	X	X							X	X	X
Boxes		X†		X		X	X	X		X	X	X	X							X	X	X
Kitchen wastes		X‡		X		X	X			X			X									
Cans				X	X	X				X		X								X	X	X
Bottles				X	X	X				X		X								X	X	X
Syringes and medicines			X		X§									X	X	X				X	X	

* Assumes on-site handling and disposal of pathological waste
† Can be stored separately
‡ Organic to sewer, inorganic in containers
§ Kept in sharps container through to ultimate disposal

FIG. 4.4.1 Red-bag wastes from hospital laboratory.

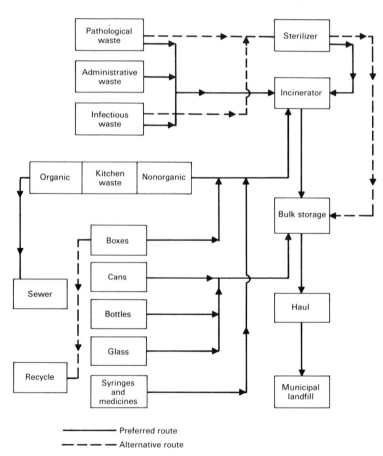

—————— Preferred route

– – – – – Alternative route

FIG. 4.4.2 Preferred method of on-site hospital solid-waste disposal.

posal of toxic and hazardous wastes. The Resource Conservation and Recovery Act (RCRA) of 1976 directs the EPA to implement a national program to regulate hazardous waste from the time it is generated to its final disposal. The RCRA defines those solid wastes which are hazardous and thus subject to regulatory controls, establishes criteria for identifying and for listing hazardous wastes, and identifies specific wastes that are excluded from the RCRA regulatory control program.

In 1984, Congress enacted amendments strengthening the Resource Conservation and Recovery Act. That legislation—the Hazardous and Solid Waste Amendments (HSWA) of 1984—made many changes in the national program that regulates hazardous waste. The most widely felt change brought about by the RCRA reauthorization, and one that affects many hospitals, is the lowering of the small-generator exemption from 1,000 kg (2,200 lb) in a calendar month to 100 kg (220 lb) in a calendar month of nonacutely toxic waste. Much of the hazardous waste managed by these smaller generators is improperly disposed of in municipal or privately owned solid-waste landfills or managed in other improper ways. When brought under the RCRA regulatory umbrella, hospitals will be required to send their waste off site to a facility authorized to receive such wastes.

Storage of hazardous wastes can be done on site, off site, or both, and is an interim waste-management technique. According to federal regulations, large-quantity generators may accumulate hazardous waste on site for 90 days or less without having to acquire a facility storage permit, provided they establish contingency plans, use proper tank and container management (including labeling), and provide personnel training according to 40 *CFR* 262.34. A generator who accumulates hazardous waste on site in tanks or containers for more than 90 days is considered an operator of a storage facility, and is subject to appropriate standards and must obtain a storage-facility permit. However, small-quantity generators may store for up to 180 days without a permit. Such on-site storage may occur without a permit being required when not more than 6,000 kg (13,216 lb) is stored for up to 270 days if the generator must ship or haul the waste over 320 km (200 mi). The sources and types of hazardous waste that may be generated at hospitals are listed in Fig. 4.4.3.

A study[3] based on a survey of some 17 hospitals indicates that solvents and chemicals are the predominant toxic or hazardous waste generated, and that these materials may be generated at a rate of some 21 mL/bed per day (0.056 lb/bed per day) (see Table 4.4.5). Based upon these data, Fig. 4.4.4 indicates that any hospital with over 125 beds may qualify as a small-quantity generator [over 100 kg/mo (220 lb/mo)] under RCRA.

4.4.3 DISPOSAL OF TOXIC AND HAZARDOUS WASTE AT HOSPITALS

Most hospitals will take a two-track program for the disposal of their hazardous wastes (see Fig. 4.4.5). Solvents and other toxic or hazardous wastes will be identified and then placed in lab packs for shipment to a treatment, storage, and disposal facility (TSDF). In some areas, several hospitals may cooperatively install a central incinerator and obtain an RCRA Part B permit. In this way, they are able to handle their hazardous wastes at this facility without going to a commercial TSDF.

Centralized incinerators are also very economically attractive for disposing of

Sources of Hazardous Wastes

Anesthesia	Nuclear Medicine
Blood bank	Nursing
Central supply services	Obstetric/gynecology
Dentistry/oral surgery	Oncology/radiation
Dialysis	Oncology
Emergency	Pathology/histology
Environmental svcs./	Pharmacy
Housekeeping/Laundry	Engineering
Food service	Print shop
Intensive care	Radiology
Clinical laboratories	Respiratory care
Materials management	Security
Morgue	Surgery

Examples of Potentially Hazardous Wastes

Acids/Caustics	Germicides
Adhesives	Heavy-metal solutions
Alcohols	Infectious waste
Ammonia	Inks/printing materials
Anesthetic gases	Insecticides
Antineoplastic drugs	Iodine
Asbestos	Mercury
Bromine	Mutagens
Carcinogens	Nitrous oxide
Chlorine	PCBs
Chromates	Pesticides
Clinical test reagents	Pharmaceutical agents
Cleaning products	Phenols
Compressed gases	Quarternary ammonium compounds
Corrosives	Radioisotopes
Photographic chemicals	Rodenticides
Disinfectants	Solvents: organic, nonchlorinated
Dyes	Solvents: organic, chlorinated
Ethylene oxide	Teratogens
Explosive gases and liquids	Toluene
Flammable gases and liquids	Water-treatment chemicals
Formaldehyde/Formalin	Xylene
Fungicides	RCRA-listed wastes

FIG. 4.4.3 Examples of potential sources and types of hazardous wastes generated by hospitals (*Hazardous Susbtance and Waste Management Research, Inc..*)

infectious wastes, and where heat recovery is involved, all hospital wastes can be disposed of efficiently in this manner.

As can be seen from this chapter, hospitals generate 4.54 to 4.08 kg/bed per day (20 lb/bed per day) of waste, 10% of which is infectious waste and must be treated prior to disposal. According to one study, hospitals also generate 21.4 mL/bed per day (0.056 lb/bed per day) of hazardous waste, which for a 200-bed

TABLE 4.4.5 Toxic-Waste Generation Rate for Hospitals

Type of material	Generation rate, mL/bed/day	Portion of waste, %
LSW*	1.24	5.80
Toluene	0.001	Negligible
Xylene	5.73	26.80
Formalin	7.48	34.95
Alcohol	6.67	31.17
Other	0.29	1.36
Total	21.4	100

*LSW = Liquid scintillation waste.
Source: Ref 3.

hospital results in (152 kg (336 lb), of hazardous waste per month. This rate [over 100 kg (220 lb) per calendar month] qualifies a hospital as a small-quantity generator of hazardous waste under federal regulations, requiring that these wastes be sent off site to an authorized treatment, storage, and disposal facility. Thus, hospitals must be aware of environmental regulations and responsible in disposing of infectious and hazardous wastes properly.

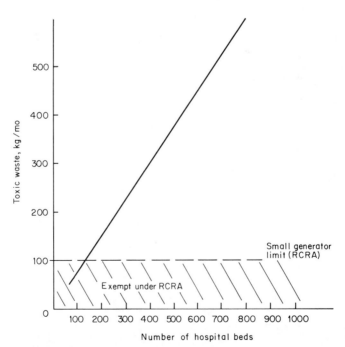

FIG. 4.4.4 Generation of toxic wastes at hospitals.

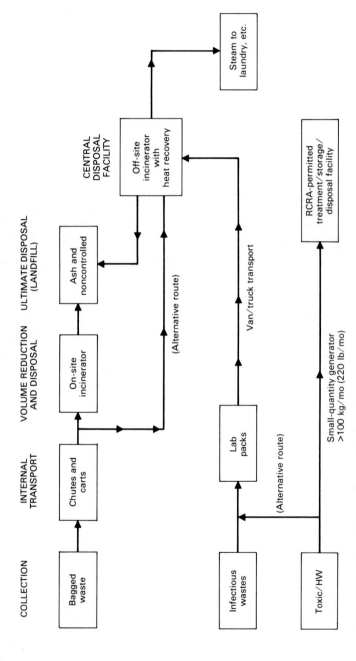

FIG. 4.4.5 Disposal of hazardous and toxic waste.

COLLECTION   INTERNAL TRANSPORT   VOLUME REDUCTION AND DISPOSAL   ULTIMATE DISPOSAL (LANDFILL)

CENTRAL DISPOSAL FACILITY

Bagged waste

Chutes and carts

On-site incinerator

Ash and noncontrolled

Off-site incinerator with heat recovery

Steam to laundry, etc.

Infectious wastes

Lab packs

Toxic/HW

RCRA-permitted treatment/storage/disposal facility

Van/truck transport

(Alternative route)

(Alternative route)

Small-quantity generator >100 kg/mo (220 lb/mo)

4.4.4 REFERENCES

1. U.S. Environmental Protection Agency, *Draft Manual for Infectious Waste Management,* U.S. EPA Office of Solid Waste, September 1982, sec. 2.5.
2. F. L. Cross and H. E. Hesketh, *Controlled Air Incineration,* Technomic Publishing, Lancaster, Pennsylvania, 1985, pp. 10 and 11.
3. *Centralized Incinerator Study for the South Florida Hospital Association,* Cross/Tessitore & Associates, P.A., Orlando, Florida, December 1985, pp. 17 and 18.

SECTION 4.5
MINING WASTES

Robert L. Hoye
Project Manager
PEI Associates, Inc.
Cincinnati, Ohio

S. Jackson Hubbard
U.S. Environmental Protection Agency
Office of Research and Development
Water Engineering Research Laboratory
Cincinnati, Ohio

Several specific industrial activities generate extremely large volumes of solid waste annually. Over the years these activities have resulted in the accumulation of billions of tons of waste on the land. These large-volume wastes are generated by the following activities:

- Extraction, beneficiation, and processing of ores and minerals
- Exploration, development, and production of crude oil, natural gas, and geothermal energy
- Disposal and utilization of wastes, including flue-gas emission-control waste, and other by-products generated from combustion of fossil fuels
- Disposal of cement-kiln dust waste

These wastes were excluded from regulation under the Resource Conservation and Recovery Act (RCRA) pending completion of studies by the U.S. Environmental Protection Agency (EPA). After completion of the study of each large-volume waste the EPA must submit a report to Congress that documents its findings and must either "determine to promulgate regulations" or "determine such regulations are unwarranted." The first report to Congress on mining waste was submitted in December 1985 .[1]

4.5.1 LEGISLATIVE AND REGULATORY HISTORY

Congress recognized the need for special consideration of the mining-waste issue when RCRA (Public Law 94-580) was first enacted in 1976. Section 8002(f) of the original act directed the EPA to conduct a "detailed and comprehensive study on the adverse effects of solid wastes from active and abandoned surface and underground mines on the environment." The study was to include an analysis of the sources and volume of mining wastes, disposal practices, potential danger, alternative disposal methods and their cost, and potential uses of the waste.

When Congress amended RCRA in the Solid Waste Disposal Act of 1980 (Public Law 96-482), the EPA was prohibited from regulating mining waste as hazardous waste until at least six months after completion of the Report to Congress required by Section 8002(f) and a new section, 8002(p). Section 8002(p) requires the EPA to conduct a comprehensive study on the disposal and utilization of mining waste in conjunction with the 8002(f) study. The amendments also required the EPA to, after holding public hearings, "determine to promulgate regulations" for mining wastes or "determine that such regulations are unwarranted." The EPA published, on November 19, 1980, an interim final rule to implement the 1980 RCRA Amendments, in which mining and mineral-processing wastes were excluded from regulation under Subtitle C (45 *Fed. Reg.* 76,618). However, EPA subsequently proposed, on October 2, 1985 (50 *Fed. Reg.* 40,292) to reinterpret the scope of this exclusion as it applies to processing wastes.[3] Under the proposed reinterpretation, only large-volume mineral-processing wastes—e.g., slag from primary metal smelters and elemental phosphorus plants, red and brown muds from bauxite refineries, and phosphogypsum from phosphoric acid plants—would be excluded with mining waste. Other mineral-processing wastes that are hazardous would be regulated under Subtitle C after promulgation of the reinterpretation. The large-volume processing wastes that remain within the exclusion would be studied and another Report to Congress prepared to complete the EPA's response to the RCRA Section 8002(p) mandate.

On July 3, 1986, the EPA published the regulatory determination for wastes from the extraction and beneficiation of ores and minerals (51 *Fed. Reg.* 24,496). The EPA found that regulation of mining wastes under Subtitle C of RCRA is not warranted; however, because of concern about actual and potential problems with these wastes, the EPA plans to develop a mining-waste program under Subtitle D. If the Subtitle D approach is not workable, it may be necessary to reexamine use of modified mining-waste standards under Subtitle C. The agency plans to complete analysis of relevant information and propose Subtitle D revisions 1988.

4.5.2 MAGNITUDE OF THE MINING INDUSTRY

The U.S. mining industry has generated up to 2 trillion kg (2 billion tons) of solid waste in a year and accounts for about 40% of the total solid waste generated annually in the United States.[2] In comparison, industrial hazardous waste for all industries other than mining amounts to about 250 billion kg (220 million tons) per year.[1] The mining industry is an important sector of the economy; the value of raw nonfuel minerals produced in the United States during 1984 was estimated to be \$23.2 billion.[3] In 1983, 389 metal mines and 5,563 nonmetal mines were in operation.

Mining activities are conducted in every state; however, most metal mines are located in the western states. Because metallic ores occur only in certain geologic formations, most mines recovering the same commodity are concentrated in a few locations. For example copper mines are located almost exclusively in Arizona, Utah, and New Mexico and most of the gold mined comes from operations in Nevada, South Dakota, and Montana.

An undetermined number of nonoperating mines (inactive and abandoned) also have an impact on the environment, ranging from unfavorable esthetics to sediment transport and acid drainage. Wastes at inactive and abandoned mines are responsible for a large portion of the total pollution resulting from mining waste. Data are inadequate to assess the total extent of their contributions to pollution problems.

4.5.3 MINING WASTES

Mining wastes include overburden, which is the barren rock that must be removed to expose an ore body in a surface mine; development waste rock, which is the barren rock removed during development of an underground mine, tailings, which result from beneficiation (i.e., froth-flotation processing) of ores; and leach residue from dump and heap leach operations. Mining and milling operations generate huge quantities of waste because (1) ore bodies are typically buried and must be uncovered and (2) the metal values recovered typically constitute only a fraction of a percent of the weight of the ore. For these reasons there are no practical means of reducing the volume of wastes generated.

The EPA has estimated that about 31 trillion kg (34 billion tons) of mine waste and 13 trillion kg (14 billion tons) of tailings were accumulated between 1910 and 1981.[4] The phosphate rock, uranium, copper, and iron ore mining segments generate over 80% of the total mine waste (overburden) and over 80% of the total tailings.[4]

4.5.4 MANAGEMENT PRACTICES

Virtually all mine wastes are disposed of on site. Because of the quantities generated, handling and hauling of waste is costly and minimized by the operator. Overburden and waste rock are usually placed in piles adjacent to the mine or backfilled into the mine. Tailings are disposed of, as a slurry, in unlined tailings impoundments. Specific management practices are site-specific and dependent upon topography, climate, hydrology, and hydrogeology. Management of mining wastes has evolved from indiscriminant dumping to a system of engineered disposal and reclamation practices regulated by a variety of state and federal agencies.

Stabilization and control technologies encompass a variety of methods for providing structural stability for tailings dams and for overburden and waste-rock piles; for preventing the evolution of excessive fugitive dust from tailings-pond slopes, inactive tailings, and overburden and waste-rock piles; for preventingboth surface and groundwater pollution by seepage and runoff from tailings ponds and overburden and waste-rock piles; and for ultimately creating a reclaimed area that is satisfactory from both the functional and esthetic standpoint.

The volume of material handled at these sites results in waste-disposal sites of extremely large proportions. Waste-rock piles cover hundreds of acres at large copper mines. Copper tailings ponds may be a square mile in surface area and each mine site may utilize several ponds. The largest single tailings pond covers over 5,000 acres. By contrast an underground lead-zinc-silver mine and flotation mill may have minimal waste rock and a tailings pond only 15 acres in size.

4.5.5 WASTE CHARACTERISTICS

In the early 1980s, the EPA sponsored waste-characterization studies involving waste sampling at 86 mines in 22 states. Mining wastes contain, with the exception of the product recovered, all constituents present in the ore. The excavation and processing of this native material can expose constituents, such as pyrites, that have acid-forming potential. Heavy metals or potentially toxic elements, such as radionuclides, formerly contained in the ore matrix may become mobile in the environment once exposed. In addition, some process reagents, particularly cyanide, may be mixed with the waste. In evaluating waste characteristics the EPA evaluated two general categories: RCRA Subtitle C hazardous waste characteristics and "other potentially hazardous characteristics." The Subtitle C characteristics of concern are corrosivity and EP toxicity. The other potentially hazardous characteristics include presence of cyanide, radioactivity, or asbestos, and acid-formation potential.

During the various mining-waste studies, 159 liquid samples were taken; only 5 were corrosive (pH2.0 or pH 12.5).[1] Some of the samples found to be corrosive were leach liquors from copper-dump leach operations. The EPA, in the regulatory determination, indicated it now considers these materials to be process solutions and not wastes. The sampling and waste characterizations also indicated that only 6% (21 of 332 samples) of the samples were EP toxic.[1] Of the 21 samples exhibiting EP toxicity, 19 were EP toxic because of lead. Industry sources often stated during these studies that the EP toxicity test, which uses acetic acid, is aggressive toward lead and that it is not indicative of actual leaching potential. Development of additional toxicity tests may be necessary to simulate the potential hazard of some of these residues.

The presence of cyanide is of concern mainly in the precious metals heap-leaching industry. An alkaline cyanide solution is sprayed over prepared ore heaps on specially constructed pads. There is essentially no data available on how much cyanide and in what forms (complexes) it remains in heap-leach residue. Cyanide is also used, in very small amounts, as a flotation reagent in the beneficiation of copper ores.

Radioactivity is associated mainly with mining waste from operations recovering uranium or phosphate. Each of the 17 uranium mines sampled had a least one waste sample with a radium-226 concentration equal to or greater than 5 pCi/g.[1] Of 13 phosphate mines sampled, 10 also had such samples. The concern with radium-226 is that it decays to gaseous radon-222 and its progeny, which can cause radiation-induced cancer.

Asbestos is of concern because of the well-documented inhalation danger that asbestos fibers pose to human health. Analysis of each of the five waste samples taken from asbestos mining and milling sites indicate that the asbestos content greatly exceeded 1% by weight; the range was 5 to 90%.[1]

The formation of acid by the oxidation of naturally occurring sulfides in mining waste is of concern because of the resulting increased potential for mobilization of toxic metals and acid drainage that can adversely affect receiving streams. Similar problems are well-documented in the coal-mining industry. Copper, gold, and silver mines are believed to have the highest acid-formation potential; however, these sites are typically located in arid areas, which lessens the potential problem.

4.5.6 FUTURE RESEARCH

The EPA's Office of Research and Development and the Office of Solid Waste are actively acquiring and evaluating information germane to the mining-waste issue. Currently research projects intended to estimate the quantities of cyanide-bearing and acid-generating wastes and to assess the potential for contaminant release are being finalized.[5] The Office of Research and Development is currently making specific assessments of the copper-dump leach and gold-silver heap-leach industries with a goal of developing conceptual alternative management practices.[6,7] Additional research and risk assessments will be conducted by the EPA to develop suitable regulatory programs for mining waste. Efforts will also begin on the other studies of large-volume wastes to fulfill the agencies' congressional mandate.

4.5.7 REFERENCES

1. U.S. Environmental Protection Agency, *Report to Congress—Wastes from the Extraction and Beneficiation of Metallic Ores, Phosphate Rock, Asbestos, Overburden from Uranium Mining, and Oil Shale,* EPA/530-SW-85-033, Office of Solid Waste and Emergency Response, December 1985.

2. *Evaluation of Management Practices for Mine Solid Waste Storage, Disposal, and Treatment,* vol. II, prepared by PEI Associates, Inc., Cincinnati, Ohio (formerly PEDCo Environmental, Inc.), for the U.S. EPA Office of Research and Development, Cincinnati, Ohio, December 1983.

3. U.S. Bureau of Mines, "Mining and Quarrying Trends in the Metal and Nonmetal Industries," preprint from the *1984 Bureau of Mines Mineral Yearbook.*

4. *Estimated Costs to the U.S. Mining Industry for Management of Hazardous Solid Wastes,* prepared by Charles Rivers Associates, for the U.S. EPA Office of Solid Waste, Washington, D.C., 1985.

5. *Quantities of Cyanide-Bearing and Acid-Generating Wastes Generated by the Mining and Beneficiating Industries and the Potentials for Contaminant Release,* prepared by Versar, Inc., for the U.S. EPA Office of Solid Waste, Washington, D.C., June 27, 1986.

6. *Gold/Silver Heap Leaching and Conceptual Alternative Management Practices,* prepared by PEI Associates, Inc., Cincinnati, Ohio, for the U.S. EPA Office of Research and Development, Cincinnati, Ohio, September, 1986.

7. *Copper Dump Leaching and Conceptual Alternative Management Practices,* prepared by PEI Associates, Inc., Cincinnati, Ohio, for the U.S. EPA Office of Research and Development, Cincinnati, Ohio, September 1986.

SECTION 4.6*

HOUSEHOLD HAZARDOUS WASTES

Gerry Dorian
Office of Solid Waste
U.S. Environmental Protection Agency
Washington, D.C.

4.6.1 DEFINITIONS

Household hazardous waste (*HHW*) has never been officially defined by the U.S. Environmental Protection Agency (EPA) in its regulations. However, a recent study on household hazardous wastes suggested the following definition:

> Household Hazardous Wastes: Solid wastes discarded from homes or similar sources as listed in 40 CFR 261.4 (b)(1) that are either hazardous wastes as listed by EPA in 40 CFR, Parts 261.33 (e) or (f), or wastes that exhibit any of the following characteristics as defined in 40 CFR Parts 261.21 through 261.24: ignitability, corrosivity, reactivity, and EP toxicity.[1]

Such a definition recognizes that there are substances and products commonly used in the home that are certainly potentially hazardous and that would be included as hazardous wastes under the Resource Conservation and Recovery Act (RCRA) if disposed of by a company or industry that generated more than the minimal amount of hazardous waste. This amount is currently 100 kg/mo.

Figure 4.6.1 presents the generic types of waste that might be considered hazardous according to the above definition.

---

*This section is extracted from the U.S. Environmental Protection Agency Office of Solid Waste and Emergency Response report, *A Survey of Household Hazardous Wastes and Related Collection Programs*, EPA/530-SW-86-038, October 1986.

I. Household cleaners
   A. Drain openers (C)
      1. Sodium hydroxide, lye, caustic soda (C)
   B. Oven cleaners (C)
      1. Sodium hydroxide, lye, caustic soda (C)
   C. Wood and metal cleaners and polishes (I)
      1. Petroleum distillates (I)
      2. Petroleum naptha (I)
      3. Turpentine (I)
      4. Isopropyl alcohol, isopropanol (I)

II. Automotive products
   A. Oil and fuel additives (I)
      1. Xylene, xylol, dimethylbenzene (I), (U239)
      2. Petroleum distillates (I)
      3. Mineral spirits (I)
      4. Methyl alcohol, methanol (I), (U154)
      5. Ethyl ether (I), (U117)
      6. Secondary butyl alcohol, secondary butanol (I)
   B. Grease and rust solvents; (I)
      1. Petroleum distillates (I)
      2. Cresylic acid, cresol (I), (U052)
   C. Carburetor and fuel-injection cleaners (I)
      1. Toluene, toluol (I), (U220)
      2. Methyl ethyl ketone, butanone (I,T), (U159)
      3. Methanol, methyl alcohol (I), (U154)
      4. Methyl chloride, chloromethane (I,T), (U045)
      5. Xylene, xylol, dimethylbenzene (I), (U239)
      6. Acetone (I), (U002)
      7. Diacetone alcohol (I)
   D. Air-conditioning refrigerants (listed)
      1. Freon 12, dichlorodifluoromethane (U075)
   E. Starter fluids (I or listed)
      1. Petroleum distillates (I)
      2. Ethyl ether (I), (U117)

III. Home maintenance and improvement products
   A. Paint thinners (I)
      1. Mineral spirits (I)
      2. Acetone (I), (U002)
      3. Petroleum distillates (I)
      4. Methanol, methyl alcohol (I), (U154)
      5. Toluene, Toluol (I), (U220)
      6. Methyl ethyl ketone, MEK, butanone (I,T), (U159)
      7. Turpentine (I)
      8. Isopropyl alcohol, isopropanol (I)
      9. Methyl isobutyl ketone, isopropylacetone (I), (U161)
   B. Paint strippers and removers (I)
      1. Acetone (I), (U002)
      2. Toluene, toluol (I), (U220)
      3. Petroleum distillates (I)
      4. Methanol, methyl alcohol (I)

FIG. 4.6.1 Household hazardous wastes and their hazardous components. RCRA hazardous waste categories: C = corrosive, I = ignitable, T = toxic, U,P = listed wastes.

C. Adhesives (I)
 1. Methyl ethyl ketone, butanone (I), (U159)
 2. Petroleum distillates (I)
 3. Acetone (I) (U002)
 4. Butyl acetate (I)
 5. Mineral spirits (I)
 6. Xylene, xylol; dimethylbenzene (I), (U239)
 7. Petroleum naptha (I)
 8. Tetrahydrofuran (I), (U213)
 9. Isobutylacetate (I)
 10. Toluene, toluol (I), (U220)
 11. Acrylic acid (I), (U008)
 12. Hexane, *n*-hexane (I)
 13. Allyl isothiocyanate, allyl isosulfocyanate, mustard oil (I)
 14. Cyclohexane (I) (U056)
 15. Formaldehyde (I), (U122)
 16. Ethylene dichloride, 1,2-dichloroethane (I), (U077)
 17. Ethylidene dichloride, 1,1,-dichloroethane (I), (U076)
IV. Lawn and garden products
 A. Herbicides
 1. 2,4-D, (D016)
 2. Silvex (U233)
 3. 2,4,5-T (U232)
 B. Pesticides
 1. Acrolein (P003)
 2. Aldicarb (P007)
 3. Aldrin (P004)
 4. Arsenic Acid (P010)
 5. Aziridine; Ethyleneimine (P054)
 6. Chlordane (U036)
 7. Creosote (U051)
 8. 2,4-D, (D016)
 9. DDD (U060)
 10. DDT (U061)
 11. Dieldrin (P037)
 12. Dimethoate (P044)
 13. Dinoseb (P020)
 14. Disulfoton (P039)
 15. Endosulfan (P050)
 16. Endrin (P051)
 17. Heptachlor (P059)
 18. Lindane (U129)
 19. Methoxychlor (U247)
 20. Methyl Parathion (P071)
 21. Parathion (P089)
 22. Pentachlorophenol (U242)
 23. Phorate (P094)
 24. Silvex (U233)
 25. 2,4,5-T (U232)
 26. Toxaphene (P123)
 27. Trichlorophenol (U230, U231)
 28. Warfarin (P001)

FIG. 4.6.1 (*Continued*)

4.6.2 PRESENCE IN THE MUNICIPAL WASTE STREAM

Although there have been many surveys of the materials that constitute the household solid waste stream, there have been very few studies designed to identify the percentage contribution of HHW. Three studies on the subject are outlined here.

Los Angeles County Sanitation Districts

The Los Angeles County Sanitation Districts conducted a limited characterization study in 1979.[2] The purpose of the study was to estimate the types and quantities of wastes, including both hazardous and nonhazardous materials, received at landfills and one transfer station in southern California. The first step in the effort was to identify these wastes. Manual sorting and characterization of 155 tons of refuse were included in the program. This program involved both a count of the number of containers associated with HHW and measurement of the quantity of their contents. More than 90% of the containers were considered empty.

The presence of hazardous wastes from all sources was found to be quite low. The Districts estimated that notably less than 1% of all refuse received at these facilities was hazardous. The majority of these wastes came from commercial and industrial sources. It was estimated that less than 20% of all hazardous waste entering the sites came from residential sources.

In a more recent effort, the Los Angeles County Sanitation Districts have been systematically sorting loads of refuse to identify hazardous waste. Since California has a zero small-quantity generator exemption, all commercial sources of hazardous waste are legally required to dispose of those wastes in permitted hazardous-waste facilities. Thus, the thrust of the program is to identify commercial violators. In 1984 and 1985, five to six refuse trucks per day were inspected for hazardous wastes. The inspections were conducted at four Los Angeles County landfills and one transfer station. The inspectors were primarily interested in identifying commercial-size quantities of hazardous wastes and, therefore, typically looked for containers larger than 1 gal in size or for entire boxes full of hazardous wastes. Nearly 3,500 loads of refuse representing some 15,000 tons of waste were inspected in 1984. Results indicated total hazardous waste to represent only 0.00147% of the total amount of waste delivered to these facilities.[3] Expressed in another way, total hazardous waste was about 15 ppm in the entire mixed municipal waste stream including commercial as well as residential sources.

University of Arizona

The Department of Anthropology at the University of Arizona in Tucson has conducted detailed residential-refuse characterization from 1973 through the present. Recently, efforts were focused on identifying HHW. The primary thrust of the characterization studies has been to count the number of hazardous-waste containers. The University of Arizona study indicates that about 100 hazardous items (containers) are discarded per household each year. Thus, approximately 120,000 households in Tucson generate approximately 11 million of these hazardous items annually.[4]

City of Albuquerque, New Mexico

Based on a 1983 survey carried out by the City of Albuquerque's Environmental Health and Energy Department, the department concluded that the city's 96,300 dwelling units yielded an estimated 800 tons of HHW annually. This represents an estimated 0.5% of the residential waste streams.[5]

4.6.3 HUMAN AND ENVIRONMENTAL IMPACTS

Impact on the Public

The presence of HHW in the residential waste stream certainly has some impact on refuse collection and disposal personnel. In the vast majority of instances, disposal of hazardous waste from homes probably does not harm anyone directly. However, there have been instances across the country in which people have been injured by HHW.

Homeowners are probably the group most often affected. When hazardous household products are misused or improperly stored or disposed, residents are the ones most likely to be exposed to the dangers, and they will be subject to the longest-term exposure. Children and pets are also often present and they may be particularly susceptible to the hazards posed by some household products.

The nature of some household hazardous wastes also increases their potential for damage to homeowners over a period of time. Materials that are used infrequently are often stored in closets, basements, or garages for long periods of time. Materials such as paint thinners, solvents, fertilizers, and others may react with their containers over the years, causing the containers to deteriorate. This further increases the potential danger to homeowners.[1]

Impact on Refuse-Collection Workers

Refuse collectors have been injured by HHW. The number and severity of accidents, injuries, or health impacts associated strictly with HHW is difficult to ascertain. The information is often vague and is not centrally located.[6] In most instances, the relationship of injuries to HHW has to be inferred. Most of the injuries occurred while the waste was being emptied or compacted. Some of the incidents that have affected refuse collectors include:

- Serious injury to one refuse collector has been directly related to HHW. A refuse worker in San Diego, California, lost his sight when hazardous waste from a residence spilled on his face. The hauling firm notified the residents of the injury and identified wastes that should not be discarded in residential waste. The San Diego Environmental Health Department was barraged with phone inquiries on how to dispose of HHW. A task force was formed that subsequently developed the city's current program.[7]

- A private firm in Lemon Grove, California, reported a number of incidents in which swimming pool chemicals splashed on collection personnel during the compaction of residential refuse. One worker lost 50% of the use of his left eye.[6]

- Used motor oil caused severe eye irritation to three disposal workers in Lemon Grove, California.[6]
- Severe eye irritation in one incident was caused by contact with paint thinner, also in Lemon Grove, California.[6]
- Some 42 incidents have been reported in Los Angeles County, California, related to HHW. Injuries to refuse collectors have been caused primarily by oil, battery acids, swimming pool chemicals, paints, and solvents.[8]
- A caustic material in residential refuse caused severe skin irritation to a refuse collector in Roscoe, Illinois.[6]

Hauling and Disposal

Household hazardous wastes can also have negative impacts during hauling and disposal of refuse. A number of these wastes, particularly automotive products, are ignitable. Therefore, it is certainly possible that some of them have caused fires in refuse-collection trucks. Fires in packer trucks are not unusual. Personnel from the City of Tempe, Arizona, report an average of five such fires in the city's residential packers each year.[9] They believe there are only two likely causes: hot ashes from barbecues or fireplaces, or chemicals (HHW). Investigations have not identified the numbers of fires attributable to each type of waste. However, if only one packer fire per year in this city of 170,000 is caused by HHW and this number is extrapolated to the entire country, household hazardous wastes are a significant source of property loss and potential injury.

Personnel at a transfer station, landfill, or other disposal site may also be injured by HHW. The activities of unloading, spreading, and compacting refuse often cause containers to rupture and contents to be sprayed into the air. Injuries and accidents at the disposal site have been reported for both haulers and landfill personnel. Again, little tabulated information is available. However, reports from disposal-site supervisory personnel indicated that incidents related to hazardous wastes occur, but rather infrequently. This relatively low level of reported incidents may be due to the perceived minor nature of these events. Only accidents that require special care such as the use of a respirator or something more than simple first aid are likely to receive recognition as being related to HHW. Furthermore, a landfill is a relatively accident-prone area and numerous cuts and burns occur that are not related to hazardous waste.

Environmental Impacts

Does household hazardous waste have any impact on the environment? While there have been no reported incidents of HHW being solely responsible for environmental deterioration, there are reasons to believe that there is an impact. Indirect evidence indicates that HHW may contribute to groundwater contamination. More than 12 former municipal waste-disposal sites are on the National Priorities (Superfund) List of sites with contamination requiring near-term corrective action. In the past, these sites received HHW and commercial and industrial wastes (probably including hazardous wastes) in addition to household refuse. The present groundwater contamination indicates that industrial (or HHW) chemicals are the source of the problem. However no direct proof exists that HHW are the only pollutants. This same situation is probably present at many active municipal solid-waste landfills.[1]

4.6.4 COLLECTION PROGRAMS

Collection programs for household hazardous waste are operated at both the local and state levels. First instituted only three or four years ago, these activities have grown rapidly in popularity.

Nearly 200 collection programs may be operated in 1986. Collection programs have many positive attributes. Education of the public and increased awareness of the presence of hazardous materials in the home are major benefits of these programs. Removal of these wastes from long-term, improper storage certainly reduces the potential for injury or accident. The disposal of these wastes in permitted hazardous-waste facilities reduces potential impact on the environment.

Participation rates in HHW collection programs have been low. Few programs can boast participation of even 1% of households in the community, and several programs report participation less than 0.2%. Quantities collected typically range from 20 to 40 lb per contributing household. This may represent several years accumulation of wastes. Thus future collection efforts in the same area may result in even lower quantities per participant household, assuming the same households participate. Unit costs for these collection programs are extremely high. Cost data are scarce and most program sponsors do not include factors for fringe benefits, overhead, and other indirect costs or donated materials or services. Relatively complete cost estimates (although based on several estimates) indicate that a well-publicized program with high participation costs more than $2 per pound of HHW collected. Programs with limited publicity and low participation may cost more than $9 per pound ($18,000 per ton).[1]

4.6.5 CONCLUSIONS

The actual degree of hazard represented by these materials, and the quantity of the materials going into the municipal landfills, are both quite uncertain. As these hazardous materials continue to accumulate in landfills the issue of a safety threshold becomes an increasingly vital public concern. A better estimate of the potential environmental damage awaits a more thorough determination of the presence of household hazardous waste as well as the types and characteristics of these wastes.

4.6.6 REFERENCES

1. *A Survey of Household Hazardous Wastes and Related Collection Programs,* EPA/530-SW-86-038, U.S. EPA Office of Solid Waste and Emergency Response, Washington, D.C., 1986.
2. Los Angeles County Sanitation Districts, Solid Waste Management Department, *Hand Sorting Fact Sheet,* Whittier, California, 1979.
3. Los Angeles County Sanitation Districts, Solid Waste Management Department, *Unannounced Search Summary 1984,* Whittier, California, 1984.
4. Department of Anthropology, University of Arizona, *Preliminary Results from Household Phase Research,* Tucson, Arizona, 1985.
5. Environmental Health and Energy Department, City of Albuquerque, *Residential Hazardous/Toxic Waste Survey,* Albuquerque, New Mexico, 1983.

6. L. J. Russel and N. Knappenberger, *Evidence of Inappropriate Disposal of Hazardous Waste From Small Quantity Generators,* Technical Memo No. 11, Association of Bay Area Governments, April 1984, p. 11.

7. Personal communication, Diane Takvorian, San Diego Department of Environmental Health, San Diego, California, November 1, 1985.

8. Golden Empire Health Planning Center, *Household Hazardous Waste: Solving the Disposal Dilemma,* Sacramento, California, 1984.

9. Personal communication, Ronald L. Ottwell, Construction and Operations Superintendent, Public Works Department, City of Tempe, Arizona, October 29, 1985.

CHAPTER 5

WASTE MINIMIZATION AND RECYCLING

SECTION 5.1
MINIMIZATION OF HAZARDOUS-WASTE GENERATION

Gary E. Hunt
Head, Technical Assistance
Pollution Prevention Pays Program
North Carolina Division of Environmental Management
Raleigh, North Carolina

Roger N. Schecter
Director
Pollution Prevention Pays Program
North Carolina Division of Environmental Management
Raleigh, North Carolina

5.1.1 INTRODUCTION TO WASTE MINIMIZATION

Liquid, solid, and/or gaseous waste materials are generated during the manufacture of any product. In addition to environmental problems, these wastes represent losses of valuable materials and energy from the production process and a significant investment in pollution control. Traditionally, pollution control has been dominated by the "end-of-the-pipe" and "out-the-back-door" viewpoints. The control of pollution in this way requires labor hours, energy, materials, and capital expenditures. Such an approach—e.g., wastewater treatment or air-pollution abatement—removes pollution from one source, but places pollutants somewhere else, e.g., a landfill.

Added regulations, higher "treatment" expenses, and increased liability costs have caused industrial and governmental leaders to begin critical examinations of end-of-pipe pollution-control measures. The value of waste minimization, reduction, and recovery has become apparent to industries taking the opportunity to look at broader environmental-management objectives rather than concentrating solely on pollution control. As will be documented throughout this chapter, waste minimization is very often economically beneficial for an industry and also results in improved environmental quality. Some of these economic advantages are highlighted in Fig. 5.1.1.

Current and future environmental regulations also provide a strong incentive to minimize waste. With more stringent control of hazardous wastes and effluent limits on discharges to air, surface waters, and municipal treatment plants, a greater emphasis on waste minimization will be required in order to develop cost-effective waste-management programs.

The Hazardous and Solid Waste Amendments of 1984 initiated new policy and regulatory responses for waste reduction, stating that "Congress declares it to be the national policy of the United States that, wherever feasible, the generation of hazardous waste is to be reduced or eliminated as expeditiously as possible." Regulations as a result of the 1984 Amendments now require a certification on waste manifest forms for off-site shipments of hazardous wastes. Waste generators must certify that they have minimized waste amounts and toxicity. In addition to manifest certification, a generator must submit a biennial report, describing efforts to minimize waste generation. Finally, permit holders for treatment, storage, and disposal facilities must certify annually that they have in place a minimization program that reduces the volume and toxicity of hazardous wastes.[1]

Waste minimization should be a key component of any comprehensive waste-management program for economic, environmental, and regulatory reasons. Such a program should take a multimedia approach so that waste material is actually reduced and not just transferred from one medium to another. Waste-reduction techniques do not have to be based on high technology or require large capital expenditures. Many techniques are just simple changes in the way materials are handled in the production process. The techniques which can be used in a management program to minimize waste generation are discussed in detail in Subsec. 5.1.2. An approach for developing and implementing a multimedia waste-minimization program using these techniques is discussed in Subsec. 5.1.3.

- Reduced on-site waste-treatment costs: capital and operational
- Reduced transportation and disposal costs for off-site wastes
- Reduced compliance costs for permits, monitoring, and enforcement
- Lower risk of spills, accidents, and emergencies
- Lower long-term environmental liability and insurance costs
- Reduced production costs through better management and efficiency
- Income derived through sale or re-use of waste
- Reduced effluent costs and assessments from local wastewater plants

---

FIG. 5.1.1 Economic incentives for waste minimization.

5.1.2 WASTE-MINIMIZATION TECHNIQUES

Waste minimization techniques can be broken down into four general classifications: inventory management, production-process modification, volume reduction, and recovery and re-use. Because the classifications are broad, there will be some overlap between the techniques discussed under each heading. In practice, waste-minimization techniques are usually used in combination so as to achieve maximum waste reduction. Each of the categories will now be discussed in detail.

Inventory Management

Proper control over the materials used in the manufacturing process is an important way to reduce waste generation. By reducing both the quantity of hazardous materials used in the process and the amount of excess raw materials in stock, the quantity of waste generated can be reduced. This can be accomplished by establishing material-purchase review and control procedures.

Developing review procedures for all material purchased is the first step in establishing an inventory-management program. Procedures should require that all material be approved prior to purchase. In the approval process all production materials are evaluated to see first if they contain hazardous constituents and then if alternative nonhazardous substitute materials are available (see under *Production-Process Modification,* on p. 5.6). This determination can be made either by one person with the necessary chemistry background or by a committee made up of people with a variety of backgrounds. Information necessary to make this determination can be obtained from the Material Safety Data Sheets provided by the chemical supplier. If these data sheets do not contain enough information, the supplier must be contacted and asked to provide the necessary information in confidence. If the chemical supplier is not able to or will not provide this information, seek a new supplier who will. Newly drawn procedures should stipulate that any material which has been approved can be ordered, while new material must first go through the approval process.

Another inventory-management procedure for waste reduction is to ensure that only the needed quantity of a material is ordered. This will require the establishment of a strict inventory-tracking system. In many instances, excess or out-of-date material is simply disposed of as hazardous waste. The disposal cost will greatly exceed the original purchase cost, often by orders of magnitude. Thus, purchase procedures must be implemented which ensure that materials are ordered only on an as-needed basis and that only the amount needed for a specific period of time is ordered.

Review procedures should also be applied during new-product development. Before a new product is produced, a thorough evaluation of the materials and processes used to make the product should be made. The use of hazardous materials should be reduced as much as possible prior to beginning production. Additionally, the proposed new production process should be evaluated for waste-reduction potential (see under *Production-Process Modification,* on p. 5.6).

Inventory control is a waste-minimization technique that is rapidly growing in popularity. It is applicable to all types of industries no matter what their size.[10,2] The great advantage of this approach is that it is not expensive or difficult to implement.

Production-Process Modification

Changes can be made in the production process which will reduce waste generation. This reduction can be accomplished by changing the materials used to make the product or by the more efficient use of input materials in the production process, or both. The potential waste-minimization techniques can be broken down into three general categories: improved operating and maintenance procedures, material change, and process-equipment modification. Each of these areas are discussed in more detail here.

Operational and Maintenance Procedures. Improvements in the operation and maintenance of process equipment can result in significant waste reduction. However, in many cases operation and maintenance is given a low priority, especially when it is related to waste management. A wide range of methods can be used to operate a production process at peak efficiency. These methods are not new or unknown and are usually very inexpensive to institute as they entail no capital costs. However, a thorough knowledge of the production process is essential.

Operational Procedures. Operating the production-process equipment more efficiently can reduce waste generation. Instituting standard operational procedures can optimize the use of raw materials in the production process and reduce the potential for materials to be lost through leaks and spills. This essentially is just fine-tuning the production process.

The first step in instituting such a program is to review all current operational procedures, or lack of procedures, and examine the production process for ways to improve its efficiency. A review would include segments of the process, from the delivery area through the production process to final product storage. Procedures for conducting such a review are presented under *Evaluation and Selection of Waste-Minimization Techniques,* on p. 5.21.

In many cases, simple operational changes can significantly reduce waste generation. For example, a manufacturer of plumbing fixtures changed the concentration of chrome in the electroplating baths to the low end of the recommended operating range. By reducing the chrome concentration from 3,700 mg/L (32 oz/gal) to 3,350 mg/L (29 oz/gal), the amount of chrome which had to be treated was reduced by 9% without affecting product quality. This not only saved raw material and treatment chemicals, but it reduced the amount of wastewater-treatment sludge generated.[10]

One important area commonly overlooked or not given proper attention in many manufacturing facilities is material-handling procedures. This includes storage of raw materials, products, and process waste, and the transfer of these items within the process and around the facility. Proper material handling will ensure that the raw material will reach the production process without loss through spills, leaks, or contamination. It also will ensure that the material is efficiently handled in the production process. Loss of raw and finished material due to improper material handling will increase both production and waste-disposal costs. Some potential sources of material loss due to improper handling procedures are shown in Table 5.1.1.

A resin manufacturer provides an example of how proper raw-material transfer procedures can reduce waste generation. It was common practice to let the residual material in the hose from the phenol delivery trucks drain into the plant's wastewater-collection system after the storage tanks were filled. The unloading procedures were changed to require that the hoses be flushed with water and the phenol–water mixture be stored in a tank for use in the production process.[19]

TABLE 5.1.1   Potential Causes of Process-Material Loss

Area	Cause or point of loss
Loading	Leaking fill-hose connections
	Draining of fill hose between filling
	Punctured, leaking, or rusting containers
	Leaking valves and piping
Storage	Overfilling of tanks
	Improper or malfunctioning overflow alarms
	Punctured, leaking, or rusted containers
	Leaking transfer pumps, valves, and pipes
	Inadequate diking or open drain valve
	Improper material-transfer procedures
	Lack of regular inspection
	Lack of training program
Process	Leaking process tanks
	Improperly operating process equipment
	Leaking valves, pipes, and pumps
	Overflow of process tanks; improper overflow controls
	Leaks or spills during material transfer
	Inadequate diking
	Open drains
	Equipment and tank cleaning

Once proper operating procedures are established they must be fully documented and become part of the employee training program. Easy to follow guidelines addressing the proper operational waste-management and emergency procedures should be set for each process step. By establishing and enforcing strict operational procedures, waste generation can be reduced.

Maintenance Program.   A strict maintenance program which stresses corrective *and* preventive maintenance can reduce waste generation caused by equipment failure. Such a program will help personnel spot potential sources of release and correct the problem before any material is lost. It should include all potential release sources identified during the operational review process. To be effective, a maintenance program should be developed and followed for each operational step in the production process, with special attention to potential leak points. A strict schedule and accurate records on all maintenance activities should be maintained. The type of information which must be collected and updated regularly in order to establish a preventive maintenance program is listed below.[17]

- A list of all plant equipment and location
- A knowledge of operating time for each item or area
- A knowledge of which items are critical to the process(es)
- A knowledge of problem equipment
- Vendor maintenance manuals
- A data base of equipment repair histories

Training.   An employee training program is a key element of any waste-reduction program. All personnel should be included who are responsible for operating and maintaining process equipment, for handling process materials, and for handling waste materials. Training should include correct operating and handling procedures, proper equipment use, recommended maintenance and inspection schedules, correct process-control specifications, and proper management of waste materials.[17] Written materials should be prepared and used in conjunction with hands-on training. This should be an ongoing process with review updates and interaction between employees and supervisors on a regular basis. One furniture manufacturer, for example, implemented a successful training program for spray-coating operators which reduced raw-material usage and waste generation while producing a better-quality product. Working closely with the operators, training personnel used video equipment to record their spray technique. The tape was later reviewed by the operator and training personnel to identify problem areas and provide corrective procedures. The operator was videotaped again later so that he or she could see the improvements. The company estimates that coating use and the associated waste-generation costs were reduced by about 10%. This represented a savings of about $60,000 per year in coating-material costs alone.[12] Savings were also realized through reduced waste from spray-booth clean-out and reduced air emissions.

Material Change.   Hazardous materials used in either a product formulation or a production process may be replaced with a less hazardous or nonhazardous material. Reformulating a product to contain less hazardous material will reduce the amount of hazardous waste generated both during the product's formulation and during its end use. Using a less hazardous material in a production process will reduce the amount of hazardous waste produced. Some examples of material change to reduce waste generation are given in Table 5.1.2.

Product reformulation is one of the more difficult waste-reduction techniques, but it is a very effective one. As more manufacturers implement inventory-management programs, pressure will increase on chemical supply companies to produce products with lower quantities of hazardous materials. Because of the proprietary nature of product formulations, specific examples of product reformulation are scarce. General examples include eliminating pigments containing heavy metals from ink and paint formulations; replacing chlorinated solvents with nonchlorinated solvents in cleaning products; replacing phenolic biocides with other, less toxic compounds in metal-working fluids; and developing new paint, ink, and adhesive formulations based on water rather than organic solvents. The applicability of this technique will be very product-specific, but with the ever increasing cost and liability associated with waste management, it is a very important waste-reduction and business strategy.

Hazardous chemicals used in the production process can also be replaced with less hazardous materials. This is a very widely used waste-reduction technique and is applicable to most manufacturing processes, as shown in Table 5.1.2. Implementation of this waste-reduction technique may require only some minor process adjustments or it may require extensive new process equipment. An electroplater replacing a cyanide plating-bath solution with a noncyanide one will have to make only minor changes in operational procedures. A circuit-board manufacturer switching to water-based flux from a solvent-based product will have to replace the solvent-vapor degreaser with a detergent parts washer. This may require only a limited capital investment. For example, one firm just used a household dishwasher to replace the solvent-vapor degreaser.[41] More extensive production changes would have to be made if a manufacturer switched from an organic-solvent coating process to an electrostatic powder-coating system. This would require extensive capital investment in new process equipment, facility modifications, and employee training, but can result in substantial cost savings.

TABLE 5.1.2 Examples of Waste Reduction through Material Change

Industry	Technique (reference)
Household appliances	Replace vapor degreaser using chlorinated solvent with an alkaline degreaser (10)
Printing	Substitute water-based ink for solvent-based ink (10)
Textile	Reduce phosphorus in wastewater by reducing use of phosphate-containing chemicals (4)
Office furniture	Use water-based paints in place of solvent-based paints (8)
Air conditioners	Replace solvent-containing adhesives with water-based products (9)
Printed circuit boards	Use water-based developing system instead of a solvent-based system (2)
Aerospace	Replace cyanide cadmium-plating bath with a noncyanide bath (13)
Ink manufacture	Remove cadmium pigments from product (2)
Plumbing fixtures	Replace hexavalent chrome-plating bath with a low-concentration trivalent chrome-plating bath (15)
Electronic components	Use ozone instead of organic biocides in cooling towers (4)
Pharmaceuticals	Replace solvent-based tablet-coating process with a water-based process (25)

Reducing or eliminating use of hazardous materials in the production process will decrease not only hazardous-waste generation but also the quantity of hazardous materials in air emissions and wastewater effluents. This will, in turn, reduce the capital investment in treatment systems needed to meet pollution discharge limits. For example, a producer of gift-wrapping paper switched from solvent to water-based inks. This saved the company $35,000 per year on hazardous-waste management costs. Furthermore, it saved the company from having to install an air-pollution control system, costing several million dollars, to control volatile organic carbon emissions.[9]

Process-Equipment Modifications. Waste generation can be significantly reduced by installing more efficient process equipment or modifying existing equipment to take advantage of better production techniques. New or updated equipment can use process materials more efficiently, producing less waste. Additionally, such efficiency reduces the number of rejected or off-specification products, thereby reducing the amount of material which has to be reworked or disposed of. The use of more efficient equipment or processes can pay for itself through higher productivity, reduced raw-material costs, and reduced waste-management costs. The necessary capital investment can usually be justified by the increased production rates, and reduced waste-management costs provide an added bonus.[2,12]

Modifying existing process equipment can be a very cost-effective method of reducing waste generation. In many cases the modifications can just be relatively simple changes in the way the materials are handled within the process to ensure that they are not wasted. This can be as simple as using an inexpensive flow-control valve on a water line to a rinse tank to reduce water consumption or re-designing parts racks to reduce drag-out in electroplating operations.[16,15]

Modifying production equipment to improve operational efficiency requires a thorough understanding of both the production process and waste-stream generation. In many cases it is best to phase in the required process modifications over a period of time to reduce any potential effects on the whole production process. Some examples of modifications to existing process equipment which can reduce waste generation are shown in Table 5.1.3.

One important factor which is sometimes overlooked in evaluating the cost effectiveness of installing new production equipment or modifying existing equipment is the cost associated with reworking or disposing of off-specification materials. This can be very expensive not only in terms of labor and materials but in

TABLE 5.1.3   Examples of Production-Process Modifications for Waste Reduction

Process step	Technique
Chemical reaction	Optimize reaction variables and reactor design
	Optimize reactant addition method
	Eliminate use of toxic catalysts
Filtration and washing	Eliminate or reduce use of filter aids and disposable filters
	Drain filter before opening
	Use countercurrent washing
	Recycle spent washwater
	Maximize sludge dewatering
	Filter only when necessary
Parts cleaning	Cover all solvent cleaning units
	Use refrigerated freeboard on vapor degreaser units
	Improve parts draining before and after cleaning
Surface finishing	Prolong process-bath life by removing contaminants
	Redesign part racks to reduce drag-out
	Re-use rinsewater
	Install spray- or fog-nozzle rinse systems
	Properly design and operate all rinse tanks
	Install drag-out recovery tanks
	Install rinsewater flow-control valves
Surface coating	Use airless air-assisted spray guns
	Use electrostatic spray-coating system
	Control coating viscosity with heat units
	Use high-solids coatings
Equipment cleaning	Use high-pressure rinse system
	Use mechanical wipers
	Use countercurrent rinse sequence
	Re-use spent rinsewater
Spill or leak control	Use bellows-sealed valves
	Install spill basins or dikes
	Use sealless pumps
	Maximize use of welded pipe joints
	Install splash guards and drip boards
	Install overflow-control devices

Source: Refs. 3, 14, 17, 28, 39.

waste-management costs. For example, in many manufacturing operations which involve coating a product, such as electroplating or painting, chemicals are used to strip off the coating from rejected products so that they can be recoated. These chemicals, which can include acids, caustics, cyanides, and chlorinated solvents, are often a hazardous waste and must be properly managed. By reducing the number of parts which have to be reworked, the quantity of waste can be significantly reduced.

In many cases the use of new, more efficient process equipment also means changing to less hazardous process materials, as discussed earlier in this section. For example, one furniture company switched from a manual staining operation using solvent-based materials to an automatic staining system using water-based stains. The new process reduced the time to stain a pallet of parts by 95% and reduced raw-material usage by 20%. The savings in labor alone paid for the new staining system in just 3 months.[12]

Examples of new, more efficient process equipment are numerous in the literature but little is usually said about reduced waste-generation and management costs. The following example shows what one company saved when new process equipment was installed:

When an automated metal electroplating system was installed to replace the manual operation, annual productivity increased and system down-time decreased from 8% to 4%. Chemical consumption decreased 25%, resulting in an annual reduction of $8,000/year in raw material costs. Water costs have been reduced by $1,100 per year, and plating wastes, including acids, caustic, and oils, have decreased from 204 kg/day to 163 kg/day (450 to 360 pounds/day). Treatment costs for the process water used in the plating operation have been reduced 25%. Annual personnel and maintenance cost savings attributable to the new system are $35,000/year. The automated system has also eliminated worker exposure to acids and caustics, which was required with the previous manual operation.[10]

Volume Reduction

Volume reduction includes those techniques that remove the hazardous portion of a waste from the nonhazardous portion. These techniques are usually used to reduce the volume, and thus the cost of disposing of, a waste material. However, once the material has been concentrated, there is a greater likelihood that the materials in the waste can be recovered. A good example of this is recovery of metals from dewatered sludge from treatment of electroplating wastewater. Some examples of waste minimization through volume reduction are shown in Table 5.1.4.

The techniques which can be used to reduce waste-stream volume can be divided into two general categories: source segregation and waste concentration. Each of these is discussed in more detail below.

Source Segregation. Segregation of wastes is in many cases a simple and economical technique for waste reduction. By segregating wastes at the source of generation and handling the hazardous waste separately from the nonhazardous portion, the volume and disposal cost of the waste will be reduced. Additionally, the uncontaminated or undiluted waste may be re-usable in the production process or it may be sent off site for recovery.

This technique is applicable to a wide variety of waste streams and industries as shown in Table 5.1.4. For example, in metal-finishing facilities, wastes con-

TABLE 5.1.4 Examples of Waste Reduction through Volume Reduction

Industry	Technique (reference)
Electronic components	Use compaction equipment to reduce volume of waste cathode-ray tubes (4)
Plumbing fixtures	Use sludge drier to further reduce volume of dewatered wastewater-treatment sludge (10)
Printed circuit boards	Use filter press to dewater sludge to 60% solids and sell sludge for metal recovery (10)
Pesticide formulation	Use separate baghouses at each process line and recycle collected dust into product (10)
Research laboratory	Segregate chlorinated and nonchlorinated solvents to allow off-site recovery (10)
Printing	Segregate waste wash solvents and re-use in ink formulations (10)
Paint formulation	Segregate tank-cleaning solvents and re-use in paint formulations (2)
Furniture	Segregate solvents used to flush spray-coating lines and pumps and re-use as coating thinner (2)

taining different types of metals can be treated separately so that the metal values in the sludge can be recovered. Keeping spent solvents or waste oils segregated from other solid or liquid waste may allow them to be recycled. Wastewater containing toxic material should be kept separate from uncontaminated process water, reducing the volume of water which must be treated.

A commonly used waste-segregation technique is to collect and store washwater or solvents used to clean process equipment (such as tanks, pipes, pumps, or printing presses) for re-use in the production process. This technique is used by paint, ink, and chemical formulators as well as by printers and metal fabricators. For example, a printing firm segregates and collects toluene used for press and roller cleanup operations. By segregating the used toluene by color and type of ink contaminant, it can be re-used later for thinning the same type and color of ink. The firm now recovers 100% of the waste toluene, totally eliminating a hazardous waste stream.[10]

Another example is collection, and re-use in the product, of dust and excess materials generated during the manufacturing process. This technique is used by one pesticide formulator to reduce the volume of hazardous waste it generates. The firm collects the dust generated during the formulation process in a separate baghouse for each process line. The collected dust is then re-used in the product as an inert filler. This has eliminated the generation of 20,412 kg/y (45,000 lb/yr) of waste, saving the company $9,000 in disposal costs and $2,000 in raw-material costs.[10] These savings paid for the necessary equipment in less than a year.

Concentration.   Various techniques are available to reduce the volume of a waste through physical treatment. Such techniques usually remove a nonhazardous portion of a waste, such as water. For example, concentration techniques are commonly used to dewater wastewater-treatment sludges and reduce the volume by as much as 90%. Available concentration methods include gravity and vacuum filtration, ultrafiltration, reverse osmosis, freeze vaporization, filter press, heat drying, and compaction. Concentration techniques are available for all types of waste streams and, as shown in Table 5.1.5, are used by a wide range of industries.

Concentration of a waste stream may also increase the likelihood that the material can be re-used or recycled. In some cases, filter presses or sludge driers can increase the concentration of metals in electroplating wastewater-treatment sludge to such a level that they become attractive purchases for metal smelters. A printed-circuit-board manufacturer dewaters its sludge using a filter press to 60% solids. The company receives $7,200 per year through the sale of the dewatered sludge for copper reclamation.[10]

Recovery and Re-use

Recovering and re-using wastes may provide a very cost-effective waste-management alternative. This technique could eliminate waste-disposal costs, reduce raw-material costs, and provide income from a salable waste. The effective use of this option will depend on the segregation of the recoverable waste from other process wastes. This will ensure that the waste is uncontaminated and the concentration of recoverable material is maximized. Waste can be recovered on site, at an off-site recovery facility, or through interindustry exchange. The selection will depend on the type of waste, production-process input materials, and economics as discussed below.

On-Site Recovery and Re-use.   In most cases the best place to recycle process wastes is within the production facility. Wastes which are simply contaminated versions of the process raw materials are good candidates for in-plant recycling. Such recovery can significantly reduce raw-material purchases and waste-disposal costs. Waste can be most efficiently recovered at the point of generation. The possibility of contamination with other waste materials is reduced, as is the risk involved in handling and transporting waste materials. Some examples of on-site waste recovery are shown in Table 5.1.5.

Some waste streams can be re-used directly. This can usually be accomplished when the waste material is not too contaminated. It may be possible to put the waste material directly back into the original production process as raw material. In the electroplating industry this can sometimes be done with the drag-out from the process bath. Alternatively, lightly contaminated waste may be re-usable in operations not requiring high-purity materials. For example, spent high-purity solvents generated during the production of microelectronics can be re-used in less critical operations, such as metal degreasing.

A waste may have to undergo some type of purification before it can be re-used. A number of physical and chemical techniques which can be used to reclaim a waste material are shown in Table 5.1.6. The method of choice will depend on the physical and chemical characteristics of the waste stream and on the recovery economics. For a number of waste streams there are economic small modular recovery units available. Some of the materials commonly recycled by these modular units are spent solvents, electroplating-process waste, and spent metal-working fluids.

As Table 5.1.5 shows, a wide range of on-site recovery methods are available and used by a variety of industries. Many of these facilities have recovered the cost of the equipment very rapidly. For example, one printed-circuit-board manufacturer uses electrolytic recovery to reclaim metals from the drag-out from the

TABLE 5.1.5   Examples of Waste Reduction through in-Plant Recycling

Industry	Technique (reference)
Printing	Use a vapor-recovery system to recover solvents (10)
Photographic processing	Recover silver, fixer, and bleach solutions (10)
Printing	Use a distillation unit to recover waste alcohol solutions for re-use (10)
Synthetic fibers	Recover laboratory solvents in a small distillation system (2)
Metal fabricator	Recover synthetic cutting fluid using a centrifuge system (2)
Paint fabricator	Use a distillation unit to recover cleaning solvents (2)
Mirror manufacturer	Recover spent xylene using a batch-distillation system (2)
Furniture	Use a batch-distillation system to recover waste cleanup solvents from coating line (12)
Printed circuit boards	An electrolytic recovery system used to recover copper and tin–lead from process wastewater (11)
Electroplating job shop	Chrome- and nickel-plating solution recovered by an evaporation unit (3)
Leather tanning	Recover chrome from tanning solutions (8)
Tape measures	Nickel-plating solution recovered using an ion-exchange unit (3)
Medical instruments	Reverse-osmosis system used to recover nickel-plating solution (3)
Power tools	Recover alkaline degreasing baths using an ultrafiltration system (10)
Chemical manufacturer	Recycling cooling water reduced wastewater generation by 50% (21)
Textile	Ultrafiltration system used to recover dye stuffs from wastewater (23)
Hosiery	Reconstitute and re-use spent dye baths (23)

copper and tin–lead plating baths. These units recover about 9 kg (20 lb) of copper metal and 4.5 kg (10 lb) of tin–lead metal each week and allow the treated wastewater to be discharged to the sewer system. The company saves over $33,200 per year in treatment and disposal costs, which covered the cost of the system in just a few months.[11] A color-portrait processor established a recovery program for all of its process chemicals. An electrolytic recovery system is used to reclaim silver from the fixer solutions. This system recovers over 46 kg/wk (2,200 troy oz/wk) of silver. Spent color-developer solution is regenerated by an ion-exchange system and re-used in the process. Finally, spent bleach solution is restandardized, allowing re-use of 90% of the bleach. This comprehensive recovery program saves the firm over $1.16 million in chemical costs each year and reduces waste generated by almost 11,360 L/d (3,000 gal/d).[10]

TABLE 5.1.6 Examples of Physical and Chemical Techniques for Waste Recovery

Sorption	Activated-carbon resins
Molecular separation	Reverse osmosis
	Ion exchange
	Ultrafiltration
	Electrolysis
Phase transition	Condensation
	Distillation
	Evaporation
	Refrigeration
	Electrolytic recovery
Chemical modifications	Cementation
	Precipitation
	Reduction
Physical separation	Filtration
	Flotation
	Liquid-liquid extraction
	Supercritical-fluid extraction
	Centrifugation
	Decanting
	Sedimentation

Source: After Ref. 35.

Off-Site Recovery. Wastes may be recovered at an off-site facility when the equipment is not available to recover on site, when not enough waste is generated to make an in-plant system cost effective, or when the recovered material cannot be re-used in the production process. Techniques used off site are similar to those used at on-site facilities. Oils, solvents, electroplating sludges, process baths, lead–acid batteries, scrap metal, and plastic scrap are commonly sent off site for recovery. For some process materials such as solvents and etchants, the supplier may provide services to pick up the material, recover it, and return it for re-use. Known as "tolling," this vendor service is an excellent way to reduce disposal and raw-material costs, as the reclaimed material is usually cheaper than virgin material. A manufacturer of airline seats relies on tolling for managing its spent solvents. The solvents are collected and recycled by the solvent vendor, saving the company over $11,000 per year in disposal costs.[5]

Interindustry Exchange. In some situations a waste may be transferred to another company for use as a raw material in its manufacturing process. Interindustry exchange can be economically advantageous to both firms as it will reduce the waste-disposal costs of the generator and reduce the raw-material costs of the user. For example, an *x*-ray film manufacturer produces a salable product from waste film stock. The company installed equipment that flakes and bales waste polyester-coated film stock, which is sold as raw-material input to another firm. Over 9 million kg (20 million lb) of film stock is exchanged each year, representing a $200,000 annual savings in collection, transport, and processing costs, and an annual profit of $150,000 from the sale of the materials.[4]

Interindustry exchange requires a strong commitment from the generator to find markets for the waste material. In some cases the waste may have to undergo some modification, such as dewatering, in order to make it a salable product. An aluminum die-casting firm developed a market for a by-product of their production process, fumed amorphous silica. After much research into uses for the product it was found to be a valuable additive to concrete. The firm marketed the waste and now sells all the fumed amorphous silica it generates to cement manufacturers. The sale of this material, along with another, has realized the company almost $1 million per year in income from sales and eliminated the cost of land disposal.[6]

A number of waste exchanges have been established as clearinghouses for information on wastes available and wanted. The service usually offered is a listing, in a catalog or computer data base, of wastes available from a generator and wanted by a user. This information is distributed throughout a specific geographic area. A company interested in a waste contacts the waste exchange, which forwards the inquiry to the listing company. Usually, actual negotiations and material transfers are handled directly by the companies. Additional information on waste exchanges can be found in Sec. 5.2.

5.1.3 DEVELOPING AND IMPLEMENTING A WASTE-MINIMIZATION PROGRAM

The development and implementation of a waste-minimization program is a key element in any environmental management program. An effective minimization program must be based on accurate and current information on waste-stream generation, and economical and technically effective waste-reduction techniques. This can be accomplished by establishing procedures to collect information, evaluate options, and identify cost-effective reduction techniques. Once identified, the techniques can be implemented and can become an established part of the facilities' management and operation. An approach to developing and implementing a waste-minimization program is summarized in Fig. 5.1.2. This approach can be used by all types and sizes of companies. The first step in developing a program is to establish a clear corporate policy. The full commitment from management of time, personnel, and financing is extremely important. Lack of this commitment is often one of the most formidable obstacles to waste reduction.

Facility Assessment

A *facility assessment,* or *audit,* provides a protocol for collecting the technical and economic information necessary to select appropriate waste-minimization techniques. Depending on the size of the facility, an audit can be done by a single person or a team. The team approach is best, as the team as a whole will be able to offer a wider range of experience, knowledge, and problem perception. An in-house team can include management and plant personnel from facilities engineering, environmental engineering, safety and health, purchasing, materials and inventory control, finance, and product quality control. The team should be selected and led by a technically competent person with sufficient authority to do the job.

Once the appropriate personnel have been selected, the next step is to conduct the facility assessment. Information should be collected on the types, quantities,

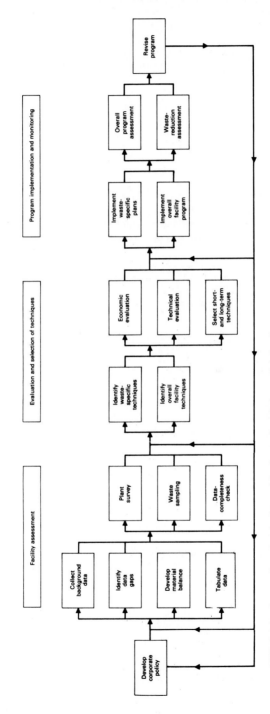

FIG. 5.1.2 Approach to developing and implementing a waste-minimization program.

5.17

compositions, and sources of all air, solid, hazardous, and wastewater waste streams. This information is obtained through a search of available background data and supplemented with detailed data from a plant survey.[44,45]

Background Information. All available background information must be collected first. This includes information on the production process, facility layout, waste-stream generation, and waste-management costs. Some sources of this information are listed in Fig. 5.1.3.

Based on the collected information, a general flow diagram or material balance for each process step can be developed. The diagram should clearly identify the source, type, quantity, and concentration of each identified waste stream (see Fig. 5.1.4). The background information can be used to develop and organize the plant survey and help identify data gaps, sampling points, problem areas, and data conflicts.

Plant Survey. After reviewing the background information and identifying additional data requirements, a survey can be conducted to (1) verify background data and fill data gaps, (2) identify additional waste streams, and (3) observe and collect data on actual operation and management practices. Each step in the man-

---

Production-Process Information

Process flow diagrams and plant layout
Sewer layout diagrams
Purchasing records
Material Safety Data Sheets
Operating manuals
Water usage rates
Plant operating schedule
Production records

Waste-Stream Information

Manifests, annual reports, and related RCRA information
Environmental-monitoring reports
Environmental permits (solid waste, hazardous waste, NPDES, pretreatment, air
    emissions)
Information on any regulatory violations
Location of all solid- and hazardous-waste collection and storage points
Diagram of air, wastewater, and/or hazardous-waste treatment units
Operating manuals for treatment units

Economic Information

Water and sewer costs
Solid- and hazardous-waste management costs
Cost of operating on-site treatment units
Waste-management contracts and billings

General Information

Current waste-minimization practices
Copies of previous environmental audits
Vendor information

FIG. 5.1.3 Background information.

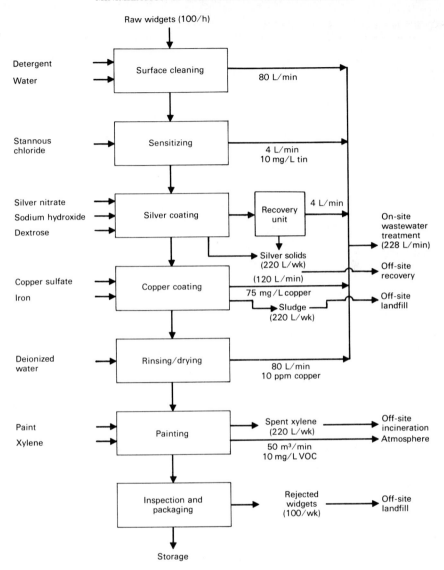

FIG. 5.1.4 Shiny widget production-process material balance.

ufacturing process from the material delivery area to final product storage must be examined. Table 5.1.7 lists some of the types of waste that can be generated in each production area. Examples of specific process information which can be collected by a survey are listed in Table 5.1.8.

If detailed or specific data on waste-stream quantity and composition are not available or cannot be calculated, then a sampling program should be included as part of the survey. Sampling points should be identified before the survey begins, based on the waste flow diagram. However, additional samples may have to be

TABLE 5.1.7 Possible Sources of Waste

Plant category	Area	Possible waste material
Material receiving	Loading docks, incoming pipelines, receiving areas	Packaging materials, off-spec materials, damaged containers, spill residue, transfer-line leaking or dumping
Raw material storage/final product storage	Tanks, silos, warehouses, drum storage yards, storerooms	Tank bottoms, off-spec and excess materials, spill residue, leaking pumps, valves, and pipes, damaged containers, empty containers
Production	Melting, curing, baking, distilling, washing, coating machinery, formulating	Washwater, solvents, still bottoms, off-spec product, catalysts, empty containers, sweepings, ductwork clean-out, additives, oil, process-solution dumps, rinsewater, excess materials, filters, leaking process tanks, spill residue, pumps, pipes, valves, hoses
Support services	Laboratories	Reagents, off-spec chemicals, samples, sample containers
	Maintenance shops	Solvents, cleaning agents, degreasing sludges, sand-blasting waste, lubes, oils, greases, scrap metal, caustics
	Garages	Oils, filters, solvents, acids, caustics, cleaning-bath sludges, batteries
	Powerhouses and boilers	Fly ash, slag, tube clean-out material, chemical additives, oil, empty containers
	Cooling towers	Chemical additives, empty containers, cooling-tower bottom sediment

Source: After Ref. 27.

taken as new waste streams are identified during the survey. Sampling of the waste streams should be done over a period of time to account for variations in production scheduling. Sampling may also have to be carried out over an extended length of time if products are produced on an irregular or seasonal basis. However, in many cases only qualitative information is needed. This usually can be calculated from the process-input composition and simple flow measurements or historical records on waste generation. Information on how to conduct a detailed sampling program can be found in Ref. 29.

TABLE 5.1.8 Examples of Information from in-Plant Survey

Area	Information
Material delivery and storage	Material-transfer and handling procedures Storage procedures Evidence of leaks or spills Inventory of materials Condition of pipes, pumps, tanks, valves, and storage/delivery area
Production process	Exact sources of all process waste Waste flow (quantity and concentration) Operational procedures Source, quantity, and concentration of intermittent waste streams (i.e. cleaning, batch dumps, etc.) Condition of all process equipment including tanks, pumps, pipes, valves, etc. Evidence of leaks or spills Maintenance procedures and schedule Potential sources of leaks and spills
Waste management	Operational procedures for waste-treatment units Quantity and concentration of all treated wastes and residues Waste-handling procedures Efficiency of waste-treatment units Waste-stream mixing

A preliminary review of the data should be performed during or just after the survey. This can help identify missing or inaccurate information. Additions and corrections should be made to the waste flow diagram and the overall data reviewed for completeness. For each waste stream the following information should be available:[30]

- Point of origin
- Subsequent handling, treatment, and disposal
- Physical and chemical characteristics
- Quantity
- Rate of generation [i.e., kilograms (pounds) per unit of product]
- Variations in generation rate
- Potential for contamination or upset
- Cost to manage and dispose

Evaluation and Selection of Waste-Minimization Techniques

Procedures used to identify, evaluate, and select applicable waste-minimization techniques will depend on the complexity of the manufacturing process, and the quantity and variety of waste generated. Successful approaches range from simple group discussions to complex computer-modeling techniques.[28,30] However, all approaches will involve the same basic steps: (1) List waste streams; (2) identify potential waste-minimization techniques for each waste stream; (3) evaluate the technical and economic aspects of each technique; (4) select the most cost-effective waste-minimization technique(s) for each waste stream. In addition to addressing specific waste streams, general recommendations on facilitywide reduction methods should also be made. These could include such areas as material handling, maintenance, and operating procedures.

There are a number of sources of information which can help identify waste-reduction techniques. One of the best sources of ideas, and one that is sometimes overlooked, is the operating personnel. Many examples of available techniques were discussed in Subsec. 5.1.2. Also, as discussed in Subsec. 5.1.4, trade associations, equipment vendors, trade journals, and state and federal agencies can provide information and further assistance.

Once the techniques for each waste stream have been identified, the technical feasibility of each technique should be evaluated. An engineering evaluation should take into account such factors as applicability, waste-reduction potential, operation and maintenance requirements, safety and health, ease of implementation, reliability, and any special design considerations. Not all of these factors are equally important for each technique. For example, an evaluation of a change in inventory management may only have to consider ease of implementation and reduction potential. At this point an engineering consultant may be employed to provide additional expertise in evaluating technical feasibility.

As part of the technical evaluation, each technique should be evaluated according to the hierarchy shown in Fig. 5.1.5. The liability associated with each step in the hierarchy increases as one goes down it. This will help during the selection process to bring the risks and liabilities associated with managing a waste into the evaluation process. As indicated, the best techniques for waste minimization are those that eliminate or reduce generation of the waste at its source. This will eliminate the risks and costs associated with management of the waste.

In addition to the technical evaluations, an economic analysis of each reduction technique should be done. Cost factors to consider include implementation costs (capital, installation, operating and maintenance) as well as cost savings from lower production costs and lower waste-management and disposal costs. Based on this information, a return on investment analysis can be done to estimate the payback period. The current waste-management cost is a very important factor which is often overlooked. These costs include not only the cost of shipping a waste off site, but also the on-site expense of labor and time required to handle, manage, track, store, treat, and manifest the waste. Other considerations are harder to quantify but are very important; they include liability insurance premiums, long-term liability and legal costs, worker health and safety, community relations, and regulatory compliance.

The completed technical and economic analyses will enable the best waste-minimization options for each waste stream to be selected. Techniques may be short-term (such as inventory control) or longer-term (such as process modification). This selection process is rather subjective and is usually based on the experience of several people who are in the decision-making process. In many cases there are just one or two technically feasible and cost-effective alternatives. In some cases several techniques may be effectively used together. For example, the first technique may be a segregation of a waste stream, which in turn enables recovery and re-use on site.

Once the minimization techniques are identified, an implementation plan should be developed for each waste stream. This should include information on the implementation schedule, equipment needs, conceptual design, implementation requirements, and management requirements, and it should also include cost estimates.

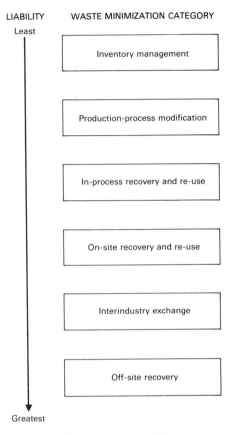

FIG. 5.1.5   Waste-minimization hierarchy.

Implementation and Monitoring of the Waste-Minimization Program

The waste-stream reduction plans, along with any general facility recommendations, will form the basis of the waste-minimization program. To ensure continued program effectiveness, procedures must be established for monitoring and evaluating the techniques once in place. The program should address review and updating procedures as well as how the program will be integrated into the management structure. In addition, the program should be dynamic in nature to allow for production changes and for development of new waste-reduction techniques.

The implementation of a waste-minimization program can be done in a phased manner. Waste streams which present management problems and those for which the investment will have a short payback period can be addressed first. Simple and low-cost techniques can also be put in quickly. In many cases this just involves improvements in inventory control, operation, and maintenance. One important factor is to keep the employees informed and involved at all steps in the development and implementation of a program. Employees will also be extremely valuable in evaluating a minimization program and in identifying ways to improve program performance.

A record-keeping system should be established to track the effectiveness of each segment of the program. Waste-generation and reduction data should be calculated in terms of product production rates (i.e., pounds of waste per pound of product, pounds of waste per unit area of product, etc.). This makes possible accurate comparison of waste-generation and reduction data over time. Associated economic data as discussed above should also be tabulated to evaluate the efficiency of the waste-stream reduction techniques. Based on the collected information, the program should undergo regular review and updating.

As discussed earlier, corporate commitment is a most important factor in the initial and continuing success of a waste-minimization program. The program must become an integral part of a firm's corporate policy, product-development procedures, operational procedures, and training programs. A senior manager or executive committee might be given the authority and resources to develop, operate, and monitor the program throughout the company. If a firm is large enough, a small staff may be needed to evaluate the success of the program and develop new waste-minimization approaches. Such a high-level management commitment will help to keep the waste-minimization program active in all parts and levels of the company, from new-product development to the maintenance staff.

Companies have taken a number of simple approaches to ensure the initial and continued success of a waste-minimization program. Many have established, at the corporate level, a waste-reduction engineering group that can provide technical assistance to all plants. Awards and financial incentives have been used to foster new ideas and innovations. Some firms also conduct annual audits of their plants to review and update existing waste-minimization programs and to identify applicable new waste-reduction methods. A companywide information-exchange program using newsletters, fact sheets, publications, and/or internal conferences to transfer ideas has been successfully used. Other companies have instituted separate capital-expenditure review procedures for waste-minimization projects that require less paperwork and have a quicker review process.

5.1.4 SOURCES OF INFORMATION AND TECHNICAL ASSISTANCE

There are a number of sources of information on waste minimization. Books, conference proceedings, and articles present general overviews of waste-reduction options for specific industrial categories (see Refs. 2, 3, 4, 10, 12, 14, 19, and 31 through 40). Additional information may be obtained from trade associations, trade journals and government research reports, regulatory agencies, and technical assistance groups.

Industrial trade associations can provide, in many cases, the most detailed and current technical information. Associations may have staff experts with extensive knowledge and experience in waste management. Through contacts with members, research projects, conferences, and trade publications, trade associations are good sources of information on what other firms are doing and what techniques are effective. Trade publications are another very good source of information. Many journals contain articles on case studies, current research, vendor information, and suggestions from industrial experts. A list of trade associations can be found in Ref. 42.

A number of federal agencies sponsor studies and research on waste minimization: they include the Environmental Protection Agency, Department of Defense, Department of Energy, and the Bureau of Mines. All agency reports are available for a fee through the National Technical Information Service (NTIS), U.S. Department of Commerce, Springfield, Virginia 22161. Identification of relevant publications can be made by conducting a computer search of the NTIS Bibliography Data Base. Access to NTIS and other helpful bibliographic data bases can be made, usually for a fee, through most university library systems and some trade associations. The research librarian at most public libraries can help with getting the necessary access to the data systems. Access to U.S. Environmental Protection Agency research reports and current research topics can be made by contacting the Hazardous Waste Engineering Research Laboratory, Office of Research and Development, U.S. Environmental Protection Agency, 26 West St. Clair Street, Cincinnati, Ohio 45268.

Another source of technical information is state regulatory agencies or technical-assistance programs. Many state regulatory programs offer technical information on waste minimization to the regulated community. Another rapidly growing source of information is technical-assistance programs, specifically on waste reduction, being established by a number of states. The level of assistance offered by these programs varies and may include on-site technical assistance, access to information data base and documents, workshops, referral services, research programs, and matching-grant programs. Currently programs are operating in California, Georgia, Illinois, Minnesota, New York, North Carolina, and Pennsylvania. Check with your respective state agencies to find out if such a program is available.[17,43]

5.1.5 REFERENCES

1. 40 *CFR* 262.41(A)(6), 262.41(a)(7), and 264.73(b)(9), respectively.

2. J. Kohl, P. Moses, and B. Triplett, *Managing and Recycling Solvents: North Carolina Practices, Facilities, and Regulations,* North Carolina Board of Science and Technology, Raleigh, North Carolina, 1984.

3. J. Kohl and B. Triplett, *Managing and Minimizing Hazardous Waste Metal Sludges: North Carolina Case Studies, Services and Regulations,* Governor's Waste Management Board and the North Carolina Hazardous Waste Management Branch, Raleigh, North Carolina, 1984.

4. G. Hunt and R. Schecter, *Accomplishments of North Carolina Industries: Case Summaries,* North Carolina Department of Natural Resources and Community Development, Raleigh, North Carolina, 1986.

5. Nominations for 1985 Governor's Award for Excellence in Waste Management, Governor's Waste Management Board, Raleigh, North Carolina, 1985.

6. T. A. Turberville, "Market Development for Fumed Amorphous Silica and Ladle Rakeout," *Pollution Prevention Pays: Waste Reduction in Alabama Proceedings, October 30, 1985,* University of Alabama Center for Urban Affairs, Birmingham, Alabama.

7. Walter B. Stumpff, "Solvent Recovery in Rubber Technology Applications," *Pollution Prevention Pays: Waste Reduction in Alabama Proceedings, October 30, 1985,* University of Alabama Center for Urban Affairs, Birmingham, AL.

8. *Outstanding Achievement in Hazardous Waste Management, State of Minnesota, April 29, 1985,* Minnesota Waste Management Board, Crystal, Minnesota, 1985.

9. *Governor's Conference on Pollution Prevention Pays, March 4–6, 1986—Awards,* by the Governor's Safe Growth Team, Nashville, Tennessee.

10. D. Huisingh, *Profits of Pollution Prevention: A Compendium of North Carolina Case*

Studies, North Carolina Board of Science and Technology, Raleigh, North Carolina, 1985.

11. T. J. Nunno and M. Arienti, "Waste Minimization Case Studies for Solvents and Metals Waste Streams," presented at the Hazardous Waste Disposal Conference, U.S. EPA, Cincinnati, Ohio, 1986.

12. J. Kohl et al., *Managing and Recycling Solvents in the Furniture Industry,* North Carolina Board of Science and Technology, Raleigh, North Carolina, 1986.

13. T. E. Higgins, *Industrial Processes to Reduce Generation of Hazardous Waste at DOD Facilities, Phase I Report, Evaluation of 40 Case Studies,* U.S. Environmental Leadership Project Office, Washington, D.C., and U.S. Army Corps of Engineers, Huntsville, Alabama, 1985.

14. M. E. Campbell and W. M. Glenn, *Profit from Pollution Prevention, A Guide to Industrial Waste Reduction & Recycling,* Pollution Probe Foundation, Toronto, Canada, 1982.

15. J. Kohl, J. Pearson, and B. Triplett, "Reducing Hazardous Waste Generation with Examples from the Electroplating Industry," *Waste Management Advisory Note,* vol. 22, Solid and Hazardous Waste Management Branch, North Carolina Department of Human Resources, Raleigh, North Carolina, 1986.

16. G. Hunt and R. Walters, "Cost-Effective Waste Management for Metal Finishing Facilities: Selected Case Studies," in *Proceedings of the 39th Purdue University Industrial Waste Conference,* Butterworth Publishers, Boston, Massachusetts, 1985, pp. 521–528.

17. *Waste Minimization Issues and Options,* prepared for the U.S. EPA Office of Solid Waste, Washington, D.C., by Versar, Inc., and Jacobs Engineering Group, 1986.

18. F. J. Hellhake, "Setting up a Preventive Maintenance Program," *Chem. Eng.,* **88**:(14)145–150 (1981).

19. D. J. Sarokin et al., *Cutting Chemical Wastes: What 29 Organic Chemical Plants Are Doing to Reduce Hazardous Wastes,* Inform, Inc., New York, 1985.

20. E. B. Jones and W. Banning, "The Role of Waste Exchanges in Waste Minimization & Reclamation Efforts," in *Hazardous & Solid Waste Minimization,* Government Institutes, Inc., Rockville, Maryland, 1986.

21. *Low- or Non-Pollution Technology Through Pollution Prevention,* prepared for the United Nations Environment Programme, Office of Industry and the Environment, by 3M Company, St. Paul, Minnesota, 1982.

22. P. A. Chubb, "Managing Waste: Critical to Competitiveness," *Wasteline—The Du Pont Report on Waste Reduction,* Wilmington, Deleware, Spring 1986.

23. *Technical Manual: Waste Abatement, Reuse, Recycle and Reduction Opportunities in Industry,* prepared by Resource Integration Systems, Inc., J. L. Richards & Associates Limited, Ontario Research Foundation, and The Proctor & Redfern Group, for Environment Canada, Toronto, Ontario, undated.

24. "Pollution Abated, Cash Crop Created," in *Ideas: A Compendium of 3P Success Stories,* 3M Company, Saint Paul, Minnesota, undated.

25. "Riker Innovation Meets Air Regulation," in *Ideas: A Compendium of 3P Success Stories,* 3M Company, Saint Paul, Minnesota, undated.

26. J. F. Meister, "Operation of the Southern Illinois University-Carbondale Hazardous Waste Program," in *Madison Seminar Proceedings: Waste Management in Universities and Colleges,* EPA 905/9-81-001, U.S. EPA Region 5, Chicago, Illinois, 1981.

27. H. W. Blakeslee and T. M. Grabowski, *A Practical Guide to Plant Environmental Audits,* Van Nostrand Reinhold, New York, 1985.

28. C. H. Fromm and M. S. Callahan, "Waste Reduction Audit Procedure—A Methodology for Identification, Assessment and Screening of Waste Minimization Options," in *Proceedings of the National Conference on Hazardous Wastes and Hazardous Materials,* Hazardous Materials Control Research Institute, Silver Springs, Maryland, 1986.

29. *Handbook for Monitoring Industrial Wastewater,* U.S. EPA Technology Transfer, Cincinnati, Ohio, 1973.

30. R. B. Pojasek, "Waste Minimization: Planning, Auditing and Implementation," in *Proceedings of Waste Reduction: The Ongoing Saga,* Tufts University Center for Environmental Management, Medford, Massachusetts, 1986.

31. M. G. Royston, *Pollution Prevention Pays,* Pergamon Press, Elmsford, New York, 1979.

32. D. Huisingh and V. Bailey, (eds.), *Making Pollution Prevention Pay: Ecology with Economy as Policy,* Pergamon Press, Elmsford, New York, 1982.

33. Illinois Department of Energy and Natural Resources, *Pollution to Profit—Reducing Industrial Waste in Illinois,* Springfield, Illinois, 1984.

34. *Massachusetts Hazardous Waste Source Reduction Conference Proceedings,* Bureau of Solid Waste Disposal, Massachusetts Department of Environmental Management, Boston, Mass., October 17, 1984.

35. K. E. Noll et al., *Recovery, Recycle and Reuse of Industrial Waste,* PB 84-127-141, Lewis Publishers, Chelsea, Michigan, 1985.

36. L. L. Tavlarides, *Process Modifications for Industrial Pollutants Source Reduction,* Lewis Publishers, Chelsea, Michigan, 1985.

37. *Economic Incentives for the Reduction of Hazardous Wastes—Final Report and Appendices,* prepared by ICF Consulting Associates Incorporated for the California Department of Health Services, Sacramento, California, 1985.

38. *Compendium of Low- and Non-Waste Technology,* vols. I and II (1981), vols. III and IV (1982), vol. V (1983) and vol. VI (1984), United Nations Economic Commission for Europe, Geneva, Switzerland.

39. U.S. Environmental Protection Agency, *U.S. EPA Summary Report—Control and Treatment Technology for the Metal Finishing Industry—In-Plant Changes,* EPA-625/8-82-008, U.S. EPA, Cincinnati, Ohio, 1982.

40. National Research Council, *Reducing Hazardous Waste Generation: An Evaluation and a Call to Action,* National Academy Press, Washington, D.C., 1985.

41. Case study files from the Pollution Prevention Pays Program, Division of Environmental Management, North Carolina Department of Natural Resources and Community Development, Raleigh, North Carolina.

42. *National Trade and Professional Associations of the U.S., 1986,* Columbia Books, Washington, D.C., 1986.

43. U.S. Environmental Protection Agency, *U.S. EPA Summary of Issues and Discussions at the Third Workshop on Implementing State Waste Reduction Programs, April 21-22, 1986,* U.S. EPA Office of Policy Planning and Evaluation, Washington, D.C.

44. U.S. Environmental Protection Agency, *Manual for Waste Minimization Opportunity Assessments,* Hazardous Waste Engineering Research Laboratory, Cincinnati, Ohio, 1988.

45. Ontario Waste Management Corporation, *Industrial Waste Audit and Reduction Manual,* Toronto, Ontario, 1987.

SECTION 5.2

WASTE EXCHANGES

Bill Quan

Chief, Environmental Assessment Unit
California Department of Health Services
Sacramento, California

5.2.1   OBJECTIVES OF WASTE EXCHANGES

The primary objective of waste exchanges is to reduce through re-use the amount of waste material that is discarded or going to landfills for disposal. The re-use of waste typically results in savings in disposal costs, conservation of natural resources and energy needed to process these resources, and the saving of limited landfill space. In recent years there has been an evolution and growth in waste exchanges in the United States. Although waste exchanges will continue to play an important role in diverting waste from landfill—one study[1] indicates that perhaps as much as 11% of the hazardous wastes now disposed off site may be recycled—it is not a panacea. Waste exchange and alternative technologies can reduce the amount of hazardous waste disposed, but for the foreseeable future, landfills will remain with us.

5.2.2   HISTORY OF WASTE EXCHANGES

Perhaps due to the comparative scarcity of land for landfills, waste exchanges appeared in Europe before they appeared in the United States. One of the earliest was established in Great Britain in 1942. The National Industrial Recovery Association was established by the government to conserve material during World War II.[2]

One of the first U.S. waste exchanges established was a company waste exchange organized to deal with the company's waste. The Investment Recovery Department, formed by the Union Carbide Corporation in 1964, initially dealt with surplus equipment and materials, but in 1969 it began to recycle chemicals, metals, and other wastes.[2]

In the early to mid-1970s a different sort of waste exchange appeared in the United States: the material exchange. These exchanges actually take physical possession of other companies' wastes and actively seek markets for them. In some cases, a material exchange like Zero Waste System in California processes the wastes to potential users' purity specifications.

It appears that the greatest beneficiary among the different types of waste exchanges since the passage of the Resource Conservation and Recovery Act (RCRA) has been the information clearinghouse. Many state governments have opted for this form of waste exchange because of its relatively low costs of operation. The first U.S. exchange of this type was formed in St. Louis, Missouri, in 1975 by the St. Louis Regional Commerce and Growth Association.

5.2.3 TYPES OF WASTE EXCHANGES

As can be inferred from the foregoing, most waste exchanges are variations of one of two distinct types: the information clearinghouse and the material exchange. The major differences between these types of exchange lie in what is transferred and the roles these exchanges play in the actual transfer of waste.[3]

Information Clearinghouse

As the name implies, what is transferred by the *information clearinghouse* is information about waste materials. This type of exchange is considered passive in the way it operates. Action is initiated either by the company with a waste to dispose of or by a company looking for wastes, either for its own use as a raw material or for marketing to industries which can re-use the waste. Information from either source is coded by the clearinghouse and published in a catalog which is periodically distributed. It was originally thought that trade secrets of companies that generated the wastes would be protected if information in the catalogs was coded. However, some clearinghouses are finding this rationale for coding to be of little validity since competitors can surreptitiously get whatever information the clearinghouse has about a particular waste simply by posing as a potential client for that waste. Regardless of coding, readers of the catalog must normally contact the clearinghouse to forward their requests for further information about a listing. Once a referral or a connection between an interested reader and a lister in the catalog has been made, the clearinghouse has completed its basic function.

Because the basic role of these clearinghouses initially was that of an intermediary, little technical expertise was required. In recent years, however, clearinghouses have come to recognize the limited value of a passive approach and have begun to pursue a more active role in the recycling of wastes. One reason for this is that state governments found that studies and manifests (documents which track waste from the site of generation to the site of disposal) show that significant quantities of recyclable wastes were still being disposed of. Eventually, some states adopted regulations to force industry to consider recycling before disposing of their waste. For example, California regulations require waste generators, under penalty of law, to respond to inquiries of the California Waste Exchange [a program run by the state's Department of Health Services (DHS)] on the disposal of waste judged economically and technically feasible to recycle. Refusal to recycle when it is feasible may lead to the doubling of a company's disposal costs by DHS.

As clearinghouses became more active in the recycling scenario, they added staff with technical expertise in industrial chemistry and chemical engineering. Technical staff of these clearinghouses become consultants to the chemical industries; they sometimes visit waste generators to identify recycling opportunities. What these field technical people have often found is that what is waste to one industry can sometimes be raw material to another industry.

Material Exchanges

This second type of waste exchange plays an entirely active role in the transfer of wastes. Unlike many of the clearinghouses, these *material exchanges* are not supported by government subsidies and they do charge for their services. The characteristic feature of the active role played by these exchanges is that the exchanges take physical possession of the waste, and they may actually analyze and process the waste to meet market needs. Because of the additional services offered, material exchanges are costlier to run than are information clearinghouses.

Table 5.2.1 shows those waste exchanges outside the United States still operational during 1983.[2,3] Figure 5.2.1 shows those U.S. waste exchanges still operational during 1985.[4,5]

5.2.4 MEASUREMENT OF THE EFFECTIVENESS OF WASTE EXCHANGES

There is little information available on the effectiveness (e.g., volume of waste recycled) of many of the waste exchanges during the early years, as they were too short-staffed to keep track of waste transfers. Another measurement problem arises when an attempt is made to compare the effectiveness of different waste exchanges. The reason is that not all waste exchanges measure the success of their programs in the same manner.

In terms of the listings in the catalog, some clearinghouses tally the volume of interest shown toward the listing, some tally the percentage of listings exchanged or recycled, some tally the volume of wastes recycled, and some tally the savings accrued. In addition, waste exchanges, particularly regional (as opposed to national) exchanges, do not usually serve the same industrial communities; they are affected by state laws and regulations. For all these reasons, it will always be difficult to directly compare the effectiveness of one exchange against that of another exchange.

TABLE 5.2.1   Waste Information Exchanges outside the United States

Exchange	Service area
CWME (Canada)	International
UKWME (U.K.)	National
NIMRA (U.K.)	National
SWE (Switzerland)	National
CIAE (Austria)	National
FWE (Austria)	National
NWE (Sweden)	International
TWME (Italy)	National
ANIC (Italy)	National
WEI (Israel)	National
MWDA (Australia)	Regional
IWES (Australia)	Regional
ANRED (France)	National
VCI (Germany)	International
DIHT (Germany)	International
VNCI (Netherlands)	National
FICB (Belgium)	National
OBEA (Belgium)	National

5.2.5 TYPES OF WASTE RECYCLED

Many waste exchanges recycle both nonhazardous and hazardous waste, but the focus in this publication is on hazardous waste. Although it is recognized that some states may be using a more stringent definition, here hazardous waste will be defined as those wastes meeting the U.S. Environmental Protection Agency's (EPA's) regulatory definition for solid waste and hazardous criteria for corrosivity, ignitability, reactivity, and toxicity.

A primary factor affecting the marketability of a hazardous waste is its purity, such as found with raw or virgin materials or by-products. An example of the latter is the lime or calcium hydroxide produced from the calcium carbide process used to manufacture acetylene gas. Little processing or treatment is necessary for these kinds of materials before they are used. In other words, they can be used "as is."

Materials less valuable and usually less pure, but traditionally re-used, are those kinds of materials considered superfluous to the manufacturing processes from which they arose. Examples of this second category of materials are solvents from pharmaceutical and paint manufacturing, slag from steel production, and rejected lead plates from lead–acid batteries. All these materials are considered unusable by the industries that generated them, but they have been considered by some other industries as suitable raw materials.

A third category of material, actually a subcategory of the second category above, is that type of material for which some processing or treatment is necessary to meet certain purity requirements. For instance, it is common to use zinc oxide dust from air-pollution control devices to manufacture zinc sulfate fertilizer by reaction of the zinc oxide dust with sulfuric acid. However, in some cases it

California Waste Exchange
Department of Health Services
Toxic Substances Control Division
714/744 P Street
Sacramento, Calif. 95814
(916) 324-1818

Colorado Waste Exchange
Colorado Association of Commerce and
Industry
1390 Logan Street
Denver, Colo. 80203
(303) 831-7411

World Association for Safe Transfer
and Exchange
130 Freight Street
Waterbury, Conn. 06702
(203) 574-2463

Southern Waste Information Exchange
P.O. Box 6437
Tallahassee, Fla. 32313
(904) 644-5516

Georgia Waste Exchange
Business Council of Georgia
181 Washington St., S.W.
Atlanta, Ga. 30303
(404) 223-2264

Industrial Material Exchange Service
IEPA-DLPC-24
2200 Churchill Road
Springfield, Ill. 62706
(217) 782-0450

Industrial Commodities Bulletin
Enkarn Corporation
P.O. Box 590
Albany, N.Y. 12210
(518) 436-9684

Northeast Industrial Waste Exchange
90 Presidential Plaza, Suite 122
Syracuse, N.Y. 13202
(315) 422-6572

Piedmont Waste Exchange
Urban Institute
University of North Carolina
Charlotte, N.C. 28223
(704) 597-2307

Tennessee Waste Exchange
Tennessee Manufacturers Association
501 Union St., Suite 601
Nashville, Tenn. 37219
(615) 256-5141

Chemical Recycle Information Program
1100 Milam Building, 25th Floor
Houston, Tex. 77002
(713) 658-2462 or 658-2459

Inter-Mountain Waste Exchange
W. S. Hatch Company
643 South 800 West
Woods Cross, Utah 84087
(801) 295-5511

Louisville Area Industrial
Waste Exchange
Louisville Chamber of Commerce
1 Riverfront Plaza
4th Floor
Louisville, Ky. 40202
(502) 566-5000

Great Lakes Regional Waste Exchange
Waste Systems Institute of Michigan,
Inc.
470 Market, SW—Suite 100A
Grand Rapids, Mich. 49505
(616) 363-7367

Midwest Industrial Waste Exchange
Rapid Commerce and Growth Association
10 Broadway
St. Louis, Mo. 63102
(314) 231-5555

New England Materials Exchange
34 North Main Street
Farmington, N.H. 03835
(603) 755-4442 or 755-9962

New Jersey State Waste Exchange
New Jersey Chamber of Commerce
5 Commerce Street
Newark, N.J. 07102
(201) 623-7070

Montana Industrial Waste Exchange
P.O. Box 1730
Helena, Mont. 59624
(217) 782-0450

FIG. 5.2.1 Major U.S. Chemical Waste Exchanges.

Western Waste Exchange
Arizona State University
Center for Environmental Studies
Krause Hall
Tempe, Ariz. 85287

Zero Waste Systems
2928 Poplar Street
Oakland, Calif. 94608
(415) 893-8257

ICM Chemical
20 Cordova Street, Suite 3
St. Augustine, Fla. 32084
(904) 824-7247

Alkem
25 Glendale Road
Summit, N.J. 07901
(201) 277-0060

Ore Corporation
2415 Woodmere Drive
Cleveland, Ohio 44106
(216) 371-4869

Techrad Industrial Waste Exchange
4619 North Santa Fe
Oklahoma City, Okla. 73118
(405) 528-7016

FIG. 5.2.1   (*Continued*)

may be necessary to first treat the zinc oxide waste to remove undesirable contaminants like heavy metals (e.g., lead).

The fourth and final category of materials that is generated by industry are materials that have not, in the past, usually been recycled. These are true waste materials and usually end up in the landfills.

The best indication of the feasibility of recycling a hazardous waste is the existence of recyclers for the particular waste type. Catalogs from the various U.S. waste exchanges show that it is both economically and technically feasible to recycle the following kinds of hazardous waste:

- Alkalis
- Acids
- Agriculturally useful materials
- Catalysts
- Containers
- Distressed, surplus, or expired-shelf-life chemicals
- Metals and metallic compounds
- Oils
- Solvents

A 1983 review[1] of California manifests shows that most of the wastes recycled off site have been solvents, oils, and aqueous metal solutions.

Since much of the tracking aspects of the spate of hazardous-waste regulations adopted to date at both the federal and state levels focuses on wastes leaving the point of generation, much more is known about the kinds and amounts of hazardous wastes recycled off site than on site. The California study cited above indicates that as much as 11% of the wastes going off site may be recyclable, and that more and more companies are recycling their wastes on site. The latter situation seems to be particularly true with the larger companies who employ chemical engineers, the kind of professionals most proficient in finding recycling opportunities if they exist.

5.2.6   ROLE OF WASTE EXCHANGES IN CLEANUP OF CONTAMINATED SITES

Although waste exchanges have traditionally been employed to solve a waste generator's problem, there is no reason why they cannot be used in the cleanup of uncontrolled waste sites.[6] An *uncontrolled waste site* is defined as having one or more of the following characteristics:

1. It is no longer used or is inactive and is no longer maintained.
2. It is a location of unpermitted or improper waste disposal.
3. It has no known owner.
4. It uses containment and monitoring techniques that are inadequate.

The commonest way of cleaning up an uncontrolled hazardous-waste site has been and continues to be excavation of contaminants and disposal at a permitted facility, a method which exacerbates an already severe scarcity of landfill disposal capacity. It appears that unless waste exchanges aggressively assert themselves, most site-remedial actions will continue to be devised without consideration of the possibility of resource recovery.

5.2.7   OBSTACLES TO WASTE TRANSFER

A number of obstacles to waste transfer arise from the generator side. Some generators do not want to reveal any information, or they reveal as little information as possible, because they fear their competition will discover the secrets of their manufacturing processes. This makes recycling difficult. Generators are also worried about potential liability if their waste is mismanaged by a recycler; they feel safer if their wastes go to a disposal site. Perhaps this is because they know where their wastes are located, which is more difficult to know if wastes are given to a recycler. The recycler may send the waste on to another party for processing or to another company for re-use without telling the generator.

However, the key barrier to successful waste transfer is economics. Although it may be technically feasible to recycle many materials or wastes, the economics are not always there. Examples of these kinds of waste materials are sludges and close-boiling liquids.

Other barriers to successful waste transfer primarily affect the recycler. There is a public stigma associated with hazardous waste which extends to recyclers of these wastes. More than one local land-use permit to construct a hazardous-waste recycling plant has been denied because of public pressure.

5.2.8   STRATEGIES FOR THE PROMOTION OF RECYCLING

To bring about increased recycling, some states have passed laws requiring waste generators to explore the potential for waste recycling; some states have gone one step further by penalizing generators for refusing to recycle when it is feasible, technically and economically.

A recent U.S. conference[7] on waste exchange has identified other promising strategies to promote recycling. Some of these follow:

1. Educate waste generators about recycling options and about liability issues. Some generators are still concerned that recycling poses a greater liability than disposal because tracking of their recycled wastes terminates at the point where the wastes are actually converted into products or re-used.
2. Educate the public about the need to recycle for no other reason than to conserve scarce natural resources.
3. Pass tax incentives for recycling.
4. Advertise successful waste exchanges and how much these exchanges saved.
5. Encourage waste exchanges to become more active.

Finally, according to a recent study by the Canadian Waste Material Exchange, waste exchanges can be even more effective if regional exchanges are strongly linked to a national exchange since waste transfers do not seem to be as limited by distance as previously thought. France has such a network of exchanges and it seems to be working effectively.

Whether the employment of such a network would result in a measurable increase in recycling in the United States is questionable because of the present differences in the hazardous-waste laws from state to state and between the states and the Federal government. This suggests that state governments may wish to modify their state hazardous-waste laws so as to promote more interstate recycling.

5.2.9 CONCLUSION

The concept of an industrial waste exchange is rather new in the United States, but in the last few years waste exchanges have become an effective tool for diverting wastes from being disposed at landfills. Working virtually in anonymity for years and considered during this time almost as an afterthought in solving the problem of hazardous waste, recycling has recently taken on a new air of respectability as evidenced by the now annual national conferences on waste exchanges.[2,7] This is a fortuitous change, for it will allow waste exchanges to share ideas and to work more closely together; as one body these exchanges will have a greater influence in the development of legislation on both a national and state level to further encourage recycling.

5.2.10 REFERENCES

1. B. Quan, "The History and Effectiveness of the California Waste Exchange," California Department of Health Services, Toxic Substances Control Division, Sacramento, California, presented at the Third International Symposium on Industrial and Hazardous Wastes, Alexandria, Egypt, July 1985.
2. R. C. Herndon (ed.), *Proceedings of the National Conference on Waste Exchange*, co-sponsored by the U.S. EPA and Florida State University, Tallahassee, March 8–9, 1983.
3. U.S. Environmental Protection Agency, *Waste Exchanges Background Information*, EPA SW-887.1, December 1980. SW-887.1.

4. Rich, *Chemical Week,* **134**(20):58 (May 16, 1984).

5. California Department of Health Services, Toxic Substances Control Division, *California Waste Exchange, A Newsletter/Catalog,* **1**(1) Sacramento, California.

6. B. Quan, "The Effective Use of Resource Recovery in the Cleanup of Uncontrolled Hazardous Waste Sites—Based on the California Experience," *Proceedings of the Conference on the Management of Uncontrolled Hazardous Waste Sites,* Hazardous Materials Control Research Institute, Silver Spring, Maryland, November 1981.

7. R. C. Herndon, (ed.), *The Second National Conference on Waste Exchange,* cosponsored by the U.S. EPA and Florida State University, Tallahassee, March 1985.

CHAPTER 6

HAZARDOUS-WASTE RECOVERY PROCESSES

SECTION 6.1

ACTIVATED-CARBON ADSORPTION

Thomas C. Voice

Department of Civil and Environmental Engineering
Michigan State University
East Lansing, Michigan

6.1.1 PROCESS SUMMARY

Activated-carbon adsorption can be used to remove a wide variety of contaminants from liquid or gaseous streams. It is most frequently used for organic compounds, although some inorganic species are also efficiently adsorbed. The process is relatively nonspecific, and thus, is widely used as a broad-spectrum treatment operation when the chemical composition of a stream is not fully understood. Common treatment applications include groundwater-treatment systems, chemical-spill response, industrial-wastewater treatment, and air-pollution control systems.

Most carbon-adsorption systems utilize granular activated carbon (GAC) in flowthrough column reactors. These systems are efficient and relatively simple to operate if properly designed. Powdered activated carbon (PAC) can be added to some existing treatment processes or used by itself in slurry reactors. The PACT® process, for example, integrates PAC adsorption into a conventional activated-sludge system.

The adsorption process is reversible. It is therefore common to remove the adsorbed contaminants after the adsorption capacity of the carbon has been exhausted. This regenerates the carbon, allowing it to be re-used. In waste-treatment applications, thermal regeneration is usually employed since organic contaminants are destroyed in the process. For small waste streams, regeneration may not be cost-effective and the carbon is simply disposed of after use. Activated carbon can also be used for chemical recovery and concentration, in which case a nondestructive regeneration process is used.

6.1.2 BACKGROUND

The use of carbon adsorption can be traced back through history to some of the earliest written records. The Ebers Papyrus of 1550 B.C. notes the medicinal benefits of wood chars, and early Sanskrit manuscripts (circa 200 B.C.) record the desirability of filtering water through charcoal. It was not until late in the eighteenth century, however, that the early experimental chemists showed that carbon did not actually destroy substances which are harmful to human health, but rather served to accumulate or absorb them. The current level of theoretical understanding originated in the late nineteenth and early twentieth centuries, when the concept evolved that it is adsorption, a surface-accumulation phenomenon, and not simple absorption that is responsible for the ability of various carbon forms to concentrate chemical compounds from gases and liquids. It was during this period that theories which built upon the foundation of the evolving science of thermodynamics were developed and refined. Many of these theories survive today.[1]

Because of the relative chemical simplicity of gaseous mixtures, most of the early theoretical efforts focused on adsorption from the gas phase, rather than from liquid solutions. The first widespread application of carbon adsorption also dealt with vapors; it was precipitated by the German introduction of gas warfare in 1915. Solution-phase applications were somewhat slower to develop. The rapid growth of the synthetic chemicals industry in the 1940s resulted in the development of adsorption-based separation and purification systems. The use of carbon to remove objectionable tastes and odors during water treatment also grew. These applications were stimulated, in part, as a result of the development of activated carbons that were much more efficient than previously used carbon forms. It was not until the late 1960s or early 1970s that one of the most promising applications of carbon adsorption—the removal of a broad spectrum of toxic organic compounds from both water and air—was recognized. One such application, the treatment of groundwater contaminated with industrial solvents, is shown in Fig. 6.1.1.

6.1.3 ADSORPTION PRINCIPLES

Mechanisms of Adsorption

A variety of attractive forces exists between fluid-phase (gas or liquid) molecules and the molecules of a solid adsorbent, all having their origin in the electromagnetic interactions of nuclei and electrons. Traditionally, three loosely defined categories have been distinguished: physical, chemical, and electrostatic interactions. *Physical adsorption* results from the action of van der Waals forces, relatively weak interactions produced by the motion of electrons in their orbitals. *Chemical adsorption,* or *chemisorption,* involves electronic interactions between specific surface sites and solute molecules, resulting in the formation of a bond

FIG. 6.1.1 Temporary carbon adsorbers, which can be shipped to a site and installed on short notice, were used to treat groundwater contaminated with industrial solvents at the Verona well field CERCLA site in Battle Creek, Michigan. These columns were removed from service upon completion of an air-stripping system that utilizes activated carbon to remove the stripped contaminants from the effluent air stream.

that can have all of the characteristics of a "true chemical bond." Chemisorption is typified by a much stronger adsorption energy than physical adsorption. While both chemical and physical adsorption result from electrostatic interactions, the term *electrostatic adsorption* is generally reserved for Coulombic attractive forces between ions and charged functional groups, and is synonymous with the term *ion-exchange*.[2,3]

In liquid solutions, consideration must also be given to the nature of the solution state, that is, the extent to which the solvent is capable of accommodating the solute. Solvophobic forces result when there is a substantial chemical incompatibility between the solute and solvent. These forces may be associated with a significant thermodynamic gradient that drives the solute out of solution. This can result in adsorption energies which are substantially higher than those which result from the surface reaction alone, and in extreme cases, are essentially independent of the characteristics of the solid. In aqueous solutions, for example, most nonpolar organic molecules will readily adsorb to any available solid surface in what is often labeled *hydrophobic bonding*.[2]

Properties of Activated Carbon

Activated carbon can be manufactured from a number of carbonaceous raw materials including coal, coke, peat, wood, and nutshells. The manufacturing process involves dehydration of the raw material; carbonization or conversion of the material to a mixture of amorphous and crystalline carbon, tars, and ash; and activation, where the tars are burned off and an activating agent such as steam or carbon dioxide serves to produce an extensive network of internal pores. The product that results from this process is primarily carbon, but it contains significant levels of other elements that were present in the raw material, in addition to a variety of oxygenated functional groups that result from the activation process. The interior of a carbon particle is highly porous, consisting of macropores, defined as having diameters greater than 1,000 Å, and micropores, defined as having diameters between 10 and 1,000 Å. The pores with diameters less than 10 Å are generally considered inaccessible to most solutes. The size distribution of the internal pores is largely a function of the manufacturing process. The total interior surface area is typically in the range of 500 to 2,000 m^2/g. The size of the carbon particles can also be controlled to produce powdered activated carbon (PAC), sometimes defined as carbon which passes a U.S. Sieve Series No. 50 sieve, and granular activated carbon (GAC).[3,4]

The adsorption properties of activated carbon are primarily a result of its highly porous structure, or equivalently, the high specific surface area of the finished product. Total surface area, such as that measured by the adsorption of nitrogen gas in the BET test, may provide an approximation of adsorption capacity. However, the actual capacity for a specific compound is more closely related to the available surface area; both molecular size and pore diameter determine the extent to which a molecule has access to the interior surface area. A number of standardized tests using adsorbates of known molecular size have been devised to more accurately reflect available surface area. Common test adsorbates include iodine, methylene blue, and molasses, which can be correlated with pores greater than 10, 15, and 28 Å, respectively. The distribution of pore sizes may also be important, since most streams to be treated by adsorption contain numerous adsorbable species, only a few of which need to be removed. A narrow range of pore diameters can serve to exclude larger molecules that might otherwise adsorb and reduce the area available for the adsorbates of interest. Adsorption capacity is theoretically independent of particle size for highly porous adsorbents, but measured capacities typically increase as particle size decreases. This may be a kinetic effect since the rate of adsorption is expected to be greater for smaller particle sizes, or this may result from the fact that small particles are likely to have fewer interior regions that are "sealed off" and unavailable to the adsorbate.[3,4]

Properties of the Adsorbate

Adsorption capacity is simply a measurement of the state of thermodynamic equilibrium for a given set of system conditions and, thus, should be related to one or more fundamental thermodynamic properties of the adsorbate. Unfortunately, the current level of understanding of the adsorption process has not produced the ability to make a priori assessments of adsorption equilibria. It is possible, however, to correlate experimental capacity measurements with other adsorbate properties and draw inferences as to relative capacity.

For liquid solutions, adsorption generally increases as the solubility of the compound in the solution decreases. It is also frequently observed that the adsorption of organic compounds usually increases as one moves up a homologous series or as molecular weight increases. These two properties are highly correlated to solubility for most classes of organics, however, and this latter "rule" can be considered as a special case of the solubility rule. It should be noted that none of these relationships can be treated as rigorous since numerous exceptions exist, especially when chemisorption is a dominant factor in the overall process. For gaseous streams, vapor pressure is analogous to solubility in liquid streams, and thus should serve reasonably well as an adsorption indicator.[3,5,6]

Adsorption Equilibria

In order to use adsorption as a waste-management process, it is necessary to understand how the equilibrium state varies as a function of system conditions. Adsorption equilibrium relationships, commonly called *isotherms,* relate the concentrations of the adsorbed compound in each of the two phases. For example, in the adsorption of phenol from an aqueous solution, the isotherm could indicate what solid-phase phenol concentration in milligrams per gram corresponds to any given solution-phase concentration in milligrams per liter. In practice, isotherms are measured experimentally by equilibrating known quantities of the adsorbent with the compound of interest and plotting the resultant concentrations. This graphical result can be expressed mathematically by fitting the experimental data to one or more mathematical relationships or isotherm models. Figure 6.1.2 shows both the mathematical expressions and graphical representations for the three most common isotherm models and their respective linear forms, which are used for curve fitting.[2,3]

An alternative to performing adsorption experiments is to utilize reported data such as that compiled by the U.S. Environmental Protection Agency (EPA) for most of the "priority pollutants" in aqueous solutions. The advantages of this approach are obvious and estimates produced may be acceptable for many situations, such as preliminary feasibility studies. In practice, however, adsorption isotherms are quite dependent upon the relative concentrations, the presence of other adsorbable species, the characteristics of the solvent, the temperature, and the methods employed to conduct the test. For these reasons, before an adsorber is designed, it is almost always necessary to perform an adsorption isotherm using the actual waste stream of interest and experimental conditions that closely reflect the actual system.[7]

Adsorption Kinetics

It is generally accepted that there are three consecutive steps in the adsorption process. First, an adsorbate molecule is transferred from the bulk fluid being treated to the exterior surface of the carbon. The terms *surface film* and *hydrodynamic boundary layer* are often used to describe the fluid immediately surrounding the particles across which the molecule must move. Second, the adsorbate is transported from the exterior of the particle through the pores to an adsorption site. (Only a small amount of the solute is adsorbed on the exterior particle surface.) Finally, at some point in this transport process (or possibly at several points) the molecule is actually adsorbed and held to the surface. The final step is probably not rate-limiting, and thus adsorption can be represented by two sequential mass-transport processes.[3]

The relative importance of film transport and internal diffusion is dependent upon the characteristics of both the adsorbate and the adsorbent. While evaluation of the mass-transport parameters is not a trivial procedure—several experi-

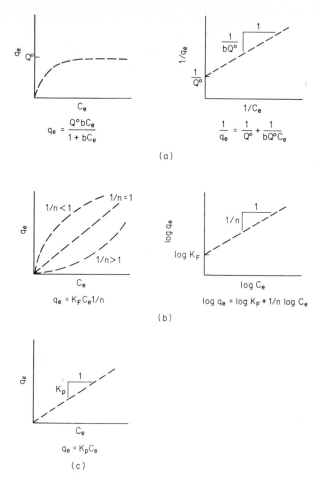

FIG. 6.1.2 Graphical and mathematical forms of three common adsorption isotherm models. (*a*) Langmuir. (*b*) Freundlich. (*c*) Linear.

ments and some sophisticated computer modeling may be involved—there are some useful insights to be gained by doing so. For example, in a system that is controlled by film transport, it may be beneficial to provide additional mixing or turbulence to the fluid phase, whereas this would not be expected to produce any difference in an internal-diffusion-controlled system. Alternatively, reducing the carbon particle size can be beneficial in internal-diffusion-controlled systems.[3]

6.1.4 REACTORS AND TREATMENT PROCESSES

From the discussion above it can be concluded that the manner in which the adsorbate is exposed to the adsorbent is an important design consideration. Several different reactor configurations are commonly employed and each has a particular set of advantages and disadvantages that extend beyond the effects resulting from mass-transfer considerations. The most common reactor schemes for adsorption from aqueous solutions are summarized in this subsection. Application of carbon adsorption to gaseous streams has utilized the fixed-bed approach almost exclusively and the material that follows applies equally well to gas or liquid streams. The most common reactor schemes are shown graphically in Fig. 6.1.3.[3,8,9]

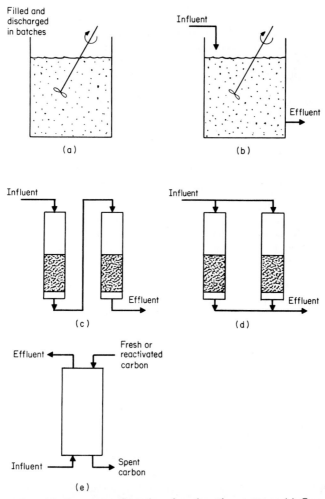

FIG. 6.1.3 Reactor configurations for adsorption systems. (*a*) Completely mixed batch (CMB). (*b*) Completely mixed flow (CMF). (*c*) Columns in series. (*d*) Columns in parallel. (*e*) Pulsed or moving bed.

Completely Mixed Reactors

Powdered activated carbon is generally used in either completely mixed batch (CMB) or completely mixed flow (CMF) reactors. In both cases, mixing is applied to the system to assure that the PAC is kept in suspension, that the slurry of PAC and wastewater is distributed uniformly throughout the tank, and that no significant spatial concentration gradients are established. In a CMB reactor the waste solution is typically introduced into a tank; PAC is added, and the slurry is held until the desired effluent concentration is reached. The rate of the reaction decreases over this time interval since the force driving the reaction, solution concentration, is decreasing. The primary advantage of this reactor configuration is that the quality of the effluent can be guaranteed—the operator needs only to hold the reactor contents longer and perhaps add more PAC to reach the goal. The primary disadvantage of this approach is that it is essentially a manual operation, although CMB systems can be automated if it is possible to continuously monitor the water quality. Alternatively, the system can be operated with a large safety margin to assure that the treatment goal is reached. CMB systems do not efficiently utilize the adsorption capacity of the carbon since the system progresses toward an equilibrium point determined by the effluent concentration of the waste, which is intentionally low.[8]

In a CMF reactor, wastewater and PAC are continuously added to a tank and the mixed contents of the tank are withdrawn at the same rate. Since the system is completely mixed, the tank contents are maintained at a steady-state concentration equal to that desired in the effluent. While this approach does not assure that a desired effluent concentration will always be reached—changes in the influent will translate to changes in the effluent—it does lend itself to unattended operation. CMF reactors also suffer from the problem that the carbon capacity is poorly utilized. An additional consideration in using a CMF reactor is that the reaction rate is also controlled by the steady-state solution concentration. The reactor must therefore be operated at a point far from equilibrium by adding more PAC than is required (at equilibrium), thereby resulting in an additional inefficiency in adsorption-capacity utilization. It should be kept in mind, however, that these inefficiencies are all relative to those obtainable with the GAC systems discussed below, and may be more than offset by the relative cost advantage of PAC.[8]

For both types of completely mixed systems it is necessary to separate the PAC from solution once an acceptable concentration is reached. It is common to use a clarifier or settling tank for this purpose. In CMB systems it may be possible to use the reaction basin itself by simply terminating the mixing and allowing the suspension to settle.

Column Reactors

Column reactors or plug flow (PF) reactors are not hydraulically feasible for PAC but have several distinct advantages over CMF reactors for GAC adsorption systems. Progression of the adsorption process in a fixed-bed column reactor is depicted in Fig. 6.1.4, where the state of the carbon bed is shown at several points along a breakthrough curve, the plot of effluent concentration as a function of time or volume treated. It can be observed from this figure that when the bed is first brought on line, only the very top layer of the bed is exposed to the influent concentration of the adsorbate since the adsorption reaction depletes the solution reaching the layers below. This depletion process establishes a finite depth in the bed, over which the concentation changes from the influent value to essentially zero. The column depth over which this occurs is called the *adsorption zone* or

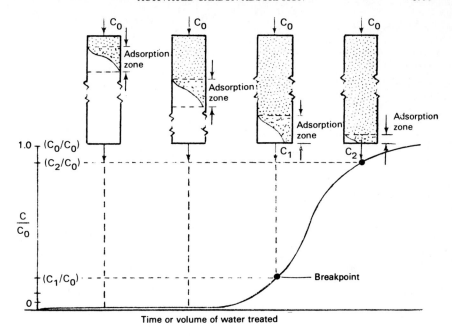

FIG. 6.1.4 Idealized breakthrough curve (*After Weber, Ref. 8*)

the *mass-transfer zone.* As the top layer continues to adsorb, it eventually reaches saturation, where the concentration in the carbon is in equilibrium with the influent, and the adsorption zone moves downward. The concentration of the effluent exiting the column will remain essentially zero until the adsorption zone reaches the bottom of the column. When the effluent reaches some predetermined concentration, termed the *breakpoint,* the adsorber is generally removed from service. The major advantage of the column reactor is its very efficient utilization of the adsorption capacity, since most of the bed is in equilibrium with the influent concentration when the carbon is removed from service. Secondarily, any given volume of carbon is exposed to an increasing concentration of solute as the reaction progresses, resulting in a much higher overall reaction rate than can be achieved in completely mixed systems. Finally, since there is always fresh (unexposed) carbon ahead of the adsorption zone, changes in influent concentration generally do not produce changes in the effluent: they simply alter the rate at which the mass-transfer zone moves.[3,8,9]

Fixed-bed adsorbers can be operated in either an upflow or downflow mode. Downflow operation allows the adsorbers to also function as filters, although both the design and operation must address the need for frequent and efficient backwashing. Upflow operation, while not tolerating heavy solids loading, generally entails fewer problems with channeling, fouling, and the development of excessive head loss. In addition, the particle size of the adsorbent can be smaller, resulting in an increase in the rate of adsorption, and thus a decrease in the adsorber size. An adsorption design may also include multiple adsorbers, in which case the columns can be operated in series or in parallel. The series arrangement is generally operated such that when the first column is removed from service, the second column is moved up to the first position and the new column (or regenerated carbon) is brought on line in the second position. The benefit of this approach is that the carbon is more fully utilized than it is in single-adsorber systems. In the parallel arrangement, the removal of the adsorbers from service is staggered and the effluent from all the in-service adsorbers is mixed, such that the system is operated using adsorbers at various points along the breakthrough curve. The result is less variation in effluent quality.[12]

A second type of column reactor is the pulsed-bed adsorber. In this process only a portion of the spent carbon is removed from service when some critical effluent concentration is reached. This is usually accomplished by operating the column in an upflow mode and periodically adding fresh carbon to the top of the column and withdrawing spent carbon from the bottom. Since only fully saturated carbon is removed from service, the pulsed-bed process is more efficient and results in lower carbon usage than fixed-bed systems. This must be balanced against higher capital costs for pulsed-bed systems, however. An additional consideration when choosing between the two types of systems is the variation in effluent quality. Whereas the fixed-bed process produces an effluent that varies from essentially zero up to some maximum level, the pulsed-bed process can be designed to produce concentration changes from some selected level below the maximum, up to the maximum, simply by controlling the size and frequency of the pulses. The result is a more constant effluent. In the extreme case of a moving-bed adsorber, the carbon is added and withdrawn continuously as in a countercurrent-flow reactor, and the effluent concentration is constant.[6,12]

Both fixed-bed and pulsed-bed adsorbers can be operated in what is termed an upflow expanded-bed mode. In this mode of operation sufficient fluid velocity is maintained to expand the bed volume by approximately 10%. This will allow suspended solids in the influent to pass through the bed, thus eliminating the need for prefiltration. A drawback to this approach is higher carbon losses as a result of abrasive wear from particle movement in the bed.[12]

Combined Adsorption and Biological Treatment Processes

In many waste streams, there will be microbiological activity on the carbon surface unless specific preventive steps are implemented. This results from the ability of activated carbon to adsorb oxygen, nutrients, and substrate material, and from the propensity of microorganisms to attach to surfaces. Whether this activity is beneficial or detrimental to the adsorption process per se is the subject of considerable debate among researchers. The major issues of contention relate to whether biological growth significantly impedes the adsorbate flux to the carbon surface; whether the carbon can serve to increase the time that organics resistant to biodegradation are held in the system, thus promoting their destruction; and whether the organisms can serve to "bioregenerate" the activated carbon.[3,8]

From a practical standpoint, it is clear that biological growth on carbon adsorbers can either enhance or degrade the overall effectiveness of the treatment process, depending upon the waste stream involved and how the system is operated. There have been several efforts designed to exploit the potential benefits of biological growth, using what are sometimes termed *bioactive carbon systems*. The most widely known is the PACT® system, a patented process in which powdered activated carbon is added to an activated-sludge system. In comparing the PACT® process to conventional activated-sludge systems, it is reported that the PACT® system can handle higher loading rates, is not as easily disrupted by influent spikes, and will produce a sludge with superior settling characteristics. This approach is extremely attractive to facilities with an existing activated-sludge system because of the savings in capital equipment costs that it permits. It is perhaps less widely known that it is common to encourage biological activity in adsorption columns by the addition of oxygen, ozone, or other oxidants prior to the column. This can have the effect of both increasing dissolved-oxygen levels in the bed and partially oxidizing organic molecules, making them more easily degraded.[13]

Modeling Reactor Performance

The efficient design of adsorption systems requires some a priori knowledge of how the system will perform. Unfortunately, the theory of activated-carbon adsorption is not developed to the point where all the process variables can be accounted for and translated into design specifications. It is therefore essential that the design process include a substantial modeling component—an expenditure of time and effort that is almost always cost-beneficial relative to the alternative: overdesign. The most efficient approach involves a combination of physical and mathematical modeling. Physical modeling of the adsorption system should include both bench-scale and pilot-scale studies. The collected data can then be analyzed and interpreted using one or more of the available mathematical models. It is important to note that these two efforts are integrally linked and cannot be conducted independently. Recently, several techniques involving "minicolumns" or "microcolumns" have been developed. These approaches use small physical models, short testing times, and innovative mathematical models to simplify the modeling process.[14]

6.1.5 DESIGN CONDSIDERATIONS

Preliminary Assessment of Adsorption Feasibility

Carbon adsorption should be considered as a potential removal process for organic contaminants that are nonpolar, of low solubility, or of high molecular weight. Table 6.1.1 shows a number of classes of organic compounds that are generally amenable to adsorption. Certain inorganic compounds will also adsorb to activated carbon, but the technology is not widely used for this application. This is because of problems in the regeneration process when there are high levels of adsorbed inorganics, and because processes that are more cost-effective are frequently available.[12]

To assess the feasibility of adsorption for a particular application, it is generally necessary to perform an adsorption isotherm on the waste stream of interest. A number of standard techniques are available in the literature. This data will provide an estimate of the adsorption capacity. When combined with data on the contaminant concentration in the waste stream, effluent goal, waste flow rate, and desired carbon regeneration (or replacement) frequency, isotherm data can be used to estimate the size and cost of an adsorber. It should be noted that this approach will provide an estimate that is frequently *lower* than the actual size and cost since it assumes equilibrium conditions and complete utilization of the carbon capacity—assumptions that are generally violated in full-scale systems. For this reason, this approach may be most valuable as a means to eliminate adsorption from consideration if an alternative process appears to be less costly in the preliminary assessment.[12]

If it is obvious that carbon adsorption will be applied to the waste stream of

TABLE 6.1.1   Classes of Organic Compounds Amenable to Adsorption on Activated Carbon

Aromatic solvents	Benzene, toluene, xylene
Polynuclear aromatics	Naphthalene, biphenyls
Chlorinated aromatics	Chlorobenzene, PCBs, Aldrin, Endrin, toxaphene, DDT
Phenolics	Phenol, cresol, resorcinol
High-molecular-weight aliphatic amines and aromatic amines	Aniline, toluene diamine
Surfactants	Alkyl benzene sulfonates
Soluble organic dyes	Methylene blue, textile dyes
Fuels	Gasoline, kerosene, oil
Chlorinated solvents	Carbon tetrachloride, perchloroethylene
Aliphatic and aromatic acids	Tar acids, benzoic acids

Source: O'Brien et al., Ref. 12.

interest, it may be advantageous to proceed directly to column studies, which are somewhat more involved and therefore more costly, in order to develop a better estimate of final costs. Alternatively, if only a very rough approximation is needed, the use of isotherm results from the literature can be considered.

Pretreatment

With many wastes it will be necessary to pretreat the stream prior to adsorption. This will, of course, have a direct impact on the cost of an adsorption system and should be considered before any evaluation beyond the feasibility study is conducted. Suspended solids, oil and grease, and unstable chemical compounds in the influent stream are the most problematic.

Suspended-solids concentrations are constrained in fixed-bed adsorbers primarily because of the development of excessive head loss and the resulting backwash requirements. A suspended-solids concentration of 50 mg/L is often considered a practical limit since this corresponds to one backwash every 24 h using 2.5% of the product water at a hydraulic loading rate of 4 gpm/ft². A limit of 5 to 10 mg/L is suggested for downflow fixed-bed columns without backwash capability and upflow fixed-bed columns. Somewhat higher limits will apply to pressurized systems. Upflow expanded beds can tolerate very high levels of suspended solids since the solids tend to pass through the bed, although a subsequent solids-removal process is generally required. Oil and grease should be limited to 10 mg/L since higher levels will tend to coat the carbon particles, thereby reducing the adsorption capacity. Similarly, inorganic species, precipitation agents, and coagulents that are unstable with respect to the wastewater will tend to foul the carbon. Hardness often produces problems of this type since it may be present in a facility's raw water supply. Precipitation and coagulation problems can often be addressed by providing effective filtration prior to adsorption, and unstable minerals can often be stabilized by pH adjustment or the addition of scale inhibitors.[4,12]

Treatability Studies

Bench-scale treatability studies should be performed once it is determined that carbon adsorption can produce an effluent of acceptable quality at a reasonable cost, on the basis of the equilibrium concentration data and carbon-usage estimates obtained from the isotherm tests. It is imperative that bench-scale tests be designed to accurately represent the intended full-scale system. A procedure that is applicable to fixed-bed columns, the most common approach, is summarized below. This procedure can be readily modified for other reactor systems.

A system of columns in series, such as that shown in Fig. 6.1.5a, should be constructed. It is recommended that the columns have a diameter of at least 30 times the average carbon-particle diameter. Hydraulic loading rates should be the same as intended in the full-scale system, typically 2 to 8 gpm/ft^2. These two parameters will allow the determination of the column lengths that are required to achieve a given set of contact times, typically 60 to 150 min for the complete series of columns. In performing the experiment, the concentration of the contaminants of interest should be determined at regular intervals for the system influent and the effluents of each of the columns. The resulting series of breakthrough curves can be plotted as shown in Fig. 6.1.5b. From this data, plus the definition of an effluent objective, the carbon-exhaustion rate as a function of contact time can be determined. This information is plotted in Fig. 6.1.5c, where it can be observed that the exhaustion rate asymptotically approaches a minimum exhaustion rate. This rate should correspond to the isotherm point determined by equilibrium with the initial concentration of the waste. The critical design information provided by this approach, and other similar approaches described in the literature, is an understanding of the relationship between contact time or bed depth and regeneration or replacement frequency.[4,12,15,16]

This bench-scale treatability test is an empirical approach to carbon-column design. As such, the data produced are representative only of the conditions used in the test. To evaluate other design variables, additional testing is normally required. This may be a tedious and lengthy process if several variables are to be optimized. An alternative is to supplement bench-scale testing with mathematical modeling as described in a previous subsection. A major advantage of this approach is that once a model system is calibrated and verified, the effect of many operational parameters can be evaluated using computer simulations.[14]

Pilot-Scale and Full-Scale Design

The details of the design process are beyond the scope of this handbook and are readily available in the literature. In general, the most efficient approach involves continuation of bench-scale and modeling activities until all design and operational variables have been optimized. It is highly recommended that pilot-scale studies be conducted as the next step, in order to verify that the design can be successfully scaled up and to fine tune the optimization of variables. With the data collected in the pilot-scale studies a full-scale design can be initiated.[3,4]

Regeneration

Adsorption is normally a reversible process; that is, under suitable conditions the materials that have accumulated in the carbon can be driven off and the carbon re-used. In waste-treatment applications where carbon usage is low, this will not be cost-effective and the spent carbon should be disposed of in a suitable manner. For larger waste-treatment systems, carbon regeneration should definitely be

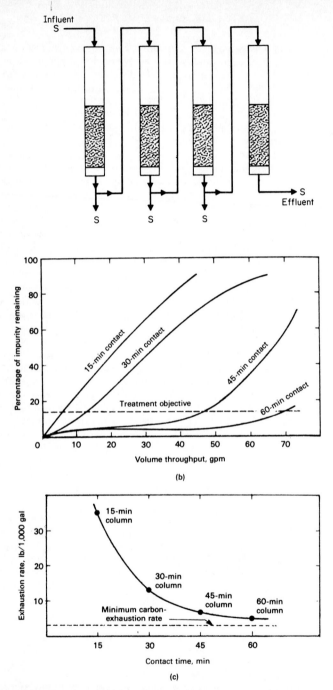

FIG. 6.1.5 Test apparatus and typical results for treatability testing. (*After O'Brien et al., Ref. 12*) (*a*) Schematic diagram of bench-scale treatability test apparatus. S = sample collection point. (*b*) Typical breakthrough curves from pilot carbon adsorbers. (*c*) Typical plot of superficial contact time versus exhaustion rate.

considered, although the critical size at which it becomes cost-effective varies with the application. The decision is essentially an economic one in which a number of factors must be considered including the cost of the carbon, the type of regeneration system required, carbon losses in regeneration, adsorption-capacity losses in regeneration, the cost of disposal of the spent carbon, and the availability of off-site regeneration services. In addition to waste-treatment systems, carbon is frequently used in product-recovery systems in which regeneration is an integral component of the process.[12]

Thermal reactivation is the most widely used regeneration technique. This is most commonly accomplished using a multiple-hearth furnace or rotary kiln operating at temperatures of 870 to 980°C (1600 to 1800°F) and a carbon residence time of approximately 30 min. Steam may be introduced to aid the reactivation process. Carbon losses of 5 to 10% can be expected. Air-pollution control equipment, including afterburners to ensure the complete oxidation of organic vapors and a particle-removal system, are normally required. With properly designed and operated control equipment the air emissions should be minimal. Some adsorbates may interfere with or serve to prohibit thermal regeneration. Included in this list are inorganic salts and minerals which are not thermally desorbed, materials which would cause air-pollution problems that cannot be contained by conventional air-pollution control equipment, explosive materials such as TNT, or strong acids which make handling the carbon difficult. In certain cases these problems can be addressed by a step such as prewashing the carbon.[4,17]

Nonthermal regeneration techniques are most commonly employed when carbon is used as a recovery medium. For example, phenol can be adsorbed from water under acidic conditions and, because of the much lower adsorption energy of phenolate, desorbed for recovery under basic conditions. A similar process has been used for the recovery of acetic acid and p-cresol. Acid desorption of basic adsorbates such as ethylene diamine is also practiced. Volatile solvents can often be adsorbed from either air or water and desorbed using steam regeneration.[6]

Costs

A generic approach to cost prediction for activated-carbon adsorbers is difficult to construct since differences between systems will radically affect the cost. A cost estimate is normally needed prior to undertaking a full-scale design, however, so the following information is provided. The data are derived primarily from municipal water and wastewater applications using fixed-bed adsorbers. It should be kept in mind that hazardous- and industrial-waste applications may be significantly different both in the equipment utilized and the manner in which it is operated.

The cost curves shown in Fig. 6.1.6 are based on material provided by O'Brien et al. Separate data are presented for the fixed-bed adsorption and thermal regeneration systems, since the sizes of these two operations are normally independent of each other. Capital costs for fixed-bed systems are shown as a function of wastewater flow rate in Fig. 6.1.6a for both backwashable and nonbackwashable adsorbers. These costs include site preparation, foundations, building, feed and backwash pumps, air compressor, electrical components, automatic controls, engineering, overhead, and profit. Operating costs are relatively minor for small, well-behaved systems, consisting primarily of power (approximately $15,000 per year for a 1 mgd system) and labor (2 to 3 checks per

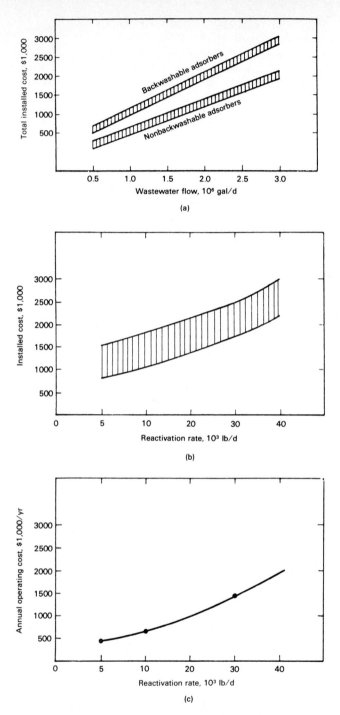

FIG. 6.1.6 Cost-estimation curves for adsorption and regeneration systems. (*After O'Brien et al., Ref. 12*) (*a*) Installed costs for fixed-bed adsorption systems. (*b*) Installed costs for reactivation systems. (*c*) Operating costs for reactivation systems.

shift). Capital costs for regeneration systems are shown in Fig. 6.1.6*b*. The data provided are for a multiple-hearth furnace or rotary kiln with an afterburner and wet scrubber, and storage tanks for spent and reactivated carbon. Operation and maintenance costs for the regeneration facility are shown in Fig. 6.1.6*c*.[4,12,18]

6.1.6 OPERATIONAL CONSIDERATIONS

Performance Monitoring

The operational demands of a carbon-adsorption system will depend largely on the size and type of system installed. Normally, an operation and maintenance plan is considered part of the design process and will be developed and instituted by the designer. Regardless of the specific details of a particular plan, however, there is a need in all systems for continuing performance monitoring of contaminant concentrations in the influent and effluent. This is required to assure that effluent objectives are being met, to determine when the carbon needs to be regenerated, and to provide a data base against which operational problems can be evaluated.[12]

Analyses of influent and effluent concentrations should be performed at regular intervals, which are determined by the regulatory requirements, the expected regeneration frequency, and the variability of flow and concentration in the influent. No general guidelines for monitoring frequency are available; rather, the frequency should be established by determining the maximum allowable period of time for which the most extreme expected deviation in system performance can be tolerated. In many cases, surrogate analytical procedures can be substituted for more costly or slower analyses, at least for some portion of the monitoring requirement. For example, total organic carbon (TOC) measurements might be substituted for biological oxygen demand (BOD) or chemical oxygen demand (COD) tests, or gas chromatograph analyses for gas chromatograph/mass spectrometry procedures. Typically such substitutions do not totally eliminate the need for the more elaborate test; they only allow the simpler technique to be used for regular monitoring as an indicator of system performance.[12]

Procedures also need to be established to monitor the performance of regeneration systems. This is most commonly accomplished by conducting adsorption isotherms on the reactivated carbon. It is most useful to measure the adsorption capacity of the carbon for the constituents of interest in the full-scale system, but surrogate tests such as iodine number can also be used.[4]

Operational Problems

Carbon adsorption is a complex process and as such there are a multitude of potential operational problems associated with it. While many of these are application-specific, many others occur with such regularity that a brief discussion is warranted. Several common problems and possible corrective actions are discussed below.

The most common problem with carbon adsorbers is the development of excessive head loss as a result of suspended-solids accumulation or biological growth in the bed, or fouling of the influent screen. Screen fouling can often be remedied by simply reversing the flow momentarily to dislodge the accumulated solids. If the problem is in the bed itself, then full backwashing with 50% bed expansion is probably warranted, although a temporary flow increase or moderated bed expansion may be sufficient for upflow beds. If problems with premature head loss continue, alternatives such as instituting a more frequent backwash cycle, increasing the scouring in the backwash process, or adding a disinfectant such as chlorine to reduce biological growth in the bed should be considered.[4]

Biological growth is also associated with hydrogen sulfide, which is a highly corrosive end product of sulfate-reducing bacteria under anaerobic conditions. Hydrogen sulfide production appears to be promoted by low concentrations of dissolved oxygen and nitrate in the influent, high concentrations of BOD and sulfate, long detention times, and low velocities. In addition to instituting steps to limit biological growth, hydrogen sulfide production may be reduced by adding oxygen or nitrate to the influent, reducing high detention times to the design specification, and increasing the removal of BOD prior to adsorption.[4]

Another common operational problem is premature exhaustion of the carbon capacity. While this may result from inadequate regeneration, it has been attributed in some cases to the presence in the influent of high-molecular-weight compounds that adsorb slowly, but with a high adsorption energy. The result is that the internal pores in the deep regions of the bed become blocked by the high-molecular-weight constituents and are unavailable to the contaminant of interest. A related group of operational problems can occur in systems with multiple adsorbable species. In this situation, a previously adsorbed compound can be displaced by a compound with a higher adsorption energy, causing an effluent spike that may exceed influent concentration levels. Similarly, a change in influent conditions such as pH can cause this type of chromatographic behavior.[3,19]

6.1.7 SOURCES OF ADDITIONAL INFORMATION

The references provided at the end of this section are intended to direct the reader to good summary and review publications on each of the major topics covered in the section. These publications are, for the most part, heavily referenced and should provide a means to locate the most significant original technical reports on almost all aspects of activated-carbon technology. For additional assistance with specific waste-management applications, several options are available. A number of private consultants, including many of the authors listed in the references, are available to help determine the applicability and feasibility of carbon adsorption for specific situations. It is particularly important, in seeking this type of expertise, to ensure that all available options are considered, including systems other than adsorption and the possibilities for waste reduction. Engineering consulting firms and the major activated-carbon manufacturers can provide assistance with adsorption-system design, construction, operation, and maintenance.

6.1.8 REFERENCES

1. W. J. Weber, Jr., "Evolution of a Technology," *Journal of Environmental Engineering,* **110**:(5)899–917 (1984).
2. T. C. Voice and W. J. Weber, Jr., "Sorption of Hydrophobic Compounds by Sedi-

ments, Soils and Suspended Solids—I: Theory and Background," *Water Research,* **17**:1,433–1,441 (1983).

3. W. J. Weber, Jr., *Physicochemical Processes for Water Quality Control,* Wiley-Interscience, New York, 1972.

4. U.S. Environmental Protection Agency, *Process Design Manual for Carbon Adsorption,* EPA 625/1-71-002a, 1973.

5. P. N. Cheremisinoff and A. C. Morresi, "Carbon Adsorption Applications," in P. N. Cheremisinoff and F. Ellerbusch (eds.), *Carbon Adsorption Handbook,* Ann Arbor Science, Ann Arbor, Michigan, 1978, pp. 1–54.

6. W. J. Lyman, "Applicability of Carbon Adsorption to the Treatment of Hazardous Industrial Wastes," in P. N. Cheremisinoff and F. Ellerbusch (eds.), *Carbon Adsorption Handbook,* Ann Arbor Science, Ann Arbor, Michigan, 1978, pp. 131–166.

7. R. A. Dobbs and J. M. Cohen, *Carbon Adsorption Isotherms for Toxic Organics,* EPA-600/8-80-023, 1980.

8. W. J. Weber, Jr., and F. E. Bernardin, Jr., "Removal of Organic Substances from Municipal Wastewaters by Physicochemical Processes," in B. B. Berger (ed.), *Control of Organic Substances in Water and Wastewater,* EPA-600/8-83-011, 1983, pp. 203–254.

9. W. J. Weber Jr., and E. H. Smith, "Activated Carbon Adsorption: The State of the Art," presented at the Fifth International Conference on Chemistry for Protection of the Environment, Leuven, Belgium, September 10–13, 1985, (1985).

10. J. L. Kovach, "Gas-Phase Adsorption and Air Purification," in P. N. Cheremisinoff and F. Ellerbusch (eds.), *Carbon Adsorption Handbook,* Ann Arbor Science, Ann Arbor, Michigan, 1978, pp. 331–358.

11. J. W. Leatherdale, "Air Pollution Control by Adsorption," in P. N. Cheremisinoff and F. Ellerbusch (eds.), *Carbon Adsorption Handbook,* Ann Arbor Science, Ann Arbor, Michigan, 1978, pp. 371–388.

12. R. P. O'Brien, J. L. Rizzo, and W. G. Schuliger, "Removal of Organic Compounds from Industrial Wastewaters Using Granular Carbon Column Processes," in B. B. Berger (ed.), *Control of Organic Substances in Water and Wastewater,* EPA-600/8-83-011, 1983, pp. 337–362.

13. D. G. Hutton, "Combined Powdered Activated Carbon–Biological Treatment," in P. N. Cheremisinoff and F. Ellerbusch (eds.), *Carbon Adsorption Handbook,* Ann Arbor Science, Ann Arbor, Michigan, 1978, pp. 389–448.

14. W. J. Weber, Jr., "Modeling of Adsorption and Mass Transport Processes in Fixed-bed Adsorbers," in A. L. Myers and G. Belfort (eds.), *Proceedings, First International Conference on the Fundamentals of Adsorption,* Engineering Foundation and AIChE, New York, 1984.

15. C. E. Adams, Jr., and W. W. Eckenfelder, Jr. (eds.), *Process Design Techniques for Industrial Waste Treatment,* Enviro Press, Nashville, Tennessee, 1974.

16. T. D. Reynolds, *Unit Operations and Processes in Environmental Engineering,* Wadsworth, Belmont, California, 1982.

17. L. A. Lombana and D. Halaby, "Carbon Regeneration Systems," in P. N. Cheremisinoff and F. Ellerbusch (eds.), *Carbon Adsorption Handbook,* Ann Arbor Science, Ann Arbor, Michigan, 1978, pp. 905–922.

18. R. H. Zanitsch and M. H. Stenzel, "Economics of Granular Activated Carbon Water and Wastewater Treatment Systems," in P. N. Cheremisinoff and F. Ellerbusch (eds.), *Carbon Adsorption Handbook,* Ann Arbor Science, Ann Arbor, Michigan, 1978, pp. 215–240.

19. W. B. Arbuckle, "Premature Exhaustion of Activated Carbon Columns," in M. J. McGuire and I. H. Suffet (eds.), *Activated Carbon Adsorption of Organics from the Aqueous Phase,* Ann Arbor Science, Ann Arbor, Michigan, 1980, pp. 237–252.

SECTION 6.2
DISTILLATION

Tony N. Rogers

Research Triangle Institute
Research Triangle Park, North Carolina

George Brant

Associated Technologies, Inc
Charlotte, North Carolina

6.2.1 DEFINITION OF DISTILLATION; APPLICATIONS OF THE PROCESS

Distillation is an operation which has been widely used throughout the chemical and petroleum industries for many years. In a significant number of situations, distillation is the only feasible method for separating components in liquid or gas streams. Distillation systems vary in complexity from simple batch processes to very sophisticated multiple-column units, with the system configuration depending upon the objectives of the operation.

Broadly defined, *distillation* is the separation of more volatile materials from less volatile materials by a process of vaporization and condensation. In engineering terminology, the separation of a liquid from a solid by vaporization is considered evaporation, and the term *distillation* is reserved for the separation of two or more liquids by vaporization and condensation.

By far the most numerous industrial applications of distillation have been for purification in chemical manufacturing and in processes involving internal solvent recycling. Distillation is also capable of recovering volatile species with little degradation, which is an important advantage in situations where the recovered species can be sold or recycled. The increasing difficulty and cost entailed in disposal of chemical wastes, combined with the rising cost of raw materials in recent years, has made distillation attractive for recovery of organic solvents from waste streams that would otherwise be discarded. Product purity of any desired level can theoretically be obtained by distillation, provided the process economics are not prohibitive.

6.2.2 GENERAL PRINCIPLES OF DISTILLATION

Basic distillation involves application of heat to a liquid mixture, vaporization of part of the mixture, and removal of heat from the vaporized portion. The resulting condensed vapor, called the *distillate,* is richer in the more volatile components, and the residual unvaporized bottoms are richer in the less volatile components. Most commercial distillations involve some form of multiple staging to obtain a greater enrichment than is possible by a single vaporization and condensation operation.

In *simple distillation,* a single equilibrium stage is used to obtain a desired separation, and the operation may be either batch or continuous. *Simple continuous distillation* (also called *flash distillation*) has a continuous feed to an equilibrium stage; the liquid and vapor leaving the stage are in equilibrium. Flash distillation is used in applications where a crude separation is adequate. The component separation in simple distillation is limited by thermodynamic partitioning constraints, and multiple staging must therefore be used to increase the separation efficiency.

Multiple staging in a column design is achieved by returning part of the condensed overhead vapors to the top of the column, thereby bringing this reflux liquid into intimate contact with the rising vapors. Either a tray or a packed column is normally used to provide the necessary gas-liquid interfacial area for mass transfer. The degree of component separation for a given system configuration is dependent upon the operating conditions, the number of stages, and the amount of reflux.

6.2.3 DESCRIPTION OF DISTILLATION PROCESSES

Distillation can be conducted in one of two possible modes: batch or continuous. Most large-scale industrial applications utilize continuous distillation;[1] however, batch distillation can prove to be advantageous in some situations.

Batch (Differential) Distillation

Simple *batch (differential) distillation* equipment consists of a heated vessel (the *still* or *pot*), a condenser to condense the overhead vapor, an accumulator (a necessity if reflux is employed), and several receiving vessels to collect the overhead and bottoms products. Operation involves charging a batch of feed to the still with subsequent boiling of the feed mixture. The overhead vapor stream is condensed and collected in an appropriate receiving vessel. *Reflux* (returning some of the condensed overhead product to the still) may or may not be employed, depending upon the desired product composition. Use of simple batch distillation is usually restricted to preliminary separations in which the products are held for additional separation at a later time.

In most batch distillation operations, the compositions of the overhead and bottoms products will vary as the distillation proceeds. In the absence of reflux or at a constant reflux ratio, the overhead product will become progressively richer in the less volatile components, while the residual liquid, or *bottoms,* becomes progressively leaner in the more volatile components. Thus, depending upon the planned ultimate fate of the products, either the overhead product composition or the bottoms product composition may dictate the extent to which the distillation process is carried out.

To obtain products within a narrower composition range, batch distillation with *rectification* may be used.[2] The components of this system (see Fig. 6.2.1) are the pot, a rectifying column, a condenser, a splitter for returning a portion of the distillate as reflux, and one or more receiving vessels for the product. In operation, a batch of feed is charged to the pot, and the system is brought to steady state under total reflux conditions. Once steady state is achieved, removal of some of the overhead product (that portion not returned to the column as reflux) is begun. Some control over the overhead product composition can be achieved by appropriate adjustment of the reflux ratio as the distillation proceeds. According to Perry and Chilton,[2] there are two primary modes of operation: (1) constant reflux ratio with varying overhead composition, and (2) constant overhead composition with varying reflux ratio.

Under the first mode, distillation is conducted at a constant, predetermined reflux ratio until the average distillate composition reaches the desired value. At this point, the overhead product is diverted to another receiving vessel, and an intermediate cut is withdrawn until the liquor remaining in the pot reaches the desired composition.

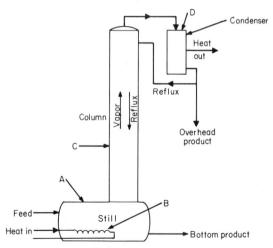

FIG. 6.2.1 Batch distillation. Still with fractionating column. A = still. B = heating coil. C = column. D = condenser.

The second mode, constant overhead composition, can be maintained by constantly increasing the reflux ratio as the run proceeds. At a point approaching total reflux, the overhead is diverted to a different receiving vessel, and the reflux ratio is decreased. An intermediate cut is then collected (as in the first mode) to achieve the desired composition in the residual liquid.

Continuous Distillation

Continuous distillation (see Fig. 6.2.2) involves the constant introduction of feed stream(s) to an appropriately designed distillation column. The feed stream (liquid, vapor, or a mixture) is fed to a column that contains plates or packing designed to provide intimate contact between vapor and liquid. According to Berkowitz et al.,[1] packing is normally used only in small-scale equipment. If only one feed stream is involved, the *cascade* (set of plates) above the feed introduc-

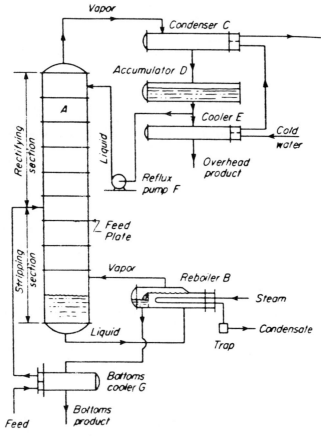

FIG. 6.2.2 Continuous distillation. Continuous fractionating column with rectifying and stripping sections.

tion point is termed the *rectifying* or *enriching section* of the column, while the cascade below the feed stage is called the *stripping section.*

Vapor rising through the rectifying section of the column is washed with liquid to remove the less volatile components, while the liquid in the stripping section of the column is stripped of the more volatile components by the rising vapor in that section. A condenser is used to condense the overhead vapor that exits the rectifying section of the column. A portion of this condensate is continuously withdrawn as the overhead product, while the remainder of the condensate is returned to the column as reflux. At the bottom of the column, liquid that exits the stripping section is collected in a reboiler, where it is heated to produce a vapor stream that is returned to the column. The portion of the liquid that is not vaporized and returned to the column is continuously withdrawn as the bottoms product.

6.2.4 LIMITATIONS OF DISTILLATION

Equipment and Operating Restrictions

Because of inherent thermodynamic constraints, distillation tends to be a rather complex operation with a number of practical limitations on its usefulness. First, equipment and auxiliaries for distillation processes are usually comparatively large. They can have heights exceeding 60 m (200 ft) and can cover large areas. Second, such equipment is expensive, and capital-recovery charges usually constitute the major portion of the product-recovery cost. Third, distillation equipment is often complex and must be operated by highly skilled personnel. Finally, recovery by distillation is energy-intensive, with nominal energy requirements being about 5.81×10^5 to 2.79×10^8 J/kg of feed (250 to 1200 Btu/lb of feed).

Applicability to Specific Waste Types

There are several important constraints on the physical form and chemical nature of wastes that can be treated by distillation. One such constraint is that the feed to a continuous-distillation column must be a free-flowing fluid with a negligible solids content. Solid materials will severely plug and foul the internals of the column. If the waste stream does contain solids or highly viscous liquids, some form of prior conditioning is required. There are, however, some free-flowing fluids that cannot be treated by distillation. These include organic peroxides, pyrophoric organics, and most inorganic wastes. Feeds that tend to polymerize can likewise cause severe operational difficulties.

The hazardous waste streams for which removal and reclamation of volatiles by distillation is most suitable are liquid organics, including organic solvents and halogenated organics. Typical industrial wastes that can be handled by distillation include the following:

- Plating wastes containing an organic component—usually the solvents are evaporated and the organic vapors distilled.
- Aqueous wastes containing phenol.
- Polyurethane waste containing methylene chloride.

- Ethylbenzene–styrene mixtures.
- Waste solvents—usually mixtures containing ketones, alcohols, and aromatics.
- Waste lubricating oils.
- Wastes containing butyl acetate that are produced via antibiotics (penicillin) manufacture.

Maintaining Product and Feed-Stream Compositions

As might be expected, continuous distillation is not well suited for wastes that exhibit large and/or frequent variations in composition. This is primarily because of the complexity of the process dynamics involved, i.e., the distillation column's ability to respond to fluctuations in feed-stream composition so as to maintain the desired overhead and bottoms compositions.

Batch-distillation designs are, on the other hand, used quite frequently in situations where mixtures of widely varying compositions are to be distilled. As previously mentioned [see under *Batch (Differential) Distillation,* on p. 6.24], careful regulation of the reflux ratio can be used to exercise control over the overhead product composition. In addition, the charge to a batch unit may have a high solids content, or it may contain tars or resins that would plug or foul a continuous unit.[2] Such materials merely settle out in the pot and are conveniently removed at the termination of the process.

Batch operation, because of its cyclic nature, involves frequent start-ups and shutdowns. Thus, an irregular supply of waste should not force undesired start-ups and shutdowns that would tend to cause operational problems. A continuous unit, by contrast, can be greatly hampered by such breaks in operation.

For specific applications, therefore, the choice between batch and continuous distillation should be studied carefully. There are many circumstances in which batch distillation may be favorable even though it is considerably more difficult to maintain the composition of the products at a constant level. Batch distillation is preferred when (1) the composition of the feed varies significantly, (2) the feed contains significant amounts of solids that would tend to foul and plug a continuous unit, and (3) excessive start-up and shutdown might be required because of an irregular supply of the feed.

6.2.5 PROCESS DESIGN CONSIDERATIONS

In distillation processes, the number of plates or stages required for a given separation is dependent on the reflux rate. As the reflux is increased, the required number of stages falls. An infinite reflux ratio would be required at the minimum number of stages, and an infinite number of stages would be required at the minimum reflux required to effect the separation. The optimum process will obviously lie somewhere between these extremes.

As the reflux rate is increased, the required number of stages decreases, resulting in a lower cost for the column, even though the column diameter must increase to maintain an acceptable pressure drop with the increased vapor flow. However, as the reflux increases, reboiler steam requirements increase proportionately. Steam is the single biggest operating cost, accounting for over 50% of the variable cost. In addition, a larger reboiler and condenser are required, increasing the cost of these components. It has been found that the optimum reflux ratio generally will be between 1.1 and 1.25 times the minimum theoretical ratio.[3]

6.2.6 SELECTION OF EQUIPMENT ITEMS

After a distillation process has been designed to achieve a specific separation for a particular feed stream, there are a number of different equipment types available for the system components. When selecting the best equipment designs for the process, consideration may be given to such factors as fouling tendency, corrosivity, throughput, versatility, and cost. The following discussion briefly outlines the many pieces of hardware required for a distillation system and the many equipment variations available to meet different process conditions. [This subsection is based on information contained in APV Equipment, Inc.'s, *Distillation Handbook* (Ref. 4).]

Column Shell

A distillation column shell can be designed either as a freestanding module or to be supported by a steel structure. A self-supporting column is generally more economical. Column fabrication in a single piece (without shell flanges) is more economical than with shell flanges, in addition to simplifying installation and eliminating potential sources of leakage. Columns over 24 m (80 ft) in length have been shipped by road without transit problems. Hazardous-waste processing may require a flange assembly to facilitate cleaning.

While columns over 1 m (3 ft) in diameter normally have been transported without trays to prevent dislodgement and possible damage, recent and more economical techniques have been devised for factory installation of trays with the tray manways omitted. Manways are added after the column has been erected, and the fitter inspects each tray. The position and number of manways are important, especially for systems which require periodic cleaning.

Packing can be installed prior to shipment in columns of 51 cm (20 in) diameter or less which use high-efficiency metal-mesh packing. Larger columns are packed on site to prevent the packing from compacting during transit and leaving voids. Random packing is almost always installed on site.

Additional requirements can include access platforms and interconnecting ladders for on-site attachment to freestanding columns.

Column Internals

During recent years, the development of sophisticated computer programs and new materials has led to many innovations in the design of trays and packings for more efficient operation of distillation columns.

Tray Devices. There are five basic types of distillation trays: sieve, valve, bubble-cap, dual-flow, and baffle trays—each with unique advantages and preferred usages. Sieve- and valve-type trays currently are most often specified for tray towers.

The hydraulic design of a tray is a most important factor. The upper operating limit generally is governed by the flood point although, in some cases, entrainment can also restrict performance. Entrainment reduces concentration gradients because some liquid flows up the column, thereby lowering efficiency. A column can also be flooded by downcomer backup. The trays fill and the pressures increase when the downcomer is unable to handle all the liquid involved. This can occur with a liquid that foams profusely. Flooding is associated with large tray-pressure drops and small tray spacings.

The lower limit of tray operations is characterized by excessive liquid weeping from one tray to the next. Unlike the upward transport of liquid via entrainment, the flow of weeping liquid is in the normal direction, and considerable amounts can be tolerated before column efficiency is significantly affected. As the vapor rate decreases, however, a point is eventually reached when all the liquid is weeping, and there is no liquid seal on the tray. This is known as the *dump point*, below which there is a severe drop in efficiency.

Sieve Tray. The *sieve tray* (see Fig. 6.2.3*a*) is a low-cost device which consists of a perforated plate that usually has holes of 0.5 to 2.5 cm (³⁄₁₆ to 1 in) in diameter, a downcomer, and an outlet weir. Although inexpensive, a correctly designed sieve tray can be comparable to other tray styles in vapor and liquid capacities, pressure drop, and efficiency. Its flexibility, however, is inferior to that of valve and bubble-cap trays, and it is sometimes unacceptable for low liquid flows when weeping has to be minimized.

It is possible, however, to increase the flexibility of a sieve-tray column for occasional low-throughput operation by maintaining a high reboil and increasing the reflux ratio. This may be economically advantageous when the low through-

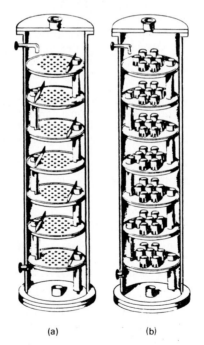

(a)   (b)

FIG. 6.2.3   (*a*) Sieve-tray column. (*b*) Bubble-tray column.

put occurs for a small fraction of the operating time. Flexibility likewise can be increased by the use of blanking plates to reduce the tray's hole area. This is particularly desirable for initial operation of a column in situations where it is envisioned that the plant capacity will be expanded after a few years. There is no evidence to suggest that blanked-off plates have inferior performance to unblanked plates of similar hole area.

Dual-Flow Tray. The *dual-flow tray* is a high-hole-area sieve tray without a downcomer. The downflowing liquid passes through the same holes as the rising vapor. Since no downcomer is used, the cost of the tray is lower than that of a conventional sieve tray.

In recent years, use of the dual-flow tray has declined somewhat because of difficulties experienced with partial bypassing of the liquid and vapor phases, particularly in large-diameter columns. Also, the dual-flow column has a very restricted operating range and a reduced efficiency because there is no cross-flow of liquid.

Valve Tray. While the *valve tray* dates back to the rivet type first used in 1922, many design improvements and innumerable valve types have been introduced in recent years. The wide selection of modern valve types (see Fig. 6.2.4) provides the following advantages:

1. Throughputs and efficiencies can be as high as for sieve or bubble-cap trays.
2. Very high flexibility can be achieved, and turndown ratios exceeding 4:1 are easily obtained, without resorting to large pressure drops at the high end of the operating range.
3. Special valve designs with venturi-shaped orifices are available for duties involving low pressure drops.
4. Although slightly more expensive than sieve trays, valve trays are very economical in view of their operating superiority.
5. Since an operating valve is continously in movement, the valve tray can be used for light to moderate fouling duties. APV Equipment, Inc., has successfully used valve trays with brewery effluents containing waste beer, yeast, and other materials with fouling tendencies.

Bubble-Cap Tray. Although many bubble-cap columns are still in operation, *bubble-cap trays* (see Fig. 6.2.3*b*) are rarely specified today because of their high

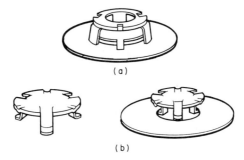

FIG. 6.2.4 Valves used in valve-tray columns. (*a*) Special two-stage valve with lightweight orifice cover for complete closing. (*b*) Two typical general-purpose valves useful for all types of services.

cost and the excellent performance of the modern valve-type tray. The bubble cap, however, does have a good turndown ratio and is suitable for low liquid flows.

Baffle Tray.   The liquid flows down the *baffle-tray* column by splashing from one baffle to the next lower baffle. The gas or vapor rises through this curtain of liquid spray. Although the baffle-type tray has a relatively low efficiency, it can be useful in treating waste flows when the liquid contains a high fraction of solids.

Column Packing.   Packing is the most economical method of contacting liquid and gas streams in distillation columns, particularly small-diameter columns. Most packings can be purchased from stock by the cubic foot. In addition, the mechanical design and fabrication of a packed column is quite simple (see Fig. 6.2.5). Packing is limited in waste-treatment applications due to its tendency to foul and its less predictable performance at low liquid flows or high column diameters because of liquid-distribution problems.

The most widely used packings are randomly dumped packings such as Raschig rings, Pall rings, and ceramic saddles (see Fig. 6.2.6). These are available in various plastics, a number of different metals, and, with the exception of Pall rings, in ceramic materials. While plastic packings have the advantage of corrosion resistance, the self-wetting ability of some plastic packings (such as those made of fluorocarbon polymers) sometimes is poor, particularly in aqueous systems. This considerably increases the height equivalent of a theoretical plate (HETP) as compared with equivalent ceramic packings.

High-efficiency metal-mesh packings have found increasing favor in industry during recent years. One type uses a woven wire mesh that becomes self-wetting because of capillary forces. This helps establish good liquid distribution as the liquid flows through the packing geometry in a zigzag pattern. If properly used,

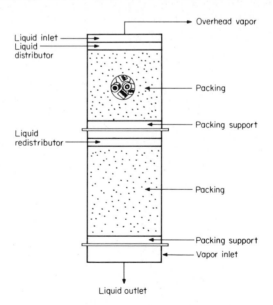

FIG. 6.2.5   Illustration of packed-column internals.

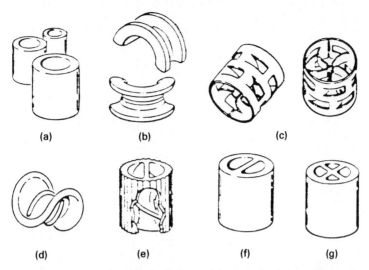

FIG. 6.2.6 Types of packing typically used in packed columns. (*a*) Raschig rings. (*b*) Intalox saddle. (*c*) Pall rings. (*d*) Cyclohelix spiral ring. (*e*) Berl saddle. (*f*) Lessing ring. (*g*) Cross-partition ring.

high-efficiency structured packings can provide HETP values in the range of 15 to 30 cm (6 to 12 in). This can reduce column heights, especially when a large number of trays is required. Such packings, however, are very expensive, and each application must be studied in great detail.

With both random and high-efficiency structured packings, considerable attention must be given to correct liquid distribution. Certain types of high-efficiency packing are extremely sensitive to liquid distribution and should not be used in columns over 0.6 m (2 ft) in diameter. Positioning of these devices and the design of liquid distribution and redistribution are important factors that should be determined only by experts.

Instrumentation

One of the most important requirements of any distillation system is the ability to maintain the correct overhead and bottoms compositions from the column by means of proper controls and instrumentation. While manual controls can be supplied, this approach is rarely used today in the United States. Manual control involves the extensive use of rotameters and thermometers, which, in turn, involves high labor costs, possible energy inefficiency, and, at times, poor quality control. Far better control is obtained through the use of pneumatic or electronic control systems.

Reboiler

Although there are many types of reboilers, the shell-and-tube thermosiphon reboiler is used most frequently. Boiling within the vertical tubes of the ex-

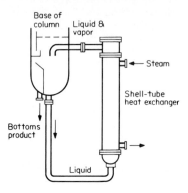

FIG. 6.2.7 Typical shell-and-tube thermo-siphon reboiler arrangement.

changer produces liquid circulation and eliminates the need for a pump. A typical arrangement is shown in Fig. 6.2.7.

For certain duties, particularly when the bottoms liquid has a tendency to foul heat-transfer surfaces, it is desirable to pump the liquid around the heat exchanger. Since boiling can be suppressed by use of an orifice plate at the outlet of the unit, fouling is reduced. The liquid being pumped is heated under pressure and then is flashed into the base of the column, where vapor is generated.

An alternative approach is the use of a plate heat exchanger as a forced-circulation reboiler. With this design, the very high liquid turbulent flow induced within the heat exchanger through the use of multiple corrugated plates holds fouling to a minimum. Meanwhile, the superior rates of heat transfer that are achieved reduce the surface area required for the reboiler.

Condenser

Since most distillation-column condensers are of shell-and-tube design, the design engineer has the option of condensing on either the shell or tube side. From the process point of view, condensation on the shell side is preferred since there is less subcooling of condensate and a lower pressure drop is required. These are important factors in vacuum duties. Furthermore, with cooling water on the tube side, any fouling can be removed more easily.

Tubeside condensation, on the other hand, can be more advantageous whenever process-fluid characteristics dictate the use of more expensive, exotic materials. In that case, capital costs of the unit may be reduced by using a carbon-steel shell.

Preheaters and Coolers

The corrosion characteristics of the waste stream dictate the selection of plate or shell-and-tube preheaters and product coolers. If the volatile organics do not excessively degrade or swell gasket materials, a plate heat exchanger is an extremely efficient preheater. Heat can be transferred from the tops and bottoms products for this purpose. The plate heat exchanger can be easily disassembled for cleaning. Heat transfer to aromatic species such as benzene and toluene normally is accomplished in a series of tubular exchangers.

Vent Condenser

It is normal practice in some distillation systems to use a vent condenser after the main condenser to serve as an air-pollution control system. Usually of the shell-and-tube type, the vent condenser will typically have one-tenth the area of the main unit. Chilled water is typically used to cool the noncondensible gases to about 7 to 10°C (45 to 50°F), although some plants use process cooling water as a backup system.

Pumps

Distillation is used to process many volatile organic fluids that are highly flammable. It is desirable that explosionproof (Class 1, Group D, Division 1) pump motors be used. Centrifugal pumps are reliable and can economically provide the required flow capacities and operating pressures. The bottoms product of the distillation unit or the reboiler may, however, contain particulate material, necessitating the use of positive-displacement pumps.

6.2.7 ECONOMIC EVALUATION OF DISTILLATION

Capital Cost

The capital cost of a distillation system is dependent upon such design variables as the size and type of reboiler, column height, column diameter, column internals, degree of automation, and materials of construction. These design considerations must be determined for the feed stream to be processed and the component separation required. Changes in component relative volatilities, feed rates, and required product purities can effect large differences in system costs.

Equipment installation costs vary with the type and size of equipment, the geographical location, and the site characteristics. Peters and Timmerhaus[5] present a general range of installation costs as a percentage of the purchased-equipment cost for various types of equipment. Taking into consideration the major equipment components of a distillation system, an installation cost of 25 to 50% of the equipment cost would be indicated. However, the APV *Distillation Handbook* refers to a typical installed system cost as being 1½ to 2 times the equipment cost.[4]

Custom-built, preassembled process systems or process modules are another option being offered by most system suppliers and this option is claimed to provide significant savings in cost and time. An overall cost savings of 25 to 40% for a preassembled unit is projected in a brochure published by Chem-Pro Equipment Corp.[6]

Operating Costs

The total operating costs will vary significantly from a small batch-distillation system to a very high throughput continuous-process system. The total cost for processing a large feed volume will be much greater than for the smaller operation, primarily because of the much greater steam usage. On a unit-cost basis, however, the steam usage becomes the predominant variable cost as the feed volume becomes large, and other costs such as labor and maintenance become less significant. In a small batch operation that may require considerable operator attention, the unit labor cost can be fairly significant (the result of low volume, high labor), while steam usage would be equivalent on a unit-cost basis to that for a high-volume process.

Economics Case Study

In a distillation-system application by APV Equipment, Inc., ethanol is recovered from an aqueous waste stream containing about 3% ethanol in addition to suspended sludge and yeast solids. One major problem is the tendency of the liquid to foul heat-transfer surfaces. A paraflow (corrugated-plate) heat-exchanger reboiler, which is less susceptible to fouling than many other types of heat-transfer equipment, is utilized to minimize this problem. The plate corrugations enhance the liquid turbulence, which assists in shearing the fouling deposits off the heat-transfer surface. In addition, the paraflow design allows easy cleaning in place and ready opening if fouling becomes excessive.

The efficiency of the column trays, however, has not been observed to decrease because of the fouling tendency of the liquid. APV Equipment, Inc., attributes this primarily to the use of valve trays as the gas-liquid contacting device. The small valves in the holes on the trays are continually in motion (with some rotation), thus helping to prevent buildup of foulant in the area around the holes. The performance and cost of the APV system are illustrated by the following process information:

Feed rate	250 L/min (70 gal/min)
Annual operation	8400 h/yr
Feed temperature	100°C (212°F)
Steam usage	0.2 lb/lb of waste processed
Ethanol concentration (feed)	3% v/v
Ethanol concentration (distillate)	95% v/v
Ethanol concentration (bottoms)	<0.02% v/v
System equipment cost (1½ times equipment cost assumed)	$1,050,000
Annualized system equipment cost (10 years, 10% interest)	$171,000
Annual operating cost	$165,300 (includes annual steam cost of $58,000)
Total annual cost	$336,300
Cost per unit volume of aqueous waste processed	$0.0025/L ($0.0095/gal)

Cost per unit volume of ethanol re- $0.084/L ($0.318/gal)
   covered

The bottoms product, containing water, solids, and no more than 0.02% v/v alcohol, can be readily utilized in related plant by-product processes.

Economic Viability

Summarizing the above discussion, distillation (either batch or continuous) tends to be energy-intensive and thus expensive unless the product can be recycled or sold. However, with sufficient by-product credits, it can sometimes compete with processes having a much lower capital-recovery cost. Berkowitz et al.[1] note that when distillation is competitive on an economic basis, it generally becomes the preferred method of recovery where salable organic species are involved.

6.2.8 ENVIRONMENTAL EFFECTS OF DISTILLATION

Air Emissions

The potential for volatile-organic emissions to the air from distillation processes includes losses through condenser vents, accumulator-tank vents, and storage-tank vents.[1]
   Emissions from the condenser vent should be primarily noncondensible gases. However, if the condenser should become overloaded for any reason, volatile organic carbon (VOC) emissions to the air could become significant. A vent condenser is typically used after the main condenser in order to minimize the amount of volatiles emitted. Emissions from operating and storage tanks would be displacement losses of equilibrium vapor from the liquid holdup.

Residuals

Waste treatment by distillation serves only to reclaim volatiles in a more concentrated form. In addition, process residuals such as still bottoms may present a waste-disposal problem. If volatiles can be recovered at desirable concentrations, it is likely that many could be sold. Otherwise, ultimate destruction or disposal that minimizes volatile emissions will be required. Solids, tars, or sludges recovered from still bottoms are normally incinerated.[1] Clearly, the ultimate fate of some distillation products may present another waste-disposal problem, i.e., that of disposing of concentrated volatile matter and/or other residuals produced by the distillation process.[7]

6.2.9 LIST OF EQUIPMENT VENDORS

A number of companies manufacture distillation equipment and provide complete package units for specific applications. Most distillation systems are custom-designed because of the large number of variable process factors that must be considered. These companies usually have computer capability for complete system design, as well as pilot plant facilities.

Most companies also provide complex distillation systems which include multicomponent units. In these, solvent extraction, carbon-adsorption, and distillation processes may be integrated in solvent-recovery or stream-purification operations. In most cases, complete package systems can be provided in preassembled modular or skid-mounted units. Such modularization usually results in reductions in installation cost and time.

Some of the major suppliers for distillation systems used in solvent-recovery and volatile-organic removal operations are listed below.

- *APV Equipment, Inc., Tonawanda, N.Y.:* Pilot-plant facilities for evaporation.
- *Aqua-Chem, Inc., Milwaukee, Wis.:* Pilot-plant facilities for evaporation.
- *Artisan Industries, Inc., Waltham, Mass.:* Pilot-plant facilities for distillation, solvent extraction.
- *Chem-Pro Corp., Fairfield, N.J.:* Pilot-plant facilities for distillation, stripping, extraction.
- *DCI Corp., Columbus, Ohio:* Provides standard package systems for distillation and live-steam stripping.
- *Glitsch, Inc., Dallas, Tex.:* Pilot-plant facilities for distillation, scrubbing, stripping, extraction.
- *The Pfaudler Company, Rochester, N.Y.:* Pilot-plant facilities for distillation.
- *Vara International, Inc., Vero Beach, Fla.:* Experts primarily in integrated adsorption and distillation systems for solvent recovery from gas streams.

This is only a representative sampling of equipment suppliers and is not intended to be a complete listing of all companies providing equipment or systems.

6.2.10 REFERENCES

1. J. B. Berkowitz et al., *Unit Operations for Treatment of Hazardous Industrial Wastes,* Arthur D. Little, Inc., Cambridge, Mass., 1978.

2. R. H. Perry and C. H. Chilton, *Chemical Engineers' Handbook,* 5th ed., McGraw-Hill, New York, 1973.

3. C. J. King, *Separation Processes,* McGraw-Hill, New York, 1977.

4. *Distillation Handbook,* 2d ed., DH-682, APV Equipment, Inc., 395 Fillmore Ave., Tonawanda, New York.

5. M. S. Peters and K. D. Timmerhaus, *Plant Design and Economics for Chemical Engineers,* 3d ed., McGraw-Hill, New York, 1980.

6. *Preassembled Process Plants, Cost and Time Effective,* Chem-Pro Equipment Corp., Fairfield, N.J. p. 200.

7. C. C. Allen, S. Simpson, and G. Brant, *Field Evaluations of Hazardous Waste Pretreatment as an Air Pollution Control Technique,* prepared for the U.S. EPA under EPA contract No. 68-02-3992, Task 9 (unpublished), 1986.

SECTION 6.3
ELECTROLYTIC RECOVERY TECHNIQUES

Aloysius A. Aguwa

General Motors Technical Center
General Motors Corporation
Warren, Michigan

Charles N. Haas

Pritzker Department of Environmental Engineering
Illinois Institute of Technology
Chicago, Illinois

Electrolytic recovery techniques are used primarily for the recovery of metals from process streams or rinsewaters. These metals must be removed or recovered from the effluent streams prior to discharge either to meet the effluent discharge limits imposed by the environmental regulations, or to recover the metals for their economic value. Although recovery techniques have been used for many years in the mining industry for refining ores, they are presently finding wide application in electroplating, rolling mills, and electronic industries.

Electrolytic recovery techniques are based on the oxidation-reduction reaction which takes place at the surface of conductive electrodes (cathode and anode). The electrodes are immersed in a chemical medium under the influence of an applied potential. At the cathode, the metal ion is reduced to its elemental form. At the same time, gaseous products such as oxygen, hydrogen, or nitrogen may evolve at the anode. The gases that are produced at the anode depend on the chemical composition of the medium. Dissolved species such as cyanide are generally oxidized at the anode.

The major process equipment consists of (1) the electrochemical reactor containing the electrodes, (2) a gas-venting system, (3) recirculation pumps, and (4) a power supply. After the metal coating or deposition at the cathode reaches the desired thickness, the metal can be removed and generally re-used or sold. In the electroplating industry, for example, either the recovered metal, which is essentially pure, is returned to the plating tank or the metal-plated cathode can now be used for an anode in the plating bath.

Electrolytic recovery techniques have been used to recover copper, nickel, zinc, silver, cadmium, gold, and other heavy metals. The capital cost and operation and maintenance costs are generally low, especially when compared with those for other processes. The process can generally pay for itself in about a year, or in a matter of weeks when precious metals such as gold are being recovered.

6.3.1 APPLICABILITY OF THE TECHNOLOGY

Electrolytic techniques have been used to recover a variety of heavy metals from process streams or rinsewaters. These processes have been demonstrated for the recovery of gold, silver, cadmium, nickel, nickel–iron alloy, copper, zinc, and other metals. One of the most common applications of the electrolytic system is the recovery of copper from sulfuric acid solutions.[1] The mining industry has used electrolytic techniques to refine ores for several years. The process is now finding wide application with electroplaters, rolling mills, printed-circuit-board manufacturers, and metal-coating firms.[2] Electrolytic techniques can also be used to recover metal from crystals and sludges and to regenerate process solutions.[1] However, electrolytic techniques are not effective at recovering nickel because of the low standard reduction potential and the high stability constants of its cyanide complexes.[3]

6.3.2 DESCRIPTION

The conventional electrochemical reactor consists of the electrodes, associated appliances such as recirculation pumps, and a power supply. The cathodes are made of stainless steel of approximately 150 mm width upon which the recovered metal is deposited.[1] After the coating or deposition of metals reaches sufficient thickness [6 mm (0.15 in.)], the metal deposited can be removed and is generally re-used or sold. The metal-plated stainless-steel cathode can then be used as an anode in the plating baths. This is the general practice in the electroplating industry. The electrolytic recovery tank is usually designed to produce high flow rates in a narrow channel.

The conventional recovery tank is very efficient at high concentration but not at low concentration due to mass-transfer-limiting conditions.[5] As a result, advanced electrolytic recovery devices have been fully developed and are commercially available. These electrochemical devices or cells may be in the form of a packed bed or fluidized bed, cells with spiral-wound electrodes, or cells with rotating cylindrical electrodes.[6]

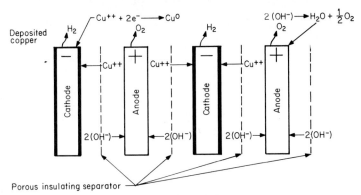

FIG. 6.3.1   Extended-surface electrolysis cells[4]

One such advanced electrolytic recovery device is the extended-surface electrolysis (ESE) recovery system. This system, unlike conventional recovery devices, recovers metal better at low concentrations than at high concentrations. However, the mechanism of metal removal is essentially the same as in the conventional types.

The ESE spiral cell is of sandwich construction, containing a fixed, "fluffy" cathode, a porous insulating separator, an anode of screenlike material, and another insulating separator. Cathode material is usually a fibrous woven stainless-steel mesh. However, the anode and cathode material may vary depending on the effluent stream to be treated. The cathode, separator material, and anode are formed into a sandwich structure which is rolled into a spiral and inserted into a pipe. This type of cell construction results in a very open structure with a void volume of 93% to 95%, which provides a low resistance to fluid flow.

A number of cells can be stacked as modules so that a large fraction of contaminant metals can be recovered from an effluent. The solution to be treated is pumped in at the top of the module and flows down through the cells, where the metals are plated out on the cathode. Figure 6.3.1 depicts this process as a copper-containing stream flows through the cell stack. The reactions take place continuously as the fluid is pumped through the various cells in the cell stack, or module. Figure 6.3.2 shows a possible placement of this recovery system in a plating line. Metal accumulated on the cathode is usually stripped out by circulating an acidic cleaner through the cell.

Another advanced electrolytic recovery device is the high-surface-area (HSA) reactor.[2] This reactor can be used to treat effluent from plating streams or other streams containing metals. The cathode achieves a high surface area by incorporating a large number of carbon-filament mats. These mats collect the materials from the still rinse bath during plant operations. When plating operations cease, vat solutions are drawn from the bath into a separate compartment of the reactor where they remove the metal from the cathode by an electrochemical or sometimes a chemical process.[2]

Other advanced electrolytic recovery devices, such as fluidized-bed electrochemical reactors, are commercially available for the removal and recovery of metals from dilute solutions.[5] This process makes recovery from dilute streams economically attractive. The electrochemical cell consists of a set of

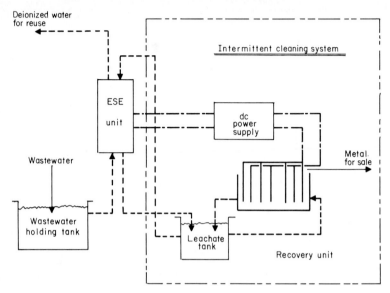

FIG. 6.3.2   Application of extended-surface electrolysis[4]

apertured, expanded-metal-mesh electrodes immersed in a bed of small glass beads. The bed is fluidized to about twice its packed depth by pumping rinsewater upward through a distributor and through the bed. The glass beads continuously scrub the surface of the electrode, thereby bringing it fresh solution when the ion concentration of the solution has been depleted near the surface. The electrodes are removed at regular intervals to remove the recovered metals. This type of electrochemical reactor is used to recover gold, silver, cadmium, nickel, nickel–iron alloy, copper, and zinc in more than 150 installations.[5]

6.3.3   ADVANTAGES AND DISADVANTAGES

The advantages and disadvantages of the electrolytic recovery techniques are summarized below. The advantages include the following:[2,3,6,7,8]

- Valuable metals are recovered, re-used, or sold.
- Discharge of toxic heavy metals is reduced or eliminated.
- Initial investment for the process is low.
- Production of toxic sludges is eliminated.
- Operating expenses are low.
- Equipment maintenance is minimal.

The disadvantages of this process, if any, may not be of any practical consequence.

6.3.4 DESIGN CONSIDERATIONS

Physical Constraints

Parameters that affect the recovery of metals using electrochemical means are the initial concentration of the metal, power supply, and characteristics of the electrodes. These parameters interact as presented in the following discussions.

As pointed out earlier, conventional electrolytic recovery is only good at high metal concentration, hence development of advanced electrolytic recovery equipment. The main problem in using conventional electrolytic recovery for dilute waters is cathodic polarization. In such systems, as the metal is plated at the electrode, the layer of solution next to the electrode that is depleted of metals forms a polarized layer that offers high resistance to metal migration to the electrode. Severe cathode polarization can lead to poor-quality deposits: the formation of dark, powdery, burned areas and "trees" that can grow across to the anodes and short out the cell.[1] This problem generally reduces the efficiency of the electrode since the electricity is used to decompose water and form hydrogen gas instead of plating out the metal.[1] The problem can be solved by agitating the solution, using low-current-density, high-surface-area electrodes, or, as in fluidized beds, by bringing the electrode in contact with fresh influent stream.[5] The effect of metal concentration (using copper as an example) on electrical efficiency is depicted in Fig. 6.3.3.

Complexation of the targeted metal in solution may also cause a problem, as is the case with nickel, which forms cyanide complexes with high stability constants. Failure of electrolytic techniques in removing nickel is also due to nickel's low standard potential.

Materials of Construction

The electrochemical reactor can be made of any material as long as no undesirable reaction would occur between the material and the chemical constituents of

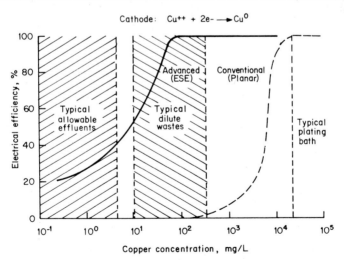

FIG. 6.3.3 Effect of concentration on electrical efficiency in metals reduction.[4] ESE = extended-surface electrolysis.

the waste. The electrodes are generally made of stainless steel. The anode and cathode material may vary with the particular effluent stream to be treated.[4]

Relation to Other Processes

Electrolytic recovery techniques can easily be adapted to other processes, especially in the electroplating industry. This is because electrolytic recovery techniques are similar to plating techniques used in the plating industry. Figure 6.3.2 shows a possible placement of an electrolytic recovery system in a plating line.

6.3.5 ENVIRONMENTAL EFFECTS

Electrolytic recovery techniques generally produce few adverse environmental effects. Use of electrolytic systems may entail a potential air-pollution problem resulting from the formation of gases such as hydrogen and chlorine at the electrodes.[2]

6.3.6 RESULTS OF CASE STUDIES

Tin and silver have been electrolytically recovered from their respective rinsewaters using conventional methods. Using a cathode area of 4.1 m^2 (45 ft^2), current density of 5 to 10 A/ft^2, and flow rate of 0.005 m ^{3}/h (1.2 gal/h), the recovery efficiency for tin was reported to be 97 to 99%, but the current efficiency was 70%.[4] A 99.8% recovery was reported for silver. In this case, the flow rate was 0.003 m^3/h (0.8 gal/h), with a cathode area of 3.15 m^2 (35 ft^2), and current density of 3 to 5 A/ft^2. A current efficiency of 25 to 50% was reported.[4]

A high-surface-area (HSA) reactor was constructed at Allied Metal Finishing in Baltimore, Maryland, in 1980 for the recovery of cadmium and destruction of cyanide.[8] The cathode is made of carbon fiber which has an enormous surface area/volume ratio. It is claimed that this reactor is capable of recovering 99.9% of the metals contained in the feed stream. The reactor can reportedly electrooxidize and destroy cyanide to below detection limits at a cost much lower than the conventional alkali-chlorination process. Since the installation, Allied Metal Finishing has saved money on water use as a result of recycling ($7,000 in the first year), reduced purchase of cadmium as metal was made available for reuse, and reduced cost associated with sludge disposal. In 1982, the estimated savings from sludge minimization were about $10,000 per year.

The GTE/Automatic Electric facility uses an electrochemical reactor to recover copper from sulfuric acid solutions.[1] It was reported that, over a five-week period, 241 kg (536 lb) of copper were recovered. The average copper concentration was approximately 5 g/L and the cathode efficiency was reported to be 90%.

An HSA system designed to recover cadmium and destroy cyanide was instituted at the X-pert Metal Finishing plant in Ontario, Canada, in 1981. It has been reported that 99.9% of the cadmium is recovered and that most of the cyanide is destroyed.[2]

These and many other case studies point to the fact that many metals can be recovered using electrolytic techniques.

6.3.7 ENGINEERING ECONOMIC EVALUATION

The capital cost of electrolytic recovery equipment depends on the flow rate, initial metal concentration, desired effluent concentration, and whether the metal is complexed. The operational cost depends on the cost of electricity, cost of pumps, and labor costs for cell operation and maintainence.

At low concentrations (generally below 10 mg/L), the ESE systems are preferred over ion exchange.[4] However, ion exchange is a more cost-effective technology at concentrations between 100 and 1,000 mg/L. Other studies have demonstrated that electrolytic recovery is more cost-effective than ion exchange in treating the same wastewater.[6] Cost of destroying cyanide using the ESE system compares favorably with costs for other methods.[9]

In general, recovery of metal such as copper or tin will pay for itself in 3 to 21 months by virtue of recovered metal values and reduced costs for chemical treatment and sludge hauling and subsequent disposal. A system recovering gold can pay for itself in less than a week.[1]

6.3.8 REFERENCES

1. C. A. Benirati and W. J. McLay, "Electrolytic Metal Recovery Comes of Age," *Plating and Surface Finishing,* March 1983, p. 26.

2. M. E. Campbell and W. M. Glenn, *A Guide to Industrial Waste Reduction and Recycling,* Pollution Probe Foundation, Toronto, Canada, 1982.

3. E. F. Hradil and G. Hradil, "Electrolytic Recovery of Precious and Common Metals," *Metal Finishing,* November 1984, p. 85.

4. U.S. Environmental Protection Agency, *Development Document for Existing Source Pretreatment Standards for the Electroplating Point Source Category,* EPA 440/1-79/003, Washington, D.C., August 1979, p. 204.

5. A. G. Tyson, "An Electrochemical Cell for Cadmium Recovery and Recycling," *Plating and Surface Finishing,* December 1984, p. 44.

6. P. M. Robertson, J. Leudolph, and H. Maurer, "Improvements in Rinsewater Treatment by Electrolysis," *Plating and Surface Finishing,* October 1983, p. 48.

7. G. E. Hunt and R. W. Walters, "Cost-Effective Waste Management for Metal Finishing Facilities: Selected Case Studies," *Proceedings of the 39th Industrial Waste Conference*, Purdue University, May 8–10, 1984, p. 521.

8. P. Horelick, "Recovery and Electrochemical Technology," paper presented at the Fourth Conference on Advanced Pollution Control for the Metal Finishing Industry, Florida, January 18–20, 1982.

9. J. K. Easton, "Electrolytic Decomposition of Concentrated Cyanide Plating Wastes," *J. Water Poll. Control,* 1967, p. 1,621.

SECTION 6.4
HYDROLYSIS

Alan Wilds
President
BDT, Incorporated
Clarence, New York

A variety of chemicals react with water. Some reactions are especially interesting because they are associated with the generation of heat or large volumes of gas, or with the production of pH extremes, or with a combination of any or all of these events. Many chemicals retain these properties when they become waste. It is not feasible to consider the direct land disposal of materials that are so reactive with water that they might ignite, explode, or produce toxic gases, or extremely acidic or basic conditions, either during the land disposal process or at a later date in the landfill. Therefore, those reactive properties should be eliminated, whenever possible, before the ultimate disposal of the materials. Some of the options available to do this are controlled incineration, open-pit burning, detonation, and chemical reaction. Not all materials that react adversely with water can be incinerated and, if they could, it might not always be desirable. Chemical reaction by hydrolysis is one good method for removing these hazardous properties.

Hydrolysis, in this context, is the use of water to destroy, decompose, or alter a chemical species. When used under controlled conditions, this process can meet the criteria for the safe disposal of reactive hazardous waste. Hydrolysis technology has been designed and is operational. An important, but not essential, feature of this technology is the use of a *hammer mill.*

6.4.1 APPLICATION

Hydrolysis can remove reactivity from organic and inorganic substances in the gas, liquid, or solid phase. The general classifications of commonly encountered reactive chemicals are as follows (with specific examples given in parentheses):

Metal alkoxides (sodium methoxide)

Metal amides (sodium amide)

Carbides (calcium carbide)

Cyanates (methylisocyanate)

Halides (benzoyl chloride, aluminum chloride)

Hydrides (lithium aluminum hydride)

Oxyhalides (phosphorus oxybromide)

Reactive metals (potassium)

Silanes (trichloromethylsilane)

Sulfides (phosphorus pentasulfide)

This list of materials conveys some idea of the variety of compounds that can be managed by hydrolysis. This is not to imply that these chemicals can be treated only by hydrolysis or that hydrolysis is always the best method. However, hydrolysis can successfully eliminate the reactive properties of these materials.

The technology enables certain specific hazardous waste streams to be handled safely that might otherwise be extremely difficult to manage. These waste streams include aerosols (contain flammable propellant such as propane under pressure), lecture bottles or cylinders (contain pressurized gases such as phosgene), certain batteries (contain reactive materials such as lithium metal and thionyl chloride), laboratory chemicals [a wide variety of water-reactive chemicals in glass or metal containers (see list above)] and a variety of specialty devices that incorporate a reactive compound (such as control devices containing sodium–potassium alloy).

6.4.2 CHEMISTRY

A variety of reactions by chemicals with water are described below:
  Metal alkoxides, e.g., sodium methoxide (water causes ignition):

$$NaOCH_3 + H_2O \rightarrow NaOH + CH_3OH$$

Sodium methoxide reacts with water to produce sodium hydroxide and methanol. Sodium hydroxide produces a solution of high pH which is caustic and corrosive. Methanol is flammable and toxic. The reaction is highly exothermic.
  Metal amides, e.g., lithium amide (ignites spontaneously on contact with water):

$$LiNH_2 + H_2O \rightarrow LiOH + NH_3$$

This reaction produces a lithium hydroxide solution and evolves ammonia gas. The solution is caustic and corrosive and the ammonia is odorous and a strong irritant. The reaction is extremely exothermic and is frequently accompanied by fire.
  Carbides, e.g., calcium carbide (produces acetylene gas on contact with moisture):

$$CaC_2 + 2H_2O \rightarrow Ca(OH)_2 + C_2H_2$$

Calcium hydroxide and acetylene gas are the products of this reaction. Calcium hydroxide is only slightly soluble in water and produces an alkaline solution.

Acetylene gas is highly flammable and forms explosive mixtures with air in a wide range of concentrations.

Cyanates, e.g., methylisocyanate: the compound is highly toxic and produces a methylamine solution and a carbonate. Under normal hydrolysis conditions, carbonic acid is not produced. The methylamine remains in solution and produces alkaline conditions.

$$CH_3NCO + 2H_2O \rightarrow CH_3NH_2 + H_2CO_3$$

Halides, e.g., aluminum trichloride:

$$AlCl_3 + 3H_2O \rightarrow Al(OH)_3 + 3HCl$$

This reaction produces aluminum hydroxide and hydrochloric acid. The acid produces solutions of low pH which are highly corrosive and very toxic.

Hydrides, e.g., lithium aluminum hydride:

$$LiAlH_4 + 4H_2O \rightarrow LiOH + Al(OH)_3 + 4H_2$$

The compound shown reacts violently with water and produces the hydroxides of lithium and aluminum, and hydrogen gas. The hydroxides produce caustic and corrosive solutions, and the hydrogen gas can form explosive mixtures in air. This reaction is often accompanied by flame and explosion.

Oxyhalides, e.g., phosphorus oxybromide:

$$POBr_3 + 3H_2O \rightarrow H_3PO_4 + 3HBr$$

Phosphorus oxybromide reacts with water to produce phosphoric acid and hydrogen bromide gas. Some of the gas goes into solution to form highly corrosive hydrobromic acid. Hydrogen bromide is very toxic. This reaction is extremely exothermic.

Alkali metals, e.g., potassium:

$$2K + 2H_2O \rightarrow 2KOH + H_2$$

The alkali metals such as potassium react extremely vigorously with water and produce caustic and corrosive potassium hydroxide and flammable and explosive hydrogen gas. This reaction is usually accompanied by flame and explosion. When potassium burns, it produces dense, white, toxic fumes of the oxide.

Silanes, e.g., trichloromethylsilane:

$$CH_3Cl_3Si + 3H_2O \rightarrow CH_3Si(OH)_3 + 3HCl$$

Hydrolysis of trichloromethylsilane produces hydrochloric acid and methylsilanol. The silanol is unstable and further reacts to a siloxane (Si-O-Si). The siloxanes precipitate out during the reaction. Hydrogen chloride produces solutions of low pH which are extremely corrosive and toxic.

Sulfides, e.g., phosphorus pentasulfide:

$$P_2S_5 + 5H_2O \rightarrow P_2O_5 + 5H_2S$$

Phosphorus pentasulfide produces hydrogen sulfide, a gas which is both odorous and toxic on contact with water, as well as the water-reactive compound phosphorus pentoxide. In the presence of excess moisture, the pentoxide further hydrolyzes to produce the phosphate. These reactions are highly exothermic.

From these reactions, it is clear that chemicals in the presence of water can produce highly undesirable conditions, such as large volumes of flammable, explosive, or toxic gases; and/or liquids or solutions which are highly corrosive because of extremes of pH. In many instances large quantities of heat may be generated. None of these conditions should be present in land disposal or uncontrolled situations.

However, the controlled hydrolysis reaction can eliminate these properties and render the materials safe for further treatment or land disposal.

6.4.3 TECHNOLOGY

Principles

Hydrolysis technology is based on the fact that water, under controlled conditions, modifies certain reactive chemical elements and compounds safely. Further, the resultant species can be handled safely. The previous examples of chemical reactions demonstrate how very hazardous chemicals, such as potassium metal and lithium amide, can be reacted to produce less hazardous substances. In the process, solutions with high or low pH may be produced, as well as toxic and flammable gases. The evolved gases must be handled appropriately and the pH of the solutions must be adjusted to more neutral values to render them safe for processing at a commercial wastewater-treatment facility. Any solids filtered from the solution can be drummed and sent to a secure landfill. The flow diagram, Fig. 6.4.1, shows how the various phases are collected and handled.

Since many chemical species react violently with water, even under ambient conditions and in relatively small quantities, it is often necessary to modify the hydrolytic process by controlling the reacting medium's pH, chemistry, temperature, and flow rate. It is always necessary to control the addition rate of the reactive compound.

The safety of personnel, the public, the environment, and the equipment is a major principle of this technology. Knowledge of the chemistry of any given reaction allows implementation of the appropriate safety measures. It is necessary to control those reactions that (1) evolve large volumes of potentially explosive

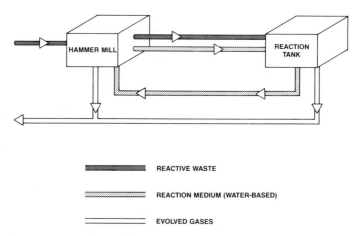

FIG. 6.4.1   Flow diagram of reactive-waste processing.

gas, (2) are exothermic, (3) result in pH in the strongly acidic or basic ranges, or (4) have high oxidation-reduction potentials. Chemical compatibility must always be considered when mixtures are being treated.

Carbides: A Case Study

To illustrate these points, consider this brief review of some work performed by the author to enable the safe reaction of carbides to be conducted by hydrolysis. Carbides react with water in varying degrees:

$$Al_4C_3 + 12H_2O \rightarrow 4Al(OH)_3 + 3CH_4 \tag{1}$$

$$CaC_2 + 2H_2O \rightarrow Ca(OH)_2 + C_2H_2 \tag{2}$$

$$Na_2C_2 + 2H_2O \rightarrow 2NaOH + C_2H_2 \tag{3}$$

$$\text{BaC} \quad \text{no reaction}$$

$$\text{SiC} \quad \text{no reaction}$$

Three of these carbide compounds (aluminum, calcium, and sodium) are water-reactive and two (barium and silicon) are not. Group 1 and group 2 cations produce acetylene, and the group 3 cation produces methane. Both gases are highly flammable and explosive, and would cause considerable problems if released in a landfill. For this discussion, calcium carbide is the example.

From reaction 2, it can be seen that calcium carbide reacts with water to produce calcium hydroxide and acetylene. The hydroxide, which is only slightly water-soluble, produces an alkaline solution, with most of the compound being precipitated. The acetylene is released as a gas.

The reaction rate of the carbides may be influenced by many factors, including

- Cation forming the carbide (e.g., lithium, calcium, aluminum, silicon)
- Physical form of the carbide (e.g., powder, granular, lumpy)
- Storage conditions of the chemical
- Reaction temperature
- Reaction-medium pH

Freshly produced calcium carbide is dark gray to black in color and is a solid that may vary considerably in particle size, from a very fine powder to large lumps. Freshly produced calcium carbide powder yields its volume of acetylene much more rapidly than an equal weight of freshly produced calcium carbide granules. Both materials produce equal volumes of gas. When stored improperly, the color might change to a light tan. This indicates that the surface material has reacted with atmospheric moisture to produce the hydroxide. The finer the particle size, the more likely the reaction will go to completion during improper storage. Tan, powdered calcium carbide is less likely to react violently with water than the black carbide. Examples have been received of laboratory powders which, on prolonged storage, have not produced any noticeable acetylene gas on contact with water.

The pH of the solution is another important factor. This is demonstrated by reference to reactions of the same material in three media: city water, and 12% and 25% sodium hydroxide solutions (see Fig. 6.4.2). It is apparent that the more

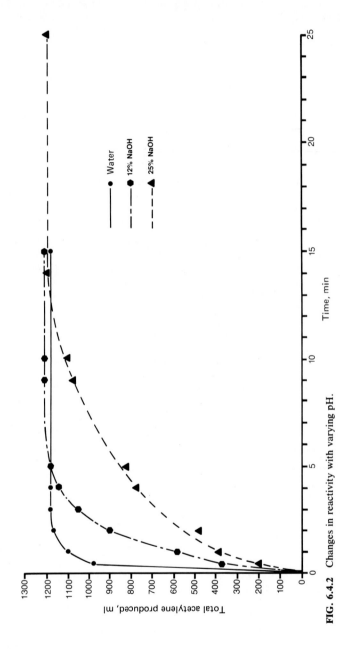

FIG. 6.4.2 Changes in reactivity with varying pH.

6.52

concentrated the alkaline solution, the more slowly the reaction proceeds. Eventually, all the acetylene is released. This indicates that under the test conditions, the reaction is completed. In city water, all the acetylene is generated in the first 3 min, whereas, in a 25% sodium hydroxide solution, only about half of the reaction has occurred in this time. It is important to understand the reaction kinetics because dangerous conditions can occur during treatment that might be extremely difficult to control. For example, in either a batch or continuous-feed process, large quantities of unreacted carbide might be reacting at the same time, thereby producing large volumes of acetylene gas. It is not safe to assume that the carbide hydrolysis reaction will go to completion in 3 min under all conditions. More generally, it is not reasonable to assume because one reaction proceeded smoothly for a particular compound, that it will always proceed this way under all conditions. Adequate testing should be performed.

Process Description

Depending on the material to be processed, the appropriate aqueous solution is prepared (most frequently alkaline with a pH greater than 10.5) and placed in the reaction tank (see Fig. 6.4.3). This solution is circulated through the hammer mill, usually at a rate of 60 m^3/h (250 gal/min) (the rate is adjustable based on the proposed reaction), and then back into the reaction tank. The reactive material is introduced into the hammer mill by a conveyor and hopper. In the hammer mill, it is reacted and then discharged to the reaction tank. Since the chemistry of the reaction medium is constantly changing as the reactive species are added, provision has been made for pH adjustment.

Reactive solids and small volumes of reactive liquids are usually received at the facility in glass bottles, metal cases or cans, or plastic bags. These are fed manually onto the conveyor (see Fig. 6.4.3) leading to the hammer mill. The rotating hammers open the containment to expose the contents, which immediately start to react with the recirculating aqueous solution.

The reacting materials and the containment remain in the hammer mill until their physical sizes have been reduced in diameter to less than 12 mm (½ in). When this has been achieved, the materials pass through a grid in the bottom of the hammer mill into the reaction tank. Most often, the reaction goes to completion in the hammer mill; however, any unreacted reactive material that passes into the reaction tank continues to react in the tank. Solids and liquids that result during the process remain in the reaction tank.

Bulk liquids and contained gases are pressure-fed and subsurface-injected into the reaction tank at controlled rates. The rate of addition is critical, especially for gases, since these must be fully reacted within the reaction tank.

(*Note:* When treating mixtures or a variety of chemicals, it is important to have a detailed knowledge of the constitution of the reaction medium to avoid major problems due to chemical incompatibilities.)

Any gases produced in the reaction are collected from the hammer mill, the reaction tank, and the reaction room, and are passed through a scrubbing system designed to remove particulates and, where appropriate, to react with the produced species (e.g., an alkaline scrubbing medium removes acid mists and an acid scrubber might remove ammonia).

When the reacting medium in the tank becomes saturated, the aqueous portion is filtered off and sent to a commercial wastewater-treatment facility. The solid phase, which consists of glass and metal fragments, shredded plastic, precipitated salts, and nonreactive species, is drummed and sent to a secure landfill.

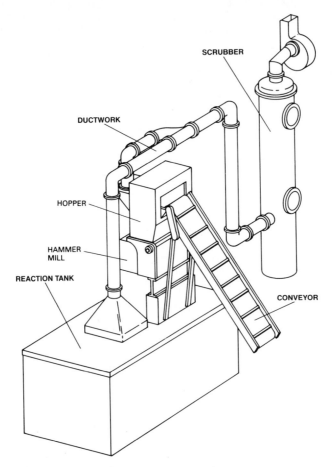

FIG. 6.4.3 Processing-equipment configuration.

Equipment Description

The equipment is housed in a dedicated room designed to handle explosions. The walls are steel-reinforced concrete set into the floor and tied together to form a rigid, five-sided box. The sixth side, the roof, is designed to lift off should an explosion occur. Personnel are not allowed in this room while material is being treated.

During processing, the reaction room is maintained at a negative pressure to the rest of the facility to reduce the risk of personnel exposure to gases produced during treatment and to ensure that all gases generated pass through the scrubber before emission from the facility. Air is withdrawn from the room through a scrubber (see Fig. 6.4.3), which is loaded with a recirculating solution appropriate for the material being treated.

Reactive wastes are loaded manually onto a single-speed conveyor. The conveyor passes through a small opening in one of the walls of the reaction room. (Before processing, it is important to determine that air flow is into the reaction room, and this should be monitored constantly during processing.) The conveyor transfers the material into the hammer mill through a small chute. The hammers and rotor are cast steel and the body of the mill is steel. The hammers rotate at 1,800 r/min and are driven by a 275 MJ (100 hp) electric motor.

The box-shaped reaction tank has a capacity of 5 m^3 (1,300 gal). The tank is constructed of carbon steel with a 9.5-mm (⅜-in) thick wall and a 13-mm (½-in) thick floor. The walls are coated internally and externally with an epoxy composition to reduce metal corrosion and erosion.

The gas-collection ductwork and the scrubber are fabricated from glass-fiber-reinforced resins and the ductwork has blowout panels to protect elbows, joints, and the scrubber should there be a sudden increase in pressure.

The plumbing is of high-density polyvinyl chloride (HPVC). The various pumps are explosion-proof, but are not corrosion- and erosion-resistant. (This design decision is based on the highly erosive and corrosive nature of the process. Replacing worn-out pumps has been shown to be more cost-effective than purchasing corrosion-resistant pumps, which has proven to only marginally increase working lifetimes.) All electrical installations, including motors, lights, fans, junction boxes, and conduit, meet the National Electrical Committee code required for acetylene gas (Division 1, Class II, Group A).

Construction Materials

Construction materials should be considered and chosen in light of the probable reactions to be carried out. Small explosions; corrosion and erosion of pumps, tanks, and other equipment; rapid pressure changes; fire; and high temperature are the major concerns in the reaction facility.

Capacities

The existing technology limits treatment rates to 0.25 kg/s (2,000 lb/h), and most materials are slower by at least an order of magnitude. The major factors to be considered in establishing limits are (1) ensuring that treatment rates do not cause explosive conditions, (2) ensuring that sufficient cooling is available, (3) ensuring that air emissions can be maintained below local standards, and (4) the physical limitations on loading of the material into the system. However, the technology could be scaled up and specific concerns could be addressed to increase treatment rates by at least a factor of 10.

6.4.4 SAFETY AND MONITORING

The requirement that a facility's management maintain a safe workplace for employees and ensure environmental safety cannot be overemphasized. During processing, reactive materials can produce explosive (methane, propane, butane, acetylene, ethylene oxide) and toxic (hydrogen cyanide, hydrogen sulfide, hydrogen bromide, phosgene, chlorine) gases. All of these materials and many others can cause fatalities or serious health problems. Consequently, all aspects of the process must be monitored and appropriate automatic shutdown procedures should be installed, especially in continuous-feed situations. The presence of any explosive or toxic gases should be monitored for continuously and reviewed on a

24-hour basis. The standards set by the Occupational Health and Safety Administration (OSHA) for time-weighted average (TWA), or short-term exposure limit (STEL) should be used as maxima. Even though areas are being monitored constantly, it is also advisable to have direct monitoring by personnel on a routine basis. Any excursions from expected behaviors should be fully investigated.

The various gases evolving during the process should be monitored at their entry into and exit from the scrubbing system to ensure that scrubbing efficiency is being maintained or exceeded. Automatic shutdown procedures should be initiated if the emissions allowance standards are surpassed.

All critical fluid flows (air, reaction solutions, cooling media) should be monitored constantly, and alarm conditions should be installed. Temperatures, pH, oxidation-reduction potentials, and liquid levels should be checked regularly. Above all, the reasons for any changes should be understood. When the monitoring systems are installed, an extensive preventive maintenance program should be initiated to ensure satisfactory operation of the complete system. Where the potential for explosion exists, the facility should be constructed as described.

All gas flows should be directed away from personnel, who should always be upwind of the source of the gas. In the specific example considered earlier, the treatment of carbides, the major hazard is the risk of explosion resulting from the achievement of sufficient levels of acetylene. Consequently, it is important to know the reaction rate of the particular source of the carbide and then to monitor the concentrations of acetylene at various locations within the facility. A flammable-gas monitor should be installed and calibrated for acetylene. Acetylene forms explosive mixtures with air in the range of concentrations from 3% to 65%. When a 3% concentration is achieved [the lower explosion limit (LEL)], the potential for an explosion exists. Detection must be provided that enables values lower than 3% concentration to be observed. Action is typically recommended at the 20% LEL and 40% LEL levels. Therefore, both audible and visible alarms should be triggered at the 20% LEL (0.6% acetylene) and the feed system should automatically shut down and the environment-purge system should be activated at the 40% LEL (1.2% acetylene).

If the air-flow system should be designed for, for example, 1 m³/s (2,000 ft³/min) through the scrubbing system, it is important to maintain this rate and alarm conditions should be triggered if the flow rate drops below this value.

Various building and fire codes, government agencies at the federal and local levels, and National Fire Protection Association (NFPA) member organizations are recommended as sources of information to help with the installation of a safe facility.

During processing, employees should be provided with the appropriate safety apparel—i.e., hard hats, eye protection, respiratory equipment, overalls, gloves, and shoes—and they should be fully trained in safe working procedures.

If an incident occurs, employees should be completely knowledgeable about their responsibilities and procedures, and an adequate supply of emergency-response equipment should be available. In the case of water-reactives, water cannot be used to help in an emergency so fire hoses and sprinkler systems must not be installed.

6.4.5 ENVIRONMENTAL CONSIDERATIONS

Reactive materials are not safe for landfilling. The potential for catastrophe is always present. Consequently, the reactivity of these materials must be eliminated before their ultimate disposal. The necessary safeguards must be installed at the treatment facility to prevent the transfer of this potential for disaster from the landfill to another location.

Because of the nature of the materials being stored and treated, the treatment plant should not be located in a known earthquake zone or in a 100-year flood zone. Under ideal conditions, the facility should be located away from heavily populated areas. Air emissions are another physical consideration: The facility must be maintained within local standards. The only problem here is to ensure that the treatment mode and its subsequent controls can achieve these requirements.

Adequate spill-control procedures must be implemented to remove the risks should materials escape from their container. It should be impossible for materials to move outside the treatment facility or for them to commingle unintentionally within the facility. Further, during and after processing, the reaction products and the aqueous phase must be prevented from leaking into the environment. The aqueous solution may have extreme pH and contain toxic species.

Secondary containment is necessary at each step of the process, from storage of incoming materials, to processing, to storage of the reaction products.

A major advantage of the hydrolysis technology is that it can be designed to be totally self-contained for solids and liquids. Consequently, the potential for contaminating local groundwater, rivers, lakes, or land is reduced to an extremely low probability. This is important because many water-reactive materials are either toxic or produce toxic compounds when processed by hydrolysis. The aqueous solution can be safely transported either to a commercial wastewater-treatment facility or to an equivalent in-house system prior to discharge. Filtered solids can be drummed and sent to a secure landfill. During processing and in storage, toxic vapors must not be released into the environment. Consequently, all air must be withdrawn from the treatment equipment and room through scrubbers designed to remove the foreseeable concentrations of the toxic species. Further, these concentrations of gases should be monitored at both their entry into and exit from the scrubbing system to ensure that required scrubbing efficiencies are being achieved and to prevent the escape of any airborne environmental hazards from the facility.

6.4.6 ADVANTAGES AND DISADVANTAGES

There is no known technology that compares with hydrolysis in the range of reactives to which it can be successfully applied. However, there are other technologies, including controlled incineration, open-pit burning, detonation, ambient exposure, and chemical reaction, which can treat some of these materials.

The disadvantage of burning, especially for organic compounds, is the potential release of species created by incomplete combustion, which may be harmful to the public and to the environment. For this reason, costly test burns and removal efficiencies to "five and six nines" (99.999 and 99.9999%) are required. Other compounds cannot be burned in an incinerator without fear of causing an explosion, with subsequent environmental problems, and, more immediately, the costly down time to return the incinerator to production.

Obviously, in open-pit burning, there are few, if any, controls, and volatile species that result from the combustion process will be released into the air.

The controlled-hydrolysis method has obvious advantages to personnel and the environment over open-pit burning, detonation, and ambient-exposure technologies, which cannot be well-controlled. Further, incorrect detonation techniques may spread the not fully deactivated and possibly toxic material over a wide area.

Water is still a relatively abundant and inexpensive resource and, in hydrolysis, with its subsequent treatment of the aqueous waste produced, the water is returned to the environment.

Cost might be considered a disadvantage of hydrolysis because the process is such that either certain chemicals have to be reacted very slowly, or very costly controls are required. Taking a short-sighted view, cost could certainly appear to be a problem. However, if landfilling is the alternative, cost should be set aside in the decision-making process, since even if no catastrophe occurs during the process of burial, it is possible that at some later date the material will have to be excavated and this has proven to be extremely costly to the initial disposer of the waste.

An advantage of this particular technology is the ability to either open packages which might not otherwise be opened, or to remotely open packages that contain very reactive materials, thereby reducing the likelihood of handling accidents to very low possibilities.

6.4.7 SUMMARY

A technology discussed has been successfully applied to the treatment of water-reactive wastes. Under the appropriate controls, this technology can be safe and environmentally attractive. In application, the current technology can be most accurately described as a working prototype. The next step must be to scale up the plant and equipment to demonstrate that, without losing its environmental- and personnel-safety attributes, it can also become cost-competitive with other methods in the short term.

With application of adequate resources, hydrolysis should fulfill its early promise of becoming an important weapon in the battle to properly dispose of hazardous wastes.

SECTION 6.5
ION EXCHANGE

Craig J. Brown, P. Eng.
Vice President and General Manager
Eco-Tec Limited
Pickering, Ontario, Canada

Ion exchange rarely represents an option for the ultimate disposal of hazardous wastes since the process generates chemical wastes which must almost invariably be further treated. Its role is usually to reduce the magnitude of a problem by converting a hazardous waste into a form in which it can be re-used, leaving behind a less toxic substance in its place, or to facilitate ultimate disposal by reducing the hydraulic flow of the stream bearing the toxic substance.

Ion exchange is nevertheless a very versatile and effective tool for the treatment of aqueous hazardous wastes. Developed more than 40 years ago as a means of producing high-purity water for such applications as boiler makeup, it is well-suited to the detoxification of large flows of wastewater containing relatively low levels of heavy-metal contaminants, such as those emanating from electroplating facilities. The purified water can then be discharged to the environment, or in some cases recycled. The contaminants are recovered from the resin by chemical regeneration with acids or alkalis, often in a form in which they can be re-used. As a result of new advances in equipment and process design, ion exchange has recently been applied to the purification of concentrated chemical solutions such as steel-pickle liquors. After ion-exchange treatment, the purified concentrate can be recycled to the process.

Several equipment designs are currently employed, ranging from fairly simple and inexpensive units similar to the domestic water softener to large, continuous moving-bed plants. The process is usually carried out cyclically, alternating between service and regeneration modes, and can be either manually or automatically controlled.

6.5.1 APPLICABILITY OF THE TECHNOLOGY

Although ion exchange has become widely accepted as a standard method of purifying water for such applications as boiler feed, pharmaceutical makeup, and semiconductor manufacture, its potential and, indeed, actual use in treating hazardous wastes has not been nearly so widely recognized. Nevertheless, the ion-exchange process has tremendous potential in these applications.

It is noteworthy that unlike many other separation processes such as evaporation and reverse osmosis that remove the water from the polluting species, ion exchange usually removes the pollutant from the water. Since the offending pollutant is often present in low concentrations, ion exchange is frequently more efficient in treating large flows of dilute hazardous waste streams than many other processes. Operating costs depend mainly on the amount of the pollutant to be removed—hydraulic flow rates affect only capital costs.

Another significant feature of the ion-exchange process is that it has the ability to separate, or *purify,* as well as to concentrate pollutants. Some exchangers are selective for certain metals and can remove low concentrations of toxic metal from a wastewater containing a high background level of a nontoxic metal such as sodium or calcium. Ion exchange can also be utilized in some applications to purify a spent chemical concentrate by removal of low-level contamination.

On the downside, the ion-exchange process has certain general limitations and disadvantages including:

1. Chemical wastes are produced if excess regenerant is required.
2. There are limitations on the concentration of what can be treated and produced.
3. Down time is required for regeneration.
4. Resins cannot yet be made which are specific to a particular substance.
5. Ion-exchange resins are prone to fouling by some organic substances.

Many of the advances made in ion-exchange resin and equipment design have been aimed at minimizing these disadvantages.

By far the greatest utility of ion exchange has been in the treatment of inorganic wastes. This is largely because most standard exchangers exhibit unfavorable kinetics of sorption for organic species from liquids nonpolar in character and because the molecular weight of many organics is too high to permit ion exchange at practical rates in polar solutions. There are a number of possible applications of ion exchange to the treatment of organic wastes—e.g., phenol removal and decolorization of kraft pulp-mill effluents—but usually the mechanism is one of adsorption by the resin as opposed to true ion exchange.

6.5.2 DESCRIPTION OF TECHNOLOGY

Ion-exchange resins can be described simply as solid, insoluble acids or bases which are capable of entering into chemical reactions in the same way as their mineral or organic acid analogs. Although ion-exchange reactions are quite complex, consideration of a few of the major factors involved in ion exchange can provide a very useful understanding of the process and its potential utility.

Resin Polymeric Structure

Although various polymeric resin matrices are employed in resin synthesis, most common resins are polystyrene-based. Divinylbenzene (DVB) is used as a cross-linking agent with the styrene. Higher DVB cross-linkage levels provide a more stable resin, but result in slower ion-exchange kinetics, necessitating lower flow rates and larger resin inventories. A somewhat newer resin structure called *macroporous* was designed to retain some of the advantages of the more highly cross-linked *gel* resin while minimizing its disadvantages.

Resin Functionality

It is the functional groups which actually participate in the ion-exchange reactions. Broadly speaking there are two types of ion-exchange resins: cation and anion. These two types are further subdivided into subclasses.

Cation Exchangers. Strong-acid cation resins have sulfonic acid groups. They exchange cations over the entire pH range. Analogous to sulfuric acid, they do not retain hydrogen tightly and consequently are somewhat difficult to regenerate with acid, usually requiring several times the theoretical or stoichiometric dosage of acid to convert them back to the hydrogen form after exhaustion.

For example, strong-acid cation-exchange resins can remove a toxic metal, such as copper, from a wastewater by exchange of hydrogen for the copper according to reaction (1):

$$Cu^{++} + 2RH \rightarrow R_2Cu + 2H^+ \tag{1}$$

Ion exchange is reversible, so that the copper can be removed from the resin by regeneration with a moderately concentrated (typically about 10%) strong mineral acid such as sulfuric or hydrochloric, according to reaction (2), to yield a more concentrated solution of the copper. In some cases it is possible to recycle the metal concentrate produced by regeneration.

$$R_2Cu + 2H^+ \rightarrow Cu^{++} + 2RH \tag{2}$$

Resins have a higher affinity for some ions than others. Generally speaking, strong-acid resins prefer cations with higher ionic charges in solution and larger hydrated radii. Table 6.5.1 qualitatively shows the relative preferences of strong-acid cation resins for various cations.

Weak-acid resins have carboxylic acid groups. Being analogous to acetic acid, they readily associate with hydrogen ions. As a result, they regenerate easily with nearly stoichiometric quantities of acid, but do not function below a pH of about 4.

TABLE 6.5.1 Affinities of Strong-Acid Cation Resins for Various Cations

Monovalent	$Ag>Cu>K>NH_4>Na>H$
Divalent	$Pb>Hg>Ca>Ni>Cd>Cu>Zn>Fe>Mg>Mn$
Trivalent	$Fe>Al$

Chelating resins are very similar to weak-acid resins. The most common chelating resins utilize an iminodiacetate functionality. The principal feature of these resins is that they exhibit a high degree of selectivity for many toxic metals such as copper, mercury, nickel, and lead.[1] The selectivity coefficients for various metals on iminodiacetate chelating resins are shown in Table 6.5.2. The selectivity coefficient is a quantitative measure of the relative preference of a resin for various ions. The relatively slow kinetics of chelating resins necessitate the use of low flow rates and large resin inventories. In addition, the chelating resins are considerably more expensive than other resins. The main disadvantage of chelating resins from an operational point of view is that they are not functional below a pH of about 3 or 4. Chelating resins utilizing a picolylamine functionality have been more recently developed.[2] These resins are highly selective for copper and operate at pHs as low as 1 and below.

Anion Exchangers.   *Strong-base anion resins* have quaternary ammonium groups which, being analogous to sodium hydroxide, do not readily associate with hydroxyl ions. As a result they function over the entire pH range, but require an excess of strong base (usually sodium hydroxide) to regenerate. Strong-base resins are capable of removing acids according to reaction (3) as well as splitting salts as shown in reaction (4).

$$ROH + HCl \rightarrow RCl + H_2O \tag{3}$$

$$ROH + NaCl \rightarrow RCl + NaOH \tag{4}$$

Aside from their ion-exchange properties, strong-base anion resins have a strong tendency to sorb strong acids. Here a distinction is made between true anion exchange of acids as shown in reaction (3) and sorption of acid according to a phenomenon known as *acid retardation*.[4] In the acid-retardation process, strong acids such as sulfuric, hydrochloric, and nitric, in their undissociated form, are sorbed by the resin. Acid sorption is reversible and the acid can be eluted from the resin simply with water. Fortunately, metal salts are not sorbed to the same extent as acids, so that a separation of acids from metal salts is possible. The difficulty in utilizing this process is in minimizing the amount of water employed in desorbing the acid from the resin. A proprietary ion-exchange technique known as Recoflo™ has been very successful in this application (see under *Short Bed,* on p. 6.64).

Weak-base anion resins, having a tertiary amine functionality which behaves similarly to ammonium hydroxide, readily associate with hydroxyl, resulting in high regeneration efficiencies. The disadvantage of weak-base resins, as one would expect, is that they function only under acidic conditions and are not capable of splitting salts.

TABLE 6.5.2   Chelating-Resin Selectivities (pH = 4)[3]

Ion	Selectivity coefficient	Ion	Selectivity coefficient
Hg^{++}	2,800	Cd^{++}	15
Cu^{++}	2,300	Co^{++}	6.7
Pb^{++}	1,200	Fe^{++}	4.0
Ni^{++}	57	Mn^{++}	1.2
Zn^{++}	15	Ca^{++}	1.0

6.5.3 DESIGN CONSIDERATIONS

Although it is possible to conduct ion-exchange reactions "batch" style in stirred tanks, it is generally more efficient to operate the resin in vertical columns in which the resin is "fixed." The resin is usually regenerated in place, either manually or automatically. Continuous ion-exchange systems have been developed; however, because of high cost and complexity, their use has been restricted to very large primary-metals (i.e., mining) applications.[5]

Several different fixed-bed ion-exchange equipment types, with different column designs and process flowsheets, are being used in the treatment of hazardous wastes. Broadly speaking, these can be grouped into four classes: cocurrent, countercurrent, short bed, and mixed bed.

Cocurrent

Although the cocurrent fixed bed is, generally speaking, the least efficient design, it is also usually the lowest-cost design. Its other advantages include simplicity of operation and a vast history of operation in the water-treatment industry, which is widely documented in the literature. Probably the most common example of this system is the domestic water softener.

The following steps are used in the cocurrent ion-exchange cycle:

1. *Service:* The polluting ions are normally removed from the feed during the service step by passing the solution downflow through the resin column, which is two-thirds filled with resin to a height of 60 to 150 cm (24 to 60 in). Service continues until leakage of the ion being removed reaches an unacceptable level. Service flow rates range from 8 to 40 bed volumes per hour.
2. *Backwash:* During this step water is passed upflow through the column at a flow sufficient to fluidize the resin bed 50 to 100%. This serves to reclassify the resin, to avoid channeling, and to remove any suspended solids which may have filtered out on the resin bed during service.
3. *Regeneration:* This is normally accomplished by passing a dilute (i.e., 1 to 5 normal) solution of either acid (for cation) or caustic (for anion) down through the column.
4. *Displacement and rinse:* Excess regenerant is displaced from the column with water at a slow flow rate. Following this, the resin is rinsed with water at high flow to remove the last trace of chemical from the resin.

A typical cocurrent ion-exchange unit is shown in Fig. 6.5.1.

Countercurrent[6]

The countercurrent method utilizes service flow in the direction opposite to the regeneration. This provides higher uptake and regeneration efficiencies and coincidentally produces less chemical waste and more concentrated recovered products upon regeneration. These features make it well suited to wastewater applications. Most of the advantages of this system accrue from the fact that the solution being treated contacts the most highly regenerated part of the ion-exchange bed as it exits from the bed. The converse occurs during regeneration.

A major consideration in the design of countercurrent systems is the method

FIG. 6.5.1 Cocurrent ion-exchange unit.

of preventing fluidization of the resin during upflow. A wide variety of patented techniques have been developed to solve this problem, which, of course, significantly increases the cost of these systems. Because of the complexity of this system, it is not recommended for the "do-it-yourselfer." In fact, it is usually commercially available only from the more engineering-oriented ion-exchange equipment suppliers.

Short Bed

The Recoflo® short-bed system was introduced in the early 1970s primarily for wastewater and chemical-recovery applications. Several hundred of these systems have been installed in the metal-finishing industry for recovery of heavy metals and waste acid. The system utilizes a number of unique features including countercurrent regeneration, fine-mesh ion-exchange resins, low exchanger loadings, fast cycle times, and short [75- to 610-mm (3- to 24-in)] resin beds. A number of significant advantages are claimed for this system,[7] including the following:

- The ability to treat and produce relatively concentrated as well as dilute solutions

- Efficient regeneration
- Compact size
- Decreased regenerant wastes
- Semicontinuous operation through extremely short off-stream times

Mixed Bed

In the mixed-bed system, two different types of resins, usually a mixture of strong-acid cation and strong-base anion, are installed in a single column. When water is passed through the column, very nearly complete deionization occurs, because the ions in solution are alternately exposed to numerous contacts with cation- and anion-exchange sites. Mixed beds are very useful for producing effluents of exceptionally high purity, although regeneration is quite inefficient.

Following the service cycle, the resins are separated by backwash. By virtue of the fact that anion-exchange resins have a lower density than cation-exchange resins, hydraulic separation into two layers occurs, enabling subsequent regeneration of each resin separately. Air agitation then remixes the resins. Alternatively, the resins can be removed from the column and regenerated externally.

Controls

The operating cycle of an ion-exchange unit can be controlled either manually or automatically. Conventional electromechanical controls utilizing timers, liquid-level controls, flow controls, and relays are most common, although microprocessor-based programmable logic controllers are finding increasing application.

Materials of Construction

A major factor to be considered in designing ion-exchange equipment is the choice of materials of construction utilized in vessels, piping, valves, and controls to withstand the aggressive attack of the regenerant chemicals as well as the feed solutions involved. Fortunately ion exchange is usually carried out at, or near, room temperature, so that inexpensive plastics such as PVC can be extensively utilized. Pressure vessels are usually rubber-lined steel, although molded fiber-reinforced plastic (FRP) vessels, being less expensive, are becoming common in diameters less than 120 cm (48 in).

6.5.4 APPLICATIONS

Electroplating Wastes

One of the industries that is most beset by hazardous-waste disposal problems is the electroplating industry. Large quantities of rinsewaters laden with heavy metals and of spent plating solutions are generated on a routine basis. This probably has been the largest area of application of ion-exchange technology for hazardous-waste treatment.

Recovery of Chromic Acid from Plating Rinsewater.   When plated parts are removed from the chrome-plating bath, the adhering film of chromic acid plating electrolyte must be rinsed off. Chromic acid recovery is a good example of the use of ion exchange to purify and concentrate a waste, converting it back into a valuable product.[8]

Rinsewater, typically containing a few hundred mg/L of chromic acid, is pumped through a strong-acid cation exchanger in the hydrogen form to remove metallic impurities (denoted as M^{+++} below) such as trivalent chrome and iron. This is necessary to avoid precipitation of metallic hydroxides in the subsequent anion-exchange bed. This exchange occurs according to reaction (5).

$$3RH + M^{+++} \rightarrow R_3M + 3H^+ \tag{5}$$

Upon exhaustion, the cation resin is regenerated with sulfuric acid.

The *decationized* chromic acid rinsewater is next directed through an anion-exchange resin in the hydroxyl form to remove the chromate. The effluent leaving the anion exchange will be deionized water, which is recycled to the final rinse tank. Uptake of chromate basically occurs according to reaction (6), although the actual mechanism is somewhat more complicated.[9]

$$2ROH + H_2CrO_4 \rightarrow R_2CrO_4 + 2H_2O \text{ (6)} \tag{6}$$

Upon exhaustion, the anion exchanger is regenerated with dilute sodium hydroxide, yielding sodium dichromate according to reaction (7).

$$R_2CrO_4 + 2NaOH \rightarrow 2ROH + Na_2CrO_4 \text{ (7)} \tag{7}$$

The sodium chromate produced in anion regeneration is passed through another cation exchanger in the hydrogen form to yield chromic acid at a concentration of 40 to 80 g/L according to reaction (8). Any excess sodium hydroxide present is also removed by the cation resin according to reaction (9). The cation resin must be subsequently regenerated with sulfuric acid.

$$2RH + Na_2CrO_4 \rightarrow 2RNa + H_2CrO_4 \text{ (8)} \tag{8}$$

$$RH + NaOH \rightarrow RNa + H_2O \text{ (9)} \tag{9}$$

A Recoflo™ chromic acid–recovery unit of this type is shown in Fig. 6.5.2. This unit, which has a service flow of 2.73 m³/h (12 gal/min), reportedly recovers in excess of 99% of the hexavalent chrome emanating from a plating operation.[10]

Recovery of Metals from Acid Copper- and Nickel-Plating Rinsewater.   The application of cation exchange to acid copper and nickel recovery from plating rinsewater was reported as early as 1952.[11] Rinsewater containing several hundred to a few thousand mg/L can be easily reduced to less than 10 mg/L by cation exchange. By carefully regenerating the resin with sulfuric acid it is possible to recover a metal concentrate containing more than 30 g/L.

The principal problem with the original ion-exchange process for nickel recovery is that an excess of sulfuric acid is required to regenerate the nickel from the resin, so that the recovered nickel sulfate product has a pH of approximately 1.[12] This cannot be recycled to a nickel-plating bath operating at a pH of 4 without adversely affecting the bath chemistry. The acid-retardation process has recently

FIG. 6.5.2 Recoflo™ ion-exchange chromic acid–recovery unit.

overcome this problem by removing the excess sulfuric acid and raising the pH of the recovered nickel sulfate concentrate without any chemical additions.

Complete deionization of nickel-plating rinsewaters to allow water recycling is problematic because of the extremely high organic content of most modern bright nickel-plating solutions.

The economics of nickel recovery by ion exchange are extremely attractive, with costs recovered through reduction in nickel salt purchases and waste-treatment expenses. As a result, nickel recovery by ion exchange has now become fairly widespread. The lower value of copper has restricted application somewhat, although the process itself is more straightforward, since deacidification of the eluate is not generally required.

Anodized aluminum is electrolytically colored in solutions containing metal salts of nickel, cobalt, or tin. Rinsewater following these processes can be treated by cation exchange in a similar fashion.[13]

Recovery of Metals from Mixed Rinsewaters. Theoretically, it is possible to combine a group of rinsewaters, deionize them, and recycle them. Even if low-cost water is in good supply, the concept is appealing, since the amount requiring final treatment is substantially reduced.

The presence of cyanide ions in the wastewater to be treated bears consideration since the cyanide may exist both as free cyanide (CN or HCN) or bound as metal cyanide complexes [e.g., $Cu(CN^{3-}_4)$].[14] The metals may be present as the free metallic cations or as the anionic complexes. The metals may therefore be exchanged by either the anion- or cation-exchange resins. Because of the high selectivity of most anion resins for some cyanide complexes (notably copper), special regeneration procedures employing both acid and caustic are sometimes required, but caution must be exercised to avoid generating toxic cyanide gas.

There are a number of potential pitfalls in recovering mixed rinsewaters:

Many cleaners and plating electolytes are heavily laden with organics such as chelating agents and surfactants, some of which may either foul the ion-exchange resin or not be removed by the resin. This may result in contamination buildup in the plant's process water which cannot be easily detected and could potentially have an adverse effect on some processes. It is important to characterize each wastewater source properly, in order to judge which wastewaters are amenable to recycling. It is also important to bear in mind that ion exchange is a chemically driven process which will ultimately result in multiplying the amount of dissolved salt in the final wastewater. Levels of such "neutral salts" are usually restricted for nonocean discharges. A trend begun in Europe some years ago toward water recycling using ion-exchange techniques is now being reversed somewhat because of the increasingly diverse character and complexity of proprietary organic addition agents used in metal-finishing solutions, as well as concerns about neutral-salt discharge.

Effluent Polishing.   Often a better way to utilize ion exchange to reduce the load on a conventional waste-treatment system is to employ resins which will selectively remove low concentrations of toxic metals from a high-flow effluent stream. Although strong-acid resins prefer divalent copper (Cu^{++}) to monovalent sodium (Na^+), it is not generally feasible to recover a few milligrams per liter Cu^{++} from a wastewater containing several thousand milligrams per liter Na^+, because the resin is not selective enough for the copper. Because of their high selectivities for metals such as copper, nickel, lead, and mercury, chelating resins have been shown to be very effective for this type of application.[3] Using an iminodiacetate resin, for example, it is possible to achieve copper effluent levels of less than 0.01 mg/L from a feed containing several hundred mg/L Cu and several thousand mg/L Na and Ca. The eluate from the ion-exchange unit, which is considerably more concentrated and lower in flow rate than the original stream, can then be more easily handled. The same resins can be used to supplement a conventional precipitation system by "polishing" the clarifier effluent to obtain extremely low pollutant concentrations.

Complexing agents such as EDTA, gluconates, quadrol, citrates, and ammonia are extensively utilized in metal-finishing solutions. These chemicals interfere with the removal of metals by the conventional hydroxide precipitation process. Chelating resins can sometimes be utilized to break the bond of these agents with the polluting metal and remove the metal. Acid regeneration of the resin yields a metal concentrate which, not containing the offending agent, can be treated by conventional hydroxide precipitation or recovered as a metal by electrolysis.[15]

Figure 6.5.3 shows how a selective ion-exchange unit can be used to treat a complex wastewater and polish the final effluent.

Cooling-Tower Blowdown

Continuous or periodic blowdown from an industrial cooling tower is necessary to prevent an excessive buildup of salts within the cooling circuit. In high-temperature cooling circuits which use chromate corrosion inhibitors, this blowdown constitutes a major source of toxic-metal pollutants. Although nonchromate inhibitors are available, they are often more expensive and less effective than chromate systems. Chromate removal by sulfate, chloride, or hydroxide exchange, using both strong- and weak-base anion resins, has been practiced for some time.[16,17] In addition to removing the chromate from solution, it is

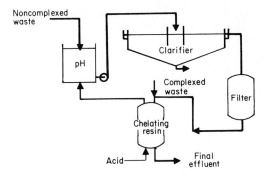

FIG. 6.5.3 Effluent polishing with a chelating resin.

possible to recycle either sodium or ammonium chromate by elution with sodium or ammonium hydroxide. Industrial operating experience is now available from several installations using ion-exchange resins for recovery of chromate from cooling-tower blowdown.

Acid Purification and Recovery

Strong mineral acids such as sulfuric, hydrochloric, nitric, and phosphoric, are used in processes such as the pickling, etching, and anodizing of a variety of metals. The acids become spent through depletion of the acid strength and buildup of dissolved metallic impurities. These solutions are hazardous because a significant quantity of free acid is usually contained in the so-called spent solution and the toxic-metal content is often very high. Although appreciable quantities of such solutions are generated, the value of the contained chemicals is usually very low, the cost of waste treatment being much more significant.

The acid sorption process has been extensively employed since 1977 to recover the free-acid content of such solutions.[18] Utilizing the Recoflo™ technique, the process has been packaged into a standard, compact, skid-mounted unit with the trademark name APU (see Fig. 6.5.4). The APU (or acid-purification unit) is normally operated on a bypass arrangement directly on the acid process bath (see Fig. 6.5.5). Contaminated acid solution is pumped from the process bath to the APU. The free acid is sorbed by the resin and the remaining metal salt solution is then directed to a neutalization and precipitation system for metals removal. In some cases valuable metals can be recovered from this "by-product" stream by electrowinning.[19] Water elution of the APU resin yields a purified acid product containing about 90% of the free acid and 10 to 60% of the original metal content. This product is recycled directly to the acid bath.

Since no significant quantities of chemicals or energy are required for the APU, operating costs for this process are extremely low, making it effective for treatment of such inexpensive chemicals as sulfuric acid. Even so, the major reasons cited for installation of such a system are reduction in neutralization costs and process stabilization, rather than acid savings.

APU systems have been successfully installed on many different processes, including the following: sulfuric and hydrochloric acid steel pickling,[20] nitric–hydrofluoric acid stainless-steel pickling,[20] nitric acid and sulfuric acid per-

FIG. 6.5.4 Acid-purification unit used for removal of iron from sulphuric acid steel-pickle liquors.

oxide etchants for brass and copper,[19] nitric acid stripping solutions for nickel,[21] hydrochloric acid zinc-stripping solutions,[22] and sulfuric acid aluminum anodizing[23] among others. Table 6.5.3 shows typical data obtained from several of these applications.

When the acid concentration is fairly dilute (i.e., $[H] \approx 1 \ N$), or when the acid is relatively weak (e.g., chromic acid or phosphoric), it is possible to re-move metal contamination by cation ex-change. Since an excess of acid (usually sulfuric or hydrochloric acid) is nor-mally required for regeneration of the cation exchanger, acid sorption can of-ten be used to recover the excess free acid for re-use by the cation-exchange unit. Applications include purification of hard chrome-plating baths,[24,25] phos-phoric acid pickling, anodizing and chemical polishing solutions,[26] hydro-chloric acid electrochemical etchants for aluminum lithographic plates and ca-pacitor foils,[27] and "sulfo" acid integral-color aluminum-anodizing baths.[13]

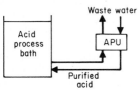

FIG. 6.5.5 Acid-purification unit (APU) installation configuration.

Radioactive Wastes

Application of ion-exchange technology in the nuclear industry is extensive, ranging from common feedwater and condensate systems to specific processes such as transuranic extractions, boron-recovery systems, spent-fuel-poolcleanup, and radioactive-waste processing.

TABLE 6.5.3 Typical Results: APU® Acid Sorption System

Application	Component	Feed, g/L	Product, g/L	By-prod., g/L
Anodizing	H_2SO_4	190	182	13
	Al	10	5.5	6
Steel pickling	H_2SO_4	127	116	10
	Fe	36	10.5	21
Steel pickling	HCl	140	142	10.3
	Fe	45.6	33.5	20.4
Rack strip	HNO_3	514	581	10
	Cu + Ni	99	47.5	70.8
Brass etch	H_2SO_4	166	157	17.2
	H_2O_2	12.4	11.6	1.7
	Cu	26.6	13.6	13
	Zn	5.9	2.9	3.3

Mixed Radioactive Waste. In nuclear power plants the various liquid streams generated from normal operations, including flows from chemical, floor, and equipment drains, are combined and concentrated by evaporation. The condensates from the evaporators are polished by ion exchange to remove the last traces of radioactivity.[28] The depleted resins are not normally regenerated but are disposed.

Cesium and Strontium Radioisotopes. Because of their stability, inorganic ion exchangers (i.e., *zeolites*) such as clinoptilolite offer significant advantages in applications involving the separation and purification of radioisotopes such as cesium and strontium. These isotopes can also be retained on the zeolites and encapsulated for long-term storage.[29]

6.5.5 RECENT ADVANCES IN ION-EXCHANGE TECHNOLOGY

Synthetic adsorbents[30]. As discussed earlier, ion exchange has found only limited application in the treatment of organic wastes. Adsorbent resins of macroporous polystyrene and acrylic show promise in the treatment of many such wastes, however. Although an extension of ion-exchange resin technology, these resins are not true ion exchangers. By controlling the pore size, porosity, surface area, and degree of hydrophilicity, it is possible to tailor adsorbents to a range of organic adsorption applications. Regeneration is accomplished with organic solvents, acids, or bases, depending on the nature of the adsorbed species. Possible applications include removal of surfactants, phenol, chlorinated pesticides, and dyes, and decolorization of kraft pulp-bleaching effluents.

Intermediate in chemical composition between activated-carbon and polymeric adsorbents is a new class of carbonaceous adsorbents. The advantage ofthese adsorbents is an even greater range of properties and possible applications.

Inclusion polymers[31]. Inclusion polymers are polymeric adsorbent resins which have been impregnated with solvents such as di-2-ethylhexyl phosphate

(DEHPA). The solvent, which remains trapped inside the pores of the resin, is capable of selectively extracting certain metals, which are free to pass in and out. These polymers can be operated in an analogous manner to ion-exchange resins to remove toxic metals from wastewater streams. As with cation resins, inclusion polymers are usually regenerated with strong acids.

Molecular-sieve zeolites[32].   Molecular-sieve zeolites are inorganic ion exchangers, some of which are naturally occurring. They exhibit a high degree of selectivity for certain ions such as ammonia. Because of this, their application is highly specialized.

Devoe-Holbein process[33].   Although not truly an ion-exchange process, the Devoe-Holbein process utilizes a new adsorbent which behaves very similarly to highly selective ion exchangers. It is reported that compositions can be tailored to remove a specific species. Like ion-exchange resins, these compositions can be regenerated with mineral acids.

Centralized ion-exchange recovery systems.   These systems have been employed in Europe for several years with good success.[34] Small portable ion-exchange columns are installed on site to treat isolated waste streams, obviating the need for other in-plant treatment systems. When exhausted, the resin columns are replaced with fresh ones and transported to a centralized regeneration facility. The economies of scale then facilitate the treatment of spent regenerant solutions and recovery of metal values where practical.

6.5.6 VENDORS

Ion-Exchange Resins

Dow Chemical U.S.A.
2020 Willard H. Dow Center
Midland, Mich. 48674
(800) 258-2436

Purolite Corp.
150 Monument Rd.
Bala Cynwyd, Pa. 19004
(215) 668-9090

Rohm and Haas
Independence Mall West
Philadelphia, Pa. 19105
(215) 592-3030

Sybron Corp.
Ionac Chemical Div.
P.O. Box 66
Birmingham, N.J. 08011
(609) 894-8211

Mobay/Bayer Chemical
Mobay Rd.
Pittsburgh, Pa. 15205-9741
(412) 777-4652

Ion-Exchange Equipment

Belco Pollution Control Corporation
119 Littleton Rd.
Parsippany, N.J. 07054
(201) 263-8900

Crane Co.
Cochrane Environmental Systems Div.
P.O. Box 191
King of Prussia, Pa. 19406

Eco-Tec Limited
925 Brock Rd. S.
Pickering, Ontario,
Canada L1W 2X9
(416) 831-3400

Graver Water
Division of The Graver Co.
2720 U.S. Highway 22
Union, N.J. 07083
(201) 964-2400

Himsley Engineering Ltd.
250 Merton St.
Toronto, Ontario,
Canada M4S 1B1
(416) 485-4441

Illinois Water Treatment Co.
Rockford, Ill. 61105

L*A Water Treatment Corporation
P.O. Box 1467
17400 E. Chestnut St.
City of Industry, Calif. 91749
(818) 912-5411

Penfield Inc.
8 West St.
Plainsville, Conn. 06479
(203) 621-9141

Water Refining Industrial, Inc.
500 N. Verity Parkway
Middletown, Ohio 45042
~~(513) 423-9421~~
708 961 5043

6.5.7 REFERENCES

1. J. Melling and D. W. West, "A Comparative Study of Some Chelating Ion Exchange Resins for Applications in Hydrometallurgy," in D. Naden and M. Streat (eds.), *Ion Exchange Technology,* Ellis Horwood, Chichester, England, 1984.

2. R. R. Grinsted, W. A. Nasutavicus, and R. M. Wheaton, "New Selective Ion Exchange Resins for Copper and Nickel," in *Extractive Metallurgy of Copper,* American Inst. Of Min., Met., and Pet. Engineers, 1976.

3. W. H. Waitz, "Ion Exchange in Heavy Metals Removal and Recovery," *Amber-hi-lite,* no. 162 (Fall 1979), Rohm and Haas, Philadelphia.

4. M. J. Hatch and J. A. Dillon, *Industrial and Engineering Chemistry Process Design and Development,* 2:(4)253 (1963).

5. F. L. D. Cloete, "Comparative Engineering and Process Features of Operating Continuous Ion Exchange Plants in Southern Africa," in D. Naden and M. Streat (eds.), *Ion Exchange Technology,* Ellis Horwood, Chichester, England, 1984.

6. I. M. Abrams, "New Developments In Counter-Current Fixed Bed Ion Exchange," *Proceedings of the 34th International Water Conference,* Pittsburgh, 1973.

7. C. J. Brown, P. J. Simmons, and I. H. Spinner, "Water and Chemical Recovery by Reciprocating Flow Ion Exchange," *Proceedings of the 40th International Water Conference,* Pittsburgh, Oct 30–Nov 1, 1979.

8. C. J. Brown, "Chromic Acid Recovery by Reciprocating Flow Ion Exchange," *Proceedings of the AES Third Continuous Strip Plating Symposium,* Annapolis, Maryland, April 1980.

9. T. V. Arden and M. Giddings, "Anion Exchange in Chromate Solution," *Journal of Applied Chemistry,* July 1961, pp. 229–235.

10. M. Dejak and T. Nadeau, "Copper, Nickel and Chromium Recovery in a Jobshop," *Plating and Surface Finishing,* 73(4) (April 1986).

11. T. J. Fadgen, "Metal Recovery by Ion Exchange," *Sewage and Industrial Wastes,* September 1952.

12. P. H. Kehoe, "Nickel Recovery from Rinsewater," *Industrial Finishing,* January 1969.

13. C. J. Brown, "Recovery and Treatment of Aluminum Finishing Wastes," *Proceedings*

of the Aluminum Finishing Seminar, Aluminum Extruders' Council, St. Louis, March 30–April 1, 1982.

14. R. Kunin, "Ion Exchange for the Metal Products Finisher," *Products Finishing,* April 1969.

15. C. I. Courduvelis, U.S. Patent 4,303,704.

16. "Recovery and Reuse of Chromates from Cooling Tower Blowdown," *Idea Exchange,* Dow Chemical Co.

17. R. Kunin, "New Technology for the Recovery of Chromates From Cooling tower Blowdown," *Amber-hi-lite,* no. 151 (May 1976), Rohm and Haas, Philadelphia.

18. C. Brown, "Acid/Metal Recovery by Recoflo Sorption," presented at the 23rd Conference of Metallurgists of the CIM, Quebec City, August 19–22, 1984.

19. J. C. Egide, "Copper Recovery System Extends Life of Bright Dip," *Products Finishing,* March 1985.

20. W. K. Munns, "Iron Removal from Pickle Liquors Using Absorption Resin Technology," presented at the International Symposium on Iron Control in Hydrometallurgy, October 19–22, 1986, Toronto.

21. W. K. Munns, "Western Forge Maintains High Quality With Purification System," *Finishers' Management,* **28**:(6)22 (June–July 1983).

22. A. K. Haines and T. H. Tunley, "The Recovery of Zinc From Pickle Liquors by Ion Exchange," *Jour. of the S. African Inst. of Min. Met.,* November 1973.

23. C. J. Brown, D. Davy, and P. J. Simmons, "Purification of Sulfuric Acid Anodizing Solutions," *Plating and Surface Finishing,* January 1979.

24. anonymous, "Ion Exchange Helps Canadian Chromium Plater," *Plating and Surface Finishing,* May 1983.

25. R. L. Costa, U.S. Patent 2,733,204.

26. C. J. Brown, "Recovery of Phosphoric Acid by Ion Exchange and Evaporation," *Proceedings of the AES 71st Annual Technical Conference,* New York (1984).

27. C. J. Brown, "Recovery of Aluminum Etch and Anodizing Solutions," *Proceedings of the Symposium on Aluminum Surface Treatment Technology* at the 169th Meeting of the Electrochemical Society, Boston, Massachusetts, May 4–9, 1986.

28. G. R. Allan, "Ion Exchange of Mixed Radwaste," *Proceedings of the International Water Conference,* Pittsburgh, Oct 24–26, 1983.

29. M. Howden and J. Pilot, "The Choice of Ion Exchanger for British Nuclear Fuels Ltd's Site Ion Exchange Effluent Plant," in D. Naden and M. Streat (eds.), *Ion Exchange Technology,* Ellis Horwood, Chichester, England, 1984.

30. E. C. Feeney, "Removal of Organic Materials from Wastewaters with Polymeric Adsorbents," in C. Calmon and H. Gold (eds.), *Ion Exchange for Pollution Control,* vol. 2, CRC Press, Boca Raton, Florida, 1979.

31. F. Vernon and H. Eccles, "Some Hydrometallurgical Applications of Hydroxy-oxime, Hydroxyquinoline and Hydroxamic Acid Solvent Impregnated Resins," *The Theory and Practice of Ion Exchange,* Society of Chemical Industry, London, 1976.

32. J. D. Sherman, "Application of Molecular Sieve Zeolites to Pollution Abatement," in C. Calmon and H. Gold (eds.), *Ion Exchange for Pollution Control,* vol. 2, CRC Press, Boca Raton, Florida, 1979.

33. D. Brener and C. W. Greer, "Novel, Synthetic Compounds for the Efficient Removal of Nickel and Zinc From Plating Effluent," *Proceedings of the 7th AESF/EPA Conference on Pollution Control for the Metal Finishing Industry,* January 27–29, 1986.

34. E. P. Reinhardt and A. J. Grussing, "Recycling is the Answer for Wastewater Treatment," *Proceedings of the 7th AESF/EPA Conference on Pollution Control for the Metal Finishing Industry,* January 27–29, 1986.

6.5.8 FURTHER READING

Calmon, C. and H. Gold (eds.), *Ion Exchange for Pollution Control,* vols. 1 and 2, CRC Press, Boca Raton, Florida, 1979.

Dorfner, K., *Ion Exchangers,* Ann Arbor Science, Ann Arbor, Michigan, 1972.

Kunin, R., *Ion Exchange Resins,* Robert E. Kreiger, Huntington, New York, 1972.

SECTION 6.6
SOLVENT EXTRACTION*

Joan B. Berkowitz

President
Risk Science International
Washington, D.C.

6.6.1   OBJECTIVES OF TREATMENT

The term *solvent extraction* is used to refer to two different types of processes. One, which is also known as *liquid-liquid extraction,* is the separation of constituents from a liquid solution by contact with another, immiscible liquid in which the constituents are more soluble. (If the constituent to be separated is ionic, the process is called *liquid ion exchange.*) The other, which is also known as *leaching,* is the separation of constituents from solids by contact with a liquid in which the constituents dissolve.[1] A commercial example of the first is extraction of uranium from nitric acid solution with tributyl phosphate dissolved in hexane.[2] A commercial example of the second is extraction of vegetable oils from flaked oil seeds with hexane.[2] In the area of hazardous-waste treatment, the term *solvent extraction* is usually taken to mean liquid-liquid extraction and, more particularly, extraction of organic components from aqueous solutions into immiscible solvents.[3] In this section, we will use a broader definition that encompasses leaching and liquid ion exchange as well as liquid-liquid extraction in the conventional sense.

---

*This section of Chap. 6 was prepared when the author was with Arthur D. Little, Cambridge, Massachusetts.

Solvent extraction is a physical separation process, used for transfer of constituents from one liquid solution (or solid matrix) to another. The constituents are unchanged chemically. Such transfers may be carried out for a variety of reasons, such as separation of a valuable constituent from impurities present in the original solution or concentration of a constituent for ease of subsequent recovery or treatment.

6.6.2 REACTIONS AND THEORETICAL CONSIDERATIONS

Four terms are used in the literature to simplify discussion of solvent extraction processes. These terms are[1]

- *Feed:* The solution, solid, or sludge whose constituents are to be separated
- *Solvent:* The liquid into which feed constituents are to be transferred*
- *Extract:* Solvent-rich solution of transferred constituents
- *Raffinate:* Residual feed from which constituents have been removed

These terms are illustrated in Fig. 6.6.1.

Solvent extraction is a thermodynamic process that depends upon the equilibrium relationships established between system components and phases when feed and solvent are brought into contact. These relationships are derived from ternary (multicomponent) phase diagrams, which are discussed at length elsewhere.[1] In the simplest instances, the constituent to be separated (C) will be completely miscible with both the solvent (S) and the feed (F), while the mutual solubilities of the pure solvent and pure feed will be very low. As illustrated in Fig. 6.6.1, when a fairly dilute solution of C in F is equilibrated with pure S, a two-phase system will result—an extract (solution of C and F in S), and a raffinate (dilute solution of C and S in F). A partition coefficient (or distribution coefficient) of C between the two phases is defined as follows:

$$K_C = \frac{X_{CS}}{X_{CF}}$$

where X_{CS} is the weight fraction of C in the extract and X_{CF} is the weight fraction of C in the raffinate. The distribution coefficient is dependent upon concentration, temperature, and pressure. The distribution coefficient defines an absolute limit on the extent of transfer of a constituent from the feed to the solvent in a single contact. To strip the feed further, multiple contacts are required. The higher the distribution coefficient, the less solvent and the lower the number of contacts that will be required to obtain efficient transfer. Other desirable properties of solvents include:[1]

- *High selectivity*—selectivity being defined as the ratio K_C/K_F. This ratio measures the relative transfer of the desired constituent and the feed matrix into the solvent. If $K_C/K_F = 1$, the composition of the extract and the raffinate will be

---

*The *solvent* may be a single liquid, a mixture of liquids, or a solution.

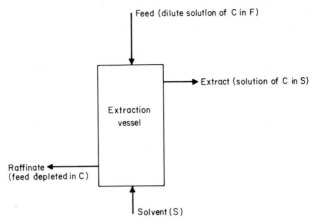

FIG. 6.6.1 Idealized solvent extraction system. (If the solvent is heavier than the feed, the extract will be in the bottoms. Otherwise, the extract will be in the overhead, as shown.)

the same; i.e., no separation will have occurred. Thus, K_C/K_F must be greater than 1, and the larger the better.

- *High saturation solubility for the constituent to be transferred*—i.e., ability of the solvent to dissolve larger quantities of the desired constituent before reaching saturation.

- *Low solubility of the solvent in the feed material*—The reason this is wanted is that the solvent generally needs to be removed from the raffinate prior to discharge for economic or environmental reasons or both (i.e., extracting solvents are often expensive and, therefore, must be recovered and/or they would contaminate the raffinate so that it would be unsuitable for sewer or direct discharge without further treatment).

6.6.3 WASTE APPLICABILITY

Solvent extraction is applicable, in principle, to the physical separation of selected waste constituents from aqueous or organic solutions (liquid-liquid extraction) or from sludge or solid mixtures (leaching). In practice, an appropriate solvent must be found that meets the criteria described in Subsec. 6.6.2. This can be difficult and there are few commercial applications of solvent extraction in use for hazardous wastes or wastewater treatment.

The principal waste-treatment application of solvent extraction, at present, is removal of phenols from wastewater effluents from petroleum-refinery, coke-oven, and phenol-resin plant operations. Suitable extraction solvents include crude oil, light oil, benzene, toluene, isopropyl ether, tricresyl phosphate, methyl isobutyl ketone, methyl chloride, and butyl acetate. The phenols transferred to the solvent phase can be recovered by subsequent extraction with caustic. Removal efficiencies are typically in the range of 90 to 98% and can exceed 99% in some applications with the use of special equipment (e.g., centrifugal and rotating-disk contactors).[3] Processes in commercial use include one developed by Lurgi around 1940 (Phenosolvan process) and one licensed by Chem-Pro Equipment Corporation.[4] The Phenosolvan process uses diisopropyl ether as the extracting solvent; the solvent is recovered from the extract by distillation and removed from the raffinate by nitrogen stripping. The Chem-Pro process reportedly uses methyl isobutyl ketone as the solvent with recovery by distillation and removal from the raffinate by steam stripping.[4] Operating costs are approximately inversely proportional to the percentage of phenol in the influent, ranging (in 1976 dollars) from $0.02 per pound of phenol recovered with a 1% phenol influent to $1.68 per pound with a 0.01% influent.[3]

Another solvent extraction process in commercial use is the recovery of acetic acid from industrial wastewaters.[14] For feeds containing up to 16% acetic acid (by weight), the solvent typically used is ethyl acetate. For feeds with higher concentrations of acetic acid, mixtures of ethyl acetate with benzene are used. Solvent is typically regenerated for the extract by azeotropic distillation, and recovered from the raffinate by stripping.

Over the past decade, a considerable amount of research and development has been done on solvent extraction processes for waste treatment. Several examples follow.

Separation of Phenol from Wastewater[5,6,17]

Phenolic compounds are major components of aqueous effluents from petroleum refineries, petrochemical plants, phenolic-resin plants and coal-processing facilities. Concentrations can be as high as several percent. As discussed earlier, there are commercial solvent extraction processes in use for recovery of phenol from wastewaters. Nonetheless, there is continuing research to develop more efficient, cheaper, and safer methods. Medic et al.[5] measured the distribution coefficients between phenol in aqueous solutions and in organic-solvent mixtures (butyl acetate and isopropyl ether; butyl acetate and octane; isopropyl ether and octanol; isopropyl ether and benzene; isopropyl ether and carbon tetrachloride; and isopropyl ether and octane). Maximum distribution coefficients were obtained with mixtures around 50% isopropyl ether and 50% 1-octanol by volume. Greminger et al.[6] compared methyl isobutyl ketone (MIBK) and diisopropyl ether (DIPE) as solvents for extracting phenol, dihydroxybenzenes, and trihydroxybenzenes from wastewater. The authors concluded that MIBK is the preferred solvent because of its high distribution coefficient and suitable physical properties for recovery of residual solvent from the raffinate by vacuum steam stripping. Capital and operating cost estimates for a system to treat 115,000 kg/h of wastewater containing 15,800 ppm of phenol are of the order of $3.3 million and $6.30 per 1,000 gal, respectively.

Drummond et al.[17] investigated the efficiency of a treatment train for wastewasters from a coal-liquefaction process. One step in the treatment process involved solvent extraction with methyl ethyl ketone in a 36-stage, continuous countercurrent extraction column. At a solvent feed rate of 0.4 L/min, and a solvent to wastewater ratio of 0.1, phenols were reduced from 9,150 ppm in the incoming stream to 40 ppm in the raffinate. Reductions in total organic carbon (TOC) and chemical oxygen demand (COD) were 93% and 92%, respectively. The raffinate was suitable for biological treatment in an activated-sludge system.

Removal of Polar Organics from Wastewater

Joshi et al.[7] investigated the suitability of several solvents for extraction of seven of the priority pollutants from wastewater. Table 6.6.1 summarizes the solvents

TABLE 6.6.1 Potentially Suitable Solvents for Extraction of Selected Pollutants from Wastewater

Potential extraction solvents	Acrolein	Acrylonitrile	N-Nitroso-dimethylamine	2-Chlorophenol	Isophorone	Bis(2-choloroethyl) ether	Bis(2-chloroethoxy) ether
n-Butyl acetate	X						
Methyl isobutyl ketone	X			X	X	X	
Tetrachloroethane	X						
Toluene	X				X		
Methylene chloride		X	X				
Tributyl phosphate		X		X	X	X	X
Hydrocarbons				X	X	X	X
Isobutyl heptyl ketone				X	X		
Diisobutyl ketone					X		X

Source: Ref. 7.

recommended for further consideration based on bench-scale measurements of equilibrium distribution coefficients of the pollutants between water and solvent, ease of recovery of the solvent from the extract by distillation, and thermal stability of the extract. Estimated operating costs (including capital charges) are in the range of $4.20 to $12.20 per 1,000 gal of wastewater treated; these costs are for a conceptual system that would include an extractor, a solvent-regeneration column, a vacuum steam stripper for removal of solvent from the raffinate, and appropriate heat exchangers.

Removal of Blasting Oil from Nitration-Plant Effluent

One step in the production of explosives by nitration of polyalcohols involves washing residual nitric acid out of the formed nitro compounds with a solution of sodium carbonate. The effluent wastewater is saturated with blasting oil (mixture of nitroglycol and nitroglycerin) and contains droplets of blasting oil in suspension. Michelson and Ostern[8] developed a solvent extraction process for removal of blasting oil from such wastewaters.

The solvent selected is dinol, a complex, commercial mixture of dinitrotoluenes, nitroxylenes, and impurities such as phenols, that is also a product of the nitration plant. A simple mixer-settler was chosen for contacting the wastewater with the dinol. The extraction is carried out at elevated temperatures (45 to 50°C) because dinol is a solid at room temperature. The density of dinol is also greater than that of water, so that the extract becomes the lower phase. A second solvent extraction with toluene can be used to remove the dinol from the raffinate. A full-scale system was put into operation in Norway in 1977 to treat a maximum flow of 1,500 L/h of wastewater. The extract solution of dinol and blasting oil is recycled to the production process. The secondary extraction to remove dinol from the raffinate was not implemented. Instead, dinol and sludge are settled out in a simple retention tank and the nitrate-containing raffinate is either used in production, applied as a fertilizer, treated biologically, or purified to produce nitrate of commercial quality.

Removal of Chlorophenols from Wood-Preserving Wastewaters

The primary contaminant in wood-preserving wastewaters is pentachlorophenol at a concentration of around 100 mg/L. Other chlorinated phenols are present at much lower concentrations. In bench-scale tests, 99% pentachlorophenol removals were achieved by extraction with a solvent mixture of No. 2 fuel oil and amyl alcohol still bottoms. The removal efficiency was reduced to 97% with No. 2 fuel oil alone as the extraction solvent.[9]

Removal of PCBs from Transformer Oils

Union Carbide Corporation investigated solvent extraction as a method for removal of PCBs from mineral and other oils.[10,11] The extraction solvent selected for pilot-plant testing was N,N-dimethylformamide (DMF) which had the highest distribution coefficient of those tested. The pilot plant consisted of two extraction columns, each 10 cm in diameter and 4.6 m long, with countercurrent flow. The concentration of PCB in the oil was reduced from 300 ppm to less than 50 ppm. Additions of water to the extract resulted in separation of an oil layer in which the PCBs were concentrated, and an aqueous layer from which DMF could be recovered by distillation. To be suitable as an extraction solvent, the DMF must contain less than 1.5% water by weight.

The Electric Power Research Institute (EPRI) developed a continuous countercurrent solvent extraction system for removal of PCBs from mineral oils.[18] The solvent used is diethylene glycol monomethyl ether (DGME). Removal efficiencies in bench-scale tests were 90 to 98%. The extract (PCBs in DGME) can be vacuum-distilled to recover the solvent, and the PCBs (reduced in volume) can be incinerated. A pilot test facility in Atlanta has been planned by EPRI and Georgia Power Company. Capital costs have been estimated at $800,000 for a 500,000 gal/yr plant.

Removal of Acetic and Formic Acids from Wood-Treatment Exhaust Liquors

The exhaust liquor remaining after treatment of wood with bisulfite can be treated by solvent extraction to recover a mixture of acetic and formic acids. Trioctylphosphine oxide in undecane has been reported to be a suitable solvent.[12]

Removal of Barium from Wastewater

A solution of dinonylnaphthalenesulfuric acid in heptane has been reported as a possible solvent for extraction of barium ions from wastewater.[12]

Removal of Water and Oils from Sludges

Oily waste sludges are produced from many metalworking, petroleum-refining and other operations. Solvent extraction has been investigated as a means for recovering the oil and drying the sludges.

In the case of mill scale, for example, removal of oil and water might allow the treated sludge to be returned to the blast furnace for metal recovery. In the case of other sludges, removal of oil and water might reduce the potential for leaching in a land disposal environment. Resource Conservation Company (RCC) of Bellevue, Washington, has developed the BEST process for separation of sludges into dry solids (4 to 5% moisture), water, and oil.[3,12] The process is based on solvent extraction of water and oil from the sludge with an aliphatic amine. Following extraction, the liquid and solid phases are separated (by centrifugation, for example). The liquid phase (extract), containing the water and oil dissolved in the amine, is further separated into an organic and an aqueous phase by heating to around 120°F. Oil is recovered from the organic phase by distillation. The amine is recovered from the aqueous phase by steam stripping. The BEST process has been tested on a variety of hazardous- and nonhazardous-waste sludges in both bench-scale and mobile (i.e., field-erectable) pilot-plant units. The mobile unit is being tested for cleanup of oil sludge lagoons at a CERCLA (Superfund) site.

Extraction of Contaminants from Soil

Solvent extraction has been used extensively for separation and concentration of metal values from ores.[13,14] The same general principles should apply to the extraction of metallic contaminants from soils, although specific details are likely to be very different. Extraction solvents are not currently available for all contaminants, and extraction efficiencies may vary for different types of soils, levels of contaminants, and site-specific parameters (i.e., interferences). A mobile system has been field-tested for extraction of lead from excavated soils with a proprietary solvent. Lead levels were reduced from 4 to 5% to less than 100 ppm. Estimated costs are in the range of $100 to $2,000 per ton of soil or sediment treated.[15]

Ellis et al.[16] investigated the use of aqueous surfactants to extract slightly hydrophilic and hydrophobic organics from soils. A low-clay soil, with a total organic carbon content of 0.12% by weight and a cation-exchange capacity of 8.7 meq/100 g, was used in a laboratory study. Soil samples were contaminated with a high-boiling crude-oil fraction containing a mixture of aliphatic and aromatic hydrocarbons at 1,000 ppm; a mixture of PCBs (Aroclor 1260) in chlorobenzenes at 100 ppm; and a mixture of chlorophenols (di-, tri-, and penta-) at 30 ppm. The commercial surfactants tested (Adsee[R]799 and Hyonic NP-90) resulted in removal of about 86% of the contaminants into 10 pore volumes of surfactant. No treatment methods were found, however, to remove the contaminants from the extract so that the surfactant could be re-used.

Extraction of Platinum and Palladium from Spent Automobile Converters

A leaching process has been tested in the laboratory for recovery of platinum and palladium from spent automobile-converter catalyst. The catalyst tested contained 300 ppm of platinum and 110 ppm of palladium in an alumina bead matrix. Leaching for 1 h with a solution of 1.5M aluminum trichloride and 0.7M nitric acid at 95°C was effective in extracting over 97% of the precious metals.[13]

Waste-Treatment Applications of Supercritical-Fluid Extraction

The critical point is the highest temperature and pressure at which liquid and gas can coexist in equilibrium.[19] At temperatures above the critical point, for example, a fluid cannot be liquefied by isothermal compression. A fluid at supercritical temperatures and pressures has a density approaching that of the liquid and a viscosity and diffusivity approaching that of the gas. The net effect is that supercritical fluids have the high-solubility-for-organics characteristics of liquids and the high-mass-transfer characteristic of gases. The basic principle of supercritical- (or near-critical-) fluid extraction is dissolution of feed to constituents into a fluid near its critical point, and release of these constituents at lower (subcritical) pressures and temperatures where their solubility is much less.

The critical points and densities of several fluids that have been tested for waste-treatment applications are given in Table 6.6.2. It may be noted that temperatures are not far from ambient, and pressures are in the 40 to 75 atm range. Critical-fluid solvent extraction processes have been designed to accept either liquid, sludge, slurry, or granular solid feeds.

The U.S. Environmental Protection Agency (EPA) has sponsored several studies on the application of critical-fluid extraction to waste treatment. One of the first applications examined was supercritical–carbon dioxide extraction, as a potential substitute for steam stripping, to regenerate spent granular activated-carbon (GAC) adsorbents.[20] In bench-scale tests, spent GAC, contaminated with volatile organic carbon components (VOCs) removed from stack gases, was regenerated in a pressure vessel by continuous countercurrent flow of CO_2 through the GAC bed at about 103 atm and 40 to 50°C. Regeneration of 436 g of GAC, contaminated with 109 to 148 g of organics, was accomplished with 100 standard L of CO_2, flowing at a rate of 15 to 20 L/min. Estimated capital and operating costs for a combined carbon-adsorption system to clean up 10,000 scfm of air

TABLE 6.6.2 Critical Points of Fluids Tested for Solvent Extraction of Organic Constituents from Wastes

Fluid	Critical temperature, °C	Critical pressure, atm	Critical density, g/cm^3	Ref.
Carbon dioxide	31.1	73.8	0.468	21
Ethylene	9.9	51.2	0.227	21
Ethane	32.3	48.8	0.203	21
Propane	96.9	42.6	0.220	21
Dichlorodifluoromethane (Solvent-12)	112	40.7	0.557	22

laden with 4,500 ppm of methyl ethyl ketone (MEK) with supercritical-CO_2 regeneration are \$530,000 and \$285,000 per year, respectively. The recovery value of MEK is estimated at \$1,386,000. (All costs are in 1980 dollars.)

Small-column tests of supercritical-CO_2 extraction were run with 7 g of GAC, contaminated with adsorbed Alachlor, Atrazine, carbaryl, diazinon, acetic acid, and phenol. For these compounds, desorption conditions were more severe (higher temperature and pressure), and larger volumes of CO_2 were required than for the VOCs discussed earlier. It was also found that regeneration is incomplete for compounds such as carbaryl and diazinon, which adsorb to GAC irreversibly.[20]

The use of supercritical-fluid CO_2 as a solvent to extract organic contaminants from wastewater is theoretically possible but as yet unproven. A standard countercurrent extractor with the dilute aqueous stream flowing down through a rising column of supercritical-CO_2 solvent should be suitable. Distribution coefficients for selected organics between water and supercritical CO_2 range from 0.1 for ethanol to 10 or more for higher-molecular-weight alcohols, esters, and aromatic hydrocarbons.[20]

Research has been conducted at Louisiana State University on supercritical-CO_2 extraction of DDT from contaminated soils. Tests were conducted on 10-g samples of finely ground soils contaminated with 500 ppm of DDT. A 350-mL batch reactor was used with CO_2 at 80 atm and 35°C. Removal efficiencies were in the range of 60 to 80% in 10 min of contact.[20]

Supercritical CO_2 has also been tested for extraction of PCBs from soil particles. In laboratory tests, 90% of the PCB contamination was extracted in less than 1 min.[23]

Krupp Research Institute in West Germany has developed a recycling process for waste oil based on supercritical-ethane extraction.[24] Feeds are waste oils contaminated with dissolved metals, water, and PCBs. A preliminary distillation step is used to remove benzene and to reduce the water content below 2.1%. The supercritical-ethane extraction is carried out at 100 atm and 42°C in a column packed with Raschig rings. The oil is extracted into the ethane, and heavier contaminants leave as column bottoms. The oil extract is treated in a three-stage purification train that yields in sequence motor-grade lube oils, machine-grade lube oils, and diesel oils; 99% pure ethane remains for recycling to the extractor. The recovered oils have been found to contain less than 50 ppm of PCBs and no metals. Krupp expects to have a pilot plant in operation in 1987 to process 100 kg/h of dry oil.

Arthur D. Little, Inc., has tested several critical-fluid extraction processes for separation of oil from mill scale, from wastes generated in processing of vegetable and special oils, and from machine-oil emulsion wastes.[21]

The recycling of mill scale from steel production to the blast furnace, via the sinter plant, is limited by the high lubricating-oil content of the scale. Bench-scale tests have been carried out to de-oil the scale by solvent extraction with propane, carbon dioxide, and dichlorofluoromethane (Solvent-12) under near-critical conditions. With carbon dioxide, de-oiling was incomplete even at 150 atm, 50°C, and a solvent/feed ratio of 17. Propane and Solvent-12 resulted in oil-removal efficiencies of 80 to 100%. Runs were made at around room temperature and pressures around 20 atm. Capital and operating cost estimates for a 90,000 t/yr plant are of the order of \$1.75 million and \$5.67 per metric ton of mill scale processed (exclusive of oil-recovery credits), respectively.[21]

Refining (and rerefining) of oils includes a clay purification step. The spent clay filter cakes typically contain 30 to 60% occluded oil, which is difficult to remove and which presents a problem for disposal. Bench-scale tests have been carried out to extract silicone and vegetable oils from spent clays with propane, Solvent-12, and Solvent-500 (an azeotropic mixture of Solvent-12 and 1,1-difluoroethane). Extractions were carried out at room temperature and pressures from 10 to 60 atm, under near-critical conditions. Oil-removal efficiencies were around 90%. Capital and operating cost estimates for a 300,000 lb/yr plant are of the order of $170,000 and $80,000 per year (exclusive of oil-recovery credits), respectively.[21]

Waste-oil lubricating emulsions are common in the metal-machining and metal-forming industries. Bench-scale tests of critical-fluid solvent extraction as an oil-recovery method were conducted on oil emulsions from aluminum-can and steel product-forming operations. The solvents tested were carbon dioxide, propane, Solvent-12, and Solvent-500. Extractions were carried out close to room temperature at pressures of 10 to 90 atm. Oil-removal efficiencies of around 56% were achieved in a single extraction with CO_2. Propane and the halocarbon solvents were less efficient and also resulted in the formation of complex emulsions which interfered with good phase separations.[21]

6.6.4 PROCESS DESIGN AND OPERATION

There are four basic components of a solvent extraction process, as follows:

- Solvent-feed contact
- Extract and raffinate separation
- Extract treatment
- Raffinate treatment

Design and operation of each of these components and of the solvent extraction process as a whole are discussed briefly below.

A wide variety of equipment is available for bringing solvent and feed into intimate contact. Systems range from simple, single-stage mixer-settlers to more elaborate multistage extraction columns, run in a crosscurrent or, more commonly, in a countercurrent mode. The extractor design will also include provision for separating the extract and raffinate phases. Specific designs and operations are discussed in detail in the open literature.[1,14,25]

The extract (contaminant-laden solvent) must normally undergo further treatment. In most instances, process economics will require purification of the solvent for recycling to the extractor. In many instances, the contaminant removed from the waste feed will also have economic value, and the recovery credit will strongly affect the choice of solvent extraction over alternative treatment and disposal methods. Separation of solvent and contaminant can be accomplished in a number of ways, depending upon the physical and chemical properties of the extract and its components. Air or steam stripping, for example, might be used to separate a volatile contaminant from a less volatile solvent. Distillation or evaporation might be used to recover a volatile solvent from a less volatile contaminant. For the removal of phenols from aromatic extraction solvents, a second solvent extraction with an aqueous alkaline solution is possible.

The raffinate from a solvent extraction process often contains a sufficiently high concentration of solvent to be worth recovering for recycling. Air or steam stripping may be applicable. Even if recovery is impractical, the raffinate may not be acceptable for sewer or direct discharge without further treatment. Possible treatment methods include carbon or resin adsorption, activated-sludge or other biodegradation methods, as well as air or steam stripping.

Many of the examples given in Subsec. 6.6.3 of solvent extraction processes applied to waste treatment include a discussion of the further treatment of both extract and raffinate.

6.6.5 ENERGY AND MATERIALS REQUIREMENTS

One of the benefits often cited for solvent extraction is the lower energy requirements compared with, for example, distillation, which often can be used to accomplish a similar purpose.[20,21,25] For solvent extraction, as an isolated unit operation, the energy requirements are generally limited to electricity for running feed and solvent pumps, stirrers, or column agitators.[3] However, as discussed in Subsec. 6.6.4, a solvent extraction system generally also includes solvent and contaminant recovery from the extract and solvent recovery or removal from the raffinate. If these secondary operations involve distillation, evaporation, or stripping, the additional steam and electrical-energy requirements can be quite significant.[3]

Materials requirements are essentially limited to makeup for solvent losses, since in most instances, solvent will be recovered for re-use from both the extract and the raffinate.

6.6.6 PROCESS EFFICIENCY

Process efficiency depends importantly on the equilibrium distribution of the contaminant of interest between the waste feed and the extracting solvent and, of course, on the kinetics of mass transfer from the feed to the solvent.[25] An increase in the solvent-to-feed ratio and multiple-stage extractions can be used, within limits, to increase net contaminant transfer. Obviously, process efficiency is critically dependent on the choice of solvent, and is also affected by temperature and pressure. The latter is a particularly significant variable in supercritical-fluid and near-critical-fluid extractions.

6.6.7 EMISSIONS AND RESIDUALS MANAGEMENT

Solvent extraction processes can be operated as totally enclosed systems. There are two output streams: the extract and the raffinate. System designs for solvent extraction invariably include provisions for these streams. The extract is normally treated to recover materials values, but in some instances may be burned as fuel or simply incinerated. The raffinate, which is often an aqueous solution, can be captured for appropriate treatment prior to discharge.

6.6.8   RECOVERY CONSIDERATIONS

In general, solvent extraction processes are most attractive relative to alternatives when the contaminant to be removed from the feed has significant material value. (See Subsec. 6.6.10.)

6.6.9   SAFETY CONSIDERATIONS

The primary health and safety considerations in solvent extraction systems are the toxicity and flammability of the solvent.[25] As in any system, the materials of construction of the extractor and auxiliary equipment must be compatible with the solvent, feed, extract, and raffinate with which they come in contact. Equipment should, of course, be diked to contain any major spills.

Supercritical-fluid or near-critical-fluid extractions have the additional safety problems associated with operations at high pressures. One of the significant benefits of critical-fluid extractions with CO_2, however, is that the "solvent" is both inert and nontoxic.

6.6.10   COSTS

Costs of several solvent extraction systems for treatment of wastes were given in Subsec. 6.6.3.

6.6.11   REGULATORY CONSIDERATIONS

Solvent extraction would be classified under physical, chemical, or biological treatment, as defined by regulations promulgated in response to the Resource Conservation and Recovery Act (RCRA). At the time of preparation of this manuscript, only interim-status (Part 265) regulations were in effect, and these were basically the same as tank-storage regulations. Final (Part 264) regulations had not yet been proposed. Technically, therefore, new applications of solvent extraction to the treatment of hazardous wastes cannot be permitted if they do not already qualify for interim status. The Hazardous and Solid Waste Amendments of 1984, however, do provide for granting of research, development, and demonstration permits.

The use of solvent extraction for treatment of wastewater would not require an RCRA permit, but would fall under the Federal Water Pollution Control Act. The latter sets effluent discharge limits, rather than equipment design and operating standards. Therefore, a discharge permit [National Pollutant Discharge Elimination System (NPDES) or State Pollutant Discharge Elimination System (SPDES)] would be required for the raffinate, prior to release to a surface water body, and solvent extraction could be used provided that the raffinate met permit limits.

6.6.12   REFERENCES

1. R. E. Treybal, "Liquid Extraction," sec. 15 in R. H. Perry and C. H. Chilton (eds.), *Chemical Engineers' Handbook*, 5th ed., McGraw-Hill, New York, 1973.

2. G. T. Austin, *Shreve's Chemical Process Industries,* 5th ed., McGraw-Hill, New York, 1984.

3. J. B. Berkowitz, J. T. Funkhouser, and J. I. Stevens, *Unit Operations for Treatment of Hazardous Industrial Wastes,* Noyes Data Corp., Park Ridge, New Jersey, 1978.

4. J. D. MacGlashar, "Extraction of Phenols from Water with Tri-Octyl Phosphine Oxide," M.S. thesis, University of California, Berkeley, LBL-19963, March 1982.

5. M. Medir, A. Anola, D. Mackay, and F. Giralt, *J. Chem. Eng. Data,* **30**:157–159 (1985).

6. D. C. Greminger, G. P. Burns, S. Lynn, D. N. Hanson, and C. J. King, *Ind. Chem. Eng. Process Des. Dev.,* **23**:748–754 (1984).

7. D. K. Joshi, J. J. Senetar, and C. J. King, *Ind. Chem. Eng. Process Des. Dev.,* **23**:748–754 (1984).

8. O. B. Michelson and S. Ostern, *Environ. Sci. and Tech.,* **13**:735–738 (1979).

9. B. K. Wallin, A. J. Condren, and R. L. Waldan, *Removal of Phenolic Compounds from Wood Preserving Wastewaters,* EPA-600/52-81-043, April 1981.

10. J. M. Napier, M. A. Travaglini, E. G. Lazgis, and M. A. Marewicz, *Evaluation and Development of Polychlorinated Biphenyl Removal Processes,* Y/DZ-1, Oak Ridge Y-12 Plant, Union Carbide Corp., Oak Ridge, Tennessee, February 5, 1982.

11. C. W. Hancher and M. B. Saunders, *Pilot Plant Studies for Solvent Extraction of Polychlorinated Biphenyl (PCB),* Y/DZ-77, Oak Ridge Y-12 Plant, Union Carbide Corp., Oak Ridge, Tennessee, December 1983.

12. S. P. Tucker and G. A. Carson, *Environ. Sci. and Tech.,* **19**:215–220 (1985).

13. "Annual Review of Extractive Metallurgy," *J. of Metals,* pp. 39–60, April 1986.

14. T. C. Lo, M. H. I. Baird, and C. Hanson, *Handbook of Solvent Extraction,* Wiley, New York, 1983.

15. *Pesticide and Toxic Chemical News,* December 18, 1985, p. 12.

16. W. D. Ellis, J. R. Payne, and G. D. McNabb, "Treatment of Contaminated Soils with Aqueous Surfactants."

17. C. J. Drummond, R. P. Noceti, R. D. Miller, T. J. Feeley III, and J. A. Cook, *Env. Prog.,* **4**:26–33 (1985).

18. *Chemical Engineering,* pp. 64–66, August 19, 1985.

19. K. E. Beth, J. S. Rowlinson, and G. Saville, *Thermodynamics for Chemical Engineers,* The MIT Press, Cambridge, Massachusetts, 1975.

20. F. R. Groves, Jr., B. Brady, and F. C. Knopf, CRC Critical Reviews in Environmental Controls, **15**:237–274 (1985).

21. R. P. deFillippi and M. E. Chung, *Laboratory Evaluation of Critical Fluid Extractions for Environmental Applications,* EPA/600/2-85/045 (NTIS No. PB 85-189843), April 1985.

22. R. C. Reid, J. M. Prausnitz, and T. K. Sherwood, *The Properties of Gases and Liquids,* McGraw-Hill, New York, 1977.

23. H. M. Freeman and R. A. Olexsey, JAPCA, **36**:67–76 (1986).

24. *Chemical Engineering,* June 24, 1985, p. 9.

25. J. L. Humphrey, J. A. Rocha, and J. R. Fair, *Chemical Engineering,* September 17, 1984, pp. 76–95.

SECTION 6.7
MEMBRANE SEPARATION TECHNOLOGIES

Janet MacNeil

Environmental Strategies Corporation
San Jose, California

Drew E. McCoy

McCoy and Associates, Inc.
Lakewood, Colorado

Membrane separation processes are destined to play an increasing role in the reduction and/or recycling of hazardous wastes. These processes include reverse osmosis, ultrafiltration, hyperfiltration, and electrodialysis, each of which separates a contaminant (solute) from a liquid phase (solvent, typically water). Membrane separation processes can function in several ways: volume reduction, recovery and/or purification of the liquid phase, and concentration and/or recovery of the contaminant or solute.

When choosing a membrane separation technique, the following basic characteristics should be taken into account:

1. *Reverse osmosis* is primarily used to separate water from a feed stream containing inorganic ions. The purity of the recovered water is relatively high, and the water is generally suitable for recycling. The maximum achievable concentration of salt in the reject stream is usually about 100,000 mg/L because of osmotic-pressure considerations.

2. *Hyperfiltration* separates ionic or organic components from water by limiting the size of membrane pores through which a contaminant can pass. It is typically used to remove contaminants having a molecular weight of 100 to 500 from water.

3. *Ultrafiltration* is primarily used to separate organic components from water according to the size (molecular weight) of the organic molecules. Membranes

are manufactured with the capability to remove contaminants with molecular weights between 500 and 1,000,000.

4. *Electrodialysis* is used to remove ionic components from water. It produces moderate-quality product water (i.e., several hundred mg/L salt), and is capable of producing concentrate streams containing up to 20% salt.

6.7.1 REVERSE OSMOSIS

When a semipermeable membrane separates two solutions of different dissolved-solids concentrations, pure water will flow through the membrane into the concentrated solution, while ions (e.g., dissolved salts) are retained behind the membrane. This process is known as *osmosis*. During *reverse osmosis,* pressure is applied to the more concentrated solution to reverse the normal osmotic flow, and pure water is forced through the semipermeable membrane into the less concentrated solution. The purified stream that passes through the membrane is called *permeate;* the concentrated stream retained by the membrane is known as *concentrate.*

Conceptually, a reverse-osmosis (RO) system has several advantages for treating wastes:

- Both the recovered solvent and the concentrated solute in some cases can be recycled to a manufacturing process, rather than requiring treatment and disposal.
- The separation process does not require an energy-intensive phase change such as is required for distillation or evaporation. Therefore, operating costs associated with energy consumption are relatively low.
- Capital costs are also relatively low. In fact, because the cost of membrane materials has been stable, some RO applications may now be economically justifiable as compared with competing technologies that have had larger price increases.
- Reverse-osmosis equipment does not require a large amount of space. Also, because RO is a straightforward mechanical process, it requires a low degree of operational skill.

Membrane Characteristics

For RO applications, membranes that have high water permeability and low salt permeability are ideal. The three most commonly used RO membrane materials are cellulose acetate, aromatic polyamides, and thin-film composites (consisting of a thin film of a salt-rejecting membrane on the surface of a porous support polymer). Temperature, pH, and other limitations of these materials are summarized in Table 6.7.1.

Module Design

After a membrane material is selected, it is fabricated into one of the following RO module designs: tubular, spiral wound, hollow fiber, or plate and frame.

TABLE 6.7.1   Physical and Chemical Limitations of Reverse-Osmosis Membranes

Membrane material	pH limits	Maximum temperature limits, °C (°F)	Other limitations
Cellulose acetate	2.5–7	29–50 (85–122)	Biologically degradable
Aromatic polyamides	4–11	35–46 (95–115)	Cannot tolerate chlorine
Thin-film composites	<1–13	46–79 (115–175)	Can tolerate moderate chlorine levels (100 ppm)

Source: McCoy and Associates, Inc.

Tubular Modules.   A typical tubular module, shown in Fig. 6.7.1, consists of a porous, membrane-lined tube that has an inside diameter of 0.63 to 3.8 cm (0.25 to 1.5 in). Feed solution is pumped through the tube, and permeate passes radially through the membrane and outer porous tube. The concentrate exits from the opposite end of the tube.

Although tubular modules have relatively small surface areas, they do have the advantage of being less susceptible to plugging by suspended solids. In many cases, it is possible to physically clean the insides of the tubes to remove any surface deposits.

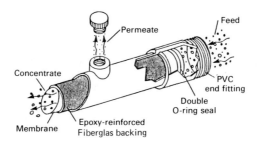

FIG. 6.7.1   Tubular reverse-osmosis module   (*U.S. EPA*)

Spiral-Wound Modules.   Spiral-wound modules consist of sandwiches of membrane and spacer materials that are wrapped like a scroll around an inner perforated tube (see Fig. 6.7.2). Two sheets of fabric-backed membrane are separated by a permeate-carrier material (typically tricot). The membranes are cemented together at the edges to seal in the permeate carrier, and the permeate carrier is in turn attached to the perforated collection tube. When the membrane is wound around the tube, two leaves of the membrane assembly are separated by a mesh spacer that serves as a flow path for the feed. The spiral-wound membrane is then inserted into a pressure shell. Feed is introduced at one end of the pressure shell and flows axially through the mesh spacer to the opposite end of the shell. Permeate is removed from the perforated collection tube.

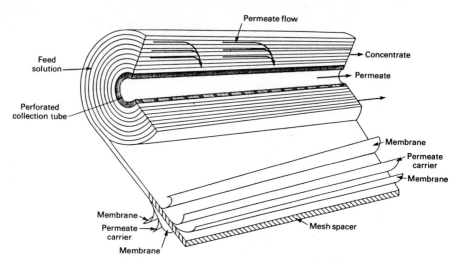

FIG. 6.7.2   Spiral-wound reverse-osmosis module [*McCoy and Associates, Inc. "Membrane Separation Technology: Applications to Waste Reduction and Recycling,"* The Hazardous Waste Consultant, *3:(3)* (May–June 1985), p. 4–6]

Hollow-Fiber Modules.   Hollow-fiber modules are available in several different fiber configurations. The Du Pont Permasep module is illustrated in Fig. 6.7.3. The fiber bundle, containing up to 4.5 million fibers, consists of doubled strands of fibers. Both ends of the fiber bundle are potted in epoxy, which is subsequently sheared to open the ends of the fibers. Each fiber bundle is then placed in a pressure vessel up to 1.2 m (4 ft) long and 10 to 25 cm (4 to 10 in) in diameter.

Feed is introduced to the module through a central distributor tube that runs the length of the module; it then flows radially outward through the fiber bundle. When the feed passes over the outside surface of the fibers, permeate flows through the fiber walls to the inner bore. Each fiber discharges its purified contents at the end of the module; concentrate is removed at the opposite end.

Other manufacturers of hollow-fiber modules have developed techniques for winding hollow fibers in a helical configuration. Feed flow across these bundles is in an axial, rather than a radial, direction.

Plate-and-Frame Modules.   Figure 6.7.4 is a schematic diagram of a plate-and-frame RO unit. In this module type, circular spacers, membranes, and sup-

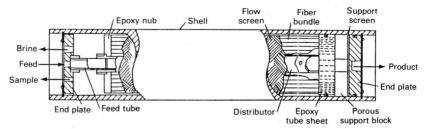

FIG. 6.7.3   Hollow-fiber reverse-osmosis module (*Lynn E. Applegate, "Membrane Separation Processes,"* Chemical Engineering, *June 11, 1984, p. 70*)

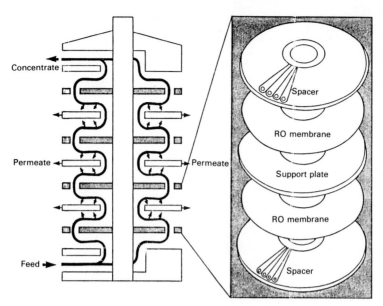

FIG. 6.7.4 Plate-and-frame reverse-osmosis module (*Pasilac, Inc.*)

port plates are assembled like a sandwich and are held together with a central bolt. Feed is introduced at the bottom of the stack at pressures ranging from 1.5 million to 7 million Pa (220 to 1,015 lb/in^2). The spacers divert flow radially across the face of an adjacent membrane. As a result of this contact, water permeates through the membrane into the porous support plate and is collected at its outer edge. Salts are concentrated as the feed passes sequentially through the stack, and are discharged from the top of the unit.

Design Considerations

The performance of an RO system is a function of membrane type, dissolved-solids concentration, feed temperature, and applied pressure. Since the composition of each specific waste stream differs, pilot tests are necessary, and are usually conducted on a 189-L (50-gal) representative wastewater sample. During pilot testing, the performance of the RO system will generally be measured by three parameters: flux, conversion, and rejection.

Flux. *Flux* is the flow rate of permeate per unit area of membrane. Typical units are m^3/(m$^2 \cdot$ d) or gal/(ft$^2 \cdot$ d) (gsfd). (For hollow-fiber units, *flux* is typically replaced by *productivity,* which is the flow rate per module.) The major factors that influence flux are the physical and chemical stability of the membrane, fouling, and concentration polarization. (For discussion of the two latter see under *Fouling and Concentration Polarization,* on p. 6.96.)

Conversion. *Conversion* (or *recovery*) is the ratio of permeate flow to feed flow, and is typically controlled by adjusting the flow rate of the reject stream leaving the RO module. When a module is operated at low conversion, the concentration of the reject stream leaving the module is not much different from the concentration of the feed stream. As conversion increases, the concentration of the concentrate will also increase; however, higher feed pressures will be required to overcome osmotic pressures.

Rejection. *Rejection,* which measures the degree to which the solute is prevented from passing through the membrane, increases with ionic size and ionic charge of materials in the feed. It is dependent on operating pressure, conversion, and feed concentration.

Fouling and Concentration Polarization

Unfortunately, the extent of membrane fouling (the most important factor that affects RO system performance) cannot be determined during pilot tests but must be analyzed during on-site "slipstream" evaluation. Fouling is typically caused by suspended materials in the feed stream or materials that precipitate during treatment, and can be inhibited by

1. Installing a feed pretreatment (e.g., coagulation and filtration) system or 5 μm prefilters to remove suspended solids
2. Prefiltering the feed with activated carbon to eliminate organic contaminants
3. Adding dispersants
4. Adjusting the pH of the feed
5. Chlorinating the feed (followed by dechlorination for aromatic polyamide membrane systems) to prevent buildup of biological organisms in the RO modules
6. Maintaining turbulent flow (promoted by the use of special membrane spacers, increasing feed flow rate, or reducing permeate production rate)
7. Using low-pressure lines made of PVC, polyethylene, or Fiberglas, and pumps and high-pressure piping made of 304 or 316 stainless steel to decrease the amount of iron-corrosion products in the feed

Concentration polarization, i.e., formation of a concentrated boundary layer of solute next to the membrane surface, can also decrease the productivity of RO systems. It can be minimized by increasing turbulence and decreasing system recovery.

Applications

Thus far, one of the major applications of reverse osmosis has been in the electroplating industry. The type and number of RO installations currently in use in the electroplating industry are summarized in Table 6.7.2. A schematic diagram of a generic plating process in which RO is utilized to recover metals for re-use in plating baths is shown in Fig. 6.7.5. Metal parts being plated pass by conveyor rack through a bath of plating chemicals and "drag out" a small amount of these chemicals to the rinse tanks. The rinse tanks are typically operated in series with the flow of rinsewater being countercurrent to the direction of parts movement. With no RO recovery, the rinsewater from the first tank would typically be dis-

TABLE 6.7.2   Reverse-Osmosis Installations Currently in Operation in the Electroplating Industry

Bath type	Type of membrane	Membrane configuration	No. of systems
Bright nickel, nickel sulfamate, and Watts nickel	Cellulose acetate	Spiral wound	150
Copper sulfate	Polyamide and cellulose triacetate	Thin-film composite, hollow fiber, and spiral wound	12
Zinc sulfate		Thin-film composite and spiral wound	1
Brass cyanide	Polyamide and cellulose triacetate	Hollow fiber	5
Copper cyanide	Polyamide	Hollow fiber	2

Source: Adapted from P. Werschulz, "New Membrane Technology in the Metal Finishing Industry," in Irwin J. Kugelman (ed.), *Toxic and Hazardous Wastes,* Technomic Publishing, Lancaster, Pennsylvania, 1985, p. 448.

posed. When RO is applied to a plating line, the most concentrated rinse stream is fed to one or more RO modules. The concentrate from the module is returned to the plating bath, and the purified water is returned to the final rinse tank.

The RO system can be operated as a closed loop if the baths operate at temperatures above 54°C (130°F) because at these temperatures water evaporation

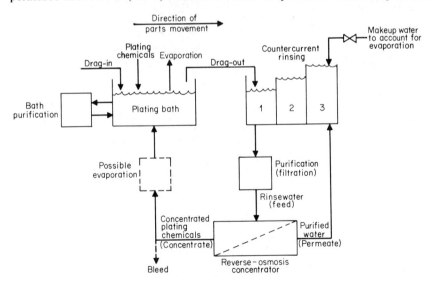

FIG. 6.7.5   Basic flow diagram for a reverse-osmosis system treating plating rinsewater (*U.S. EPA*)

from the bath offsets the inflow of water from the RO system. Otherwise, an additional step (e.g., evaporation) or a bleed stream in the concentrate return line may be needed to remove excess water from the RO concentrate.

Economic Considerations

Capital Costs.   In addition to the costs of membrane modules, significant capital costs are likely to be incurred for pumps, piping, instrumentation, pretreatment, and storage facilities. In fact, membrane modules often account for less than 30% of the cost of the complete system.

Operating Costs.   Probably the most important operating cost will be for membrane replacement. In many cases, this factor alone will determine the economic viability of a specific application. Although membranes in desalination service may have lifetimes of over five years, membrane life in waste-treatment applications is likely to be much shorter.

Additional operating costs that must be considered include power costs for pumps, labor for operation and maintenance, replacement filtration cartridges and pretreatment chemicals, and membrane-washing chemicals. In closed-loop systems where buildup of minor contaminants may occur, the cost of providing high-grade raw materials, such as deionized water for electroplating rinse tanks, should also be considered.

Credits.   Several credits are frequently obtained when membrane technologies are applied to waste streams; these include credits for chemicals recycled to the process, energy credits for the thermal-energy content of recycled streams, savings for makeup water and associated sewer charges, and credits for reduced waste-treatment and disposal costs.

6.7.2 ULTRAFILTRATION AND HYPERFILTRATION

Ultrafiltration (*UF*) and *hyperfiltration* (*HF*) utilize pressure and a semipermeable membrane to separate nonionic materials from a solvent (such as water). These membrane separation techniques are particularly effective for the removal of suspended solids, oil and grease, large organic molecules, and complexed heavy metals from wastewater streams.

In ultrafiltration and hyperfiltration systems, the membrane retains materials based solely on size, shape, and molecule flexibility. As a feed solution is pumped through a membrane module, the membrane acts as a sieve to retain dissolved and suspended materials that are physically too large to pass through its pores. The retained materials (concentrate) then exit the module separately from the purified solvent or permeate.

The major difference between hyperfiltration and ultrafiltration is that hyperfiltration typically removes species having a molecular weight of 100 to 500; ultrafiltration removes species having a molecular weight greater than 500. The two membrane separation methods utilize identical operating principles. (Although reverse osmosis has also been called hyperfiltration, this is a misnomer because the two processes have different operating principles and are generally used in different applications.)

Membrane Characteristics

UF and HF membranes have an asymmetric structure designed to maximize productivity per unit surface area. They are composed of a thin (0.1 to 1.0 μm), selective, surface layer supported by a porous, spongy layer about 100 μm thick. Membrane pore sizes typically range from 10 to 1,000 Å.

Two common membrane materials are polysulfone and cellulose acetate. Polysulfone is the most versatile because it can tolerate temperatures between 0 and 79°C (32 and 175°F) and pH from <1 to 13. It also can be cleaned with a wide array of cleaning agents. Cellulose acetate is also a popular membrane material; however, it can only be used at pH 2.5 to 7 and temperatures from 0 to 50°C (32 to 122°F).

The retention capability of UF and HF membranes is characterized by *molecular-weight (MW) cutoff* (the molecular weight of the solute that is too large to pass through the pores). UF membranes are available that retain molecular weights from 500 to 1,000,000. However, since there is no uniform method of measuring the MW cutoff of membranes, MW cutoffs specified by different manufacturers are not necessarily comparable.

Module Design

Several types of UF and HF membrane configurations are available that differ in cost, membrane area/volume ratio, and resistance to fouling: (1) tubular, (2) spiral wound, (3) hollow fiber, and (4) plate and frame. Table 6.7.3 lists advantages and disadvantages of these four common module types. Tubular and spiral-wound UF and HF modules are identical to those used for reverse osmosis (see under *Tubular Modules,* on p. 6.93, and *Spiral-Wound Modules,* on p. 6.93). Therefore, only UF and HF hollow-fiber and plate-and-frame modules will be discussed here.

UF and HF Hollow-Fiber Modules. UF and HF hollow-fiber modules differ in design and operation from those used in reverse osmosis. The UF and HF hollow-fiber membrane consists of acrylic copolymer and has a 0.1-μm skin on the inside supported by a spongy outer structure. The fibers, which are significantly larger than those used in RO, may have inside diameters of 1,100 μm (high-fouling applications) or 500 μm (low-fouling applications).

As shown in Fig. 6.7.6, the fibers are aligned in a parallel fashion and are potted in epoxy on either end inside a low-pressure vessel. The feed enters the vessel at one end and flows through the interior of the membrane fibers. The low-molecular-weight solutes and water permeate the inner membrane skin and are removed through a product port. Reject material continues to flow through the fibers and is removed at the other end of the vessel.

If fouling occurs, the fiber bundle may be backwashed and flushed by recycling the feed or other solution through the fiber bores while the permeate port is closed.

UF Plate-and-Frame Modules. UF plate-and-frame modules consist of membrane-covered support plates stacked horizontally in a frame apparatus (see

TABLE 6.7.3 Comparison of Ultrafiltration and Hyperfiltration Module Configurations

Module type	Advantages	Disadvantages
Tubular	Easily cleaned chemically or mechanically if membranes become fouled Can process high-suspended-solids feed with minimal pretreatment Good hydrodynamic control Individual tubes can be replaced	Relatively high volume required per unit membrane area Relatively expensive
Spiral wound	Compact, good membrane surface/volume ratio Less expensive than tubular modules	Susceptible to plugging by particulates Badly fouled membranes are difficult to clean
Hollow fiber	Compact, very good membrane surface/volume ratio Economical	Very susceptible to plugging by particulates Potentially difficult to clean
Plate and frame	Good membrane surface/volume ratio Well-developed equipment	Susceptible to plugging at flow-stagnation points Potentially difficult to clean Expensive

Source: U.S. EPA.

Fig. 6.7.7). When the plates are compressed hydraulically (at pressures up to 1 million Pa, or 145 lb/in^2), holes in the membrane-covered plates form flow channels for the feed and concentrate. As the feed flows between the membrane-covered plates, permeate exits via the support plates and is collected at the top of the module.

The remaining liquid continues to flow through the module for further treatment. Concentrate leaves the module end opposite the feed intake. Internal flow within the module may be arranged in a combination of parallel and series flow patterns by using section plates.

FIG. 6.7.6 Hollow-fiber ultrafiltration module (*Adapted by special permission from Lynn E. Applegate, "Membrane Separation Processes," Chemical Engineering, June 11, 1984, p. 86*)

Design Considerations

Concentration polarization is more of a problem in ultrafiltration systems than it is in RO applications. This is due to the smaller diffusion constants of organic materials, which limit diffusion of materials retained by the membrane back into the liquid. The effects of concentration polarization may be reduced

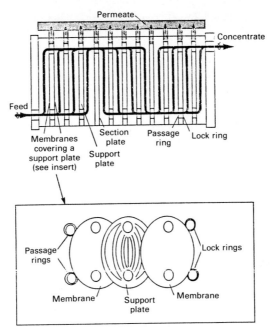

FIG. 6.7.7 Plate-and-frame ultrafiltration module (*Pasilac, Inc.*)

by operating at higher feed velocities or increasing the treatment temperature. Fluxes usually double for a 15° to 25°C (59° to 77°F) temperature increase.

Applications

Ultrafiltration and hyperfiltration are currently being used in various pollution-control applications to

1. Remove complexed toxic metals from metal-finishing wastewater
2. Concentrate oily wastes from metal-finishing rinsewaters, can-forming rinsewaters, aluminum- and steel-coil-cleaning rinsewaters, and detergent rinsewaters from automotive- and aircraft-chassis cleaning
3. Concentrate electrodeposition paint baths to remove water and contaminants
4. Recover oil from waste-oil emulsions in metal-machining and rolling operations
5. Remove dyes from textile-industry effluents

6.7.3 ELECTRODIALYSIS

Electrodialysis relies on ion-exchange membranes in a direct-current electrical field to separate ionic species from solution. Like RO membranes, electrodialysis membranes are sensitive to fouling. This sensitivity has limited waste-treatment applications of the technology, although the development of the electrodialysis reversal process has significantly reduced complications due to fouling.

Conventional Electrodialysis

Figure 6.7.8 shows a schematic diagram of an electrodialysis cell. The cell consists of an anode and a cathode separated by two cation-selective membranes, three liquid flow chambers, and two anion-selective membranes. Each of the electrodes is rinsed to remove localized buildup of hydrogen at the cathode and oxygen at the anode.

Feed is introduced to the flow path between the anionic and cationic membranes. The cationic membrane will pass only cations, and the anionic membrane will pass only anions. When a direct-current charge is applied to the cell, cations are attracted to the cathode (negatively charged) and anions are attracted to the anode (positively charged). Ions in the feed will pass through the appropriate membranes according to their charge, thus "desalting" the feed stream.

When an ion from the feed stream encounters a membrane that is selective for ions of the opposite charge, its migration toward the appropriate electrode is halted. The ions are therefore depleted from one flow passage and concentrated in an adjacent flow passage. By manifolding the appropriate passages, product water and brine are collected.

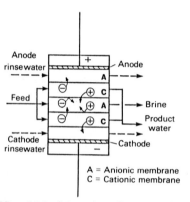

A = Anionic membrane
C = Cationic membrane

FIG. 6.7.8 Schematic diagram of an electrodialysis cell [*McCoy and Associates, Inc., "Membrane Separation Technology: Applications to Waste Reduction and Recycling,"* The Hazardous Waste Consultant, *3:(3) (May–June 1985)*, p. 4–21]

The Reversible Electrodialysis Process

In order to reduce fouling problems, nearly all new electrodialysis (ED) installations utilize a reversible electrodialysis process, which periodically reverses the polarity of electrodialysis cells. This is accomplished by using membranes that can function in either the anion- or cation-selective mode, and platinum-coated titanium electrodes that can function either as cathodes or anodes. By reversing the polarity of electrodialysis cells, flow to and from the concentrating and depleting chambers is reversed. Precipitates and surface films tend to redissolve or are physically purged by the flow reversal. In order to accomplish this flow reversal, automatic valve systems are needed to accommodate the changes in component functions. For a system that cycles every 20 min, each flow reversal is typically accompanied by a 1- to 2-min purge that disposes of any potentially contaminated product.

Electrodialysis System Configuration

As shown in Fig. 6.7.9, electrodialysis stacks may contain hundreds of individual cells in a horizontal or vertical configuration. The membranes typically measure

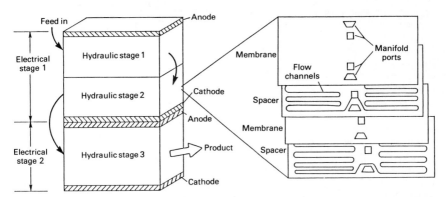

FIG. 6.7.9 Typical electrodialysis stack [*McCoy and Associates, "Membrane Separation Technology: Applications to Waste Reduction and Recycling," The Hazardous Waste Consultant, 3:(3) (May–June 1985), p. 4–22*]

about 0.5 mm in thickness, and the flow paths consist of 1-mm thick spacers having die-cut channels to distribute the liquid over the surface of the membranes. Alternating spacers contain the appropriate manifold ports for fluid flow: feed-solution-in–demineralized-water-out; or brine-water-in–brine-water-out. When operated as a single stage, all the alternating compartments operate in parallel off the manifolds.

A single pass through an electrodialysis cell will typically produce a demineralization ratio of 30 to 60%. If a greater degree of demineralization is desired, the feed must be routed through a second stage in series.

Depending on the volume of water being treated, an electrodialysis stack may contain multiple hydraulic and electrical stages. Thus, feedwater might pass through two stages having common cathodes and anodes, and then through a final stage with separate cathodes and anodes, all in the same stack. One manufacturer (Ionics) utilizes a single stage per stack for flow rates in excess of 546 m^3/d (100 gal/min), and multiple stages per stack for desired output flows of <546 m^3/d (100 gal/min).

Applications

Since electrodialysis is suited only to the removal or concentration of ionic species, the metal-finishing and electroplating industry is probably the greatest potential market for these systems.

Figure 6.7.10 is a flow diagram of an electrodialysis system being used to treat rinsewater from a nickel-plating line. Water containing nickel salts from the No. 1 rinse tank is pumped through a pretreatment filter. If organic chemicals are used in the process, this may be an activated-carbon filter; otherwise, its primary purpose is to remove suspended solids that might foul the electrodialysis stack.

The electrodialysis stack typically operates at 275,800 to 413,700 Pa (40 to 60 lb/in²), and brine leaving the stack is recycled to the feed until the desired concentration of nickel in the brine is reached. Chemicals from the concentrate tank are recycled to the plating bath, and product water is recycled to the rinse tanks.

Overall conversion in the ED system is 90%—i.e., 90% of the feed stream is recovered as product water. Since some leakage of ions between the brine and

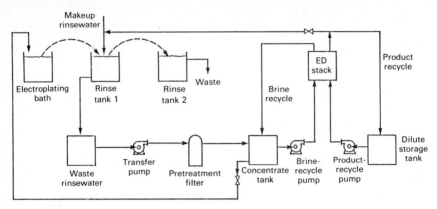

FIG. 6.7.10 Electrodialysis system for treating plating rinsewater (*Lynn E. Applegate, "Membrane Separation Processes,"* Chemical Engineering, *June 11, 1984, p. 83*)

product streams occurs, the concentration of nickel salts in the product water is approximately 10% of the concentration in the feed stream. The salt concentrations in the feed, concentrate, and product-water streams were 14,280 mg/L, 163,100 mg/L, and 1,309 mg/L, respectively.

In general, membrane-area requirements (hence, capital costs) and energy consumption increase with increases in total dissolved-solids (TDS) concentrations. As a result, unless a valuable concentrate stream is being produced, ED is generally limited to feed streams with moderate TDS contents. Most existing systems are used to treat brackish water having TDS concentrations less than 5,000 mg/L. Where a valuable concentrate is being produced, salts may be concentrated to 20% or more. This is significantly beyond the feasible operating range of reverse-osmosis systems.

6.7.4 GUIDE TO MEMBRANE MANUFACTURERS

Membrane manufacturers with products that may be suitable for hazardous-waste treatment are listed in Table 6.7.4.

TABLE 6.7.4 Membrane Manufacturers

Manufacturer	Product description
Aquatech Systems 7 Powderhorn Dr. Warren, N.J. 07060 (201) 560-1750	Electrodialysis systems
Basic Technologies, Inc. 3879 W. Industrial Way Riviera Beach, Fla. 33404 (305) 848-1717	Reverse-osmosis and ultrafiltration systems utilizing spiral-wound, tubular, and hollow-fiber modules

TABLE 6.7.4 Membrand Manufacturers (*Continued*)

Manufacturer	Product description
Desalination Systems, Inc. 1238 Simpson Way Escondido, Calif. 92025 (619) 746-8141	Reverse-osmosis and ultrafiltration systems, primarily spiral-wound modules
Envirogenics Systems Co. 9255 Telstar Ave. El Monte, Calif. 91731 (818) 573-9220	Reverse-osmosis systems
HPD Incorporated 1717 North Naper Blvd. Naperville, Ill. 60566 (312) 357-7330	Electrodialysis and dialysis systems for use in the electronics and metal-finishing industries to recover acids and metal salts
Medro Systems Inc. 1305 Summit Ave. Plano, Tex. 75074 (214) 423-8488	Reverse-osmosis and ultrafiltration systems, primarily using spiral-wound modules
Osmonics, Inc. 5951 Clearwater Dr. Minnetonka, Minn. 55343 (612) 933-2277	Reverse-osmosis, ultrafiltration, and hyperfiltration systems for metal-finishing applications and treatment of oily aqueous wastes
Pasilac, Inc. 660 Taft St. N.E. Minneapolis, Minn. 55413 (612) 331-7710	Reverse-osmosis and ultrafiltration systems utilizing spiral-wound and plate-and-frame modules
Polymer Research Corp. of America 2186 Mill Ave. Brooklyn, N.Y. 11234 (718) 444-4300	Specialized membranes for reverse osmosis and ultrafiltration
Resources Conservation Co. 3101 N.E. Northrup Way Bellevue, Wash. 98004 (206) 828-2400	Reverse-osmosis systems specially designed to tolerate high-scaling or precipitate-forming feeds
Vaponics, Inc. Cordage Park Plymouth, Mass. 02360 (617) 746-7555	Reverse-osmosis and ultrafiltration systems

Source: "Membrane Separation Technology: Applications to Waste Reduction and Recycling," McCoy and Associates, Inc., *The Hazardous Waste Consultant,* 3:(3) 4-25 and 4-26 (May–June 1985).

6.7.5 FURTHER READING

Applegate, L. E., "Membrane Separation Processes," *Chemical Engineering,* **91**:64–89, June 11, 1984.

Basta, N., "Use Electrodialytic Membranes for Waste Recovery," *Chemical Engineering,* **93**:(5)42–43 (March 3, 1986).

Cartwright, P. S., "An Update on Reverse Osmosis for Metal Finishing," *Plating and Surface Finishing,* **71**:62 (April 1984).

————, "Effluent Treatment with Membranes," presented at the 1984 Membrane Technology/Planning Conference, Cambridge, Massachusetts, October 30–31, 1984.

————, "Reverse Osmosis and Ultrafiltration in the Plating Shop," *Plating and Surface Finishing,* April 1981.

Cooper, A. R., "Ultrafiltration," *Chemistry in Britain,* September 1984, pp. 814–818.

Coplan, M. J., et al., "Implications to the CPI of a New High Performance Reverse Osmosis Membrane," presented at the American Institute of Chemical Engineers 1984 Annual Meeting, San Francisco, California, November 25–30, 1984.

Finkenbiner, K. T., "Reverse Osmosis for Industrial Waste Treatment," presented at the 7th Annual American Electroplaters and Surface Finishers Society/EPA Conference in Pollution Control for the Metal Finishing Industry, Kissimmee, Florida, January 27–29, 1986.

Gooding, C. H., "Reverse Osmosis and Ultrafiltration Solve Separation Problems," *Chemical Engineering,* January 7, 1985, pp. 56–62.

Katz, W. E., "State-of-the-Art of the Electrodialysis Reversal (EDR) Process," *Industrial Water Engineering,* **21**:(1)12–20.

Kim, B. M., "Membrane-Based Solvent Extraction for Selective Removal and Recovery of Metals," presented at the American Institute of Chemical Engineers Summer National Meeting, Denver, Colorado, August 28–31, 1983.

"Membrane Separation Technology: Applications to Waste Reduction and Recycling," McCoy and Associates, Inc., in *The Hazardous Waste Consultant,* **3**:(3)4-1–4-27 (May–June 1985).

Roush, P. H., "Ultrafiltration Treatment of a Combined Electroplating and Metal Finishing Wastewater," in *Proceedings of HazMat Southwest,* published by Tower Conference Management Company, Wheaton, Illinois, 1984, pp. 281–297.

U.S. Environmental Protection Agency, *New Membranes for Treating Metal Finishing Effluents by Reverse Osmosis,* EPA 600/2-76-197, U.S. EPA, October 1976.

U.S. Environmental Protection Agency, *Treatment of Electroplating Wastes by Reverse Osmosis,* EPA 600/2-76-261, U.S. EPA, September 1976.

Werschulz, P. "New Membrane Technology in the Metal Finishing Industry," in Irwin J. Kugelman (ed.), *Toxic and Hazardous Wastes,* Technomic Publishing, Lancaster, Pennsylvania, 1985, pp. 444–454.

SECTION 6.8
AIR STRIPPING AND STEAM STRIPPING

Joan V. Boegel
Polaroid Corporation
Waltham, Massachusetts

Stripping is a physical unit operation in which dissolved molecules are transferred from a liquid into a flowing gas or vapor stream. The driving force for mass transfer is provided by the concentration gradient between the liquid and gas phases, with solute molecules moving from the liquid to the gas until equilibrium is reached.

In *air stripping,* the moving gas is air, usually at ambient temperature and pressure, and the governing equilibrium relationship is Henry's law. As applied to hazardous-waste treatment, air stripping is used to remove relatively volatile dissolved organic contaminants, such as toluene and trichloroethylene, from water or aqueous waste. Practical feed concentrations are limited to about 100 mg/L organics. A properly designed and operated packed-tower air stripper can achieve greater than 99% removal of volatile organics from water. Residuals from an air-stripping process include the treated water, often suitable for re-use, and the contaminated off gas.

Steam stripping uses live steam as the gas phase. In this case, the vapor-liquid equilibrium between water and the organic compound(s) is the key equilibrium relationship. Like air stripping, steam stripping has been successfully applied to the removal of hazardous organics from aqueous waste. Steam stripping is more widely applicable in that it can effectively remove less volatile or more soluble compounds not easily removed by air stripping. Examples are methyl ethyl ketone and pentachlorophenol. Higher concentrations—up to several weight percent solvent in water—can be handled. Again, better than 99% removal is to be expected. Steam stripping has also been used to remove recyclable, water-immiscible solvents from such nonvolatile impurities as oil and grease, paint solids, and polymeric resins. Steam stripping results in a treated bottoms stream and an overhead vapor. The overhead vapor is condensed for further treatment or recovery.

This section will provide information about both air and steam stripping as they are currently used in hazardous-waste treatment.

6.8.1 AIR STRIPPING

Applicability

Air stripping is useful for the removal of volatile organic compounds (VOCs) from water or aqueous waste streams. VOCs having Henry's law constants above 10 atm are readily air-strippable at ambient temperatures. These include chlorinated solvents like methylene chloride, tetrachloroethylene, and 1,1,1-trichloroethane; aromatic solvents like benzene and toluene; and trihalomethanes.

Air stripping has found its widest application in the remediation of aquifers contaminated with solvents. The recent literature includes many reports of successful use of air stripping to clean up groundwaters contaminated with low levels of volatile solvents.[1,2,3] Contamination has resulted from leaking underground storage tanks and/or past solvent-disposal practices.

Air stripping may also be used for the removal of volatiles from industrial aqueous wastes containing traces of dissolved solvents. Recent data from commercial treatment, storage, and disposal facilities operating with interim-status permits under the Resource Conservation and Recovery Act (RCRA)[4,5] indicate that many nominally inorganic hazardous aqueous wastes such as metal-laden rinsewaters and spent acids contain part per million levels of such common solvents as methylene chloride, toluene, and trichloroethylene. As new RCRA regulations restrict the levels of VOCs acceptable in the residuals from treatment of such wastes, air stripping may be considered as one feasible pretreatment option.

Restrictive Waste Characteristics

Air stripping applications are limited to dilute liquid aqueous wastes with volatile-organic concentrations less than about 100 mg/L. Suspended solids in the waste stream should be removed prior to air stripping. Naturally occurring dissolved ferrous iron in contaminated groundwater would be oxidized on contact with air in the stripping tower. The resulting ferric hydroxide would plate out on the packing, increasing the pressure drop through the tower and reducing the surface area available for air-water contact. To avoid this problem, iron should be oxidized and removed by filtration upstream of the air stripper.

Process Description

Although transfer of volatiles from water to air can be achieved in aerated tanks, sprays, or spray towers, air stripping is most efficiently accomplished in a packed tower with countercurrent flow of air and water. Such a device is shown schematically in Fig. 6.8.1.

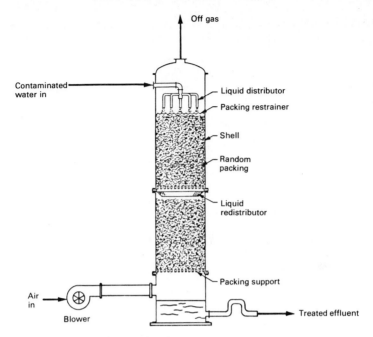

FIG. 6.8.1 Packed-tower air stripper.

Contaminated water is pumped to the top of the tower and distributed uniformly across the packing. It flows downward in a film layer along the packing surfaces. Air is blown into the base of the tower and flows upward, contacting the water. Packed-tower operation provides a high level of turbulence and a very large surface area for mass transfer. Volatile organics are transferred from the water to the air and carried out the top of the column.

Theory and Design Equations

Packed-tower air stripping is one example of a countercurrent gas-liquid desorption process, the theory of which has been well-developed in the chemical engineering literature. As practiced for the treatment of contaminated groundwater and aqueous wastes, air stripping can be adequately described by the relatively simple case of desorption from very dilute solution where the gas-liquid equilibrium relationship can be described by a straight line.

The straight-line equilibrium relationship for air stripping is Henry's law, which states that the partial pressure of a gas or volatile compound in the air above a dilute aqueous solution is directly proportional to its concentration in the solution. Mathematically,

$$P_A = (H_A)(x_A) \tag{1}$$

Where P_A = partial pressure of compound A in air, atm
H_A = Henry's law constant for compound A, atm
x_A = mole fraction of A in water, dimensionless

If both sides of Eq. (1) are divided by the total system pressure, one can then determine the equilibrium concentration of the compound in air.

$$Y_A = \frac{P_A}{P_T} = \left(\frac{H_A}{P_T}\right)\left(x_A\right)$$

(2)

Where Y_A = equilibrium mole fraction of compound A in air
P_T = total system pressure, atm

Henry's constants for selected volatile organic compounds are presented in Table 6.8.1.

TABLE 6.8.1 Henry's Constants for Volatile Organic Compounds

Compound	H, atm*
Vinyl chloride	3.55×10^5
Carbon tetrachloride	1.29×10^3
Tetrachloroethylene	1.1×10^3
Trichloroethylene	5.5×10^2
1,1,1-Trichloroethane	4.0×10^2
Toluene	3.4×10^2 (25°C)
Benzene	2.4×10^2
Chloroform	1.7×10^2
1,2-Dichloroethane	61
Bromoform	35
Methyl ethyl ketone	5.0 (24°C)
Acetone	2.0 (25°C)
Pentachlorophenol	0.12

*At 20°C unless otherwise noted.

The depth of packing required for removal of VOCs in a countercurrent tower is a function of the required removal, the air-to-water ratio, the liquid loading rate, the packing characteristics, and the water temperature. If continuous steady-state operation is assumed, Henry's law is applicable, and the incoming air is free of organic contaminants, one can write a mass balance around a differential section of the tower and integrate to obtain the following design equation:

$$Z = \frac{L}{(K_L a)C_o} \cdot \frac{R}{R - 1} \ln\left[\frac{C_{in} / C_{out} (R - 1) + 1}{R}\right]$$

(3)

Where Z = packing height, m
L = liquid loading rate, kmol/(s · m^2)
$K_L a$ = product of overall liquid mass-transfer coefficient K_L (m/s) and the specific interfacial area a (m^2/m^3) in the packed tower
C_o = molar density of water, 55.6 kmol/m^3 at 20°C

$$R = \text{stripping factor (dimensionless)} = \left(\frac{H_A}{P_t}\right)\left(\frac{G}{L}\right)$$

C_{in} = influent concentration of the compound to be removed, µg/L
C_{out} = effluent concentration of the compound to be removed, µg/L

Equation (3) is often simplified by grouping the terms into two dimensionless factors referred to as the height of a transfer unit HTU and the number of transfer units NTU. Thus,

$$Z = (HTU)\,(NTU) \tag{4}$$

Where $HTU = \dfrac{L}{(K_L a)\,C_o}$

$$NTU = \frac{R}{R-1}\,\ln\frac{C_{in}/C_{out}\,(R-1)+1}{R}$$

The overall liquid mass-transfer coefficient K_L includes the diffusional resistance to mass transfer in both the liquid and gas phases. It is related to the local gas and liquid mass-transfer coefficients k_g and k_l by the expression:

$$\frac{1}{K_L} = \frac{1}{k_l} + \frac{C_o}{k_g H_A} \tag{5}$$

Where k_l = local liquid-phase mass-transfer coefficient, m/sec
k_g = local gas-phase mass-transfer coefficient, kmol/(s · atm · m²)

For very high values of the Henry's constant H_A, $1/K_L \simeq 1/k_L$. This is generally the case with the compounds presented in Table 6.8.1. For most of these compounds, the liquid-phase resistance contributes more than 90% of the total resistance. For Henry's constants below about 150 atm, the local gas-phase resistance becomes increasingly important.

$K_L a$, the product of the overall mass-transfer coefficient and the interfacial area, depends on many factors including the physical and chemical properties of the volatile contaminant, the packing type and material, the temperature, and the liquid loading rate. Values for $K_L a$ are best determined experimentally with a pilot plant treating the actual water or waste stream of interest and using the type of packing intended for the full-scale towers. In a well-designed pilot plant, the tower diameter should be at least 10 times the packing diameter to avoid "wall effects"—inefficient mass transfer resulting from channeling of liquid flow along the tower walls. In addition, "end effects"—removal occurring outside the packing—must be corrected for by obtaining data at several different packing heights. An excellent description of air-stripper pilot-plant design, operation, and data evaluation is presented by Ball et al.[6]

In the absence of pilot-plant data, $K_L a$ may be estimated from empirically derived correlations. Two correlations reported in the mass-transfer literature—that of Sherwood and Holloway[7] and the Onda correlation[8] are most frequently employed for air-stripper design. A recently published comparison of these two correlations[9] concluded that the Onda correlation more accurately predicts the

$K_L a$ for typical air-stripping operations. This is probably because Onda's model includes both gas- and liquid-phase resistances to mass transfer, whereas the Sherwood-Holloway equation makes the simplifying assumption of liquid-phase control.

Design Considerations

A packed-tower air stripper consists of the tower shell itself; packing material; tower internals including the liquid distributor and redistributors, packing support plate, and demister pad; pumps and piping for the water; and a blower to provide air to the base of the tower.

Typical materials of construction for the tower shell are fiber-reinforced plastic (FRP), aluminum, coated carbon steel, and stainless steel. Its corrosion resistance and low cost make FRP the preferred material, but it may not be suitable for very tall outdoor towers subjected to high winds or in situations in which the air stripper is to be operated at elevated temperatures. Plastic packing, usually polyethylene or polypropylene, is light, corrosion-resistant, and inexpensive. Ceramic or steel packing might be specified for high-temperature operation.

Establishing and maintaining uniform flow of the water throughout the packed tower is critical to efficient operation. Feed distributors may be of several types—trays, trough-and-weir arrangements, header-lateral piping, and spray nozzles.

In tall towers, liquid redistributors—trays or rings—may be placed at intervals of about 10 ft (3.048 m) to redirect any water channeling along the tower walls back into the packing.

As is clear from Eq. (3), the key design variables in air stripping are the liquid loading rate, the air-to-water ratio, the packing height, and the characteristics of the tower packing. Typical towers range from 1 to 12 ft (0.3 to 3.66 m) in diameter and have packing heights of from 5 to 50 ft (1.5 to 15.2 m). Liquid loading rates are generally from 5 to 30 gpm/ft^2 [3.4 to 20.4 L/(s · m^2)], and volumetric air-to-water ratios may be as low as 10:1 or as high as 300:1. These ranges of practical

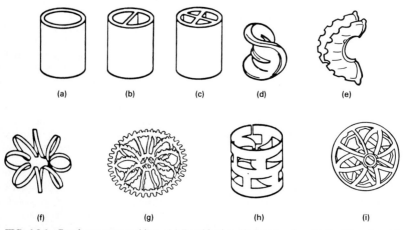

FIG. 6.8.2   Random tower packings. (*a*) Raschig ring. (*b*) Lessing ring. (*c*) Partition ring. (*d*) Berl saddle. (*e*) Intalox saddle. (*f*) Tellerette. (*g*) Ceilcote Type K Tellerite. (*h*) Pall ring. (*i*) Jaeger Tri-Pack.

operation are established by gas-pressure drop and flooding considerations as well as by mass-transfer requirements.

Various types of plastic packings from simple Raschig rings to more complex Tri-Packs are available in a range of sizes from about 0.5 to 3.5 in (1.3 to 8.9 cm). Figure 6.8.2 illustrates several packing shapes. As an alternative to these random or dumped packings, stacked or structured packings are sometimes used for air-stripping towers.

Environmental Impact

In its simplest form, air stripping converts a water-pollution problem to an air-pollution problem. The off gas leaving the top of the tower contains all toxic volatile organic compounds removed from the aqueous waste. Because of dilution with the stripping air and dispersion in the atmosphere, the magnitude of this air-pollution problem and of the potential related health risks may be considered negligible by the appropriate regulatory authorities. This has sometimes been the case for groundwater cleanup projects where initial concentrations of contaminants were very low and the operation was to be temporary.

However, when atmospheric discharge of toxic organics is not acceptable, air-pollution control measures can be implemented. Two applicable technologies are vapor-phase carbon adsorption and fume incineration. Discussion of these technologies is beyond the scope of this section.

Current Research and Recent Advances

Current research in the technology of air stripping includes (1) the development and demonstration of improved packings and (2) experimental work on high-temperature air stripping.

New plastic packings have been designed with open structures and many edges or points to minimize pressure drop while increasing the wetted surface area available for air-water contact and mass transfer. Compared to traditional rings and saddles, these packings should result in shorter towers and therefore in lower operating costs since energy requirements for air blowers and water pumps will be reduced. Preliminary experimental work and definition of packing characteristics has been done, of course, by the packing manufacturers themselves. Jaeger has published information[10] on its Tri-Packs, which have been marketed in the United States since 1980. Ceilcote has recently released data[11] on its Type K Tellerite, which was introduced in 1985.

Air stripping is a temperature-dependent process, largely because the volatility and thus the Henry's coefficient of an organic compound increases with increasing temperature. In addition, the viscosity of water decreases at high temperature, reducing liquid-phase resistance to mass transfer. Therefore, the effectiveness of an air stripper can be enhanced by heating the contaminated influent water. Higher concentrations of strippable compounds may be reduced to acceptable levels with reasonable tower heights. Compounds with low Henry's coefficients that are not readily strippable at typical groundwater temperatures [50 to 70°F (10 to 21°C)] may become strippable at temperatures from 100 to 180°F (38 to 82°C). Recently published studies have demonstrated the use of high-temperature air stripping to remove methyl ethyl ketone,[12] and isopropanol and tetrahydrofuran[13] from contaminated groundwater.

6.8.2 STEAM STRIPPING

Applicability

Like air stripping, steam stripping can be used to remove VOCs from water or aqueous waste streams. However, steam stripping is more broadly applicable, in that it can treat

1. Aqueous wastes contaminated with more soluble, less volatile compounds, not readily air-strippable, including acetone, methanol, and pentachlorophenol
2. Higher concentrations, up to several percent by weight, of volatile organic compounds in an aqueous waste
3. Nonaqueous wastes such as spent solvents contaminated with nonvolatile impurities

This subsection is limited to discussion of steam stripping for treatment of aqueous wastes. The reader interested in steam stripping and the related process of steam distillation as applied to treatment of nonaqueous wastes and to solvent recovery is referred to Ellerbe.[14]

Due to higher capital and operating costs, steam stripping is unlikely to be considered for groundwater-decontamination projects where air stripping is adequate. However, steam stripping has been successfully demonstrated for the decontamination of a groundwater containing ketones and alcohols as well as chlorinated solvents at 1,000 ppm.[15] Steam stripping may be employed for the treatment of industrial wastewaters, particularly where the organic contaminants can be recovered for re-use or concentrated for more efficient destruction on site. Examples are the recovery of solvent from an aqueous raffinate stream for recycling to an extraction process;[16] the stripping of methanol, turpentine, phenolics, and organosulfur compounds from paper mill wastewaters;[17] and the removal of such high-boiling materials as pesticides (e.g., Aldrin), heat-transfer fluids (e.g., biphenyl), and antimicrobials (e.g., pentachlorophenol) from industrial-process wastewaters.[18]

Restrictive Waste Characteristics

Suspended solids should be removed from a waste stream prior to steam stripping in order to avoid fouling the packing or plugging tray towers. Steam stripping would not be appropriate for treatment of aqueous wastes if the organic contaminant were a noncondensible gas such as vinyl chloride. The gaseous contaminant would leave the tower with the overhead vapor, but it would not be liquefied in the condenser. Instead, the gas would probably be vented to the atmosphere, resulting in an air-pollution problem.

Process Description. Steam stripping is a continuous fractional-distillation process without rectification. Figure 6.8.3 is a schematic diagram of a steam-stripping operation. The feed stream enters the top of a packed or tray tower. There is no reboiler; instead, live steam is injected directly into the bottom of the tower. When aqueous waste is steam-stripped, the bottoms product is relatively clean water. The overhead vapor—consisting of a mixture of organic vapors and steam—is condensed in a water-cooled heat exchanger and collected in an accu-

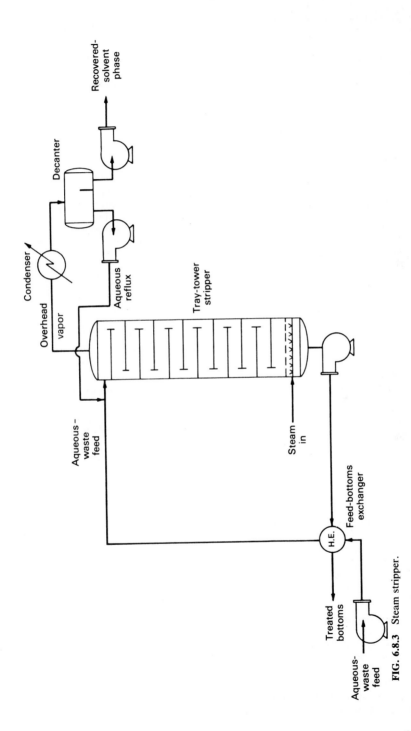

FIG. 6.8.3 Steam stripper.

6.115

mulator tank where immiscible organic and aqueous phases gravity-separate. The aqueous phase, saturated with the organic contaminant(s), might be combined with the feed and returned to the top of the stripper. The organic phase might be recovered for re-use as a solvent or a fuel. In the case of an organic contaminant completely miscible with water (e.g., acetone) there will be no phase separation. The condensed overhead product is a concentrated aqueous solution which may be further treated or recycled.

Theory and Design Equations

Steam stripping is essentially a distillation process. Therefore, the design equations and methodology developed for continuous fractional distillation will apply. (See Sec. 6.2 for a discussion of distillation.) Vapor-liquid equilibrium relationships for water and the organic contaminant(s) must be known empirically or estimated. In many cases, the mole fraction of organic contaminant in water is so low that a straight-line equilibrium relationship, or constant relative volatility, may be assumed.

Design Considerations

Process equipment for steam stripping includes a packed or tray tower; a boiler or steam generator; a water-cooled condenser and distillate decanter; pumps and piping for the aqueous feed, the bottoms, the aqueous reflux stream, and the condensed solvent product; and a bottoms-to-feed heat exchanger. Most steam strippers operate at atmospheric pressure, but vacuum operation may be employed to minimize steam requirements.

Since a steam stripper operates near the boiling point of water [100°C (212°F)], plastics are generally inappropriate. Stainless-steel columns and internals are commonly used. If packing rather than trays is specified, the packing may be steel or ceramic.

Unlike air stripping, steam stripping is an energy-intensive separation process. The economics of the process can be improved by various energy-conservation techniques. The use of a bottoms-to-feed heat exchanger to preheat the feed stream while cooling the treated bottoms for discharge is an obvious energy-saving step which can be implemented in any steam-stripping system. When steam stripping is used for continuous on-site treatment of an industrial-process waste stream, creative engineering approaches involving other on-site processes can result in further cost savings. For example, paper mills using steam strippers to remove methanol, turpentine, and organosulfur compounds from aqueous wastes save energy by using steam from existing multiple-effect evaporators as the stripping fluid.[19,20]

Environmental Impact

Steam stripping produces a treated bottoms stream suitable for appropriate discharge to a municipal sewage-treatment system or into surface water, or for re-use in on-site industrial processes. The nature of the overhead distillate stream depends on the solubility characteristics of the organic contaminants. If the organic is miscible or highly soluble in water, the distillate is a concentrated aqueous stream probably requiring further treatment, but possibly suitable for recycling to an on-site industrial process. In such cases, steam stripping may be viewed as a physical waste-minimization or concentration technique by which toxic organic contaminants in a dilute wastewater are concentrated into a waste stream of much smaller volume that is more readily amenable to treatment by chemical, biological, or thermal methods. If the organic contaminant is only slightly soluble in water, the condensed overhead stream will gravity-separate into aqueous and organic phases. The aqueous phase, saturated with the organic contaminant, is returned to the stripper with the feed stream. The organic phase, if it consists of a single solvent component, may be suitable for re-use, perhaps after a drying operation to remove residual water. Where the organic phase is a mixture of various compounds, destruction of this residual waste by incineration or use as a fuel substitute is appropriate.

The steam stripper's condenser would normally be vented to the atmosphere. This could have an adverse environmental impact if noncondensible toxic organics (e.g., vinyl chloride) are stripped, or in the case of a process upset where inadequate cooling could result in the release of organic vapors. Air-pollution control may be required.

6.8.3 REFERENCES

1. K. Sullivan and F. Lenzo, "Strip Those Organics Away," *WATER/Engineering and Management,* December 1983, pp. 46–47.

2. W. D. Byers and C. M. Morton, "Removing VOC from Groundwater; Pilot, Scale-up and Operating Experience," *Environmental Progress,* 4:(2)112–118 (May 1985).

3. R. L. Gross and S. G. TerMaath, "Packed Tower Aeration Strips Trichloroethylene from Groundwater," *Environmental Progress,* 4:(2)119–121 (May 1985).

4. *Facility Test Report for Environmental Waste Enterprises,* prepared for the U.S. EPA Office of Research and Development, by Metcalf & Eddy, Inc., Woburn, Massachusetts, 1986.

5. *Facility Test Report for U.S. Pollution Control Inc.,* prepared for the U.S. EPA Office of Research and Development, by Metcalf & Eddy, Inc., Woburn, Massachusetts, 1986 (RCRA Confidential Business Information).

6. W. P. Ball et al., "Mass Transfer of Volatile Organic Compounds in Packed Tower Aeration," *J. Water Pollut. Control Fed.,* 56:(2)127–136 (February, 1984).

7. T. K. Sherwood and F. A. L. Halloway, *Trans. Am. Inst. Chem. Eng., 1940,* 36:39–70.

8. K. Onda et al., *Chem. Eng. Jpn., 1968,* (1):56–62.

9. P. V. Roberts et al., "Evaluating Two-Resistance Models for Air Stripping of Volatile Organic Contaminants in a Counter-current, Packed Column," *Environ. Sci. Technol.,* 19:(2)164–173 (1985).

10. Technical Bulletin JTP-600, Jaeger Tri-packs, Inc.

11. Ceilcote (a division of General Signal) Bulletin 12-20.

12. T. Johnson et al., "Raising Stripper Temperature Raises MEK Removal," *Pollution Engineering,* September 1985, pp. 34–35.

13. B. L. Lamarre et al., "Design, Operation and Results of a Pilot Plant for Removal of Contaminants from Ground Water," *Proceedings of the Third National Symposium on Aquifer Restoration and Ground-Water Monitoring,* May 25–27, 1983, pp. 113–122.

14. R. W. Ellerbe, "Steam Distillation/Stripping," sec. 1.4 in Philip A. Schweitzer (ed.), *Handbook of Separation Techniques for Chemical Engineers,* McGraw-Hill, New York, 1979, pp. 1–169 to 1–178.

15. D. V. Nakles and J. E. Bratina, "Technology for Remediation of Groundwater Decontamination," *Proceedings of the National Conference on Hazardous Wastes and Hazardous Materials,* Atlanta, Georgia, March 4–6, 1986, pp. 133–135.

16. M. F. Nathan, "Choosing a Process for Chloride Removal," *Chemical Engineering,* **58**:(3)93–100 (January 30, 1978).

17. K. E. McCance and H. G. Burke, "Contaminated Condensate Stripping—an Industry Survey," *Pulp & Paper Canada,* **81**:(11)78–81 (November 1980).

18. L. A. Robbins, The Dow Chemical Company, *Method of Removing Contaminants from Water,* U.S. Patent 4,236,973.

19. T. Burgess and D. Voigt, "Nekoosa Cleans Condensates with Steam Distillation," *Pulp & Paper Canada,* **79**:(8)72–74 (August 1978).

20. D. A. Donovan et al., "Steam Stripping of Foul Condensates at Great Lakes," *Pulp & Paper Canada,* **81**:(2)28–33 (February 1980).

SECTION 6.9
THIN-FILM EVAPORATION

C. Clark Allen, Ph.D.
Research Triangle Institute
Research Triangle Park, North Carolina

Steve I. Simpson
Associated Technologies, Inc.
Charlotte, North Carolina

In the selection of potential pretreatment processes for removing volatile organics from hazardous wastes, agitated thin-film evaporation must be considered for those wastes that would foul or plug conventional evaporators or which require low operating pressure during evaporation. Certain slurries, sludges, or viscous wastes which cannot be handled by flash evaporators, forced-circulation evaporators, distillation columns, falling-film evaporators, or other conventional equipment can be successfully processed or recovered in agitated thin-film evaporators.

The unique feature of this equipment is not the thin film itself (falling- and rising-film evaporators use thin liquid layers) but rather the mechanical agitation device for producing and agitating the film. This mechanical agitator permits the processing of high-viscosity liquids and liquids with suspended solids.[1] The agitation at the heat-transfer surface not only promotes heat transfer but also maintains precipitated or crystallized solids in manageable suspension and prevents or minimizes fouling of the heat-transfer surface.

6.9.1 APPLICABILITY OF THE TECHNOLOGY

Generic Waste Streams

Thin-film evaporators can be used to remove or recover organic components from waste streams. The bottoms of the evaporator will have smaller amounts of volatiles than the feed, either because of the reduction in concentration or the reduction in the volume (or both). For mixed (water and organics) waste streams, a thin-film evaporator can selectively remove the water (the organics are high-boiling) to improve the Btu value of the bottoms so that they can be more easily incinerated, or selectively remove the organics (the organics are low-boiling) so that the organics can be recycled or burned. Thin-film evaporators can be used to recover high-boiling water-soluble solvents, such as ethylene glycol, without adding water to the recovered product.

Specific Hazardous Wastes

Hazardous wastes which can be treated in a thin-film evaporator include waste solvents from the chemical, plastics, electronics, and pharmaceutical industries. Both halogenated and nonhalogenated organics can be recovered, together with alcohols, ketones, esters, glycols, ethers, aromatic hydrocarbons, petroleum napthas, freons, and specialty solvents. Still bottoms, spent lubricating oil, coating residues, obsolete paints, and inks can be treated in a thin-film evaporator.

Physical Form of the Waste Stream

Thin-film evaporators are used to treat liquids and sludges. Products handled in falling-film evaporators usually have low viscosities, while agitated thin-film evaporators readily process liquids with viscosities ranging up to 10,000 cP under operating shear (equivalent to 100,000 to 400,000 cP without subjected shear, as measured by standard instruments in the laboratory).[1]

Waste Streams for Which It Does Not Work

Thin-film evaporators do not work well for wastes which react immediately when heated, or for extremely viscous materials. Filtering is required if the solids in the waste are comparable in size to the clearance of the rotors.

6.9.2   DESCRIPTION OF THE TECHNOLOGY

Chemical and Mechanical Principles

Agitated thin-film evaporators are designed to spread and continuously agitate and renew a thin layer, or film of liquid on one side of a metallic surface, with heat supplied to the other side. Heat can be supplied by either steam or heated oil; heated oils are typically used to heat the waste to temperatures higher than can be achieved with 1.0 mPa or 150 psig saturated steam (>185°C or 360°F), although higher steam pressure can be employed.

A volatile component has to be transported within the film to the interface, then vaporized. Transport of the volatile component within the film is accomplished by molecular diffusion or by eddy diffusion. Molecular diffusion, the only possibility in nonagitated laminar flow, is extremely slow and decreases with increasing viscosity of the film liquid. Eddy diffusion can be influenced and increased by adding turbulence to the film. Values of diffusivities in agitated thin-film evaporators are on the order of 10^{-6} m^2/s, or 1,000 to 10,000 times greater than the molecular diffusivities achieved in nonagitated evaporators.[1]

In the removal of volatiles from aqueous or mixed hazardous wastes, the efficiency of the removal and the residual concentration of volatiles will depend upon waste viscosity and concentration, the boiling points of the volatiles, and evaporator's operating pressure and temperature. For complete separation of close-boiling components by distillation, a fractionation column of adequate design can be added to an agitated thin-film evaporator, which serves as a reboiler. A thin-film evaporator with vapors flowing countercurrent to the thin liquid film can be expected to have a fractionation effect of 1.25 to 1.5 theoretical plates, as opposed to the single-plate maximum efficiency of a conventional still-pot reboiler.[2]

Deodorization is an extreme stripping operation where only traces of a low-boiling odor impurity in a feedstock must be removed. Again, it is the diffusion process, not the thermal load, that determines the required surface. Deodorization can be further facilitated by using a stripping gas (or steam) to lower the partial pressure of the low-boiling impurities.[2] Residual volatiles concentrations of less than 1,000 ppm have been achieved routinely in similar applications, and less than 100 ppm is possible if conditions are optimal.

Description of the Process

There are two general configurations of mechanically agitated thin-film evaporators: horizontal and vertical. A typical unit consists of a motor-driven rotor with longitudinal blades which rotate concentrically within a heated cylinder.

In the vertical design (Fig. 6.9.1), product enters the feed nozzle above the heated zone and is mechanically transported by the rotor and gravity down a helical path on the inner heat-transfer surface. The evaporator does not operate full of product; the liquid or slurry forms a thin film or annular ring of product from the feed nozzle to the product outlet nozzle. Holdup, or inventory, of product in a thin-film evaporator is very low—typically about 2 kg/m^2 (½ lb/ft^2) of heat-transfer surface.[3]

With typical tip speeds of 900 to 1,200 cm/s (30 to 40 ft/s),[1] centrifugal forces distribute the liquid feed as a thin film on the heated cylinder wall, and the wave action produced by the rotating blades provides rapid mixing and frequent surface regeneration of the thin, turbulent liquid layer on the transfer surface.

As illustrated in Fig. 6.9.2, the rotor may be one of several *zero-clearance* designs, a rigid *fixed-clearance* type; or, in the case of tapered rotors, an adjustable-clearance construction may be used.[3] The clearance is the space between the shell and the periphery of the circle described by the rotor blade tips.[5] One vertical design includes an optional residence time-control ring at the end of the thermal surface to hold back liquid (and thus build up the film thickness),[3] for cases involving a very low bottoms product rate.

Equipment Configurations

The vertical design illustrated in Fig. 6.9.1 is manufactured by Luwa Corporation (Charlotte, North Carolina) and incorporates a cylindrical thermal zone. Some manufacturers utilize a tapered thermal zone, as illustrated in Fig. 6.9.3.

Advantages and Disadvantages

A major advantage of thin-film evaporators is that organic compounds can be recovered from viscous liquids. Disadvantages include the need to leave a fraction

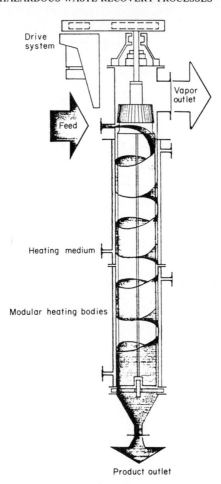

FIG. 6.9.1   Flow path for thin-film evaporator.

of the organics in some sludges to maintain fluidity. This is necessary to prevent high maintenance costs and reduced heat transfer that can result from bake-on on the heat-exchange surfaces. The capital cost of a thin-film evaporator is generally higher than that for steam-stripping equipment; however, energy efficiency may be higher, and distillate is not contaminated with stripping-steam condensate.

6.9.3   DESIGN CONSIDERATIONS

Capacities, Flows, and Other Design Specifications

Typical operating characteristics for various applications of thin-film evaporators are listed in Table 6.9.1.

Luwa manufactures vertical thin-film evaporators with cylindrical heating bodies (see Fig. 6.9.1). The body of each vertical thin-film evaporator is made in

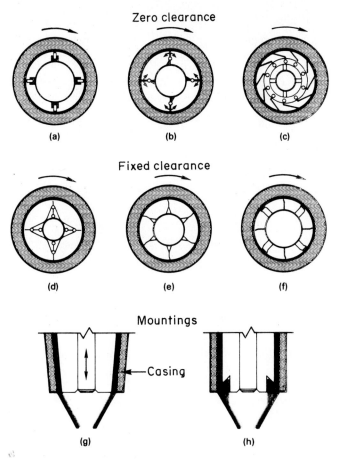

FIG. 6.9.2 Rotors for thin-film evaporators. (*a*) With carbon or teflon wipers. (*b*) With pendulum-hinged blades. (*c*) With scraping hinged blades. (*d*) Low-viscosity. (*e*) Medium-viscosity. (*f*) High-viscosity. (*g*) Adjustable-clearance mounting. (*h*) Fixed-clearance mounting with residence-time control ring.

sections, including one or more jacketed thermal sections, a top or vapor head, and a bottom discharge cone. The thermal body sections must have a uniform bore for precise rotor clearance. The sections are rolled, machined, and sometimes honed (or bored), and then are assembled as a unit.[3]

The general design specifications for standard thin-film evaporators are:[3]

- *Pressure:* Full vacuum to about 0.1 mPa (15 psig)
- *Temperature:* To 340°C (650°F)
- *Heating:* Steam to 1.0 MPa (150 psig)
  Dowtherm to 0.7 MPa (100 psig)
- *Sizes:* To 40 m² (430 ft²) of surface

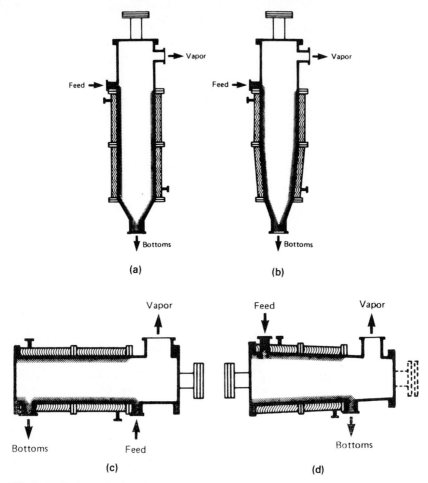

FIG. 6.9.3   Horizontal and vertical thin-film evaporation. (*a*) Vertical cylindrical thermal zone. (*b*) Vertical tapered thermal zone. (*c*) Horizontal cylindrical thermal zone. (*d*) Horizontal tapered thermal zone.

Possible Efficiencies, Capabilities

In most situations, the heat-transfer rate between the liquid and the wall (the equipment's U-value) determines the size and effectiveness of thin-film equipment. Figure 6.9.4 shows overall U-values of several dozen products, differing in latent heat of vaporization, heat conductivity, viscosity, boiling-point rise, and surface tension. These data were developed using saturated-steam or high-temperature heating mediums. The U-values include the resistance of a 0.63-cm (¼-in) stainless-steel wall or a 1.3-cm (½-in) stainless-steel-clad wall.[1] A U-value can be used to calculate the temperature of the heating fluid necessary to achieve a specified flow of overhead product.

The heat-transfer coefficients are grouped into main applications to facilitate

TABLE 6.9.1 Typical Operating Characteristics for Thin-Film Units

Size of heat-transfer surface	$0.1–40$ m^2 [$1–430$ ft^2]
Capacity	
Steam heated	
Water evaporation	60 kW/m^2 [$50{,}000$ Btu/(h $\cdot$ ft^2)]
Organics distillation	63 kW/m^2 [$20{,}000$ Btu/(h $\cdot$ ft^2)]
Hot-oil heated	
Organics distillation	25 kW/m^2 [$8{,}000$ Btu/(h $\cdot$ ft^2)]
Operating pressure	
Standard	Full vacuum to atmospheric pressure
Special	Any positive pressure
Heating steam in jackets	1.0 MPa [150 psig]
Maximum hot-oil temperature	$350°C$ [$650°F$]
Liquid throughput	Up to $900–1100$ kg/m^2 [$200–250$ lb/(h $\cdot$ ft^2)]
Pressure drop (vapor flow)	0.5 mmHg
Overhead-to-bottoms splits	Up to $100{:}1$
Residence time	
Uncontrolled	$3–10$ s
Controlled	$3–100$ s
Product viscosities at operating conditions	$10{,}000$ cP
Blade-tip speed	
Nonscraping blades	$9–12$ m/s [$30–40$ ft/s]
Scraping blades	$1.5–3$ m/s [$5–10$ ft/s]
Recommended maintenance	Twice a year; more often when processing extreme products

Source: Ref. 5.

preliminary evaluations. U-values for concentration or distillation show the characteristic increase with increased heat flux, resulting from the additional circulation of the formed vapor bubbles (nucleate boiling usually takes place in thin-film evaporators).[1]

In many applications of the agitated thin-film evaporator, mass transfer (not heat transfer) determines the size and capacity of the equipment. Deodorization, the removal of low boilers, and dehydration are mass-transfer-controlled processes.[1] The stripping of volatile organics from hazardous waste falls within this category if the aim is to remove the volatiles to low residual levels.

Materials of Construction[5]

Mechanical construction of units with a fixed-clearance rotor requires machining of both the inside diameter of the shell and the outside diameter of the rotor to assure concentricity and dynamic balancing of the rotor, and consideration of the effects of differential expansion of rotor and shell. Although clearances may vary in a narrow range determined by viscosity, surface tension, and thermal conductivity of the material handled, they usually fall in the range of 0.08 to 0.25

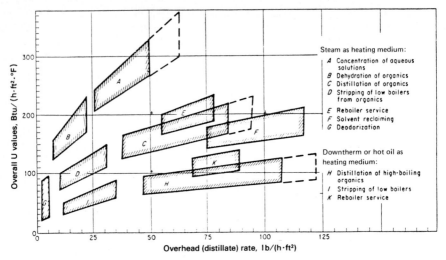

FIG. 6.9.4 Heat transfer in agitated thin-film equipment for various services.

cm (0.03 to 0.10 in). Such small clearances emphasize the importance of the machining operations.

Adjustable clearance is available in the tapered-shell units by moving the rotor in or out with respect to the fixed position of the shell. This offers some advantage in a pilot plant or commercial unit where a wide variety of materials will be processed.

Zero-clearance, or wiping, blades are spring-loaded or free-floating and are forced against the wall by centrifugal force as the agitator rotates. This design is primarily for materials that remain in a liquid state at the final concentration. Luwa and Kontro, however, have designs suitable for materials that will be evaporated to the solid state.

Maintenance Considerations[3]

At the drive end of the rotor, a double mechanical seal is most often supplied, although single mechanical seals can be used satisfactorily. A coaxial design like the Crane 151 seal, which is more durable and accommodates more vibration, has a typical life expectancy of about four years. Because of torque and some axial loading by the multiple belt drive, a self-aligning roller bearing is recommended; it also has a life of about four years. The bearing and seal can be inspected and serviced without removing the top cover or rotor.

Reasons for premature failure include failure of the double mechanical seal and the self-aligning roller bearing. The double mechanical seal can fail as a result of the loss of cooling fluid, the presence of solids or abrasives in the seal coolant, chemical attack on elastomeric O-rings, or abnormal vibration. The self-aligning roller bearing can fail as a result of inadequate lubrication or mixing of lubricant types; sporadic lubrication, which can permit water or other fluids into the bearing; or abnormal vibration.

The internal bearing most commonly used at the discharge end of vertical machines is a non-load-bearing design. The full weight of the rotor is supported at the top, and the function of the spider-type bearing assembly is to center the rotor. A hardened pin on the rotor fits into a carbon cup bushing in the center of the spider. Often this bearing design is preferred to an external bearing to eliminate a mechanical seal and bearing at the dirty end of the machine. This design often employs a small amount of lubricant: oil, process liquid, water, steam, or air. Typical life of the carbon bushing is two years. The bearing can be inspected or serviced by simply removing the bottom cone.

Premature failure of the bearing can be caused by the loss of lubricant, poor lubricating properties of lubricant, or process upsets with liquid backing up into rotor area.

Original-equipment rotors last typically 10 to 30 years with little or no maintenance.

Relation to Other Processes

Thin-film evaporators can be used to separate volatile compounds from the solids in sludges. These volatiles can be purified by dehydration, or the individual compounds can be separated by distillation or other processes.

Utility Support

Utilities required to support a thin-film evaporator include loading and handling facilities, a screening process to remove large solids, a source of steam or heating oil, electricity, and cooling water for the condenser. A vacuum pump is needed to facilitate the separation of high-boiling compounds.

6.9.4 ENVIRONMENTAL EFFECTS OF PROCESS

Air and Water

A thin-film evaporation system has three streams which exit the system—the distillate which was evaporated and then condensed back to the liquid phase, the concentrates stream which was not evaporated, and the noncondensible gases which are vented from the condenser.

In the case of volatile-organics removal from hazardous waste, the distillate will contain the organics which have been removed from the waste. Depending upon water content and miscibility, it may be possible to separate the organics and water by decanting. The organics can be incinerated or re-used, depending upon purity. In cases where fractionation of the organics into individual components is desired, vapors from the thin-film evaporator can be used as feed for a fractionation column. In some cases, it may be desirable to pass the distillate of the thin-film evaporator through activated carbon to remove the organics by adsorption.

The exhaust vent of the condenser contains noncondensible gases (primarily air) and, potentially, a small amount of organic vapors which may pass through the condenser without condensing. This exhaust can be further processed by passing it through activated carbon to remove the organics by adsorption. The exhaust vent of the condenser is connected to a vacuum pump when the evaporation is under vacuum. The vacuum-pump vent is then a source of potential emissions. At start-up, the vent emissions are expected to be greater than during normal operation.

Residuals

The treated waste exits the bottom of the evaporator, and may have a substantially reduced volatile-organics content. Often, however, only 85 to 90% of the organics are recovered. The residual sludge from a thin-film evaporator has been landfilled and burned for its fuel content.

Recycling Possibilities

The thin-film evaporator is frequently used for recovering organic materials from wastes at hazardous-waste management facilities as well as for in-plant recycling or waste-minimization applications at the point of generation.

6.9.5 ENGINEERING ECONOMIC EVALUATIONS

Capital Costs

1984 budget prices for four standard evaporators with 316L stainless-steel wetted parts are listed in Table 6.9.2.

TABLE 6.9.2 Budget Prices: Thin-Film Evaporators

Heating surface m^2 (ft^2)	Budget price, 1984
0.1 (1)	$ 16,000
1.0 (11)	45,000
5.0 (54)	120,000
15.0(160)	210,000

Table 6.9.3 compares the cost of the evaporator alone to the installed-system cost for two cases. In the case designated *simple,* the heating medium is assumed to be steam, and the system is controlled manually. In the case designated *sophisticated,* the heating medium is a recirculating thermal fluid and the system is completely automated.[5] The cost of a heat generator (steam boiler) is not included.

Table 6.9.3 gives a very generalized example, based, however, on actual projects involving agitated thin-film evaporators. From Table 6.9.3 we can say that installed cost could range from twice the evaporator-only cost to as high as four times, depending upon the degree of sophistication.[5]

Operating and Maintenance Costs

As an example of operating costs, assume that a hazardous waste containing 2% by weight volatile organics, 88% by weight water, and 10% by weight sol-

TABLE 6.9.3 Cost Distribution of Agitated Thin-Film Units, $000

	Simple	Sophisticated
Evaporator	100	100
Components as % of evaporator		
Main auxiliaries (condenser, pumps, vacuum system, controls)	30	150
Piping, fittings (materials only)	10	20
Structural frame	5	5
Installation (foundation, erection, piping, wiring, insulation)	100	100
Total installed cost	245	375
Evaporator as % of installed cost	41	27

Source: Ref. 5.

ids is to be processed in a steam-heated thin-film evaporator to remove the volatile organics. Assume that the limiting factor in the degree of separation is a solids content of 50% by weight, above which the concentrates will not flow from the evaporator.

In this situation the residual level of organics in the concentrates will depend

TABLE 6.9.4 Estimated Operating Costs for Volatiles Removal from Hazardous Waste

Assumptions	
Feed rate	40 L/min (10 gal/min) continuous
Feed composition	10% solids, 2% organics, 88% water
Feed temperature	42°C (60°F)
Bottoms concentration	50% solids, less than 1,000 ppm organics
Steam pressure	1 MPa (150 psig), saturated
Heat-transfer coefficient	4 MJ/(h)(m^2)(°C) [200 Btu/(h)(ft^2)(°F)]
Steam cost	$4.4/Mg ($2/1,000 lb)
Electricity cost	$0.05/kWh
Labor cost (including overhead)	$20/labor-hour
Operators required	0.25 (automated system)
Operating period	8,760 h/yr

Operating Costs*	
Steam	$85,000/yr
Electricity	$7,500/yr
Operating labor	$45,000/yr
Spare parts	$10,000/yr
Maintenance	$10,000/yr
Total	$157,500/yr

Unit Costs	
$/L waste treated	0.0075
$/L VOC removed	0.37
$/Mg VOC removed†	462

*Does not include laboratory analyses and residual, disposal costs.
†Assumes a density of 0.8 g/cm^3.

TABLE 6.9.5   Major Manufacturers of Agitated Thin-Film Evaporators

Blaw-Knox Company (Buflovak Division)	Buffalo, N.Y.
Cherry Burrell (Chemetron Division)	Louisville, Ky.
Luwa Corporation	Charlotte, N.C.
Kontro Company, Inc.	Orange, Mass.
Pflaudler Company, Inc.	Rochester, N.Y.
Artisan Industries, Inc.	Waltham, Mass.

upon the vapor pressure of the organics, the viscosity of the concentrates, the speed of the evaporator rotor, the heating temperature, and the operating pressure; but residual levels of 1,000 ppm and less could be expected for low-boiling organics. Assume a continuous feed flow rate of 40 L/min (10 gal/min). Table 6.9.4 lists estimated operating costs and assumptions.

6.9.6 LIST OF EQUIPMENT SUPPLIERS

There are six major suppliers of agitated thin-film equipment in the United States: Buflovak Division of Blaw-Knox, Chemetron Division of Cherry Burrell, Luwa, Kontro, Pflaudler, and Artisan. The locations of these companies are presented in Table 6.9.5. The Luwa concern had the first production units onstream in 1946 in Switzerland. The Pflaudler design is based on development work by Arthur Smith. Buflovak introduced its design with a bottom drive in the mid-fifties. The tapered-shell designs of Kontro and Luwa (Fig. 6.9.3) evolved from the standard cylindrical designs. Basic design configurations furnished by the six suppliers are shown in Table 6.9.6. This table also relates liquid-vapor flow characteristics to design and location of the vapor separator.[5] Countercurrent flow occurs when the vapor is removed from the top of the evaporator and the bottoms are removed

TABLE 6.9.6   Configurations of Commercial Agitated Thin-Film Units

Configuration	Manufacturer	Liquid-vapor flow
Vertical, cylindrical shell, integral separator	Luwa, Chemetron	Countercurrent
Vertical, cylindrical shell, external separator	Buflovak	Cocurrent
Vertical, cylindrical shell, internal separator and condenser	Pflaudler	Separated
Horizontal, tapered shell, integral separator	Kontro	Co- or countercurrent
Vertical, tapered shell, integral separator	Kontro	Co- or countercurrent
Horizontal, cylindrical shell, external separator	Artisan	Co- or countercurrent

Source: Ref. 5.

from the bottom; both the liquid and vapors flow in the same direction in cocurrent flow. In separated flow, the vapors are condensed in an internal condenser.

6.9.7 REFERENCES

1. A. B. Mutzenberg, "Agitated Thin-Film Evaporators: Part 1—Thin-Film Technology," *Chemical Engineering,* McGraw-Hill, New York, September 13, 1965.
2. R. Fischer, "Agitated Thin-Film Evaporators: Part 3—Process Applications," *Chemical Engineering,* McGraw-Hill, New York, September 13, 1965.
3. H. L. Freese and W. T. Gregory, "Volume Reduction of Liquid Radioactive Wastes Using Mechanically Agitated Thin-Film Evaporators," prepared for National Meeting of American Institute of Chemical Engineers, June 4–8, 1978, Philadelphia, Pennsylvania.
4. *Luwa Thin-Film Evaporation Technology,* Brochure no. EV-24, Luwa Corporation, Charlotte, North Carolina, 1982.
5. N. Parker, "Agitated Thin-Film Evaporators: Part 2—Equipment and Economics," *Chemical Engineering,* McGraw-Hill, New York, September 13, 1965.

6.9.8 FURTHER READING

McCabe, W. L., and J. C. Smith, *Unit Operations of Chemical Engineering,* 2d ed., McGraw-Hill, New York, 1967, p. 433.
Perry, R., and C. H. Chilton (eds.), *Chemical Engineers' Handbook,* 5th ed., McGraw-Hill, New York, 1973, pp. 11-27 to 11-38.

SECTION 6.10
FREEZE-CRYSTALLIZATION

James A. Heist, P.E.
Vice President, Research and Development
Heist Engineering Corporation
Raleigh, North Carolina

A freeze-crystallization process separates water from solutions (including hazardous wastes) by cooling the solution until ice crystals begin to form. Ice, with few exceptions, crystallizes as a pure material. The ice is separated from the remaining liquid and impurities, washed, and melted to produce a purified stream. The dissolved components of the stream are concentrated into a reduced volume. The cost is typically $0.025 to $0.15 per gallon of purified water, including equipment amortization.

Hazardous components of liquid wastes are concentrated into a smaller volume, which can be more economically destroyed by thermal oxidation or solidified.

Freeze-crystallization is a flexible process that can adjust to the needs of the application, and thus always operate at high efficiency. An explanation of the process' flexibility and variety is presented in Subsec. 6.10.2.

6.10.1 HAZARDOUS-WASTE APPLICATIONS

Generic Applications

The freeze-crystallization process is inherently insensitive to the composition of the waste stream it is processing. The process adapts automatically to changes in influent composition. It can handle a wide variety of waste-stream components without pretreatment:

- Acids and bases are concentrated. The melted ice has a corresponding change in pH of 3 to 4 units toward neutral, regardless of the volatility of the solute.
- Volatile organics remain in the concentrate.
- Salts, when at saturation, crystallize and are removed from the process as a sludge.

The process has been tested on a variety of hazardous wastes and industrial waste streams:

- The process is planned for remediation of hazardous-waste lagoons.
- A program is continuing to recover process materials for recycling from an ammunition plant's wastewater.
- Testing has demonstrated the ability of the process to concentrate pickle liquors for recycling.
- By-product recovery from an organic-chemical waste stream is being developed for implementation in 1988–1989.
- The Navy is pilot testing metal-finishing wastewaters for recovery of the process materials and water re-use.

Limitations

The ability of freeze-crystallization to process waste streams is constantly being expanded. Past limitations that have been resolved include:

- *Freezing-point depression:* Early versions of the process encountered difficulty when the freezing-point depression (FPD) exceeded 8 to 12°C (15 to 22°F). Process control and modifications have been developed that allow freeze-crystallization to function to at least 40°C (72°F) FPD.
- *Eutectic conditions:* A *eutectic* is the occurrence of more than one material crystallizing from the solution at the same time. Incorporation of fractionation processes to separate these solids has been successfully demonstrated in a number of pilot tests.
- *Immiscible-liquid formation:* In many hazardous wastes there are organic chemicals at saturation that form second liquid phases during processing. Several methods have been devised to isolate this liquid phase.
- *Refrigerant emulsification:* In secondary-refrigerant freeze processes (see Subsec. 6.10.2 for descriptions of the freeze-process variations) some hazardous wastes emulsify the refrigerant in contact with the wastewater. These have been successfully broken with modest doses of surfactants.

6.10.2 TECHNICAL DESCRIPTION

Process Description

There are four basic steps in the freeze-crystallization process, as shown in Fig. 6.10.1:

1. Crystallize one or more materials from the solution.

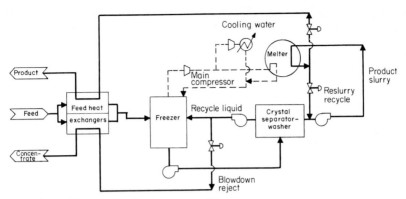

FIG. 6.10.1 Freeze-crystallization process diagram.

2. Separate the crystals from the solution, and from each other.

3. Melt those crystals that have fusion temperatures below ambient.

4. Using a heat pump, remove heat from the crystallizer and transfer it to the melter.

Slurry from the crystallizer is pumped to the separation device, where the crystals are separated from the concentrated liquid. The concentrated liquid is recycled to the crystallizer, where further crystallization takes place; any amount of recovery of water from the feed can be effected. The ice is melted, and discharged from the system through a refrigerant stripper and cross heat exchanger.

The concentrate consists of any slurry of precipitated materials that is bled from the system and any blowdown of concentrate. These flows may or may not be cross-exchanged with the feed to recover the cold value.

Each of the operations can be implemented in several ways. The principal components of the process and the equipment variations that have been devised include:

Crystallizers. Crystallization can be induced by either of two methods: (1) by removing heat through a heat-transfer surface, or (2) by direct contact of a boiling refrigerant with the process fluid. The direct-contact refrigerant can be either the solvent (operating at its triple point) or a secondary refrigerant chosen for its operating properties.

Crystal Separators. Ice can be separated from the concentrate by filtration, by centrifugation, or by wash columns. Each of these separation devices must wash the ice to remove adhering concentrate. The wash column allows countercurrent washing, and is therefore more efficient than the other devices.

Crystal Melters. Any crystal that melts at a temperature below ambient can be used to condense the refrigerant and improve the energy efficiency of the process. These crystals can be melted either by direct contact with the condensing refrigerant or in a heat exchanger.

Refrigeration System. The purpose of the refrigeration system is to remove heat in the crystallizer and reject it to a convenient sink. The sink may be either melting crystals or cooling water. Conventional refrigeration equipment has been adapted for this purpose.

Development Status

Variations of the freeze-crystallization process have been in development for over 30 years. In the 1950s the process was commercialized for fractionating *p*-xylene from its isomers. The U.S. Department of Interior for a period of over 20 years developed at least six different versions for seawater desalting. Only one of these was ever commercially implemented.

Crystallization is still as much art as science. New applications usually require some pilot testing. Kinetics and equipment design determine crystal size, which dictates the design of the rest of the plant. Refrigeration equipment, heat exchangers, solids/liquid separators, control schemes, and pumping equipment are well-developed for the application. Other than the crystallizer the only process component that requires testing is the solids-fractionation equipment needed when more than one material crystallizes.

Resource recovery and the cost of disposal of residuals are creating a much better market condition for this process. The use of freeze processing for reclaiming materials and cleaning up wastewaters is not wide at this time, but a good deal of testing has been done. As the process is implemented in more and varied applications the ability to design installations without pilot testing will develop.

Process Advantages and Limitations

The freeze-crystallization process' concentrated form allows more economical final disposition of hazardous materials. The economics of thermal destruction methods are particularly enhanced by preconcentration.

Freeze-crystallization is unique in its ability to concentrate to high levels of solutes, to produce a highly purified product, and to work with a wide variety of wastes. Characteristics of the process that enhance its ability to process hazardous wastes include:

- The process is energy-efficient because the heat of fusion of water is only 15% of the heat of vaporization. A heat pump is also used to reduce energy use.
- Water is converted to ice, which is mechanically separated from the concentrate; the concentrate retains volatiles that are in the waste.
- Pretreatment is usually not necessary. The process is not affected by saturated salts or organic materials, which just form additional solid and liquid phases as they are concentrated by the removal of water.
- The process is dynamically stable because of the large holdup-to-throughput ratios.
- Product qualities are stable and can be monitored automatically to control discharges. Since the removal of all contaminants is relatively uniform, one "tracer" can be used to inexpensively monitor and control discharges.
- All contaminants are reduced proportionally in the treated water; the melt contains from 0.01 to 0.1% of the concentration of contaminants in the feed.
- Phase equilibria can be altered chemically and physically with the process to

promote selective precipitation, allowing, in some applications, resource recovery from the waste stream.

The main disadvantages of freeze-crystallization arise from the infancy of the technology. The process characteristics dictate the conditions where it is most applicable. The factors that limit use of this technology include:

- Pilot testing is usually required because the data base on the process is limited.
- The process has rather high capital costs. However, lower operating costs often result in a lower life-cycle cost than with other treatment technologies.
- Long-term operating and maintenance (O&M) costs and reliability are unproven.

Integration with Other Processes

The flexibility of the freeze-crystallization process enhances its integration with both the generating and the disposal processes. The process can contribute to integration by adapting to changes in the following ways:

- Waste sources often produce wastes of variable flows and compositions. A freeze plant maintains constant flows of either product (clean water) or concentrate, and over short periods of both. This is done by automatically controlling product-concentrate split, by varying the amount of ice being produced, and by allowing the inventory of material within the process to change.
- The freeze process can adapt quickly to the needs of other processes if the final disposal equipment has variable capacity based on waste characteristics or process variables. Variable flow of concentrate from the process has little effect on the operations of the freeze process, due to the comparatively large holdup in the system.
- The freeze plant can be turned down to a fraction of its capacity quickly, and brought to full capacity on demand.

6.10.3 DESIGN CONSIDERATIONS

Design Variables

The design variables for freeze-crystallization equipment fall into three categories: (1) process characteristics, dictated by the application; (2) production objectives, which are often an optimization of the total costs of freeze-crystallization and final disposal; and (3) equipment and process options.

Important waste characteristics that dictate the design of the freeze process include:

- Physical properties that dictate process design and efficiency include freezing-point depression (FPD), viscosity, densities, and interfacial tensions.
- Phase equilibria define when multiphase separators will be needed to ensure effluent quality.

Production objectives and specifications define the ultimate concentration to which the freeze process must perform. This concentration dictates the freezing-point depression, amount of eutectic crystals formed, and the physical properties of the concentrate.

Once the performance specifications have been defined and the physical properties and phase equilibria determined, equipment variables include:

- The type of crystallizer, direct or indirect, and if direct, primary or secondary refrigerant.
- The type of crystal separator and eutectic fractionator. The design reflects the crystal fraction and the physical properties of concentrate and crystals.
- The melter can use direct or indirect contact with the hot refrigerant.
- The refrigeration system can use either an open or closed cycle. Refrigerants are usually chosen so that the operating pressure in the crystallizer is less than 15 psi above ambient.

Flow Sheets and Layouts

The flow sheet in Fig. 6.10.2 reflects an application with the following characteristics:

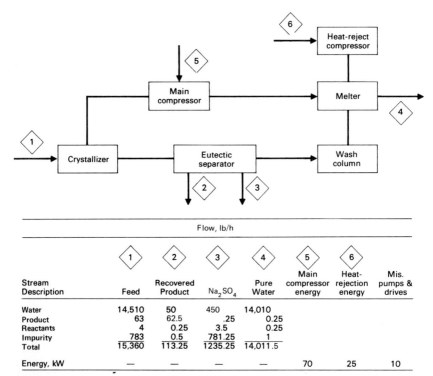

Flow, lb/h							
	1	2	3	4	5	6	
Stream Description	Feed	Recovered Product	Na$_2$SO$_4$	Pure Water	Main compressor energy	Heat-rejection energy	Mis. pumps & drives
Water	14,510	50	450	14,010			
Product	63	62.5	.25	0.25			
Reactants	4	0.25	3.5	0.25			
Impurity	783	0.5	781.25	1			
Total	15,360	113.25	1235.25	14,011.5			
Energy, kW	—	—	—	—	70	25	10

FIG. 6.10.2   Flow diagram for freeze-crystallization application: recovery of materials in chemical reactor blowdown.

- *Source:*   Chemical reactor blowdown
- *Flow:*   6.8 m³/h (30 gal/min)
- *Composition:*   Product   0.004 wt fraction
                       Reactants   0.001
                       Impurities   0.051
                       Water   0.944

Both the product and by-product become less soluble as the temperature is lowered, so the product begins to crystallize immediately. The by-product doesn't crystallize until a significant amount of water is removed, but when it does crystallize it must be removed from the wastewater stream to prevent buildup in the reaction loop. The water balance is satisfied by the removal of ice from the system, the by-product level is maintained at a satisfactory level in the reactor by the bleed of concentrate from the freeze system, and the product is recycled to the reactor for recovery.

The mass balance is based on laboratory testing of actual wastewaters. The energy and utility consumption is an estimate from test results and a preliminary design. The layout, Fig. 6.10.3, is based on vessel sizing from the preliminary design.

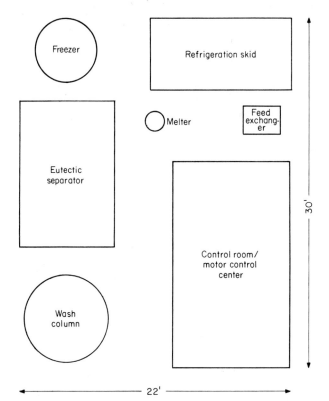

FIG. 6.10.3 Freeze-crystallization applied to reactor blowdown in a chemical plant: preliminary facility design.

Materials of Construction

In general materials of construction for freeze-crystallization plants are less expensive than for other processes. First, corrosion is much reduced at the lower temperatures. The process operates in the absence of oxygen, so oxidation processes are reduced by both reactant availability and temperature.

The process is designed to operate at low pressures. Often a plastic material of construction is feasible, as is lined carbon steel.

6.10.4   ENVIRONMENTAL EFFECTS

Air Emissions

Emissions from this process are virtually nonexistent for two reasons:

1. This is a closed process—none of the process is open to the atmosphere.
2. At the operating temperatures of the process the vapor pressure of the hazardous components in the waste are reduced by one to two orders of magnitude.

Water Effluents

Each component of a wastewater purified by freezing is reduced in concentration by a factor of 1,000 to 10,000:1 (ratio of concentration in the feed to concentration in the melted ice). This is usually sufficient to meet the requirements of even the most stringent discharge permit. If more stringent requirements are imposed, the melted-ice stream can be polished at very little additional cost.

The stability of the process, as well as the slow and even response to shock loading, allows automatic monitoring. The response time of the monitor can be built into the discharge-line residence time, so that if the effluent water goes out of specification it can be recycled back into the feed to the plant.

Solid Discharges

The solids discharged from the process can be directly landfilled in suitable facilities. The advantage of the freeze process is that the solids and the small amount of concentrated liquid waste are much more amenable to conventional destruction and stabilization technologies.

Recovery of materials from wastes is a possibility with crystallization processes. With freezing, conditions can be varied to selectively crystallize specific solutes. For instance, phenols can be crystallized from water at concentrations of less than 5 wt % , and melted to produce a high-quality by-product.

6.10.5   CURRENT RESEARCH

State of the Art

The state of the art in freeze crystallization is typified by the following statements:

- A wide variety of freeze-crystallization processes have been defined and developed.
- Several commercial equipment options are available on the market.
- The components of freeze-crystallization processes are off-the-shelf commercial items.
- Few applications have been studied or tested; there will be no commercial installations on hazardous wastes until at least early 1989.
- New applications currently require pilot testing and full-scale design efforts for new installations.

Areas of Current Development (See Fig. 6.10.4)

Given the state of the art, the current efforts for wide-scale implementation of this process center on such areas of applications engineering as:

- Laboratory and pilot-plant testing for a variety of hazardous wastes
- Developing equipment to separate eutectic crystal mixtures and to achieve greater ice purification
- Reducing capital requirements and improving reliability and operating efficiencies through commercial-scale equipment designs
- Continuing to define process cycles that will further reduce energy consumption and operating costs

FIG. 6.10.4 25,000 gal/d absorption freeze vapor compression (AFVC) pilot plant operated by the U.S. Department of the Interior Office of Water Research and Technology at the Wrightsville Beach Test Facility from 1979 through 1981. (*U.S. Department of Interior*)

6.10.6 ENGINEERING ECONOMICS

Capital Costs

Capital costs for freeze-crystallization are widely variable and highly dependent on the application. Still, the two main variables affecting capital costs are size and materials of construction. The basic cost for a freeze-crystallization plant can be estimated using the following formula:

$$\text{Cost}_2 = \text{Cost}_1 \times [(m^3/h)/2.27]^{.60}$$

Here Cost_1 is based on purified effluent recovered, and varies with the process design. The base cost for a 2.27 m^3/h (10 gal/min) unit varies from \$700,000 for a direct-contact process to \$1.5 million for an indirect process. The cost basis is carbon-steel construction.

Operating and Maintenance Costs

The primary O&M costs are for electricity to operate the process and labor for operations and maintenance. Electric consumption is, for a "standard" application, approximately 16 kWh/m^3 (60 kWh/1,000 gal) for direct-contact processes and 33 kWh/m^3 (125 kWh/1,000 gal) for indirect-freeze processes. The only factor that affects energy consumption to any degree is the freezing-point depression in the concentrate. Correction of energy estimates for this factor is done with the following equation:

$$E = 7 + (8.5) \cdot [10^{(.02241 \cdot -T)}]$$

Here E is the energy consumption in kilowatt hours per cubic meter, and T is the crystallizing temperature in degrees Centigrade.

Labor requirements will reflect individual operating philosophies and conditions (i.e., stand-alone plant or integrated with other production facilities). Costs will reflect local wage and benefit levels. Smaller facilities have commensurately larger costs per unit of throughput if they don't have other activities with which to share operating personnel.

Other costs vary with the size and throughput of the plant, and to some extent with the fractional conversion, the materials of construction, design philosophy, and refrigerant choice. All of these lumped together range from \$1.40 per cubic meter (\$0.005 per gallon) in larger—i.e., 20 m^3/h (100 gal/min)—systems to no more than \$5.50 per cubic meter (\$0.025 per gallon) for smaller systems.

6.10.7 FURTHER READING

There are no texts on freeze-crystallization processes. However, the technology can be understood from the basic chemistry, physics, and unit operations involved. The titles listed in this subsection contain pertinent texts, monographs, symposia proceedings, and technical reports that will give a better understanding of the process.

Phase equilibria and physical and thermodynamic properties can be estimated with conventional models. Unit operations involved include heat transfer, crystallization, mass transfer, refrigeration, hydraulics, and momentum transfer.

A great deal of the work done under Department of Interior sponsorship is available in Research and Development Progress Reports published by the Office of Saline Water and its successor organizations. A bibliography of reports by that agency is titled *Office of Water Research and Technology, Research and Development Progress Reports, 1975.* Patent literature is also a fertile source of information on both the process technology and applications for it. Early work in freeze-crystallization involved chemical-fractionation applications. More recent patents have been split between water-purification and chemical-fractionation applications.

Unit operations and relevant theory and modeling:

Holman, J. P., *Heat Transfer,* 5th ed., McGraw-Hill, New York, 1981.

McCabe, W. L., and J. C. Smith, *Unit Operations of Chemical Engineering,* 3d ed., McGraw-Hill, New York, 1975.

Prausnitz, J. M., *Molecular Thermodynamics of Fluid-Phase Equilibria,* Prentice-Hall, Englewood Cliffs, New Jersey, 1969.

Stoecker, W. F., *Refrigeration and Air Conditioning,* McGraw-Hill, New York, 1958.
Crystallization:

Mullin, J. (ed.), *Industrial Crystallization,* Plenum Press, New York, 1976.

Randolph, A. D., and M. A. Larson, *Theory of Particulate Processes,* Academic Press, New York, 1971.

Youngquist, G. R. (ed.), *Advances in Crystallization from Solutions,* AIChE Symposium Series, vol. 80 no. 240, AIChE, New York, 1984.
Note: Other AIChE Symposium Series publications dealing with crystallization include nos. 110, 193, and 215.

General freeze-crystallization literature:

Barron, T. S., "Use of Freeze Crystallization Systems for Concentration of Liquid Hazardous Wastes," Hazardous Materials Management Conference and Exposition, Anaheim, California, 1986.

Heist, J. A., "Freeze Crystallization," *Chem. Eng.,* **86**:(10) (March 7, 1979), McGraw-Hill, New York.

Heist, J. A., and T. S. Barron, "Freeze Separation Processes—Energy Efficiency by Flexibility," Fifth Industrial Energy Conservation Technology Conference and Exhibition, Houston, (1983), Texas Industrial Commission (Austin), 1983.

Klarman, A. F. and J. A. Heist, "Hazardous Waste Minimization by Reuse and Recycling at a Naval Aviation Depot," 24th Annual Aerospace/Airline Plating and Metal Finishing Forum and Exposition, Phoenix, Ariz., April 4–7, 1988, SAE.

Modifications to and Testing of the Absorption Freezing Vapor Compression Pilot Plant at the Wrightsville Beach Test Facility, prepared by the North Carolina State University Water Process Development Laboratory for the U.S. DI Office of Water Research and Technology, under Contract 14-34-0001-9530, December 1981.

Robinson, R. S., *Developmental Testing of a Secondary Refrigerant Freezing Desalination Pilot Plant at the Wrightsville Beach Test Facility,* NTIS No. PB296-811, U.S. DI Office of Water Research and Technology, Washington, D.C., 1979.

Wrobel, P. J., and T. S. Barron, "Separation Process Economics," Seventh Industrial Energy Conservation Technology Conference and Exhibition, Houston, Texas Industrial Commission, Austin, 1985.

CHAPTER 7

PHYSICAL AND CHEMICAL TREATMENT

SECTION 7.1
FILTRATION AND SEPARATION

Li-Shiang Liang, Ph.D.
Director of Process and Engineering Development
Memtek Corporation
Billerica, Massachusetts

Filtration is a method for separating solid particles from a fluid through the use of a porous medium. The driving force in filtration is a pressure gradient, caused by gravity, centrifugal force, vacuum, or higher than atmospheric pressure. In this chapter, the discussion of filtration is limited to separation of solids from an aqueous medium. See Refs. 1 to 3 for broader reviews of filtration processes.

The relationship of various filtration processes to each other and to other solid-liquid separation processes can be illustrated as shown in Fig. 7.1.1.

7.1.1 APPLICABILITY OF THE TECHNOLOGY

Generic Waste Streams

The applications of filtration to treatment of hazardous waste fall into two categories:

1. Clarification, in which suspended solid particles in concentrations of typically less than 100 ppm are removed from an aqueous stream. This is usually accomplished by depth filtration or cross-flow filtration. The primary aim is to produce a clear aqueous effluent which can either be discharged directly or further processed. The suspended solids are concentrated in a reject stream.

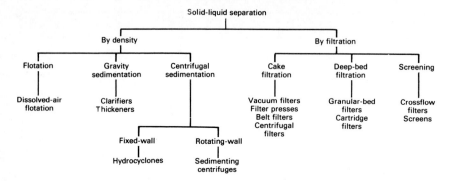

FIG. 7.1.1   Relationship of filtration processes to other solid-liquid separation processes.

2. Dewatering of slurries of typically 1 to 30% solids by weight. The aim is to concentrate the solids into a paste or solid form for disposal or further processing. This is usually accomplished by cake filtration.

Specific Hazardous Wastes

Example 1.   Neutralization of strong acids with lime or limestone, followed by settling and thickening of the resulting precipitated solids as a slurry. The slurry can be dewatered by cake filtration, and the effluent from the settling step can be filtered by depth filtration prior to discharge.

Example 2.   Precipitation of dissolved heavy metals as hydroxides, carbonates, or sulfides, followed by settling and thickening. Filtration applications similar to Example 1.

Example 3.   Treatment of oily wastewaters by breaking of emulsions followed by dissolved-air flotation. The float is dewatered prior to incineration or disposal.

Example 4.   Destruction of toxic organic compounds by biological methods such as activated-sludge treatment followed by aerobic or anaerobic digestion of the resulting sludge. The sludge is dewatered prior to incineration.

Wastes for Which Filtration Will Not Work

Filtration will not remove substances which are dissolved. Waste streams containing soluble toxic organics, for example, cannot be treated directly by filtration.

Filtration processes, except for membrane filtration (see Sec. 6.7), cannot selectively separate chemical components in the same phase.[4]

Filtration processes require the flow of the liquid phase through a porous medium. Sludges or tars which are highly viscous cannot be filtered.

Filtration is not applicable to treatment of solid wastes.

7.1.2 DESCRIPTION OF THE TECHNOLOGY

Chemical and Mechanical Principles

Depth Filtration.   In *depth filtration,* liquid containing typically less than 100 ppm of suspended solid is forced through a porous medium. Various physical and chemical forces cause the solids to be captured and retained inside the medium, while the liquid flows through the bed. Depth filtration is used in granular-media filters, in screen filters with precoat, and in disposable cartridge filters.

Depth filtration is usually a batch process. As the solid particles deposit in the porous medium, the number and sizes of the pores decrease so that the pressure drop through the medium increases. The number of available retention sites also decreases so that at some time an increasing fraction of the suspended particles in the feed is not retained in the medium and is carried through in the effluent. The filtration cycle is stopped when either the pressure drop or the solids concentration in the effluent exceeds established limits. The filter medium must then be either replaced or cleaned.

Depth filtration can be a continuous process if the filter medium is continuously cleaned. One example of a continuous sand filter is discussed later.

Theoretical investigations of depth filtration are reviewed in Refs. 5 to 7. Most take a mechanistic approach in developing models for the transport and attachment of particles to the porous medium and the consequent decrease in the permeability of the medium. The resulting equations are complex and usually must be solved by finite differences on a computer. To date, the theoretical results have not been useful for practical design and operation of depth filters.

In practice, semiempirical equations are used to interpret test data and to design and size granular-bed filters.[6,8,9] One example is Eq. (1).

$$t = \frac{KH}{Cq} \tag{1}$$

Where t = total run time
H = operating head available (beyond the head loss through a clean bed)
C = concentration of suspended solids in the feed
q = flow velocity
k = a constant, to be empirically determined

The total pressure drop through the filtration system must also include the pressure drop for a clean medium and the pressure losses through associated piping, valves, and flow-distribution manifolds.

Cake Filtration.   In *cake filtration* the solid particles are retained at the surface of a porous medium as the liquid flows through the medium. The solids accumulate as a cake of increasing thickness. Solids in the feed slurry are thereafter retained on the surface of the cake; in effect the cake becomes the filter medium. The flow resistance in the cake increases with cake thickness. At the end of the filtration cycle the cake can be dewatered by mechanical compression or by air drying. The cake can also be washed with clean water or solvent prior to or after the dewatering step to remove the original liquid. The cake is then removed from the filter medium, and the medium is washed and re-used.

Most cakes formed from wastewater slurries are compressible, so that the filtration rate is not proportional to the applied pressure difference; increasing the pressure will further compress the cake and increase its flow resistance. In the case of an extremely compressible cake, the cake may form a dense skin at the interface with the filter medium so that the filtration rate becomes insensitive to the applied pressure. Smaller particles in the slurry may not be retained at the surface of the cake but may be deposited within the cake or the filter medium (as in depth filtration). The increase in flow resistance with time is then greater than that due to the cake buildup alone.

Recent theoretical models of the cake-filtration process have focused on the relationship of flow resistance and porosity within the cake to the applied pressure gradient.[10,11] Most of the models rely on the use of numerical integration to solve the resulting equations; they offer valuable insight into the fundamental mechanisms of the process but as yet no widely applicable design methods.

Semiempirical equations are used for interpretation of test data and equipment design. Equation (2), for example, based on the approximations of negligible flow resistance in the filter medium and constant pressure differential, is used for rotary vacuum filters.

$$Y = k \left[\frac{P^{(1-s)}}{\mu \alpha_0} \right]^{1/2} \frac{C^m}{t^n} \tag{2}$$

Where Y = cake yield, in weight of solids (on dry basis) per unit filter area per unit time

P = pressure differential across filter cake

C = solids concentration by weight in feed slurry

μ = liquid viscosity

s = compressibility coefficient, ranging from 0.2 to 0.6 for most industrial-waste sludges

α_0 = a constant which is a measure of the "specific resistance" in the cake

t = filtration time

k,m,n = constants, to be empirically determined

In classical filtration theory, $k = \sqrt{2}$ and $m = n = 0.5$.

Equations based on assumptions of constant flow and of variable flow and pressure can be found in Ref. 10.

Cross-flow Filtration. The most common design for a *cross-flow filter* is a porous tube. The suspension is recirculated at high velocity through the tube by a pump and concentrated as the filtrate flows through the porous medium. The turbulent flow prevents the formation of a filter cake which would otherwise decrease the filtrate flow rate. In a continuous operation, a small fraction of the recirculating stream, called the *reject,* is bled off to maintain constant the solids concentration as fresh feed wastewater is introduced into the recirculation loop. Cross-flow filtration is most effective in clarification of suspensions containing less than 0.5% suspended solids.

Description of Processes

Granular-Bed Filters.   Depth filtration of industrial wastewaters is usually carried out in *granular-media filters*. The filters consist of a tank, with an underdrain covered by a porous grid which supports a bed of granular materials of typically 1 to 2 m (3 to 6 ft) in depth. The bed may be composed of a single material (uniform bed) or it may be composed of stratified layers of different materials (multiple bed). Above the bed is a feed-distribution manifold which feeds the wastewater uniformly to the top of the bed at typically 0.08 to 0.40 $m^3/(m^2 \cdot min)$ [2 to 10 gal/(min $\cdot$ ft^2)]. The top of the tank may be open if gravity is the driving force, or it may be closed for pressurized feed.

During the filtration run, the top surface of the medium can be scoured with high-pressure water jets to prevent retained solids from agglomerating into "mudballs" which will increase the flow resistance. At the end of the filtration run, the bed must be cleaned. The most common method is to agitate the bed with air to dislodge the trapped particles and then backwash with water to flush the solids out of the bed. The total backwash volume will normally be 1 to 5% of the wastewater volume treated per filtration run. Typical air-scouring rates are 0.9 to 2.4 $m^3/(m^2 \cdot min)$ [3 to 8 standard $ft^3/(min \cdot ft^2)$] for 2 to 5 min, and backwash rates are 0.4 to 1.6 $m^3/(m^2 \cdot min)$ [10 to 40 gal/(min $\cdot$ ft^2)] for 3 to 10 min. Backwash techniques are discussed in detail in Refs. 9 and 12.

Backwashing will classify the bed media by size and density. In a multiple bed, the layers of media must have progressively smaller particle sizes in the flow direction; to prevent reverse stratification, the density of each layer must be higher than that above. A common combination is coarse anthracite above fine sand. In the case of a three-medium bed, the bottommost layer is composed of fine garnet particles. Typical particle sizes are 0.8 to 2.2 mm for anthracite, 0.4 to 1.4 mm for sand, and 0.3 mm for garnet.

To achieve continuous operation with granular-media filters operated in a batch mode as described above, more than one filter must be installed in parallel, so that one can be backwashed while the other(s) are operating.

One design of granular-medium filter can be operated in a continuous mode. The feed wastewater flows upward through a sand bed, typically 1 m (40 in) deep. The sand at the bottom of the bed is continuously removed by an air lift, air-scoured and separated from the deposited solids, and then replaced on top of the filter bed. The bed is therefore continuously cleaned.[13]

Rotary Vacuum Belt Filters.   Rotary vacuum filtration is the most widely used method of mechanical sludge dewatering. In the belt filter version, the filter medium is a continuous belt of fabric which winds around a horizontal rotating drum and several rollers (see Fig. 7.1.2). The drum is partially immersed in a shallow tank of the feed slurry. The circumference of the drum is perforated and divided into shallow, tray-shaped sections covered by grids which support the filter belt. A multiport valve connects each section to the vacuum source. Vacuum is applied to each section as it is immersed into the feed slurry. Filtrate flows through the filter medium from the outside of the drum and is collected in the trays and directed through internal piping to a filtrate receiver. Cake formation proceeds as the section is rotated through the slurry. The section emerges from the slurry and the cake is dewatered by the applied vacuum. Air enters the cake as the residual liquid is drawn out to the receiver. The vacuum is then released as the belt is pulled away from the drum and over a roller, where the sharp change in curvature causes the cake to be discharged. The belt is spray-washed before being reimmersed into the slurry tank.

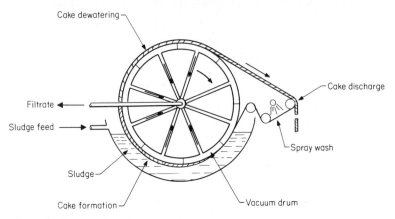

FIG. 7.1.2　Cross section of a rotary vacuum belt filter.

The rotary vacuum belt filter is effective in dewatering sludges which will form a thick and dry enough cake to be discharged by a roller. The advantage of this method of filtration is that it is a continuous process which requires relatively little operator attention. The disadvantages are that the maximum pressure difference is limited by vacuum, that thin, gelatinous sludges cannot be dewatered, and that the power consumption by the vacuum pump is relatively high.

Rotary Vacuum Precoat Filters.　The fabric filter medium in this version of the rotary vacuum filter covers the entire drum face. Before the filtration cycle, the slurry tank is filled with a suspension of a precoat material such as diatomaceous earth. Vacuum is applied so that a layer of up to 0.15 m (6 in) of the precoat is formed on the drum. The precoat suspension is then replaced with the feed slurry. Filtration and dewatering proceed as described earlier except that after the dewatering stage the cake is removed with a scraper knife. The knife is set so that typically 0.08 to 0.8 mm (0.003 to 0.03 in) of the precoat is removed every revolution of the drum along with the cake. The precoat serves several purposes: it acts as the filter medium which will retain the larger particles by cake filtration and the smaller particles by depth filtration, and its presence permits complete cake removal since part of the precoat is also removed. Furthermore, fresh precoat is exposed as the filtration section reenters the slurry tank.

Advantages and disadvantages of the vacuum precoat filter are similar to those of the vacuum belt filter except that, because of the precoat, the former can clarify and dewater sludges which contain "fines" and which therefore form a thin, fragile cake. One disadvantage of the filter is that up to an hour is required for application of the precoat, so that its operation is not truly continuous.

Other Vacuum Filters.　The rotary drum coil filter is similar to the belt filter except that the filter medium consists of two layers of metal coils. It is most suitable for dewatering sludges which are fibrous or can be flocculated.[14] The filtrate quality is poor, typically 500 to 2,000 mg/L of suspended solids.

The vacuum disk filter consists of disk-shaped grids covered on both sides by filter cloth and mounted on a horizontal axis. The disks are rotated through a tank of feed sludge, and cake formation and dewatering proceed as with rotary drum filters. The cakes are discharged by a scraper. The particular advantage of the disk filter is that a large filtration area can be provided in a given space.

Filter Press. Filter presses used in sludge dewatering are mostly of the recessed-plate type. A press consists of a number of plates stacked either horizontally or vertically. A horizontal filter press is shown in Fig. 7.1.3. Each plate has a recessed cavity on each side which is lined with a filter cloth. The surface of the recessed area is corrugated or grooved to allow free drainage of the filtrate through the filter medium to the discharge ports. During a filtration cycle, the plates are pressed together either hydraulically or mechanically and the feed slurry is pumped into the filtration chambers formed by the plates. Cake filtration proceeds on the fabric medium until the entire chamber is filled with the cake. More feed slurry is pumped into the chambers to dewater the cake by displacing the water with sludge solids. Consolidation of the cake occurs both in the direction perpendicular to the filter medium and parallel to the medium toward the edges of the plate. At the end of the dewatering stage, the pressure holding the plates together is released and the plates are separated individually. Ideally the cake in each chamber is dry enough so that it drops freely into a hopper below the press. The filter cloth is then washed and the press is closed for the next cycle. Vertical filter presses are described in Ref. 15.

The feed slurry is usually pressurized by positive-displacement pumps such as progressive-cavity or diaphragm pumps to gauge pressures up to 1,550 kPa (225 lb/in^2). To prevent premature compression of the cake and the resulting decrease in cake permeability, the feed pressure can be increased in stages. During the initial part of the cycle, when the cake thickness is low and the permeability is high, a low feed pressure is used. As the cake thickness increases and the permeability decreases, the pressure is increased. Finally maximum pressure is applied during the dewatering stage to compress the cake. Typical maximum gauge pressure is 690 kPa (100 lb/in^2), although gauge pressure as high as 1,550 kPa (225 lb/in^2) is used.

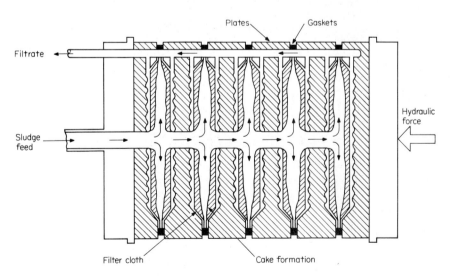

FIG. 7.1.3 Cross section of a filter press.

To further dewater the cake, a movable, impermeable diaphragm can be installed between the filter cloth and the plate. Compressed air is released into the space between the diaphragm and the plate to mechanically squeeze and dewater the cake.

The advantage of the filter press in most waste-sludge applications is that it will produce the driest cake of all dewatering devices. The disadvantages are that it must be operated in a batch mode and that the process is labor-intensive.

Belt Filter Presses.   A *belt filter press* is a continuous sludge-dewatering device which combines cake formation under gravity with cake dewatering by mechanical compression between two moving belts of cloth. Belt filters have been used mostly for dewatering of sludges from biological treatment of wastewaters. Most belt filters include a flocculation chamber where polymer is added to the feed sludge to agglomerate the particles. Proper flocculation of the sludge is important to create solids which will dewater readily under gravity and form a cake with sufficient mechanical strength to be squeezed without being extruded from the belts. The sludge is fed onto the top of the first moving belt (see Fig. 7.1.4), where it is thickened as liquid filters through the belt. The belt then makes a sharp turn over a roller, and the sludge is discharged onto the second moving belt. The two belts then advance together and slowly converge to apply pressure to form a cake and begin the dewatering portion of the process. Sufficient water must be removed in this low-pressure dewatering step to create a sludge cake which is firm enough that it does not squeeze out between the belts. The sludge cake is subjected to progressively higher mechanical compression by running the two belts together under tension over small rollers. The repeated changes in belt curvature shear the cake and promote dewatering. A final compression step using a hydraulically applied set of rollers may be used before the belts diverge and the cake is discharged. The belts are then washed and realigned before returning to the initial step.

The advantages of a belt filter press are that it permits a continuous process with relatively low labor and power requirements. It produces a cake with a solids content intermediate between those from a filter press and a vacuum filter. The disadvantages are that the filtrate quality is poor (200 to 500 ppm from the

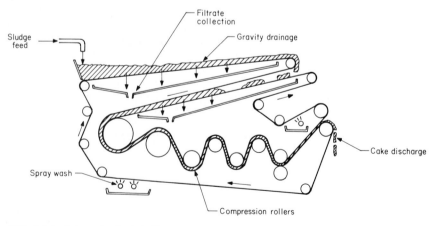

FIG. 7.1.4   Schematic of a belt filter press.

gravity stage and 500 to 2,000 ppm from the high-pressure stage),[16] and that the belt washwater rate can be as high as 70 to 100% of the sludge feed rate. The result is that a combined filtrate and backwash stream with a flow rate of typically 85 to 110% of the feed must be reprocessed for removal of suspended solids.

7.1.3 DESIGN CONSIDERATIONS

Design Constraints

The selection of filtration equipment for any specific application must take into account the following:

1. The flow rate, composition, temperature, and rheological properties of the waste stream to be filtered. Average values and fluctuations should be obtained; the fluctuations may be smoothed out by equalization or further pretreatment.
2. The objective of the filtration; e.g.,
   a. To remove suspended solids from a wastewater stream and produce a clear effluent.
   b. To concentrate a dilute suspension to a slurry form for further dewatering.
   c. To dewater a slurry for either recovery or disposal of the solids.
3. The number of operating hours per day for the filter, and whether holdup volume is available in the event of shutdown for repairs or maintenance.
4. The availability of space, chemicals, filter aids, and supporting utilities. This is particularily important in the case of remedial action on a site.
5. Economic constraints.

Capacities and Capabilities

The capacity of a filtration device depends on the properties of the material to be filtered, the filter medium, the type and size of the equipment, and the operating conditions. Table 7.1.1 gives some guidelines on the capacities and capabilities of typical filters.

With regard to the solids content achievable in dewatering equipment, keep in mind the actual properties of the cake that is required for the recovery or disposal option; for example, some sludges can be rigid and handleable at 20% solids, while others will still flow at 50% solids and may be thixotropic.

The capacities given are for standard equipment supplied by sample vendors. One vendor suggests that for installations operating less than 7 h/d, 5 d/wk, one large machine is more cost-effective than two smaller ones; for installations operated 16 h or more per day, two or more machines are necessary to provide redundancy.[14]

Design Methods

In practice, the design and sizing of filtration devices is still based on previous experience and/or pilot studies. Semiempirical equations are useful in interpreta-

TABLE 7.1.1 Typical Filter Capacities and Capabilities

	Rotary drum vacuum belt filters	Belt filter presses	Filter presses	Granular-bed depth filters
Sizes	1–70 m^2 (10–750 ft^2)	1.0–2.0 m (3.3–6.6ft) beltwidth	0.02–16 m^3 (0.7–565 ft^3)	0.2–10.5m^2 (2–115 ft^2)
Solids in feed, %	2–5	2–8	——	——
Solids in cake, %	15–20	15–25	28–40	
Loading	10–30) kg/($m^2 \cdot$ h) [2–6 lb/($ft^2 \cdot$ h)]	190–270 kg/(m $\cdot$ h) [127–181 lb/(ft $\cdot$ h)]	——	0.12–0.49 m^3/($m^2 \cdot$ min) [3–12 gal/($ft^2 \cdot$ min)]
Cycle time, h	——	——	2–4	——
Influent TSS,* mg/L	——	——	——	20–400
Effluent TSS, mg/L	——	——	——	2–20

*TSS = total suspended solids.

tion and extrapolation of data. Pilot-scale test equipment for several types of filters is described here.

Granular Depth Filters. The same size and packing of the filter media as are to be used in the full-scale device must be used in the model filter,[6] which is usually a cylindrical column. Wall effects can be neglected if the ratio of column diameter to grain size is greater than 50. Sample ports can be located at several locations along the height of the column.

Pilot tests will provide data on pressure drop and effluent quality as functions of feed wastewater characteristics, media types, size and depth, and flow velocity. Backwash and air-scouring techniques can be tested.

Rotary vacuum filters. Bench-scale tests require a test leaf, which is a disk-shaped device which models the surface of a rotary-drum vacuum filter. The surface area of the leaf is 0.01 m^2 (0.1 ft^2). A complete filtration cycle can be simulated with the leaf, and effects of operating parameters (such as filter medium, vacuum level, cake formation, and dewatering times) on filtration rates and cake yield can be investigated using samples of the wastewater sludge.

Pilot-scale tests provide more accurate data which take into account actual operating conditions such as variations in the feed composition. Scaled-down versions of actual equipment are used as pilots. A typical pilot-scale drum filter is 0.9 m in diameter and 0.3 m wide (3 ft by 1 ft).

Filter Presses. Bench-scale and pilot-scale filter presses are available from vendors. Typically a 20-L (5-gal) sample of the sludge is required per bench test. Test parameters include chemical preconditioning of the sludge, filter medium and aids, and operating pressure.

Sizing of a full-scale press is straightforward given data on cycle time and on the solids content and density of the cake.

Materials of Construction

Filtration equipment is available in a wide range of materials. Wetted surfaces in particular, such as the drum in a rotary vacuum belt filter, may be constructed from corrosion-resistant materials such as stainless steel and Fiberglas-reinforced plastics. Polypropylene plates are increasingly being used in filter presses for wastewater applications. The equipment manufacturers should be consulted when selecting materials of construction both compatible with the wastewater or sludge to be processed and cost-effective over the expected lifetime of the equipment.

Filter fabrics are available in numerous combinations of materials, construction, and weave. The selection of a fabric medium is currently more an art than a science, and is usually based on bench-scale and pilot tests and on experience. The selection criteria include

- Compatibility with the waste to be filtered
- Permeability to the liquid
- Ability to retain solids and produce a clear filtrate
- Ease with which the cake can be discharged
- Ease with which the fabric can be washed
- Expected life and cost of replacement

The reader should consult Refs. 2 and 17 for detailed discussions of fabric media.

The design and materials of construction of auxiliary equipment must not be overlooked. Failure of a sludge feed pump or vacuum pump, for example, may shut down the filter.

Environmental effects are important: these include solar radiation, rainfall, wind, dust, and freezing.

Relation to Other Processes

In hazardous-waste applications, filtration is usually only a step in an integrated waste-treatment system. A wastewater-lagoon cleanup system, for example, may consist of the following process steps:

1. Coarse screening.
2. Air or steam stripping to remove organics.
3. Chemical reduction or oxidation, followed by pH adjustment to precipitate dissolved heavy metals.
4. Coagulation and flocculation of the suspended solid particles.
5. Density-driven separation, such as sedimentation or dissolved-air flotation, to concentrate the suspended solids in a sludge form.
6. Filtration of the effluent from the above separation processes to remove remaining suspended solids (granular-media filters, precoat filters, or cross-flow filters), followed by carbon adsorption to remove remaining organics prior to discharge.
7. Dewatering of the sludge from step 5 to a cake dry enough for chemical stabilization and disposal. Addition of flocculating agents and filter aids may be necessary before dewatering.
8. Recycling of the reject from step 6 and the filtrate from step 7 back to step 4.

7.1.4 ENVIRONMENTAL EFFECTS

Air and Water

Air emissions from filtration processes exposed to the atmosphere can occur if there are volatile components in the wastewater or sludges. Volatile organics, for example, may separate from the filtrate in the receiver of a vacuum filter and be discharged to the atmosphere. Exposure of the operator to hazardous gases must be considered in selecting a filtration process.

Most filtration processes will generate a liquid reject stream which must be reprocessed, usually by recycling of the stream to pretreatment steps ahead of the filtration step or by a further and finer filtration step. Examples of these streams are:

Process	Reject stream
Granular-media filter	Backwash
Vacuum belt filter	Filtrate, vacuum-pump seal water, and filter-cloth washwater
Filter press	Filtrate and filter-cloth washwater
Belt filter press	Filtrate and filter-cloth washwater

Solid Residues

The filter media in most filtration processes are re-usable, the exception being the disposable-cartridge-type filters.

In dewatering processes such as those employing vacuum filters and filter presses, the main objective of the process is to generate a solid residue which can be either reprocessed, incinerated, or disposed of.

Recycling Possibilities

The filtrate from granular-media filters, vacuum precoat filters, and cross-flow filters may be clean enough to be recycled as service water in the waste-treatment plant, for example, or as spray water for dust control or recharge water for a contaminated aquifer in site-remediation programs.

Metals in filter cakes from dewatering processes may be recovered economically, depending on the composition of the cake. Recovery methods are based on acid digestion, electrochemical purification, electromagnetic separation, and other processes.

7.1.5 CASE STUDIES

Manufacturer of Organic Chemicals, Plastics, and Fibers[18]

Belt filter presses are used to dewater aerobic digested sludge from an activated-sludge process. Average waste-sludge production is 36,600 kg/d (72,000 lb/d) dry basis at 1.6 to 2.6% solids. Eight belt-press units with 1.5-m (5-ft) wide woven polyester fiber belts are installed.

Sludge is conditioned with 75 to 225 kg of bentonite and 7.5 to 22.5 kg of polymer per dry metric ton of sludge (150 to 450 lb and 15 to 45 lb per dry ton, respectively). The feed flow rate per unit is up to 0.41 m³/min (110 gal/min), with sludge loading rate of up to 12.7 dry t/d (14 dry tons/d), or 353 kg/(m · h) [236 lb/(ft · h)]. Final solids content of the cake is 13 to 17%, and overall solids capture is 96 to 99%.

Metal-Forming and Plating Operations[19]

Gravity sand filters are used to polish the effluent from settlers prior to discharge to stream. Wastewater-treatment steps before filtration include chrome reduction, lime precipitation of heavy metals, and sedimentation.

Eight filter cells are installed, with seven on line at maximum flow of 20.8 m³/min (5,500 gal/min). The filter medium is anthracite, 1.2 to 1.4 mm effective size, 1.2 m (4 ft) deep. The design loading is 0.2 m³/(m² · min) [5 gal/(min · ft²)]. After a run length of 12 h, the bed is cleaned with a combination of air scouring and water backwash. Backwash rate is 0.4 m³/(m² · min) [10 gal/(min · ft²)] and air rate is 0.9 m³/(m² · min) [3 standard ft³/(min · ft²)].

Data for a year are shown in Table 7.1.2.

Arsenic-Sludge Dewatering[20]

Leachate from an abandoned gold smelter is collected and treated before discharge to a river. Ferric chloride and hydrated lime are added to precipitate the 300 to 600 mg/L of arsenic as a calcium ferrocyanate complex. Polymer is added to flocculate the precipitate as a gelatinous hydrous sludge, which settles to 5% total solids in a clarifier.

Bench-scale tests with mechanical dewatering equipment showed that only the filter press could dewater the sludge to 20 to 25% solids or higher. Subsequent pilot-scale tests with a 300-mm filter press showed that the addition of cement-kiln dust improves the cake solids and the filtration time. Table 7.1.3 shows typical results.

TABLE 7.1.2 Annual Data for Filtration of Effluent from Metal-Forming and Plating Operations

	Current limits		Concentrations, mg/L					
			Feed			Effluent		
	Max.	Ave.	Min.	Max.	Ave.	Min.	Max.	Ave.
Al	1.0	0.1	1.4	0.5	0.1	0.7	0.3	
Cu	0.6	0.4	0.03	0.21	0.08	0.02	0.11	0.05
Pb	0.5	0.09	1.14	0.32	0.01	0.14	0.08	
Zn	0.7	0.5	0.33	2.37	0.89	0.03	0.38	0.14
F	15	3.2	17.7	10.6	4.6	15.4	10.4	
TSS*	20	5	23	11	3	14	9	

*TSS = total suspended solids.

TABLE 7.1.3   Typical Results of Filter-Press Dewatering with Lime Dust Added

Solids in feed, %	Kiln dust, %	Combined feed solids, %	Cake solids, %	Run time, min
3.7	5	9.5	32.0	140
7.7	5	10.3	32.5	140
4.5	10	12.7	37.5	70
5.5	20	12.6	36.3	35

The estimated capital cost for dewatering 400 t (dry basis) of 5% sludge was $780,000 (1984), including building and engineering, and the estimated annual cost was $38,000.

Dewatering of Textile-Waste Activated Sludge[21]

Pilot-scale tests showed that only the rotary vacuum precoat filter and the pressure filter were able to produce a sludge cake with solids content above 20%. The rotary filter was selected on the basis of "lower estimated capital cost, ease of operation and unusual ability to produce a 'dry-handleable' sludge cake over a wide range of operating conditions."

The effects of operating variables (such as drum speed and chemical-conditioning dosages) on filtration rate and cake yield were investigated on bench, pilot, and plant scale. The best operating conditions for a full-scale, 2.43-m diameter by 4.88-m long (8 ft by 16 ft) rotary vacuum precoat filter were as follows:

Drum speed	2 min/r
Vacuum	635 mm (25 in) Hg
Precoat material	Celatom FW-18
Knife advance	0.13 mm (0.005 in)/r
Precoat consumption	312 kg/t (625 lb/ton)*
Ferric chloride usage	5 kg/t (10 lb/ton)*
Polymer usage	0.5 kg/t (1.0 lb/ton)*
Filter productivity	180 kg/h (400 lb/h)
Cake solids	20 to 22% by wt
Filtrate suspended solids	10 mg/L

Capital cost was $506,000, of which $145,000 was for equipment and installation, and operating cost was $80.56 per metric ton (dry basis), of which $49.06 was for the precoat.

7.1.6   CURRENT RESEARCH AND ADVANCES IN THE STATE OF THE ART

Research

Depth Filtration.   Theoretical research to date has focused on the mechanisms of particle deposition in a granular, porous medium and the resulting decrease in liquid permeability, from both a mechanistic and a statistical approach.[6,7] The results have been useful in the interpretation of data and formulation of semiempirical equations used in design and operation of depth filters.

Cake Filtration.   One area of current research covers several physical phenomena which occur simultaneously during cake filtration (in addition to cake formation); for example, convection and deposition of fine particles in the cake, as in depth filtration, and compression and consolidation of the solid matrix in the cake during filtration[10,11] and during mechanical compression and dewatering.[22] Classical filtration theory does not address these phenomena and is therefore of limited use in correlation of data on dewatering of most industrial sludges.

Another focus of research is on methods for improving the performance of current sludge-dewatering devices; for example, cake formation during filtration can be limited by using mechanical means to minimize the flow resistance through the cake.[23,24,25] The filter then performs as a sludge-thickening device. Cake formation and dewatering must be carried out separately.

State of the Art

The basic designs of granular deep-bed filters, rotary vacuum filters, and filter presses have not changed in more than 50 years. Improvements are constantly being proposed, as evidenced by published papers and patents.

The major recent advances are in the development of continuous-filtration devices such as moving-bed sand filters, belt filter presses, rotary filter presses (cake thickness limited by rotating scrapers) and cross-flow filters. These devices offer the advantages of relative compactness and low labor costs.

Recommended areas for future design improvements include:

- *Filtrate quality and consistency:*   With effluent discharge limits less than 1 ppm for most toxic substances, the solids-removal efficiency of liquid filters must be improved and made more insensitive to process upsets. Membrane filtration devices (see Sec. 6.7) have become cost-effective in the current environment of tightening effluent limits and escalating costs for hauling and disposal of waste liquids.

- *Solids content in dewatered sludge cakes:*   The incentives for improvements are again the increasing cost of hauling and disposal.

- *Productivity:*   With compressible waste-sludge cakes, methods other than increasing the pressure are necessary to increase the sludge dewatering rates.

- *Flexibility of design and operating parameters to allow treatment of several types of waste streams:*   This is particularily important for filtration systems installed in centralized waste-treatment and disposal facilities and for mobile systems to be used in site-remediation programs.

7.1.7   LIST OF EQUIPMENT VENDORS

The following is only a partial list of vendors of filtration equipment for hazardous-waste applications. More extensive lists with addresses and telephone numbers can be found in Refs. 26 and 27.

Granular-Bed Filters	*Belt Filters*
Parkson	EIMCO
Serck Baker	Komline-Sanderson
Ecodyne	Clow Corp.
Duriron	Parkson
EIMCO	Carter Ralph

Rotary Belt Filters	*Rotary Precoat Filters*
Komline-Sanderson	Komline-Sanderson
Dorr-Oliver	Serck Baker
Westech	Westech
EIMCO	EIMCO

Filter Presses	*Cross-Flow Filters*
EIMCO	Mott Corp.
Hoesch Industries	Memtek Corp.
JWI Inc.	
Bethlehem Corp.	
Netzsch, Inc.	

7.1.8 REFERENCES

1. D. B. Purchas, *Solid-Liquid Separation Technology,* Uplands Press, Croydon, England, 1985.

2. R. H. Perry and D. Green, *Perry's Chemical Engineers' Handbook,* 6th ed., McGraw-Hill, New York, 1984.

3. L. Ricci (ed.), *Separation Techniques 2: Gas/Liquid/Solid Systems, Chem. Eng.,* McGraw-Hill, New York, 1980.

4. E. H. Dohnert, "Filtration," in D. J. DeRenzo, (ed.), *Unit Operations for Treatment of Hazardous Industrial Wastes,* Noyes Data Corp., Park Ridge, New Jersey, 1978.

5. J. P. Herzig et al., "Flow of Suspensions Through Porous Media," *Flow Through Porous Media,* Am. Chem. Soc. Pub., Washington D.C., 1970, pp. 130–157.

6. K. J. Ives, "Deep Bed Filtration: Theory and Practice," *Filtn. and Sepn.,* **17**:(2)157–166 (March–April 1980).

7. C. Tien, "Dynamics of Deep Bed Filtration," *Proc. 3rd World Filtration Conf.,* 1982, Uplands Press, Croydon, England, pp. 186–195.

8. T. D. Lekkas, "A Modified Filterability Index for Granular Bed Water Filters," *Filtn. and Sepn.,* **18**:(3)214–216 (May–June, 1981).

9. A. V. Metzer, "The Design and Sizing of Granular Media Filtration Systems for Industrial Waste Treatment," *Proc. 33rd Ind. Waste Conf.,* 1978, Purdue University, West Lafayette, Indiana, pp. 32–36.

10. F. M. Tiller et al., "Filtration Theory in Its Historical Perspective—A Revised Approach with Surprises," *Proc. 2nd World Filtration Conf.,* 1979, Uplands Press, Croydon, England, pp. 1–13.

11. M. S. Willis et al., "A Rigorous Filtration Theory and a New Characterization of Filter Cakes," *Proc. 2nd World Filtration Conf.,* 1979, Uplands Press, Croydon, England.

12. J. E. Gay, "Removing the Limits on Deep-Bed Filtration Through New Backwashing Techniques," *Proc. 33rd Ind. Waste Conf.,* 1978, Purdue University, West Lafayette, Indiana, pp. 776–786.

13. R. M. Shimokubo, "Continuous Sand Filtration: An Innovative Approach to Rapid Sand Filtration," *Filtn. and Sepn.,* **20**:(5)376–380 (September–October 1983).

14. *Komline-Sanderson Coil Filter Sludge Dewatering Practice,* Bulletin 122-7508, Komline-Sanderson Co., Peapack, New Jersey.

15. C. A. Jahreis, "Filtration: Advances and Guideline. 1. Recent Developments in Pressure Filtration," *Chem. Eng.,* **83**:(4)80–83 (February 16, 1976).

16. K. C. Goel et al., "Selection of Belt Presses for Sludge Dewatering," in C. P. Huang (ed.), *Ind. Waste, Proc. 13th Mid-Atlantic Conf.,* 1981, Ann Arbor Science, Ann Arbor, Michigan, pp. 341–349.

17. D. B. Purchas, "Art, Science and Filter Media," *Proc. 2nd World Filtration Conf.,* 1979, Uplands Press, Croydon, England, pp. 479–494.

18. R. A. Poduska, and R. C. Stroupe, "Belt-Filter Press Dewatering Studies, Implementation and Operation at the Tennessee Eastman Company Industrial Activated Sludge Wastewater Treatment System," *Proc. 35th Ind. Waste Conf.,* 1980, Purdue University, West Lafayette, Indiana, pp. 437–455.

19. R. D. Knapp and E. G. Paulson, "Gravity Filtration for Reduction of Suspended Metals in a Lime Precipitation Treatment System," *Proc. 37th Ind. Waste Conf.,* 1982, Purdue University, West Lafayette, Indiana, pp. 95–104.

20. R. W. Wilson et al., "Arsenic Sludge Dewatering," *Proc. 39th Ind. Waste Conf.,* 1984, Purdue University, West Lafayette, Indiana, pp. 343–351.

21. R. E. Komoski et al., "Precoat Filtration of Waste Activated Sludge," *Proc. 32nd Ind. Waste Conf.,* 1977, Purdue University, West Lafayette, Indiana, pp. 603–610.

22. M. Shirato et al., "Expression Theory and Its Practical Utilization," *Proc. 2nd World Filtration Conf.,* 1979, Uplands Press, Croydon, England, pp. 280–287.

23. F. M. Tiller and J. R. Crump, "How to Increase Filtration Rates in Continuous Filters," *Chem. Eng.,* **84**:(12)183–187 (June 6, 1977).

24. T. Toda, "Recent Advances in the Application of the Rotary Filter Press," *Filtn. and Sepn.,* **18**:(2)118–122 (March–April 1981).

25. A. Bagdasarian et al., "Bench Scale Filter for Studying Thin-Cake Filtration," *Filtn. and Sepn.,* **20**:(1)32–37 (January–February 1983).

26. *Chem. Eng. Equip. Buyers' Guide Issue, 1987,* supplement to *Chem. Eng.,* **93**:(15) (August 4, 1986), McGraw-Hill, New York.

27. *Poll. Eng.,* **18**:(10)65–256 (October 1986).

SECTION 7.2
CHEMICAL PRECIPITATION

Neville K. Chung
Metcalf & Eddy, Inc.
Wakefield, Massachusetts

Chemical precipitation is a process by which a soluble substance is converted to an insoluble form either by a chemical reaction or by changes in the composition of the solvent to diminish the solubility of the substance in it. The precipitated solids can then be removed by settling and/or filtration. Precipitation is commonly used to reduce the hardness of water by removing calcium and magnesium. In the treatment of hazardous waste, the process has wide applicability to the removal of toxic metals from aqueous wastes.

7.2.1   APPLICABILITY TO HAZARDOUS WASTES

Chemical precipitation is applicable to the treatment of aqueous hazardous wastes containing toxic constituents that may be converted to an insoluble form. This includes wastes containing the metals arsenic, barium, cadmium, chromium, copper, lead, mercury, nickel, selenium, silver, thallium, and zinc.

Specific hazardous wastes identified under the Resource Conservation and Recovery Act (RCRA) to which precipitation is applicable are listed in Table 7.2.1.[1]

Other aqueous wastes identified under RCRA that commonly contain metals that are removable by precipitation are corrosive wastes (D002) and spent pickle liquor from steel-finishing operations in the iron and steel industry (K062). The "California List" of metal-containing wastes that were specified for prohibition from land disposal by the 1984 RCRA Amendments, i.e., liquid wastes containing arsenic, cadmium, hexavalent chromium, lead, nickel, selenium, and thallium at specified concentrations, are treatable by chemical precipitation.

Major industries that are sources of metal-containing wastes are metal plating and polishing, steel and nonferrous metals, inorganic pigments, mining, and the

TABLE 7.2.1   RCRA-Listed Wastes Containing Metals

EPA hazardous-waste no.	Metal contaminant
D004	Arsenic
D005	Barium
D006	Cadmium
D007	Chromium
D008	Lead
D009	Mercury
D010	Selenium
D011	Silver

electronics industry. Hazardous wastes containing metals are also generated from cleanup of uncontrolled hazardous-waste sites, e.g., as leachate or contaminated groundwater.

7.2.2   PROCESS DESCRIPTION

The chemical precipitation process for heavy-metal removal is implemented typically as illustrated in Fig. 7.2.1. A chemical precipitant is added to the metal-containing aqueous waste in a stirred reaction vessel. The dissolved metals are converted to an insoluble form by a chemical reaction between the soluble metal compounds and the precipitant. The resultant suspended solids are separated out by settling in a clarifier. Flocculation, with or without a chemical coagulant or settling aid, may be used to enhance the removal of suspended solids.

Several different chemical precipitants have been shown to be effective in removing heavy metals from aqueous wastes. Hydroxide precipitation using lime as the precipitant is by far the most widely used method. Most metals can also be

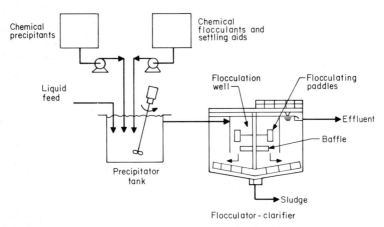

FIG. 7.2.1   Chemical precipitation.

precipitated as sulfides, and certain metals as carbonates. Sodium borohydride, a reducing agent, can reduce and precipitate metals as elemental metal. Descriptions of these precipitation processes are given in this subsection.

Hydroxide Precipitation

Hydroxide precipitation involves the use of calcium hydroxide (lime) or sodium hydroxide (caustic) as the precipitant to remove metals as insoluble metal hydroxides. The reaction is illustrated by the following equation for precipitation of a divalent metal using lime:

$$M^{++} + Ca(OH)_2 \rightarrow M(OH)_2 + Ca^{++} \tag{1}$$

The effluent concentration levels attainable by hydroxide precipitation are dependent on the metals present; the precipitant used; the reaction conditions, especially pH; and the presence of other materials which may inhibit precipitation. Effluent metal concentrations of less than 1.0 mg/L, and sometimes less than 0.1 mg/L, approaching theoretical solubilities, are achievable.

The theoretical solubilities of several metal hydroxides are shown in Fig. 7.2.2. As indicated by the solubility curves, the metal hydroxides are amphoteric; i.e., they are increasingly soluble at both low and high pH, and the point of minimum solubility (optimum pH for precipitation) occurs at a different pH value for every metal. At a pH at which the solubility of one metal hydroxide may be minimized the solubility of another may be relatively high. In most cases a pH between 9 and 11, selected on the basis of jar tests or operating experience with the waste, produces acceptable effluent quality. For a waste containing several metals, however, more than one precipitation stage with different pH control points may be required to remove all metals of concern to desired levels. Otherwise an alternative precipitant may be required.

The precipitation of chromium as chromic hydroxide requires that the chromium be present in the trivalent form; hexavalent chromium cannot be removed directly by hydroxide precipitation. Pretreatment to reduce hexavalent chromium to the trivalent state is accomplished by lowering the pH to between 2 and 3 and adding a reducing agent such as sulfur dioxide, sodium bisulfite, sodium metabisulfite, or ferrous sulfate. The process is illustrated by the following reaction in which sulfur dioxide is the reducing agent:

$$3SO_2 + 3H_2O \rightarrow 3H_2SO_3 \tag{2}$$

$$3H_2SO_3 + H_2Cr_2O_7 \rightarrow Cr_2(SO_4)_3 + 4H_2O \tag{3}$$

The pretreated chrome waste can then be treated, either alone or mixed with other metal-bearing wastes, by lime or caustic to raise the pH to above 8 to precipitate chromic and other metal hydroxides.

Sulfide Precipitation

Sulfide precipitation has several potential advantages as an alternative to hydroxide precipitation. The solubilities of metal sulfides, shown in Fig. 7.2.3., are lower than those of corresponding metal hydroxides;[2] the metal sulfides are not

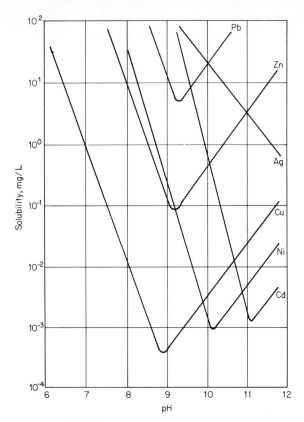

FIG. 7.2.2   Solubilities of metal hydroxides as a function of pH.[2]

amphoteric and hexavalent chromium can be precipitated in the process without a separate reduction step. The process involves the combination of a heavy-metal ion with a sulfide radical, as illustrated by the following reaction of a divalent metal with ferrous sulfide:

$$M^{++} + FeS \rightarrow MS + Fe^{++} \tag{4}$$

The removal of hexavalent chromium is accomplished through reduction and precipitation of the chromium as chromic hydroxide in a single step as follows:

$$Cr_2O_7^= + 2FeS + 7H_2O \rightarrow 2Fe(OH)_3 + 2Cr(OH)_3 + 2S + 2OH^- \tag{5}$$

Several sources of sulfides have been used including sodium sulfide (Na_2S) or sodium hydrosulfide (NaHS), which are soluble, and ferrous sulfide (FeS), which is only slightly soluble.

A disadvantage of sulfide precipitation is the potential of generating toxic hydrogen sulfide gas. Care must be taken to maintain the pH above 8 to prevent evolution of hydrogen sulfide. Even at alkaline pH, hydrogen sulfide may be emitted through rapid hydrolysis where a soluble sulfide salt is added to water. A

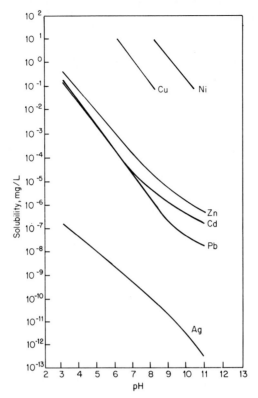

FIG. 7.2.3 Solubilities of metal sulfides as a function of pH.[2]

second disadvantage of sulfide precipitation is the potential for release of excessive sulfide in the effluent, which may then require posttreatment. An excess of sulfide is needed to drive the reaction to completion.

To minimize the potential hydrogen sulfide hazard and excess sulfide release, the addition of soluble sulfide salt should be done in a closed system with adequate ventilation, and sulfide dosage carefully controlled, e.g., by the use of a sulfide-specific ion electrode to measure free sulfide. An excess of 0.5 mg/L free sulfide can be successfully maintained by this method.

Ferrous sulfide has been found to be practicable as a sulfide precipitant because it is soluble enough to precipitate other heavy metals yet its own solubility is low enough to maintain a very low free sulfide concentration, virtually eliminating the problem of hydrogen sulfide evolution in the reaction tank, and assuring a low concentration of soluble sulfide in the effluent. Because ferrous sulfide is reactive and therefore unstable, however, it is necessary to generate ferrous sulfide on site. This is accomplished by reacting ferrous sulfate, a soluble sulfide such as sodium hydrosulfide, and lime. There is therefore still a potential for hydrogen sulfide generation.

The use of ferrous sulfide, along with other features, is incorporated in the proprietary Sulfex process for sulfide precipitation.

Carbonate Precipitation

For certain metals, i.e., cadmium and lead, carbonate precipitation may produce effluent metal concentrations comparable with those achievable by hydroxide precipitation, with the benefits of lower operating pH and denser, more filterable sludge. A pH of 10 or greater is required for effective precipitation of cadmium and lead hydroxide, but these metals can be precipitated as carbonates at pH of 7.5 to 8.5.

The precipitation of metals by sodium carbonate (soda ash) is accomplished as illustrated by the following example, where M represents a divalent metal:

$$Na_2CO_3 + M^{++} \rightarrow MCO_3 + 2Na^+ \tag{6}$$

The process is not effective for all metals. Tests conducted by Patterson et al.[3] indicated no improvement in effluent quality, operating pH, or sludge characteristics over hydroxide precipitation for zinc and nickel; and in fact the kinetics of the process appear to be such that zinc and nickel carbonates will not form within typical treatment-plant detention times.

Sodium Borohydride Precipitation

Sodium borohydride is a reducing agent that can be used to precipitate metals from solution as the insoluble elemental metal. The pH-dependent reaction for a divalent metal is illustrated by the following simplified equations:

$$4M^{++} + NaBH_4 + 2H_2O \rightarrow NaBO_2 + 4M + 8H^+ \tag{7}$$

$$4M^{++} + NaBH_4 + 8OH^- \rightarrow NaBO_2 + 4M + 6H_2O \tag{8}$$

The process is usually carried out in the pH range 8 to 11 to assure efficient utilization of borohydride. At pH below 8, borohydride consumption increases as a result of hydrolysis of the borohydride, while at pH greater than 11 the rate of reaction decreases. The optimum pH is determined by testing to balance borohydride usage against reaction time and effluent quality.

Sodium borohydride may be especially advantageous for recovery of metals from waste solutions, and is effective for removing lead, mercury, nickel, copper, cadmium, and precious metals such as gold, silver, and platinum. Sludge volume reduction of 50% or more has been reported for the sodium borohydride process over lime precipitation. Sodium borohydride is available as a free-flowing active powder or as a stabilized solution of sodium borohydride in caustic soda. The caustic solution is generally used, because it is easy to handle with standard chemical metering and storage equipment.

Cementation

Cementation for the purposes of this section is the displacement and precipitation of a metal from solution by a metal higher in the electromotive series. Strictly speaking, it is an electrochemical rather than purely chemical precipitation process. It is included here as a viable alternative to other precipitation processes. The process can be used to remove and recover reducible metallic ions from aqueous waste streams. Examples are the precipitation of silver from photographic-processing solutions, the precipitation of copper from printed-circuit etching solutions or brass-mill effluents, and the reduction of hexavalent chromium in waste streams for subsequent precipitation by lime. The use of cementation offers significant economic advantages over other methods in situations in which a valuable metal such as gold, silver, or copper can be recovered using scrap reductant material.

Other Precipitation Processes

There are other precipitation processes that may be considered for treating wastes containing specific metallic compounds, especially where the wastes contain a single metal, where metal recovery is desired, or both. Trivalent cations such as iron, aluminum, and chromium can be selectively removed from solutions containing divalent and monovalent cations by phosphate precipitation. This may be a potentially effective method of separating and recovering a trivalent metal such as chromium from a mixed waste solution.

Barium can be precipitated as barium sulfate, a very stable, insoluble material that is nonhazardous. Selenium can be recovered from waste solutions by reaction with sulfur dioxide to precipitate elemental selenium; selenides may be removed by precipitation as the insoluble iron selenide salt; and silver may be precipitated from solution as insoluble silver chloride.

7.2.3 DESIGN CONSIDERATIONS

The following design considerations and alternatives related to the use of precipitation for treatment of hazardous wastes are associated either with the precipitation processes themselves or with the nature of the hazardous metal-bearing wastes to be treated. In many cases hazardous metal-bearing wastes are relatively concentrated solutions, and contain mixtures of several constituents at high concentrations. Examples are spent plating, cleaning, and pickling baths. Treatment of these wastes by precipitation often requires special design and operating procedures.

Waste Segregation and Pretreatment

In designing a precipitation system for wastes which originate from a variety of different sources, more cost-effective treatment may be achieved if certain wastes are segregated. Many metal-containing wastes that are treatable by precipitation also contain cyanide that must be removed not only to meet disposal requirements but also because it acts as a complexing agent which inhibits precipitation. Wastes which contain hexavalent chromium cannot be treated directly by hydroxide precipitation. The chromium must first be reduced to the trivalent state.

Cyanide can be pretreated for destruction to carbon dioxide and nitrogen by alkaline chlorination. Hexavalent chromium can be reduced to trivalent chromium by a reducing agent such as sulfur dioxide at low pH. Although these processes may be applied to a combined waste stream, segregation of the cyanide and hexavalent chromium wastes will allow pretreatment of smaller waste streams. This will require smaller reaction tanks and chemical feed equipment, and reduced chemical usage for pH adjustment as well as for chlorine and reducing agent since there will be less material that exerts demand for these chemicals.

Jar Testing

The highly variable nature of hazardous wastes and the substantial effect that many common waste constituents may have on precipitation processes make testing a necessity for the proper selection of process design parameters for treatment of hazardous wastes by precipitation. Bench-scale "jar tests" on waste samples can be run simply and quickly to determine precipitant dosages, optimum pH, settleability of precipitates, settling-aid requirements, and residual metal concentrations achievable. Jar testing may also be used to compare alternative precipitation processes. In the batch treatment of concentrated wastes, jar tests are commonly conducted to custom-design the treatment procedure for each different batch of waste.

Lime vs. Caustic

Either lime or caustic can be used as the source of hydroxide ions for precipitation of metal hydroxides. There are several important differences which must be considered by the design engineer in selecting a system for a particular application. Caustic is more expensive than lime. However, it is easier to handle and feed, requiring relatively simple storage, pumping, and metering equipment. Lime must be slaked (if purchased as quicklime), slurried, and stored in an agitated tank. To prevent plugging of lines and valves, lime slurry is usually fed from a loop through which the slurry is continuously pumped from the slurry tank to the point of application and back to the tank. A short feed line and a pinch valve are used to minimize plugging. Sometimes rubber hoses are used so that caking in the lines can be more easily cleaned out. Because of these requirements, the cost of chemical feed systems for lime slurry can be substantially greater than for caustic.

Offsetting the higher feed-system cost for lime, the sizing and cost of downstream solids-separation and dewatering equipment can be greater for caustic precipitation than for lime precipitation. Precipitates formed by caustic precipitation usually do not settle and dewater as well as sludges from lime precipitation systems. Consequently, larger clarifiers and dewatering equipment may be required to achieve the same effluent quality and dewatering results achieved with lime. If the waste has a high sulfate content, such as from sulfuric acid, precipitation of calcium sulfate will increase the sludge volume from lime precipitation. On the other hand, this material will also enhance the dewaterability of the sludge. Because of these considerations, the choice between lime and caustic for hydroxide precipitation should not be made on the basis of chemical and feed-system costs only, especially for larger systems. Tests should be conducted to assess differences in sludge settleability and dewaterability and the effect on total system costs. The advantages and disadvantages of lime and caustic precipitation are summarized in Table 7.2.2.

TABLE 7.2.2   Advantages and Disadvantages of Lime and Caustic Precipitation

	Precipitant	
	Lime	Caustic
Advantages	Low chemical cost More settleable and filterable precipitate	More reactive Easy to handle
Disadvantages	More difficult to handle than caustic Higher feed-equipment cost Higher maintenance cost More sludge if high sulfates are present	High chemical cost Less settleable and filterable precipitate Higher clarification and dewatering cost Higher dissolved solids in effluent

Heat of Reaction

Many metal-containing hazardous wastes which are treatable by precipitation are relatively concentrated acid solutions. Treatment of these wastes by precipitation usually requires neutralization to a pH greater than 8. The heat of the exothermic neutralization and precipitation reaction can result in substantial temperature rise in the reaction tank, and may even cause localized splashing and emission of volatile materials at the point of addition of the reactant. It is important that this factor be adequately considered in developing design and operating procedures for precipitation systems in order to ensure safe operation and protection of equipment. The heat of reaction can be dealt with in one or more of the following ways:

1. Dilution of concentrated wastes, perhaps by bleeding into dilute metal-containing wastes.
2. Control of neutralization rate, i.e., reagent feed rate, to avoid excessive temperature rise in the reaction tank. Maximum operating temperature for common plastic tank materials is about 150°F (65°C).
3. Addition of lime slurry (and dilution water if necessary) to the reaction tank prior to adding acid waste. This coats the tank and protects it from strong acid, and allows dissipation of the heat of reaction throughout the mass of slurry and dilution water in the tank.
4. Cooling of the reaction tank externally or by aeration.
5. Adequate ventilation of emissions.
6. Use of a surface impoundment with large surface area for neutralization.

Precipitation of Complexed Metals

There are many chemical compounds which act as complexing agents, inhibiting or preventing the use of conventional precipitation methods for waste treatment. Among the more common complexing agents encountered in metal bearing wastes are ammonia, cyanide, and ethylenediaminetetraacetic acid (EDTA).

Pretreatment of ammonia by air or steam stripping, and cyanide by alkaline chlorination, can be employed to remove or destroy these materials prior to hydroxide precipitation; EDTA can be precipitated as the free acid at low pH.

The treatment of complexed metal-containing wastes by chemical reduction and precipitation has been reported for wastes containing copper, tin, and lead. The pH of the waste stream is lowered to break the various metal complexes, followed by addition of a reducing agent to reduce the metals to their lowest oxidation state, which permits precipitation of the metals. Lime is then added to precipitate the metals out of solution. Another method that has been investigated and reported to be effective for certain copper and nickel complexes is precipitation at high pH (11.6 to 12.5) with lime. The drastic increase in pH is believed to prompt a shift in the complex dissociation equilibrium to produce uncomplexed metal ions which can then be precipitated.[2]

7.2.4 REFERENCES

1. *Resource Conservation and Recovery Act of 1976*, Public Law 94-580, *Environmental Reporter*, Bureau of National Affairs, Inc., Washington, D.C., **71**:3101 (1984).
2. *Development Document for Effluent Limitations Guidelines and Standards for the Metal Finishing Point Source Category*, EPA 440/1-83/091, June 1983.
3. J. W. Patterson et al., "Carbonate Precipitation for Heavy Metals Pollutants," *J. Water Pollut. Control Fed.*, December 1977.

7.2.5 FURTHER READING

Bowers, A. R., et al., "Predicting the Performance of a Lime Neutralization/Precipitation Process for the Treatment of Some Heavy Metal-Laden-Industrial Wastewaters," *Proc. 13th Mid-Atlantic Ind. Waste Conf.*, 1981, p. 51.

Brantner, K. A., and E. J. Cichon, "Heavy Metals Removal: Comparison of Alternate Precipitation Processes," *Proc. 13th Mid-Atlantic Ind. Waste Conf.*, 1981, p. 43.

Christensen, E. R., and J. T. Delwiche, "Removal of Heavy Metals from Electroplating Rinsewaters by Precipitation Flocculation, and Ultrafiltration." *Water Research (G.B.)*, **16**:729 (1982).

Cook, M. H. and J. A. Lauder, "Sodium Borohydride Controls Heavy Metal Discharge," *Pollut. Eng.*, **13**:36 (1981).

Croy, L. K., and W. R. Knocke, "Comparative Metals Precipitation Techniques," *Proc. Amer. Soc. Civil Engr, Environ. Eng. Spec. Conf.*, 1980, p. 534.

Korchin, S. R., "Performance Report on a New Heavy Metals Removal System Via Sulfide Precipitation," *Proc. 26th Ontario Ind. Waste Conf.*, Ontario Water Resources Comm., Toronto, Ontario, Chem. Abs., 168567p, 1980, p. 92.

McAnally, S., et al., "Nickel Removal from a Synthetic Nickel-Plating Wastewater Using Sulfide and Carbonate for Precipitation and Coprecipitation," *Sep. Sci. Technol.*, **19**:191 (1984).

Peters, R. W., et al., "Crystal Size Distribution of Sulfide Precipitation of Heavy Metals," *Ind. Crystal. 84 Proc. (Netherlands)*, 1984, p. 111.

Peters, R. W., and D. Bhattacharyya, "The Effect of Chelating Agents on the Removal of Heavy Metals by Sulfide Precipitation," *Proc. 16th Mid-Atlantic Ind. Waste Conf.*, 1984, p. 289.

Peters, R. W., and Y. Ku, "Removal of Heavy Metals from Industrial Plating Wastewaters by Sulfide Precipitation," *Proc. of the Ind. Wastes Symp., Water Pollut. Control Fed.,* 1984, p. 279.

Pugsley, E. B., "Commentary on Soluble-Sulfide-Precipitation for Heavy Metals Removal from Wastewaters," *Env. Prog.,* 1:(3)A8 (1982).

Resta, J. J., et al., "Soluble-Sulfide Precipitation of Metal Finishing Wastewater," *Proc. 16th Mid-Atlantic Waste Conf.,* 265, 1984.

U.S. Environmental Protection Agency, *Control and Treatment Technology for the Metal Finishing Industry Series: Sulfide Precipitation,* Technology Transfer Summary Report, EPA 625/8-8-80-003.

Whang, J. S., et al., "Design of Soluble Sulfide Precipitation System for Heavy Metals Removal," *Proc. 13th Mid-Atlantic Ind. Waste Conf.,* 64, 1981.

SECTION 7.3
PHOTOLYSIS

Philip C. Kearney
Natural Resources Institute
Agricultural Research Service
U.S. Department of Agriculture
Beltsville, Maryland

Paul H. Mazzocchi
Chairman, Department of Chemistry and Biochemistry
University of Maryland
College Park, Maryland

7.3.1 PHOTOCHEMICAL PRINCIPLES

The process of photochemistry occurs continuously in the earth's atmosphere and on its surface. It is a natural and important environmental process that occurs daily, degrading numerous organic compounds on the soil surface, in the atmosphere, and in various bodies of water. The ultraviolet radiation that reaches the surface of the earth is more or less limited to wavelengths longer than 295 nm.

The purpose of this section is to describe photochemical processes that can be effectively employed to photodegrade selected hazardous wastes. Various techniques like the use of sensitizers and commercial high-intensity light sources, which effectively remove the wavelength limitation of sunlight, make the large-scale photodegradation of hazardous wastes a viable process.

Definitions and Rules

In a photochemical reaction, only the light absorbed is effective in producing a photochemical change.[1,2] Therefore, there must be an overlap between the absorption spectrum of the reacting species and the emission spectrum of the light source to realize a direct photochemical reaction. In cases where there is no ab-

7.33

sorption or the excited state required is unavailable via direct irradiation, a process called *sensitization* may be employed.[3] The reactant of interest is not irradiated directly, but a secondary material called a *sensitizer* is employed. A typical sensitizer is a molecule whose excited-state properties are 4 to 5 kcal/mol higher in energy than that of the substrate. Energy transfer is exergonic, it will take place at every collision, and its rate is diffusion-controlled. The use of sensitizers is a convenient way to achieve the triplet state of molecules such as alkenes, whose triplet states are unavailable via direct irradiation.

The electronically excited state, a state of high potential energy, may undergo a variety of processes in the course of deexcitation. The process we are mainly interested in is reaction and, in this context, competing processes are undesirable since they essentially waste light, resulting in a decrease in efficiency. Some of the competing processes are fluorescence and phosphorescence—i.e., emission of light from the excited singlet and triplet states, respectively; internal conversion—i.e., conversion of potential energy to thermal energy; and intersystem crossing—i.e., conversion of an excited state of one multiplicity to another by electronic-spin inversion (e.g., singlet-triplet). These processes and their energy relationships are outlined in Jablonski diagrams which are presented in several books.[1,2,4]

An important parameter in determining the efficiency of a photochemical reaction is the quantum yield. *Quantum yield* is defined as the ratio of the number of molecules of material undergoing that process to the number of quanta of light absorbed. In a given reaction one can usually define a number of different quantum yields, e.g., the quantum yields for product formation, fluorescence, and internal conversion. A primary process is defined as an event that follows directly after the absorption of a photon and thus the sum of all of the primary quantum yields must equal unity. Quantum yields for product formation may vary from very small measurable quantities to very large values in photoinitiated thermal chain processes like photochlorination and photopolymerization.

Equipment

Photochemical equipment generally consists of two components: a vessel to contain the material being irradiated, and the light source. The number of vessel designs that have been used over the years is too numerous to mention. Vessels generally have been made from glass or fused silica and are of two basic designs: those in which the light source is immersed in the reaction medium and those in which the light source is external to the medium. Characteristics of the three main types of mercury resonance lamps that are used as light sources are listed in Table 7.3.1.

For most purposes low- or medium-pressure mercury lamps are most convenient for large-scale reactions. Medium-pressure lamps are often used when an immersion geometry is desired.

7.3.2 APPLICATIONS

Pesticides

Modern pesticides are divided into various use categories, the primary categories being the insecticides, herbicides, and fungicides. The chlorinated hydrocarbon

TABLE 7.3.1   Operating Characteristics of Mercury Resonance Lamps

Light source	Low pressure	Medium pressure	High pressure
Operating mercury pressure	10 m	30 cm	100 atm
Expected lifetime mercury	2,000–10,000 h	1,000 h	200 h
Principal emission	254 nm	254, 265, 297, 302, 313, 366 nm	Continuum with self-absorption hole at 254 nm

insecticides (DDT, Endrin, Dieldrin, Aldrin, and toxaphene) that once dominated the market have been banned over the last two decades. The organic herbicides (Atrazine, Alachlor, 2,4-D, and others) now represent 70% of the U.S. sales of pesticides. The extensive use of certain herbicides, coupled with their persistence and mobility, has led to their presence in groundwater. This finding has stimulated research on destruction methods, including photolysis.

Perhaps one of the most important reactions in the area of pesticide photodegradation is the photodehalogenation reaction; this is because so many of the materials of interest are halogenated and so many of those are recalcitrant— i.e., they are resistant to environmental biodegradation or hydrolysis.

$$\text{Aryl-Cl} + \text{RH} \xrightarrow{\ h\nu\ } \text{Aryl-H} + \text{RCl}$$

The general reaction involves irradiation of a halide in the presence of a hydrogen source resulting in the dehalogenation of the substrate. There are apparently two mechanisms involved in photodehalogenation reactions. The first is simple excitation followed by homolytic carbon-halogen bond cleavage to give the aryl and halogen radicals. This is the preferred process in chlorobenzene[5] and

in alkyl halides[6] but, in the latter case, the initially formed radicals probably undergo electron transfer to give the alkylcarbenium ion and halide, which proceed to products. The second process is quite different,

proceeding by initial electron transfer to give the aryl radical anion, which expels halide ion, forming the aryl radical which then proceeds to give product through normal processes.

The requirement for an electron donor has been shown[7,8] and in the case of pure 1-chloronaphthalene there is a dramatic increase in reduction efficiency with increasing concentration, suggesting that a charge-transfer exciplex is involved.[7] The reverse relationship occurs with chlorobenzene. It appears that electron transfer is the major process in 1-chloronaphthalene, whereas direct C—Cl bond homolysis is the major process with 1-chlorobenzene. In any case, the potential exists to dramatically increase the efficiency of many photodehalogenation reactions by the addition of suitable electron donors to promote electron-transfer-initiated dehalogenation.

Atrazine [2-chloro-4-(ethylamino)-6-(isopropylamino)-s-triazine] is one of the most widely used herbicides in the United States. A mobile UV unit containing 66 low-pressure mercury resonance lamps was used to determine the feasibility of on-site photolysis.[9] A product study revealed a series of dehalogenated s-triazine molecules lacking one or both alkyl side chains (Fig. 7.3.1). A survey of nine herbicides and two insecticides at three concentrations (10, 100, and 1,000 ppm) in the same 66-lamp unit has recently been completed. The rate of photolysis decreased as the concentrations increased, probably because of the opaque nature of the formulating solvents required to keep the pesticides in solution. The herbicide paraquat (1,1'-dimethyl-4,4'-bipyridinium ion) was photolyzed very slowly at concentrations in the range of 1,500 ppm.[10] Addition of acetone as a photosensitizer accelerated fragmentation of the parent compound

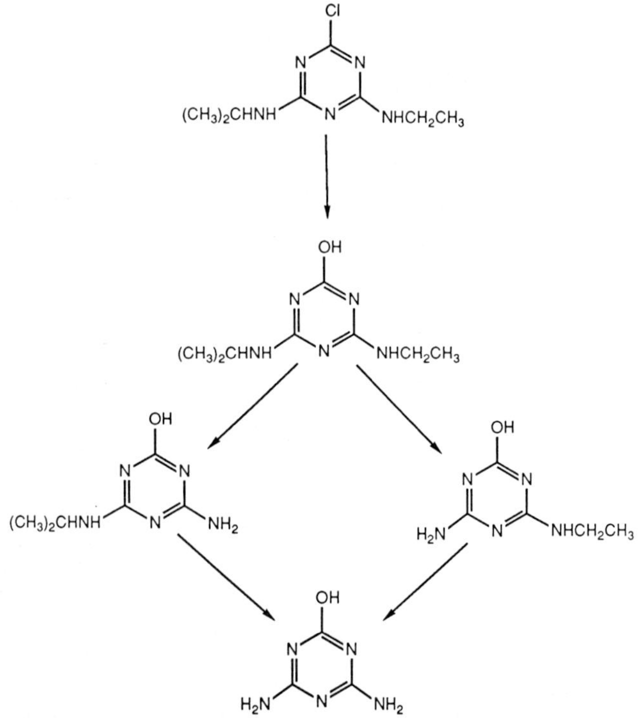

FIG. 7.3.1 Products of photolysis from Atrazine.[9]

and yielded as products 4-carboxy-1-methylpyridinium ion; 4,4'-bipyridyl, 4-picolinic, hydroxy-4-picolinic acid; succinic acid; N-formylglycine; malic acid; and oxalic acid.

Dioxins

One of the most successful large-scale photodegradation processes was carried out on tetrachlorodibenzodioxin.[11] The 2,3,7,8-tetrachlorodibenzo-p-dioxin (2,3,7,8-TCDD) is a potent teratogen, is acutely toxic, and causes chloracne on exposure. This compound arises during the production of trichlorophenol by treatment of tetrachlorobenzene with sodium hydroxide at 150 to 200°C (302 to 392°F). Both hexachlorophene and 2,4,5-T (a component of Agent Orange) were produced from trichlorophenol. In 1974 TCDD was confirmed in a waste-storage tank at a chemical plant in Verona, Missouri. A solvent extraction system (hexane, 99.9% efficient) was developed to remove the TCDD from the waste and the hexane-TCDD solution was subjected to photolysis, which resulted in dissociation of the carbon-halogen bond. The photoreduction products of TCDD are shown in Fig. 7.3.2. Addition of isopropyl alcohol to the hexane facilitated the destruction process. A 7,600-m³ (2,000-gal) glass-lined reactor was present at Verona for the removal of the TCDD from about 17,480 m³ (4,600 gal) of waste. The waste contained 7 kg TCDD. Calculations showed that eight 10-kW lamps could reduce a 6,600-L batch of hexane extract containing 40 mg/L of TCDD by 99.79% in 22 h. The extraction-photolysis process for destroying TCDD was implemented in full-scale plant operation within 19 months of initial laboratory experiments. Because of the extreme toxicity of TCDD, great care had to be exercised in the movement and overall processing of the waste.

Munitions

A major problem faced by the military is the safe disposal of unused propellents. Specifically the migration of TNT (2,4,6-trinitrotoluene) into surface waters near shell-loading and handling operations has been a major concern due to the formation of "pink water" by natural photochemical processes. Pink water is caused by the formation of primarily 1,3,5-trinitrobenzene, as well as 2,4,6-trinitrobenzonitrile, 4,6-dinitroanthanil, and several azoxybenzene compounds, some of which are suspected or known carcinogens.

Since TNT in aqueous solution is rapidly photolyzed, a number of studies

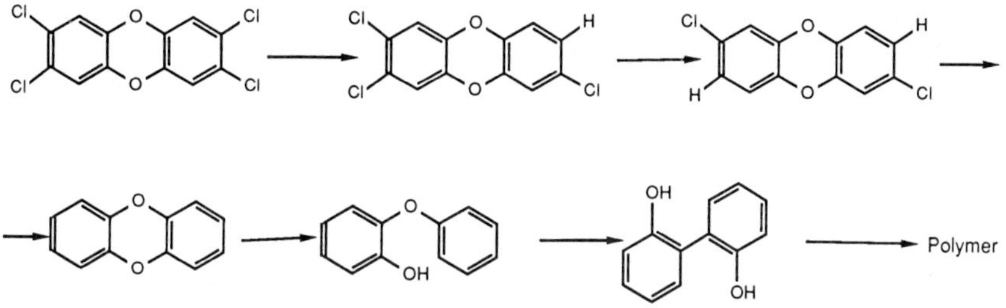

FIG. 7.3.2 Products of photolysis from 2,3,7,8-tetrachlorodibenzo-p-dioxin (2,3,7,8-TCDD).[11]

have been directed at the intentional destruction of TNT by UV irradiation, usually in the presence of a strong oxidant. An excellent review of these studies has been published.[12]

Nitroaromatics are known to undergo photoreduction to anilines via a hydrogen-abstraction process.[13] The intramolecular analog of this hydrogen-abstraction process, which we will see is very important in subsequent sections, was one of the first photoreactions discovered.[14]

In aqueous hydrochloric acid, however, it has been shown that the conversion of nitrobenzene to trichloroaniline is an electron-transfer process[15,16] with chloride ion acting as the electron donor. The nitrosobenzenes, formed by dehydration of the dihydroxy intermediates, are converted to the anilines in dark processes.

In a recent study[17] aqueous solutions containing 1 and 10 ppm ring ^{14}C-TNT, the ^{14}C-labeled TNT was rapidly degraded with respect to aromatic substituents with a slower fragmentation of the ring. An analysis of the photoproducts by GC mass spectrometry showed nine compounds (Fig. 7.3.3), including TNT (I), 2,4,6-trinitrobenzaldehyde (II), 1,3,5-trinitrobenzene (III), 3,5-dinitrophenol (IV), 3,5-dinitrocatechol (V), 2,6-dinitrohydroquinone (VI), 5-nitroresorcinol (VII), 1,3,5-trihydrohydroxybenzene (VIII), and oxalic acid (IX).

7.3.3 PHOTOLYSIS IN COMBINED DISPOSAL PROCESSES

Photolysis of many aromatic compounds leads to incomplete ring fragmentation. Some of the pesticide products of photolysis may pose some potential toxicological problems even though they are less toxic than the starting material. Consequently it may be advantageous to have a secondary treatment process that mineralizes the photoproducts to CO_2, H_2O, and Cl. Microbial metabolism by indigenous, selected, or engineered microorganisms holds considerable promise for effecting complete destruction of some photoproducts generated by UV. The photoproducts from TNT were rapidly metabolized to CO_2 in soils.[17] Conversely ring rupture from ring ^{14}C-TNT via UV photolysis was very slow and ring cleavage of TNT by soil microorganisms has not been reported. As in the case of TNT, many photoproducts are more polar and biodegradable than the parent com-

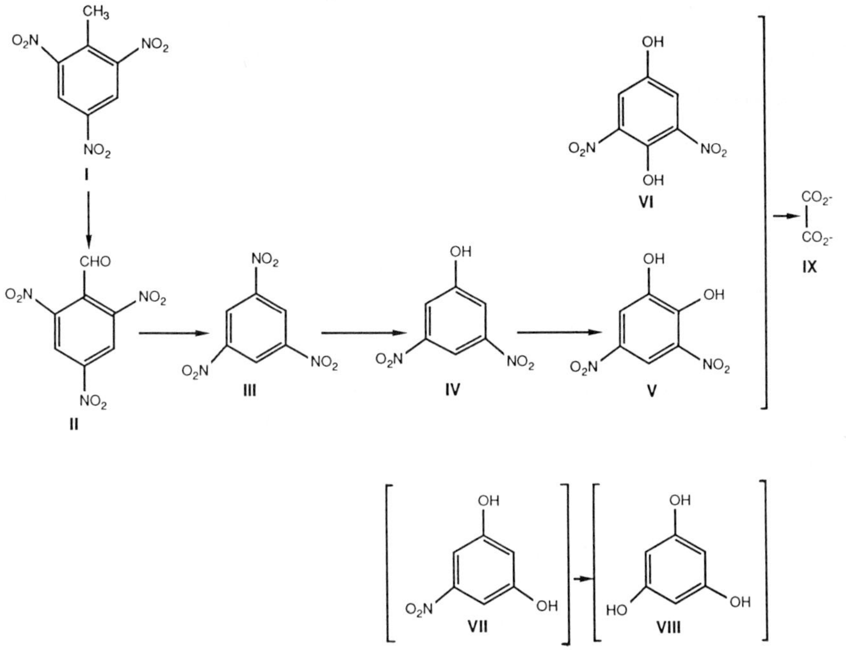

FIG. 7.3.3 Products of photolysis from TNT.[9]

pound, and therefore more susceptible to microbial metabolism. A two-stage dis-posal process that offers the advantages of both processes may provide destruc-tion of a toxic substance, whereas each process independently may not achieve the same result.

7.3.4 ECONOMICS

There is very limited information on the costs of destroying toxic wastes by photolysis. The capital costs can be substantial depending on the equipment used. The operating costs for destroying nine herbicides and two insecticides in a 66-lamp mobile UV unit have been estimated.[18] Three concentrations were examined (10, 100, 1,000 ppm) and the cost per gallon calculated based on the time required for 90% disappearance of the parent pesticide. The en-ergy use of the 66-lamp UV unit is about 1.5 kW/h and the cost of electricity is roughly 5 cents per kilowatt hour. As anticipated, the cost was dependent on concentration and increased as the concentration increased. The average cost of destroying 90% of the 11 pesticides at 10 ppm was 0.3 cents per gallon; 100 ppm, 0.7 cents per gallon; and 1,000 ppm, 4.9 cents per gallon. Efforts to increase the quantum yields of the various processes by the use of sensitizers and other techniques to optimize the system could be very effective in lower-ing the program costs.

7.3.5 REFERENCES

1. J. G. Calvert and J. N. Pitts, Jr., *Photochemistry*, Wiley, New York, 1966.
2. N. J. Turro, *Modern Molecular Photochemistry*, Benjamin/Cummings Publishing, California, 1978.
3. S. L. Murov, *Handbook of Photochemistry*, Marcel Dekker, New York, 1973.
4. A. J. Gordon and R. A. Ford, *The Chemist's Companion*, Wiley, New York, 1972.
5. N. J. Bunce and L. N. Ravenal, *J. Am. Chem. Soc.*, **99**:4,151 (1977).
6. P. J. Kropp, G. S. Poindexter, N. J. Pienta, and D. C. Hamilton, *J. Am. Chem. Soc.*, **98**:8,135 (1976); S. A. McNeely and P. J. Kropp, *J. Am. Chem. Soc.*, **98**:4,319 (1976).
7. N. J. Bunce, P. Pilon, O. Ruzol, and D. J. Sturch, *J. Org. Chem.* **41**:3,023 (1976).
8. M. Ohashi, K. Tsujimoto, and K. J. Seki, Chem. Soc. Chem. Commun. 384, 1973.
9. P. C. Kearney, Q. Zeng, and J. M. Ruth, *Treatment and Disposal of Pesticide Wastes*, ACS Sympos. Ser. 259, 1984, pp. 195–209.
10. P. C. Kearney, J. M. Ruth, Q. Zeng, and P. H. Mazzocchi, *J. Agric. Food Chem.*, **33**:953 (1985).
11. J. H. Exner, J. D. Johnson, O. D. Ivins, M. N. Wass, and R. A. Miller, "Process for Destroying Tetrachlorodibenzo-*p*-dioxins in a Hazardous Waste," Chap. 17 in *Detoxication of Hazardous Waste*, Ann Arbor Science, The Butterworth Group, Ann Arbor, Michigan, 1982.
12. R. J. Spanggord, T. Mill, T. W. Chore, W. R. Mabey, J. H. Smith, and S. Lee, *Environmental Fate Studies on Certain Munition Wastewater Constituents. I. Literature Review*, Final Report on Contract No. DAMD17-78-8081, U.S. Army Medical Research and Development Command.
13. J. Faure, J. P. Fouassier, D. J. Lougnot, and R. Salvin, *Nova. J. Chim.*, **1**:15 (1977); G. Beck, Dobrowolski, J. Kiwi, and W. Schnabel, *Macromol.*, **8**:9 (1975).
14. G. Ciamician and P. Selber, *Ber.*, **34**:1040 (1901).
15. G. G. Wubbels and R. L. Letsinger, *J. Am. Chem. Soc.*, **96**:6,690 (1974).
16. R. Hurley and A. C. Testa, *J. Am. Chem. Soc.*, **89**:6,917 (1967); G. G. Wubbels, J. W. Jordan, and N. S. Mills, *J. Am. Chem. Soc.*, **95**:1,281 (1973).
17. P. C. Kearney, Q. Zeng, and J. M. Ruth, *Chemosphere*, **12**:1,583 (1983).
18. P. C. Kearney and M. T. Muldoon, personal communication, 1986.

CHEMICAL OXIDATION AND REDUCTION

Edward G. Fochtman
Chemical Waste Management, Inc.
Riverdale, Illinois

Oxidation-reduction reactions are those reactions in which the oxidation state of one reactant is raised while the oxidation state of the other reactant is lowered. When electrons are removed from an ion, atom, or molecule, the substance is oxidized; when electrons are added to a substance, it is reduced. When metal atoms (e.g., Zn^0) are converted to metal ions (e.g., Zn^{2+}) they lose electrons, that is, they are oxidized. If the metal ions (i.e., Zn^{2+}) take on electrons, they are converted to the metal (Zn^0) and the ions are reduced.

Oxidation-reduction or "redox" reactions have an important role in treatment of wastes. Such reactions are used in treatment of metal-bearing wastes and of inorganic toxic wastes such as metal-containing wastes, sulfides, cyanides, and chromium, and also in the treatment of many organic wastes such as phenols, pesticides, and sulfur-containing compounds.

Since these treatment processes involve chemical reactions, generally both reactants are in solution. However, in some cases a solution is reacted with a slightly soluble solid or with a slightly soluble gas.

7.4.1 OXIDATION-REDUCTION TECHNOLOGY

Principles of Oxidation-Reduction

Chemical oxidation-reduction reactions are those reactions in which the oxidation state of one reactant is increased and the oxidation state of another reactant is decreased. Two very common reactions are:

Oxidation $\qquad$ $NaCN + H_2O_2 \rightarrow NaCNO + H_2O$ $\qquad$ (1)

Reduction $\qquad$ $2H_2CrO_4 + 3SO_2 \rightarrow Cr_2(SO_4)_3 + 2H_2O$ $\qquad$ (2)

In reaction (1) the cyanide ion (CN) oxidation state of nitrogen is (formally) increased from -1 to $+1$, and the oxidation state of oxygen is decreased from -1 to -2.

In reaction (2) the oxidation state of chromium is reduced from $+6$ to $+3$, and the oxidation state of sulfur is increased from $+4$ to $+6$.

Oxidation

Chemical oxidation is widely used to treat both hazardous and nonhazardous wastes. The technology is well-established and represents a safe means of waste treatment that is easily monitored and controlled. While chemical oxidation is most suited to treatment of liquids it can be used for slurries and sludges. Since the oxidizing agents tend to be nonselective and represent a major portion of the treatment cost, this type of treatment is more suited to wastes with low organic content. A variety of oxidizing agents are available. Costs for treating a concentrated cyanide waste have been estimated at about $0.23 per gallon ($0.06 per liter).[1] Typical oxidation systems for cyanide are listed in Table 7.4.1.[1]

Organic wastes that have been treated by chemical oxidation include phenols, amines, mercaptans, and chlorophenols. However, some organics are resistant to oxidation by most oxidation agents under ambient temperatures and pressures and may require increased temperature, use of a catalyst, or use of ultraviolet light.

TABLE 7.4.1 Oxidation Agents for Treatment of Cyanide

Agent	Oxidation reaction
Sodium hypochlorite	$2NaCN + 5NaOCl + H_2O \rightarrow N_2 + 2NaHCO_3 + 5NaCl$
Calcium hypochlorite	$4NaCN + 5Ca(OCl)_2 + 2H_2O \rightarrow 2N_2 + 2Ca(HCO_3)_2 + 3CaCl_2 + 4NaCl$
Chlorine	$2NaCN + 5Cl_2 + 12NaOH \rightarrow N_2 + 2Na_2CO_3 + 10NaCl + 6H_2O$
Calcium polysulfide	$NaCN + CaS_x \rightarrow NaCNS + CaS_{x-1}$
Potassium permanganate	$NaCN + 2KMnO_4 + 2KOH \rightarrow 2K_2MnO_4 + NaCNO + H_2O$
Hydrogen peroxide	$NaCN + H_2O_2 \rightarrow NaCNO + H_2O$
Ozone	$NaCN + O_3 + H_2O \rightarrow NaCNO + O_2$

Reduction

Chemical reduction can also be used to treat a hazardous-waste constituent. The most prevalent use of reduction in current waste treatment is the reduction of chromium(VI) to chromium(III). The Cr^{3+} is much less toxic than Cr^{6+} and can be precipitated as the relatively insoluble $Cr(OH)_3$ for removal.

Although chemical reduction can reduce metals to their elemental state for recovery, this process has found only limited application. It appears that reduction to the elemental metal will be more widely used as pretreatment standards further restrict discharges of metals to publicly owned wastewater-treatment plants.

Sulfur dioxide and ferrous sulfate are generally used for chromium reduction. The reactions are shown in Table 7.4.2.[1]

TABLE 7.4.2   Agents Used for Reduction of Chromium(VI)

Agent	Reduction reaction
Sulfur dioxide	$3SO_2 + 3H_2O \rightarrow 3H_2SO_3$
	$2CrO_3 + 3H_2SO_3 \rightarrow Cr_2(SO_4)_3 + 3H_2O$
Bisulfite	$4CrO_3 + 6NaHSO_3 + 3H_2SO_4 \rightarrow 2Cr_2(SO_4)_3 + 3Na_2SO_4 + 6H_2O$
Ferrous sulfate	$2CrO_3 + 6FeSO_4 + 6H_2SO_4 \rightarrow 3Fe_2(SO_4)_3 + Cr_2(SO_4)_3 + 6H_2O$

7.4.2   OXIDATION-REDUCTION PROCESSES

Treatment of waste by oxidation utilizes equipment very similar since both generally involve the mixing of two aqueous liquids, the waste and the treatment chemical, or contacting an aqueous solution with a gas. Both batch and continuous processing systems are used.[2] Some reactions are rapid, on the order of 1 to 2 s, and can be conducted in a pipeline reactor. Oxidation of sodium bisulfite by sodium hypochlorite is one such reaction. Mixing of the reactants in a pipeline results in a rapid reaction which can be monitored by temperature rise and oxidation-reduction potential. Oxidation of some organics by hypochlorite may require several hours at an elevated temperature, and it may be more feasible to conduct such reactions in batches or, at a minimum, in three, four, or five continuous stirred reactors. Reaction rates are increased at elevated temperatures. However, this requires increased energy and, at temperatures above 100°C (212°F), pressure-rated equipment. These changes in processing conditions to increase reaction rates must be weighed against the increased capital and operating costs.

7.4.3   OXIDATION REAGENTS

While there are many chemicals which are oxidizing agents, relatively few of them are used for waste treatment. These oxidants vary in oxidation potential, convenience, cost, and by-product production, and each new application requires careful consideration, often with bench- or pilot-scale evaluations before selection of the oxidizing reagent and equipment. Commonly used oxidizing agents are discussed in this subsection.

Sodium Hypochlorite

Sodium hypochlorite is probably the most widely used oxidant. It is readily available, and the aqueous solution is easily transported, stored, and metered into the reacting system.

The term *available chlorine* is frequently used in referring to hypochlorites and hypochlorite solutions. This usage sometimes leads to confusion concerning reagent concentrations. The available chlorine in a substance is a measure of the oxidizing power of its active chlorine compared to that of elemental chlorine. For NaOCl, one hypochlorite ion is equivalent to one Cl_2 molecule. Thus pure NaOCl has an available chlorine content of $1 \times MW\ Cl_2/MW$ NaOCl, or $(1 \times 70.90/74.46) \times 100\%$, or 95.2%. A 5.25 wt % solution of NaOCl, specific gravity at 20°C of 1.08, contains $5.25 \times 0.952 = 5.00$ wt % available chlorine or $5.00 \times 1.08 \times 10,000 = 54,000$ mg/L available chlorine.

One of the major uses of chlorine or hypochlorite solutions is for the treatment of cyanide-containing wastes from ore extraction, synthetic organic-chemical manufacture, and metal finishing. Metal-finishing wastes are the most numerous and result from rinsewaters, spent processing solution, and spills. The cyanide is first oxidized to cyanate at high pH, typically 11. Reduction of the pH to 6 converts the cyanate to bicarbonate and nitrogen. While the theoretical requirement is 6.8 units of chlorine per unit of cyanide, the actual requirement is higher due to the presence of oxidizable organic compounds.

Capital costs for batch cyanide-treatment units rated at 0.76 m³/d (200 gpd) are on the order of $10,000 plus installation.[3] Continuous units with capacities of 0.076 m³/min (20 gpm) cost $50,000 plus installation. Treatment costs range from $2.50 to $7 per pound ($5.50 to $15.40 per kg) of cyanide for on-site treatment and from $3 to $4.50 for off-site treatment. Costs are quite dependent upon chemical consumption.

Hydrogen Peroxide

Hydrogen peroxide is sold as a colorless water solution at concentrations of 30 to 70%. While not a hazardous substance, hydrogen peroxide requires proper handling. It is a strong oxidizing agent that releases oxygen and heat during decomposition. Solutions of high concentration require some special safety precautions. However, suppliers readily supply storage and handling information.[4]

Most industrial applications utilize hydrogen peroxide in concentrations of 35 to 70 wt %. Physical properties of a 50 wt % solution are shown in Table 7.4.3. It has been used for oxidation of phenolic wastewater, and for treatment of paper mill effluent, drilling mud, and other types of organic wastewaters.

Special inhibitors are added to hydrogen peroxide during manufacture to inhibit decomposition during storage. Small amounts of some metals may be catalytic to the decomposition reaction. Properly inhibited hydrogen peroxide has a decomposition rate in large tanks of less than 1% per year at ambient temperatures and in drums of less than 2% per year. The material should be stored only in containers that have been properly designed and thoroughly conditioned.

TABLE 7.4.3   Properties of 50 wt % H_2O_2 in Water

Active oxygen content, wt %	23.5
H_2O_2 per liter, g	600
Specific gravity at 20°C	1.20
Pounds/gal	10.0
Boiling point, °C (°F)	114 (237)
Freezing point, °C (°F)	−52 (−62)

High-purity aluminum alloy is recommended for storage tanks and the manufacturer should be consulted for proper design and fabrication specifications for these tanks. The tanks must be vented and preferably should be stored in a relatively cool place.

Hydrogen peroxide can be delivered in 30- or 55-gal drums or in tank trucks carrying up to 15 m^3 (4,000 gal). There are special aluminum railroad tank cars that have capacities up to 30 m^3 (8,000 gal).

Protective clothing and safety goggles should be worn when handling hydrogen peroxide. Although it is not a systemic poison (a 3% solution is often used as a gargle or mouthwash), care should be taken to avoid contact with the skin. Any exposed areas should be washed immediately with water. Clothing or other organic material that has been exposed to the hydrogen peroxide should be washed thoroughly before it is allowed to dry since the peroxide may concentrate and cause a fire. Hydrogen peroxide may form a shock-sensitive material when mixed with organic substances. One should avoid contamination of the material with heavy metals such as silver, lead, copper, chromium, cobalt, nickel, mercury, iron, and iron oxide, since such contamination can cause rapid decomposition and overpressuring of a vessel.

The hydrogen peroxide solutions can be assayed by reaction with potassium permanganate.

Hydrogen peroxide is a relatively safe oxidant and has uses in the treatment of many industrial wastes including cyanides, formaldehyde, hydrogen sulfide, hydroquinone, mercaptans, phenols, and sulfites. In addition, it has been used as an additive to scrubber solutions that are used to remove hydrogen sulfide, sulphur dioxide, nitric oxide, and nitrogen dioxide.

Calcium Hypochlorite

Calcium hypochlorite, which contains approximately 65% available chlorine, is sold in granular or tablet form. It is easy to use and to handle and may offer some advantages over other oxidizers in this regard.

The usual precautions in handling an oxidizer should be observed when handling calcium hypochlorite. These include the wearing of goggles, coveralls, rubber gloves and boots, and a respirator if there is a possibility of dusting.

The chemical should be stored in cool, dry, well-ventilated places and away from combustible materials. It does not lose its strength rapidly during storage. However, extensive heating will cause decomposition and may cause drums to rupture.

At temperatures above 177°C (351°F), calcium hypochlorite will decompose rapidly, with evolution of oxygen and heat.[5] Temperatures over 57°C (135°F), if sustained for several days, may cause decomposition. Since heat and oxygen do evolve during decomposition, contamination with an organic substance may start a chemical reaction that will result in a fire of great intensity.

All oxidizers should be stored separately from materials that react with them, such as acids, propellents, explosives, solvents, pesticides, paint products, household products, oils, beverages, and soaps and detergents, as well as organic materials such as rags or other debris.

Calcium hypochlorite can be purchased in 50-, 75-, and 100-lb fiber drums, smaller plastic pails, and even smaller polyethylene bottles that are often used for treatment of swimming-pool water.

Calcium hypochlorite has been used for the treatment of several industrial wastes. In the treatment of cyanide solutions, particularly copper or nickel cyanides, it requires about 4.5 to 5.5 lb per 1 lb of free cyanide to produce the cyanate. Calcium hypochlorite has also found use in the food industry for sanitizing equipment, in papermaking for bleaching of paper dyes, in the treatment of industrial cooling water, in the treatment of swimming pools and lobster storage ponds, and in several food-preparation areas such as the handling of eggs, fish, nuts, sugar, meats, and dairy products.

Potassium Permanganate

Potassium permanganate is available as crystals or granules which are dark purple with a metallic appearance. An additive is sometimes used to enhance the flow properties. This additive gives the material a gray appearance. Potassium permanganate is a relatively easy oxidizer to handle and presents little health hazard during handling or storage. Dust masks should be worn to avoid damage to the respiratory system, and eye protection should be worn.

It is recommended that potassium permanganate be stored in a cool, dry place. Concrete floors are, of course, preferred. The product should be stored away from acids, organic materials, oils and combustible materials, since these may be a source of fire or explosion.

Potassium permanganate is used as an aqueous solution. The solubility is:[7]

Temperature, °C	Solubility, g/L
0	28
20	65
40	125
60	230
80	350
100	500

Although oxidation reactions conducted in aqueous solutions are generally regarded as safe, there are violent reactions that can occur under unfavorable conditions: for example, when hot, concentrated solutions of permanganate and reducing agents are combined or when dry potassium permanganate is brought in contact with liquid organic compounds. Explosions can occur if solid potassium permanganate is brought in contact with concentrated sulfuric acid or hydrogen peroxide. Potassium permanganate can also react violently with metallic powders, elemental sulfur, phosphorus, carbon, hydrochloric acid, hydrazine, and metal hydrides. Contact with all organic materials should be avoided.

Permanganate may be stored indefinitely in a cool, dry area in closed containers. Dry potassium permanganate should not be exposed to intense heat. Fires should be extinguished with water. Potassium permanganate can cause brown stains on wood, plastics, fabric, ceramics, and skin. These can be easily removed by treatment with a 2% sodium bisulfite solution in water or even by lemon juice followed by a water rinse.

Potassium permanganate has been used to control phenols and other industrial pollutants, in addition to organic and inorganic material in wastewater streams. It has been used for the removal of soluble iron and manganese found in acid mine waters; it destroys many of the odor-producing compounds in rendering plants and sewage plants. It has also been used to oxidize cyanides.

Ozone

Ozone, which boils at minus 112°C at atmospheric pressure, is an unstable gas with a characteristic penetrating odor that is readily detected at concentrations as low as 0.01 to 0.05 ppm. The gas is water-soluble to the extent of about 20 mg/L at 0°C and 8.9 mg/L at 20°C. This compares to oxygen solubility of 6.9 mg/L at 0°C and 4.3 mg/L at 20°C.[9] In general it is used by dispersing the gas in aqueous media.

Ozone itself is unstable, having a half-life of 20 to 30 min in distilled water at 20°C; the half-life in ambient atmosphere is on the order of 12 h.

Ozone is generated on site, generally with an electrical discharge, in concentrations of 1 to 3% in air and 2 to 6% in oxygen. At these concentrations there is no explosive hazard.

Ozone is generated by passing oxygen or air through a discharge gap between two electrodes, one of which has a dielectric.[8] The corona in the discharge gap breaks the oxygen molecules, forming nascent oxygen, which reacts with other oxygen molecules to form the ozone. The preferred generation method utilizes pure oxygen, although ambient air can be used to form ozone. If ambient air is used, the yield is reduced because of moisture present; also, the discharge can form nitric acid, which can result in severe corrosion of downstream equipment. Generation of ozone by electrical discharge produces heat, which tends to cause decomposition of the product gas. It is important that the ozone generator be operated as cool as possible. The efficiency of conversion of electrical energy to ozone has increased during the last 50 years from about 33 kWh/lb (72.6 kWh/kg) of ozone to a current 10.6 kWh/lb (23.3 kWh/kg) of ozone.

Since ozone is a relatively expensive oxidant, it must be utilized efficiently. It is only slightly soluble in aqueous media, and it is important that the mass transfer of the ozone from the oxygen or air into the water be effected as efficiently as possible. The equipment employed to effect this contacting is generally of one of the following types: spray towers, packed beds, bubble-plate or sieve-plate towers, or diffuser plates. A considerable effort has been expended in the recent past to increase the efficiency of ozone transfer into aqueous media. This is generally accomplished either by introducing the gas as small bubbles to provide a maximum contact surface and time, or by introducing the gas as a large bubble or gas stream which is then broken up in the fluid. Typically, this is done by diffuser plates or injector contactors. Not all of the ozone is absorbed by aqueous media and it may be necessary to treat the off gas to reduce the concentration to discharge limits.

Ozone is most widely used for the treatment of drinking water and has been used in Europe for almost 100 years. In 1916 there were 49 plants in Europe, 26 of them in France. By 1940 the number of drinking-water treatment plants throughout the world using ozone had risen to 119, and by 1977 there were 1,043 plants used for drinking-water treatment.[9]

The capital cost for a plant using 4,600 lb (2090 kg) ozone per day has been estimated as $1,860,000, with an annual operating cost of $296,495. In some cases this is competitive with other treatment processes such as activated carbon.[9]

Automatic instrumentation that is convenient for process control is available for continuous on-line monitoring of ozone in gases at concentrations of a few parts per million and 1 to 6%. Measurement of the ozone in water is somewhat more of a problem and is generally done by the iodometric method.[8] There are interferences with this method if other oxidants are present.

Although ozone is a powerful oxidant, there are a number of refractory chemicals that are oxidized slowly if at all by ozone. Oxidation of some of these chemicals can be enhanced with ultraviolet light. For instance, cyanide complexes of iron, which react very slowly with ozone, are rapidly decomposed with the combination of ozone and radiation. Similarly, solutions of polychlorinated biphenyls, which are stable to oxidation, are destroyed rapidly by the UV-ozone combination.[10,11,12] The amount of UV radiation applied to increase the effectiveness of ozone ranges from 0.04 to 1.3 W/L at ambient temperatures. This type of system has been used for halogenated solvents such as methylene chloride, trichloroethylene, carbon tetrachloride, and vinyl chloride. Packaged commercial equipment is available for this treatment.

7.4.4 REDUCTION REAGENTS

Reducing agents are used to treat wastes containing hexavalent chromium, mercury, organometallic compounds, and chelated metals. The largest single use is for the treatment of hexavalent chromium, which is reduced to the less toxic trivalent state. The trivalent chromium can be removed from aqueous solutions by precipitation of the relatively insoluble hydroxide.

Several chemicals can be used for reduction.

Sulfur Dioxide

Sulfur dioxide is widely used for chemical reduction. It is supplied in various container sizes up to 1 ton. It has a boiling point of $-10°C$ ($14°F$) and a pressure of 48 psi at ambient temperature. The sulfur dioxide is vaporized and fed to the process through a regulator and meter. At 1 atm pressure, the solubility in water is 18.5% at $0°C$ ($32°F$) and 5.1% at $40°C$ ($104°F$).[13]

Reduction is conducted at low pH, and the process can be controlled by simple pH and oxidation-reduction meters.

Sulfur dioxide has found widespread use in chromium treatment because it enjoys the advantage of low amounts being required, as illustrated in Table 7.4.4 for a specific chromium(VI)-containing waste.

Dry sulfur dioxide does not represent a corrosion problem. Vessels for treating aqueous wastes are often of steel lined with hard rubber, which is satisfactory at moderate temperatures.

Sulfur dioxide concentrations in air can be determined by drawing a known volume through a hydrogen peroxide solution, adding isopropyl alcohol, and adjusting the pH with perchloric acid. The sulfate ion content is then determined

TABLE 7.4.4   Relative Consumption in Treating Chromium Waste

Chemical	Amount required, lb/lb or kg/kg
Sulfur dioxide	1.9
Sodium bisulfite	3.0
Sodium metabisulfite	2.8
Ferrous sulfate	8.8

using perchlorate.[13] The concentration in water is determined by distilling the sulfur dioxide into hydrogen peroxide and titrating the acid formed.

Cunningham[1] prepared an estimate of the cost to treat 2,000 gpd ($8.8m^3/d$) of a chromium waste from a plating operation: the estimated capital cost was $230,000 (1976 dollars—this would be about $400,000 in 1987). The total treatment cost was about $0.25 per gallon ($0.06 per liter) (1987 cost).

Sodium Borohydride

Use of sodium borohydride, $NaBH_4$, is increasing as environmental concerns and regulations require more complete removal of metals. Sodium borohydride is sold as a 12% solution in 40% sodium hydroxide.[14] It can be used directly or diluted with tap water to the desired concentration. It can be stored in the same type of equipment used for 50% caustic solutions. Decomposition in the absence of contaminants is slow, about 0.1% per year. Storage temperature should be below 60°C (140°F); it should be used above 18°C (65°F) since it becomes viscous at low temperatures. Storage containers should be vented.

A 12% sodium borohydride in 40% sodium hydroxide solution boils at 100°C (212°F) and freezes at 18°C (64°F). The specific gravity is 1.4 g/mL. Procedures used for handling strong caustic should be followed. The solution is a corrosive liquid under U.S. Department of Transportation (DOT) regulations and is packaged in 5-gal and 55-gal containers with 10% head space.

Sodium borohydride is being used to treat wastes. There are currently more than 45 installations for the treatment of heavy-metal wastes.

Sodium borohydride reacts with hydroxyl ions according to

$$NaBH_4 + 8OH^- \rightarrow NaBO_2 + 6H_2O + 8e^-$$

which indicates that 1 mol of $NaBH_4$ will reduce 8 mol of a monovalent metal ion. Table 7.4.5.[14] shows the theoretical weight ratio of metal reduced per weight of

TABLE 7.4.5   Weight Ratios: Reduced Metal Obtainable from Ionic Species/Theoretical Amount of Sodium Borohydride

Metal	Valence	Weight ratio*
Lead	Pb^{2+}	22
Cadmium	Cd^{2+}	12
Mercury	Hg^{2+}	21
	Hg^+	42
Copper	Cu^{2+}	7
Silver	Ag^+	22
Gold	Au^{3+}	14
Nickel	Ni^{2+}	6
Palladium	Pd^{2+}	11
Platinum	Pt^{4+}	10
Cobalt	Co^{2+}	6
Rhodium	Rh^{3+}	7
Iridium	Ir^{4+}	10

*Weight ratio = maximum kilograms of metal reduced per kilogram of sodium borohydride.

reagent. These ratios are seldom obtained in practice; however, they do indicate the potential effectiveness of this reducing agent.

$NaBH_4$ has been used to treat lead at 500 to 3,500 mg/L, mercury at 10 to 50 mg/L, silver at 10 to 120 mg/L, and cadmium at 5 to 60 mg/L.[15,16]

$NaBH_4$ has also been used for the reduction of organic compounds, using water and low-molecular-weight alcohols as solvents. It reacts with ketones, organic acids, and amides, and has been proposed for dehalogenation of organic contaminants.

Sodium borohydride can be assayed by hydrogen evolution, by reduction of nicotinamide adenine dinucleotide to a UV-absorbing species, or colorimetrically by reducing phosphomolybdic acid to a blue color. Hydrogen evolution is most accurate for high concentrations; several other methods are used for trace concentrations.

Water containing several grams per liter of metals can be treated for between \$0.15 and \$0.25 per gallon. Use of sodium borohydride is rapidly increasing as regulations require more complete removal of metals from aqueous discharges.

7.4.5 POTENTIAL FOR WASTE TREATMENT

Chemical oxidation and reduction are currently utilized for waste treatment, and this use is expected to increase as regulations require more complete removal of contaminants discharged to municipal wastewater-treatment systems, to surface waters, or to deep wells.

In general, chemical treatment costs are highly influenced by the chemical cost. Thus, oxidation and reduction treatment tends to be most suitable for low concentrations (less than 1%) in wastes. This restriction is expected to continue. However, there will be an increase in the need to treat such wastes, and an expansion of this type of treatment in the future.

7.4.6 REFERENCES

1. J. B. Berkowitz, J. T. Funkhouser, and J. I. Stevens, *Unit Operations for Treatment of Hazardous Industrial Wastes,* Noyes Data Corp., Park Ridge, New Jersey, 1978.

2. Product Catalog Nos. 4101, 4201, 1101, Lancy International, Inc., Zelienople, Pennsylvania.

3. P. E. Radimsky et al., *Alternative Technology for Recycling and Treatment of Hazardous Wastes,* Second Biennial Report, California Department of Health Services, Sacramento, California, July 1984.

4. *Industrial Waste Treatment with Hydrogen Peroxide,* Company Bulletin, FMC Corporation, Princeton, New Jersey, 1981.

5. A. K. Chowdhury and L. W. Ross, "Catalytic Wet Oxidation of Strong Waste Waters," *Water,* AIChE Symposium Series, no. 166, 1976, pp. 46–58.

6. *Product Data and Application Information for HTH,* Olin Chemicals, Stamford, Connecticut.

7. Product Data Sheets, Potassium Permanganate, Carus Chemical Company, LaSalle, Illinois.

8. F. L. Evans III, *Ozone in Water and Wastewater Treatment,* Ann Arbor Science, Ann Arbor, Michigan, 1977.

9. R. G. Rice and M. E. Browning, *Ozone Treatment of Industrial Wastewater,* Noyes Data Corp., Park Ridge, New Jersey, 1981.

10. I. D. Zeff, *New Developments in Equipment for Detoxifying Halogenated Hydrocarbons in Water and Air,* Internal Document, Ultrox International, Culver City, California, September 1985.

11. J. D. Zeff, E. Leitis, and J. A. Harris, "Chemistry and Application of Ozone and Ultraviolet Light for Water Reuse—Pilot Plant Demonstration," 38th Annual Purdue Industrial Waste Conference, West Lafayette, Indiana, May 1983.

12. G. R. Peyton and W. H. Glaze, "The Mechanism of Photolytic Ozonation," ACS Symposium Series, *Aquatic Photochemistry,* April 1986.

13. Kirk-Othmer, *Encyclopedia of Chemical Technology,* 3d ed., Wiley, New York, 1983.

14. M. J. Lindsay and M. E. Hackman, "Sodium Borohydride Reduces Hazardous Waste," 40th Annual Purdue Industrial Waste Conference, West Lafayette, Indiana, 1985.

15. M. M. Cook, J. A. Lander, and D. S. Littlehale, *Case Histories—Use of Sodium Borohydride for Control of Heavy Metal Discharge in Industrial Waste Waters,* Internal Document, Morton Thiokol, Inc., Ventron Division, Danvers, Massachusetts.

16. M. Fleming, *Reduction of Nickel (II) complexes in Spent Electroless Plating Baths,* Internal Paper, Morton Thiokol, Inc., Ventron Division, presented at Electroless Nickel Conference, Chicago, Illinois, April 1985.

SECTION 7.5
DEHALOGENATION

Alfred Kornel, Ph.D.
U.S. Environmental Protection Agency
Hazardous Waste Engineering Research Laboratory
Cincinnati, Ohio

The application of chemical dehalogenation reactions for the purpose of detoxification or destruction of hazardous compounds is a relatively recent (5 to 7 years) phenomenon. For more than 100 years organic chemistry has concerned itself with creation of new haloorganics and has applied a large variety of reactions toward this goal. Roughly during this same time period, organic dehalogenation reactions have been researched primarily to create new useful intermediates or products. An excellent example of this is the chemical destruction of ethylene dibromide (EDB) by alcoholic potassium hydroxide to yield vinyl bromide and acetylene gas.[1] This process, described in the chemical literature in 1878, is still being used with modification today. Thus, if we review the literature available on chemical dehalogenation, we can find a multitude of applications designed for production of new compounds from pure starting materials. However, very little literature is available for the specific chemical destruction of haloorganics in dilute, complex material such as soils, sediments, sludges, or various aqueous and mixed waste streams. The reason for this is simply that chemists have not been asked until recently to look at the environmental problems of widely dispersed toxic or hazardous halogenated species. Until now, it has been the engineer who has directed disposal practices, whether they were landfill, containment, or some form of thermal destruction. The limitations of these methods, whether real or perceived, have been, and will certainly continue to be, debated especially when incineration of a matrix such as dioxin-contaminated soil is considered.

When dealing with dehalogenation of organic compounds, we can classify them as being either haloaliphatic or haloaromatic. This classification implies that the halogen,—i.e., fluorine, chlorine, bromine, or iodine—is covalently bonded to either an aromatic or aliphatic carbon atom. The nature of this chemical bond directs the type of chemistry required for breaking the bond. Reaction conditions used for dehalogenation of a chlorobenzene (an aromatic) generally differ from those used for dechlorination of chloroethylene (an aliphatic). A review of dehalogenation chemistry can be found in any good-quality college-level chemistry text. Those wishing further information are referred to the works of Fieser and Fieser[2] or Morrison and Boyd.[3]

Other chemical methods for dehalogenation of organic compounds include molten-salt chemistry, ultraviolet and other radiation sources, wet oxidation, elevated temperature (300 to 600°C) catalytic reduction, and incineration. Further, microbial dehalogenation and indeed even the dehalogenation of the haloorganics which occur in a large variety of organic life forms are considered "chemistry" although on an enzymological level. These subjects are not considered here. This information is covered elsewhere in this book (see Chapter 9).

7.5.1 CURRENT APPLICATIONS OF THE TECHNOLOGY

Haloaromatics

Perhaps the best-known problem compounds of current times are the polychlorinated biphenyls (PCBs), their mixtures, and the polychlorinated dibenzodioxins (PCDDs), of which 2,3,7,8-tetrachlorodibenzo-p-dioxin (2,3,7,8-TCDD) is the greatest environmental problem. Other hazardous compounds of this class which are beginning to emerge as problematic include the halogenated benzo- and dibenzofurans, and other halogenated complex aromatic chemicals. Of these compounds, the PCBs were products of commerce introduced by the Monsanto Company in 1929 and were widely applied in dielectric heat-exchange and hydraulic media. The PCDDs and related haloaromatics were by-products of various chemical processes, industrial accidents, or transformer (PCB) fires. For this reason, the PCBs greatly outweigh the dioxins or furans in total mass in the environment. It is estimated that 1.253 billion lb of PCB fluids were sold in the United States from 1930 to 1975. Roughly 77% (965 million lb) of this was used for electrical equipment.[4] A considerable quantity of this PCB fluid has entered the environment. In the late 1980s, PCB contamination exists in soils, sediments, and sludges, and in both fresh and salt waters. The level of contamination can be from the few parts per billion range up to parts per thousand by weight. However, dioxins usually occur in the parts per billion range in the environment. For these reasons, dehalogenation chemistry has recently focused on the PCB problem and then more so on PCB fluids still in use either in a relatively pure form, or in a simple matrix. Haloaromatics which occur in complex matrices, such as soils or sediments, are more difficult to detoxify by dehalogenation chemistry which was designed for pure chemical streams. This is not to imply that dehalogenation chemistry will not function under these conditions. It is to state that more research is required in this area of chemistry.

Commercial Dehalogenation of Haloaromatics

The commercial processes for decontamination of PCB-laden dielectric, heat-exchange, or hydraulic fluids represent a classical case of application of chemical dehalogenation. Incineration or land disposal of these fluids represents economic as well as environmental concerns. Organic chemistry was applied and has succeeded in the elimination of the problem (PCB) and made available a recycled product, namely a valuable oil.

Let us trace the history of these commercial PCB-dehalogenation chemistries. The scientific literature shows Ligget[5] to have demonstrated the use of an organic reagent, sodium biphenyl, for the analytical determination of organic halogen in 1954. Modifications of this reagent were made by Oku et al. in 1978[6] and Smith and Bubbar in 1979.[7] These two recent articles deal with the dechlorination, or chemical destruction, of PCBs with a sodium napthalene reagent. We have now seen how an early chemical application designed for analytical purposes has been further enhanced and applied to a current chemical problem.

The works referenced above were employed as a basis for the commercialization of a PCB-disposal process by the Goodyear Tire and Rubber Company in 1980. This process employed a sodium napthalene reagent in tetrahydrofuran. The nature of this reagent's reactivity also enables it to react with a wide variety of other haloorganics such as polybrominated biphenyls, halogenated pesticides, herbicides, and PCDD. With regard to PCB dechlorination, the process is capable of dehalogenation of from approximately 100 to 300 ppm of the haloorganic in hydraulic or heat-transfer oils. Certain drawbacks of the process include the handling of metallic sodium and of tetrahydrofuran, the latter a flammable ether, and also the necessity of having to work under a nitrogen blanket. Despite these technical problems the process and variations have been adapted by the Acurex Corporation, General Electric Co., and PPM Inc. for PCB disposal in various organic liquid media. The SunOhio PCB-removal process was developed independently and its chemistry is unrelated. Of these companies only three have retained commercial interest. They are SunOhio, PPM, and Acurex, with the Acurex process recently sold to Chemical Waste Management, Inc.

A generalized comparison of the SunOhio, Acurex, and PPM processes is given in Table 7.5.1. The information was obtained from the respective companies and interested persons are urged to contact them for further information.[8,9,10]

As can be seen from Table 7.5.1, all three vendors possess mobile capabilities; however, apparently only the SunOhio process is continuous. There is an advantage to a continuous process when dealing with transformer oil; the core of the device is constantly flushed, allowing for more complete cleaning. In a typical application of the SunOhio process, the PCB levels of a transformer oil 90 days after treatment rose to between 1 and 3% of the initial ppm level. A resampling of batch-cleaned transformer oil may indicate greater PCB levels than for a continuous process, indicating that further cleanup may be necessary. Information on resampling for PCB levels in transformers after dehalogenation by the PPM or Acurex processes was not available.

Both SunOhio and PPM give a guarantee or certificate of destruction; it is not yet known how Chemical Waste Management will deal with the recently aquired Acurex process. In the SunOhio process, final oil-quality tests are conducted with regard to PCB levels, acid, dielectric, power factor, and interfacial tension. This is followed by a formal written report within eight weeks of completion. The PPM literature reveals that a certificate of destruction is issued to the customer within 60 days of completion; this company makes no mention as to the quality of the processed oil except that PCB levels are less than 2 ppm or nondetectable limits.

Volatile emissions are controlled in all three of the processes so that no adverse health or environmental effects occur. Further, there are no water dis-

Table 7.5.1   Commercial PCB-Dehalogenation Processes

Company	Max. level of PCB in oil, ppm	Final PCB level in oil, ppm	Types of fluids	Process description	Quantities of oil processed	Cost range $ per gallon of oil
SunOhio	2,600	<2.0	Transformer and heat-transfer fluids	Batch or continuous processing optionally on energized transformers up to 69-kV; mobile capabilities	>6 million gal	†
Acurex	1,000*	<2.0	Transformer oil	Batch processing; mobile capabilities	N.A.‡	N.A.
PPM Inc.	10,700	<2.0	Transformer oil, heat transfer and hydraulic fluids, lubricants and fuel oils	Batch processing; mobile capabilities on Annex lll facility§	>8.5 million gallons	1.75–2.00

*Example given in company literature.
†Cost varies with type of installation, PCB level and other contaminants, etc.
‡N.A. = not available
§Most of the PPM Inc. processing is at four sites: Atlanta, Georgia; Kansas City, Missouri; Grayback Mount, Utah; Philadelphia, Pennsylvania.

charges from the processes. However, there are, as in any chemical process, re-action products arising from the dehalogenation of the PCBs and eventual quenching of the reagent. The solid by-products which are contained consist primarily of a polyphenyl polymer, sodium salts, some entrapped oil, and other nontoxic material. The volatile by-products generally consist of nitrogen, hydrogen, water vapor, and some hydrocarbons. Further solid disposable materials that these processes could generate are absorbent and filter cakes, gloves, and other processor safety equipment.

Presented here are three commercial processes for dechlorination of PCBs in a variety of oils. All three make use of proprietary formulations, two of which are similar to those originally commercialized by the Goodyear Tire and Rubber Company. All three processes produce essentially the same desired result; lowering amounts of, or removing, PCBs from useful oils or fluids, and at least one process permits re-use of these materials. This is the first example of chemical dehalogenation providing viable alternatives to landfilling or incineration of otherwise useful materials. When considering selection of one of these processes for application to a PCB contamination problem, we suggest the responsible party contact the companies. This author cannot recommend one process over another as the choice for each application will depend on specific information as to type of oil, PCB levels, water content, and other chemical content which may interfere with the reaction.

7.5.2 EXPERIMENTAL DEHALOGENATION TECHNIQUES FOR HALOAROMATICS

As seen in Subsec. 7.5.1., there exist commercial processes for PCB dehalogenation in oils. These processes have major limitations when their use is attempted in any other matrix—e.g., soils or sediments which contain water or are exposed to atmospheric oxygen. As the commercial processes rely on active sodium metal they cannot be safely used on wet materials. This has prompted research into the chemistry of dehalogenation such that reactions which can tolerate water and can still effectively dehalogenate the target pollutant have come about. Work in this area has been sponsored since 1981 by the U.S. Environmental Protection Agency (EPA), Hazardous Waste Engineering Research Laboratory (HWERL), Cincinnati. One of the first groups to research this area was the Franklin Research Institute, who have applied a new dehalogenation reagent, NaPEG[™], for possible in situ dehalogenation of PCBs in soils. The NaPEG reagent is formed by the reaction of sodium or its hydroxide salt with a polyglycol or polyglycol monoalkylether in the presence of oxygen.[11,12] This NaPEG reagent has been tested by both the EPA and Franklin Institute personnel with respect to dehalogenation of PCBs in both solution and soil applications. Initial solution studies presented very promising results with regards to oxygen stability and dehalogenation effectiveness of these NaPEG reagents. Soil studies employing these reagents under laboratory conditions also showed promise for the reagents' ability to perform under in situ conditions. However, field studies revealed a greater susceptibility to water deactivation than had previously been expected.

The Franklin Research Institute's NaPEG reagent is not the only polyethylene glycol–based dehalogenation reagent. Others have been developed by General Electric Co.,[13] Vertac Chemical Corporation,[14] and Galson Research Corporation.[15] All these processes employ alkaline polyethylene glycolates (APEGs) as the reactive species. And initially these processes were designed for application to chloroaromatics in hydrocarbon or nonpolar oils. It has been only in the past three to four years that research has been directed toward the in situ approach to dehalogenation of haloaromatics in complex matrices. Of these groups apparently only the Galson Research Corporation is actively pursuing an APEG dehalogenation of PCBs and PCDDs in soils.

7.5.3 APEG CHEMISTRY

The origin of the concept of APEG reagents as dehalogenation agents can be traced back to the work of Starks,[16] who in 1971 introduced the term *phasetransfer catalysis*. This term describes the process of heterogeneous reactions involving two distinct solution phases, namely aqueous and organic. The reaction is initiated and sustained by small amounts of a suitable catalytic material. This catalyst, as its name implies, permits transfer of the substance to be dehalogenated or otherwise reacted from one phase to the other. Primarily, investigations centered on exchange of anions such as CN− or OH− from an aqueous phase to substitute a halogen (Cl, Br) on an aliphatic compound residing in the organic phase. The catalysts were primarily of the tetraalkyl ammonium or phosphonium structure. Further applications and improvements of these systems were described by Makosza,[17] Lehmkuhl et al.,[18] and Regen.[19] The groups of Makosza and Regen both concentrated on catalytic routes involving functionality of the previously mentioned tetraalkyl ammonium species; however, Regen has carried the process further by formulating a triphase catalytic system employing a solid-phase catalyst at the organic-aqueous interface. Lehmkuhl et al. were apparently one of the first groups to employ open-chain polyethylene glycol dialkyl ethers complexed with alkali metal salts as phase-transfer agents. This group also drew the analogy between these complexed substances and the macrocyclic crown ethers which had been employed in nucleophilic substitution reactions. Work by Lehmkuhl was primarily concerned with substitution reactions on the aliphatic halogen of benzyl bromide. Further investigation of the use of both polyethylene glycols and their monoalkyl ethers by Gibson[20] and Kimura and Regen[21] revealed that the glycol-alkali moiety not only behaves as its own phase-transfer agent or catalyst, but also is actually incorporated in the substitution reaction. Gibson employed alkaline polyethylene glycolates in phase-transfer-catalyzed Williamson ether synthesis, whereas Kimura and Regen revealed the extraordinary effect of these reagents in two-phase dehydrodehalogenation reactions, employing alkylbromo compounds as reactants.

Comparisons of the alkali salt polyethylene glycol complexes with modified crown ether catalysts have been carried out by Stott et al.[22] based on the earlier works of Lehmkuhl et al.[18] and Gokel and Garcia.[23] The results of the Stott group indicated that the onium salts and polyethylene glycols were cost-effective catalysts of choice when compared with a series of crown ethers based on 18-crown-6; preference for the salt was based on the high price of the macrocyclic crown ethers.

In an excellent article by Brunelle and Singleton[24] dealing with dechlorination of PCBs by use of polyethylene glycol potassium hydroxide (KPEG) it was revealed that the PEGs and the polyethylene glycol monoalkyl ethers (PEGMs) not only act as phase-transfer catalysts, but also are functional nucleophiles under alkaline conditions. In a more recent study Brunelle et al.[25] have used this APEG or APEGM reagent to dehalogenate PCBs in transformer oil. These articles also allude to the products which are formed by these reagents in haloaromatic systems. Further it should be stressed that the reactions employing PEG or PEGM formulations are tolerant of oxygen, water, and other contaminants, are found to be of low toxicity, and are economically attractive. Thus they may have application for in situ decontamination of soils, sediments, and sludges which have become contaminated by either aryl or alkyl haloorganics.

In a recent article by Kornel and Rogers,[26] the APEG reagents were evaluated with regard to the effect of both water and solvent dilution on dehalogenation efficiency. The effects of water and other diluents could play an important role for in situ application of these reagents. It was confirmed that if the proprietary solvent used by Peterson[15] was incorporated into an APEG formulation that is diluted by water, the dehalogenation reaction proceeds even at room temperature. At elevated temperatures, e.g., 50 to 100°C (122 to 212°F), the dehalogenation reactions are complete in relatively short times (from 10 min to 2 h) when dealing with Aroclor 1260 in solution. This has prompted the EPA to investigate in situ and slurry methods of application of these reagents to contaminated soil systems.

In these systems artificial heating of the material by either conventional or radio-frequency (RF) methods is under investigation.[27]

As stated, APEG reagents are being applied to both in situ and slurry soil-decontamination systems. The *in situ method* would be essentially thus: apply a water–APEG reagent solution directly to the contaminated area, heat the area with RF or other techniques while containing volatiles, and thus decontaminate the area. The *slurry method,* which could be termed the *on site method,* requires the contaminated soil to be excavated, placed into a mobile APEG reactor system, and decontaminated. The reactor is heated by conventional methods and volatiles are contained. This system is being designed by Galson Research Corporation and should be evaluated in conjunction with the EPA in 1987 under actual field conditions.

7.5.4 OTHER EMERGING DEHALOGENATION TECHNIQUES

There are a number of other groups who are investigating innovative and novel dehalogenation techniques not yet discussed. These novel techniques can also be termed chemical methods. They include catalytic (low-temperature) and electrochemical processes and are all directed at dehalogenation of PCBs, though they may be applied toward other haloorganics.

An electrochemical dehalogenation process proposed and under development by Massey and Walsh[28] is designed to address the PCB problem in fluids and soils. Current work has dealt only with PCBs in oils such as mineral oil. As the process incorporates a proprietary organic solvent which acts as conducting medium, we cannot discuss the chemistry. However, we can give a brief description of the process. Essentially, the PCB-contaminated oil is mixed with the proprietary conducting medium in a circulating electrochemical cell. A current of low voltage (5 to 15 V) is passed through the mixture at a temperature from ambient to 95°C (203°F). The current flow causes dechlorination of the PCBs and results in substituted biphenyls and chloride salts. The reaction is stated to be rapid, to be highly specific to haloorganics, and to proceed to the detection limits of PCBs in organics, e.g., 2 ppm. According to the authors of the process, the equipment is easily moved, fits in a small space, and could treat contaminated oil for approximately $1 per gallon. The companies performing this work expect commercialization of the technology in the late 1980s.

A recent catalytic dechlorination scheme has been described by Chu and Vick of Union Carbide Corporation.[29] Their catalytic process, which employs a proprietary nickel catalyst, is claimed to dehalogenate PCB with efficiencies approaching 100%. Primarily the PCB either in pure form or as a solution in oils is mixed with dimethyl formamide and isopropanol. Zinc metal as well as the nickel catalyst is added and reaction proceeds. The major dechlorinated product is biphenyl with some chloro- and dichlorobiphenyl formed, which if left in contact with the catalyst also undergoes dechlorination to biphenyl. The other major product is a concentrated metal halide solution.

Advantages of this process are similar to those of all the other PCB-in-oil destruction technologies in that the process can be mobile, contains the reaction products, and recycles the transformer oil. A further advantage may be this method's ability to dehalogenate "pure" PCB fluids more cost-effectively than the sodium treatments. The nickel catalyst is claimed to be inexpensive and recyclable, as only zinc is consumed by the reaction forming zinc chloride. The solvents are also said to be recyclable. No cost evaluation or date for commercialization has been presented at this time.

7.5.5 *DEHALOGENATION OF HALOALIPHATICS*

Less emphasis has been placed on the dehalogenation of haloaliphatics than haloaromatics. This is especially so when dealing with these halogenated aliphatic compounds in complex matrices such as soil, sediment, or sludges. Some concern is now being given to the presence of aliphatic halogenated solvents in ground and drinking water. In these cases, the general process is to treat the water supply for removal of the problem chemicals. This still leaves the problem of eventual disposal of haloaliphatic concentrates. Other industries such as dry cleaning and commercial degreasers are faced with similar problems in disposing of concentrates and still bottoms containing mixtures of oils, sediments, and halogenated solvents.

As chemists have not until recently been asked to look at these problems, there are no specific solutions. There are, however, applications of existing technologies which could, as in the case of PCBs in oils, render a solution to these problems.

A catalytic destruction of brominated organics which is applicable to debromination of both haloaromatic and aliphatic compounds was patented in 1975.[30] Though the process requires elevated temperatures and hydrogen gas for formation of hydrogen bromide (HBr) from the precursor materials, it does produce a salable product, namely, the HBr. Problems are encountered when the bromo compounds are in oily wastes as these tend to foul the catalysts. Perhaps others will investigate the application of lower-temperature catalytic methods for dehalogenating aliphatic compounds. There is certainly going to be a demand for this type of chemistry as regulations evolve and emerge.

One of the earliest methods for dehalogenation was that referred to at the beginning of this section,[1] namely, the use of ethanolic potassium hydroxide. Another early and still used dehalogenation technique for aliphatics is the use of zinc metal to effect dehydrohalogenations. An excellent review of this chemistry can be found in *Reagents for Organic Synthesis*,[31] as well as in the general organic chemistry books previously mentioned.[2,3]

One article dealing with *Entgiftung* (detoxification) of aliphatic compounds reports on the phase-transfer catalytic properties of benzyltributylammonium chloride.[32] The compounds dehalogenated in this study were ethylene dibromide and dichloride. Another article also deals with the dehalogenation of vicinal dihalogenated aliphatics and employs glycol dimethyl ethers such as tetraethylene glycol dimethyl ether.[33] Recently a patent was granted to the USEPA, which covers the chemical destruction of halogenated aliphatic hydrocarbons.[34] Application of these as well as creation of other organic dehalogenation reactions may well solve the potential problem of haloaliphatic wastes, both in concentrated contained form or in other matrices in the environment. No commercial processes are yet available so no cost comparisons can be made.

7.5.6 *CONCLUSION*

Chemical dehalogenation has been, and will continue to be, a viable alternative for the destruction of certain types of hazardous toxic materials. There are sufficient processes available for commercial- and laboratory-level dehalogenation of "neat," or pure, haloorganic materials if contained in a simple or suitable matrix. The commercial PCB-detoxification processes are good examples of this. What is not available are cost-effective, environmentally sound methods for decontamination of hazardous halogenated compounds in dilute or trace quantities in complex matrices. There are as yet (late 1980s) no commercialized field or site chemical dehalogenation processes either for in situ or on-site applications. Such processes are under development, and are expected to be field-tested. What is required are more innovative chemical dehalogenation processes which can be applied to the problem of environmental pollution caused by haloorganics. These haloorganics are present in soils, fresh and salt water, sediments, and aquifers. This area of research, namely, application of chemical dehalogenation for these environmental problems, will certainly be of great importance for the future.

7.5.7 ACKNOWLEDGMENT

The author wishes to thank Donald L. Wilson of the U.S. Environmental Protection Agency's Hazardous Waste Engineering Research Laboratory, Cincinnati, Ohio, for his effort in researching and obtaining the information on the commercialized PCB-dehalogenation processes.

7.5.8 REFERENCES

1. Justus Liebig's *Annalen der Chemie,* Band 191, 1878, p. 368.
2. L. F. Fieser and M. Fieser, *Organic Chemistry,* 2d ed., D.C. Heath, Boston, 1950, pp. 59, 60, 78, 151, 157, 288, 362, 665, and 708.
3. R. T. Morrison and R. N. Boyd, *Organic Chemistry,* 2d ed., Allyn and Bacon, Boston, 1966, pp. 466, 483, 541, 588, and 733.
4. J. H. Exner, *Detoxication of Hazardous Waste,* Ann Arbor Science, Ann Arbor, Michigan, 1982, p. 131.
5. L. M. Liggett, "Determination of Organic Halogen with Sodium Biphenyl Reagent," *Analytical Chemistry,* **26**:(4)748 (April 1954).
6. A. Oku, K. Yasufuku, and H. Kataoka, "A complete dechlorination of polychlorinated biphenyl by sodium naphthalene," *Chemistry and Industry,* **4**:841 (November 1978).
7. J. G. Smith and G. L. Bubbar, "The Chemical Destruction of Polychlorinated Biphenyls by Sodium Naphthalenide," *Journal of Chemical Technology and Biotechnology,* **30**:620 (1980).
8. SunOhio, 1700 Gateway Blvd., SE, Canton, Ohio 44707, (216) 452-0837. Personal communication with EPA.
9. PPM Inc., 1875 Forge Street, Tucker, Georgia 30084. Personal communication with EPA.
10. Acurex Corporation, 485 Clyde Avenue, Mountain View, California 94042. Chemical Waste Management, Inc., 150 West 137th Street, Riverdale, Illinois 60627. Personal communication with EPA.
11. L. L. Pytlewski and K. Krevitz, *Reagent and Method for Decomposing Halogenated Organic Compounds,* U.S. Patent 4,337,368.

12. —— *Method for Decomposition of Halogenated Organic Compounds*, U.S. Patent 4,400,552.

13. D. J. Brunelle, *Method for Removing Polyhalogenated Hydrocarbons from Nonpolar Organic Solvent Solutions*, U.S. Patent 4,353,793.

14. K. J. Howard and A. E. Sidwell, *Chemical Detoxification of Toxic Chlorinated Aromatic Compounds*, U.S. Patent 4,327,027.

15. R. L. Peterson, *Method for Reducing Content of Halogenated Aromatics in Hydrocarbon Solutions*, U.S. Patent 4,532,028.

16. C. M. Starks, "Phase-Transfer Catalysis. I. Heterogeneous Reactions Involving an Ion Transfer by Quaternary Ammonium and Phosphonium Salts, *Journal of the American Chemical Society*, **93**:195 (1971.)

17. M. Makosza, "Two-Phase Reactions in the Chemistry of Carbanions and Halocarbenes—A Useful Tool in Organic Chemistry," *Pure and Applied Chemistry*, **43**:(3–4)439 (1975).

18. H. Lehmkuhl, F. Rabet and K. Hauschild, "Phasentransferkatalyse durch offenkettige Polyathylenglykol-Dreivate; I. Substitutions reaktionen von Benzylbromid mit Kaliumsalzen," *Synthesis*, March 1977, p. 184.

19. S. L. Regen, "Triphase Catalysts," *Angewandte Chemie*, **18**:(6)421 (June, 1979).

20. T. Gibson, "Phase-Transfer Synthesis of Monoalkyl Ethers of Oligoethylene Glycols," *Journal of Organic Chemistry*, **45**:1,095 (1980).

21. Y. Kimura and S. L. Regen, "Poly (ethylene glycols) Are Extraordinary Catalysts in Liquid-Liquid Two-Phase Dehydrohalogenation," *Journal of Organic Chemistry*, **47**:2,493 (1982).

22. P. E. Stott, J. S. Bradshaw, and W. W. Parish, "Modified Crown Ether Catalysts. 3. Structural Parameters Affecting Phase Transfer Catalysts by Crown Ethers and a Comparison of Effectiveness of Crown Ethers to That of Other Phase Transfer Catalysts," *Journal of the American Chemical Society*, **102**:4,810 (1980).

23. G. W. Gokel and B. J. Garcia, "Crown-Cation Complex Effects. III. Chemistry and Complexes of Monoza-18-Crown-6," *Tetrahedron Letters*, **4**:317 (1977).

24. D. J. Brunelle and D. A. Singleton, "Destruction/Removal of Polychlorinated Biphenyls from Non-Polar Media. Reaction of PCB with Poly (Ethylene Glycol)/KOH," *Chemosphere*, **12**:(2)183 (1983).

25. D. J. Brunelle, A. K. Mediratta, and D. A. Singleton, "Reaction/Removal of Polychlorinated Biphenyls from Transformer Oil: Treatment of Contaminated Oil with Poly (ethylene glycol)/KOH," *Environmental Science and Technology*, **19**:(8)740 (1985).

26. A. Kornel and C. Rogers, "PCB Destruction: A Novel Dehalogenation Reagent," *Journal of Hazardous Materials*, **12**:161 (1985).

27. H. Dev, "Radio Frequency Enhanced In-Situ Decontamination of Soils Contaminated with Halogenated Hydrocarbons," *Proceedings of the 12th Annual Hazardous Waste Research Symposium*, EPA 600/9-86/022, Cincinnati, Ohio, April 20, 1986.

28. M. J. Massey and F. M. Walsh, "An Electrochemical Process for Decontaminating PCB Containing Transformer Coolants," *Proceedings EPRI PCB Seminar*, Seattle, Washington, October 22–25, 1985; M. J. Massey, Environmental Research and Technology, Inc., 696 Virginia Rd., Concord, Massachusetts 01742; F. M. Walsh, Trace Technologies, Inc., 225 Needham St., Newton, Massachusetts, 02164.

29. N. S. Chu and S. C. Vick, "Chemical Destruction of Polychlorinated Biphenyls," *Proceedings EPRI PCB Seminar*, Seattle, Washington, October 22–25, 1985; S. C. Vick, Union Carbide Corp., Old Saw Mill River Rd., Tarrytown, NY.

30. R. A. Davis, *Recovery of Bromine from Organic Bromides*, U.S. Patent 3,875,293.

31. L. F. Fieser and M. Fieser, *Reagents for Organic Synthesis,* Wiley, New York, 1967, p. 1,276.

32. D. Martinetz, "Entgiftung aliphatischer Dihalogenverbindungen durch festflussig phasentransfer-katalysierte Diacetatbilding," *Zeitschrift Chemie,* **22**:257 (1982).

33. C. Tarchini, T. DinkAn, Gerald-Jan, and M. Schlosser, "65. Hydroxidbewirkte Halogenwasenstaff-Abspaltungen ans Vic-Dihalogeniden unter heterogenen Bedingungen: Reaktionssteuerung dank überraschender Solvens and Metall-Einflusse," *Helvetica Chimica Acta,* **62**:635 (1979).

34. C. J. Rogers and A. Kornel, Chemical Destruction of Halogenated Aliphatic Hydrocarbons, U.S. Patent 4,675,464.

SECTION 7.6
OZONATION

Frederick C. Novak

Director of Marketing, H2M Group
Melville, New York

7.6.1 APPLICABILITY OF THE TECHNOLOGY

Ozone Generation and How It Is Used

Ozone is generated when a high voltage is projected across a discharge gap in the presence of oxygen, or an oxygen-containing gas, such as air. A major advantage of ozone is that it can be generated on site from air or oxygen and used immediately. This avoids the storage and handling problems usually associated with conventional oxidants.

Ozone is used as a gas in the normal range of concentrations as follows:

Carrier gas	O_3, wt %	g/m^3	lb/ft^3
Air	1–2	12–24	.00074–.00148
Oxygen	2–5	24–60	.00148–.00370

Ozone Applications in Water, Wastewater, and Air

Ozone has been used for more than 80 years, primarily for disinfection of municipal drinking water. The first ozone water plants in the United States were installed 45 years ago to remove color, odor, and taste. The first wastewater plant to use ozone for disinfection was operated in 1973. At present there are more than 1,000 water plants using ozone for disinfection and color and odor removal as well as 50 wastewater plants using ozone worldwide. In addition, ozone plus UV and/or carbon systems for the prevention of trihalomethane (THM) formation and the removal of halogenated organic compounds from drinking water are operating in many countries.

New processes, equipment, techniques, and monitoring devices have been developed in recent years to aid in compliance with effluent guidelines for industrial and municipal plants. Ozone's oxidation power means that it can react rapidly with a large number of organic compounds and that it can destroy bacteria and viruses.

Ozone odor-control systems have been used mainly to treat hydrogen sulfide and mercaptans from municipal and industrial plants. Most odors are from organic compounds, many of which are oxidizable under controlled conditions. The ozone can be applied directly to the gas stream (dry oxidation) if sufficient mixing (ozone with odor molecules) and reaction time (more than 5 s) are used. Wet-oxidation systems combine ozone with scrubbers, utilizing pH control (acid and/or base) to remove a variety of organic odors. Monitoring or control equipment is used to sense ozone presence in the treated off gas or automatically adjust ozone input to the system. Many systems use heat, a catalytic chamber, or carbon as the final off-gas treatment to ensure that no ozone is discharged to the immediate environment.

Industrial and Hazardous-Waste Applications

Industrial-waste categories containing a wide variety of wastes, including hazardous materials, for which ozone treatment has been used include:[1]

Aquaculture	Mining
Breweries	Organic chemicals
Biofouling control	Paint and varnish
Cooling towers	Petroleum refineries
Cyanide and cyanates	Pharmaceuticals
Electroplating	Phenols
Food and kindred products	Photoprocessing
Hospitals	Plastic and resins
Inorganic manufacturers	Pulp and paper
Iron and steel	Soaps and detergents
Leather tanneries	Textiles

Detailed reports and data are not always available on ozone applications for industrial processes or wastewater treatment because of the competitive and proprietary nature of the industrial field.

Organic and Inorganic Reactions

Some common chemical oxidation treatments utilizing ozone are those for the decomposition of cyanide, phenol, organic acids, cresol, xylol, aldehydes, mercaptans, hydrogen sulfide, nitride, iron, and manganese. Several pharmaceutical and chemical manufacturers use ozone to oxidize organics to create new products.

Complete oxidation of organic contaminants to nitrogen and carbon dioxide gas can be obtained with some compounds if sufficient ozone doses and reaction time are used. However, most compounds form intermediate compounds that compete for the ozone and in some cases are difficult to oxidize. Therefore, chemical kinetics for various compounds affect the oxidation rate. Because of the complexity of organic hazardous wastes, laboratory feasibility tests and/or pilot-plant studies should be conducted prior to designing an ozone oxidation system.

Ozone and UV radiation have been used to detoxify industrial organic wastewater containing aromatic and aliphatic polychlorinated compounds, aliphatic amines, ketones, and alcohols.[2] Greater than 99% apparent detoxification of four pesticides (DDT, PCP, PCB, and Malathion) was demonstrated using ozone and UV.[3]

The decomposition of aqueous ozone produces a hydroxide radical that reacts with dissolved organic and inorganic substances. This concept is used to predict reaction rates and pathways of ozone reactions for well-defined systems.[4] Forty rate constants related to reactions of ozone with inorganic aqueous solutes are reported.[5] Ozone rapidly reacts with oxidizable inorganics such as ferrous iron, hydrogen sulfide, iodide, and nitrite ions. Some inorganics react under modified conditions (pH adjustments) or at slower rates (manganese). Theoretically, 0.4 and 2.2 mg O_3 are required to react with 1.0 mg iron and manganese, respectively. Systems designed to treat these metals normally use 0.9 and 5.0 mg O_3 per 1.0 mg of metal, respectively.

7.6.2 OZONE TECHNOLOGY

Oxidation Potential

Ozone is a relatively unstable gas consisting of three oxygen atoms per molecule (O_3) and is one of the strongest oxidizing agents known. It can be substituted for conventional oxidants such as chlorine, hydrogen peroxide, and potassium permanganate.

Relative oxidizing strengths of ozone, hypochlorous acid, and chlorine are 2.07, 1.50, and 1.40 V, respectively. These oxidants have been used for treating both organic and inorganic hazardous contaminants.

Flow Diagram

Hazardous wastewaters are normally complex; therefore laboratory and/or pilot-plant studies are needed to determine if ozone is feasible for use with a specific waste.

A typical pilot plant is shown in Fig. 7.6.1. Batch or flowthrough pilot-plant studies are normally conducted after laboratory feasibility tests. This flow diagram was for a cyanide-destruction study.[6] The purpose of the study was to collect data for final design. Some key data include ozone dose rates and concentration, gas-to liquid ratios, ozone utilization, ozone transfer rate, and effect of pH and temperature reaction rates.

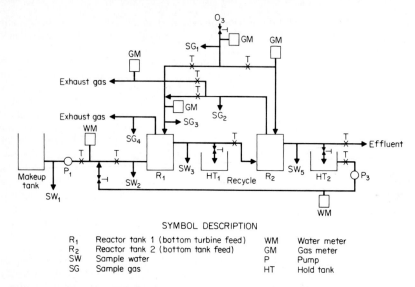

FIG. 7.6.1   Pilot-plant flow diagram.

Equipment

Ozone generators are essentially inefficient since most of the energy is lost as heat, light, and sound. Inefficient cooling and moist air are detrimental to generating ozone as well as corrosive to the equipment. Figure 7.6.2 depicts the normal stages of the ozone-generation process. Intake air is filtered and dried to a dew point of at least $-45°C$ ($-50°F$). One dryer regenerates (4 to 8 h) while the second

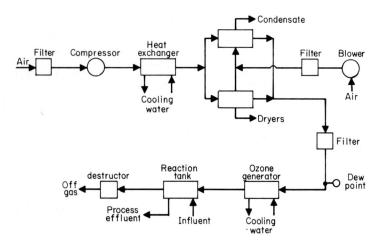

FIG. 7.6.2   Ozone generator system.

air dryer is in operation. The air is filtered again to less than one 1 μm before entering the ozone generator. Air for dryer regeneration is also filtered.

Relationship of Energy Requirements to Ozone Output and Concentration

Figure 7.6.3 shows the data obtained for an ozonator designed to generate 480 g/h (25 lb/d) at a concentration of 1.5% by weight. Energy requirements increase as the concentration increases for specific ozone outputs, or the concentration is held constant and the ozone output is increased. Changes in these variables are obtained by changes in gas flow or power input. Ozone output for an oxygen feed gas is approximately 2 times greater than for an air feed at the same power and gas flow. Power requirements for oxygen generation must be considered when comparing operating costs. Approximately 16 to 19 Wh/g (7.3 to 8.6 kWh/lb) is needed to generate ozone (average power including air drying) from an air feed.

Ozone Injection and Mixing Systems

Ozone utilization, or transfer efficiency, is determined by a number of factors such as wastewater characteristics, kinetics, pH, and temperature. Higher efficiencies are obtained with mass-transfer-type reactions, lower gas-to-liquid ratios (less than 1.0 is recommended), higher ozone concentrations, and smaller bubble size, which latter results in more surface area for a given volume of gas.

Use of porous-type diffusers (ceramic or stainless steel) in tanks at depths of 3 to 6 m (10 to 20 ft) with countercurrent flow is the most common ozone-injection technique. Deeper tanks normally increase ozone transfer. Turbine-type injectors provide higher and more rapid ozone transfer due to the greater turbulence and generally require smaller tanks (less depth), but more operating energy. Packed

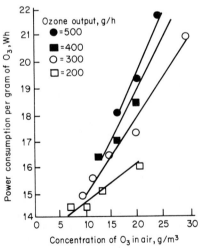

FIG. 7.6.3 Operating data from ozone generator.

columns, venturi tubes, and orifice devices have also been used. High-temperature, catalytic, or combination thermal-catalytic ozone destructors have been used to treat off gases containing excess ozone.

Safety Considerations

At high concentrations, ozone is a toxic gas. Since ozone is generated on site and used immediately, the supply of ozone can be stopped by turning off the electrical power.

Exposure to ozone concentrations of 1 to 75 ppm by volume for a period of 10 min is considered harmless. At this range of concentration, most people will begin to feel irritated nasal passages, thereby providing a warning well below the toxic level. Ozone-monitoring equipment with alarms is available and is used with most ozone systems.

Advantages and Disadvantages

Some of the advantages of using ozone include:

- Generated on site from air; used immediately; no storage or handling of strong oxidants; stop generating by turning off power; low maintenance.
- Very strong oxidant; reacts with a large variety of organics; reacts with inorganics including metals; does not form chlorinated organics; residuals react with constituents or revert back to oxygen; short reaction time if mixed properly (less process space); lower dose rates than other oxidants; makes some organics more biodegradable.
- Temperature and pH less critical than with other oxidants; treated effluents are normally oxygen rich.

Some disadvantages of using ozone are:

- Higher capital cost than other oxidant systems; operating cost dependent on electrical cost.
- Will not degrade low-molecular-weight chlorinated organics; must treat off gas if ozone present.
- Residual oxidant (ozone) essentially nonexistent.

7.6.3 DESIGN CONSIDERATIONS

Background Information on Ozone Equipment

Essentially all new ozone generators are the tube type described in Subsec. 7.6.2 (see under *Equipment,* on p. 7.68). An Otto Plate–type ozonator was first developed in the early 1900s. It is a series of water-cooled aluminum blocks, dielectrics, and stainless-steel electrodes. Air is passed between the dielectrics. A Lowther Plate–type ozonator was developed later and is air-cooled, using air or oxygen to generate ozone.

Materials of Construction

Good engineering practices should be followed in the design of all ozone systems. Parts that will not contact the ozone should be designed to normal standards. The ozone generators and outlet pipes are normally made from stainless steel, although some parts can be made from PVC (Type II) if only an air feed system is to be used. Dielectrics are usually made of special high-dielectric glass and electrodes can be of stainless steel or special aluminum alloys. Seals can be made of Teflon and ethylene propylene or silicone rubber. Reaction tanks are normally constructed of stainless steel, concrete, Fiberglas, or PVC. Some specific materials to avoid are natural rubber, PVC Type I, copper, brass, bronze, and galvanized steel.

Pressure, Temperature, and Dew Point

Following certain design criteria will increase ozone-generator output:

- Very dry input gas: $-45°C$ $(-50°F)$ dew point
- Low-temperature input gas: below $4°C$ $(40°F)$
- Sufficient cooling-water flow: 2 L/g O_3 (240 gal/lb O_3) at $16°C$ $(60°F)$ or lower
- High-dielectric-constant material of minimum wall thickness
- A critical ratio of discharge-gap width to pressure ratio
- High voltage and frequency

Trade-offs must be made to optimize the ozone output. High voltage and frequency increase the temperature, which can cause dielectric fractures.

Monitoring and Control

Monitoring equipment is used to provide operating information related to efficiency and reliability. These items include power meters, ozone-concentration meter for the gas stream, and monitors for cooling-water flow and input temperature, and for dew point. Control equipment is used to adjust or turn off the ozone system to ensure protection of the generators, environment, and plant operators. Control equipment is normally recommended by the manufacturer.

7.6.4 ENVIRONMENTAL EFFECTS

Air

All ozone systems should include some type of ozone-destruction equipment to prevent excess ozone from entering the environment. Air systems designed for odor control must provide sufficient ozone to react with the contaminant as well as turbulent mixing and adequate reaction time. It is not practical to continuously add stoichiometric amounts of ozone resulting in ozone-free and/or odor-free off gas. Monitoring and/or control equipment is used to verify off-gas safety.

Excessive concentrations of ozone discharged to the environment are corrosive to electrical components and metals previously identified, as well as detrimental to foliage.

Water and Residuals

Ozone will react with almost all oxidizable materials in an aqueous system. Therefore a slight overdose of ozone is normally injected to aqueous systems so as to result in a residual amount of ozone. If excessive ozone is added, the off gas from the reactor will contain ozone. Unreacted residual ozone will revert back to oxygen in a relatively short period of time.

7.6.5 CASE STUDIES

General Motors: Cyanide Destruction

Ozone is a more powerful oxidizing agent than chlorine. Only 2 to 3 parts of ozone (depending on the metal content) is required to rapidly oxidize 1 part of cyanide. The ozone is generated on site from air or oxygen and used immediately. This rapid cyanide destruction by ozonation indicates the reaction is mass-transfer-oriented. A second reaction, which proceeds slowly because it is kinetically oriented, is the oxidation of the cyanate ion to bicarbonate, ammonia, nitrogen, and oxygen.

Similar cyanide-degradation curves were obtained during an extensive series of laboratory and flowthrough pilot-plant studies on plating wastewaters. Figure 7.6.4 shows that approximately 80 mg O_3/L (absorbed) was required to decrease the total cyanide (TCN) to 2.85 mg/L and free cyanide (FCN) to 0.0 mg/L. There-

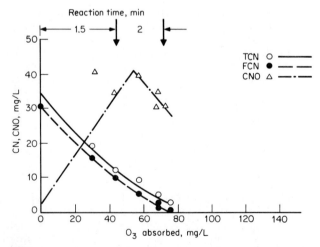

FIG. 7.6.4 Cyanide degradation with single ozone injection. TCN = total cyanide. FCN = free cyanide. CNO = cyanate.

fore 2.5 mg O_3/mg FCN, or 1.3 mol O_3/mol FCN is required to oxidize the FCN to zero. Theoretically 1.85 mg O_3/mg FCN is required for complete oxidation.

Theoretically 52.6 mg cyanate (1.3 mol CNO) should be formed as a result of the cyanide destruction. However, some of the ozone is reacting with the cyanate before all the FCN is oxidized. The slopes of the cyanide-degradation curves (speed of reaction) can be altered by changing the rate of ozone absorption in the wastewater.

Ozone-to-cyanide ratios and reaction rates will change with varying cyanide concentration as well as with differences in the amounts of metals (copper, iron, cadmium, zinc) contained in plating wastes. Other factors that affect the rate of reaction are pH, temperature, and ozone concentration.[7]

The difference between the FCN and TCN after ozonation (Fig. 7.6.4) can be related to the presence of heavy-metal complexes. After the FCN has been oxidized, additional ozone injection oxidizes the cyanates and precipitates the heavy metals.

Settling and/or filtration can remove the final amounts of TCN and minimize the heavy-metal concentrations in the treated effluent.

Textile Wastewaters

Because they contain a variety of organic dyes and chemical additives, textile-dye wastewaters are extremely complicated and difficult to treat. Unit treatment processes for textile wastewaters include biological, sedimentation, chemical coagulation, polymer flocculation, activated carbon, and more recently, ozone. One or more of these unit processes can be used to treat textile wastewaters. Process selection depends on the type and concentration of the dye waste as well as the quality of wastewater needed for acceptable discharge.

A mobile pilot plant was assembled to treat textile wastewater at several manufacturing plants. At one plant, ozone and/or chemical coagulation was used to obtain various degrees of effluent quality—from opaque to crystal clear. The object was to recycle rather than discharge the wastewater. Twenty test runs were completed with eight different dyes and 100% wastewater recycling. Ozone was the only treatment. Freshwater makeup for evaporation was added to the closed-loop water system. Chemical additions during dye-batch changeovers were reduced by 75%. Energy savings were obtained as a result of treating the water and returning it to the system at 60 to 66°C (140 to 150°F) whereas the fresh-water supply was at 20°C (68°F). Dyed products were acceptable.[6]

Hazardous-Waste Sites

Abandoned and active hazardous-waste sites, hazardous plumes without identified sites, and municipal landfills containing hazardous wastes are constantly being identified. Since most landfills do not have liners and since hazardous materials are disposed of at these sites by a variety of methods, the leachates, surrounding soils, and waters (surface and/or groundwater) contain a wide variety of hazardous compounds.

Standard and innovative technology can be used to detoxify wastes from these hazardous-waste sites. One identifiable process is oxidation. Since ozone is one of the most powerful oxidants and can be generated on site, greater interest in its application is expected. This is especially true if in situ or on-site treatment is selected. On-site treatment is expected to be less costly than transferring the hazardous waste to another location.

Over a period of years, petroleum products were spilled in an area near the Durlacher Wald water-treatment plant in Germany. This resulted in contamination of the groundwater wells. An abandoned chemical dump site was leaching cyanide from the opposite side of the landfill. Ozone is being used to oxidize the biorefractory organic pollutants so that a natural underground biological activity can purify the water supply.[8]

7.6.6 CURRENT RESEARCH AND ADVANCES IN THE STATE OF THE ART

IOA Journal

The International Ozone Association (IOA) publishes its *Science and Engineering Journal* several times per year. This journal presents worldwide technical papers related to a wide variety of ozone applications. The IOA has regional committee meetings at least twice a year and an international conference every two years. The present IOA Headquarters is in Zurich, Switzerland, and the Pan American Committee is based in Norwalk, Connecticut.

Papers in Science and Engineering Journal

Some topics related to hazardous-waste treatment with ozone that are covered in recent technical papers in the *Science and Engineering Journal* include ozonation of aromatic compounds, waste-gas purification plants using ozone, high temperature wastes from manufacturing TNT and RDX (cyclonite) using ozone and ultrasound, oxidation of organic compounds using ozone and hydrogen peroxide, ozone schemes for destruction of organophosphorus pesticides, PAC- and ozone-assisted activated-sludge treatment of toxic organic compounds, and ozonation of aniline and anilinium.

U.S. Environmental Protection Agency

The U.S. Environmental Protection Agency (EPA) prepared a design manual for the standardization of terms and of specifications for ozone systems for ozone disinfection of wastewater. In addition to equipment-design standards, the manual includes sections on ozone properties and terminology, process control, safety, and associated costs. This type of information could be applied to ozone systems for hazardous-waste treatment.

7.6.7 ENGINEERING ECONOMIC EVALUATION

Capital Costs

It is difficult to obtain specific capital costs for ozone systems; this is because of the highly competitive nature of the industry as well as the individual manufacturers custom-design to meet clients' specifications. Capital costs can vary as much as 50% depending on the interpretation of ozone-generator performance specifications. Major factors that must be evaluated include maximum ozone output at a specific concentration, variable ozone output at a specific concentration, auxiliary equipment (air preparation, reaction tanks), monitoring and control instrumentation, spare parts, installation costs, start-up and training, and performance warranties.

Some budgetary costs (average prices received from several manufacturers) for ozone generated from air at 1% O_3 by weight, including air preparation but no reaction tanks are 19 g/h (1 lb/d), \$10,000; 95 g/h (5 lb/d), \$22,000; 500 g/h (26 lb/d), \$50,000; 1,000 g/h (52 lb/d), \$76,000; 5,000 g/h (260 lb/d), \$275,000; 10,000 g/hr (520 lb/d), \$680,000.

Operating Costs

Ozone-generator operating costs depend primarily on the cost of electrical energy. Other factors affecting these costs include peak electric demand for the dryers, ozone concentrations used, blowers and compressors, and types of injectors used. Power consumption will range from 15 to 26 Wh/g (6.8 to 11.8 kWh/lb) for an air feed system at 1.5% O_3 by weight including air preparation and ozone injection.

High frequency (2,000 kHz) and oxygen feed will reduce these energy costs by approximately 50%. Operations and maintenance costs are relatively low (approximately 20 to 30%) of ozone systems.

7.6.8 EQUIPMENT SUPPLIERS

- Wellsbach Division of Polymetrics, Inc., Sunnyvale, Calif.
- Emery Industries, Inc., Cincinnati, Ohio
- Griffin Technics, Inc., Lodi, N.J.
- Infilco Degremont, Richmond, Va.
- Mitsubishi Electronic Sales Medama, Rancho Dominguez, Calif.
- PCI Ozone Corporation, West Caldwell, N.J.
- Hankin Environmental Systems, Ontario MIR4G3, Canada
- Trailigaz, F-95140 Garges-les-Gonesse, France
- Mannesman Handel AG, 4030 Ratingen 1, Germany
- Messer Griesheim GmbH, 4000 Dusseldorf, Germany
- Schmidding-Werke GmbH & Co., 5000 Koln 60, Germany
- Ebara Jitsugyo Co., Ltd., Tokyo, Japan

7.6.9 REFERENCES

1. F. Novak, "Ozone for Industrial Wastewater Treatment," Eighth Annual Industrial Pollution Conference of WWEMA, Houston, Texas, June 6, 1980.

2. E. Leitis, J. Zeff, and L. Cole, "Progress In The UV-Ozonation of Industrial Wastewaters," Sixth World Congress—International Ozone Association, May 1983.

3. *Briefing-Technologies Applicable to Hazardous Waste*, sec. 2.20, *Chemical Oxidation*, prepared by Metcalf & Eddy, Inc., for the U.S. EPA, Cincinnati, Ohio, May 1985.

4. J. Staehelin, Jr., and H. Hoigne, "Decomposition of Ozone in Water in the Presence of Organic Solutes Acting as Promoters and Inhibitors of Radical Chain Reactions," *Environmental Science and Technology*, **19**:206 (December 1985).

5. H. Hoigne, H. Bader, W. Haag, and J. Staehelin, "Rate Constants of Reactions of Ozone With Organic and Inorganic Compounds In Water," *Water Research*, **19**:(8)993–1004 (1985).

6. F. Novak, "Destruction of Cyanide Wastewater By Ozonation," Third International Symposium on Ozone, 1979, International Ozone Association.

7. G. Sukes, R. Pordon, and K. Gupta, "The Destruction of Cyanide in Wastewater with Ozone," Water Pollution Control Federation Conference, October 1984, New Orleans.

8. R. Rice, "Purification and Recycling of Groundwater Contaminated with Petroleum Products and Cyanides—The Karlsruhe (Federal Republic of Germany) Drinking Water Treatment Plant," *Management of Uncontrolled Hazardous Waste Sites, Proceedings of Fifth National Conference*, November 1984, pp. 600–603.

7.6.10 FURTHER READING

The Journal of the International Ozone Association (IOA), and *Science and Engineering Journal*, Pan American Committees, 83 Oakwood Avenue, Norwalk, Connecticut 06850. Also available from the IOA: a series of books related to ozone applications for organic reactions and odor control, and the proceedings from six international symposia.

American Chemical Society, *Ozone Chemistry and Technology*, Am. Chem. Soc., Washington, D.C., 1959.

Evans, F., III, *Ozone in Water and Wastewater Treatment*, Ann Arbor Science, Ann Arbor, Michigan, 1972.

Rice, R., and M. Browning, *Ozone treatment of Industrial Wastewater*, Noyes Data Corp., Park Ridge, New Jersey, 1981.

SECTION 7.7
EVAPORATION

B. Tod Delaney
Ground/Water Technology Inc.
Rockaway, New Jersey

Ronald J. Turner
U.S. Environmental Protection Agency
Hazardous Waste Engineering Research Laboratory
Cincinnati, Ohio

Evaporation has long been used as a unit operation in the manufacture of various products in the chemical-process industries. In addition, it is currently being used for the treatment of hazardous wastes such as radioactive liquids and sludges, metal-plating wastes, and other organic and inorganic wastes. In industries which generate large volumes of aqueous wastes with dilute hazardous constituents, evaporation can be used to concentrate the waste stream for additional treatment or shipment off site. For this reason, much of the subsequent discussion in this section deals with evaporation as a concentration process which is used primarily to reduce the volume of an aqueous waste. The "solvent" is water and the "solute" is an inorganic or organic solid with very low vapor pressure at the evaporator temperatures.

If an aqueous hazardous waste contains moderately volatile organic constituents, treatment after evaporation may be required to separate these organics from the water phase. Final disposal of the concentrated residue is also required.

The major disadvantages of evaporation as a waste-treatment technology are high capital and operating costs, and high energy requirements. Potential operational problems include salt buildup on heat-exchange surfaces, foaming, and solids decomposition.

7.7.1 APPLICATION TO HAZARDOUS WASTES

The application of evaporation technology to the treatment of hazardous waste is limited by the character of many hazardous waste streams and by limitations inherent in evaporator design. Design choice is dependent on the liquid to be evaporated. This means that the waste stream must be completely characterized and relatively homogeneous in nature. This is often not the case with waste streams, particularly those treated off site.

Despite this, evaporation is used in the treatment of hazardous wastes. The process equipment is quite flexible and can handle wastes in various forms: aqueous, nonaqueous, slurries, sludges, and tars. Evaporation is commonly used as a pretreatment method to decrease quantities of material for final treatment. It is also used in cases where no other treatment method has been found to be practical, such as in the concentration of trinitrotoluene for subsequent incineration. Along with its use in industry, it is also used by several independent operators of hazardous-waste treatment or disposal facilities.

One particular site utilizes 101 hectares (250 acres) of surface ponds for holding aqueous wastes and rainwater runoff. This site also uses sprays to enhance evaporation and estimates that 12 million gal of water is evaporated per month.

The most effective evaporation technique for treatment of hazardous waste is solar evaporation. Because of the simplicity of design, many of the limitations of other evaporator technologies do not apply.

Traditionally, solar evaporation has been used as a method for sludge dewatering. Recently, it has been used quite extensively in the handling of hazardous wastes, specifically pesticides (from pesticides spraying) and fluoride wastes. Use of solar evaporation has many advantages: It conserves energy and resources; the required capital investment is low; nonhomogeneous waste streams can be treated; and it causes minimal release of pollutants to the environment. However, it also has several disadvantages. It is greatly dependent upon climatic conditions (humidity, precipitation, wind velocity, and solar-radiation intensity), requires utilization of large amounts of land, is attractive to waterfowl, and is largely dependent upon the integrity of the liner used. State and federal requirements may limit the use of solar evaporation in the future.

The most important factor to consider is the rate of precipitation versus the rate of evaporation. Most favorable areas in the United States are primarily in the southwest, where evaporation exceeds rainfall by a factor of 2. Solar evaporation can be employed in other geographical areas if closed systems are used to prevent rainfall from accumulating in evaporation ponds or lagoons. In those areas where annual precipitation equals or exceeds annual evaporation, the effectiveness of transparent materials (to allow maximum penetration of solar radiation) as roofing materials for evaporation has been studied. Results indicate that a reduction in effective evaporation of 25 to 30% from unroofed ponds should be incorporated in designing roofed evaporation lagoons.

7.7.2 *EVAPORATION PROCESS*

Evaporation is defined as the conversion of a liquid into vapor. Specifically, it involves the vaporization of a liquid from a solution or a slurry. The objective of evaporation is to concentrate a solution which is composed of (1) a volatile solvent and (2) a solute which is not appreciably volatile; concentration is accomplished by driving off the solvent as a vapor. Unlike the residue produced by a *drying* process, which is generally a solid, the residue produced by evaporation is usually a more viscous liquid. Evaporation differs from *distillation* in that no attempt is made to fractionate the vapor.

All evaporation systems require the transfer of sufficient heat from a heating medium to the process fluid to vaporize the volatile solvent. The amount of heat required to cause the transformation is called the *latent heat of vaporization.*

The processes generally used for evaporation are called *boiling heat transfer* and *flash vaporization.* Boiling heat transfer is the most commonly used. Here, liquor is transformed to a vapor on the surface of an evaporator body by condensing steam or other source of heat. Flash vaporization occurs as the result of flow into a region where the pressure is below the saturation vapor pressure of the liquid. In this case, the heat required for evaporation comes from the sensible heat released during cooling of the liquid. Flash vaporization takes place within an enclosed vessel.

7.7.3 *EVAPORATOR TYPES*

The function of evaporation equipment is (1) to provide the means to transfer heat to the liquid, and (2) to allow the vaporization process to occur. The evaporation process has been accomplished commercially with a variety of equipment designs. The three most common types are the rising-film, falling-film, and forced-circulation evaporators. The first two rely on boiling heat transfer and the latter relies on flash vaporization.

Rising-Film Evaporator

In a *rising-film evaporator,* a film of liquor is evaporated as it passes upward in a film on the inside of heated tubes. The liquor enters the bottom of the tubes, usually at less than boiling point. As the liquor begins to boil, vapor is generated which moves upward carrying liquor with it, resulting in an upward-flowing film (rising film) on the interior of the tubes.

The nature of this flow is similar to the familiar "percolator" action of a coffee pot. A deflector plate mounted above the tubes provides primary separation of the vapor and liquid exiting the tubes. The vapor enters a reservoir mounted above the heating element, where it undergoes additional entrainment separation. The liquor is collected in an annular space at the lower end of the vapor body and flows from there to the next vessel in the evaporator train. The vapor is passed through a heat exchanger which provides heat for the tubes in the next evaporator vessel.

The principal parts of the body of a rising-film evaporator are the heating element, vapor body, liquor box, and deflector plate. A typical configuration is shown in Fig. 7.7.1.

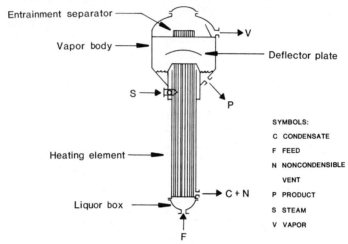

FIG. 7.7.1   Body of a rising-film evaporator.

Falling-Film Evaporator

In a *falling-film* evaporator, a film of liquor is evaporated as it flows downward on the inside of heated tubes or the outside of heated plates. The liquor is pumped to the top of the heating element, where it is distributed in a film onto the heating-element surface. As the liquor flows downward it is heated to boiling (if not already at the boiling point) and vapor is driven off.

A plate design utilizes hollow plates formed from two sheets of metal as the heating surface. Plates are arranged into a bundle to form a heating element. Steam flows through the inside of the plates through a common header. Removal of condensate and steam is also through headers. Liquor is continuously recirculated to the top of the heating element and distributed in a film onto the outside of the vertical plates. The film of liquor flows by gravity down the outside of the heated plates and into a reservoir at the bottom of the vapor body. The vapor produced flows out the top through an entrainment separator.

In the case of a tubular design, there is a continuous flow of liquor and vapor down the interior of heated pipes. A pump continuously circulates the liquor from the bottom of the tubes to the top. This also keeps the vapor generated in the tubes under pressure, assuring its downward flow. A reservoir mounted at the lower end of the heating element provides for separation of the vapor–liquid mixture. Primary separation is achieved by scrubbing of the vapor as it flows through the liquid curtain at the bottom of the tubes and centrifugal action as the vapor turns 180° to flow out the upper part of the vapor body.

The principal parts of the body of a falling-film evaporator are the heating element, vapor body, recirculation pump, and distributor, the latter used to uniformly film the liquor onto the heating surface. A vertical-tube, falling-film body is shown in Fig. 7.7.2.

Forced-Circulation Evaporator

In a *forced-circulation evaporator,* liquor is evaporated by flash vaporization resulting from an increase in heat achieved by flow through heated tubes. The li-

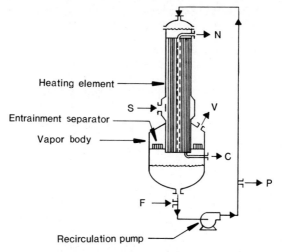

FIG. 7.7.2 Body of a falling-film evaporator.

quor is continuously recirculated from a reservoir through a tubular heating element and back to the reservoir by means of a pump. Sufficient pressure is maintained within the heater to allow heating of the liquor without boiling in the tubes. Flash vaporization (evaporation) occurs as the liquor reenters the reservoir, which operates at a pressure slightly lower than the saturation vapor pressure of the returning liquor.

The principal parts of a forced-circulation evaporator are the heating element, vapor body, and recirculation pump. A typical configuration is shown in Fig. 7.7.3.

Additional types of evaporators that are commonly used are solar (previously discussed), batch-pan, and natural-circulation evaporators.

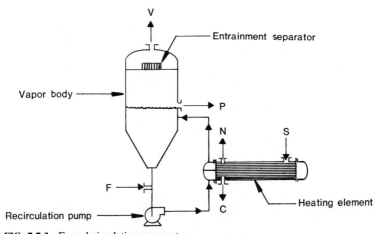

FIG. 7.7.3 Forced-circulation evaporator.

Batch pans are the simplest mechanical evaporation devices. Heat is generally supplied by use of a jacketed vessel or a vessel with internal coils. Evaporation occurs in batches as opposed to a flowthrough system.

Natural-circulation evaporators are similar in design to shell-and-tube heat exchangers. They are generally vertical in design with the fluid making a single pass through the system. The flow of liquid through the tubes is driven by the gradient in densities between the liquid in the central liquid-return tube and the two-phase mixture in the evaporation tubes.

7.7.4 AUXILIARY EQUIPMENT

Heat exchangers, flash tanks, pumps, and ejectors are common auxiliary equipment items incorporated with evaporator bodies to complete an evaporator system. Since the performance of heat exchangers and flash tanks is included in a heat and mass balance, a brief description of those items will be given.

Heat exchangers are used to (1) preheat liquor and (2) condense vapors from the last effect. The heat exchangers are typically of the tube-and-shell design with liquid on the tube side and vapor on the shell side. High heat-transfer performance is achieved by maintaining adequate liquid velocities in the tubes. The principal parts of a heat exchanger are a heating element and liquid reservoir. Liquor preheaters can either stand alone or be integrated into the evaporator.

Liquor and condensate will flash as they flow to regions where the pressure is below the saturation vapor pressure of the liquid. Evaporation occurs, converting sensible heat to latent heat as the liquid cools. In some cases, the flashing can occur within an evaporator reservoir, making it unnecessary to add special equipment. However, when it is necessary to keep streams separated, separate flash tanks can be used. Liquid streams are generally flashed to achieve an improved steam economy by putting heat into the system where it can accomplish the most additional evaporation.

A flash tank is usually a vertical, cylindrical vessel enclosed at both ends. The liquid is normally introduced tangentially to the cylinder near the bottom. An orifice plate is placed in the feed line to the vessel to cause the flashing to occur beyond that point. Primary vapor/liquid separation is achieved by (1) centrifugal flow of the liquid, and (2) a low upward velocity of the vapor to promote settling of entrained droplets. The vapor discharges from the top of the vessel and liquid from the bottom.

7.7.5 SELECTION OF EVAPORATORS

Properties of the liquid to be evaporated are critical in final selection of an appropriate evaporator system. Since operating costs are a significant factor in overall cost, heat-transfer characteristics and energy requirements are important considerations. Other considerations are the potential for corrosion, foaming, entraining, salting, and scaling. Some of these concerns can be largely eliminated by use of proper construction material. Some can be controlled through design choice. Each type of evaporator design has advantages and disadvantages. However, the appropriate design is specific to the liquid to be treated and must be determined through the use of pilot units. Properties of liquids which are critical to determination of final design include the following:

- *Heat capacity:* *Heat capacity* is the amount of heat required to raise the temperature of a unit mass of material one degree or the amount of heat released by a one degree temperature decrease of a unit mass. This type of heat is known as *sensible heat.*

- *Heat of vaporization:* *Heat of vaporization* is the amount of heat needed to vaporize a unit mass of liquid at constant temperature and pressure. This type of heat is encountered with a change in phase and is termed *latent heat.*

- *Density:* *Density* is the ratio of the mass of a material to its volume.

- *Thermal conductivity:* *Thermal conductivity* is a measure of the rate of heat transfer by conduction through a material per unit area and per unit temperature gradient.

- *Boiling-point rise:* *Boiling-point rise* is the difference between the boiling point of liquor and that of pure water at the same pressure.

- *Heat-transfer coefficient:* The overall *heat-transfer coefficient* is a measure of the rate of heat transfer between the steam and liquid per unit surface and per unit temperature difference.

7.7.6 ENVIRONMENTAL CONSIDERATIONS

Evaporation is an expensive technology, both in terms of capital costs and operating costs. Additionally, mechanical evaporation produces a condensate and a bottoms stream, one or both of which may require further processing or disposal.

Solar evaporation is often less expensive, but the cost of land for evaporation may be high. In addition lagoons used for solar evaporation often require bottom liners to prevent the migration of hazardous constituents into the groundwater and will need to meet other pertinent federal and state environmental regulatory standards. For instance, the State of California requires that a solar evaporation pond must be emptied once every 12 months to be considered as a treatment process rather than a disposal practice.

7.7.7 FURTHER READING

Dierks, R. D., and W. F. Bonner, "Putting Evaporators to Work: Wiped Film Evaporator for High Level Wastes," *Chemical Engineering Progress,* 72(4):61-62 (April 1976).

Fosberg, T. M., and H. L. Claussen, "Falling-Film Evaporators Recover Chemicals Efficiently," *The Journal of the Technical Association of the Pulp and Paper Industry,* 65(8):63–66 (August 1982).

Goodlett, C. B., "Putting Evaporators to Work: Concentration of Aqueous Radioactive Waste," *Chemical Engineering Progress,* 72(4):63–64 (April 1976).

Mehra, D. K., "Selecting Evaporators," *Chemical Engineering,* 93(3):56–72 (February 3, 1986).

Petrie, J. C., et al., "Putting Evaporators to Work: Vacuum Evaporator-Crystallizer Handles Radioactive Waste," *Chemical Engineering Progress,* 72(4):65–71 (April 1976).

Shuckrow, A. J., et al., *Concentration Technologies for Hazardous Aqueous Waste Treatment,* Contract No. 68-03-2766, U.S. EPA Water Engineering Research Laboratory, Cincinnati, Ohio, 1980.

Stickney, W. W., and T. M. Fosberg, "Treating Chemical Wastes by Evaporation," *Chemical Engineering Progress,* **72**(4):41–46 (April 1976).

Wakamiya, W., "Shale Oil Wastewater Treatment by Evaporation," *Chemical Engineering Progress,* **77**(5):54–60 (May 1981).

Welty, R. K., "Solar Evaporation of Fluoride Wastes," *Chemical Engineering Progress,* **72**(3):54–57 (March 1976).

SECTION 7.8

SOLIDIFICATION AND STABILIZATION TECHNOLOGY

Carlton C. Wiles

U.S. Environmental Protection Agency
Hazardous Waste Engineering Research Laboratory
Cincinnati, Ohio

In hazardous-waste management, *solidification and stabilization (S/S)* is a term normally used to designate a technology employing additives to reduce the mobility of pollutants, thereby making the waste acceptable under current land disposal requirements. The use of this technology to treat hazardous waste may become more important as regulations restrict the use of land for disposing of hazardous waste. This section of Chap. 7 reviews the technology and provides information to help assess its potential role in managing hazardous waste. Information is provided to assist in the proper selection, use, and evaluation of S/S technologies. Regulatory factors affecting their use are also discussed.

7.8.1 DEFINING SOLIDIFICATION, STABILIZATION, AND ENCAPSULATION

Solidification and *stabilization* are treatment processes designed to either improve waste-handling and physical characteristics, decrease surface area across which pollutants can transfer or leach, or limit the solubility of or detoxify the hazardous constituents. *Solidification* implies that these results are obtained primarily by production of a monolithic block of treated waste with high structural integrity. *Stabilization* describes processes which limit the solubility of or detoxify the contaminate; the physical characteristics may or may not be improved or changed. The term *fixation* is used to mean either solidification or stabilization. *Surface encapsulation* is defined as a technique to isolate the waste by placing a jacket or membrane of impermeable material between the waste and the environment.[1]

It is important that one understand what is meant when discussing the technology. The following definitions are being used in this section:

- *Solidification:* A process in which materials are added to the waste to produce a solid is referred to as *solidification*. It may or may not involve a chemical bonding between the toxic contaminant and the additive.
- *Stabilization:* *Stabilization* refers to a process by which a waste is converted to a more chemically stable form. The term includes solidification, but also includes use of a chemical reaction to transform the toxic component to a new, nontoxic compound or substance. Biological processes, however, are not considered.
- *Chemical fixation:* *Chemical fixation* implies the transformation of toxic contaminants to a new, nontoxic form. The term has been misused to describe processes which did not involve chemical bonding of the contaminate to the binder.
- *Encapsulation:* *Encapsulation* is a process involving the complete coating or enclosure of a toxic particle or waste agglomerate with a new substance, e.g., the S/S additive or binder. *Microencapsulation* is the encapsulation of individual particles. *Macroencapsulation* is the encapsulation of an agglomeration of waste particles or microencapsulated materials.

7.8.2 ADVANTAGES AND DISADVANTAGES

Advantages and disadvantages of S/S will vary with the process, the binders, the waste, the site conditions, and other unique factors. As an example, processes using reactions of the pozzolan-cementation type are relatively low in cost and easy to use. However, these processes will increase the total volume of end material which must be managed. In many cases, the volume increases can be significant (even double). In the case of encapsulation with polymeric materials, volume increase can be very small and in some cases product performance significantly increased, but sometimes at an increased cost in dollars and difficulty of processing. Specific conditions must be carefully considered when comparing the advantages and disadvantages of S/S processes. More information is provided on each process in the following subsections.

7.8.3 CHARACTERIZING S/S TECHNOLOGIES

Solidification and stabilization technology may be characterized by the binder used or by the binding or containment mechanism; or it may be categorized by process type.

Binders

Binder systems can be placed in two broad categories: inorganic and organic. Most inorganic binding systems in use include varying combinations of hydraulic cements, lime, pozzolans, gypsum, and silicates. Organic binders used or experimented with include epoxy, polyesters, asphalt, polyolefins (primarily polyethylene and polyethylene-polybutadiene), and urea-formaldehyde. Combinations of inorganic and organic binder systems have been used. These include diatomaceous earth with cement and polystyrene; polyurethane and cement, and polymer gels with silicate and lime cement.[2]

Knowledge of the binders currently in use for waste solidification and stabilization can be important in assessing and selecting a technology. Such knowledge provides insights into processing requirements, waste-pretreatment requirements, waste-binder interactions, and expected product performance.

Pollutant Binding Mechanisms

Solidification and stabilization systems may also be classified or identified by waste-containment mechanism or reaction type.[3] These include:

Sorption.   *Sorption* involves adding a solid to take up any free liquid in a waste. Examples are activated carbon, anhydrous sodium silicate, gypsum, clays, and similar particulate materials. Most sorption processes merely remove the liquid onto the surface of the solid (similar to a sponge soaking up water), and do not reduce contaminate leaching potential. Sorption has been used to remove free liquids from waste in order to satisfy the no-liquids-in-landfill requirements. However, under provisions of the Hazardous and Solid Waste Amendments (HSWA) the use of sorbents that merely soak up liquids is no longer permitted. Selected sorbents, however, can be used to enhance the performance of solidification and stabilization processes.

Pozzolan Reactions.   This process uses a fine, noncrystalline silica in fly ash and the calcium in lime to produce low-strength cementation. Physical trapping of the contaminant in the cured pozzolan-concrete matrix is the primary containment mechanism. Water is removed in hydrating the lime-pozzolan concrete. Some characteristics of lime–fly ash pozzolanic processes are:

- Requires that waste be blended with reactive fly ash (or other pozzolanic material) to a pasty consistency. Free water is required for reactions.
- Lime is blended into the waste–fly ash mixture. Typically 20 to 30% lime is required to produce acceptable strength, but this varies with the type of fly ash used.
- Bonding depends on formation of calcium silicate and aluminate hydrates.
- System is highly alkaline. With certain wastes this can cause release of undesirable gas or leachate.
- Materials such as sodium borate, carbohydrates, and potassium bichromate may chemically interfere with the bonding reactions. Oil and grease may physically interfere to reduce contaminate containment.
- Without special additives or waste pretreatment, hazardous waste treated by lime–fly ash will probably still be classified as hazardous based upon the EPA's toxicity-leaching procedure (EP).
- System is relatively inexpensive.
- Cured composites of lime–fly ash and waste may not be as durable or control contaminate leaching as well as portland cement composites.

Pozzolan–Portland Cement Reactions.   In this process portland cement is combined with fly ash or other pozzolans to produce a relatively high-strength waste-and-concrete matrix. Waste containment is primarily by entrapment of waste particles. Soluble silicates may be added to aid processing and to assist in containment of metals through the formation of silicate gels. Water is removed in the hydration of the portland cement. In variations of this technology, gypsum or aluminous cement may be used with or instead of portland cement. Characteristics of this process are:

1. Pozzolan products added to portland cement react with free calcium hydroxide to improve strength and chemical resistance of solidified product.
2. Portland cement types are available and can be selected to enhance a specific desirable reaction. Type I portland cement has been most widely used.
3. A variety of materials (e.g., soluble silicate, hydrated silica gels, clays) have been added to increase or change a desired performance. In some cases, selected sorbents are required to help contain the pollutant in the rather porous solid matrix. Many of the additives are considered proprietary by the vendors. Some of the more common additives and their intended functions are described in general terms here:
   a. Soluble silicates, such as sodium silicate or potassium silicate. These agents will generally "flash set" portland cement to produce a low-strength concrete. These materials are beneficial in reducing the interference from metal ions in the waste solution.
   b. Selected clays to absorb liquid and bind specific anions or cations.
   c. Emulsifiers and surfactants to allow the incorporation of immiscible organic liquids. Research in the nuclear-waste field has indicated that waste turbine oil and greases can be mixed into cement blends if dispersing agents are used and the proper mixing system is employed.
   d. Proprietary absorbents that selectively bind specific wastes. These materials include carbon, silicates, zeolitic materials, and cellulosic sorbents; they hold toxic constituents and are encapsulated with the waste.
   e. Lime (CaO) can be used to raise the pH and the reaction temperature to improve setting characteristics.
4. Cement-based processes are more versatile than lime–fly ash processes, can be formulated for exceptional strength, and have been found to retain selected contaminants effectively.
5. Certain waste components may interfere with the setting and stability of cement-based processes. These include such materials as borates, calcium sulfate, phenol, oil, and grease. Acids can react with the concrete and destroy the matrix after setting, thus abetting the release of pollutants to the environment.

Thermoplastic Microencapsulation.   This process blends waste particulates with melted asphalt or similar materials. Physical entrapment is the primary containment mechanism for both liquids and solids. Additional information on this process is given here:

- The waste is dried and mixed with a heated plastic material, such as asphalt, polyethylene, or wax.
- The process can be effective for some soluble and toxic materials not readily treated by other processes.
- Compatibility of waste with the plastic binder is more important than in other

processes because some mixes of reducers and oxidizers are potentially reactive at elevated temperatures. For instance, greases can soften asphalt, and xylene and toluene will diffuse through asphalt. Thermoplastic processes are sensitive to high salt concentrations.

• The waste may require more pretreatment for some thermoplastic processes than for pozzolan-cementation processes. Pollutant containment, however, may be significantly improved for selected compatible waste.

• Processing is more difficult (e.g., higher temperatures, specialized equipment) than for pozzolan and pozzolan–portland cements and therefore more expensive.

Macroencapsulation.   This process isolates a large volume of waste by jacketing with any acceptable material. A 55-gal drum is a simple example. More sophisticated and more effective macroencapsulation processes employing polyethylenes and similar resins have been investigated.

Although macroencapsulation, except for overpacking, has not been widely used, it is considered in this chapter in order to provide insight into its potential for helping to overcome contaminate-retention problems which may occur with other processes. The approach involves the use of a generally applicable S/S process (e.g., pozzolanic, lime–fly ash, pozzolan-cement) to microencapsulate the contaminate; to compensate for potential incomplete contaminate isolation, this is followed by macroencapsulation of the matrix containing the contaminant.

Several alternatives can be considered for macrocapsulation.[4] Figure 7.8.1 depicts three different schemes. Overpacking (scheme 3) with 55-gal metal drums has been a method of choice in many operations, but this technique suffers from potential relatively quick deterioration of the container. Also leakage from lids and bung holes is a problem. Improvements (e.g., plastic liners, new closure devices) have been made. Plastic containers with heat-sealed or friction-sealed lids have also been introduced.

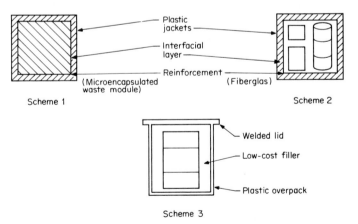

FIG. 7.8.1   Three schemes for macroencapsulating hazardous waste: scheme 1, polymeric resin fused onto surface of microencapsulated waste; scheme 2, polymeric resin fused onto or sprayed onto Fiberglas substrate holding containers of waste; scheme 3 overpacking of 55-gal waste containers with welded plastic containers.

Three methods can be used for macroencapsulation by the resin fusion process of schemes 1 and 2: (1) microencapsulation of waste particulates with polybutadiene and subsequent macroencapsulation of the matrix with high-density polyethylene (HDPE), (2) use of low-cost cementing agents to form a waste block onto which the HDPE jacket is fused, and (3) use of a preformed Fiberglas–thermosetting resin matrix as a substrate for the HDPE jacket. These are depicted in Fig. 7.8.2.

In the first method, 1,2-polybutadiene is mixed with the particulated waste; after solvent evaporation, thus yields free-flowing, dry, resin-coated waste particulates. Further processing involves the formation of a block of the polybutadiene-waste mixture by the application of moderate heat and pressures in a mold. The final encapsulated module is produced by fusing a ¼-in thick HDPE jacket onto the block.

The second method involves the use of low-cost pozzolans or pozzolan–portland cement to form the waste block onto which the HDPE jacket is fused. This modification provides the opportunity to process sludges without having to form dry particulates and to treat the limited number of wastes that are not compatible or will not agglomerate with polybutadiene.

The third method employs a Fiberglas–thermosetting resin matrix as a substrate for the HDPE jacket. Powdered HDPE is fused onto Fiberglas; this yields receivers to contain hazardous waste, which is usually containerized. The receivers are then sealed with additional Fiberglas (for the lid) and the HDPE.

Equipment and processes for methods 1 and 2 have been demonstrated for managing low-level radioactive waste.[5] The discussions of particular macroencapsulation processes in this section emphasize particular materials for illustrative reasons and because research has concentrated on them. Other materials may also be applicable. The cost of this resin fusion type of macroencapsulation process will vary with the materials used and the degree of waste pretreatment required. Performance advantages over pozzolan and pozzolan-cement and similar S/S processes, however, may be realized in increased volume ratios of waste to binder and in improved pollutant containment.

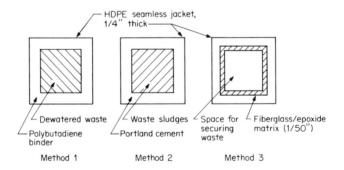

FIG. 7.8.2 Three methods that can be used for macroencapsulation by the resin fusion process of schemes 1 and 2: method 1, microencapsulation of particulated waste with polybutadiene and subsequent macroencapsulation with HDPE; method 2, use of low-cost cementing agents to form a waste block onto which HDPE jacket is fused; method 3, use of Fiberglas-epoxide matrix as substrate for HDPE jacket.

Process Types

There are several S/S processing schemes available for consideration. These include

- *In-drum processing:*   In this process, the S/S binders are added to the waste contained in a drum or other container. After mixing and setting, the waste-binder matrix is normally disposed of in the drum.

- *In-plant processing:*   *In-plant processing* refers to a plant and/or process specifically designed for solidification and stabilization of bulk waste material. The process may be conducted within a plant to manage the waste from an internal industrial operation or a plant may be specifically designed and operated to solidify and stabilize waste from external sources.

- *Mobile-plant processing:*   *Mobile-plant processing* refers to S/S processes and equipment which either are mobile or can be easily transported from, and set up, site to site.

- *In situ processing:*   The addition of binders directly to a lagoon or the injection of solidifying or stabilizing materials to the soil subsurface, etc., to promote the solidification and stabilization of the contaminated sludge and/or soil are referred to as in situ processing. This type of S/S processing may have considerable application to remediation of contaminated soils and surface impoundments at Superfund sites.

7.8.4   EVALUATING THE PERFORMANCE OF S/S

At present there are no established standards and protocols for testing and predicting performance of S/S products for treating nonradioactive hazardous waste. Tests described in the Resource Conservation and Recovery Act (RCRA) regulations—the Extraction Procedure (EP) and a modification of the EP, the Multiple Extraction Procedure (MEP)—are classification tests (e.g., tests that are used to classify waste as to whether or not it is hazardous).[6] However, the leaching procedures are used to support waste-delisting decisions. Therefore they can be used to determine delisting potential of a waste that has been solidified and stabilized. If the S/S product does not pass, it is still considered a hazardous waste and must be managed accordingly. A new proposed procedure, the toxicity-characteristics leaching procedure (TCLP), is designed to accommodate wastes containing organics.[7] The TCLP can require leaching be done at a lower pH than in the EP depending upon specific conditions encountered during the test. Without care in formulating and processing to achieve products with better performance characteristics, waste materials that have been solidified and stabilized may have a more difficult time meeting leaching requirements for delisting.

Most procedures used to test the physical properties of solidified waste involve standard testing procedures for concrete, such as confined and unconfined compressive strength tests; the wet-dry, freeze-thaw durability test; and similar procedures to help determine structural integrity and durability. These procedures and modifications would appear to be applicable to testing of solidified waste products; however, at this time regulatory product-performance and acceptance criteria for physical properties have not been established. Therefore, it is difficult to predict expected product performance from testing results although comparisons between different processes and binders are possible.

The U.S. Nuclear Regulatory Commission (NRC) has developed a position on characteristics which solidified waste must have to be acceptable. Because the NRC waste is somewhat unique, (i.e., low-level radioactive), and requirements differ, the standard may not be entirely applicable to nonradioactive hazardous waste. Even so, the NRC position on standards which S/S products must meet provides a basis to which processors can work. Such criteria is needed for non-radioactive S/S hazardous waste.

7.8.5 FACTORS AFFECTING SELECTION AND PERFORMANCE OF S/S TECHNOLOGIES

Factors which will affect the selection, design, implementation, and performance of S/S processes and products are

- Treatment objective
- Waste characteristics (chemical and physical)
- Process type and processing requirements
- S/S waste-management requirement
- Regulatory requirements
- Economics

These and other site-specific factors (i.e., location conditions, climate, hydrogeology, etc.) must be carefully considered to assure acceptable performance.

Treatment Objectives

In the evaluation and selection of S/S techniques, both the waste-treatment objective and the criteria used to determine how well the process performs to meet that objective must be considered. When using S/S to treat hazardous waste, there are three levels to which one can attempt to treat. In level I, the objective is to remove free liquids from the waste so that the waste will pass the Paint Filter Liquids Test (PFLT) and therefore can be disposed into the landfill. Solidification rather than sorption is required. Therefore if the solidification is not apparent, the product may be subject to a compressive test as proof that solidification was accomplished.

For level II treatment, the objective is to make the waste acceptable for land disposal. At the current time this means only that the waste not contain free liquids. Therefore, the treatment objective is to remove free liquids and the acceptance criteria is the same as for level I. However, provisions of HSWA require the EPA to review the list of hazardous wastes and determine which are acceptable for land disposal. If it is determined that a waste is not acceptable it will be banned from land disposal unless treated to remove the unacceptable characteristic. In this process new criteria will be developed and used to determine whether the waste, either prior to treatment or after treatment, is acceptable for land disposal. Therefore waste treated by S/S processes may be required to meet these criteria before it can be disposed of in the land. It is possible that a standard based on the level of toxic constituents will be used and the toxicity-characteristics leaching procedure (TCLP) or some modification will be used to determine whether the toxic constituent(s) of concern leach from the waste at unacceptable levels. In this case the S/S treatment objective will be to contain and prevent the toxic constituents from entering the environment at concentrations greater than those established as protective to human health and the environment.

For level III treatment the objective is to treat the waste so it can be delisted (classified as nonhazardous) and therefore acceptable for disposal in nonhazardous-waste disposal facilities. Current requirements are that the treated waste must pass the EP Tox and meet other case-specific criteria. Under the waste-banning provisions and associated activities, these criteria may change and become more strict and therefore more difficult to meet.

Waste Characteristics

Waste characteristics are among the most important factors affecting waste solidification and stabilization. Generally the waste must be compatible and must be hazardous based on toxicity only. Small amounts of some compounds can seriously reduce the strength and containment characteristics of binder-waste mixes. Impurities can affect the strength, durability, and permeability of portland cement and asphalt mixtures.[8,9,10] Since cement plays a major role in current waste-S/S technologies these same effects can be expected. Even minor quantities of some waste compounds act as accelerators or retarders and can cause poor performance in S/S products. (Table 7.8.1)

Selected organics have been shown to affect the unconfined compressive strengths and leaching characteristics of fly ash–lime S/S formulations.[10,11] Adypic acid adversely affected the unconfined compressive strengths. Methanol retarded the setting time of the formulations. Benzene and xylene also acted as retarders but to a lesser extent. Methanol, xylene, and benzene increased the concentrations of toxic constituents in leachate from the solidified and stabilized samples.

Since investigators have concluded that a significant correlation exists between the effects of organic compounds on lime–fly ash pozzolanic systems and on the hydration of portland cement the information concerning additives and interferences in using portland cement should be applicable to S/S systems using pozzolanic reactions. Even so, some organic and inorganic wastes contaminated with organics are probably acceptable for solidification and stabilization using pozzolans. Wastes such as rolling-mill sludges, electroplating residues, or oily sludges from petroleum refineries have been treated. However, organic wastes containing hydroxyl or carboxylic acid functional groups, such as biological wastes, paint sludges, and some solvents, can be expected to delay or completely inhibit the pozzolanic or portland cement-based reactions responsible for solidification.[9]

Table 7.8.2 lists selected chemicals that exert adverse effects on portland and pozzolan-based S/S processes. In addition to the chemical effect, temperature and humidity conditions during setting are important. Temperatures below 0°C (32°F) will cause retardation of set while those over 30°C (86°F) will accelerate setting. Temperatures over 66°C (151°F) may completely destroy the reactions. High humidity can also accelerate setting. Extensive mixing, especially after the gel-formation phase, may destroy the solids and result in an extremely low-strength product.

The waste-treatment industry can be a potentially important source of infor-

TABLE 7.8.1   Effects of Selected Chemicals on Cement-Based Pozzolanic Processes[15]

Chemical or material	Important functions							
	Flocculant	Dispersant	Wetting agent	Chelating agent	Matrix disruptor	Retarder	Accelerator	Destroys reaction
Carboxylic acids		X				X		*
Carbonyls		X				X		*
Amides			X				Allows for better mixing	
Amines	X						X	*
Alcohols			X					*
Sulfonates		X				X		*
Glucose/sugar				X		X		
Chlorinated hydrocarbons					X	X		X
Oil								>25–30%§
Calcium chloride					>4%‡		<2%‡	>4%§
Iron†	X				X		X	
Tin					X	X		
Lead					X	X		
Borates					X	X		*
Magnesium	X				X	X		
Gypsum (hydrate)						X		
Gypsum (anhydrate)							X	
Silica	X					X‡		*

*At high concentrations.
†Ratio of Fe^{2+} to Fe^{3+} important.
‡Only in certain forms.
§By weight.

mation to better understand and judge the capability to handle waste. While it is difficult to accurately ascertain the amount of research the industry has conducted, their experiences can be valuable. Comments from vendors and others on factors important in S/S can be found in Ref. 12.

Inorganics generally are easier to successfully solidify and stabilize than organics with currently available technology (see Tables 7.8.2 and 7.8.3). In addition, it may be easier to pretreat the waste stream to accommodate hard to handle inorganics. In most cases organic wastes apparently do not enter into chemical reactions to form new organic-inorganic compounds or complexes which can chemically bind organic contaminants. Organics are probably held by physical entrapment in available pores, although research is indicating that in selected cases, the organics may be present in cement gel phases.[13,14]

Physical characteristics of the waste and the binder are also important. Particle size and shape in the waste and of the hardened binder can play an important role in the performance of treatment processes in the field. Viscosity of mixes can change with particle size and shape and affect the amount of water available for reactions. Proper water/binder ratios are important in producing mixes which will yield acceptable strength. Overmixing or undermixing can adversely affect the strength of the final product or even prevent an initial set. The more information available concerning waste and binder characteristics as they relate to proper formulations, the better the opportunity to assure acceptable S/S products.

Retardation of set may or may not be detrimental. If the set is retarded to the point that unacceptable strength or pollutant containment is reached, an inferior product will result that will not perform satisfactorily. However, if retardation of the set is merely a delay in the time to reach an acceptable strength, then it is not a significant problem. In this situation, economics and the processing schedule become controlling factors.

Process Type and Processing Requirements

The type of S/S process required (i.e., in drum, in plant, etc.) and specific processing conditions (e.g., waste modification; mixing modes; transportation, placement, and storage of treated wastes) are important factors to be considered in evaluating and selecting S/S technology. It is easier to control and provide proper mixing of the binder with the waste matrix in a drum or in a plant process than when solidifying a pit, pond, or lagoon. Special processing requirements such as treatment to remove interfering agents, use of thermosetting binders, and macroencapsulation, will also affect the evaluation and choice of S/S technology.

S/S Product Management

The waste-management objective for the treated product (i.e., disposal in landfill, storage, transportation, etc.) will also be important in the selection of an S/S technology. Depending upon regulatory requirements, placement in an RCRA hazardous-waste landfill may not require the same degree of solidification and stabilization as is required for delisting. Delisting requires that the material no longer be hazardous, while for placement in an RCRA hazardous-waste landfill it generally does not have to meet that same criterion. Other placement schemes are currently proposed or used—e.g., placement in drums (plastic or metal) for

TABLE 7.8.2 Vendors' Identification of Factors Important in the Solidification and Stabilization of Hazardous Waste[15]

Company	Waste type (Liquid, solid, sludge)	Important parameters			
		Reaction mechanism	Critical factors	Organics	Accelerators/retarders
A	All	Cement with silicaceous material	pH, alkalinity, saturated-solids content; particle surface area; viscosity	Chelators—negative, nonpolar oils treatable up to 20%	Gels retard
B	All	Cement with silicaceous material	Colloidal gel formation, pH (alkalinity); particle size; valence state	N/A	Gels retard
C	All	Cement with silicaceous material	Like to work at 40% solids or pumpable; particle size; colloidal gel ≤540 mesh; free water; alkalinity, 20–40% as CaCO$_3$	BOD: 3,000–4,000 ppm; Organic N: 5,000 ppm; TOC: 30,000 ppm—all treatable	N/A
D	All	Self-cementing and others	Solids content 10–40%, but <20% need volume reduction; viscosity—use a slump test to measure correct mix consistency; tin, lead may be problems; pH; temperature	Up to 25% oil treatable	Gels?
E	All—prefer sludge, solid	Cement and pozzolan	Particle size—smaller size and random distribution preferred; percent solids (only from economic view); mixing, colloids have effect; alkalinity seems more important than pH	N/A	Gels—must be "conditioned" (mixed, destroyed) prior to processing; add drying agent or other materials

F	All—prefer liquid, sludge	Pozzolan, lime, cement	Gypsum—helpful in making tendrils; flat particles—better product; solids content; smaller particles need more reactants; avoid high-speed mixing; pH	Organic coating a problem—overcome by adding more reactant	Gels work against reactions, i.e., Al, Fe; large amounts of sulfates retard set
G	All—matter of economics only	Pozzolan, cement, silicates, and 4 or 5 other reactive ingredients; can use up to 30 reagents; all inorganic	Hydrophobic materials need more reagents; mixing has effect; water glass decreases set time; alkalinity; valence state	Better results with high-MW material; solvents tend to inhibit set; too much oil retards	Sodium inhibits set—overcome by more reagents; chrome appears to be accelerator
H	All—except no biological waste	Cement-silicate	Particle size, shape, and wettability; alkalinity; temperature-limited, evaporation a problem; use mixing/vibration technique	Must disperse oil if present	Silicates and sulfates—accelerators; glucose and borates—retarders (borate can be treated with sodium metasilicate); gels—difficult to treat, attempt to treat by forming emulsion

TABLE 7.8.3 Compatibility* of Selected Waste Categories with Waste Solidification and Stabilization Techniques

Waste component	S/S Treatment type			
	Cement-based	Lime-based	Thermoplastic solidification	Organic polymer (UF)†
Organics				
1. Organic solvents and oils	Many impede setting, may escape as vapor	Many impede setting, may escape as vapor	Organics may vaporize on heating	May retard set of polymers
2. Solid organics (e.g., plastics, resins, tars)	Good—often increases durability	Good—often increases durability	Possible use as binding agent	May retard set of polymers
Inorganics				
1. Acid wastes	Cement will neutralize acids	Compatible	Can be neutralized before incorporation	Compatible
2. Oxidizers	Compatible	Compatible	May cause matrix breakdown, fire	May cause matrix breakdown
3. Sulfates	May retard setting and cause spalling unless special cement is used	Compatible	May dehydrate and rehydrate causing splitting	Compatible
4. Halides	Easily leached from cement; may retard setting	May retard set; most are easily leached	May dehydrate	Compatible
5. Heavy metals	Compatible	Compatible	Compatible	Acid pH solubilizes metal hydroxides
6. Radioactive materials	Compatible	Compatible	Compatible	Compatible

*Compatible indicates that the S/S process can generally be successfully applied to the indicated waste component. Exceptions to this may arise dependent upon regulatory and situation-specific factors.
†UF = urea-formaldehyde resin.
Source: Ref. 1.

storage in warehouses or underground mines, in situ injection into mined cavities—and in each case the choice as to final placement will affect evaluation and selection of an S/S system. Knowledge, understanding, and consideration of these and other factors such as economics—and of the interactions of the various factors—are important in selecting the S/S technique or system best suited for the given situation.

Regulatory Requirement

Regulatory factors can be expected to play a major role in the use of S/S technologies for managing hazardous waste. With the possible exception of the EP, MEP, or TCLP there are currently no set performance criteria which S/S products must meet. For economic and related reasons (e.g., volume increases which reduce available landfill space), a processor will normally produce an S/S waste product which will meet minimum requirements necessary to remove free liquids and/or produce a solid with a structural integrity sufficient to meet their specific processing, transport, and placement requirements.

Theoretically, almost any waste can be solidified and/or stabilized. Additions of large quantities of binders can overcome most problems that might make a waste difficult to solidify. Most processors reject wastes that require uneconomical amounts of binder. However, the important point is that systems and processes can be altered to meet different performance criteria. The criteria must be established before attempting to select a particular S/S technique. Whether or not S/S technology becomes an important technology for treating hazardous waste will be dependent upon regulatory requirements and the ability of the technology to meet these requirements.

Economics

The cost of solidification and stabilization has generally been considered low compared with those for other treatment techniques. The reasons for this are the availability of rather cheap raw products (e.g., fly ash, cements, lime) used in the more popular processes, simple processing requirements, and the use of readily available equipment from the concrete and related construction industries. In addition, earlier treatment objectives were often driven by a need only to produce a more manageable waste (e.g., removal of liquid) rather than to produce a product to meet a more stringent regulatory requirement. The latter will most likely require more additives or more expensive processing and will therefore increase cost.

It is impossible to provide accurate costs for stabilization and solidification. Final costs will be dependent upon site-specific conditions. Important factors include those listed here.

Waste Characteristics. The physical form and chemical makeup of the waste undergoing S/S treatment will have an important effect on the cost. If pretreatment of the waste is required to remove excess liquids or to remove and/or alter interfering constituents, costs will be increased.

Transportation. Requirements for transporting raw materials to the plant or site and transporting finished products to disposal will affect costs.

Process.   The S/S process selected will affect costs. Cements, fly ash, and so forth, are cheaper raw materials than are polyolefins and similar materials. Processing requirements for the latter may also make their use more expensive. This however, may be balanced by better performance of the products. Increased volume generated by processes of the cement and fly ash type may also result in added transportation and disposal costs. Process type (e.g., in drum, in plant) also affects economics.

Other Factors.   Special health and safety requirements will affect costs, as will any special regulatory requirements. Quality assurance and quality control (QA/QC) and associated analytical costs may be a cost factor and must be carefully considered in estimating costs. Regulatory factors will probably play an increasingly important role in the costs of solidification and stabilization as processing and product-performance requirements may become more exacting.

7.8.6 CONCLUSIONS

The role that S/S technology eventually has in management of hazardous waste depends upon regulatory actions and subsequent judicial interpretation and on the ability of the technology to meet performance criteria which may be developed. As restrictions on landfilling become stronger and wastes are banned from land disposal, S/S technology could potentially be an important alternative technology with a major use being to treat waste in order to make it acceptable for land disposal. Lower permeability, lower contaminate leaching rates, and similar characteristics may make hazardous wastes acceptable for land disposal after stabilization. Depending upon the technical requirements of any developed performance criteria and/or the willingness of processors to meet them, S/S technology has the potential for making a major contribution as one of the alternatives for managing hazardous waste.

7.8.7 REFERENCES

1. P. G. Malone, L. W. Jones, and R. J. Larson, *Guide to the Disposal of Chemically Stabilized and Solidified Waste,* SW-872, U.S. EPA Office of Water and Waste Management, Washington, D.C., September 1982.

2. R. W. Telles, M. J. Carr, H. R. Lubowitz, and S. L. Unger, *Review of Fixation Processes to Manage Hazardous Organic Waste,* unpublished report, U.S. EPA Hazardous Waste Engineering Research Laboratory, Cincinnati, Ohio, 1984. (Attn: Carlton C. Wiles)

3. M. J. Cullinane, Jr., L. W. Jones, and P. G. Malone, *Handbook for Stabilization/Solidification of Hazardous Waste,* U.S. EPA Hazardous Waste Engineering Research Laboratory, Cincinnati, Ohio, EPA 540/2-86/001, June 1986.

4. H. R. Lubowitz and C. C. Wiles, "Management of Hazardous Waste by Unique Encapsulation Processes," *Land Disposal of Hazardous Waste, Proceedings of the Seventh Annual Research Symposium,* EPA-60019-81-0021, U.S. EPA, Cincinnati, Ohio, March 1981.

5. S. L. Unger and R. W. Telles, "Surface Encapsulation Process for Managing Low-level Radioactive Wastes," *Proceedings, Waste Management '86,* Tucson, Arizona, March 1986.

6. 40 *CFR* Part 261, "Appendix II—EP Toxicity Test Procedure," *Federal Register,* Vol. 45, No. 98, p. 3,317, 1980.

7. 40 *CFR* Parts 261, 271, and 302, "Hazardous Waste Management Systems; Identification and Listing of Hazardous Waste; Notification Requirements; Proposed Rule," *Federal Register,* Vol. 51, No. 114, June 13, 1986.

8. J. N. Jones, R. M. Bricka, T. E. Myers, and D. E. Thompson, "Factors Affecting Stabilization/Solidification of Hazardous Waste," *Land Disposal of Hazardous Waste, Proceedings of the Eleventh Annual Research Symposium,* EPA/600/9-85/013, U.S. EPA, Cincinnati, Ohio, April 1985, pp. 353–359.

9. J. N. Jones, L. W. Jones, and N. R. Francinques, *Interference Mechanisms in Waste Stabilization/Solidification Process: Literature Review,* unpublished report, U.S. EPA Hazardous Waste Engineering Research Laboratory, Cincinnati, Ohio. (Attn: Carlton C. Wiles)

10. B. K. Roberts, "The Effect of Volatile Organics on Strength Development in Lime Stabilized Fly Ash Compositions," M.S. thesis, University of Pennsylvania, Philadelphia, 1978.

11. R. L. Smith, *The Effect of Organic Compounds on Pozzolanic Reactions,* Report No. 57, Project 145, I. U. Conversion Systems, Horsham, Pennsylvania.

12. *Critical Characteristics and Properties of Hazardous Waste Solidification/ Stabilization,* unpublished report, prepared by JACA Corporation for the U.S. EPA Hazardous Waste Engineering Research Laboratory, Cincinnati, Ohio, 1985. (Attn: Carlton C. Wiles)

13. D. Chalasani, F. K. Cartledge, H. C. Eaton, M. E. Tittlebaum, and M. B. Walsh, "The Effects of Ethylene glycol on a Cement Based Solidification Process," *Hazardous Waste and Hazardous Materials,* 3:(2)169–173 (1986).

14. M. B. Walsh, H. C. Eaton, M. E. Tittelbaum, F. C. Cartledge, and D. Chalasani, "The Effect of Two Organic Compounds on a Portland Cement Based Stabilization Matrix," *Hazardous Waste and Hazardous Materials,* 3:1,111–123 (1986).

CHAPTER 8

THERMAL PROCESSES

SECTION 8.1

LIQUID-INJECTION INCINERATORS

Joseph J. Santoleri

Principal Consultant
Four Nines, Inc.
Plymouth Meeting, Pennsylvania

8.1.1 OVERVIEW OF THE TECHNOLOGY AND ITS APPLICATIONS

In the field of hazardous-waste incineration, there is more experience with liquid-injection (LI) incinerators than all other types combined. In a survey conducted in 1981, approximately 64% of the total number (219) of incinerators in service at the time were of the LI type.[1] Various types of incinerators are used to handle liquid waste as well as wastes in other forms, i.e., solids, sludges, slurries, and fumes. In this section, we will confine the discussion to the liquid-injection type, which, as the name implies, is applicable almost exclusively to pumpable liquid waste. The waste is burned directly in a burner (combustor) or injected into the flame zone or combustion zone of the incinerator chamber (furnace) via nozzles. The heating value of the waste is the primary determining factor for nozzle location.

Waste liquids in the process industries are numerous in kind and therefore defy easy definition. Disposal of these waste liquids has become a serious problem for the plant operator. Regulations today cover wastes once considered wastewater streams as well as the organic streams known to be made up of hazardous compounds. Liquid-injection incinerators are available today to handle the various liquid streams that are being generated by the process industries. These units may be on site (at the location of the generator) or off site (at a sister facility or at a commercial disposal operation). This section will attempt to explore the various designs available and the basic components necessary to provide an adequate liquid-injection system.

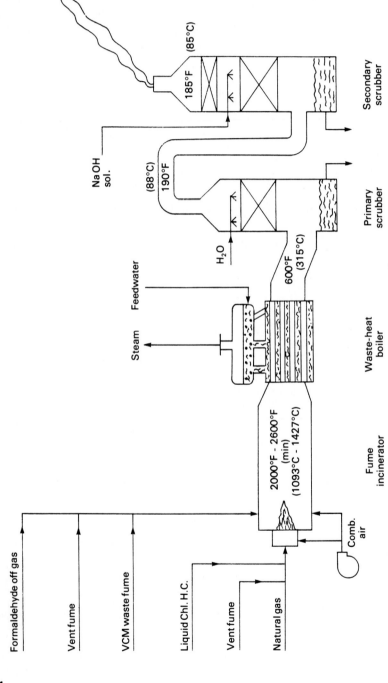

FIG. 8.1.1 Liquid-injection incinerator system. VCM = vinyl chloride monomer.[7]

Liquid-Injection Incinerators

Liquid-injection incinerators are usually refractory-lined chambers (horizontal, vertical up or down), generally cylindrical and equipped with a primary combustor (waste and auxiliary-fuel fired) and often secondary combustors or injection nozzles for low-calorific-value materials (aqueous wastes containing either organic or inorganic compounds, or both). (See Fig. 8.1.1.)

These units operate at temperature levels from 1000°C (1832°F) to 1700°C (3092°F). Residence time in the chamber may vary from milliseconds to 2.5 seconds. The viscosity determines whether the material being incinerated is considered a liquid, slurry, or sludge. Liquid units today are capable of burning high-viscosity materials of 4,500 SSU or less. Critical to the operation of the unit would be the atomizing nozzle used to convert the liquid stream into finely atomized droplets.

When considering the liquid-injection incinerator, one must be concerned about the ability of the entire system, i.e., the design, components, controls, and pollution-control system, to meet the requirements of the federal and state regulations. All units handling hazardous compounds (as listed in Sec. VIII Par. 261.30) must meet the basic requirements of 40 *CFR* Subpart 0—Incinerators—as follows:

1. 99.99% destruction and removal efficiency (DRE) of the principal organic hazardous constituents (POHC)
2. 99% removal of hydrogen chloride [1.8 kg/h (4 lb/h)]
3. Stack emissions not to exceed 180 mg per dry standard cubic meter (0.08 gr per dry standard cubic foot).[2]

Waste-Liquid Data

The physical, chemical, and thermodynamic properties of the waste must be considered in the basic design requirements of the total incinerator system. This includes storage tanks, mixers, pumps, control valves, piping, atomizers, combustors, refractory, heat recovery, quench system, and air-pollution control equipment. The types of data needed by the designers to properly engineer the total system and the individual components are listed in Table 8.1.1.

TABLE 8.1.1 Waste-Liquid Fuel Data Needed for Incinerator System Design

---

Chemical composition (including halides, sulfur, nitrogen, etc.)
Specific gravity
Heat of combustion
Viscosity
Corrosivity
Ignitability
Reactivity
Polymerization
Solids content (type, percentage, physical data)
Metals
Slagging properties (temperature, eutectic data)

8.1.2 LIQUID-INCINERATOR SYSTEM DESIGN

Storage

In order to design the storage system to permit a feed of uniform mixture and heating value, proper mixing in the storage tank is required. Layering often occurs in large tanks, especially where wastes are introduced from various sources.[3] Mixing may be accomplished by internal mixers, external recycling pumps, or air or steam spargers. Venting of the tank is also necessary to prevent pressure buildup. Vent gases should be exhausted to the incinerator chamber (see Fig. 8.1.2).

Materials of construction of all components in the storage and feed system—tanks, pumps, mixers, piping, valves, and nozzles—will be determined by chemical composition, viscosity, corrosivity, and solids content of the waste stream(s). If solids are known to be inert materials, filters should be used to prevent unnecessary carry-over of these materials into the incinerator and the air-pollution control system. If, however, these solids are organic, they must be included as part of the waste stream and design should provide for moving this material directly to the incinerator.

Transport

High-viscosity materials (slurries and sludges) and liquids containing solids are a serious consideration in the design of the feed system to the incinerator. Various pump types are available and their selection and use is predicated on the properties of the waste stream. Progressive-cavity pumps have been used successfully, provided the materials of the rotor and stator are selected properly. Gear pumps

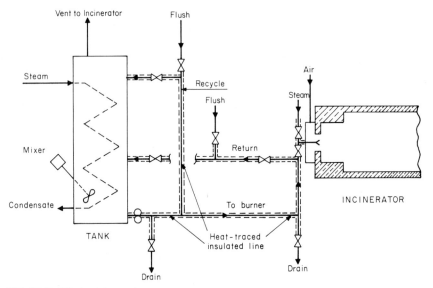

FIG. 8.1.2 Waste-piping system.

have been the workhorse for most lower-viscosity materials (equivalent to No. 6 fuel oil). However, in this case, the combination of viscosity and speed are critical for long-life, low-maintenance operation. In systems where a variety of waste streams must be pumped into the incinerator, the operator must be instructed about the proper feed system to use with each waste. In some cases, pumps may not be practical because of the high maintenance costs and shutdowns that result. Pressurizing with an inert gas (nitrogen) may be the safest and most economical method. Diaphragm pumps are in use today at many installations where other pumps have failed.

The atomizer design usually dictates the pressure requirements of the transport system. Where a hydraulic (mechanical) nozzle is used, requiring high pressures, a pump design needing close clearances between the impeller and the housing is also usually indicated. High-viscosity materials and liquids containing solids will score the surfaces of the pump, causing excessive wear and rapid deterioration of pump pressure. This results in a reduction of flow as well as poor atomization. Materials transport is critical in the proper design of an incinerator. Certain liquid wastes and sludges must be maintained at temperature levels sufficiently high to permit pumping. If cooled, the material will solidify in the pipeline. However, if heating is not controlled, the material may polymerize if heated to too high a temperature. Once polymerized, it is impossible to liquefy the material. As a result, the piping with the polymer presents a solid-waste disposal problem.[3]

Atomizers

The method of injection of the liquid into the burner or the incinerator furnace is one of the most important features of a well-designed system. The main purposes for injecting the liquid as a spray are

1. To break up the liquid into fine droplets
2. To place the liquid droplets in a specific zone with a specific pattern with sufficient penetration and kinetic energy
3. To control the rate of flow of the liquid discharged to the nozzle

Organic and aqueous liquids pass through three phases before actual oxidation is initiated. Hydrocarbons ignite at temperatures as low as 230°C (446°F) and as high as 650°C (1022°F). They are heated, vaporized, and superheated to ignition temperature (see Fig. 8.1.3). These droplets must be exposed to high-temperature surroundings so as to receive heat by radiation and convection as rapidly as ispractical. At the same time, they must also be in intimate contact with oxygen. If the droplet diameter is large, fewer droplets will be produced and the total surface available for heat transfer will be minimized.

An example is shown below:

$$\text{Ratio of number of droplets} \quad \frac{N\ 125\ \mu m}{N\ 300\ \mu m} = 13.8$$

$$\text{Ratio of surface areas} \quad \frac{A\ 125\ \mu m}{A\ 300\ \mu m} = 0.1736$$

For an equal flow rate of liquids, the total surface available for heat transfer is as follows:

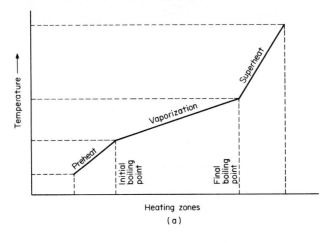

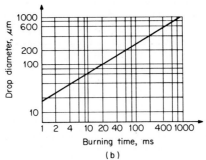

FIG. 8.1.3 (*a*) Hydrocarbon state as a function of temperature. (*b*) Burning time as a function of droplet diameter.

$$\frac{A_{t125}}{A_{t300}} = \frac{N_{125}}{N_{300}} \times \frac{A_{125}}{A_{300}}$$

$$= 13.8 \times 0.1736$$

$$= 2.396$$

Where A_t = total surface area
       N = number of droplets
       A = surface area of droplet
      125 = 125-μm droplet
      300 = 300-μm droplet

Therefore, the rate of heat transfer to the 125-μm droplets will be at least 2.4 times as great as that for the 300-μm droplets. Because of their smaller diameter and volume, the 125-μm droplets will be vaporized at an even higher rate. The fact that there are almost 14 times more droplets means turbulence will be generated, with flames initiated at their surfaces. (Note on Fig. 8.1.3, 150 ms versus 30 ms, which is 5 times more.)

    With high-viscosity materials, the nozzle orifices have to be made larger to

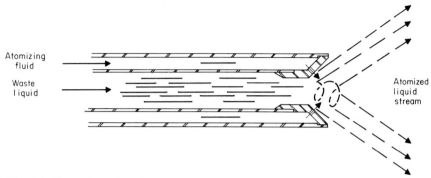

FIG. 8.1.4   External atomizer tip.

minimize pressure drop, erosion, and blockage. Therefore, nozzles must be designed to cause a shearing action of the liquid to break it into many smaller-diameter particles. Most nozzles designed to atomize viscous liquids use a pneumatic fluid (steam or compressed air) to properly break up the viscous stream into droplets which can be carried into the combustion zone (see Figs. 8.1.4 and 8.1.5).

Combustion-Air System

Proper mixing of the combustion air with the liquid droplets is necessary. As the liquid is vaporized and superheated to ignition temperature, oxygen reacts with the hydrocarbon vapor to allow release of the energy. As this occurs, there is a sudden rise in temperature. This increases the velocity of the gases in the zones surrounding the droplet, causes increased mixing, and finalizes the oxidation reaction. With low-boiling-point hydrocarbons, this reaction occurs rapidly at the initial boiling-point temperature, and the rapid rise to reaction temperatures (1000 to 1700°C or 1832 to 3092°F) will provide the necessary heat sink to the incoming colder liquid droplets (see Fig. 8.1.3).

As the viscosity of the liquid increases, droplet sizes tend to get larger and also the 90% boiling point of the hydrocarbon is at a much higher level. To completely vaporize and superheat the droplet, more time is needed. Increased turbulence created by high-intensity burners permits this reaction to be achieved

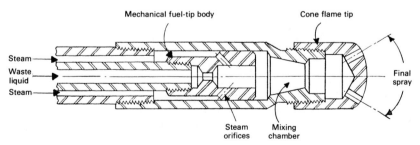

FIG. 8.1.5   Internal-mix atomizer.

rapidly. Energy is imparted to the combustion air of the combustor and this, in turn, permits more rapid mixing of the air with the fuel droplet. In many burners, this turbulence permits an internal recycling of the hot products of combustion, which transfer heat into the atomized droplets and raise them to the ignition point.

As the superheated liquid reaches ignition temperature, oxygen must be available to complete the oxidation reaction. Many problems result if the vaporized hydrocarbon contacts a low-oxygen-level stream (see Fig. 8.1.6). Pyrolysis results, with a cracking of the hydrocarbon into carbon, carbon monoxide, and hydrogen, and time for completion of combustion is increased.

A secondary reaction will occur when these combustion gases eventually reach the proper oxygen supply. In a true thermal oxidizer, the most rapid oxidation occurs when single-stage combustion is designed into the system.

If the vaporized liquid contains solids, the incinerator design must allow the particles to be carried into the gas stream without agglomeration. A high-swirl or a cyclonic design may cause the solids to be reagglomerated into larger particles which become more difficult to burn. Proper design for the air-mixing device and the nozzle location are therefore important. Proper oxygen concentration at the surface of these solids is needed to ensure that gradual oxidation will occur. Since they are solid, the particles will burn at the surface and heat will be transferred inward to the core of the droplet. Sufficient time should be provided to permit complete burnout of the solids in suspension.

If inert materials are carried with the liquid as it is vaporized, these tend to be carried into the gas stream as particulates. Depending on the type of atomizer, the composition of the solid, and the temperature of the oxidizer, a percentage will become submicron in size and be carried with the gas stream. The heavier particulates will become molten and agglomerate. The combustor design should be such as to collect these without plugging the flow passages of the system.

A slanted horizontal design or a vertical orientation is most often used for these types of waste (see Fig. 8.1.7).

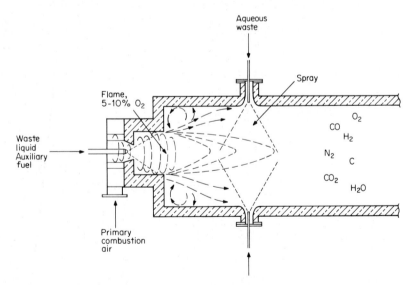

FIG. 8.1.6 Aqueous-waste incinerator without secondary air.

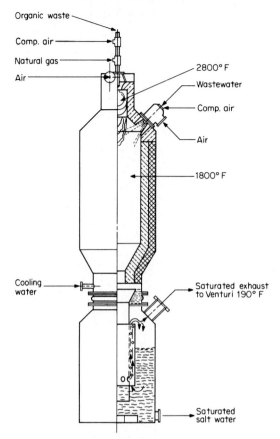

FIG. 8.1.7   Sub-X incinerator (T-Thermal).

Combustors

Primary and secondary combustion units are utilized in the liquid-injection incinerator systems. Primary units are utilized for those wastes with sufficient heating value to supply the necessary heat input to the system without need for auxiliary fuel. With nozzle-type burners, wastes with heating values from approximately 2500 kcal/kg (4500 Btu/lb) and above can be burned satisfactorily. The type of burner design, air mixing, turbulence, and so forth, determine the minimum heating value for burning without auxiliary fuel.

Figure 8.1.8—"Gross heating value, Btu/lb, vs. adiabatic temperature"—provides a guide to the ability of a combustor to burn a particular waste. Low-intensity or laminar-flame burners (air-pressure drop varies from 50 to 150 mm) (2 to 6 in w.c.) utilize high excess-air levels (25 to 60%) to provide the mixing of air with the waste fuel. These usually demonstrate low turbulence. High-intensity burners operate with combustion-air pressures of 200 to 500 mm w.c. (8 to 20 in w.c.). As a result, the increased turbulence resulting from the energy imparted by the air allows operation at much lower excess-air levels (0 to 20%). The ideal minimum combustion temperature for burning a fuel is approximately 1200° to

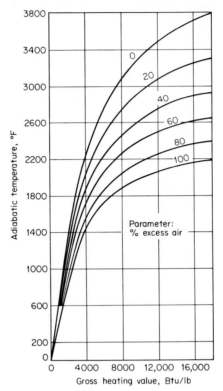

FIG. 8.1.8 Gross heating value, Btu/lb, vs. adiabatic temperature.

1315°C (2200 to 2400°F). A review of the curve (Fig. 8.1.8) indicates that a high-intensity burner would be able to oxidize a 2500 kcal/kg (4500 Btu/lb) waste material at 1200°C and 10% excess air. A low-intensity burner operating at 40% excess air and 1200°C (2200°F) would be limited to a minimum heating value of 3220 kcal/kg (5800 Btu/lb). This burner would require auxiliary fuel to properly burn a 2500 kcal/kg (4500 Btu/lb) material.

For example, if the waste to be burned had a heating value of 2770 kcal/kg (5000 Btu/lb) and 554 kcal/kg (1,000 Btu/lb) had to be disposed of, a high-intensity burner rated at 5×10^6 Btu/h would be required. The burner would be designed to light off with auxiliary fuel at approximately 2×10^6 Btu/h for initial warm-up. Once the chamber temperature reached a level of 1093°C (2000°F) plus, the waste could be introduced.

The low-intensity burner requires a minimum heating value of 3220 kcal/kg (5800 Btu/lb) to operate. In order to burn the 2770 kcal/kg material, auxiliary fuel will be required to elevate the heating value of the fuel mixture to 3220 kcal/kg. Assuming auxiliary fuel is available with a heating value of 10,000 kcal/kg (18,000 Btu/lb), the amount of fuel required and the size of burner needed can be determined as follows:

$$5,000(x) + 18,000(1 - x) = 5,800$$

Where x = % of LHV waste
  $1 - x$ = % of HHV waste

Solving for x, $x = 93.846\%$ and $1 - x = 6.1547\%$

Total fuel flow = 1,000/0.93846 = 1,065 lb/h
1065 − 1000 = 65 lb/h of aux. fuel
Total heat input = (5,000) (1,000) + 18,000(65) = 6.179×10^6 Btu/h
Additional aux. fuel input = 1.179×10^6 Btu/h
Burner must be rated at @ 6.18×10^6 Btu/h vs. the 5.0×10^6 Btu/h for the
  high-intensity burner

8.1.3 WASTE CHARACTERISTICS DETERMINE INCINERATOR DESIGN

In the process industries, waste liquids are generated which are numerous in kind and defy easy definition. The wastes can be categorized as organic, aqueous organic, and aqueous organic-inorganic (see Table 8.1.2). Each of these can again be divided into those that are considered hazardous and those considered nonhazardous. Appendix VIII to Title 40 of the *Code of Federal Regulations* lists those compounds and waste mixtures that are considered hazardous by the U.S. Environmental Protection Agency (EPA) and must be treated accordingly.

Those organic wastes listed in Table 8.1.2 will in most cases sustain combustion and would be injected into the primary combustor. Many are utilized as a source of fuel in boilers, process heaters, incinerator systems, etc. Note that the list also contains halogenated wastes. This subject will be covered in a little more detail later.

Aqueous Waste Systems with Ash

In reviewing the aqueous wastes, note the distinction between those wastes with and without ash. Non-ash-bearing wastes are easily handled in either a horizontal or vertical (up-oriented) incinerator system. The waste is usually injected downstream of the burner but into the flame of the main burner. Care must be taken to design for proper sizing, location, number, and direction of the nozzles for injecting the aqueous waste.

Since a highly aqueous waste stream is not viscous, it may be atomized by

TABLE 8.1.2 Organic Waste Liquors

Aqueous		Organic (nonaqueous)
Ash	Without ash	
Sodium glutamate	Phthalic anhydride	Solvents
Molasses ferment	Phenol	Waste Oils, sludges
Synthetic phenols	Oil/Water	Halogenated hydrocarbons
Polyesters	Maleic anhydride	Organic acids
Caprolactum		Organometal components
Agricultural-process		Aromatics
wastes		

hydraulic pressure alone in a mechanical atomizing nozzle or a dual-fluid nozzle (steam or compressed air) may be utilized. Use of compressed air for these nozzles has been very successful. The air serves as combustion air and reduces the heat load on the system that would be required with steam. The design and controls should provide the proper ratio to ensure adequate spray pattern and droplet size through the entire range of flow.

Nozzles should also be protected in the event of loss of waste flow by maintaining the flow of the atomizing fluid. In many units, these nozzles are designed to be physically removed from the incinerator chamber to prevent burnout of the nozzle tip.

Provision should also be made to provide sufficient air to oxidize the organics contained in the waste stream. Incinerators have been known to produce soot as a result of a pyrolysis condition at the entry of the aqueous organic waste (see Fig. 8.1.6, Fig. 8.1.9).

Aqueous wastes containing organics and inorganic material have been produced by the agricultural-process industries in the manufacture of herbicides, pesticides, and other products. The pharmaceutical industry also produces a high-water-content waste containing organics with phosphorous-, chlorine-, and sulfur-bearing compounds. Aqueous solutions have no self-burning capabilities when the water content is greater than 75%. If the wastewater contains ash (salts), consideration must be given to the maximum operating temperature to minimize refractory problems. These wastes are normally injected into the incinerator downstream of the primary combustor in a lower-temperature zone rather than directly into the high-temperature refractory zone.

Incinerator designs for the oxidation of those wastes with high ash content are vertically oriented. This allows gravity discharge of the molten ash material into the quench zone (quench pot with sprays, submerged-exhaust sub-X, or hot tap port). Critical to the design and operation of incinerators for this waste type is the selection and design of the refractory. A reaction occurs at these elevated temperatures between the metallic oxides resulting from oxidation and the com-

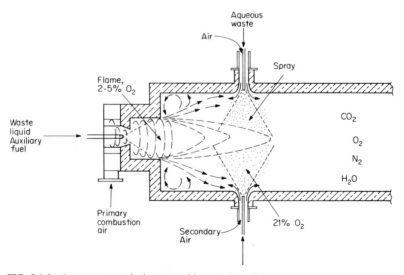

FIG. 8.1.9 Aqueous-waste incinerator with secondary air.

pounds that make up the refractory lining (typically alumina and silica). Severe spalling and excessive maintenance and down time have resulted from improper selection of refractories. Testing of various refractory types with the slag that will result from the incineration process at the operating temperatures expected is extremely important. This is covered in more detail in Refs. 4 and 5.

Aqueous Wastes Incinerated with Organic Wastes Used as Auxiliary Fuel

When handling both organic wastes as auxiliary fuel and aqueous wastes containing ash as well as organics, we must remember that the primary purpose of the incineration process is to completely oxidize the organics. Therefore, conditions must be established to provide the necessary atomization, mixing, temperature, and time for the various types of wastes being incinerated. As indicated above, certain waste compositions may limit the operating temperature, e.g., aqueous wastes with salts. Therefore, it is important to determine how and where each waste may be introduced into the incinerator. Organic wastes are usually introduced into the primary combustor. These provide the necessary heat input and temperature needed for the low-calorific-value waste streams. If these streams will not cause a problem with refractories at high temperatures, they may be introduced with the HHV waste in the primary combustor. Remember, however, that the temperature in this zone must be maintained at 1200°C (2192°F) or greater.

Halogenated Hydrocarbons

With the increase in production in recent years in the plastics industry, the generation of wastes from vinyl chloride and polyvinyl chloride processes has expanded rapidly. In many of the larger facilities where an increasing amount of product is produced, an increase in toxic gaseous vents and viscous liquid waste material results. Since these materials have been determined to be carcinogenic, it is necessary to dispose of them properly. Several schemes have been used for the on-site disposal of wastes generated from these plants. They have included carbon adsorption, chemical fixation, solvent absorption, and incineration. Industry has gained a considerable amount of expertise in these areas over the past 20 years. Incineration has resulted in the ultimate answer to the disposal of these halogenated waste materials. There are incineration systems operating today in which chlorinated waste materials are being disposed of properly. (See Table 8.1.3 for a list of chlorinated waste materials and their origins.)

The objective in incinerating residues containing chlorine is to convert as much as possible of the chlorine content to hydrogen chloride. HCl can be absorbed in water from the reaction gas mixture. In this way, re-usable hydrochloric acid is recovered in an environmentally safe way. The quantity of chlorine contained in the waste is often the measure of the heating value of the waste material. High chlorine percentages are typical of wastes produced in the VCM process (see Fig. 8.1.10). Heating values in the range of 3500 to 8000 Btu/lb (1944 to 4444 kcal/kg) are typical. Chlorinated hydrocarbon fuels tend to be slower-burning and soot-forming. As a result, many systems have operated with high excess air to promote mixing of the oxygen with the hydrocarbons to minimize soot formation. This has required an addition of auxiliary fuel to overcome the drop in flame temperature with high excess air. This also increases the operating horsepower and the size of downstream pollution-control equipment. High excess air

TABLE 8.1.3 Sources of Chlorinated Organics

Process or product	Waste products
Vinyl chloride monomer	Ethylene dichloride
	Ethyl chloride
	Vinyl chloride
Propylene glycol	Dichloroisopropyl ether
Epichlorohydrin propyl oxide	Dichloropropylene
Insecticides	Benzene hexachloride
	Hexachlorobutadiene
	Octochlorocyclopentene
Herbicides	Carbon tetrachloride
	Tetrachloroethylene
Chlorinated elastomers	Dichloropropylene

also promotes the formation of a much higher level of free chlorine.[6] These may not appear as problems in the incinerator since the results will indicate proper destruction of the POHCs; however, problems will occur in the waste-heat recovery equipment as well as in the downstream scrubbers. Free chlorine is much more aggressive than hydrogen chloride in attacking metals. Free chlorine requires caustic in the final tails tower for proper pollution control when scrubbing, while HCl is freely absorbed in water. It is also important that the combustor be designed for thorough mixing of air and wastes, thereby achieving maximum combustion efficiency. This promotes maximum temperature levels achievable under equilibrium conditions. This will also require proper selection of refractories so that these will withstand the temperatures which will be reached in high-efficiency combustors. Water and steam cooling are often used for the purpose of safe temperature control. In one situation a waste fuel having a heating value of 8000 Btu/lb had a theoretical flame temperature of 3800°F. Steam cooling prevented overheating of the refractory and also provided an excess of hydrogen to

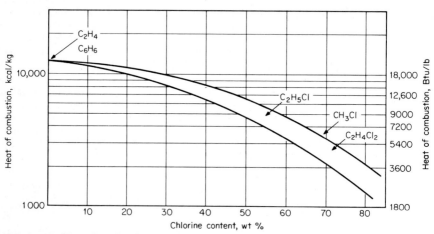

FIG. 8.1.10 Heat of combustion vs. chlorine content.

minimize chlorine formation. The incineration section should be designed with the following goals:

1. Minimize auxiliary fuel requirements.
2. Operate at minimum excess air.
3. Minimize free chlorine.
4. Provide a highly turbulent combustion section to ensure maximum flame temperatures in a minimum chamber volume.
5. Minimize soot formation.

Systems have been designed for operation at very low heat-release rates, that is, in the range of 1×10^6 to 2×10^6 Btu/h, where heat recovery is not critical, and systems have been designed for heat-release rates as high as 70×10^6 to 80×10^6 Btu/h. These systems include not only waste-heat recovery in the form of a steam boiler but also recovery of acid by-product. In most plants where VCM is manufactured, hydrochloric acid is required as a raw material. This aids the plant operator in investing in a waste-disposal system that will be economical in terms of the return on investment.[7] (See Fig. 8.1.1.)

8.1.4 LIQUID-INJECTION INCINERATOR VENDORS[1]

As mentioned above, the number of liquid-injection incinerators in use today is much greater than the number of all other types combined. However, many of the other incinerator designs include a liquid-injection burner or a secondary chamber which is very similar to a liquid-injection incinerator. Figure 8.1.11 lists

Bigelow-Liptak	John Zink	Process
Brule	Kelly	Combustion
Burn-Zol	McGill	Sure-Lite
Coen	Met-Pro	T-Thermal
Entech	Peabody	United
Hirt	Prenco	

FIG. 8.1.11 Liquid-injection incinerator vendors.

those organizations who are able to supply a liquid-injection incinerator as a separate system with or without heat recovery but with the necessary pollution-control systems. In many systems today, waste liquids serve as a source of auxiliary fuel. The system would therefore require a properly designed liquid-injection burner or nozzle to be able to properly combust this material.

8.1.5 ENVIRONMENTAL EFFECTS

The amount of effort that has been directed toward the proper design of air-pollution control and wastewater-control systems associated with liquid-injection incinerators is probably greater than that for all other areas of incineration. Since most liquid-injection incinerators do not generate solids or ash, the air-pollution control problem is minimized. In most cases, it is a matter of providing the proper scrubber for absorption of the acid gases generated, i.e., HCl, SO_2, NO_X, and others. The technology for the air-pollution control systems in this area is well-established.

In systems burning liquid wastes that contain solids or inert materials, particulates which form may be submicron in size because of the high temperatures generated in the combustion process. It is very important that pilot tests be conducted to evaluate the particulate loading and size distribution. This will affect the selection of an air-pollution control system. Wastes containing phosphorous have required venturi scrubbers with pressure drops at levels of 100 in w.c. (186.8 mmHg). An alternative would be either an ionizing wet scrubber, an electrostatic precipitator (ESP), a collision scrubber, or a baghouse.

Water

The wastewater discharge is usually dependent on the acid gases formed in the incinerator, the ash composition and size, and the scrubber system employed. Many of the larger systems today are incorporating spray dryers for acid-gas removal followed by ESPs or baghouses. Where packed-tower scrubbers or venturi scrubbers are used in HCl scrubbing, a lean acid solution is generated. This is then delivered to a lagoon where it is neutralized with lime (much less expensive than caustic) and then discharged to the plant's wastewater-treatment system.

8.1.6 REFERENCES

1. I. Frankel et al., "Survey of the Incinerator Manufacturing Industry," *Chemical Engineering Progress,* March 1983, pp. 44–55.

2. *Resource Conservation and Recovery Act,* "Standards for Owners and Operators of Waste Facilities: Incinerators," 40 *CFR* 264, RCRA 3004, January 25, 1981, Revised July 21, 1982.

3. J. J. Santoleri, "Design and Operating Problems of Hazardous Waste Incinerators," *Environmental Progress,* 4(4):246–251 (November 1985).

4. J. S. Caprio and H. E. Wolfe, "Refractories for Hazardous Waste Incineration—An Overview," ASME Solid Waste Processing Division Conference, May 1982.

5. H. Tsuruta, "Waste Liquid Incineration and Valuable Product Recovery in the Chemical Industry," *Chemical Economy and Engineering Review,* 4(12):45–51 (December 1972).

6. K. Mitachi et al., "Treatment of Liquid Waste Derived from Chlorinated Hydrocarbon Plants," *Chemical Engineering,* Oct. 1978, p. 73.

7. J. J. Santoleri, "Optimum Energy and By-Product Recovery in Chlorinated Hydrocarbon Disposal Systems," ASME—1982 National Waste Processing Conference.

SECTION 8.2
ROTARY KILNS

Charles F. Schaefer
Manager, Pyro Processing Development
Minerals Systems Division
Boliden Allis, Inc.
Milwaukee, Wisconsin

Anthony A. Albert
Manager, Project Sales
Minerals Systems Division
Boliden Allis, Inc.
Milwaukee, Wisconsin

8.2.1 OVERVIEW OF THE TECHNOLOGY AND ITS APPLICATIONS

With the increasing awareness of the problems associated with the disposal of hazardous wastes, attention has focused on incineration as potentially the best method of disposing of certain wastes. It is assumed that approximately 60% of all hazardous waste can be successfully incinerated.[1] Two of the most important operating conditions for proper incineration are temperature and residence time. These conditions, along with the chemical and physical form of the waste, determine the size of the incinerator. The desire to have the waste mixed during incineration favors rotary-kiln technology.

Rotary kilns provide a number of functions necessary for incineration. They provide for the conveyance and mixing of solids, provide a mechanism for heat ex-

FIG. 8.2.1 Rotary-kiln and cooler installation.

change, serve as host vessel for chemical reactions, and provide a means of ducting the gases for further processing. Rotary kilns are equally applicable to solids, sludges, and slurries and are capable of receiving and processing liquids and solids simultaneously. A typical rotary-kiln installation is shown in Fig. 8.2.1.

The processing of raw materials in the solid, or semisolid, state at high temperatures in rotary kilns, without substantial fusion of the charge, is routinely accomplished in the cement, lime, clay, phosphate, iron ore, aggregates, and coal industries, to cite a few examples.[2] Kilns designed for processing temperatures of 1370°C (2500°F) with air flow of hundreds of thousands of actual cubic feet per minute are common in most of these industries.

The processing of hazardous wastes with temperatures in the 1100°C (2000°F) range is readily applicable to the kiln technology.

In the processing of solids, such as iron ore and limestone, the material is uniformly fed to the kiln by either gravity or mechanical means. The particle-size distribution of the feed determines the actual load point, as the feed system is designed to minimize entrainment of fine particles into the gas stream. If the feed material is of a large size, the design has to take into account the impact of the feed on the refractory lining. Where the feed is a sludge, fines entrainment is of minimal concern at the load point. The design then minimizes the potential sweeping action of the gas stream.

8.2.2 PROCESS CONSIDERATIONS

When designing a kiln to process solid hazardous wastes, it is first necessary to characterize the material to be processed.

A chemical analysis of the material will indicate whether or not any endothermic or exothermic reactions will occur during processing. This will have a direct impact on the quantity of fuel required and the volume of process gases generated.

The specific heat of the material must be known to calculate the heat required to elevate its temperature from ambient to the required processing temperature.

Having characterized the material to be processed, the specific application can be addressed. The design production rate is established by the specific application. The processing temperature and required residence time may be known at this point. If they are not, then bench- and batch-scale testing are required to establish these process parameters.

Kilns can be designed for loading of 5 to 10% of their internal volume, in which case residence time is relatively short (1 h or less). In these cases no internal dams are provided. Heat transfer is enhanced in a lightly loaded kiln. Heat transfer in the processing zone is primarily by radiation from the kiln wall. The ratio of kiln wall to bed surface increases with decreased kiln loading. In addition, a lightly loaded kiln results in improved bed turnover, which promotes even heating of the bed.

If the required residence time is relatively long (greater than 1 h), a dam is provided at the material-discharge end of the kiln. A dam can significantly increase the residence time in a given kiln. When dams are employed, kiln loadings increase. From a practical standpoint kiln loadings should not exceed 15 to 20% of the internal kiln volume.

A particle-size analysis of the material is required to assess the potential for dust entrainment in the process-gas stream. Material size also impacts directly on the required residence time at processing temperatures.

The material's calorific value and moisture are factors which bear upon fuel quantities and process-gas volumes. Moisture content also affects how the material behaves in the early stages of pyroprocessing. Moisture that results in a pasty consistency can lead to agglomeration of the material, which is undesirable. In these situations the use of chains in the kiln inhibits agglomerate formation and breaks down any agglomerates formed.

For a given desired production rate, the size of the kiln and its loading are a function of the material's bulk density as it undergoes processing.

The dynamic angle of repose of the material affects the time required to transport the material through the kiln and, therefore, affects kiln residence time.

To summarize, the following characteristics of the material to be processed in a rotary kiln must be established before a kiln can be designed.

- Chemical composition
- Specific heat
- Size consistency (distribution)
- Calorific value
- Moisture content
- Bulk density
- Dynamic angle of repose

8.2.3 DESIGN CRITERIA

The flow of material through a rotary kiln is determined by the kiln's slope and rotational speed, as well as by the characteristics of the material being processed. The kiln is installed on a slight slope so that the bed of solids advances through the kiln by the force of gravity. As the kiln rotates, the material follows the rotation until it breaks the surface of the bed and tumbles down the sloping surface.

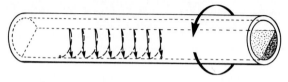

FIG. 8.2.2   Axial movement through the reactor.

The slope of the kiln produces a slight forward movement toward the discharge end. (See Fig. 8.2.2.)

Figure 8.2.3 illustrates some of the common terminology used to describe kiln solids flow systems. Kiln residence time is expressed by the formula[3]

$$T = \frac{1.77 \sqrt{\Theta}\, F}{SDN}$$

Where T = residence time, min
Θ = dynamic angle of repose, degrees from horizontal
L = kiln length, ft
S = kiln slope, degrees from horizontal
N = rotational speed of kiln, r/min
F = factor; 1.0 for undammed kilns; >1.0 for dammed kilns
D = inside diameter, ft

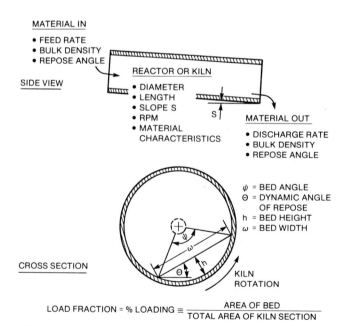

FIG. 8.2.3   The kiln solids flow system—definition of terms.

TABLE 8.2.1 Some Common Factors in Rotary-Kiln Design

Slope, in/ft	S	Θ	$\sqrt{\Theta}$	$1.77\sqrt{\Theta}$
1/4	1.192	15	3.873	6.855
5/16	1.491	20	4.472	7.915
3/8	1.790	25	5.000	8.850
7/16	2.087	30	5.477	9.694
1/2	2.385	35	5.916	10.471
9/16	2.684	40	6.324	11.193
5/8	2.980	45	6.708	11.873

Some common factors are shown in Table 8.2.1.

The slope of the kiln normally ranges from 0.02 m/m to 0.04 m/m (0.25 in/ft to 0.5 in/ft). Increased slopes enhance initial transport of material in kilns. Steeper slopes are generally employed in cases where the material decreases in unit volume as it undergoes processing.

Rotational speed usually ranges from 0.5 r/min to 1.5 r/min for heavily loaded kilns and 1 to 3 r/min for lightly loaded kilns.

As can be seen, there is an infinite number of combinations of lengths and diameters that will result in the required residence time. Generally, however, for solid hazardous-waste applications L/D ratios range from 3:1 to 10:1.

Kiln loading is expressed by the formula

$$L = \frac{Q}{VA}\ (100)$$

Where L = loading, %
  Q = material volumetric rate, ft³/min
  V = material transport rate, ft/min
   = shell length, ft, divided by retention time, min
  A = kiln's internal cross-sectional area, ft²

The depth of the bed and, therefore, the percent loading can be increased by installing a dam at the discharge end of the kiln. This dam affects the depth of the bed along the entire length of the kiln as illustrated in Fig. 8.2.4.

After a kiln size is selected, the process-gas velocity in the kiln is calculated. Depending upon the particle-size analysis of the material being processed, process-gas velocities should fall within the range of 165 to 1,000 m/min (500 to 3,000 ft/min); lower velocities corresponding to finer particles. The length/diameter ratio should be adjusted to give the desired process-gas velocity.

The quantity of process gas results from combusting the fuel, evaporating water, and any resultant gaseous products of chemical reactions.

The fuel required is established by the heat required to elevate the material to the required processing temperature, evaporate the water, and satisfy any endothermic chemical reactions; minus the heat supplied by the material and that supplied by any exothermic reactions. The radiant heat loss from the kiln shell and the sensible heat contained in the process-gas stream exiting the kiln must also be compensated for.

Two modes of kiln operation can be employed in processing solid hazardous wastes:

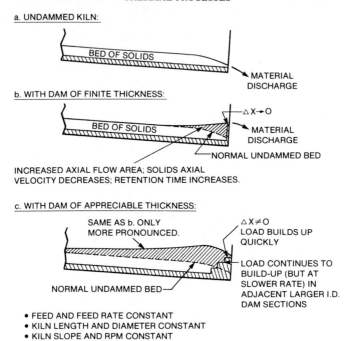

a. UNDAMMED KILN:

BED OF SOLIDS

MATERIAL
DISCHARGE

b. WITH DAM OF FINITE THICKNESS:

△X→0

BED OF SOLIDS

MATERIAL
DISCHARGE

NORMAL UNDAMMED BED

INCREASED AXIAL FLOW AREA; SOLIDS AXIAL
VELOCITY DECREASES; RETENTION TIME INCREASES.

c. WITH DAM OF APPRECIABLE THICKNESS:

SAME AS b. ONLY
MORE PRONOUNCED.

△X≠0
LOAD BUILDS UP
QUICKLY

LOAD CONTINUES TO
BUILD-UP (BUT AT
SLOWER RATE) IN
ADJACENT LARGER I.D.
DAM SECTIONS

NORMAL UNDAMMED BED

- FEED AND FEED RATE CONSTANT
- KILN LENGTH AND DIAMETER CONSTANT
- KILN SLOPE AND RPM CONSTANT

FIG. 8.2.4 Kiln-bed profiles with and without dams.

Parallel flow
Counterflow

If the principal objective is to volatilize organics for subsequent destruction in the gas stream, the material to be decontaminated is fed at the same end of the kiln as the kiln is fired. Solids and gases travel in the same direction, i.e., parallel flow.

Parallel-flow operation is also recommended when the material to be processed has fuel value or when exothermic chemical reactions occur. In this case, more gas-stream residence time results to destroy the organics and provide for better utilization of heat released.

In situations where significant quantities of water are to be evaporated and endothermic chemical reactions occur, a counterflow operation is recommended, because it is more fuel-efficient. The kiln is fed at the end opposite to the one where it is fired. Solids and gases travel in opposite directions.

8.2.4 MECHANICAL CONSIDERATIONS

In addition to process considerations, there are mechanical considerations in kiln design that require attention.

The length-to-diameter ratio of the kiln determines how many supports are required. The rotary kiln is supported on riding rings by support rollers. As the length-to-diameter ratio increases, additional supports may be required. Generally, for solid hazardous-waste applications, two supports are sufficient.

Specifications for kiln shell thickness, riding rings, support rollers, and drive sizes are determined by the total loading of the kiln. A major component of the total load is the refractory lining in the kiln.

Refractory lining is selected to optimize capital and operating costs. The functions of kiln refractories are to protect the mild-steel kiln shell and to minimize radiant heat loss.

The total horsepower required to turn the kiln is equal to the friction horsepower plus load horsepower.

$$\text{Friction hp} = \frac{W \times bd \times td \times N \times F \times 0.0000092}{rd}$$

Where W = total vertical load on all roller shaft bearings, lb
bd = roller shaft bearing diameter, in
rd = roller diameter, in
td = tire (riding ring) diameter, in
N = rotational speed of kiln, r/min
F = Coefficient of friction of support bearings:
0.018 for oil-lubricated bearings
0.016 for grease-lubricated bearings

$$\text{Load hp} = (D \times \sin \Theta)^3 \times N \times L \times K$$

Where D = inside diameter of lining or shell, ft
N = rotational speed of kiln, r/min
L = length of shell, ft
K = 0.00092 for material with 40° angle of repose
0.00076 for material with 35° angle of repose
Sin Θ is based upon percent loaded cross-sectional area of kiln

8.2.5 HEAT TRANSFER

Mixing of the solids in the bed occurs only above the shear line in the cascading layer of material (see Fig. 8.2.5). If the load has extremes in particle size with a high (15 to 25%) level of fines, those fines could collect in a "cold kidney" and the resultant mass would be less pervious to penetrating gases.

Heat transfer appears to be proportional to the square of the kiln's internal diameter. The rules of thumb which attribute 60 to 80% of kiln heat transfer to radiation are presently under investigation. Numerous tests and demonstrations indicate that the bulk of the heat transfer may well be attributable to convective heat transfer. The heat transfer through the plane surfaces (chord and arc segments) accounts for about one-half the total. Bed permeability and mass exchange in the cascading load appear to account for the majority of the rest. The mean free path for radiation increases linearly with diameter and, therefore, the kiln gases are more emissive.[4] Theoretically, the inside-wall temperatures may be lower with larger kilns, as the flame is more loosely coupled convectively to the wall. This may then lessen the underbed heat trans-

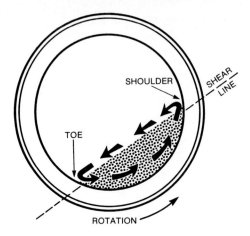

FIG. 8.2.5 Solids flow in cross section.

fer. Practical experience with kilns as large as 7.6 m (25 ft) has shown these effects to be minimal.

The radiant heat for the rotary kiln is provided by standard auxiliary-fueled burners.

The single-burner concept, either parallel (cocurrent) flow or countercurrent flow, provides a flame for about 20 to 30% of the kiln length (see Fig. 8.2.6). The burner can be designed to be movable, allowing some flexibility in locating the burning zone. The heat for the remaining 70% of the kiln length is derived from the sensible heat of the hot gases.

In either case, the gas-temperature profile can be raised (flattened) by increasing the gas-to-solids ratio. The associated penalty for doing so is the handling of larger quantities of hot waste gas at higher velocities. This, in turn, increases fuel consumption, produces an increased pressure drop across the kiln, and results in a need for larger equipment for off-gas handling and in increased particulate entrainment.

The axial-burner concept incorporates a multiplicity of primary burners to provide heat release for almost the full length of the kiln (see Fig. 8.2.7). The

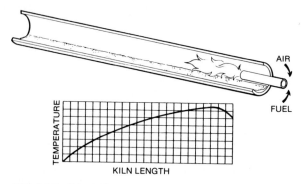

FIG. 8.2.6 Primary-burner concept.

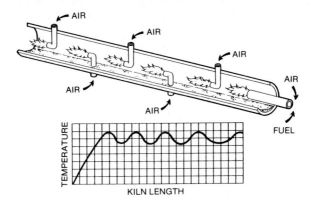

FIG. 8.2.7 Axial-burner concept.

spacing of the burners is limited by practical considerations to avoid having the flame from one burner impinge on and overheat the adjacent burner tube.

In kilns of this type, the gas temperatures are erratic as the admitted air is heated, mixed with the fuel, combusted, then cooled as the products of combustion give up the sensible heat, then heated again. The cycle continues for the length of the kiln. These erratic temperatures can cause control problems. If the kiln atmosphere does not have enough excess fuel to support combustion, the axial burners may require an auxiliary fuel supply as well.

The ported-kiln concept incorporates multiple air ports that allow heat release over the full range of the active zone (see Fig. 8.2.8). The multiplicity of ports results in almost a continuum of air for combustion. The resulting sheet of flame has no discontinuities and the temperatures are well-controlled.

The radial momentum of incoming air keeps the hottest portion of the flame from impinging on the refractory. As a result, the temperature profile is easily controlled, which provides the ultimate flexibility for process regulation.

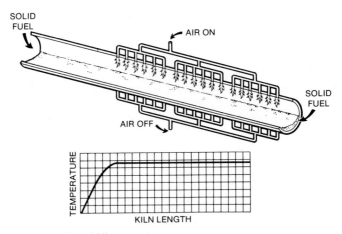

FIG. 8.2.8 Ported-kiln concept.

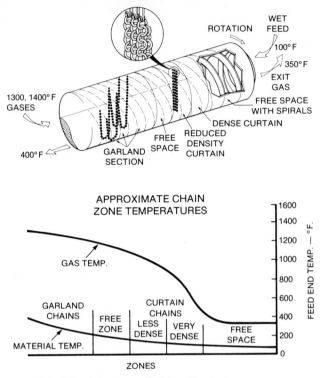

FIG. 8.2.9 Kiln-chain system (*Boliden Allis, Inc.*)

In countercurrent kiln operations, the use of kiln chain toward the feed end of the kiln serves three functions:

- Enhances heat transfer from the process-gas stream to the material being processed
- Inhibits formation of agglomerates and breaks down any agglomerates formed
- Suppresses dust emissions from the kiln

Basically, two configurations of chain are employed over a given length of kiln—garlands and curtains. The chain is hung from hangers secured to the kiln shell. Figure 8.2.9 illustrates an arrangement of kiln chain.

Kiln chain adds to the total kiln load; therefore, its weight must be factored into the mechanical design.

8.2.6 TYPES OF INSTALLATIONS

There are two categories of plants for processing solid hazardous wastes—permanent installations and transportable plants.

Permanent installations are at the sites of generators of solid hazardous

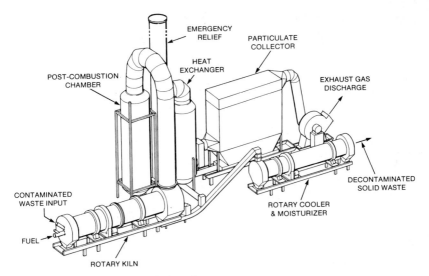

FIG. 8.2.10 Hazardous-waste incineration system.

wastes, or are installed as regional facilities accepting wastes from multiple sources.

Transportable facilities can be constructed in skid- and/or wheel-mounted modules. This type of plant is most likely to be used to clean up existing abandoned dump sites.

For permanent installations large kilns can be employed. The kilns would be shipped in sections and welded in the field. However, note that the size of transportable kiln modules is limited by highway load limits.

A generic rotary-kiln-based incineration system is shown in Fig. 8.2.10. Even though the volume of material to be processed and the type of installation would dictate the equipment sizing, the process concepts would be as shown.

The rotary kiln would volatilize the hazardous constituents, which would then be combusted in the kiln gas stream. A postcombustion chamber, equipped with independent burners and fans, ensures complete destruction of the hazardous constituents. An emergency-relief stack located downstream of the postcombustion chamber would provide a means to vent process gases if the heat exchanger experienced a shutdown.

The heat exchanger (e.g., waste-heat boiler, quencher, or other such device) reduces the off-gas temperature to protect the particulate collector. The induced-draft (ID) fan provides process-gas flow control for the system. Cleaned gases are exhausted to the atmosphere.

The incinerated solids are discharged to a high-temperature conveying device which transports the material to a cooler-moisturizer device. A rotary cooler provides the reverse heat-transfer mechanism described for the kiln. Additional cooling with the added benefit of moisturizing can be accomplished by adding water sprays to the cooler. The material is discharged at less than 65°C (150°F) for further handling by belt conveyor.

The process, design, and mechanical considerations for rotary coolers are similar to those used for rotary kilns. The heat-transfer consideration is simply reversed.

The incineration of hazardous wastes in a rotary kiln has become popular because the designs, concepts, and theories are well-established and proven in many solids-processing industries.

8.2.7 REFERENCES

1. A. P. Dillon, "Hazardous Waste Incineration Engineering," *Pollution Technology Review No. 88,* Noyes Data Corporation, Park Ridge, New Jersey, 1981, chap. 2.1, p. 3.
2. Ibid., chap. 4, p. 94.
3. U.S. Bureau of Mines, Technical Paper No. 384, U.S. Department of the Interior, 1927.
4. R. Pierce, *Computer Simulation of Solids Flow in Heavily Loaded Rotary Kilns,* Allis-Chalmers Publication 22B 10880.

8.2.8 FURTHER READING

Harris, J., et al., "Combustion of Hazardous Wastes," *Sampling and Analysis Methods, Pollution Technology Review No. 117,* Noyes Data Corporation, Park Ridge, New Jersey, 1985.

Kilns-flow of material, Allis-Chalmers Publication 22B1212-2, Allis-Chalmers Corporation, Milwaukee, Wisconsin.

SECTION 8.3
FLUIDIZED-BED THERMAL OXIDATION

George F. Rasmussen
Vice President of Engineering
Waste-Tech Services, Inc.
Golden, Colorado

Robert W. Benedict
Manager, Engineering and Development
Waste-Tech Services, Inc.
Golden, Colorado

Craig M. Young
Manager, Quality Control and Analysis
Waste-Tech Services, Inc.
Golden, Colorado

8.3.1 HISTORICAL BACKGROUND

Beginning with C. E. Robinson's 1879 patent of an ore-roasting furnace, bubbling[1]-bed type *fluidized beds* have become the standard process reactor for coal gasification, catalytic cracking of heavy oils, ore roasting, calcining, cooling, drying, sizing, and combustion. The versatility of the fluidized-bed process is due to its excellent gas-solids contacting, stable temperature control, and the ability to control residence time.

In the early 1920s the *circulating-fluids system* was developed by increasing the velocity of the fluidizing gas stream to the point where the waste and bed solids particles become entrained and elutriate out of the vessel at essentially the same rate as the fluidizing air. The primary application of circulating-fluids systems has been to biomass-fueled power boilers.

8.3.2 DESCRIPTION OF THE TECHNOLOGY

Systems for fluidized-bed thermal oxidation of hazardous waste generally consist of the principal elements shown in Fig. 8.3.1, which may be segmented into the air fluidization system, fluidized-bed vessel, feed systems, and off-gas cleanup equipment. Energy-recovery systems in the form of a steam generator or air preheater are often added when economically justified.

Air from the fluidizing-air fan is distributed across the bottom of the bed via an air-distribution system. As the velocity of the air increases, the granular bed material becomes suspended in a churning gas–solids mixture having physical properties similar to a fluid.

Liquids and sludges with solids up to approximately 1.25 mm (½ in) can be fed through a series of pneumatically assisted injection nozzles, directly into the fluidized bed. The nozzles may be liquid- or gas-cooled to minimize buildup of carbonized waste at the nozzle tip.

Solids can be fed above or into the bed. Above-bed feeding allows for simpli-

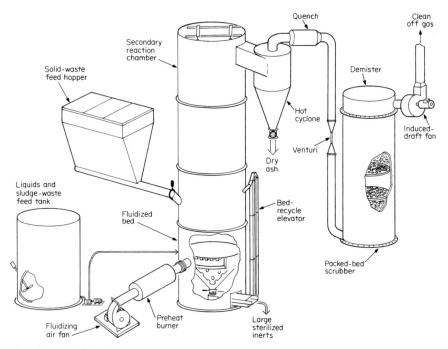

FIG. 8.3.1 Fluidized-bed system for combustion of hazardous waste.

fication of the feed system because the need to seal against the bed pressure is eliminated. However, above-bed feeding may result in incomplete combustion due to fines carry-over. Solid wastes may require size reduction to meet vendor limitations. Typical maximum inert-particle sizes are 7.6 cm (3 in) for units with parallel-header air distributors and 2.5 cm (1 in) for units with plate-type air distributors. Unlike fluidized beds, circulating beds inject solid wastes and finely ground limestone through a common feed nozzle into the solids recycling line.[2] This requires that the waste be reduced to a typical size of 2.5 cm (1 in) to permit proper mixing and entrainment.

Off-gas cleanup is accomplished with the same types of wet and dry equipment utilized by other thermal oxidation technologies.

Fluidized beds have several inherent parameters that make them uniquely suited for safe destruction of hazardous and toxic wastes, including

- Ability to handle waste streams having widely varying heating values (results from thermal "inertia" of large mass of bed material)
- Low susceptibility to over-pressure emissions (results from sealed-system design)
- Elimination of hot and cold spots (results from superior in-bed turbulence and mixing)

8.3.3 DESIGN CONSIDERATIONS

A variety of bubbling-type fluidized-bed designs exist. An understanding of their operating parameters and components is necessary for a comparison. This subsection provides a brief background on the important fluidized-bed design features and terms. For additional information the reader may wish to consult one of the references on fluidized beds listed at the end of this section.[3,4,5]

Operating temperatures vary depending on the particular application. For nonhazardous-waste sludges, operating temperatures of bubbling-type fluidized beds range from 650 to 1200°C (1200 to 2192°F).[6] Refinery wastes are combusted in fluidized beds that operate between 700 and 815°C (1300 and 1500°F).[7] For chlorinated solvents, destruction efficiencies greater than 99.99% are achievable in fluidized beds operated at 775°C (1427°F)[8] or as low as 743°C (1370°F) with the secondary reaction chamber at 1200°C (2192°F).[9] Circulating-fluids systems utilize limestone, which reacts with halogen gases in the combustion chamber to form dry salt compounds. Due to the low ash-melting point of the calcium salt compounds created, the circulating bed has an upper limit on operating temperature of 772°C (1421°F) when burning highly halogenated wastes. Multiple passes through the combustor via the solids recycling system provide the residence time required to overcome the lower operating temperature of the circulatory bed and accomplish destruction of hazardous waste.

The superficial-air velocity in the bed region is called the *fluidizing velocity*. It is limited by the minimum fluidization velocity and by the air rate where solids entrainment becomes excessive.

At minimum fluidizing velocity the bed-particle drag forces in the upwardly moving gases exceed the weight of the bed particles. At this point the bed is described as an *incipiently fluidized bed*. The minimum fluidizing velocity is affected by the bed-material particle size, particle size distribution, and particle density.

Above the minimum fluidizing velocity, bed turbulence increases. A bed in this operating range is considered a *bubbling bed,* or *dense-phase fluidized bed.* For waste combustion, bubbling beds are typically run at 2 to 10 times the minimum fluidizing velocity to assure uniform suspension of the bed media.

Further increases in fluidizing velocity produce a *dilute-phase bed* where solids elutriation becomes excessive and no distinct boundary exists between the bed region and the gas region above the bed. When the solids are recovered from the gas and reinjected, a *circulating fluidized bed* exists.

These different regions can be seen by plotting the pressure drop across the bed versus fluidizing velocity, as in Fig. 8.3.2.

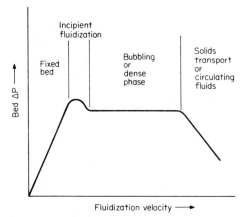

FIG. 8.3.2 Fluidizing regimes.

Bubbling beds are typically divided between shallow beds (<0.9 m or 3 ft) and deep beds (>⅛.¼ M OR ½ FT(. Since the bed depth is proportional to the pressure drop across the bed, this parameter affects the required fan power for the air-supply system. For this reason, shallow beds are typically used for waste combustion and deep beds are typically used for catalytic processes.

The function of the air distributor is to supply the air uniformly across the cross section of the bed and prevent solids flow back into the air-supply system. Two common types of air distributors are the perforated plate and the parallel header (see Fig. 8.3.3). The perforated plate consists of a plate with bubble caps or tuyeres over a plenum area. The parallel-header type of distributor uses rectangular channels for the plenum with bubble caps mounted on these channels. By sloping the holes in the bubble caps, solids backflow into the air system is minimized. The parallel-header distributor design allows for easy removal of noncombustible material up to 7.6 mm (3 in) via the on-line bed-removal system. The parallel-header design also eliminates the potential for warpage, which is common with a perforated-plate distributor.

To assure good fluidization the pressure drop across the distributor at minimum flow should be at least 35% of the bed pressure drop.

The above-bed secondary reaction chamber can be run at higher gas temperatures and provides additional residence time as well as providing an area where solids can disengage from the gas stream.

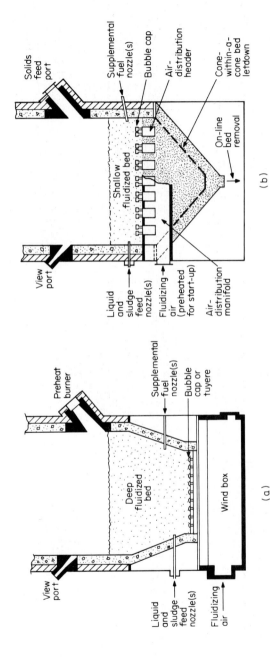

FIG. 8.3.3 Methods of fluidizing-air distribution. (*a*) Plate design. (*b*) Header design.

8.35

8.3.4 PHYSICAL CONSTRAINTS

The heating value of the waste and the presence of low-melting-point compounds can limit fluidized-bed operation. Wastes with higher heating values in excess of 5.27 MJ/kg (5000 Btu/lb) can be combusted without supplemental fuels. If the fluidizing air is preheated with the hot off gases, autogenous combustion can occur with wastes having higher heating values as low as 3.68 MJ/kg (3500 Btu/lb).

Of greater concern is the presence of compounds which may cause bed agglomeration. These compounds are typically from two possible sources: bed media or the wastes. The use of silica sand as bed material at operating conditions above 900°C (1652°F) results in bed fusion. Crushed refractory having an ash-softening point above 1300°C (2372°F) easily eliminates this concern. The most common problem compounds in wastes are low-melting alkali materials. Also, many of the halide salts have melting points below 1000°C (1832°F). Many different methods have been used to avoid bed agglomeration, including lower operating temperature or the use of additives which will react with the alkali material to form compounds that melt at temperatures above 1200°C (2192°F). Operation at 1200°C (2192°F) with wastes containing up to 1.7 wt % sodium has been demonstrated in the presence of additives.

8.3.5 ENVIRONMENTAL IMPACTS

The primary environmental impacts of fluidized-bed combustion are air emissions and ash generation. Hydrogen chloride, sulfur dioxides, and particulates are controlled with commonly available air-pollution control equipment. Nitrogen oxide emissions are typically lower than with other combustion processes because of the lower operating temperatures of fluidized beds. Carbon monoxide and hydrocarbon emissions are low because the bed is a well-mixed reaction zone. The emission values vary greatly with type of waste and fuel firing but range between 35 and 100 ppm CO, 60 and 250 ppm NO_x, and 0 and 5 ppm hydrocarbon. Ash-generation volumes are waste-dependent. Heavy metals in the waste are typically retained in the ash. Hydrocarbon levels in ash have been measured in the trace range between 0 and 10 ppm.

8.3.6 CASE STUDIES OF FLUIDIZED-BED APPLICATIONS

Fluidized beds are ideally suited for the destruction of solids, liquids, and sludges, separately or together. Over 1,000 fluidized-bed combustion units are operating worldwide with over 160 of these, sized between 35 and 2460 MJ/min (2 and 140 million Btu/h), destroying sludges from municipal waste-treatment facilities.[10] More than 20 systems are in operation worldwide destroying hazardous and toxic wastes. Figure 8.3.4 shows a transportable 26.4 MJ/min (1.5 million Btu/h) fluidized-bed system suitable for small on-site cleanups. Figure 8.3.5 is a 386 MJ/min (22 million Btu/h) fluidized-bed facility destroying chlorinated chemical plant solid, liquid, and sludge wastes.

A large chlorinated ethane production facility recently installed a fluidized-bed thermal oxidation system for destruction of highly chlorinated sludge and

FIG. 8.3.4 Transportable fluidized-bed thermal oxidation system.

FIG. 8.3.5 386 MJ/min (22×10^6 Btu/h) fluidized-bed incineration system for chemical plant wastes.

solidwastes   containing   dichloroethane,   perchloroethylene,   tri-   and
tetrachloroethanes, carbon tetrachloride, and chloroform. Destruction and removal efficiencies (DREs), greater than 99.99% were attained at bed temperatures of 950°C (1742°F) with 2-s residence time.

A fluidized-bed thermal oxidation system located in Franklin, Ohio, having
a 7.6-m (25-ft) diameter secondary reaction chamber, burned an aqueous solution of methyl methacrylate monomer and an aqueous phenol-creosol
sludge. In both cases, the DRE for the waste constituents was greater than
99.999% with temperatures of 732 to 900°C (1350 to 1650°F) for the phenol-
creosol and 788 to 843°C (1450 to 1550°F) for the methacrylate, with residence
times of approximately 11 s.

Another commercial fluidized-bed thermal oxidation system, sampled by the
U.S. Environmental Protection Agency (EPA), demonstrated a destruction and
removal efficiency of 99.99% when burning mixtures containing 1,1,2-trichloro-
1,2,2-trifluoroethane (3.0%), 1,1,1-trichloroethane (3.4%), trichloroethylene
(2.4%), and tetrachloroethylene (3.3%) at typical bed temperatures of 743°C
(1370°F) and temperatures in the secondary reaction chamber of 1180°C (2160°F).
Emissions of 0.18 g/dscm (0.08 gr/dscf) and HCl-removal efficiencies of 99.73%
were obtained with an off-gas cleanup system consisting of a lime slurry–water
quench, high-efficiency cyclone, and cross-flow wet scrubber containing
Tellerette packing. Small amounts of NaOH were added to the cross-flow scrubber for final HCl removal.

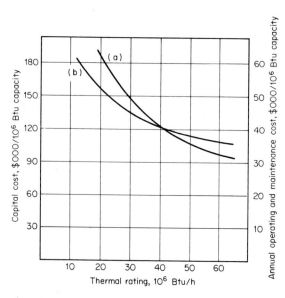

FIG. 8.3.6 Capital and operating and maintenance costs for
fluidized-bed systems. (*a*) Capital cost. (*b*) Annual operating
and maintenance cost.

8.3.7 ECONOMICS

Fluidized beds have a lower capital cost than rotary types of incinerators and are more expensive than liquid-injection incinerators. Figure 8.3.6a shows the installed capital cost of fluidized-bed systems for a range of thermal ratings. Cost data is for basic systems installed on a prepared site and excludes permits, taxes, and freight. Capital costs are approximately 10% greater for circulating-fluids systems in the 350 to 890 MJ/min (20 to 50 million Btu/h) range normally associated with hazardous waste. Operating costs of fluidized beds are comparable to those for rotary technologies, with power costs fractionally higher due to the fluidizing-air fan (see Fig. 8.3.6b). Maintenance costs are lower than for rotary technologies and comparable to those for liquid-injection systems. This is due primarily to longer refractory lives: fluidized beds, having no moving parts within the combustion zone, minimize refractory wear and mechanical fatigue. Operating costs are based on round-the-clock operation 5 days per week, 50 weeks per year and include operating and maintenance labor, utilities and consumables, property taxes, and maintenance materials.

8.3.8 NEW DEVELOPMENT AREAS

A major development area is the use of active beds rather than inert beds. A two-stage bed has been used with catalytic and dry off-gas cleanup applicability.[11] The EPA is currently testing ways to improve metal retention on bed particles.[12] A combustion process which captures phosphorus compounds on bed media has also been developed.[13] These are just a few examples of how fluidized beds are being improved.

8.3.9 SUMMARY

Fluidized beds offer an economical means for the safe thermal destruction of hazardous and toxic wastes. The technology is the choice in many cases where thermal oxidation is to be applied to many hazardous and toxic wastes. The reasons are (1) its superior combustion environment created by the bed turbulence, (2) the "thermal inertia" associated with the mass of the bed, (3) the process simplicity resulting from the lack of moving parts associated with the high-temperature vessel, and (4) the versatility of the technology, with capability to destroy solids, liquids, and sludges.

8.3.10 LIST OF HAZARDOUS-WASTE EQUIPMENT AND SERVICES VENDORS

On-Site Services

Waste-Tech Services, Inc., 18400 W. Tenth Avenue, Golden, Colorado 80401, (303) 279-9712

NOTE: On-site services include design, permitting, installing, ownership, and operation on a generator's site as a service.

Equipment Suppliers

Fluidized-Bed Combustion Systems

Keeler Dorr-Oliver, 77 Havemeyer Lane, P. O. Box 9312, Stamford, Conn. 06994-9312, (203) 358-3741

Thermal Processes, Inc., 20000 Governors Drive, Olympia Fields, Ill. 60461-1974, (203) 358-3741

The International Boiler Works Co., P. O. Box 498, East Stroudsburg, Pa. 18301-0498, (717) 421-5100

Waste-Tech Services, Inc., 18400 W. Tenth Avenue, Golden, Colo. 80401, (303) 279-9712

Circulating Fluidized Combustion Systems

Ogden Environmental, Inc., P. O. Box 85608, San Diego, Calif. 92138

8.3.11 REFERENCES

1. C. E. Robinson, *Furnace for Roasting Ore*, U. S. Patent 212,308, February 18, 1879.
2. D. A. Vrable and D. R. Engler, "Transportable Circulating Bed Combustor (CBC) For the Incineration of Hazardous Waste," Haz Pro '86, Baltimore, Maryland, April 1–3, 1986.
3. Kuni and Levenspiel, *Fluidization Engineering*, Robert F. Krieger Publishing, New York, 1977.
4. F. A. Zenz and D. F. Othmer, *Fluidization and Fluid Particle Systems*, Reinhold, New York, 1960.
5. J. R. Howard, *Fluidized Beds, Combustion and Applications*, Applied Science Publishers, 1983.
6. P. B. Liao, "Fluidized-Bed Sludge Incinerator Design," *Journal WPCF*, (48):(8)1899–1913 (1974).
7. F. Rubel, *Incineration of Solid Waste*, Noyes Data Corp., New Jersey, 1974, pp. 178–193.
8. G. P. Rasmussen and J. N. McFee, "Fluidized Bed Incineration Systems for the Ultimate Disposal of Toxic and Hazardous Materials," *Proceedings of the 1st Annual Hazardous Materials Management Conference*, Philadelphia, Pa., July 1983.
9. R. R. Hall, G. T. Hunt, M. M. McCabe, and J. O. Milliken, "Fluidized-Bed Incinerator Performance Evaluation," *Proceedings of the Ninth Annual Research Symposium, Incineration and Treatment of Hazardous Waste*, EPA 600-9-84-015, Cincinnati, Ohio, July 1984.
10. J. F. Mullen, "Fluidized Bed Combustion—An Economic, Proven Method of Industrial Waste Disposal," LICA General Seminar 1985, Dearborn, Michigan, September 4–6, 1985.
11. L. J. Meile, F. G. Meyer, A. J. Johnson, and D. L. Ziegler, *Rocky Flats Plant Fluidized Bed Incinerator*, Rockwell International, Energy Systems Group, Golden, Colorado, March 8, 1982.
12. R. D. Litt, and T. L. Tewksbury, *Trace Metal Retention When Firing Hazardous Waste in a Fluidized-Bed Incinerator*, EPA Research & Development, EPA-600/S2-84-198, January 1985.
13. N. F. Baston, *Fluidized Bed Incineration of Waste*, U.S. Patent 4,359,005, 1982.

SECTION 8.4

HAZARDOUS WASTE AS FUEL FOR BOILERS

Howard B. Mason
Manager, Energy Engineering Department
Acurex Corporation
Mountain View, California

Robert A. Olexsey
U. S. Environmental Protection Agency
Hazardous Waste Engineering Research Laboratory
Cincinnati, Ohio

8.4.1 APPLICABILITY OF WASTE FIRING

Thermal destruction of hazardous waste through cofiring with conventional fuels in boilers is widely practiced in industry. It can be an economical and reliable method for on-site waste disposal while recovering energy from the waste heat of combustion. From a hardware standpoint, cofiring is broadly applicable to the boiler design types and fuel types used in waste-generating industries. Cofiring has been used with firetube and watertube boiler designs and with gas, oil, and coal as the conventional fuel. Waste materials are usually fired as an alternate fuel with independent burners or atomizer guns rather than blended with the conventional fuel. In that context, the modifications for waste cofiring are not significantly more complicated than modifying a boiler for dual-fuel firing. Although some boilers derive the total heat input from waste firing, waste inputs from 10 to 25% of total input are more typical. Wastes with heats of combustion greater than about 11,600 kJ/kg (5000 Btu/lb) are candidates for cofiring. Constraints on the quantity of wastes to be cofired may arise because certain constituents, such as chlorine, heavy metals, or suspended solids, can promote corrosion and fouling of boiler tubes or erosion of waste-delivery system components. These operational effects are the most significant consideration when assessing cofiring feasibility. From an environmental perspective, numerous field tests have shown boilers to have waste-destruction efficiencies comparable to those of incinerators when the boiler is well-operated.

8.4.2 WASTE-COFIRING TECHNOLOGY

The technology for hazardous-waste cofiring predates the regulation of hazardous waste, going back to the time when waste was fired for heating-value credits. The additional capital costs and operating costs associated with waste as an alternate fuel, and with the lower quality of the waste fuel, were usually offset by the energy value recovered. In 1983, approximately 700,000 t of hazardous waste was fired in boilers.[1] The majority of these wastes were waste organic liquids generated by the chemical industry and fired on site.[1,2,3] Table 8.4.1 shows that cofiring technology is broadly applied over the installed boiler types in the chemical industry.[3] The corresponding conventional fuel for these boilers is 46% natural gas, 26% coal, 24% oil, and 4% others. The boilers at the lower end of the capacity spectrum, below about 9 MW (30 × 10^6 Btu/h) heat input, are predominantly single-burner firetube boilers fired almost entirely on gas and oil. The midcapacity range of boilers, from 9 to 73 MW (30 × 10^6 to 250 × 10^6 Btu/h) are mostly single-burner packaged watertube designs for gas and oil firing, and watertube stokers for coal firing. Some older units in this capacity range are multiburner field-erected designs, with several units converted from coal firing still in use. The largest units [>73 MW (>250 Btu/h)] are multiburner field-erected designs. For coal firing, as the design capacity increases, the units are increasingly suspension-fired with pulverized coal, rather than stoker-fired.

For most older cofiring installations, the cofiring capability was retrofitted after the boiler unit was operational. Increasingly for newer units, the cofiring requirement is included in the procurement specifications for the boiler. Figure 8.4.1 shows a schematic of a typical waste-fuel retrofit for a multiburner boiler.[4] Specific design details of the retrofit are discussed in Subsec. 8.4.3. Nearly all liquid-waste-fired boilers use a similar system, in which the waste is introduced to the boiler separately from the conventional fuel, either through a separate burner for multiple-burner units, or through a separate atomizer for

TABLE 8.4.1 Waste Firing in the Chemical Industry Distributed by Boiler Capacity

Boiler capacity		Waste fired in capacity range, %
MW	(10^6 Btu/h)	
<2.9	(<10)	0.2
2.9–14.6	(10–50)	11
14.6–29.2	(50–100)	6
29.2–73	(100–250)	24
>250	(>73)	58

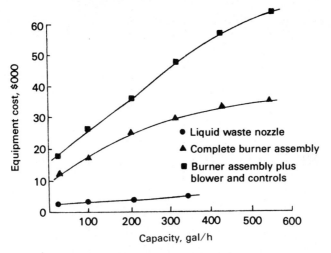

FIG. 8.4.1 Typical waste-storage and delivery system.[4]

a single-burner design. The heating values of wastes fired in the chemical industry are summarized in Table 8.4.2. Industry practice largely conforms to the probable form of impending regulations for boiler cofiring, which will simplify permitting if the wastes are highly combustible. The regulations will also encourage holding the waste-firing rate to a minor fraction of the total boiler heat input. This also conforms to industry practice: it has been reported that 65% of waste fired in boilers contributes 25% or less of the heat input to the boiler in which it is fired.[3]

Although industry statistics show cofiring to be commonly practiced, current applications are only a fraction of the total waste-destruction potential. Waste destruction in 1983 of 700,000 t is less than 0.5% of total waste generation of 264 million t.[5] Combined with incineration, which accounts for about 1% of generated waste,[5] thermal destruction currently handles less than 2% of waste generation. An estimated 15% of generated waste has sufficient heating value to be thermally destroyed,[6] so total thermal destruction by boilers or incinerators could, in principle, be expanded by more than an order of magnitude. Spare capacity in incinerators is estimated at 1 million t, or less than 0.5% of total waste generated.[5] By contrast, the capacity of industrial boilers for cofiring is virtually untapped, with current waste-heat input accounting for less than 0.5% of the installed capacity in

TABLE 8.4.2 Heating Value of Wastes Fired in the Chemical Industry

Heat content of waste		Percent of total waste burned
kJ/kg	(Btu/lb)	
<4,640	(<2,000)	<0.1
4,640–11,600	(2,000–5,000)	0.5
11,600–23,200	(5,000–10,000)	13
23,200–34,800	(10,000–15,000)	55
>34,000	(>15,000)	31

the chemical industry.[7] The primary considerations bearing on the practicality of using the available boiler capacity for waste cofiring are reviewed in the Subsec. 8.4.3.

8.4.3 DESIGN AND OPERATING CONSIDERATIONS

The hardware and operational considerations in retrofitting an existing boiler, or specifying a new boiler for waste cofiring, are

- Waste storage and transport
- Burner modifications
- Boiler operation
- Safety

Air-pollution control devices may also need to be added or modified for solid-waste firing, or if the wastes are unusually high in chlorine, metals, or suspended solids. The specific design considerations, and the degree to which these considerations impose a constraint on waste-firing rate, or waste-firing economics, depend on the chemical and physical property ranges of the wastes. These waste properties must be evaluated for the specific boiler design and plant operational practice to assess compatibility of the waste with the boiler, and potential adverse impacts of cofiring. Generalizations and absolute limits as to acceptable operating practice are usually not appropriate, and each situation needs to be reviewed with the burner or boiler vendor.

Waste Storage and Transport

The storage and transport system shown in Fig. 8.4.1 is typical of industrial cofiring applications although many plants also use a smaller day tank near the boiler room, particularly if several units are serviced by the central storage tank. The design considerations for the central tank and the day tank focus on maintaining a pumpable mixture of fairly uniform composition which can be atomized by the burner. If the raw waste contains large percentages of free water or undissolved solids, the waste should be treated by settling and decanting prior to introduction to the waste-cofiring system. Most wastes are nonhomogeneous to some degree and the central tank and day tank usually are continually mixed by agitators or recirculation pumps. Where continual recirculation is used, strainers are usually installed to reduce solids, thus protecting pumps and downstream burner and boiler components. This is particularly important if the waste contains abrasive solids or metals which have high fouling and corrosion potential. Strainers are usually effective in minimizing solids greater than about 40 μm. Additionally, if the waste is viscous, steam-heat coils may be required in the tanks to maintain pumpability and atomization, particularly during cold weather.

Burners

Burner design and operation for cofiring is directed at achieving atomization and mixing to obtain a stable flame with good carbon burnout and high waste destruction. For industrial boilers, a load turndown capability of at least 5:1 is an essential operational requirement. This turndown capability can strongly affect cofiring application. Since the waste streams are usually of variable physical properties and heating value, they are less easily controlled at low-load, high-turndown conditions. Most operators cut waste firing at loads below 20 to 30% of design capacity. When stopping waste firing, the waste gun should be pulled to avoid coking. Waste-firing turndown can be improved if the supply from the day tank has a return recirculating loop so that only a fraction of the pumped waste goes to the burner. This practice also suppresses surges and rapid variations in waste flow, which can cause smoking and flameouts in a straight-through system.

Whether waste is fired as a dual fuel from a single burner, or through a separate waste burner, dual waste guns are strongly recommended so that the atomizers can be cleaned without interrupting firing. Waste nozzles are usually air- or steam-atomized, with steam more common for lower-quality wastes with suspended solids. If abrasive, erosive solids are not filtered in the supply lines, erosion-resistant carbide atomizer tips may be needed. For viscous wastes, heated lines, similar to residual-oil systems, may be needed to achieve good atomization.

Even with appropriate design of the atomizer and supply system, the waste flame is typically more ragged and variable due to nonhomogeneities. To accommodate this, two modifications may be needed to the boiler's control systems. Because the waste-flame variations may trip the flame safeguard system for multiple-burner units, the safety interlock logic for shutting the boiler down during flameout may be modified to require a flameout in at least one of the burners firing conventional fuel. The waste-fuel flow control, particularly during load changes, is operated on manual at many plants with the response to a load demand met initially by the conventional fuel. This reduces the potential for the smoking and flameouts which can occur if the waste flow is varied rapidly. The problems with waste-flame control and stability are usually less severe if the boiler has combustion-air preheat, or if the waste is atomized through the conventional flame.

Boiler Operation

The principal operating concerns are to minimize the effects of waste firing on boiler materials, maintenance, and unit efficiency. These concerns, more than any other factor, dictate the maximum levels of wastes which are feasible for long-term cofiring. The impacts of cofiring on boiler operation can be much less of a concern with new units having materials and configurations designed for wastes than with existing units that are being retrofitted.

Materials and maintenance problems arise mainly from inorganic materials, such as ash, metals, and chlorine, contained in the waste. Ash and metals deposit and/or condense on waterwalls in the furnace section and on convective section tubes. The resultant slagging and fouling reduce heat-transfer efficiency and can promote corrosion, particularly if the waste or fuel is high in sulfur or chlorine. Fouling can be a special concern for gas-fired designs where the convective-tube passes are tightly spaced. Refractory surfaces are also susceptible to chemical attack by slag deposits, particularly alkali metals. Mitigative measures include more frequent soot blowing, better filtration of the waste, or dilution of the waste with cleaner waste compounds. It is also essential that the waste flame not impinge on the side wall or back wall over the load range of concern. For new units, the watertube and refractory materials and convective design can be selected to reduce chemical attack and fouling impacts. Boiler design concepts for dealing with chemical process wastes are well-established.[8,9,10]

The most common corrosion concern related to cofiring is chlorine attack, which can occur when halogenated wastes are fired. Composite chlorine levels in waste fuels that are in excess of 1.5 to 2.5% can cause watertube corrosion unless special materials and operational procedures are used.[11,12,13] This threshold level is usually avoided in industry by firing wastes with chlorine levels below 5%,[3] which, typically, gives a composite chlorine level below 2%. A further precautionary trend is promoted by impending boiler regulations which will probably facilitate permitting of cofiring facilities if the chlorine content is low, possibly of the order of the upper range of conventional fuels. Even when fireside corrosion is avoided by constraining chlorine levels, it is still important to maintain temperatures at the exit of the convective section above the acid dew point of the HCl-H_2O combustion products to avoid cold-end corrosion. Nonmetallic stacks or stack liners are also recommended.

The primary impact of cofiring on efficiency results from the higher levels of excess air which are necessary with the waste to maintain a stable flame and ensure carbon or carbon monoxide burnout. The excess-air requirements are highest with lower-quality, nonhomogeneous wastes and lowest with clean, homogeneous, volatile wastes. Indeed, some volatile organic solvents can be fired cleaner than heavy oil or coal. Higher excess-air levels can also be required to achieve higher convective-section temperatures. This practice may be used to reduce cold-end corrosion during operation at low loads.

Other efficiency impacts result from free water in the waste or a higher hydrogen-to-carbon ratio, which increase H_2O losses out the stack. An altered H/C ratio can also alter the flame luminosity and gaseous radiative heat-transfer properties of the flame; this can shift the heat-absorption load in the furnace and thereby affect stack losses. The overall effect of cofiring is generally minimal if the waste has similar properties to the conventional fuel.

Safety

Hazardous-waste cofiring may require special safety precautions in addition to those normally accorded addition of an alternate fuel. Those wastes classified RCRA) as ignitable may contain highly volatile compounds, and the increased explosion hazard needs to be considered in designing tank vents, and in specifying the waste-atomization guns to suppress prevolatilization in the delivery lines. Wastes listed for their toxic characteristics require stringent safety procedures in the event of spills or leaks. Procedures for routine maintenance such as changing strainers, cleaning atomizer tips, and cleaning the boiler's heat-transfer surfaces need to be modified to protect plant personnel. Containment or capture of fugitive emissions from tank vents, pump seals, and valves also needs to be practiced.

8.4.4 ENVIRONMENTAL EFFECTS

The overriding environmental concern with hazardous-waste cofiring is that the boiler's thermal environment produce a sufficiently high destruction and removal efficiency (DRE) of the principal organic hazardous constituents (POHCs) being fired. The DRE (in percent) is defined for each POHC as

$$DRE = \left(1 - \frac{M_{out}}{M_{in}} \right) \times 100$$

where M_{out} is the mass flow rate of a specific POHC in the stack and any other relevant ash or scrubber effluent streams, and M_{in} is the mass flow rate of the POHC into the boiler through the waste feed stream. A composite DRE, mass-weighted for all POHCs fired in the boiler, is also used for an overall determination of the boiler's thermal-destruction capability. Other environmental concerns are the potential for formation of hazardous products of incomplete combustion (PICs) from the intermediary combustion products of the POHCs and possibly the conventional fuel. Also, there is a risk that the addition of the waste flame will degrade the thermal environment, thus aggravating normally acceptable emissions from the conventional flame, and possibly inducing carbon monoxide or unburned-carbon emissions from the waste or conventional fuel.

A sizable data base addressing the above concerns was developed between 1981 and 1986 by the U.S. Environmental Protection Agency (EPA) with cooperation from industrial boiler operators. Preliminary surveys of industrial cofiring practice,[14,15] showed that the boiler population was very diverse in terms of fuels and wastes fired, firing configuration, capacity, configuration of heat-exchange surfaces, and operating practice. Each of these parameters could, in principle, influence the "time-temperature-turbulence" environment in the boiler which determines DRE. An initial series of 42 tests was run on 11 different boilers to obtain a profile of boiler thermal-destruction capabilities and industry experience over the spectrum of designs, wastes, and fuels. Table 8.4.3 lists the boiler specifications, the wastes cofired, and general operating conditions.[16,17,18,19] Overall, the boilers and wastes selected spanned the following ranges:

- *Conventional fuels:* Gas, oil, coal, and wood
- *Wastes:* Volatile and semivolatile wastes ranging from aqueous mixtures to flammable solvents
- *Boiler designs:* Firetubes, packaged watertubes, field-erected watertubes; stokers
- *Load:* 0.35 to 32 kg/s steam (2,500 to 250,000 lb/h)
- *Residence time (furnace)::* 0.3 to 2 s

For these initial tests, no attempt was made to span the ranges of boiler operational settings commonly used. Rather, tests at nominally steady load, excess air, and waste-to-fuel ratio were used. For a typical test, the boiler was first tested in a baseline mode without waste firing, and then sampled for at least three replicate runs at a common condition. Volatile and semivolatile POHCs were quantified by drawing a specified quantity of flue gas over sorbents, which were then extracted or desorbed to a gas chromatograph mass spectrometer (GC/MS). Periodic samples of the inlet waste stream were also analyzed by GC/MS to permit calculation of DRE.

The test results showed that for the spectrum of industrial boilers and wastes tested, thermal destruction by cofiring was generally very effective. Very few DRE values less than 99.99% were experienced, and most were in the 99.999% vicinity. The values less than 99.99% were commonly those POHCs with the low-

TABLE 8.4.3 Summary of Test-Site Boilers

Site	Boiler type	Steam capacity kgls (10³ lb/h)		Furnace volume m³ (ft³)		Furnace waterwall surface m² (ft²)		Primary fuel	Number of burners (injection ports)	Typical waste fuels	Injection mechanism	Control device	Typical operation
A	Watertube stoker	1.3	(10)	17.4	(613)	106	(1,144)	Wood waste	2	Creosote sludge	Mixed with wood	Multicyclone	Fluctuating loads, combustion air, and waste feed
B	Packaged firetube	1.1	(8.5)	1.1	(39)	8.0	(83)	Natural gas	1	Alkyd wastewater	Air-atomized oil gun	None	Low boiler load. Maximum waste fire rate of 42 mL/s (40 gph)
C	Field-erected watertube	29	(230)	322	(11,400)	170	(1,800)	Natural gas or oil	6	Phenolic waste	One or two steam-atomized burners	None	Low load with reduced number of burners; high excess air
D	Field-erected converted watertube stoker	11.4	(90)	62	(2,200)	140	(1,520)	No. 6 oil	4	Methanol and toluene wastes with chlorinated organics	One of the lower-level steam-atomized burners	None	About 50% capacity with 3 or 4 burners in service
E	Packaged watertube	13.9	(110)	42	(1,480)	665	(7,160)	No. 6 oil	1	Methylmethacrylate by-product wastes	Two steam-atomized waste guns in main burner throat	None	Part load with maximum 250 mL/s (240 gph) waste-firing rate for loads above 50%

TABLE 8.4.3 Summary of Test-Site Boilers (*Continued*)

Site	Boiler type	Steam capacity kg/s (10³ lb/h)		Furnace volume m³ (ft³)		Furnace waterwall surface m² (ft²)		Primary fuel	Number of burners (injection ports)	Typical waste fuels	Injection mechanism	Control device	Typical operation
F	Field-erected converted watertube	7.6	(60)	96	(3,390)	100	(1,100)	No. 6 oil, gas, or propane	2	Paint solvents	Lower steam-atomized oil burner	None	Part load with maximum 190 mL/s (180 gph) waste-firing rate for loads above 50%
G	Modified packaged firetube	5.0	(40)	6.4	(226)	20	(220)	None	1	Highly chlorinated organics	Available air-atomized oil gun	2 scrubbers in series	Part load with start-up on natural gas; total chlorine up to 80% of waste fuel.
H	Field-erected, tangentially fired watertube	32	(250)	520	(18,400)	515	(5,540)	Pulverized coal	12 coal, 6 oil	Methyl acetate waste fuel	One or two steam-atomized oil burners	ESP*	At boiler capacity with maximum 440 mL/s (420 gph) waste-firing rate
I	Packaged watertube	7.8	(62)	41	(1,430)	76	(820)	Natural gas	2	Aniline waste high in nitrate organics	Either upper or lower steam-atomized burner	None	Staged combustion for low NO_x with maximum 130 mL/s (120 gph) waste flow
J	Packaged firetube	1.3	(10)	1.5	(51)	2.6	(91)	None	1	Artificially blended fuels	Available oil burner	None	Typical excess air of 17%
K	Packaged watertube	7.6	(60)	65	(2,270)	47	(508)	No. 6 oil	1	Blended waste with light oil	Mixed with heavy oil	None	Typical 70/30% heavy and light oil mixture

ESP = electrostatic precipitator
Source: Refs. 16–19.

est inlet concentrations, thus indicating a possible sampling or analysis artifice. Table 8.4.4 summarizes the DRE values for the volatile POHCs.[16] All sites except F yielded DREs mass-averaged across all POHCs of greater than 99.99%. Apart from site F, which exhibited some burner malfunctions, there was no discernible effect of boiler design configuration or fuel on DRE. Similarly, the DREs for a specific POHC, mass-weighted across all boilers, were all greater than 99.99%. There was no apparent physical or chemical property of the wastes which correlated with the observed changes in DRE for the various POHCs. A number of potential correlations of DRE with boiler design and operating parameters, and emissions of other species, such as carbon monoxide and oxides of nitrogen, were examined. No consistent, strong correlation was observed for the boilers and POHCs tested. Although not shown, the results for the semivolatile POHCs were similar.

Although there is no regulation for products of incomplete combustion, the volatile and semivolatile PICs emitted were evaluated, particularly for chlorinated compounds, where detection is simplified. Significant quantities were observed, with a mass-average emission rate of volatile chlorinated PICs on the order of 10 times that for the volatile chlorinated POHCs. By contrast, the same type of comparison for incinerators showed volatile PICs of the same magnitude as volatile POHCs.[20] Although a conclusive data base does not exist, there may be a higher potential for PIC emissions from boilers than from incinerators.

The data base established for nominally steady-state operation was extended by the EPA with three additional tests studying the effects of nonsteady and off-design operation on DRE.[21] The boilers tested were a 13.9 kg/s (110,000 lb/h) gas- and oil-fired package watertube, a 44.2 kg/s (350,000 lb/h) gas-fired field-erected watertube, and an 18.9 kg/s (150,000 lb/h) coal-fired stoker. Transients and off-design settings of excess air, load, and waste-to-fuel ratio were tested along with induced malfunctions in waste feed. Test results showed DRE values comparable to the earlier steady-state tests, indicating the boilers to be fairly effective for thermal destruction, even under some upset conditions. As earlier, no useful correlations of DRE with boiler or waste characteristics were observed.

Both the steady-state and nonsteady test series showed that from an operational and maintenance standpoint, cofiring was accommodated by the sites without major disruption. Except for the case where the waste was a wastewater, efficiency losses were generally less than 1%. These losses were largely caused by increased excess air needed to maintain a stable flame and complete carbon burnout when the waste was fired. Boiler controls were usually complicated by waste firing, which typically was operated manually while the conventional fuel was left on automatic. Maintenance impacts included frequent soot blowing and boiler-tube cleaning.

8.4.5 CURRENT RESEARCH AND DEVELOPMENT

Since waste cofiring is widely used as alternate fuel firing, past and current research and development has focused on environmental effects rather than on refining the process itself. The major environmental R&D effort was the boiler field-test program by the EPA, summarized in Subsec. 8.4.4, which was conducted to support the boiler regulatory effort by the EPA's Office of Solid Waste

TABLE 8.4.4 Summary of Test-Average DREs for Volatile POHCs*

POHC	Site B	Site D	Site E	Site F	Site G	Site H	Site I	Site J	Site K	Range	Weighted average
Carbon tetrachloride			99.9990–99.9998 (99.9996)†	99.98–99.9990 (99.995)	99.995–99.9990 (99.998)	99.97–99.9994 (99.98)	99.9990–99.9993 (99.9993)	99.997–99.9998 (99.9990)	99.9998	99.97–99.9998	99.9992
Trichloroethylene			99.994–99.9995 (99.998)	99.98–99.998 (99.996)			99.99990–99.99992 (99.99991)	99.998–99.99993 (99.9996)	99.99990	99.98–99.99993	99.9994
1,1,1-Trichloroethane						99.97–99.9996 (99.994)				99.97–99.9996	99.994
Chlorobenzene			99.995–99.99990 (99.998)	99.96–99.992 (99.98)		99.990–99.997 (99.992)	99.997–99.9990 (99.998)	99.8–99.97 (99.95)	99.99992	99.8–99.99992	99.992
Benzene							99.97–99.98 (99.97)		99.996	99.97–99.996	99.990
Toluene	99.991		99.997	99.90–99.97 (99.95)						99.90–99.99996	99.998
Tetrachloroethylene		99.9992–99.99990 (99.9996)					99.998	99.9990–99.9997 (99.9990)	99.99996	99.994–99.9992	99.998
Methylmethacrylate		99.994–99.9992 (99.998)	99.95–99.997 (99.991)							99.95–99.995	99.991
Mass-weighted average	99.991	99.994–99.99990 (99.998)	99.95–99.99990 (99.995)	99.90–99.9990 (99.98)	99.995–99.9990 (99.998)	99.97–99.9996 (99.991)	99.97–99.99992 (99.998)	99.8–99.99993 (99.9990)	99.996–99.99996 (99.9997)	99.8–99.99996	99.998

*Each test-average DRE is generally based on the weighted average of triplicate measurements.
†Numbers in parentheses represent the site-specific POHC average DRE.

and Emergency Response. This field effort is largely complete. Continuing R&D is directed at laboratory testing and engineering analysis to help interpret and generalize the field data. Subscale testing is underway at both EPA and contractor facilities to evaluate formation and fate mechanisms for a variety of flame configurations and waste types. Specific topics include characterizing the species and quantities of products of incomplete combustion compared to POHC emissions for various POHC types, quantifying the partitioning of heavy metals across particle size ranges and determining their fate with air-pollution control devices, and determining flame upset conditions which will cause POHC breakthrough through the flame and cause DRE to drop below 99.99%. The engineering analysis effort is directed at correlating the test data, determining theoretical bases for observed cause-effect relationships, and developing semiquantitative procedures for evaluating the relative effectiveness of a boiler or incinerator for thermal destruction.

8.4.6 ENGINEERING ECONOMIC EVALUATION

Boiler cofiring capital and operating cost considerations have been reviewed by McCormick and Weitzman for the EPA.[22,23] Cost estimates or cost algorithms were developed for on-site liquid organic wastes with heating values between 18,500 and 23,200 kJ/kg (8,000 to 10,000 Btu/lb) fired in gas- or oil-fired boilers where wastes supply 10 to 50% of heat input. Three boiler classes were used for cost estimation:

- Single-burner firetube firing gas, distillate oil, or combined gas and oil, with a boiler capacity range of 1.3 to 6.3 kg/s (10,000 to 50,000 lb/h) of steam
- Packaged watertube firing gas, distillate, or residual oil, with a boiler capacity of 2.6 to 12.6 kg/s (20,000 to 100,000 lb/h) of steam
- Multiburner watertube firing gas or residual oil, with a boiler capacity of 12.6 to 25.2 (100,000 to 250,000 Btu/lb)

Primary capital-cost categories are storage and transport equipment, boiler modifications, and air-pollution control equipment. Storage tanks for 5,000, 10,000, and 20,000 gal of liquid hazardous wastes have an installed cost (1982 basis) of $34,000, $46,000, and $61,000, respectively. Feed pumps cost in the range of $1,000 to $1,500 for redundant systems. Piping and filters can cost from $2,000 to $7,000 or more depending on flow rate, solids loading, and solids size distribution.

The primary boiler modification is retrofit of the existing burner(s) for waste firing. Figure 8.4.2 shows three equipment cost options. The upper curve is for a complete burner system, including forced-draft blowers and burner controls for regulation of the fuel–air mixture and flame safeguard. The middle curve is for complete waste-oil-gas dual-fuel burner assemblies, including pilot, spark ignition, and dampers. The lower curve is for the simplest case, where only waste-atomizer nozzles are required. Total installed costs for the options on Fig. 8.4.2 are typically 1.5 to 2 times the equipment costs.[22]

For gas- and oil-fired boilers firing liquid wastes, retrofit of air-pollution control devices is not typically required. Wastes which conform to the probable form of impending cofiring regulations should be sufficiently low in chlorine and in metals or solids to be feasible for cofiring without scrubbers or electrostatic pre-

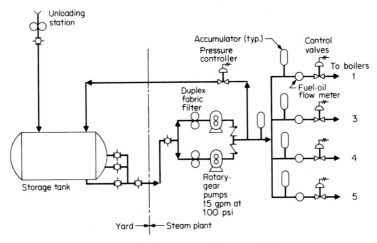

FIG. 8.4.2 Equipment costs for burner retrofit for waste firing.

cipitators. With solid wastes, or highly chlorinated wastes, retrofits may be necessary; Reference 23 presents cost algorithms for several options.

Operation and maintenance costs are highly dependent on the specific boiler design and waste characteristics and the compatibility thereof, as well as specific plant engineering and safety practices enforced at the site. In a typical situation, the primary cost is labor for cleaning strainers in the waste feed-line, cleaning atomizers, performing maintenance on pumps and agitators, and manually controlling waste feed rates to the boiler. Other maintenance costs could include more frequent soot blowing and boiler-tube washes during scheduled outages. These are also labor-intensive.

The key factor which usually dominates the economics of boiler cofiring is the trade-off between conventional fuel displacement, boiler-efficiency impacts, and eliminating the need for alternate waste disposal. The net fuel savings from cofiring is

$$Q_f = Q_w - Q_s \left(\frac{1}{n_2} - \frac{1}{n_1} \right)$$

Where Q_f = net fuel savings
  Q_w = heat input with waste, Btu/lb
  Q_s = constant heat output, Btu/lb
  n_1 = boiler efficiency during firing with conventional fuel
  n_2 = boiler efficiency during cofiring

The corresponding fuel mass savings is

$$F_f = \frac{Q_f}{HV_f}$$

Where F_f = mass (or volumetric) fuel conservation rate
  HV_f = fuel heating value

The fuel cost savings can then be calculated given local data on unit prices of fuels.

Savings from elimination of alternate waste-disposal options can vary tremendously depending on waste composition, the degree of hazard, local logistics, and disposal options available. Typical costs are $0.04 to $0.05 per pound, although costs for specific wastes can be five times that, or more.

8.4.7 REFERENCES

1. U.S. Environmental Protection Agency, *Assessment of Incineration as a Treatment Method for Liquid Organic Hazardous Wastes, Background Report III: Assessment of the Commercial Hazardous Waste Incineration Market,* U.S. EPA Office of Policy, Planning and Evaluation, March 1985.

2. U.S. Environmental Protection Agency, *National Survey of Hazardous Waste Generators and Treatment, Storage, and Disposal Facilities Regulated under RCRA in 1981, Preliminary Highlights of Findings,* U.S. EPA Office of Solid Waste and Emergency Response, August 30, 1983.

3. W. P. Moore, "Chemical Industry's Position on Regulatory Control of Burning Hazardous Waste in Boilers," Paper 85-78.4 presented at the 78th Annual Meeting of the Air Pollution Control Association, Detroit, Michigan, June 1985.

4. E. T. Murphy, "Combustion of Alternate Liquid Fuels in High Efficiency Boilers," Paper 83-50.1 presented at the 76th Annual Meeting of the Air Pollution Control Association, Atlanta, Georgia, June 1983.

5. G. A. Vogel et al., "Incinerator and Cement Kiln Capacity for Hazardous Waste Treatment," presented at the Twelfth Annual Research Symposium: Land Disposal, Remedial Action, Incineration and Treatment of Hazardous Waste, Cincinnati, Ohio, April 1986.

6. "Hazardous Waste: Potential for Incineration Market; Capacity Will Not Meet Demand," *Environment Reporter,* June 21, 1985, p. 301.

7. R. A. Olexsey, "Incineration of Hazardous Waste in Power Boilers: Emissions Performance Study Rationale and Test Site Matrix," *Proceedings of the Tenth Annual Research Symposium: Incineration and Treatment of Hazardous Waste,* EPA-600/9-84-022, September 1984.

8. *Steam, Its Generation and Use,* Babcock and Wilcox, New York, 1978.

9. O. de Lorenzi, *Combustion Engineering,* Combustion Engineering Inc., New York, 1955.

10. R. G. Schwieger, "Power from Waste," *Power,* February 1975, p. s-1.

11. H. H. Krause et al., "Corrosion and Deposits from Combustion of Solid Wastes," *Journal of Engineering for Power,* **95**:45 (January 1973).

12. A. J. Cutler et al., "The Role of Chlorides in the Corrosion Caused by Flue Gases and Their Deposits," *Journal of Engineering for Power,* **93**:307 (July 1971).

13. H. H. Krause et al., "Corrosion and Deposits from Combustion of Solid Waste, Part 2—Chloride Effects on Boiler Tube and Scrubber Metals," *Journal of Engineering for Power,* **96**:216 (July 1974).

14. J. W. Harrison et al., *Assessment of Hazard Potential from Combustion of Wastes in Industrial Boilers,* EPA-600/7-81-108, July 1981.

15. C. Castaldini et al., *A Technical Overview of the Concept of Disposing of Hazardous Wastes in Industrial Boilers,* Final Report on U.S. EPA Contract No. 68-03-2567, Acurex Corporation, Mountain View, California, June 1984.

16. C. Castaldini et al., *Engineering Assessment Report: Hazardous Waste Cofiring in Industrial Boilers,* Final Report TR-84-159/EE on U.S. EPA Contract No. 68-02-3188, Acurex Corporation, Mountain View, California, June 1984.

17. C. Castaldini et al., "Field Tests of Industrial Boilers Cofiring Hazardous Wastes," in *Proceedings of the Tenth Annual Research Symposium: Incineration and Treatment of Hazardous Wastes,* EPA-600/9-84-022, September 1984, p. 57.

18. R. C. Adams, "Field Tests of Industrial Boilers and Industrial Processes Disposing of Hazardous Wastes," in *Proceedings of the Tenth Annual Research Symposium: Incineration and Treatment of Hazardous Wastes,* EPA-600/9-84-022, September 1984, p. 62.

19. J. T. Chehaske and G. M. Higgins, "Summary of Field Tests for an Industrial Boiler Disposing of Hazardous Wastes," in *Proceedings of the Tenth Annual Research Symposium: Incineration and Treatment of Hazardous Wastes,* EPA-600/9-84-022, September 1984, p. 76.

20. R. A. Olexsey et al., "Emission and Control of By-Product Emissions from Hazardous Waste Combustion Processes," in *Proceedings of the Eleventh Annual Research Symposium: Incineration and Treatment of Hazardous Waste,* EPA/600/9-85/028, September 1985, p. 8.

21. H. I. Lips and C. Castaldini, *Engineering Assessment Report: Hazardous Waste Cofiring in Industrial Boilers Under Nonsteady Operating Conditions,* Technical Report TR-86-103/ESD, on EPA Contract No. 68-03-3211, Acurex Corporation, Mountain View, California, August 1986.

22. R. J. McCormick and L. Weitzman, "Preliminary Assessment of Costs and Credits for Hazardous Waste Co-Firing in Industrial Boilers," in *Proceedings of the Tenth Annual Research Symposium: Incineration and Treatment of Hazardous Waste,* EPA-600/9-84-022, September 1984, p. 96.

23. R. J. McCormick and L. Weitzman, *Preliminary Assessment of Costs and Credits for Hazardous Waste Co-Firing in Industrial Boilers,* Final Report on U.S. EPA Contract No. 68-02-3176, Acurex Corporation, Mountain View, California, October, 1983.

SECTION 8.5
CEMENT KILNS

John F. Chadbourne, Ph.D.
Director of Environmental Services
General Portland Inc.
Dallas, Texas

The portland-cement industry is widely distributed throughout the world. Portland cement is produced in very high temperature furnaces by heating calcium, silicon, aluminum, and iron oxides to temperatures exceeding 1425°C (2600°F). A generalized schematic diagram of a cement pyroprocessing system is shown in Fig. 8.5.1. Pyroprocessing is discussed in Subsec. 8.5.3. This energy-intensive industry annually consumes the equivalent of over 100 million t of coal. Cement kilns can accommodate a wide variety of fuels, and their very high operating temperatures assure virtually complete destruction of the organic constituents in the fuels. Replacement of fossil fuels, such as coal, petroleum coke, oil, and natural gas, with high-energy wastes, such as still bottoms, solvents, tars, tires, and refuse, has become increasingly common as energy costs have increased and environmental regulations have become more comprehensive. The cement industry represents a superior technology for the management of hazardous and other types of waste worldwide.

In addition to combustible waste, cement kilns can be used to treat many other types of waste. Wet-process kilns can accommodate brines and aqueous metal-bearing wastes. Acidic waste can be neutralized and used as kiln feed. Solid sludges, such as lime-alum sludge from water-treatment plants, can be used to produce cement. Metal-bearing wastes from many sources can also be used to provide the iron, aluminum, and silicon oxides required in production. Relatively large quantities of heavy metals can be safely treated in cement kilns because these metals are immobilized as they become part of the finished cement or cement-kiln dust. Halogenated materials, which may be difficult to treat through alternative technologies, are useful for removing excess sodium and potassium which may be present in the raw materials from which cement is manufactured. The key factor is an acceptable economic return, which is essential to create the necessary incentives to solve material-handling and operating problems.

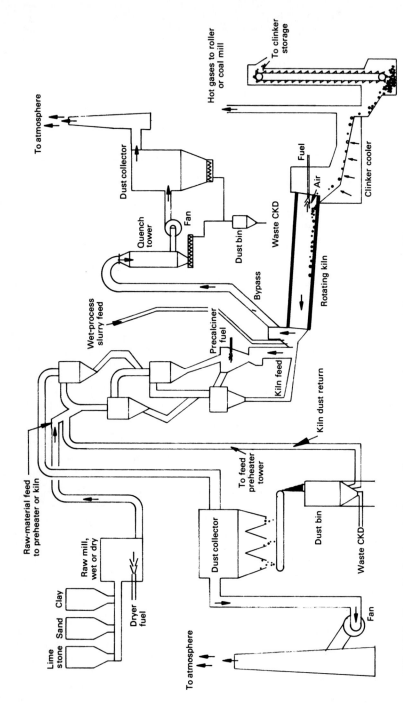

FIG. 8.5.1 Generalized schematic diagram of cement pyroprocessing system. CKD = cement-kiln dust.

8.58

8.5.1 ACCEPTABLE MATERIALS AND LIMITATIONS

Types of Waste

Cement kilns are a versatile alternative for the management of a wide variety of hazardous wastes. The raw materials used to produce cement often contain trace quantities of virtually every natural element, including alkali chlorides and sulfates; heavy metals, such as lead, cadmium, chromium, and arsenic; and organic materials. Many of these constituents are also contained in fossil fuels, such as coal, oil, and petroleum coke, and in the water used to prepare slurry for wet-process kilns. Materials present in the feed, fuels, or hazardous waste introduced into the kiln become part of the cement product, become part of a waste material known as cement-kiln dust, or are converted to other forms within the kiln. With few exceptions, materials introduced into cement kilns will be oxidized and stabilized, requiring no further treatment.

Hazardous-Waste Fuels

The use of combustible hazardous waste as fuel in cement kilns has become increasingly common during the last 10 years. Prior to 1970, natural gas and oil were preferred fuels due to the ease of handling, consistency, and cost. In response to extreme escalation of the cost of these fuels in the early 1970s, the cement industry converted to coal and began seeking low-cost alternative fuels. Petroleum coke, coal tailings, shredded or whole automobile tires, and refuse-derived fuel (RDF) have been successfully used as fuel for cement kilns.

Liquid-Waste Fuels. Prior to the advent of the Resource Conservation and Recovery Act (RCRA), spent solvents, and still bottoms from the recovery of spent solvents, were often discarded in landfills. The low cost of the landfill-disposal option made these materials largely unavailable for use as fuel in industrial furnaces and boilers.

Appropriate waste recovery, segregation, and analysis, as well as safe handling, are important factors which must be addressed in order to successfully use combustible liquid wastes as alternative fuels for cement kilns. Paint thinners, degreasing solvents, solvent washes from the ink and printing industries, chemical by-products from pharmaceutical and chemical manufacturing, waste oils, and other flammable, readily pumpable materials can be used to prepare high-quality fuels for cement kilns. The primary constituents of typical hazardous-waste fuel (HWF) from these sources are xylene, toluene, mixed aliphatic hydrocarbons, acetone, and methyl ethyl ketone.[1] A wide variety of other organic compounds, including aldehydes, esters, alcohols, ketones, phthalates, alcohol ethers (such as cellosolve), aromatic compounds, phenols, amines, amides, ethers, nitriles, freons, and other halogenated organic compounds, are commonly found in HWF mixtures. More than 250 organic compounds have been approved for use in HWF.[2]

The choice of acceptable compounds depends primarily on potential occupational exposures and material-safety criteria. Virtually any organic compound can be completely destroyed at the elevated temperatures in a properly operating cement kiln. Process operating constraints, as discussed at Subsec. 8.5.3, may

TABLE 8.5.1 Specifications for Supplemental Fuel: General Portland Waste-Fuel Program

Heat content	10,000 Btu/lb minimum
Odor	Characteristic of Solvents Reference ASTM 1296-69
Suspended solids	30% maximum
Sulfur	3% maximum
Nitrogen	3% maximum
Halogens	5% maximum
Inorganic acids and bases	Extractable pH 4 to 11
Water	1% maximum as separated phase
Ash content	10% maximum
Barium	5,000 ppm maximum
Chromium	1,500 ppm maximum
Lead	2,500 ppm maximum
PCB	None detected (<50 ppm)
Benzene	OSHA exclusion level

also become significant when certain compounds are present in excessive quantities. However, cement kilns can effectively use very "dirty" fuels and produce quality cement. For example, chlorinated solvents, resins, and waste containing large quantities of lead and other heavy metals can be effectively used as fuel for cement kilns without significant adverse environmental effects. However, the relatively low cost of primary fuels plays a significant role in the partitioning of available HWF between industrial furnaces, such as cement kilns, and industrial boilers, which require "cleaner" fuels. In practice, pulverized coal, petroleum coke, and shredded tires are more difficult to burn than pumpable HWF, provided adequate limitations on heat content, halogens, viscosity, ash, and certain heavy metals have been established. Specifications for HWF used in the General Portland HWF program are presented in Table 8.5.1.

Solid-Waste Fuels. A limited amount of work has been completed to evaluate the use of solid waste as fuel for cement kilns. In particular, spent pot lining (SPL) from the aluminum industry has proved to be an effective fuel when used to replace 5 to 10% of the primary fuel.[3] The SPL contains significant amounts of fluoride and some cyanide in addition to sodium, potassium, aluminum, and other metals. The cyanide is effectively destroyed when SPL is used as fuel in cement kilns. In addition, small amounts of fluoride and aluminum may be beneficial to the process (see under *Cement Physiochemistry,* on p. 8.62). For many kilns, the sodium and potassium content of SPL is acceptably low. It is expected that similar materials will be "discovered" in other industries in the future and that the cement industry will play an important role in recovering waste in the form of solid fuels.

Aqueous Wastes

Large quantities of aqueous hazardous waste are currently being managed through deep-well injection. Many of these wastes are strong brine solutions; materials which contain heavy metals; or inorganic acids, such as hydrofluoric, sulfuric, or hydrochloric acid. At the present time, it is more economical for most generators to use deep-well injection than to consider alternatives, such as cement kilns. However, many of these materials can be effectively stabilized in the cement process.

Aqueous solutions containing sodium fluoride have been effectively used to replace 10% of the water required to prepare slurry for a wet kiln operated by General Portland. A single 625 t/d wet-process kiln evaporates over 378,000 L/d (100,000 gal/d) of water fed into the kiln with the slurry feed. At a 10% substitution rate, such kilns could consume more than 11.4 million L/yr (3 million gal/yr) of aqueous waste. Metals present in the waste are stabilized in the cement clinker or kiln dust, requiring no further treatment. Substitutions considerably in excess of 10% are also possible, depending upon supply, economics, and process chemistry. In addition, strong inorganic acids can be effectively neutralized with limestone in a reactor and used to prepare wet-process slurry feed. However, this technology has not yet been demonstrated because the economic incentive has been marginal at best.

The most significant limitation on the use of aqueous waste in wet-process kilns is the presence of organic contaminants in the aqueous waste stream. Many aqueous wastes contain small amounts of organic contaminants, introduced either through the process or through poor work practices that result in commingling of waste streams. Organic contaminants present in aqueous waste which is used to prepare slurry are expected to be volatilized when they are fed into the kiln. Organic contaminants present at a level of 0.1% (1,000 ppm) in an aqueous waste used to prepare 10% of the slurry for a 625 t/d kiln would result in an increase in organic emissions of approximately 1.6 kg/h (3.5 lb/h). Thus, this relatively small amount of contaminants would result in organic emissions in excess of 8.9 t/yr. Proper segregation, separation, and pretreatment of aqueous waste streams may be necessary to control potential emissions of organic compounds when these materials are used. One alternative may be preprocessing with activated carbon to remove organic materials, and combustion of spent activated carbon as fuel in the combustion zone of the kiln.

Solid Wastes

It is also possible to use solid hazardous waste as feed components in cement kilns. Iron-bearing slag and residue from acid plants and other industrial processes have been successfully used for cement plants with iron-poor mix. These iron-bearing by-products are often disposed as waste and may contain substantial quantities of other metals. Lime sludge from water-treatment plants typically contains aluminum and calcium hydroxide which can be effectively used to produce cement. However, transportation and handling costs must be weighed against the costs of other resource-recovery or disposal alternatives. Organic solid wastes, such as tars, sludges, and plastics, can be effectively destroyed through pyrolysis, using the pyrolyzer gas as fuel for the cement kiln. The ash can become part of the kiln feed, depending on the chemical composition of the ash. Cement kilns can accommodate a wide variety of metal-bearing sludges and hazardous solid waste.

Unacceptable Materials

Although cement kilns can accommodate a variety of materials both as fuel and as feed in the cement process, there are limitations on the use of hazardous-waste materials and there are materials which are not desirable in the cement process. Unacceptable materials can be divided into two categories: those which are technologically unacceptable because of process constraints, and those which are unacceptable because of public opinion. The amount of sodium, potassium, chlorides, sulfates, chromium, lead, and other metals will be limited depending on the quantities of these materials present in the feed and other fuels at each particular plant. However, it is important to understand that excessive amounts of any "contaminant" can be expected to result in adverse operating or production effects, such as unstable kiln operation, poor-quality cement, refractory damage, objectionable cement color, or unacceptable safety risks. For the most part, these factors must be evaluated on a case by case basis. It is also possible to produce significant adverse environmental effects through excessive emissions of toxic metals.[4] In addition, both the need for proper management of residual kiln dust and the nature of certain potential uses of the dust may result in technological limitations on the acceptability of certain hazardous wastes intended to be used in cement kilns.

Public acceptance and worker safety should be carefully considered in accepting hazardous waste for use in cement kilns. For example, polychlorinated biphenyls (PCBs) can be effectively destroyed in cement kilns.[5,6] However, public reaction to these controversial substances and the perceived potential for "disaster" play an important role in gaining approval to manage such hazardous wastes at each cement plant. Similarly, pesticides and extremely hazardous materials can be effectively destroyed when they are burned in a cement kiln. Again, however, the material-handling system, including transportation, transfer, and storage of these materials, is an important consideration. It is much easier to gain acceptance from the public and the work force for resource-recovery projects designed to effectively utilize other people's waste than it is to gain approval to simply destroy extremely hazardous or controversial wastes. Moreover, the necessity for adequate employee training can become a significant obstacle when extremely hazardous or controversial materials are introduced. Objectionable odors and the emission of organic compounds impose significant limitations on the acceptability of waste materials. For example, because of the odor problem, mercaptans and sulfides would require special handling that could adversely affect economic return. The use of sewage sludge and other materials containing organic compounds in the cement-kiln feed is similarly unacceptable because of increased organic emissions and objectionable odors.[7,8]

8.5.2 CEMENT PYROPROCESSING

Cement kilns are manufactured in a variety of designs and sizes. A small kiln may produce only 8.9 t/h as compared with large kilns which may produce 175 t/h. Most older kilns use the wet-process technology, which is generally simpler, but more energy-intensive. Modern dry-process kilns incorporate designs to improve energy efficiency and to accommodate variations in raw materials. However, the high temperatures and the chemistry of cement production are essentially the same in all cement kilns from the burning zone to the calcining zone.

Cement Physiochemistry

Cement is produced by heating limestone mixed with sand, and shale or clay, to prepare an intermediate product called *clinker,* which is then interground with gypsum. Cement kilns produce clinker from feed materials which contain approximately 80% calcium carbonate ($CaCO_3$), 15% silicon dioxide (SiO_2), 3% aluminum oxide (Al_2O_3), and 2% ferric oxide (Fe_2O_3). The feed is heated in countercurrent heat exchangers, where water is evaporated from the feed, followed by calcination or CO_2 evolution from carbonates at 500 to 900°C (930 to 1650°F). The calcium oxide (CaO) formed reacts with SiO_2, forming dicalcium silicate ($2CaO \cdot SiO_2$) between 850 and 1250°C (1560 to 2280°F). Tricalcium aluminate ($3CaO \cdot Al_2O_3$) and tetracalcium aluminoferrite ($4CaO \cdot Al_2O_3 \cdot Fe_2O_3$) are formed at temperatures above 1200°C (2190°F), producing a liquid phase in which dicalcium silicate and calcium oxide dissolve. The temperature at which this liquid phase will form depends on eutectic mixtures which are affected by trace constituents. For example, calcium fluoride at levels of 0.1% can decrease the temperature required to form the liquid phase by about 25°C (45°F). The exothermic reaction between dicalcium silicate and calcium oxide to form tricalcium silicate occurs in the liquid phase and raises the temperature to approximately 1450°C (2640°F), depending on the relative quantities of aluminum, iron, silicon, calcium oxide, and other materials which may be present in the mix. In order to achieve clinkering temperatures, combustion gases in the firing zone of cement kilns must exceed 1650°C (3000°F). A generalized schematic diagram of a cement pyroprocessing system is shown in Fig. 8.5.1.

Volatilization and Condensation. At cement-kiln temperatures, materials which are ordinarily considered nonvolatile not only melt, but boil. This is illustrated in the following tabulation of melting and boiling points for materials of interest that is shown below.

Substance	Melting Point, °C	Boiling Point, °C
Potassium chloride (KCl)	790	1407
Sodium chloride (NaCl)	800	1465
Sodium sulfate (Na_2SO_4)	884	1500
Lead oxide (PbO)	890	1472

It is easy to understand why sodium and potassium in the feed or fuel will be volatilized when there is sufficient chloride or sulfate available in stoichiometric proportions. Similarly, lead compounds introduced with HWF will be volatilized in the burning zone and carried with exhaust gases to cooler regions of the kiln. For most cement kilns the exhaust gases remain above 1100°C (2000°F) for approximately 3 to 5 s as the gases approach the calcining zone. The exhaust gases rapidly cool as heat is transferred to the feed material in the calcining zone. Exhaust gases exiting the kiln stack are typically at temperatures between 150°C (300°F) and 250°C (480°F). Consequently, potassium chloride, lead oxide, or other materials volatilized in the burning zone will condense onto particle surfaces rather than being emitted with the stack gases.

Materials which are volatilized in the burning zone become gaseous particles smaller than 0.1 µm in size. Small particles have a large surface-to-mass ratio. They exhibit Brownian motion and collide with other particles, where they condense at cooler temperatures. Consequently, materials volatilized in the burning zone will tend to accumulate onto the surfaces of smaller particles of feed material. As the feed material, which has trapped, for example, volatilized chlorides, moves into hotter regions of the kiln, the volatile components once again evaporate and recirculate within the kiln system. Enrichment of alkali chlorides and other volatile materials occurs in the kiln feed, typically in a zone with temperatures of approximately 800°C (1475°F), unless these alkali chlorides are removed from the kiln system. Alkali chlorides and other undesirable volatile constituents can be removed from the cement process using electrostatic precipitators or baghouses. Small particles are more readily entrained in the kiln exhaust gases. Consequently, the smallest particles move into the electrostatic precipitator or baghouse to be collected rather than emitted to the atmosphere. For precipitators, this particle-size segregation continues for each sequential electrical field, such that dust collected in the last fields tends to be much smaller in size and enriched in alkali chlorides and sulfates. Depending upon the process and the chemistry of the raw materials and fuel, a portion of the kiln dust is returned to the process to minimize wasted feed. For electrostatic precipitators, it may be possible to return dust from the first fields and waste high-alkali dust from the last fields. Understandably, it is extremely difficult to conduct a material balance for volatile materials present in trace quantities in feed and fuel. For recirculation of volatile materials under dynamic operating conditions, long-term averages must be used to determine partitioning coefficients for the volatile materials between clinker and kiln dust. However, emissions to the atmosphere are extremely small and can be readily determined through actual emission testing of the kiln stack.

Wet-Process Cement Kilns

The primary difference between kiln designs is in the preparation of feed materials prior to calcination. In wet-process kilns, feed is prepared in the form of a slurry containing 30 to 35% water. The limestone, sand, and clay are interground in wet mills to produce a uniform feed. Blending of the slurry produces a very uniform feed which results in more stable operation for most wet-process kilns. Aqueous waste and certain hazardous solid waste may be added to the raw mill in preparing slurry.

The kiln slurry is pumped into the elevated end of a rotary kiln which is inclined at an angle of approximately 5°. As the slurry enters the kiln, it begins moving toward the burning end under gravitational force. Exhaust gases are pulled from the burning zone toward the feed end of the kiln using induced-draft fans. The countercurrent exchange of feed materials and exhaust gas transfers heat from the gas as water evaporates and carbon dioxide is evolved. Wet-process kilns typically use hundreds of tons of chain, hung inside the kiln near the inlet of the feed end, to enhance the transfer of heat from the exhaust gases to the feed material. This is commonly known as the *chain zone,* where exhaust temperatures drop from approximately 1000 to 300°C (1830 to 570°F) and material temperatures increase from 25°C to approximately 500°C (75 to 930°F). Water and volatile organic compounds in the slurry are evaporated in the chain zone. Similarly, materials which were volatilized in the burning zone tend to recondense in the chain zone. Dust which is entrained in the exhaust gas from the kiln or chain zone is conveyed to an electrostatic precipitator or baghouse and returned to the kiln or removed as kiln dust.

Dry-Process Cement Kilns

Modern cement kilns typically use dry-process technology which eliminates the energy penalty from evaporation of slurry water. An efficient dry-process kiln will consume about 60% of the energy required to produce a ton of cement in a typical wet-process kiln. However, raw grinding and homogenization of the feed materials is more difficult and the design of the modern preheater kiln design is more complicated, resulting in increased process-control challenges.

There are many dry-process kiln designs. Kilns of the simplest design type, known as *long dry kilns,* are very similar to wet-process kilns utilizing chains at the feed end for heat transfer and very long kiln lengths [typically 300 m (500 ft)] to calcine and prepare the feed. Modern dry-process kilns use the suspension preheater or Leopold grate preheaters to prepare and calcine the feed before it enters the rotary kiln. Approximately 30 to 60% of the kiln fuel may be burned in precalciner kilns where up to 90% of the carbon dioxide may be removed from the limestone prior to entering the kiln.

Dry-process kilns also use electrostatic precipitators and baghouses to collect entrained feed and materials which have been volatilized and recondensed as fine particles. In modern preheater systems, most, if not all, of the material collected in the dust collector is returned as feed. Consequently, an alternative for removing volatilized alkali chlorides and sulfates is included in the design of most dry-process kilns. These are commonly known as *alkali bypass systems,* and they operate by removing gases from the kiln at a point prior to recondensation of volatile materials onto the surfaces of the feed. The bypass gas is then quenched using ambient air and water sprays, followed by removal of entrained particles using a separate electrostatic precipitator or fabric filter. Preheater kilns which are equipped with a bypass can tolerate high levels of sodium, potassium, sulfate, and chloride in the feed and fuel materials. However, there is an energy penalty associated with the increased operation of the bypass necessary to remove additional amounts of alkali chlorides and sulfates. Preheater kilns which are not equipped with a bypass are severely limited with respect to alkali and chloride tolerance because volatilized alkali chlorides recondense and accumulate when they are not removed, plugging the preheater cyclones.

8.5.3 HAZARDOUS-WASTE FUEL FACILITIES

Receiving and Storage

In the United States, HWF is subject to requirements governing hazardous-waste storage facilities as set forth in the *Code of Federal Regulations* at Title 40, Parts 264 and 265. These federal regulations require development of procedures to prevent hazards, contingency and emergency plans, personnel training, waste analysis, inspections, record keeping and manifesting of waste, and adequate spill containment to prevent or control releases to the environment.[9] These requirements are satisfied through a permit-review process, which provides a mechanism for regulatory agencies and the public to ensure that adequate standards are maintained. Ordinarily, state and local permitting requirements must be satisfied in addition to federal requirements.

In the General Portland HWF program, rigorous analysis of each shipment of waste is required to assure worker safety and provide information to regulatory agencies. Prior to arranging for deliveries to permitted transportation, storage, and disposal facilities (TSDFs), each generator's waste must be prequalified. Prequalification includes comprehensive analyses to determine organic constituents using gas chromatography equipped with flame-ionization and electron-capture detectors. The RCRA metals are determined using atomic-absorption spectrometry. Heat content, ash, chlorides, sulfur, and nitrogen are determined using bomb calorimetry and ion chromatography. In addition, the viscosity, flashpoint, and compatibility with other HWF components are determined during prequalification.

Each shipment of HWF is sampled when the transport vehicle arrives at the plant; this sampling, to determine organic constituents, metals, heat content, chlorides, and other parameters of interest, is performed at an on-site laboratory prior to authorization to unload the vehicle into a transfer tank. The 95,000-L (25,000-gal) receiving tanks are used to blend materials to specifications. Following fuel blending a final analysis is performed on the tank contents prior to accepting the blended HWF for transfer to the fuel-storage and burn tanks.

Construction and operation of a properly managed TSDF entails considerable capital and operating costs. A typical HWF facility may consist of four 95,000-L (25,000-gal) tanks for receiving and blending, a 568,000-L (150,000-gal) tank for fuel storage, and a 95,000-L (25,000-gal) tank for stormwater management. The tanks should be equipped with agitators to suspend solids and prepare homogeneous fuels. Volatile organic compounds may be controlled through installation of internal floating covers on fixed-roof tanks. Concrete foundations and spill-containment structures are essential. In addition, off-loading pads, pumps, strainers, and transfer pumps and filters should be included in the facility design. Decontamination facilities and an on-site laboratory complete the facility. In 1986 the cost for constructing such a facility was approximately $1 million. The HWF laboratory will typically be staffed by four to six professional and technical people. Operating costs for salaries, equipment, training, medical surveillance, occupational-exposure monitoring, and maintenance are approximately $500,000 per year. The permit-review process typically takes two years or more for a hazardous-waste TSDF. The cost of obtaining permits and the time required depend on the degree of regulation and the public reaction.[10]

Residual Waste Materials

Strainer and Filter Sludge.   The processing of hazardous waste into HWF results in the generation of residual materials. Hazardous wastes pumped from delivery vehicles typically contain miscellaneous debris including rubber gloves, rags, plastics, chunks of rubber, glass, mop strings, fibers, and "paint skin" covered with solvents. These materials must be removed from the HWF using strainers or grinders to protect pumps and other equipment from damage and plugging. In addition, General Portland uses a final filter to remove materials greater than 0.32 cm (⅛ in) in diameter prior to delivery of the fuel for combustion in the kiln. The strainers and filters must be cleaned periodically and the residual waste must be properly managed as hazardous waste. The sludges are mixed with kiln dust or other absorbent material and placed in drums, which are sealed prior to transportation to a hazardous-waste disposal facility. Similarly, small spills which occur during hose decoupling and other operations are absorbed with kiln dust and placed in sealed drums. These drums must be properly labeled, stored, and manifested. Best management practice for these materials is combustion in a rotary-kiln incinerator. Typically, less than 1,000 kg (2,200 lb) per month will be generated for a facility processing 3 million L (10 million gal) per year.

Stormwater.   The hazardous-waste storage facility must be protected from stormwater runoff and run-on. Stormwater which falls directly into the spill-containment area or onto the off-loading pad may become contaminated with small amounts of hazardous-waste constituents. Although this material could be discharged under a National Pollutant Discharge Elimination System (NPDES) permit, the expense of analysis and administrative procedures dictates that this material be consumed in the process. For wet-process kilns this can be readily accomplished by using the stormwater to prepare feed slurry for the kiln. A small quantity of organic constituents may be emitted to the atmosphere as a result of use of stormwater in the kiln. However, since the quantity of materials is extremely small (typically less than 100 ppm organics) and the substitution rate in preparing slurry is extremely small, there is no significant increase in organic emissions from the kiln. For dry-process kilns the stormwater may be used for gas conditioning in the water spray towers before the bypass dust collector or the raw-grinding mills.

Cement-Kiln Dust. The cement industry has traditionally produced a by-product known as cement-kiln dust (CKD), which is returned in part to the process, but is primarily stored as solid waste. It has uses in soil-stabilization projects and may be used to "fix" liquid hazardous waste for land disposal. As previously explained (see under *Cement Physiochemistry*, on p. 8.62) CKD is enriched in sodium and potassium chlorides and sulfates, as well as volatile metal oxides, such as lead oxide. Exhaust gases leaving the cement kiln tend to contain a large proportion of small particles because these are more easily entrained in the gas stream. For this reason, CKD typically has a mass median diameter on the order of 10 μm. Cement-kiln dust is composed primarily of finely ground particles of calcium carbonate, silicon dioxide, calcium oxide, sodium and potassium chlorides and sulfates, metal oxides, and other salts, depending on the raw materials and the fuel. The U.S. Bureau of Mines has prepared a summary of CKD characteristics from 103 samples collected throughout the United States.[11] The average data from the Bureau of Mines report can be rearranged to provide an approximate composition of CKD as shown in Table 8.5.2.

Cement-kiln dust has a strong affinity for water. When water is added to CKD, hydration products are formed which cause the CKD to harden, or "set." Typical unconfined compressive strength in excess of 422,000 kg/m^2 (600 psi) results when CKD is cured for seven days in a closed container with free water.

TABLE 8.5.2 Approximate Composition of Cement-Kiln Dust

Constituent	Wt %
$CaCO_3$	55.5
SiO_2	13.6
CaO	8.1
K_2SO_4	7.6
$CaSO_4$	5.2
Al_2O_3	4.5
Fe_2O_3	2.1
MgO	1.3
NaCl	1.1
KF	0.4
Other	0.6
Total	100.0

The addition of approximately 10 to 15% water by weight causes CKD to become "sticky," effectively eliminating fugitive emissions from windblown dust. Because of the fine particle size, there is a large surface area available within CKD. As water moves into the pore space between these particles, surface interactions between the water and CKD cause the water to be physically bound. As with many clays, kiln dust will hold from 40 to 50% of its weight in water before reaching saturation. This water is held within the dust by surface tension, resulting in very low permeabilities for saturated CKD. Permeabilities on the order of 10^{-6} to 10^{-5} cm/s (1 to 10 ft/yr) are typical. Consequently, it is extremely difficult to move water through "set" CKD.

The sources and fate of lead in raw materials, fuels, and CKD have been extensively studied at the General Portland Los Robles cement plant.[12] It is important to understand the relative contributions from feed, fuel, clinker, and kiln dust in the overall material balance. The Los Robles kiln is a long, dry-process system which can be nominally described as follows:

Constituent	Throughput, t/h
Feed to kiln	116
Clinker produced	67
Coal or coke burned	8
Kiln dust produced	4
Hazardous-waste fuel (30% of energy)	3.6

On a weight basis, 10 ppm of lead in the feed would be equal to 150 ppm in coal or coke, 330 ppm in HWF, and 300 ppm in CKD. Long-term average data for the material balance, established through daily composite sampling for over a year, demonstrated that virtually all of the lead from HWF was found in CKD at this particular facility. Different partitioning of lead and other metals between clinker and CKD may occur in other kiln systems. However, most metal oxides which are volatile at burning-zone temperatures will tend to accumulate in CKD. Apparently anomalous results reported in the literature typically result from sampling errors, products of incomplete combustion (PICs), and differing kiln designs.[13,14] For example, most of the lead must necessarily be found in the clinker in the case of a preheater kiln which is not equipped with a bypass dust collector.

The solubility and mobility of lead and other metals in CKD is extremely important in considering the long-term use or land storage of CKD as a waste byproduct. Lead is perhaps the best example of a heavy metal in CKD because it is typically found in concentrations exceeding 100 ppm in HWF. This results in a significant increase in the lead content of CKD from cement kilns burning HWF. Theoretically, the soluble lead concentration for a given solution can be determined from thermodynamic principles.[15] In a simple system containing lead ions and water the equilibrium concentration of soluble lead is extremely small in near-neutral and slightly alkaline solutions because insoluble lead oxide (PbO) will form as a precipitate. In acidic solutions (below pH 5), lead oxide can be dissolved, releasing lead ions into solution. Similarly, in very alkaline solutions (pH above 12), lead ions can enter solution as aqueous lead hydroxide complexes, such as $[Pb(OH)_3]^-$. Between pH 5 and 12 substantial quantities of lead carbonate ($PbCO_3$) and lead carbonate-hydroxide $[Pb_3(CO_3)_2(OH)_2]$ will be present as precipitated insoluble forms of lead. Since kiln dust contains calcium oxide, solutions of water with CKD will be alkaline. The presence of these alkaline constituents, and the high level of carbonates, provide kiln dust with a very large buffer capacity. The strongly buffered carbonate and pH system minimizes the solubility of lead in the CKD environment.

Figure 8.5.2 shows the relationship between soluble lead and pH as a function of total lead concentration in samples of CKD generated at the General Portland Los Robles plant. Virtually all of the lead in Los Robles CKD is available to participate in aqueous equilibria because the lead has been condensed onto particle surfaces. For CKD containing lead which originated in the feed, the soluble lead fraction would be lower, but still related to the total lead in the CKD. Nevertheless, Los Robles CKD containing 165 ppm of total lead exhibited a lead solubility of less than 0.05 μg/mL (safe drinking-water standard) over a pH range from 6.2 to 12.8. Soluble lead increases as the total lead increases. The minimum lead solubility occurs at a pH of approximately 10.

Soluble lead ions in contact with CKD are not typically released into the environment because the mobility of lead is severely restricted by surface absorption of the lead ions onto soils and other particles. Data on absorption of lead onto clays, carbonates, and hydrated metal oxides in soils and river sediments indicate that typically 90% of the lead potentially present in solution is adsorbed onto solid surfaces.[16]

Lead which dissolves in water in contact with CKD will be removed from solution as the water moves away from the CKD through surface adsorption and

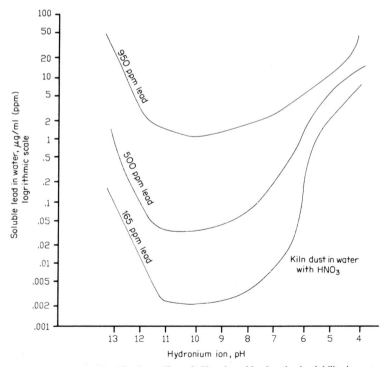

FIG. 8.5.2 Los Robles kiln dust: effect of pH and total lead on lead solubility in water.

precipitation of insoluble lead oxide and carbonate salts. The release of lead from CKD can be controlled by minimizing contact with excessive amounts of water and by employing proper waste-management practices. Lead stabilized in kiln dust as insoluble oxides and carbonates is only sparingly soluble in water. Kiln dust placed above the water table and segregated from stormwater run-on will remain dry because of the extreme affinity of CKD for water. Lead ions which enter solution will be removed by absorption onto surfaces of soils adjacent to the CKD. Consequently, the mobility of lead stabilized in CKD is severely limited. This combination of factors illustrate the effectiveness of cement kilns for stabilizing heavy metals in the CKD matrix.

8.5.4 ENVIRONMENTAL ASSESSMENT

Tests of Kiln Emissions

Extensive studies of air emissions from cement kilns burning hazardous waste have been conducted in the United States and in other countries.[17,18,19,20] Many of these studies have included evaluation of conventional air contaminants, such as particulate matter, sulfur dioxide, and nitrogen oxides, in addition to heavy metals and organic contaminants. Comparative emission tests on kilns burning conventional fuels and HWF show that there is no significant change in air emissions when HWF is properly used to replace conventional fuels. There is a potential for reduction in emissions on kilns burning HWF. This potential results from lower sulfur in the HWF and typically easier combustion of HWF as compared with conventional fuels. Natural variations in operating conditions within the kiln system and the emission-control systems typically cause more variation in kiln emissions than does the choice of fuels. Moreover, changes in raw materials and variability in fossil fuels can produce more change in kiln emissions than the use of HWF.

In removing alkalies, chlorides, and sulfates, particle size is of paramount importance. Kilns equipped with electrostatic precipitators can selectively enrich CKD in volatile components because materials collected in the final fields tend to be finer than particles collected in the first fields. Removing the last fields and returning dust from the first fields minimizes the CKD wasted while still removing undesirable alkali chlorides and sulfates.

Changes in the nature of CKD collected in the last fields may affect resistivity or cause the material to stick to the collecting electrodes. Consequently, some adjustment to the operating conditions of the electrostatic precipitator may be necessary to maintain compliance with emission limits. However, a kiln which is routinely compliant burning conventional fuels is expected to remain compliant burning HWF.

Destruction and Removal Efficiency and Heavy Metals. Emissions of very small quantities of organic contaminants and heavy metals have been the focal point of most research on cement kilns burning hazardous waste. For more than 10 years it has been known that cement kilns can effectively destroy polychlorinated biphenyls (PCBs) to levels below the detection limit in stack gas.[5,21] Similarly, tests on refractory materials, such as carbon tetrachloride, perchloroethylene, chlorobenzenes, and sulfur hexafluoride, have demonstrated that cement kilns can achieve a destruction and removal efficiency (DRE) in excess of 99.99%.[17,18,19] This is not surprising, since laboratory studies, as well as thermodynamic and kinetic principles, indicate that virtually all organic materials will be destroyed at temperatures exceeding 1000°C for a reaction time as long as 2 s.[22,23]

However, there is at least one study which indicates a cement-kiln DRE as low as 91%.[24] In addition, reported results do not always exceed 99.99%. In the case of the extremely low DRE result, the kiln was in an abnormal condition, such that burning-zone temperatures were too low to produce clinker. Virtually all of the other reported DREs less than 99.99% are believed to be due to contamination PICs and analytical errors. This is easy to understand given the extremely low level of organic contaminants which must be measured and the numerous sources of sample contamination and analytical error.

Because of the large variation in kiln design and energy efficiency, emissions from cement kilns may range from approximately 100 g·mol/kg clinker (200 lb·mol/ton clinker) to over 225 g·mol/kg clinker (450 lb·mol/ton clinker) of exhaust. A "typical" kiln producing 89 t/h of clinker may have emissions of 100 standard m^3/s (3,500 ft^3/s) consuming 369 billion J/h (350 million Btu/h). This is equivalent to 14,060 kg/h (31,000 lb/h) of fuel having a heat content of 25 million J/kg (11,250 Btu/lb) or 25 million J/L (90,000 Btu/gal), which is typical of HWF. If we assume that this kiln is burning 100% HWF (25 million gal/yr), one can calculate the allowable emissions to achieve 99.99% DRE for a principal organic hazardous constituent (POHC) present at 10%, 1%, or even 0.5% levels.

POHC level in HWF, %	Allowable emissions to achieve 99.99% DRE,	
	kg/h	g/m^3
10	0.141	390
1	0.0141	39
0.1	0.00141	3.9

The sample volume for a typical DRE test is approximately 20 L (0.71 ft^3). Consequently, contamination of a few micrograms per sample can easily lead to apparently low DRE results. Sample contamination is a chronic problem for materials commonly found in laboratories, such as methylene chloride and methyl ethyl ketone, and it is imperative that these materials be analyzed prior to use in DRE tests. Storage of collected samples in a refrigerator or laboratory environment for longer than a few days will typically result in contamination.

The threshold limit value (TLV) established for safe workplace exposures for most hazardous materials found in the workplace is considerably higher than the stack-gas concentration. The TLV for relatively toxic materials, such as lead (200 $\mu g/m^3$), ozone (200 $\mu g/m^3$), the pesticide Endrin (100 $\mu g/m^3$), methylisocyanate (50 $\mu g/m^3$), arsine (50 $\mu g/m^3$), and mercury (50 $\mu g/m^3$) are higher than the allowable stack-gas concentration for a POHC present at 1% in hazardous waste burned for 100% of the energy in a typical cement kiln.

In order to ensure 99.99% DRE during normal operation, regulatory agencies have required that use of HWF be automatically discontinued under adverse operating conditions.[25,26,27] These conditions typically include

1. Low oxygen
2. High carbon monoxide

3. Low temperatures

4. Positive pressure in the burning zone

Permit conditions requiring automatic cutoffs and prohibiting the use of HWF during start-up and shutdown provide assurance to the public that deleterious emissions will not result from the use of HWF during adverse operating conditions.

Storage-Tank and Fugitive Emissions

HWF typically contains volatile solvents, such as methyl ethyl ketone, acetone, and methanol. Typically, the vapor pressure will exceed 1,050 kg/m² (1.5 psi) at ambient tank temperatures. When liquids are pumped into the tanks, the vapors are displaced. Filling a 95,000-L (25,000-gal) tank with hazardous waste is expected to result in potential emissions of approximately 23 kg (50 lb) per loading cycle. Installation of an internal floating cover, as proposed under storage tank New Source Performance Standards,[28] can effectively control these potential emissions. However, the use of activated carbon is not recommended due to significant fire hazards which result from heat released when ketones, aldehydes, and some alcohols are absorbed onto activated carbon.

Fugitive emissions of volatile organic compounds must also be considered in the design and operation of hazardous-waste storage facilities. During the transfer, storage, and handling operations, emissions are expected from pumps, agitators, valves, flanges, spills, and hose-draining operations. The magnitude of these potential emissions can be minimized through facility design and work-management practices.

Air-Quality Impact

Cement kilns are elevated point sources which exhibit excellent dispersion in the atmosphere. On an annual basis, ground-level concentrations for a typical cement kiln will be less than one-millionth of the average stack concentration. This high level of dispersion occurs because most cement kilns are equipped with stacks exceeding 61 m (200 ft) in height and the relatively high exhaust temperatures and momentum increase the effective stack height well beyond the actual stack height. This is illustrated in Table 8.5.3, which summarizes the maximum impact calculated from hourly meteorology over a one-year period using the U.S. Envi-

TABLE 8.5.3   CRSTER Environmental Impact Analysis for General Portland Inc., Paulding Cement Kiln—Nominal Emissions of 1.0 mg/s

Averaging time	Maximum impact		Dispersion factor
	Distance from stack, km	POHCs found, mg/m³	
Annual	4.0	0.000009	1,319,430
3-month	4.0	0.000027	439,810
24-hour	2.5	0.000119	100,135
3-hour	1.5	0.000559	21,240
1-hour	1.5	0.001143	10,380

ronmental Protection Agency's (EPA's) CRSTER dispersion model. Even for relatively short averaging times, such as 1 h, stack gases are effectively diluted in the atmosphere by a factor of over 10,000. Consequently, stack gases containing 100 $\mu g/m^3$ of, for example, Endrin, would be diluted to a concentration of less 0.01 $\mu g/m^3$ at ground level. The potential risk to health or the environment from PICs or residual organic contaminants in properly operated industrial furnaces is much smaller than risks willingly accepted by society in day to day life.[29]

In contrast to emissions from elevated point sources, storage-tank and fugitive emissions occur at or close to ground level and there is no momentum or buoyant plume rise to enhance dispersion. During calm conditions and atmospheric inversions, storage-tank and fugitive emissions are expected to result in much higher workplace concentrations than would occur ordinarily. Occupational-exposure monitoring and the use of proper protective equipment, such as organic-vapor respirators, is recommended during vehicle sampling, filter cleaning, spill cleanup, and at other times when workers are in contact with volatile hazardous-waste liquids. It is also essential that proper employee protection in the form of goggles and impermeable clothing be provided to insure against occupational exposures.

8.5.5 SUMMARY

Cement kilns can be used to treat a wide variety of hazardous wastes. Properly managed, this can be accomplished with no significant increase in emissions or environmental impact. Many of the hazardous wastes also have value as alternative sources of raw materials or fuel, resulting in bona fide resource recovery rather than simple waste destruction or stabilization. Changes can typically be made to existing kiln systems to accommodate hazardous wastes without extraordinary capital expenditures. In addition, changes to an existing source do not create a new source of emissions, as would result from the construction of a new hazardous-waste incinerator. For these reasons, the use of cement kilns to manage hazardous wastes is expected to increase dramatically over the next decade.

8.5.6 REFERENCES

1. J. L. Hurt, *Hazardous Waste Facility Annual Report—Los Robles ID No. CAT 080031628,* State of California Department of Health Services, Sacramento, California, March 1986.

2. J. F. Chadbourne, *Supplemental Fuel Program Operation Plan—General Portland Inc., Demopolis, Alabama,* Hazardous Waste Permit Application No. ALD981019045, Alabama Department of Environmental Management, Montgomery, Alabama, June 1986.

3. R. C. Dickie, "Spent Potlining as a Fuel Supplement in a Cement Kiln," *Journal of Metals,* **36:**(7)22 (July 1984).

4. P. G. Deane, "The Determination of Thallium in Cement Kiln Precipitator Dusts and Iron Oxides by Differential Pulse Anodic Stripping Voltommetry," *Zement-Kalk-Gips,* no. 12/82, December 1982, pp. 285–287 (translated from German).

5. L. P. MacDonald, D. J. Skinner, F. J. Hopton, and G. H. Thomas, "Burning Waste Chlorinated Hydrocarbons in a Cement Kiln," Fisheries and Environment Canada, EPS 4-WP-77-2 and EPA/530/SW-147C, April 1977.

6. M. W. Black and J. R. Swanson, "Destruction of PCB's in Cement Kilns," *Pollution Engineering,* June 1983, pp. 50–53.

7. S. McClure, "A Real Stinker: City's Sludge Burning Project Runs Afoul of Wrinkled Noses," *Detroit Free Press,* October 6, 1982.

8. H. Warren, "Burning Sludge Raises Big Stink," *Detroit News,* October 6, 1982.

9. "Hazardous Waste Management System: Burning of Waste Fuel and Used Oil Fuel in Boilers and Industrial Furnaces," *Federal Register,* vol. 50, no. 230, pp. 49,163–49,270, November 29, 1985.

10. B. L. Krag, "Hazardous Wastes and Their Management," *Hazardous Waste and Hazardous Materials,* 2:(3)251–308, Mary Ann Liebert, Inc., 1985.

11. B. W. Haynes and G. Kramer, *Characterization of U.S. Cement Kiln Dust,* Bureau of Mines Circular, IC No. 8885, Avondale, Maryland, 1982.

12. J. F. Chadbourne and E. F. Bouse, *Los Robles Cement Plant Cement Kiln Dust Waste Classification Report,* California Department of Health Services, August 1985.

13. E. E. Berry, L. P. MacDonald, and D. J. Skinner, *Experimental Burning of Waste Oil as a Fuel in Cement Manufacturing,* Environment Canada, EPS 4-WP-75-1, June 1976.

14. B. Dellinger, M. D. Graham, and D. A. Tirey, "Predicting Emissions from the Thermal Processing of Hazardous Wastes," *Hazardous Waste and Hazardous Materials,* 3:(3)293–307, Mary Ann Liebert, Inc., 1985.

15. W. Stumm and J. J. Morgan, *Aquatic Chemistry: An Introduction Emphasizing Chemical Equilibria in Natural Waters,* Wiley-Interscience, New York, 1970, pp. 321–326.

16. J. D. Hem, "Geochemical Controls on Lead Concentrations in Stream Water and Sediments," *Journal of the Geochemical Society and the Meteoritical Society,* 40:(6)599–609, Pergamon Press, New York, June 1976.

17. R. E. Mournighan et al., in *Proceedings of the International Conference on New Frontiers for Hazardous Waste Management,* EPA-600/9-85/025, U. S. EPA, Washington, D.C., 1985, pp. 533–549.

18. R. E. Mournighan, J. A. Peters, and M. R. Branscome, *Effects of Burning Hazardous Wastes in Cement Kilns on Conventional Pollutant Emissions,* Air Pollution Control Association, Pittsburgh, Pennsylvania, June 1985.

19. E. T. Oppelt, "Hazardous Waste Destruction," *Environmental Science and Technology,* 20:(4)312–318 (April 1986).

20. D. C. Simeroth, et al., *Evaluation of Combustion Processes for Destruction of Organic Wastes—General Portland,* Report No. C-84-020, State of California Air Resources Board, October 15, 1985.

21. B. Ahling, *Combustion Test with Chlorinated Hydrocarbons in a Cement Kiln at Stora Vika Test Center,* Swedish Water and Air Pollution Research Institute, Stockholm, Sweden, March 1978.

22. B. Dellinger and D. L. Hall, "Surrogate Compounds for Monitoring the Effectiveness of Incineration System," *Journal of the Air Pollution Control Association,* 36:(2)179–183 (February 1986).

23. B. Dellinger et al., in *Proceedings of the Eleventh Annual Research Symposium: Incineration and Treatment of Hazardous Waste,* EPA-600/9-85-028, U.S. EPA, Cincinnati, Ohio, 1985, pp. 160–170.

24. J. A. Peters, T. W. Hughes, J. R. McKendree, L. A. Cox, and B. M. Hughes, *Industrial Kilns Processing Hazardous Wastes—San Juan Cement Company Demonstration Program,* EPA-600/52-84, U.S. EPA, Cincinnati, Ohio, 1984.

25. P. P. Walling, *Permit to Operate an Air Contaminant Source,* No. 0363000002P016, State of Ohio, Environmental Protection Agency, Columbus, Ohio, November 30, 1984.

26. K. W. Kizer, *Resource Recovery Facility Permit,* No. CAT 080031628, State of California Department of Health Services, Sacramento, California, April 10, 1986.

27. J. W. Warr, Air Permit No. 105-0002-Z004, Alabama Department of Environmental Management, Montgomery, Alabama, May 22, 1986.

28. "Standards of Performance for New Stationary Sources: Volatile Organic Liquid Storage Vessels Constructed After July, 1984," *Federal Register,* vol. 49, no. 142, pp. 29,698–29,718, July 23, 1984.

29. K. E. Kelly, "Methodologies for Assessing the Health Risks of Hazardous Waste Incinerator Stack Emissions to Surrounding Populations," *Hazardous Waste,* **1:**(4)507–531, Mary Ann Liebert, Inc., 1984.

SECTION 8.6
WET OXIDATION

W. M. Copa
Vice President, Technical Services
Zimpro Inc.
Rothschild, Wisconsin

W. B. Gitchel
Rothschild, Wisconsin

8.6.1 GENERAL DESCRIPTION

Wet oxidation is an aqueous-phase oxidation process brought about when an organic and/or oxidizable inorganic-containing liquid is mixed thoroughly with a gaseous source of oxygen (usually air) at temperatures of 150 to 325°C (302 to 617°F). Gauge pressures of 2,069 to 20,690 kPa (300 to 3,000 lb/in^2) are maintained to promote reaction and control evaporation.

In waste treatment, wet oxidation in the temperature range of 150 to 200°C (302 to 392°F) improves sludge dewaterability. Intermediate temperatures of 200 to 280°C (392 to 536°F) are used in such applications as spent activated-carbon regeneration and conversion of refractories to biodegradable substances. Still higher temperatures [280 to 325°C (536 to 617°F)] provide essentially complete oxidation.

The process may be carried out in a batch reactor, but usually a continuous-flow system is used. The elements of the systems are a high-pressure liquid-feed pump, oxygen source (air compressor or liquid-oxygen vaporizer), reactor, heat exchangers, and process regulators. A basic flow diagram is shown in Fig. 8.6.1.

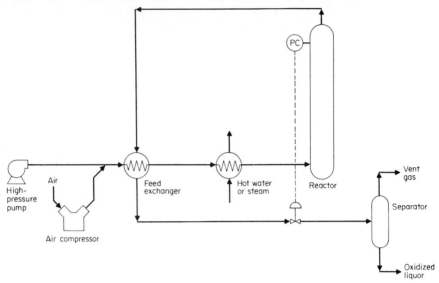

FIG. 8.6.1   Wet-oxidation flow diagram.

8.6.2   APPLICABILITY OF WET OXIDATION TO HAZARDOUS WASTES

The reactions of the wet-oxidation process occur under the conditions of the general description irrespective of the state of dispersion of the oxidizable substance—e.g., in soluble, colloidal, or fine to coarse suspensions. The main criterion is that the oxygen utilization, or oxygen demand (OD), range from about 20 to 200 g/L (0.167 to 1.67 lb/gal) of water fed.

If the waste is found in any of the above forms with the oxygen demand in the applicable range, the mixture may be fed directly. Highly concentrated mixtures may be diluted to form a single feed stream of the required OD or may be delivered as a side stream into the aqueous stream at a ratio conforming to OD requirements. The latter technique has been successfully applied to toxic oils, chemical-process scrubbing liquors, and propellents, as well as to such industrial processes as extractive metallurgy and wood pulping. Examples of hazardous-waste oxidation are described in Subsec. 8.6.6.

Almost all applications of wet oxidation in hazardous-waste treatment are effective. Most organic compounds are stoichiometrically oxidized, the carbon going to carbon dioxide, the hydrogen to water, any halogen to halides, any sulfur to sulfate, any phosphorus to phosphate, and any nitrogen to ammonia or elemental nitrogen. The most resistant compounds are the halogenated aromatics. For example, mono- and dichlorobenzene were reported about 70% destroyed, compared to better than 99% for a wide variety of other compounds.[1]

At or above 85% of stoichiometric OD utilized, the remaining organic matter is comprised of low-molecular-weight compounds—predominantly carboxylic acids—and as such is readily treated biologically. Tests have shown that detoxification of hazardous substances is essentially complete in these cases.[2]

8.6.3 DEVELOPMENT OF WET-OXIDATION TECHNOLOGY

History

A major impetus for the development of wet oxidation as a commercial process came in the early 1960s with applications to the recovery of pulping chemicals from waste liquors.[3] This application required the upper range of temperature, 300 to 327°C (572 to 620°F). At about the same time, wet oxidation was applied for the complete oxidation of sewage sludge.[4]

The work with sewage sludge led to the discovery of a new application of wet oxidation—that of conditioning the sludge for easy dewatering, utilizing the lower range of temperature, 177 to 204°C (350 to 400°F). Within 10 years, the utilization of intermediate temperatures, 200 to 240°C (392 to 464°F), was adapted to such applications as regeneration of spent activated carbon.[5]

Commercial applications to the treatment of hazardous waste appeared in 1983 and are reported in more detail in Subsec. 8.6.5.

Theoretical Aspects of Wet Oxidation

The formation of intermediate products in the course of the wet-oxidation reaction was recognized in early developments and has been confirmed by ongoing research.[6,7,8,9] Since certain intermediate products resist further oxidation under initially imposed conditions, they tend to accumulate. As a consequence, if the course of the wet oxidation is followed by a test for a non-compound-specific parameter such as chemical oxygen demand (COD) or total oxygen demand (TOD), successive phases of decreasing reaction rates will be apparent.

Reports on the adaptation of classical kinetic theory to wet oxidation generally show that a model of the first-order type fits data quite well. That is,

$$\frac{dC}{dt} = - kC$$

where C is the concentration of oxidizable substance. In such treatment it is assumed that oxygen partial pressure and gas-liquid mixing are sufficient to provide unlimited oxygen supply, thus assuring pseudo-first-order kinetics with respect to the oxidizable substance.

When the data from the nonspecific analysis are treated by the first-order model, two or more rate constants, k, $k_2 \ldots k_l$, of decreasing value may be identified. The concept is illustrated by Fig. 8.6.2.

The effect of temperature on wet oxidation rates is such that for most organic compounds a temperature in the range of 200 to 280°C (392 to 536°F) can be found which will result in better than 99% elimination of the compound in 15 to 60 min. This result is essentially accomplished in the initial high-rate phase and corresponds to first-order rate constants, k_1, of 0.1 to 1.0 min^{-1}.

For more difficult compounds, higher temperatures may be resorted to. A number of homogeneous and heterogeneous catalysts have also been disclosed which are quite effective.[10,11] Adjustment of pH, especially to lower values, can accelerate rates.

Wet-oxidation reaction rates are not as sensitive to pressure as they are to

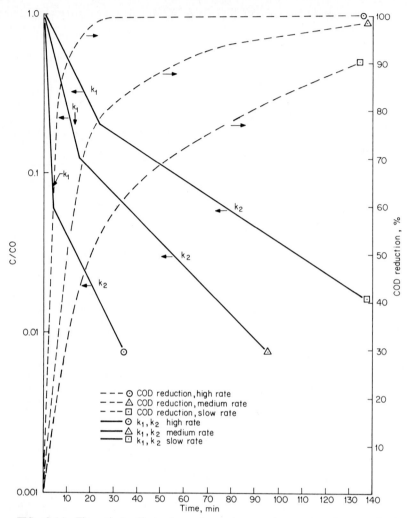

FIG. 8.6.2 Theoretical effect of reaction rates on the extent of wet oxidation. COD = chemical oxygen demand.

temperature. The stoichiometry of oxygen supply and the pressure necessary to control evaporation are of first concern.[3] Some investigators have reported rate models including the ½ power of oxygen partial pressure as a term.[12]

The question of oxygen transfer is certainly of concern but is beyond the scope of this section. From practical experience the authors find that entirely adequate oxygen transfer can be obtained even at inputs of 200 g/L OD and 95 to 99% reaction in 15 to 60 min.[3] The great enhancement of oxygen solubility and oxygen-transfer coefficients provided by the reaction temperature explain the observation.[6]

Chain-reaction mechanisms involving such free radicals as hydroxyl, hydroperoxyl, and oxygen have been proposed. Such mechanisms may explain the relatively rapid complete oxidation of organics and account for the reference to water as a "catalyst." It is noted that substances such as copper salts, which are proven effective in oxidations with hydrogen peroxide or ozone, also catalyze many wet oxidations.

At system conditions the water, fixed gases, and solids or solutes adjust to theoretical vapor-liquid equilibria. The pressure can be regulated to effect a distribution of water anywhere from near 100% liquid to near complete evaporation. In practice the distribution is adjusted to retain a liquid phase sufficient to provide transport for solids or solutes, and to allow sufficient evaporation for removal of heat of reaction. The heat thus released can be harnessed to produce steam or hot water.

8.6.4 WET-OXIDATION EQUIPMENT

Wet-oxidation equipment is available in four basic configurations for application of this technology to hazardous-waste treatment. These configurations differ mainly in reactor design, which can be categorized as follows.

1. *Aboveground vertical-column reactor:* The technology for wet oxidation utilizing an aboveground vertical-column reactor is owned by Zimpro Inc., Rothschild, Wisconsin. Zimpro Inc. wet-oxidation technology is by far the most widely applied in the hazardous-waste market.[13] The vertical-column reactor that Zimpro uses is essentially a cocurrent bubble column which may or may not be baffled depending on the desired mixing conditions.

2. *Stirred-tank reactor:* The technology for wet oxidation utilizing a stirred-tank reactor, commonly referred to as the Wetox™ system, is also owned by Zimpro Inc. The Wetox™ system reactor essentially consists of a series of completely stirred tank reactors.[14]

3. *Supercritical water:* The supercritical-water wet-oxidation technology is owned by MODAR, Inc., Natick, Massachusetts. The configuration of the supercritical reactor is similar to Zimpro's vertical column but the reactor is operated at extreme conditions of temperature and gauge pressure, e.g., greater than 374°C (705°F) and 22,048 kPa (3,200 lb/in^2).[15]

4. *Underground vertical-tube reactor:* The technology for wet oxidation utilizing an underground vertical-tube reactor is marketed by VerTech Treatment Systems, Westminster, Colorado, among others. The underground vertical-tube reactor utilizes conventional oil-field drilling technology to provide a drilled, cased, and cemented well as the wet-oxidation reactor. The reactor extends to a depth of approximately 1 mi and is reported to operate at temperatures of 288°C (550°F).[16]

8.6.5 ENVIRONMENTAL IMPACT OF WET-OXIDATION SYSTEMS

Hazardous wastes are treated by wet oxidation for the purpose of detoxification. Ultimately two effluent streams, off gas and oxidized liquid effluent, exit from the wet-oxidation process. The liquid effluent can be further separated in some cases into a liquid fraction, with all soluble components, and insoluble solids. The environmental impact of these effluent phases—gas, liquid, and solid—must be addressed when considering wet oxidation for hazardous-waste treatment.

Off-Gas Components and Treatment Systems

Wet-oxidation treatment of hazardous wastes is usually carried out to the extent that the toxic organic compounds are nearly completely destroyed. However, the oxidation products are not entirely carbon dioxide and water. Some low-molecular-weight organic compounds, such as acetaldehyde, acetone, acetic acid, and methanol, are also formed as breakdown products. These low-molecular-weight compounds are more volatile than the parent compounds and are distributed between the process off-gas phase and the oxidized liquid phase. The concentration of these low-molecular-weight compounds in the process off gas is dependent on the concentrations of these same components in the liquid phase, which in turn is determined by the degree of oxidation accomplished, the type of waste being treated, and the concentration of organics in the influent waste. Typically, the concentration of these low-molecular-weight compounds in the off gas from the wet-oxidation process ranges from 10 to 1,000 ppm (measured as total hydrocarbons expressed as methane).

The volatile organic components that are emitted from the wet-oxidation process in the process off gas can be controlled by a variety of technologies including scrubbing techniques, granular-carbon adsorption, and fume incineration. The specific technology that is used is selected on a case by case basis.

Composition and Disposal of Liquid Effluents

The liquid-phase effluent from the wet-oxidation treatment of hazardous wastes usually contains measurable concentrations of low-molecular-weight compounds, predominantly carboxylic acids and other carbonyl-group compounds. These soluble organic compounds are readily treated by biological wastewater-treatment processes, or a combination of biological and physical adsorption processes. The combination of wet oxidation followed by biological and physical adsorption treatment can produce a high-quality effluent water that is suitable for direct discharge or re-use. The overall treatment capabilities of these combined processes are discussed in Subsec. 8.6.6.

Solid Residuals

The effluent from a wet-oxidation system will contain suspended solids if the influent waste has a high metal content, an ash component, or dissolved salts which are precipitated during the wet-oxidation process. Typically, the level of wet oxidation required to detoxify hazardous wastes will produce an effluent that contains an insoluble ash. The ash material is usually comprised of metal oxides and other insoluble salts such as sulfates, phosphates, and silicates. The insoluble ash can usually be dewatered and subsequently landfilled.

8.6.6 CASE HISTORIES

General Application to Hazardous-Waste Treatment

Wet oxidation has been proven to be effective in treating a variety of hazardous wastes. In general, wastes containing inorganic and organic cyanides and sulfides can be easily treated by wet oxidation at temperatures less than 250°C (482°F) and gauge pressures below 13,800 kPa (2,000 lb/in^2). Wastes containing aliphatic hydrocarbons, chlorinated aliphatic organics, phenols, aromatic hydrocarbons, and heterocyclic compounds containing oxygen, nitrogen, or sulfur atoms are also easily oxidized under similar wet-oxidation conditions. Wastes containing halogenated aromatic compounds with at least one nonhalogen functional group, e.g., pentachlorophenol or 2,4,6-trichloroaniline, are effectively treated by wet oxidation at temperatures of 250°C (482°F) to 320°C (608°F) and gauge pressures of up to 20,700 kPa (3,000 lb/in^2). Wastes containing halogenated aromatic compounds, e.g., hexachlorobenzene, dichlorobenzene, and PCBs, require extremely high temperatures [greater than 320°C (608°F)] and gauge pressures of 20,700 kPa ($<$3,000 lb/in^2) for effective treatment by wet oxidation.

In most cases where wet oxidation is applied to hazardous wastes, the treatment objectives are the detoxification of the wastewater for subsequent biological treatment, which can be accomplished in a publicly owned treatment works (POTW) or an industrial treatment plant. To meet this detoxification requirement, a reduction of specific toxic organic compounds to low milligrams per liter concentration levels is usually required. Overall destruction levels of specific organic compounds that can be accomplished by wet oxidation are usually dependent on the structure of the compound, the waste matrix, and the processing temperature employed.

Specific Installations

Casmalia Resources, Santa Barbara County, California. Casmalia Resources operates a Class I hazardous-waste disposal facility in Casmalia, California. Casmalia Resources installed a 37.8 L/min (10 gal/min) wet-oxidation unit in 1983. The wet-oxidation unit was provided by Zimpro Inc., Rothschild, Wisconsin. The installation is pictured in Fig. 8.6.3. The wet-air oxidation unit was permitted by the State of California Department of Health Services and the Santa Barbara County Air Pollution Control District to treat the following classes of hazardous wastes: cyanide wastes, sulfide wastes, phenolic wastes, nonhalogenated pesticide wastes, solvent still bottoms, and general organic wastewaters.

Effectiveness of the wet-air oxidation treatment for each of these categories of wastes was determined by a series of demonstration tests. Each demonstration test consisted of a continuous 8-h wet-air oxidation run. During each run, composite samples of waste feed and oxidized effluent were obtained and analyzed to determine the destruction levels for the specific hazardous component.[17] These results are summarized in Table 8.6.1.

The liquid wet-oxidation effluents at the Casmalia Resources site are discharged to evaporation ponds for posttreatment. After evaporation, the remaining residue of metal oxides and salts is landfilled.

Casmalia Resources continues to use the wet-oxidation system for commercial

FIG. 8.6.3   Wet-oxidation installation (Casmalia Resources).

treatment of a variety of hazardous wastes, including pesticide wastewater, wastes from petrochemical processes, propellent wastewaters, solvent-recycling wastes, cyanide-containing metal-plating wastes, spent caustic wastewaters containing sulfides and phenols, and general organic-chemical production wastewaters.

Bofors-Nobel Inc., Muskegon, Michigan.   Bofors-Nobel, through Environmental Systems Corporation, a joint-venture company formed specifically to treat wastes from chemical manufacturing and an abandoned landfill site, uses wet oxidation as an integral part of its hazardous-waste treatment system. A flow diagram of this system is shown in Fig. 8.6.4. Wet oxidation is used to detoxify several chemical-production wastewater streams which contain toxic organic components in the concentration range of 600 to 1,200 mg/L and total organic concentrations in excess of 50 g/L (total volatile solids.)[18] Typical wet-oxidation treatment results are summarized in Table 8.6.2.

Effluents from the wet oxidation of some of the process wastewater contain substantial amounts of ammonia. These oxidized effluents are directed to a neutralization process where crystalline ammonium sulfate is recovered.

A second wet-oxidation system is employed by Bofors-Nobel to regenerate powdered activated carbon from the powdered activated-carbon treatment (PACT™) unit. The PACT™ unit involves the controlled addition of powdered

Table 8.6.1 Summary of Casmalia Resources Wet-Oxidation Treatment Performance

	Wastewater class									
	Phenolic and organic sulfur		General organic		Cyanide		Pesticide		Solvent still bottoms	
Sample description	Influent	Effluent	Influent	Effluent	Influent	Effluent	Influent	Effluent	Influent	Effluent
COD, g/L	108.1	11.6	76.0	2.5	37.4	4.2	110.	5.15	166.	45.9
DOC, g/L	—	3.68	20.83	0.69	14.71	1.71	26.60	1.04	15.20	7.99
BOD$_5$, mg/L	—	—	29,680	880	—	603	41,000	2,530	83,000	21,900
pH	13.0	8.3	1.9	3.3	12.6	9.0	2.0	4.4	12.5	8.2
Total phenols, mg/L	15,510	36	—	—	—	—	—	—	—	—
Organic sulfur, mg/L*	3,010	180	—	—	—	—	—	—	—	—
Cyanide, mg/L	—	—	—	—	25,390	82	—	—	—	—
Pesticides, mg/L†	—	—	—	—	—	—	169.0	0.93	—	—
Total solids, g/L	88.6	59.7	10.0	3.1	135.1	91.2	8.1	0.7	38.6	30.0
Total ash, g/L	57.1	50.2	0.7	1.1	112.9	77.4	1.1	0.4	25.5	19.8
Volatile solids, g/L	31.5	9.5	9.3	2.0	22.4	13.8	7.0	0.3	13.1	10.2

*Organic sulfur assumed to be equal to total sulfur minus sulfate sulfur (inorganic sulfide concentration is nil)
†Sum of Dinoseb, methoxychlor, carbaryl, and malathion concentrations

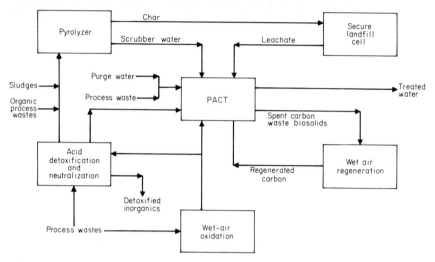

FIG. 8.6.4 Flow diagram of the hazardous-waste treatment system at Bofors-Nobel, Muskegon, Michigan.

activated carbon to the aeration basin of a biological wastewater-treatment system. This latter treatment system is used to process the wet-oxidation effluents, certain low-strength process wastewaters, and groundwater from the contaminated site. Both wet-oxidation systems at Bofors-Nobel are 37.8 L/min (10 gal/min) skid-mounted units supplied by Zimpro Inc. Both units are fabricated from titanium for maximum compatibility with a variety of production wastes. Each system is designed to operate at a maximum temperature of 280°C (536°F) and a maximum gauge pressure of 15,200 kPa (2,200 lb/in²).

Dominion Foundries and Steel Company, Hamilton, Ontario, Canada. Dominion Foundries and Steel Company (DOFASCO) uses wet oxidation to detoxify spent coke-oven gas (COG) scrubbing liquors and recovers salable products from the treated stream.[19] The COG liquor contains cyanides, phenols, thiocyanates, thiosulfates, and other organic constituents. In wet oxidation, cyanides are converted to carbon dioxide and ammonia, thiocyanates are converted to sulfates and ammonia, and thiosulfates are converted to sulfates. In the DOFASCO wet-oxidation process, hot gases and vapors are separated from the liquid stream in the reactor and are used to preheat the incoming waste. The oxidized liquid brine is concentrated by this hot separation and contains approximately 35% ammonium sulfate and excess sulfuric acid. Typical analyses of waste feed and oxidized brine are reported in Table 8.6.3. Crystals of ammonium sulfate can be re-

TABLE 8.6.2 Wet-Oxidation Treatment at Bofors-Nobel

	Toxic component	COD
Feed waste	600–1,200 ppm	70–80 g/L
Wet-oxidation effluent	1–9 ppm	30–40 g/L
Percent reduction	99 +	50 +

TABLE 8.6.3 Wet Oxidation of COG Liquor at DOFASCO*

Component	Waste feed	Oxidized brine
pH	8.4	1.9
COD, g/L	100.1	1.0
NH_4SCN, g/L	94.8	0.008
$(NH_4)_2S_2O_3$, g/L	24.7	—
Total S, g/L	44.7	130.6
$SO_4^{- -}$ − S, g/L	0.9	130.3
NH_3, g/L	25.2	116.9

*Oxidation temperature = 270°C (518°F). Reactor gauge pressure = 9,050 kPa (1,300 lb/in²). Waste feed flow = 69.0 L/min (18.25 gal/min).

covered from the wet oxidized brine and are sold as fertilizer.

The wet-oxidation system at DOFASCO was supplied by Zimpro Inc. It was designed to treat a process flow stream of 56.7 L/min (15 gal/min) and to operate at a maximum temperature of 280°C (536°F) and a maximum gauge pressure of 11,000 kPa (1,600 lb/in²). The wet-oxidation unit is fabricated from various grades of unalloyed titanium.

Northern Petrochemical Company, Morris, Illinois.   Northern Petrochemical Company (NPC) has used wet oxidation since 1981 for treatment of spent caustic scrubbing liquors from an ethylene production facility. The spent caustic scrubbing liquor typically contains cyanides, sulfides, and organic hydrocarbons.[20] Typical oxidation results are summarized in Table 8.6.4

TABLE 8.6.4 Wet Oxidation of Spent Caustic Scrubbing Liquor at NPC*

Component	Waste liquor	Oxidized liquor
pH	1.3	8.1
COD, g/L	32.6	0.7
Sulfide, g/L	12.3	0.00069

*Oxidation temperature = 304°C (580°F) Reactor gauge pressure = 13,000 kPa (1,900 lb/in²). Waste liquor flow = 30.2 L/min (8.0 gal/min).

The oxidized liquor from the NPC wet-oxidation system is directed to an on-site biological treatment plant for final disposal.

The wet-oxidation system at NPC was supplied by Zimpro Inc. The system was designed to treat a waste-liquor flow of 47.2 L/min (12.5 gal/min) and to operate at a maximum temperature of 320°C (608°F) and a maximum gauge pressure of 15,900 kPa (2,300 lb/in²). The wet-oxidation system is fabricated of 304L and 316L stainless steel.

Uniroyal Chemicals Ltd, Elmira, Ontario, Canada.   Uniroyal Chemical uses wet oxidation to detoxify a variety of chemical-production wastewater streams. The detoxified wet-oxidation effluents are then directed to a biological treatment facility for final disposal. The wet-oxidation system was supplied by WetCom Engineering, a firm that previously licensed the Wetox™ system. This particular installation used the Wetox™ wet-oxidation system configuration, which is based on a cascade of completely stirred tank reactors.[21] The system is designed to treat a waste-stream flow of 37.8 L/min (10 gal/min) at a maximum temperature of 250°C (482°F) and a maximum gauge pressure of 16,600 kPa (2,400 lb/in²). The reactor and heat exchangers are fabricated from various materials and alloys, including titanium and Hastelloy.

8.6.7 ECONOMICS OF WET OXIDATION

Capital Cost

Capital costs for wet oxidation depend on several factors such as capacity of the system, oxygen demand of the wastewater, severity of oxidation required to meet the treatment objectives, and materials of construction of the wet-oxidation system. A range of installed capital costs versus wet-oxidation capacity is shown in Fig. 8.6.5. In preparing this cost range, it was assumed that the wet-oxidation unit would operate at 280°C (536°F) and 13,800 kPa (2000 lb/in²), and would accomplish a 50 g/L COD reduction. This cost does not include site preparation, a building, or any ancillary storage tankage.

A detailed comparison of capital costs (1979 dollars) of wet oxidation versus liquid incineration showed capital costs of wet oxidation to be approximately 50% higher.[22] A similar difference is valid at the present time (1988).

Operating Cost

Operating costs for wet oxidation as a function of the capacity of the unit are shown in Fig. 8.6.6. These estimates of operating costs are made using the as-

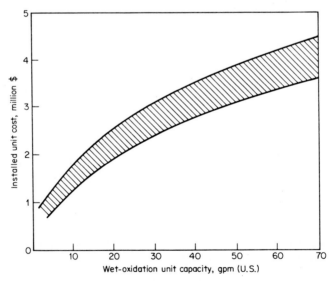

FIG. 8.6.5 Wet oxidation of hazardous waste: installed capital costs vs. unit capacity.

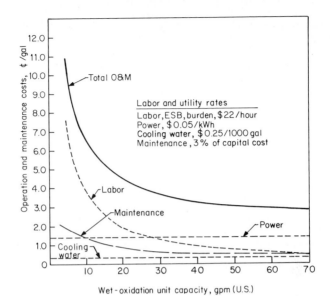

FIG. 8.6.6 Wet oxidation of hazardous waste: O&M costs vs. flow rate.

sumptions that the wet-oxidation unit will operate at 280°C (536°F) and 13,800 kPa (2,000 lb/in² gauge), and will accomplish a 50 g/L COD reduction. The operating cost is greatly affected by the labor element at low unit capacities. Operating experience at hazardous-waste treatment facilities indicates that one dedicated operator is required for the operation of a single wet-oxidation system within the capacity range of 37.8 to 264.6 L/min (10 to 70 gal/min).

Operating costs for wet oxidation and liquid incineration are greatly influenced by the external-energy requirements. With liquid incineration of hazardous wastewaters, it is necessary to supply sensible heat, heat of vaporization for the water, and heat to elevate the water vapor, combustion products, and excess air to the required combustion temperature of 1100 to 1400°C (2000 to 2500°F). This external-heat requirement may be as high as 5580 MJ/m³ (20,000 Btu/gal) for liquid incineration. In contrast, the only energy requirement for wet oxidation is the difference in enthalpy between influent and effluent streams, which is typically 139.5 MJ/m³ (500 Btu/gal). This latter energy requirement is provided by the exothermic heat of oxidation when the oxygen uptake exceeds 15 to 20 g/L COD.[23]

A comparison of operating costs of wet oxidation and liquid incineration indicated that the operating cost for wet oxidation was approximately 30% of the operating cost of liquid incineration.[22] Similar operating costs are valid at the present time (1986), especially for wastewaters of less than 250 g/L COD, which require fuel for liquid incineration but are processed autothermally in wet oxidation.

8.6.8 REFERENCES

1. M. J. Dietrich, T. L. Randall, and P. J. Canney, "Wet Air Oxidation of Hazardous Organics in Wastewater," *Environmental Progress,* **4:**(3)171–177 (1985).

2. T. L. Randall and P. V. Knopp, "Detoxification of Specific Organic Substances by Wet Oxidation," *Journal of the Water Pollution Control Federation,* **52**:(8)2,117–2,129 (1980).

3. J. E. Morgan and C. M. Saul, "The Zimmermann Process in a Soda Pulp Mill Recovery System. Development of a Commercial Process," *APPITA,* **22**(3):60–75 (1968).

4. F. J. Zimmermann, "New Waste Disposal Process," *Chemical Engineering,* August 1958, pp. 117–120.

5. P. V. Knopp and W. B. Gitchel, "Wastewater Treatment with Powdered Activated Carbon Regenerated by Wet Air Oxidation," 25th Purdue Industrial Waste Conference, May 1970.

6. P. N. Cheremisinoff and F. Ellerbusch (eds.), *Carbon Adsorption Handbook,* Ann Arbor Science, Ann Arbor, Michigan, 1978, pp. 547–553.

7. C. R. Baillod and B. M. Faith, "Wet Oxidation and Ozonation of Specific Organic Pollutants," NTIS PB83254060, August 1983.

8. H. R. Devlin and I. J. Harris, "Mechanism of the Oxidation of Aqueous Phenol with Dissolved Oxygen," *Ind. Eng. Chem. Fund.,* **23**(4):387–392 (1984).

9. M. K. Conditt and R. E. Sievers, "Microanalysis of Reaction Products in Sealed Tube Wet Air Oxidation by Capillary Gas Chromatography," *Anal. Chem.,* **56**(13):2,620–2,622 (1984).

10. R. B. Ely, E. M. Pogainis, and C. A. Hoffman, *Catalyzed Process and Catalyst Recovery,* U.S. Patent 3,912,626, October 1975.

11. T. L. Randall, "Wet Oxidation of Toxic and Hazardous Compounds," in *Proceedings of the Thirteenth Mid-Atlantic Industrial Waste Conference,* Newark, Delaware, June 1981, pp. 501–508.

12. J. J. A. Ploos Van Amstel, "The Oxidation of Sewage Sludge in the Liquid Water Phase at Elevated Temperatures and Pressures," doctoral thesis, Technische Hogeschool te Eindhoven, Netherlands, 1971.

13. J. A. Heimbuch and A. R. Wilhelmi, "Wet Air Oxidation—A Treatment Means for Aqueous Hazardous Waste Streams," *Journal of Hazardous Materials,* **12**:187–200 (1985).

14. C. R. Baillod, R. A. Lamparter, and B. A. Barna, "Wet Oxidation for Industrial Waste Treatment," *Chemical Engineering Progress,* March 1985.

15. S. H. Timberlake et al., "Supercritical Waste Oxidation for Wastewater Treatment: Preliminary Study of Urea Destruction," Twelfth Intersociety Conference on Environmental Systems, San Diego, California, July 1982.

16. G. C. Rappe, "New Sludge Destruction Process," *Environmental Progress,* **4** (1985).

17. J. L. McBride and J. A. Heimbuch, "Skid Mounted System Gives California Hazardous Wastes Hot Time," *Pollution Engineering,* July 1982.

18. L. A. Zadonick and A. R. Wilhelmi, "Multi-faceted Approach to Successful Hazardous Waste Treatment," *Pollution Engineering,* April 1984.

19. "Helping Steel Kick the Coke Oven Gas Habit," *The Reactor,* no. 41, Zimpro/Passavant Inc., Rothschild, Wisconsin, February 1979.

20. D. J. DeAngelo and A. R. Wilhelmi, "Wet Air Oxidation of NPC Spent Caustic Liquors," 1982 Spring National AIChE Meeting, Anaheim, California, June 1982.

21. "New Processes Steel Show at Canadian Gathering," *Chemical Engineering,* November 1981, p. 69.

22. A. R. Wilhelmi and P. V. Knopp, "Wet Air Oxidation—An Alternative to Incineration," *Chemical Engineering Progress,* August 1979.

23. P. T. Schaefer, "Consider Wet Oxidation," *Hydrocarbon Processing,* October 1981.

SECTION 8.7
PYROLYSIS PROCESSES

J. K. Shah
Surface Combustion Division
Midland-Ross Corporation
Toledo, Ohio

T. J. Schultz
Surface Combustion Division
Midland-Ross Corporation
Toledo, Ohio

V. R. Daiga
Surface Combustion Division
Midland-Ross Corporation
Toledo, Ohio

Pyrolysis is defined as the chemical decomposition or change brought about by heating in the absence of oxygen. Application of pyrolysis to hazardous-waste treatment leads to a two-step process for disposal. In the first step, wastes are heated, separating the volatile components (combustible gases, water vapor, etc.) from nonvolatile char and ash. In the second step, volatile components are burned under the proper conditions to assure incineration of all hazardous components.

There are several variations of the process that include energy recovery and recovery of valuable components from the volatiles and/or solid residual.

Pyrolysis is especially applicable to hazardous-waste treatment because it provides precise control of the combustion process. Traditional excess-air incineration is an *exothermic* process, operating at temperatures of 800 to 1370°C (1500 to 2500°F). The temperature is controlled by adjusting the waste feed rate and the amount of excess combustion air. This control is very difficult and often results in temperature excursions out of the desirable operating range.

The first step of the pyrolysis treatment of hazardous waste is *endothermic* and generally done at 425 to 760°C (800 to 1400°F). The heating chamber is called the *pyrolyzer*. The endothermic property makes the process much easier to control. Hazardous organic compounds can be volatilized at this low temperature, leaving a clean residue. The volatiles are burned (second step) in a fume incinerator at conditions to achieve destruction efficiency of more than 99.9999%. Separating the process into two very controllable steps allows precise temperature control and makes it possible to build smaller equipment.

The unique features of pyrolysis make it especially applicable for processing the following waste types:

1. Waste stored in containers or drums that cannot be easily drained.
2. Sludges or liquids that contain:
   a. High ash content
   b. Volatile inorganics, e.g., NaCl, $FeCl_2$, Zn, and Pb
   c. High levels of chlorine, sulfur, and/or nitrogen

The thermal treatment during pyrolysis renders most of the solid-waste residues from the pyrolyzer nonhazardous. Additional thermal oxidation to burn out carbonaceous residue may be desirable in some applications. The volume of the solid residue will be considerably less than the volume of the feed material.

8.7.1 PROCESS DESCRIPTION

Pyrolysis includes a broad category of reactions. Sometimes pyrolysis is described as *thermal decomposition* or *destructive distillation;* however, these terms refer to only a part of the sequence of reactions in a pyrolysis process.

Pyrolysis is a thermal process for transformation of solid and liquid carbonaceous materials into gaseous components and a solid residue containing fixed carbon and ash. Pure pyrolysis is accomplished by "indirect heating" to perform the reaction in the absence of air or flue gases. Modified forms of pyrolysis can be used with the direct firing of air and fuel at stoichiometric ratio in a furnace. The selection of indirect or direct heating depends upon the intended use of the flue gases in a heat-recovery process. The gaseous and solid products formed during pyrolysis, gasification, and incineration are summarized in Table 8.7.1. A typical laboratory facility for evaluating the pyrolysis process for waste disposal is shown in Fig. 8.7.1. The radiant tubes are for indirect heating and burners are for direct heating.

When indirect heating is used to drive the pyrolysis process, flue gases from the heating process are exhausted separately from the volatile gases of the pyrolysis reaction. This method may be used to produce a gaseous fuel for use in a separate heat-recovery system. Electrically heated radiation panels, using resistance heaters or induction-heated metallic walls, may also be used to successfully heat the reaction zone without contamination of hydrocarbon gases from the burner's flue products.

Modified pyrolysis with direct firing of the furnace is effective where the heat-recovery unit can be directly coupled to the pyrolysis chamber. Direct firing with external fuel and air or partial oxidation of the hydrocarbons generated is used to furnish the heat for pyrolysis reactions. The sensible heat from both the heating gases and the ultimate burning of the hydrocarbon gases evolved from the pyrolysis process can be recovered in the downstream unit.

TABLE 8.7.1 Products of Pyrolysis and Incineration Processes

Process	Gaseous phase to afterburner	Solid-phase residue	Use
Classic pyrolysis (indirectly heated)	Volatilized & decomposed feed material	Char/Ash	Product recovery
Direct-fired pyrolysis	Volatilized & decomposed feed materials with burner flue products	Char/Ash	Energy recovery and destruction of hazardous waste
Starved-air gasification	Partially oxidized, volatilized, & decomposed feed materials plus burner flue products	Char/Ash	Retard combustion rate
Excess-air combustion	Mostly burned combustion products	Ash	Direct incineration

A number of commercially available incinerators claim to operate in a "pyrolysis mode." This usually amounts to starved-air combustion or incineration with insufficient quantities of air to stoichiometrically combust the organic feed. The term *pyrolysis* can be loosely fitted into this context because some of the feed is decomposed by the action of heat, but much of the feed is incinerated to provide that heat. Hence, starved-air combustion does not offer all the benefits inherent in a true pyrolysis system.

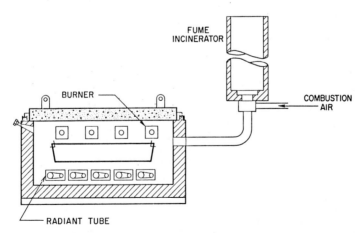

FIG. 8.7.1 Laboratory facility for pyrolysis of solid waste (*Courtesy of Midland-Ross Corporation*)

A two-step approach is generally advantageous for thermal destruction of hazardous wastes. Pyrolysis as one of the steps is followed by fume incineration (see Fig. 8.7.2). Separation of the pyrolysis or endothermic destructive distillation step from the fume incineration or exothermic oxidation reaction makes the system controllable.

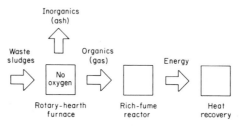

FIG. 8.7.2 Pyrolytic incineration—a two-step process.

Chemical Reactions

The chemical reactions during a pyrolysis process involve thermal decomposition, rearrangement of atoms in a molecule, and polymerization of smaller molecules. These reactions are very complex and depend upon several factors such as the reaction time and temperature, catalytic effect of the container wall, and composition of the waste material. In general, the reaction can be expressed as follows:

$$\text{organic waste} \xrightarrow{\Delta H} \text{volatiles (organics and } H_2O) + \text{ash} \\ + \text{char (nonvolatile carbonaceous)}$$

The progress of the heterogeneous reactions during the pyrolysis process can be followed by use of a thermal gravimetric furnace as shown in Fig. 8.7.3.

Operating Conditions

The thermal gravimetric type of apparatus permits determination of time-temperature requirements for the pyrolysis process. The degree of decomposition as a function of time at 427°C (800°F) for various materials, as measured by a thermal gravimetric furnace, is presented in Fig. 8.7.4. The rate of thermal decomposition increases with increase in temperature, as shown in Fig. 8.7.5 for polypropylene decomposition.

The operating conditions for a pyrolyzer are based on a trade-off analysis of capital and operating costs. Time and temperature requirements are also governed by the thickness of the waste-material bed. The equipment size depends upon processing rate, bed thickness, and residence-time requirements. Operating cost, as measured by fuel consumption, increases with increasing temperature, but capital cost decreases, in many cases, because smaller equipment is needed to operate at higher temperature.

Operating temperatures in a pyrolyzer can range from 427 to 748°C (800 to 1400°F) for different wastes. Residence times can vary from a few minutes to a

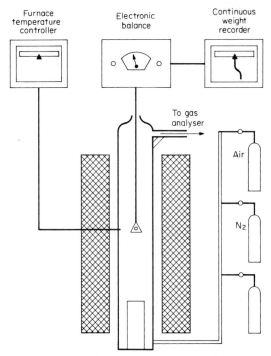

FIG. 8.7.3   Thermal gravimetric furnace to study heterogeneous reactions (*Courtesy of Midland-Ross Corporation*)

few hours, depending upon bed thickness. The pressure in a pyrolyzer is typically negative (about 0.05 to 0.1 in w.c.) to avoid leakage from the pyrolyzer. The system's oxygen concentration should be near zero and should be controlled so as not to exceed a 2% level.

Types of Wastes

The pyrolysis process can be applied to solids, sludges, and liquid wastes. Wastes with the following characteristics are especially amenable to pyrolysis:

- Sludge material which is either too viscous, too abrasive, or that varies too much in consistency to be atomized in a liquid incinerator.
- Wastes which undergo partial or complete phase changes during thermal processing, such as plastics.
- High-residue materials, such as high-ash liquids and sludges, with light, easily entrained solids that would generally require substantial stack-gas cleanup.
- Materials that contain salts or metals which melt and volatilize at normal incineration temperatures. Materials like NaCl, zinc, and lead, when incinerated, may cause refractory spalling, fouling of the heat-exchanger surface, and submicron aerosol emissions.

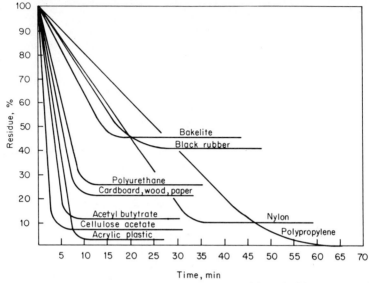

FIG. 8.7.4 Thermal decomposition of waste material at 800°F (*Courtesy of Midland-Ross Corporation*)

Effluent Treatment

Pyrolysis is an intermediate process step in thermal treatment of hazardous waste. It generates a gaseous stream and a solid residue. Environmental protection regulations require that all effluents from a hazardous-waste treatment facility be safely disposed.

The gaseous effluents from a pyrolyzer should be combusted in a fume incinerator with a destruction efficiency required for the particular waste. The products of incomplete combustion (PICs) and principal organic hazardous constitu-

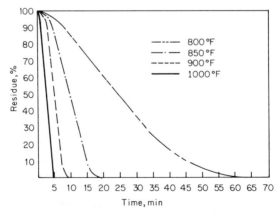

FIG. 8.7.5 Thermal decomposition of polypropylene at various temperatures (*Courtesy of Midland-Ross Corporation*)

ents (POHCs) in the effluent from the fume incinerator should be limited to concentrations below the allowable emission levels. This can be achieved by selecting proper operating temperature and residence times in the fume incinerator, with the proper oxygen concentration and efficient mixing of fumes and air.

Particulate emissions from pyrolysis systems are normally expected to be very low, and particulate emission-control equipment is often not necessary. Particulate-loading and plume-opacity measurements should be carried out during test burns to assure that no particulate emission-control equipment is necessary. If particulate-control equipment is installed, then the solids collected should be disposed of according to environmental regulations.

The effluent gases from a fume incinerator should be chemically scrubbed if they contain acid gases such as SO_2, NO_X, or HCl in concentrations above the permitted levels. The spent scrubbing media should be disposed of according to environmental regulations.

The solid residue from the pyrolyzer should be analyzed for toxicity. If it is determined to be nontoxic, it can be disposed of in a conventional landfill. Toxic solid residue should be disposed of in a properly regulated landfill.

8.7.2   EQUIPMENT DESIGN

All pyrolysis equipment should be designed to maintain an oxygen-deficient atmosphere. Good seals are key to minimizing leakage of air into the pyrolyzers. Operating with positive pressure in a pyrolyzer to prevent infiltration of air is normally unacceptable since it could result in leakage of fumes into the ambient environment. Sealing of doors, rotary joints, and feed inlets is important in pyrolyzers. Water seals in horizontal planes and ceramic-fiber seals coupled with inflatable elastomeric seals are some examples of good seal design in practice today.

Pyrolyzers heated by direct firing should be equipped with burners capable of operating under stoichiometric conditions from full firing to lowest firing range during process operation.

Types of Equipment

Pyrolysis processes can be carried out in a variety of furnace designs, such as:

1. Rotary hearth
2. Rotary kiln
3. Roller hearth
4. Roller rail
5. Car bottom

Pyrolyzer furnaces can either be direct-fired or they can be indirectly heated by radiant tubes or electrical heaters. Induction heating of a metallic cylindrical furnace can also be used for indirectly heated pyrolyzers. A comparison of direct and indirect heating is presented in Table 8.7.2.

Rotary Hearth.   The *rotary-hearth* pyrolyzer is normally used for processing of sludges or granular solids. It has a doughnut-shaped hearth rotating in a stationary furnace chamber. Waste material is continuously fed in a cold zone. A sta-

TABLE 8.7.2 Comparison of Direct and Indirect Heating Modes

Direct	Indirect
Overall heat-transfer mechanisms are more efficient.	Intermediate heat-transfer steps are required.
Cannot independently vary pyrolyzer/temperature in different locations. Limited to single operations.	Can independently vary pyrolyzer temperatures in several locations. Utilizes multizone operation.
Wall conduction not required for heat transfer. Can insulate on ID* and operate at high temperature (about 1193°C or 2200°F max.).	Wall conduction required for heat transfer. Walls cannot be insulated. Pyrolyzer temperature limited by wall alloy selection (about 748°C or 1400°F max.).
Insulated metal wall will not undergo significant thermal expansion.	Hot metal walls will exhibit significant thermal expansion and will require both axial and radial accommodations.
Cool metal wall will retain its strength. It is not subject to significant thermal stress.	High temperature will degrade strength of metal wall. Heavier wall construction to enhance strength may aggravate thermal stress.
Pyrolyzer exhaust includes all burner products of combustion and evolved volatiles.	Pyrolyzer exhaust consists of evolved volatiles only.

*ID = inside diameter.

tionary spreader allows feed material to be distributed uniformly in a radial direction as the hearth rotates under the spreader. The feed is passed through a hot zone as the hearth rotates, while solid residue is continuously discharged by a screw conveyor, as shown in Fig. 8.7.6.

Rotary-hearth pyrolyzers may be operated in a semicontinuous mode. Small trays with waste material are fed through air locks and processed trays are discharged through air locks. Rotary-hearth pyrolyzers are simple to operate and are low-maintenance operations. Floor-space requirements may be somewhat higher than for other continuous operations because of circular shape and horizontal orientation.

Rotary Kiln.   *Rotary-kiln pyrolyzers* consist primarily of a horizontal rotating cylinder with the load to be heated tumbling on the inside, as shown in Fig. 8.7.7. Waste material to be processed is introduced at one end of the kiln and rolls slowly down the length of the kiln after many rotations.

Transport of the work, usually loose solids or sludges, can be assisted by angling the axis of rotation of the kiln a few degrees so that the charge end is higher than the discharge end. Work may be conveyed through the kiln at a closely controlled rate by installing a helical auger on the kiln's inside diameter, transforming the kiln into a sort of Archimedes' screw.

Rotary kilns can be directly heated by firing gas or oil burners into the interior or indirectly heated by heating the kiln's outside diameter with burners, radiant elements, or induction coils. The burners of a direct-fired rotary kiln can be positioned at either end to provide cocurrent (same direction as work) or countercurrent (opposite direction of work) gas flow. A comparison of these two modes of operation is summarized as follows:

Countercurrent firing	Cocurrent firing
Work heated almost to firing-zone temperature.	Work heats almost to exhaust-zone temperature.
Fines and volatiles exhaust out charge end with little residence time.	Fines and volatiles exhaust out discharge end ensuring longer residence time.

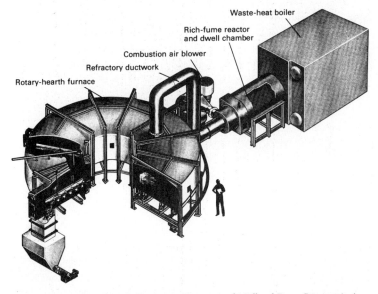

FIG. 8.7.6 Rotary-hearth pyrolyzer (*Courtesy of Midland-Ross Corporation*)

Rotary-kiln pyrolyzers are normally operated in a countercurrent mode. They are used for continuous processing of sludges and granular as well as chunky materials. Liquids may be processed in a rotary kiln with metallic walls and internal flights. Conventional seal design for a rotary kiln may require regular maintenance to assure oxygen-deficient atmosphere.

Roller Hearth. The *roller-hearth* pyrolyzer is used for processing of containerized feed. Feed can be sludges, solids, or liquids in containers. Discrete feed enters the furnace at one end at regular time intervals, and processed material discharges at the other end (see Fig. 8.7.8). Roller-hearth furnaces have rollers instead of rails. The rollers may be free or driven. Free rollers require a separate advancing mechanism, such as a pusher mechanism. Externally driven rollers move the load through the furnace. Containers should be opened to allow fumes to escape. Sealing of doors and rollers is an important feature of this type of furnace for pyrolysis.

Roller-hearth pyrolyzers have many moving parts inside the furnace. These parts are of metallic construction and therefore may have higher corrosion and wear rates than refractory materials. The roller-hearth furnaces require regular maintenance to maintain proper integrity for pyrolysis application. The advantage of a roller hearth is its length. They can be made longer for increased production rates.

FIG. 8.7.7   Rotary-kiln pyrolyzer (*Courtesy of Midland-Ross Corporation*)

FIG. 8.7.8   Roller-Hearth pyrolyzer for thermal disposal of containerized toxic material.

Roller Rail.   The length of the *roller-rail* pyrolyzer is fixed by material handling, which is achieved from charge- and discharge-end fixtures external to the furnace. This feature eliminates driven rollers and, therefore, maintenance of the operation is significantly reduced. Processing of containerized feed is in a batch mode. A door seal separating the furnace from the environment is important.

Car Bottom.   Batch processing of containerized feed in pyrolysis mode can be accomplished in a *car-bottom furnace,* as shown in Fig. 8.7.9. The car with a load is moved inside the furnace chamber and is lifted to seal against the furnace wall. Loading and unloading is done through the same door. A hood can be provided in front of the furnace to ventilate vapors from the load when containers are opened. The ventilated air can be used as combustion air in the fume incinerator.

Capacities and Constraints

Pyrolyzers are normally designed for capacities as shown in Table 8.7.3. Maximum capacity in a single process unit is sometimes constrained by equipment-manufacturing limitations. Doughnut-shaped rotary-hearth furnaces are designed for a maximum outside diameter of 7.5 m to 9 m (25 to 30 ft). Rotary kilns for pyrolysis processes are normally limited to diameters up to 1.8 m (6 ft). Roller-hearth furnaces may be longer (as an example: 12 to 15.5 m or 40 to 50 ft in length), but roller-rail furnaces are limited to 3.6 to 4.5 m or 12 to 15 ft in length.

The pyrolyzers are connected to fume incinerators to combust and destroy the gaseous products formed during pyrolysis. The capacity of the fume incinerator depends on the mode of operation of the pyrolyzer. If the pyrolyzer is operated in a continuous mode, then the needed capacity of the fume incinerator is determined by the average feed rate of waste to the pyrolyzer. The capacity of the incinerator is fully utilized when the pyrolyzer is operated in a continuous mode at its full capacity. A pyrolyzer operated in a batch mode (discrete feed at regular time interval of several minutes or hours) generates fumes with varying amount of energy content over the entire cycle for the batch load. Fume incinerators are designed for the peak loading to assure that the fumes generated in the pyrolyzer will be adequately combusted in the fume incinerator. The time interval for peak loading of fumes may be only a fraction of the total cycle time of a batch process. In this case, the fume incinerator is not utilized to its capacity over the entire cycle time.

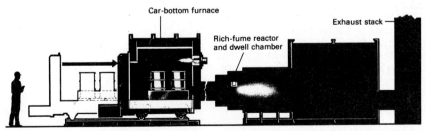

FIG. 8.7.9   Car-bottom pyrolyzer for batch-type thermal disposal process (*Courtesy of Midland-Ross Corporation*)

TABLE 8.7.3 Capacities of Various Types of Pyrolyzers

Type of pyrolyzer	Capacity range	
	kg/h	(tons/h)
Rotary-hearth	450–4,500	(0.5–5)
Rotary-kiln	225–2,700	(0.25–3)
Roller-hearth	900–4,500	(1–5)
Roller-rail	225–900	(0.25–1)
Car-bottom	90–900	(0.1–1)

It is reasonable to operate multiple batch pyrolyzers in parallel connected to a single fume incinerator for increased utilization of the fume incinerator's capacity. This method involves staggering the operation of multiple batch pyrolyzers.

Utility Requirements

Pyrolysis is an endothermic process, so energy input is necessary to heat the waste to the operating temperature. The fuel requirement varies with type of waste and process temperature. Approximate fuel-energy requirements for wastes with increasing amount of water content are summarized in Table 8.7.4. A

TABLE 8.7.4 Fuel-Energy Requirements for Wastes as Related to Water Content

Water in waste, wt %	Fuel input per weight unit of waste	
	$J/kg \times 10^6$	(Btu/lb)
10	1.3	(550–700)
25	1.7	(750–950)
50	2.4	(1050–1200)

heat-recovery system such as a waste-heat boiler can recover most of the sensible heat from both the heating gases in the pyrolyzer and the ultimate burning of the hydrocarbon gases generated by the pyrolysis process. Recovered energy, in many cases, is more than the energy input to the system.

Materials of Construction

Hazardous-waste materials may contain sulfur, chlorine, alkali metals, and other components which are corrosive to metallic and refractory materials used in furnace construction. Materials of construction should be selected to provide reasonable life expectancy. Past experience under similar environmental conditions is the best basis for selection of material. Corrosion testing with coupons of potential materials under simulated conditions can be used for material selection when operating data are limited.

Pyrolyzers are operated under reducing conditions. It is advisable to use corrosion-resistant, high-temperature alloy steels for metallic components in contact with the corrosive, reducing (oxygen-deficient) atmosphere in a pyrolyzer.

8.7.3 ENGINEERING ECONOMICS

Pyrolytic systems are normally custom-designed to treat a specific waste. Capital cost and operating cost can vary with the type of waste to be processed and the degree of process sophistication required. Computerized control and remote material handling are some features being used in current systems for highly toxic wastes.

Operating costs normally decrease with increasing capacity, as labor requirements are generally fixed for a given range of operation. The operating cost for labor, utilities, and maintenance items should be considered in the economic analysis. The operating cost will be higher if a scrubber system is required.

8.7.4 ADVANTAGES AND DISADVANTAGES

Pyrolysis processes for hazardous-waste disposal systems have attained attention due to several advantages. These advantages may be listed as follows:

1. Pyrolysis is a low-temperature process (compared to incineration), thereby increasing refractory life and reducing maintenance.
2. Entrainment of particulates is reduced and the need for particulate emission-control equipment is thereby lessened.
3. Pyrolysis is controllable because of its endothermic nature.
4. Heterogeneous solid or liquid waste can be homogenized into a high-heating-value gaseous stream by pyrolysis for controlled burning.
5. Recoverable constituents are concentrated in the solid residue or char for recovery.
6. Volume of waste is significantly reduced.
7. Condensible vapors with economic value can be recovered.
8. Noncondensible combustible vapors can be used as a source of energy.

There are some disadvantages to the pyrolysis processes, as noted:

- Some of the energy content in a waste feed may be retained by the char.
- Fume incineration is required to destroy PICs, POHCs, or carcinogens present in the waste.

8.7.5 FURTHER READING

Deneau, K. S., *Pyrolytic Destruction of Hazardous Waste*, Midland-Ross Corp., Toledo, Ohio, July 1981.

Eggen, A., and O. A. Powell, "Experience in Slagging Pyrolysis Systems," *Incinerator and Solid Waste Technology 1962–75,* ASME, New York, 1975.

Elghossain, M., and T. R. Parr, *Pyrolysis of Waste Plastics to Useful Energy,* Final report for the period August 29, 1983 to December 31, 1984; December 1984.

Frankel, I., N. Sanders, and G. Vogel, "Survey of the Incinerator Manufacturing Industry," *Chemical Engineering Progress,* March 1983, p. 44.

Hammond, V. L., et al., *Pyrolysis—Incineration Process for Solid Waste Disposal,* Battelle, Richland, Washington, December 1972.

Hemsath, K. H., et al., *Experimental Investigations of Liquid and Solid Waste Incineration Parameters,* Surface Combustion Division, Midland-Ross Corp., Toledo, Ohio, February 1971.

Milkovich, J. J., "What's Happening to Pyrolysis?" *Pollution Engineering,* March–April 1972, p. 44.

Schultz, T. J., *Disposal of Highly Toxic Substances Using Advanced Technology,* Midland-Ross Corp., Toledo, Ohio, June 1980.

"Solid Waste Disposal," *Chemical Engineering,* Deskbook Issue, June 21, 1971, p. 155.

Weismantel, G. E., Sludge Pyrolysis Schemes Now Head for Tryouts, *Chemical Engineering,* December 8, 1975, p. 90.

Witt, P. A., Jr., "Disposal of Solid Wastes," *Chemical Engineering,* October 4, 1971, p. 62.

SECTION 8.8
OCEANIC INCINERATION

Donald G. Ackerman, Jr.
Scientific Director
SITEX Environmental Inc.
Research Triangle Park, North Carolina

Ronald A. Venezia
Vice President
SITEX Environmental Inc.
Research Triangle Park, North Carolina

Incineration of hazardous wastes on land is a relatively familiar process. It is less widely known that hazardous wastes are incinerated at sea. Oceanic incineration is a routine practice in European waters. In recent years, about 100,000 t of liquid hazardous wastes, mostly organochlorines, have been incinerated annually by several vessels.[1] One American company operates two incinerator ships and has applied for research and operating permits in U.S. waters. Another American company, Tacoma Boatbuilding Co., started construction of two incinerator ships to be operated under the U.S. flag. Neither ship has been finished and there are no plans to do so in the near term. Routine operation in U.S. waters is not occurring because the U.S. Environmental Protection Agency (EPA), which has lead federal agency regulatory authority, has not promulgated final regulations and issued operating permits.

Oceanic incineration of liquids is a straightforward adaptation of land-based technology. All oceanic incineration systems to date have been vertical liquid-injection units, but there have been several conceptual designs for horizontal liquid-injection units. One vessel was designed to and did burn liquids and solids. However, the basic design was not very effective and the vessel never became operational. Several conceptual designs have involved placing rotary kilns on vessels for incinerating solids.

A large amount of testing has been performed on incinerator ships in U.S. and European waters. The four nations that have sponsored such tests [Federal Republic of Germany (F.R.G.), France, The Netherlands, and the United States] have found the technology to be without adverse environmental effects and an acceptable method for disposing of hazardous wastes.

This section describes oceanic incineration: suitable wastes, the technology, environmental effects, results of testing, and costs.

8.8.1 APPLICABILITY

Suitable Wastes

Hazardous wastes were classified on the basis of elemental composition,[2] and examination of chemical waste streams showed that those suitable for incineration fell into four classes of generalized compounds, each containing the following: (1) —CH and —CHO, (2) —CHN and —CHON, (3) —CHCl and —CHOCl, and (4) —CHNCl, —CHF, —CHS, —CHX (X = halide), —CHOP, and —CHOSi.

Wastes burned at sea have been almost exclusively organochlorines because they are more costly to dispose of on land.[3] Specific types of wastes burned at sea have included distillation ends from manufacture of vinyl chloride, Herbicide Orange, and mineral-oil dielectric fluid containing polychlorinated biphenyls (PCBs). Other classes of compounds such as hydrocarbons or oxygenated organics (e.g., alcohols, acids, or ketones) have usually been disposed of on land. However, as stated above, there is no technical reason to exclude wastes containing these compounds from incineration at sea.

Physical Forms

Currently, only pumpable liquid wastes are suited for oceanic incineration, as existing vessels are not designed to burn solids. Organic solids may be suitable if they are soluble or can be dispersed in liquids. Wastes may be "wet," emulsions in water, or even contain layers of water. In order to maintain adequate combustion temperatures, wastes of very low or negligible heat content, such as contaminated water or perchlorinated organic compounds, must be cofired or blended with wastes of high heat content or with fuel oil.

8.8.2 DESCRIPTION OF THE TECHNOLOGY

Maritime Technology

Several features of maritime technology are important, and all are reflected in international and federal regulations. First, incinerator ships are treated as chemical tankers, and they are required by the International Maritime Organization (IMO) to be of Type II construction (double hull, double bottom, and subdivided), which provides significant protection against the breaching of cargo tanks during collisions or groundings and reduces the size of a potential spill from such an accident. Second, each tank must be equipped with level indicators and safety devices in order to prevent overfilling. During loading, provisions are made to prevent escape of vapors by connecting the tank vents either to the storage facility or to adsorbent cartridges containing activated carbon. Third, vessels are prohibited from carrying their own pumping equipment for loading tanks, so that loading or unloading can be accomplished only when a vessel is in port. Fourth, the layout and interconnection of cargo piping must be such that wastes taken on board can be disposed of only through the incinerators.

Incineration Technology

Design. Oceanic incinerators have been straightforward adaptations of land-based technology. The principal difference between oceanic and land-based incinerators is that mounting an incinerator in a ship requires compactness in both envelope and height. Space on a ship is much more limited than on land, and tall, heavy stacks could cause problems with stability. These additional design constraints are not difficult to meet. Another major difference is that scrubbers are not required on oceanic incinerators because the sea has a buffering capacity more than adequate to neutralize acidic emissions and because incineration sites are located far enough from land that the emissions will not degrade buildings or other structures or adversely affect human health and welfare.[4]

There have been five operational oceanic incineration ships, all equipped with vertical liquid-injection incinerators. There was a sixth vessel, the Matthias III, which was designed to burn liquids and solids. However, it did not pass a certification test, and was scrapped. As might be expected, different owners have opted for different numbers, sizes, and configurations of incinerators and burners. Two of these vessels, Matthias I and II, are no longer operational. The other three, Vesta and Vulcanus I and II were operational as of 1988, and extensive descriptions of the Vulcanus I and II are given in Ref. 5. Table 8.8.1 presents data on the latter three vessels and the nearly complete Apollo I and II.

Figure 8.8.1 is a photograph of the Vulcanus II, which represents the state of the art in maritime incineration technology. The small white discharge above the stacks is ammonium chloride, which is formed by the reaction of HCl in the stack gas with ammonia that is injected near the exit plane of one stack. The plume from the stacks is not usually visible. Consequently, ammonia is injected periodically in order to make the plume visible. During several tests sponsored by the EPA, it has been necessary that the plume be visible so that airplanes or research vessels could sample it. Periodically injecting ammonia was considered to be a cost-effective and environmentally inoffensive technique.

Much of the discussion on technology is derived from the authors' experience and published technical reports on the Vulcanus I and II. There are few technical reports on the Vesta and, of course, none on the Apollo I and II. Except as otherwise noted, the following discussion is based on the Vulcanus II.

The incinerators on the Vesta and Vulcanus I and II are virtually identical in all dimensions, having been designed and built by Saacke AG, Bremen, F.R.G. Figure 8.8.2 is a schematic of an incinerator on the Vulcanus II.[5] Each unit has two main sections, a combustion chamber and a stack, which are connected by a converging section. As is typical for most high-intensity combustion systems, the stacks are narrower than the combustion chambers. This design effectively fills the entire volume of the combustion chamber with flame, increasing both the time that combusting wastes remain in the flame zone and the velocity of the exit

TABLE 8.8.1 Technical Data on Existing Incinerator Ships

Parameter	Vesta	Vulcanus I	Vulcanus II	Apollo I & II
Vessels				
Owner/operator	Lehnkering AG	Ocean Combustion Service, BV	Ocean Combustion Service, BV	—
Number of tanks	9	8	8	12
Max. volume, m^3	1,100	3,176	3,161	5,600
Max. capacity, t	1,430	4,128	4,110	7,280
Length, m	72	102	94	112
Width, m	11	14.4	16	18
Year constructed	1979	1972*/1983†	1982	1985‡
Per incinerator				
Type	Bottle	Bottle	Bottle	Bottle
Number on ship	1	2	3	2
Overall height, m	10.74	10.5	10.98	11.28
OD,§ combustion chamber, m	5.97	5.5	5.97	6.40
OD, stack, m	3.47	3.77	3.47	3.66
Height of stack, m	4.0	3.69	4.0	2.90
Number of burners	3	3	3	4
Air feed rate, max., m/h	90,000	90,000	90,000	64,000
Max. capacity, t/h	12	12	12	12.5
Manufacturer	Saacke, AG	Saacke, AG	Saacke, AG	Coen, Inc.

*The year when the Vulcanus was converted to an incinerator ship.
†The year when the cargo section was replaced.
‡The Apollo I is about 95% complete and the Apollo II about 65%.
§OD = outside diameter.

gas, and thus effectively eliminating the possibility of entrainment of air into the stack and the effects of wind on the incineration process. This design, narrowing to the exit, has been termed "bottle" to distinguish it from the "cup" type used on the Matthias I, II, and III, which was wider at the top than the bottom.[6]

Each incinerator has three rotary-cup burners, located at the same level on the periphery and near the base of the combustion chamber. Combustion air is supplied to each burner and also aids the atomization process. These burners are designed to produce relatively short and very turbulent flames. The burners are directed tangentially at a circle about the vertical axis of the incinerator to impart swirling motion and better mixing.

Each burner can be fed with waste or fuel oil but not both simultaneously. Fuel oil is used to heat the incinerators to operating temperature and to cool them in a controlled manner and also may be burned through one or two burners to incinerate a waste of very low heat content. The feed rate of waste is varied in order to maintain an approximately constant heat release rate.

FIG. 8.8.1 Photograph of the Vulcanus II.

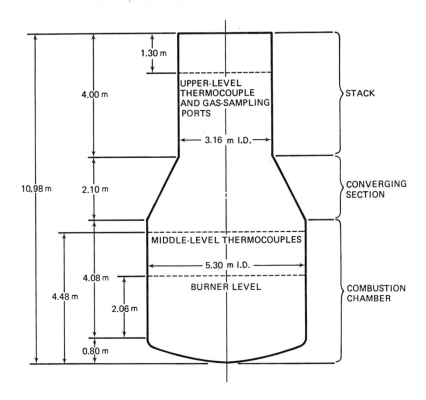

FIG. 8.8.2 Schematic of an incinerator on the Vulcanus II.

The incinerators on the Apollo I and II are also identical, having been designed and built by the U.S. firm Coen Company, Inc. They also are of the "bottle" type, and each has four burners. Atomization is provided by high-pressure air, and secondary combustion air is provided around each burner.

Each of the Matthias vessels mounted one incinerator of the "cup" type. This feature of the design made them less efficient than those of the Saacke design because the wide exit allowed greater radiant heat loss, reducing the combustion temperature and the velocity of the exiting gas, and thus allowing outside air to be entrained into the stack and the wind to affect the process. The burners also used high-pressure air to atomize the waste, and secondary combustion air was fed around each burner. The number of burners ranged from 8 on one level on the Matthias I to 10 on each of two levels on the Matthias III.

Materials of Construction. The incinerators designed by Saacke AG are lined with a high-alumina-content refractory suited to withstand temperatures up to 1700°C (3100°F) and attack from the acidic gases. They are mounted on rollers in order to prevent damage to the refractory from motion of the ship and to allow an entire incinerator to be lifted out of a ship for easy maintenance. All cargo tanks, pumps, and piping are made of low-carbon steel.

Process Monitoring and Control. International regulations require that the following operational parameters be measured and recorded: status of operation of burners and pumps, feed rates of waste and air, temperature, and concentrations of O_2, CO_2, and CO. These records must be permanent and tamper-proof. Time and date, position of the ship, and speed and direction of the wind are recorded manually. All records are turned over to the cognizant regulatory authority at the end of each voyage.

Feed rate of waste is measured by a flow meter on each incinerator. Control is manual. Feed rate of combustion air is monitored by meters in each air supply duct. Control of air feed rate is automatic (Vulcanus II) or manual (Vesta and Vulcanus I). Temperatures at several points in each incinerator are measured by thermocouples mounted in the walls. Control of temperature is effected by thermostats which, as required by international regulations, allow waste to be fed only when wall temperatures are above 1200°C (2192°F) and which shut off the flow of waste if wall temperatures drop below 1100°C (2012°F). Based on past testing, flame temperatures will be about 275°C (495°F) higher than wall temperatures at the level of the burners.[7,8] Concentrations of O_2, CO, and CO_2 in the stack gas are measured by continuous monitors. International technical guidelines require that a minimum of 3% excess oxygen be maintained to assure complete combustion. Combustion efficiency can be determined continuously from measurements of CO and CO_2 and is a useful surrogate for destruction efficiency, which cannot be measured directly during operations. Each of the monitoring devices mentioned provides a signal to a control panel and data-logging system.

Physically, the life expectancy of an incinerator ship is not much different from that of other vessels. Although stack gases from incinerator ships are not scrubbed, these acidic emissions do not create an on-board corrosion problem because the vessels are maneuvered to keep the exhaust plumes from impinging on them. The Vulcanus I has been in operation since 1972, when she was constructed by modifying an existing chemical tanker. Apart from routine maintenance on the incinerators and routine maritime maintenance, the only major work on the Vulcanus I occurred in 1978 to rearrange the cargo piping, remove residues of Herbicide Orange, and replace each plating; and in 1982, when the cargo section of the vessel was replaced. Replacement of cargo tanks every 10 years or so is normal for package chemical tankers engaged in the bulk-chemical trade.[9]

Other Technologies. As of mid-1986 two firms have proposed to enter the field of oceanic incineration and employ somewhat different technologies from that of existing ships.[10] Only very general information was available to the authors because of the preliminary nature of these proposals. SeaBurn Inc. of Greenwich, Connecticut, proposed to use a barge that can hold up to 144 standard intermodal tank containers (5,000-gal tanks mounted in metal frames). Environmental Oceanic Services (EOS), Seattle, Washington, proposed to lease a ship onto which it will load up to 16 intermodal tank containers. Tanks would be filled with liquid wastes at the point of generation, moved to the port by truck or rail, and then loaded by crane.

Both organizations proposed to use horizontal liquid-injection incinerators designed by John Zink Company, Tulsa, Oklahoma. SeaBurn Inc.'s barge will mount four units and EOS's vessel one. The incinerators will use air-atomization burners and have residence times of about 2 s. Each will be equipped with a water-injection section at the exit, designed to eliminate the thermal lift of the plume. These are not designed to be scrubbers.

The feasibility of incinerating solids on-board ships was investigated, and it was concluded that only the rotary kiln was readily adaptable to shipboard use.[11,12] Special mountings and seals necessary to adapt rotary kilns to shipboard use were not considered to present a difficult engineering problem.

8.8.3 ENVIRONMENTAL EFFECTS

Oceanic incineration has been shown not to cause measurable or adverse environmental effects when conducted in accordance with international regulations. The literature on the testing of oceanic incineration is extensive.[5,6] American[13,14,15] and European[16] tests of the marine environment have failed to measure changes in water quality (e.g., pH, chlorinity, metals), unburned wastes in the water (e.g., organochlorines, PCBs), effects on biota exposed to emissions at the point of impact of the plume, or unburned wastes in the exposed biota.

Environmental impact statements have been prepared for designating the Gulf Coast[17] and North Atlantic[18] Oceanic Incineration Sites, adopting amendments to Annexes I and II of the London Dumping Convention,[19] a U.S.-flagged incinerator-ship program,[20] an incineration system mounted on an offshore oil platform,[21] and the oceanic incineration of Herbicide Orange.[22] Comparative environmental and cost assessments have been made for oceanic and land-based incineration of organochlorine compounds.[31,32] An environmental assessment of the proposed oceanic incineration of Silvex herbicides was made.[24] In all of these studies, calculations using plume-dispersion models and results of stack tests have shown that concentrations of emitted species will be below levels that cause adverse effects. Recently, the EPA conducted a detailed risk analysis which showed that the risk of spills and loss of cargo from collision, groundings, and so forth was extremely small. The cumulative risk of loss of some part of the cargo from three or more tanks of an incinerator ship in all locations in the Gulf of Mexico was estimated to be once in 1,200 years of 14 voyages per year.[25]

8.8.4 RESULTS OF TESTING

Organochlorines on the Vulcanus II

In order to obtain certification to operate internationally, Vulcanus II was tested in European waters. The first tests were performed in January 1983 by The Netherlands Organization for Applied Scientific Research (TNO) on heavy distillation ends from the manufacture of vinyl chloride monomer; Tests were performed by TNO using their standard method[26] and EPA-approved methods[27,28] for semivolatile organic compounds. Results of these tests met all international and U.S. federal requirements and resulted in certification by The Netherlands. Averages of operational data are given in Table 8.8.2. Table 8.8.3 presents the composition of the waste and the destruction efficiencies measured.

Chemical Waste Management, Inc., contracted with TRW Inc., Redondo Beach, California, to perform additional tests using EPA-approved methods in February of 1983, while the Vulcanus II was still in European waters.[29] Light and heavy distillation ends resulting from the manufacture of vinyl chloride monomer were incinerated. A volatile-organic sampling train (VOST) was used to sample the effluent gas for selected volatile organochlorine compounds.[30] Samples were drawn at a fixed point 40 cm past the inner wall.

Averages of operational data are given in Table 8.8.2. Five compounds present in the waste were selected as principal organic hazardous constituents (POHCs),[36] and destruction efficiencies for them are listed in Table 8.8.4. The destruction efficiencies met international and U.S. federal requirements. Carbon tetrachloride does not support combustion and is ranked number 4 in the EPA's Hierarchy of Incinerability.[31] Therefore, its presence at 20.5% in the waste presented a considerable challenge to the incineration system; this challenge was more than adequately met.

TABLE 8.8.2 Average Values of Operational Parameters on Incinerator Ships

Operational parameter	Reference number					
	26	29	33	8	7	37
Combustion, %	99.98	99.99	99.99	99.99	99.97	99.98
Feed Rate, t/h*	13.1	15.8	6.22	7.3	11	
O_2, %	10.3	10.6	10.1	8.9	10.1	10.2
CO_2, %	9.2	9.6	9.1	10.3	9.2	9.2
CO, ppm	4	22	8	10	25	21
Total hydrocarbons, ppm	—	—	—	25	15	7
NO_X, ppm	—	—	—	—	99	—
Excess air, %	98	103	95	94	94	96
Heat content, kcal/kg	—	3130	6030	5550	—	—
Ash content, %	—	0.15	0.10	—	0.15	—
Wall temperature, °C†	1143	1166	1303	1273	1038	1241
Outlet temperature, °C	833	1038	—	—	—	—

*Per incinerator.
†Thermocouples are located in different places in different vessels.
Note: Comparisons between vessels are not meaningful.

TABLE 8.8.3 Results of First Tests of the Vulcanus II

Compound	Concentration in waste, wt %	Destruction efficiency, %	
		TNO's method	EPA's method
Carbon tetrachloride	0.1	>99.9966	—
1,1,2-trichloroethane	38.2	99.9996	99.9995
Tetrachloroethane	0.6	99.9999	99.9934
1,1,1,2-tetrachloroethane	0.6	99.9998	99.9983
1,1,2,2-tetrachlorethane	1.8	99.9982	99.9983
1,2,4-trichlorobenzene	0.3	99.9905	—
Hexachlorobenzene	0.003	99.8321	—

Source: Ref. 2.8.

Incineration of PCBs on the Vulcanus I: 1981–1982

The first oceanic incineration of waste containing polychlorinated biphenyls (PCBs) occurred in U.S. waters during December 1981 and January 1982. The wastes were incinerated in the Gulf Coast Oceanic Incineration Site by the Vulcanus I under a research permit (HQ-81-002) issued by the EPA. Stack-gas emissions were tested by TNO.[32] Several accidents occurred in handling the samples from these tests, causing the loss of two of three sets. Consequently, the EPA decided that insufficient valid data resulted from the test program and authorized a second research burn of PCB-containing wastes under the same permit.

During the second burn, which took place during August of 1982, the EPA sponsored a major research program involving testing both on the ship and in the marine environment. Testing on the ship was performed by TRW Inc.[33] Samples of stack gas were taken using the standard sampling train and procedures,[34] except that sampling was at a fixed point. Fixed-point sampling was used because it had been shown during earlier tests that the stack gases from this ship were well-mixed across the diameter and that traversing according to EPA Method 1 was not necessary.[7,8,13]

Samples of stack gas and feedstock from six of the ten tests were analyzed for PCBs, chlorobenzenes (CBs), tetrachlorodibenzo-*p*-dioxins (TCDDs), tetrachlorodibenzofurans (TCDFs), 2,3,7,8-TCDD, and 2,3,7,8-TCDF. No PCB, CB, TCDD, or TCDF was found in any of the stack-gas samples. Traces of TCDFs were found in the feedstock, as expected. Destruction efficiencies were calculated in a conservative manner using detection limits. Results of these calculations are given in Table 8.8.5. Calculation of destruction efficiency for

TABLE 8.8.4 Results of the Second Test of the Vulcanus II

Compound	Concentration in waste, wt %	Destruction efficiency, %
1,1-dichloroethane	6.8	99.99994
1,2-dichloroethane	7.8	99.99996
1,1,2-trichloroethane	38.9	>99.999995
Chloroform	26.1	99.9996
Carbon tetrachloride	20.5	99.998

TABLE 8.8.5 Results of Incineration of PCBs on the Vulcanus I

Compounds	Concentration in waste, wt %	Concentration in stack gas, $\mu g/m^3$	Destruction efficiency, %
PCBs	27.49	<25.3	>99.9999
CBs	6.87	<4	>99.9999
TCDFs	0.0000048	<0.00014	>99.96
TCDDs	ND*	ND	NM*

*ND and NM mean not detected and not meaningful, respectively.

TCDDs was not meaningful. Averages of operational data are given in Table 8.8.2.

Testing in the marine environment involved sampling ambient air, sampling seawater, and exposing marine organisms in control (no exposure to the plume) and exposed areas of ocean. Results were as follows: Samples of ambient air in control areas showed no detectable PCBs or other organochlorine compounds and only traces of nonchlorinated compounds. Samples of air from the plume showed no detectable PCBs or other organochlorine compounds.[14] Acidities and salinities of water samples were the same in control and plume-impact areas. Polychlorinated biphenyls were not detected in samples from control and plume-impact areas of surface water, in collectors placed on cages holding animal specimens, in samples of neuston, or in one of the exposed species of fish. Polychlorinated biphenyls were detected in the other species of fish, but all fish of that species, which had been collected in Galveston Bay, were contaminated with PCBs before the testing started. The three animal species also were analyzed for changes in metabolic activity resulting from exposure to the plume, but no effects were found. The investigators concluded that no environmental effects of the incineration were detected.[15]

Herbicide Orange on the Vulcanus I

During the period of July to September 1977, the Vulcanus I burned three loads of Herbicide Orange, a total of 10,400 t or 2.31 million gal.[8] Herbicide Orange was an equivolume mixture of the *n*-butyl esters of 2,4-dichlorophenoxyacetic acid (2,4-D) and 2,4,5-trichlorophenoxyacetic acid (2,4,5-T). It was contaminated with an average of about 2 ppm of 2,3,7,8-TCDD. The operation took place under permits issued by the EPA in an EPA-designated incineration site in the Pacific Ocean about 120 mi (193 km) west of Johnston Atoll.

Each of the three burns was extensively tested and monitored in accordance with a test plan approved by the EPA and the U.S. Air Force.[35] Stack-sampling operations utilized a benzene-filled impinger train and a modified EPA Method 5 (MM5) train and standard procedures. A total of 52 samples of stack gas was taken during the three burns. Stack samples were acquired by a remotely activated, water-cooled, stainless-steel probe capable of traversing the starboard stack diameter of 3.4 m. Ten points along one diameter of the starboard stack were selected by EPA Method 1. Although the probe could only reach the eight points closest to the port, the EPA determined that this deviation from the procedure would have a negligible effect on the quality of the data.

On-line monitoring data taken during a traverse across the starboard stack

TABLE 8.8.6 Results of Incineration of Herbicide Orange

Test	Emission concentration of 2,4-D or 2,4,5-T, $\mu g/m^3$	Destruction efficiency, %	Emission Concentration of 2,3,7,8-TCDD, ng/m^3	Destruction efficiency, %
1	<7	>99.9999	<24	>99.9
2	<3	>99.9999	<112	>99.9
3	<130	>99.9999	<106	>99.9

showed that the composition (i.e., total hydrocarbons, O_2, CO, and CO_2) was invariant across the stack at distances greater than 10 cm from the inside wall. This result demonstrated that sampling of effluent gas could be determined using a fixed-position probe.

Samples taken on board the ship were analyzed extensively in the laboratory using standard methods.[36] Neither 2,4-D, 2,4,5-T, nor TCDD was detected in the samples. Table 8.8.6 presents the destruction efficiencies. Averages of operational data are given in Table 8.8.2.

Other Tests

Tests of the Vesta in 1979. After being commissioned, the Vesta was tested by the Umveldtbundesamt, the cognizant environmental agency in the F.R.G., during September, 1979.[37] The tests showed that combustion and destruction efficiencies were in accordance with international requirements. Average percentage destruction efficiencies for the following compounds were: benzene, >99.95; chlorobenzene, >99.98; dichlorobenzenes, >99.99; hexachlorobenzene, >99.96; and hexachlorobutadiene, >99.98. Because chlorobenzenes were present in the waste, the stack-gas samples were analyzed for TCDDs, but they were not detected. Averages of operational data are given in Table 8.8.2.

Tests on Shell Chemical Company Waste in 1974–1975. The first officially sanctioned oceanic incineration in U.S. waters was performed on the Vulcanus I between October 1974 and January 1975 in what is now the Gulf Coast Incineration Site. Wastes from Shell Chemical Company's manufacturing complex at Deer Park, Texas, were burned. They were a mixture of organochlorine compounds consisting chiefly of trichloropropane, trichloroethane, and dichloroethane. The EPA coordinated an extensive sampling effort on stack-gas emissions and the marine environment.

Destruction efficiencies of the major compounds averaged 99.95% and ranged from 99.92 to 99.98%. Samples of water taken in the vicinity of plume touchdown showed no significant differences in acidity, chlorinity, or copper (the most abundant heavy metal in the waste) from samples taken in control areas. Organochlorine residues were not detected in samples of seawater taken in the vicinity of plume touchdown. Analyses of samples of phytoplankton and zooplankton showed no obvious or subtle adverse effects, and the investigators concluded that no adverse effects had been detected.[13]

Tests on Shell Chemical Company Waste in 1977. The second oceanic incineration in U.S. waters occurred during March and April of 1977.[7] Again, organochlorine wastes from Shell Chemical Company's plant in Deer Park, Texas, were incinerated by the Vulcanus I. The first burn (about 4,100 t of waste) was monitored under the direction of the EPA. Averages of operational data are given in Table 8.8.2. The overall destruction efficiency for the waste was calculated to be greater than 99.99%.

References 38 through 42 describe other test projects on incinerator ships.

8.8.5 ENGINEERING AND ECONOMIC EVALUATIONS

Engineering. There have been a substantial number of engineering evaluations of oceanic incineration. Most of these have been sponsored by the EPA,[2,3,11,23,31] although other governmental agencies also have sponsored studies.[12,20] The EPA has sponsored several comparative evaluations of oceanic and land-based incineration.[3,23,25] Also, several of the evaluations of land-based incineration made by the EPA are applicable to oceanic incineration.[2,43–47]

Economic. There is a lack of consistent and reliable information on unit costs of the various waste-management practices. Unit costs vary widely and depend on factors related to the specific waste, and to packaging, transportation, density of generators, regulatory programs that are still evolving and evolving at different rates for each management practice, and competition between suppliers of waste-management services. Reliable cost estimates for oceanic incineration are further complicated because the EPA's applicable regulatory program is less advanced than those for other management practices and because a commercial market has not yet been established.

The EPA and other governmental agencies have sponsored a number of individual and comparative economic evaluations of oceanic and land-based incineration.[3,20,48] All cost analyses to date have shown that the costs of oceanic incineration are lower than those for land-based incineration. The latest such evaluation[48] also includes other waste-management practices. Table 8.8.7 shows estimated costs per metric ton for the various waste-management technologies. The estimates in Table 8.8.7 show a considerable overlap between the costs of the two modes of incineration and they show that incineration is considerably

TABLE 8.8.7 Estimated Costs of Several Disposal Technologies

Management technique	Type/form of waste	Price, 1981, $/t
Oceanic incineration	Organochlorines	200–250
	PCBs	500–800
Land-based incineration	Liquids	53–237
	Solids, toxic liquids	395–791
Landfill	Drummed	168–240
	Bulk	55–83
Deep-well injection	Oily wastewater	16–40
	Toxic rinsewater	132–264

more expensive than the other management practices. However, there is now a general recognition that the "true" cost of the other management practices is not factored into current prices. For example, the current price for landfilling a drum of waste does not include the future cost associated with that drum resulting from remediating that landfill if it fails.

8.8.6 SUPPLIERS

As of 1988, only two firms offer oceanic incineration services. Waste Management, Inc., Oak Brook, Illinois, operates the Vulcanus I and Vulcanus II through its subsidiary Ocean Combustion Service BV, Rotterdam, The Netherlands. Both vessels are certified for international operations, although the EPA has not yet granted permits for operation out of U.S. ports. Lehnkering AG, Duisburg, F.R.G., operates the Vesta, which also is certified internationally, but which currently operates only in European waters.

8.8.7 REFERENCES

1. "Incineration at Sea: History, State of the Art, and Outlook," paper presented by the European Council of Chemical Manufacturers' Federations (CEFIC) for consideration of the Scientific Group of the London Dumping Convention at the March, 1985, meeting, March, 1985.

2. R. S. Ottinger, J. L. Blumenthal, D. D. Dalporto, G. Gruber, M. Santy, and C. C. Shih, *Recommended Methods of Reduction, Neutralization, Recovery, or Disposal of Hazardous Waste*, vol. III, *Disposal Processes Descriptions, Ultimate Disposal, Incineration, and Pyrolysis Processes*, U.S. EPA, Cincinnati, Ohio, PB 224 582, August 1973.

3. C. C. Shih, J. E. Cotter, D. D. Dean, S. F. Paige, E. P. Pulaski, and C. F. Thorne, *Comparative Cost Analysis and Environmental Assessment for Disposal of Organochlorine Wastes*, EPA-600/-2-78-190, U.S. EPA, Washington, D.C., 1978.

4. R. A. Venezia, "Incineration At-Sea of Organochlorine Wastes," paper presented December 8, 1978, at the Second Conference of the Environment, Paris, France, U.S. EPA, Office of Research and Development, Research Triangle Park, North Carolina, 1978.

5. D. G. Ackerman, J. F. Metzger, and L. L. Scinto, *History of Environmental Testing of the Chemical Waste Incinerator Ships M/T VULCANUS and I/V VULCANUS II*, TRW Inc., Redondo Beach, California, June 1983.

6. H. Compaan, *Incineration of Chemical Wastes at Sea. A Short Review*, 2d ed., Report No. CL 82/83, Netherlands Organization for Applied Scientific Research, Central Research Organization—TNO, Delft, The Netherlands, 1982.

7. J. F. Clausen, H. J. Fisher, R. J. Johnson, E. L. Moon, C. C. Shih, R. F. Tobias, and C. A. Zee, *At-Sea Incineration of Organochlorine Wastes Onboard the M/T Vulcanus*, EPA-600/2-77-196, U.S. EPA, Research Triangle Park, North Carolina, 1977.

8. D. G. Ackerman, H. J. Fisher, R. J. Johnson, R. F. Maddalone, B. J. Matthews, E. L. Moon, K. H. Scheyer, C. C. Shih, and R. F. Tobias, *At-Sea Incineration of Herbicide Orange Onboard the M/T Vulcanus*, EPA-600/2-78-086, U.S. EPA, Research Triangle Park, North Carolina, 1978.

9. Personal Communication with M. L. Neighbors, Diversified Maritime Services, Inc., Washington, D.C., October 15, 1985.

10. P. S. Zurer, "Incineration of Hazardous Wastes at Sea. Going Nowhere Fast," *Chem-*

ical & Engineering News, American Chemical Society, Washington, D.C., December 9, 1985, pp. 24–42.

11. R. J. Johnson, D. G. Ackerman, J. L. Anastasi, C. L. Crawford, B. Jackson, and C. A. Zee, *Design Recommendations for a Shipboard At-Sea Hazardous Waste Incineration System,* U.S. EPA Cincinnati, Ohio, Contract No. 68-03-2560, Work Directive No. T5017, June 1980.

12. U.S. Environmental Protection Agency, U.S. Department of Commerce (Maritime Administration), U.S. Department of Transportation (Coast Guard), U.S. Department of Commerce (National Bureau of Standards), *Report of the Ad Hoc Work Group for the Chemical Waste Incinerator Ship Program,* NTIS Report No. PB-81-112849, September 1980.

13. T. A. Wastler, C. K. Offutt, C. K. Fitzsimmons, and P. E. DesRosiers, *Disposal of Organochlorine Wastes by Incineration At-Sea,* EPA-430/9-75-014, U.S. EPA, Washington, D.C., 1975.

14. M. A. Guttman, N. W. Flynn, and R. F. Shokes, *Ambient Air Monitoring of the August 1982 M/T VULCANUS PCB Incineration at the Gulf of Mexico Designated Site,* U.S. EPA Office of Water Regulations and Standards, Washington, D.C., 1983.

15. *Biological Monitoring of PCB Incineration in the Gulf of Mexico,* prepared by TerEco Corp. for the U.S. EPA Office of Water Regulations and Standards, Washington, D.C., 1982.

16. French Ministry of the Quality of Environmental Life, *Incineration at Sea of Industrial Wastes Containing Chlorine—Report on Experiments Relating to the Activity of the Incinerator Ship MATTHIAS II Operated by INCIMER, Inc.,* Neuilly-sur-Seine, France, 1976.

17. U.S. Environmental Protection Agency, *Final Environmental Impact Statement, Designation of a Site in the Gulf of Mexico for Incineration of Chemical Wastes,* Washington, D.C., July 1976.

18. U.S. Environmental Protection Agency, *Final Environmental Impact Statement for North Atlantic Incineration Site Designation,* EPA-440/5-82-025, U.S. EPA Office of Water Regulations and Standards, Washington, D.C., December 1981.

19. U.S. Department of State and U.S. Environmental Protection Agency, *Final Environmental Impact Statement for the Incineration of Wastes At Sea Under the 1972 Dumping Convention,* Office of Environmental Affairs, U.S. Department of State, Washington, D.C., February 1979.

20. U.S. Department of Commerce, *Final Environmental Impact Statement: Maritime Administration Chemical Waste Incinerator Ship Project,* vol. 1, MA-EIS-7302-76-041F, U.S. Maritime Administration, Washington, D.C., 1976.

21. J. D. Andis, C. A. Atkinson, M. E. Drehsen, D. A. Keefer, T. L. Sarro, S. J. Sheffield, G. R. VanDyke, P. J. Weller, and S. D. Wolf, *Final Environmental Impact Statement for the Offshore Platform Hazardous Waste Incineration Facility,* vol. 1, U.S. EPA Office of Research and Development, Washington, D.C., September 1982.

22. U.S. Air Force, *Final Environmental Statement for Disposition of Orange Herbicide by Incineration,* U.S. Department of the Air Force, Washington, D.C., November 1974.

23. S. F. Paige, L. B. Baboolal, H. J. Fisher, K. H. Scheyer, A. M. Shaug, R. L. Tan, and C. F. Thorne, *Environmental Assessment: At-Sea and Land-Based Incineration of Organochlorine Wastes,* EPA-600/2-78-087, U.S. EPA Office of Water Regulations and Standards, Washington, D.C., 1978.

24. D. G. Ackerman, L. L. Scinto, R. J. Johnson, T. L. Sarro, and R. Scofield, *Preliminary Environmental Assessment of the At-Sea Incineration of Liquid Silvex,* U.S. EPA Industrial Environmental Research Laboratory, Research Triangle Park, North Carolina, Contract No. 68-02-3174, Work Assignment No. 2, 1980.

25. U.S. Environmental Protection Agency, *Assessment of Incineration as a Treatment for*

Liquid Organic Hazardous Wastes, vol. IV, *Comparison of Risks from Land-Based and Ocean-Based Incineration,* U.S. EPA Office of Policy, Planning and Evaluation, Washington, D.C., 1985.

26. J. W. J. Gielen, and H. Compaan, *Monitoring of Combustion Efficiency, Destruction Efficiency and the Emission of Some Toxic Substances During the Certification Voyage of the Incineration Vessel "VULCANUS II,"* Report No. R 83/136, Central Research Organization—TNO, Delft, The Netherlands, 1983.

27. D. G. Ackerman, R. G. Beimer, and J. F. McGaughey, *Plan for Sampling and Analysis of Volatile Organics During a Certification Voyage of the M/T VULCANUS II,* TRW Inc., Redondo Beach, California, 1983.

28. J. W. J. Gielen and H. Compaan, *Monitoring of Combustion Efficiency and Destruction Efficiency During the Certification Voyage of the Incinerator Vessel 'Vulcanus II'* *January 1983,* Report No. R 83/53, Central Research Organization—TNO, Delft, The Netherlands, 1983.

29. D. G. Ackerman, R. G. Beimer, and J. F. McGaughey, *Incineration of Volatile Organic Compounds by the M/T VULCANUS II,* TRW Inc., Redondo Beach, California, 1983.

30. G. Jungclaus, and P. Gorman, *Evaluation of a Volatile Organic Sampling Train (VOST),* report to the U.S. EPA Office of Solid Waste, Washington, D.C., and Office of Research and Development, Research Triangle Park, North Carolina, Contract No. 68-01-5915, 1982.

31. U.S. Environmental Protection Agency, *Guidance Manual for Hazardous Waste Incinerator Permits,* SW-966, U.S. EPA Office of Solid Waste, Washington, D.C., 1983.

32. H. Compaan, *Monitoring of Combustion Efficiency, Destruction Efficiency, and Safety During the Test Incineration of PCB Waste; Part I: Combustion and Destruction,* Report No. CL 82/122, Central Research Organization—TNO, Delft, The Netherlands, 1982.

33. D. G. Ackerman, J. F. McGaughey, and D. E. Wagoner, *At-Sea Incineration of PCB-Containing Wastes Onboard the M/T VULCANUS,* EPA-600/7-83-024, U.S. EPA, Office of Environmental Engineering and Technology, Research Triangle Park, North Carolina, 1983.

34. J. H. Beard III and J. Schaum, *Sampling Methods and Analytical Procedures Manual for PCB Disposal: Interim Report, Revision O,* Attachment E, U.S. EPA Office of Solid Waste, Washington, D.C., February 10, 1978.

35. D. G. Ackerman, and R. F. Maddalone, *Monitoring, Sampling and Analysis Plan for the Incineration of Herbicide Orange Onboard the M/T VULCANUS,* U.S. EPA, Office of Research and Development, Research Triangle Park, North Carolina, Contract No. 68-01-2966, 1977.

36. J. W. Hamersma, D. G. Ackerman, M. M. Yamada, C. A. Zee, C. Y. Ung, K. T. McGregor, J. F. Clausen, M. L. Kraft, J. S. Shapiro, and E. L. Moon, *Emissions Assessment of Conventional Stationary Combustion Systems: Methods and Procedures Manual for Sampling and Analysis,* EPA-600/7-79-029a, U.S. EPA Office of Research and Development, Research Triangle Park, North Carolina, 1979.

37. K. Lutzke, W. Guse, and K. D. Hoffmann, *Research Into the Combustion Efficiency and Destruction Efficiency During the Incineration of Selected Wastes on an Incineration Vessel,* SACSA VIII/3/10-E, presented by the Federal Republic of Germany to the Eighth Meeting of the Standing Advisory Committee for Scientific Advice, Antibes, France, November 24–28 1980.

38. H. Compaan, *On the Occurrence of Organic Chlorides in The Combustion Products of an EDC Tar Burnt by Incinerator Ship Vulcanus; A Preliminary Investigation,* Report No. CL 74/93, Central Laboratory—TNO, Delft, The Netherlands, 1974.

39. A. A. Van der Berg, P. M. Houpt, A. G. Keyzer, K. Koopmans, A. C. Lakwijk, P. Van Leeuwen, L. Pot, S. J. Spijk, E. Talman, and B. J. Van Woudt, *On the Occurrence of Organic Chlorides in the Combustion Products of an EDC Tar Waste Burnt by*

the Incinerator Ship "VULCANUS." A Preliminary Investigation, Report No. CL 74/95, Central Research Organization—TNO, Delft, The Netherlands, 1975.

40. H. Compaan, and W. F. M. Hesseling, *Measurements of Emissions on Board the Incinerator Vessel "MATTHIAS II,"* Report No. CL 79/76, Central Research Organization—TNO, Delft, The Netherlands, 1979.

41. H. Compaan, A. A. Van der Berg, J. M. Timmer, and P. van Leeuwen, *Emission Measurements on Board the Incineration Ship "Vulcanus" During the Incineration of Organochlorine and Organofluorine Containing Wastes,* Report No. CL 81/108, Central Research Organization—TNO, Delft, The Netherlands, 1982.

42. D. G. Ackerman, R. J. Johnson, E. L. Moon, A. E. Samsonov, and K. H. Scheyer, *At-Sea Incineration: Evaluation of Waste Flow and Combustion Gas Monitoring Instrumentation Onboard the M/T VULCANUS,* EPA-600/2-79-137, U.S. EPA, Research Triangle Park, North Carolina, 1979.

43. U.S. Environmental Protection Agency *Engineering Handbook for Hazardous Waste Incineration,* U.S. EPA Office of Research and Development, Washington, D.C., Report No. SW-889, NTIS PB81-248163, 1981.

44. D. G. Ackerman, L. L. Scinto, P. S. Bakshi, D. L. Anderson, R. G. Delumyea, R. J. Johnson, G. Richard, and A. M. Takata, *Guidelines for the Disposal of PCBs and PCB Items by Thermal Destruction,* EPA-600/2-81-022, U.S. EPA Office of Research and Development, Washington, D.C., 1981.

45. L. Mason, and S. L. Unger, *Hazardous Material Incinerator Design Criteria,* U.S. EPA, Cincinnati, Ohio, NTIS PB80-131964, 1979.

46. *Review of Proposed Action to Dispose of Orange Herbicide by Incineration,* prepared by Arthur D. Little, Inc., for the U.S. Air Force Environmental Health Laboratory, Kelly Air Force Base, Texas, Contract F33615-74-C-5116, 1974.

47. P. J. Weller, D. G. Ackerman, and J. D. Andis, *An Assessment of the Ability of the M/T VULCANUS to Incinerate Wastes Containing PCBs,* Report to the U.S. EPA, Industrial Environmental Research Laboratory, Cincinnati, Ohio, Contract No. 68-02-3174, Work Assignment No. 82, 1982.

48. U.S. Environmental Protection Agency, *Assessment of Incineration as a Treatment Method for Liquid Organic Hazardous Wastes, Summary and Conclusions,* U.S. EPA Office of Policy, Planning and Evaluation, Washington, D.C., 1985.

SECTION 8.9
MOLTEN-GLASS PROCESSES

J. L. Buelt
Pacific Northwest Laboratory
Richland, Washington

8.9.1 GENERAL PROCESS DESCRIPTION

Various molten-glass processes are becoming commercially available for the destruction and/or immobilization of hazardous wastes. The processes can destroy combustible and some toxic portions of hazardous waste and simultaneously incorporate residuals, such as ash and nonvolatile heavy metals, into a stable glass form. The final product is reduced in volume and mass by driving moisture from the waste permanently, destroying portions of the waste thermally, and consolidating the residuals into a dense glass and crystalline product.

Molten-glass processes operate by the principle of *joule heating*. Electrodes placed in the molten glass apply an electrical voltage to the molten glass, passing electrical current through alkaline ionic components in the glass.[1] The electrical resistance of the molten glass creates heat within the confines of the electrodes when voltage is applied. The heat is distributed evenly within the molten glass by natural, convective currents in the fluid.

Molten glass processes fall into the following three basic categories:

1. In situ vitrification (ISV)
2. Vertical joule-heated glass melters
3. Horizontal joule-heated glass melters

Each of the basic process types is potentially amenable to a variety of hazardous wastes.

8.9.2 PRIMARY FUNCTIONS

The primary functions of all three basic types of molten-glass processes are to destroy the organic and nitrate chemical components in the waste, immobilize the remaining portions of the waste into a geologically stable glass material, and reduce the overall volume of the waste. Even when glass-former additions are necessary the consolidation of the waste form and mass reductions from drying, combustion, and decomposition account for the volume reduction. Molten-glass processes are viewed as permanent treatment measures producing a waste form that is potentially delistable and relatively safe for transport. The processes reduce the volume of the waste by factors of between 2 and 100, depending on the composition, destructibility, and density of the waste. Volume reduction generally reduces the overall disposal cost significantly.

8.9.3 PROCESS DESIGN AND OPERATION

In Situ Vitrification

In situ vitrification is a thermal treatment process that converts contaminated soil into a chemically inert and stable glass and crystalline product. Figure 8.9.1 illustrates the major components and the operating sequence of the vitrification. A square array of four molybdenum and graphite electrodes is inserted into the ground to the desired treatment depth. Because soil is not electrically conductive when the moisture has been driven off, a conductive mixture of flaked graphite and glass frit is placed among the electrodes to act as a starter path. An electrical potential is applied to the electrodes to establish an electrical current in the starter path. The resultant power heats the starter path and surrounding soil initially to temperatures above 2000°C, well above the initial soil-melting temperatures of 1100°C to 1400°C. The graphite starter path is eventually consumed by oxidation and the current is transferred to the molten soil, which is electrically conductive. At this stage, temperatures have been measured between 1450 and 1600°C. As the molten or vitrified zone grows, it incorporates nonvolatile and semivolatile hazardous elements (such as heavy metals) and radionuclides that may be present (including radon-generating isotopes). The high temperature of the process destroys organic components by pyrolysis. The pyrolyzed by-products migrate to the surface of the vitrified zone, where they combust in the presence of oxygen. A hood placed over the area being vitrified directs the gaseous effluents to an off-gas treatment system if necessary.

In some cases where volume reduction attained by the process is high and the potential for occupational hazards are low, as with some hazardous-waste sludges, it may be more prudent to consolidate the materials into a single setting by conveying sludge or contaminated soil to the electrodes. This is especially true when the contamination depth is shallow (i.e., less than 3 m). In these cases, feeding sludges to the electrodes would eliminate the amount of down time necessary to move process equipment from setting to setting. Feeding sludges to the electrodes would also consolidate all of the sludge at a disposal site into a much smaller area, leaving the remainder of the disposal site available for future operations. The location at which to consolidate contaminated soils could be selected to be in proximity to current plant operations. This could improve process effi-

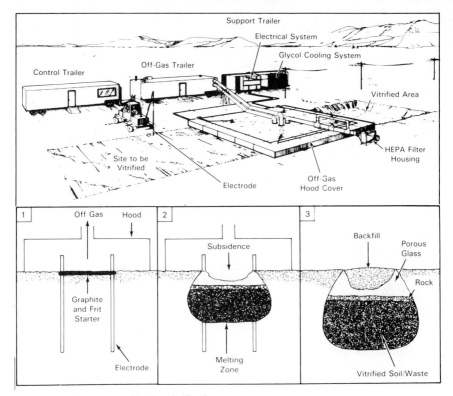

FIG. 8.9.1 The process of in situ vitrification.

ciency by permitting sharing of process equipment, operations, and monitoring with existing functions at the plant.

The large-scale process equipment for ISV currently undergoing testing by Pacific Northwest Laboratory (PNL)* is depicted in Fig. 8.9.1. Controlled electrical power is distributed to the electrodes and special equipment contains and treats the gaseous effluents. The process equipment required to perform these functions can be described most easily by dividing it into six major components:

- Electrical power supply
- Off-gas hood
- Off-gas treatment system
- Glycol cooling system
- Process-control station
- Off-gas support equipment

Except for the off-gas hood, all the components are contained in three transportable trailers. They consist of an off-gas trailer, a process control trailer, and

*PNL is operated by Battelle Memorial Institute for the U.S. Department of Energy (U.S. DOE) under Contract DE-AC06-76RLO 1830.

a support trailer. All three trailers are transportable to accommodate a move to any site over a compacted ground surface. The off-gas hood and off-gas line, which are installed at the site for collection of gaseous effluents, are partially dismantled and placed on a flatbed trailer for transport.

With the exception of the electrical power supply, the components of the ISV process are devoted primarily to the containment and treatment of process off gases. The magnitude and complexity of the off-gas system components are dependent entirely on the type of waste being treated. The degree of destruction and/or retention of potentially toxic components in comparison with regulatory requirements determines the need for an off-gas system. To help identify the need for and complexity of an off-gas system for the ISV process for a specific waste, the applications, limitations, and overall behavior of the ISV process must be understood. They are discussed in Subsec. 8.9.4.

The existing large-scale prototype is capable of treating wet, contaminated soils and sludges at greater than 4,000 kg/h. The large-scale process is indicative of the size of the process necessary to treat most contaminated soils and sludges economically, but no limitation of scale has been identified. The large-scale equipment and capabilities are described in greater detail by Buelt and Carter.[3]

The ISV process has been broadly patented in the United States, Canada, Great Britain, and France. Battelle holds a partially exclusive license to those patents; Battelle has exclusive worldwide rights to all ISV technology except for radioactive-waste applications.

Vertical Glass Melter

The *vertical glass melter* is designed to reduce the volume of both noncombustible and combustible wastes and incorporate residual solids into a stable glass matrix. It is adapted from the Vermel-type melters used in the bottle and plate-glass industry.[4] The wastes are fed to a pool of molten glass, which is kept between 1000°C and 1200°C by an electrical current between Inconel 690® electrodes* at opposite ends of the melting cavity.[5,6] The glass is contained within the melting cavity, which is lined by a fused-cast refractory known as Monofrax K3.®† This refractory is backed by various composites of insulating refractory, all encased in an airtight steel shell, shown in Fig. 8.9.2. As electrical potential is applied to the electrodes, an electrical current is established through the conductive molten glass, thereby introducing power to the melt to sustain operation.

The two different types of waste (combustible and noncombustible) require two different feeding techniques. A cylindrical drop tube, a mechanism located in the lid of the melter (Fig. 8.9.2), is inserted 10 cm into the melt during operation with combustible wastes. As the waste is fed into the top of the drop tube, the waste dries and combusts, and the combustion gases are forced into contact with the molten glass. This creates a scrubbing action by the molten glass on the particulates in the combustion gas. Air cooling is provided at the bottom of the drop tube to maintain a viscous, relatively noncorrosive layer between the drop tube and the molten glass.

When feeding noncombustible slurries and solid wastes, the drop tube is

---

*Inconel 690® is a registered trademark of the Huntington Alloys Corporation, Huntington, Alabama.

†Monofrax K3® is a registered trademark of the Carborundum Co., Niagara Falls, New York.

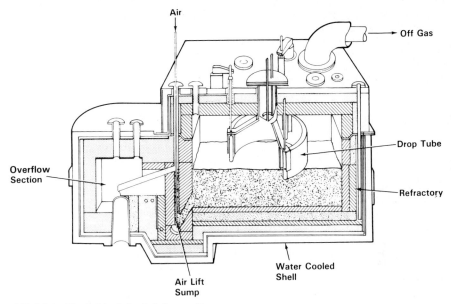

FIG. 8.9.2 Vertical joule-heated glass melter.

raised above the glass melt to allow for cold-cap formation. The noncombustible waste is premixed with glass formers, if necessary, before being introduced to the surface of the glass melt. The heat from the glass pool evaporates the moisture and establishes a cold cap in which the glass-formation reactions take place. The nonvolatile waste constituents flux with the glass formers, making a molten-glass composition.

As molten glass is accumulated, it is drained through an overflow and flows by gravity into a receiving drum. The glass-pouring operation is controlled by an air lift in the riser of the overflow.[7] The air lift can drain glass continuously or intermittently during operation. The riser and the overflow are heated by silicon carbide heating elements to maintain the fluidity of the glass while it is being poured.

The process is easily adapted to be transportable to treat hazardous wastes at the sites of a variety of waste generators. Consequently, the reduction in volume and stabilization of the waste make shipment and burial or on-site storage much more economical. Buelt[8] provides a preconceptual design for such a mobile system.

The vertical glass melter may be scaled to a variety of process rates. For noncombustible slurries, the melter can generally be designed to accommodate a 90 L/h rate for every square meter of exposed glass surface area in the melting cavity.[8] The design rate for vitrification of high-level nuclear waste has been extensively demonstrated by PNL at 30 to 100 L/(h · m²). The upper range is easily achievable with most hazardous wastes because high-level nuclear wastes generally contain many more refractory components than are expected for hazardous wastes. Therefore, the desired production rate can be accommodated by selecting the appropriate surface area of molten glass within the confines of the refractory. Systems have been designed up to 200 L/h (~%240 kg/h), but larger systems are conceivable.

The process rate for combustible wastes is dependent on the diameter of the drop tube. Generally, 0.8 kg/h of combustible wastes can be fed into the melter for every centimeter of drop-tube diameter. Scaling the processing rate linearly with the drop-tube diameter provides adequate contact time within the glass to ensure proper scrubbing efficiency of particulates. The drop tube, however, must not occupy more than 50% of the surface area of the molten glass.

Horizontal Glass Melters

The *horizontal glass melter* is an electrically heated glass melter that is also a tunnel incinerator with a pool of molten glass in its basin.[9] Like the vertical melter, the horizontal melter is equipped with feed and off-gas systems designed to provide flexibility for the various types of wastes capable of being treated in the melter.[10] It consists of a relatively long, shallow chamber of refractory linings, half-filled with molten glass, which are encased in an outer skin of stainless steel (see Fig. 8.9.3). The melter is designed to provide a long residence time of the gaseous effluents at high temperature. The glass exposes the combustion gases to a constant 1200°C infrared flux independent of the rate of burning of the combustible material being fed to one end of the chamber. Exposure of the combustion gases to this temperature as they pass over the glass before being exhausted at the other end of the chamber helps ensure complete combustion. Particulates that do not settle onto the molten-glass surface can be collected by a series of filters at the off-gas exit and recycled back to the glass once the filters are loaded. The process was patented by H. Larry Penberthy in 1981.[11]

The glass is heated by passing an electrical current through an array of electrodes, in the glass bath, and protruding through the side walls. High-purity iron electrodes have been used with this type of melter.[10] Power is controlled by thermocouples that sense the temperature above the molten-glass pool.[12] When high-energy materials are being treated in the melter, power to the electrodes is automatically reduced. When wet sludges are treated, power requirements increase.

Waste materials and glass formers are added at one end of the chamber through a ceiling port. Combustible components ignite immediately, while the inorganic portions and particulates fall to the surface of the molten glass, where they react with and are incorporated into the molten glass in the melter. When needed, combustion air is added to the melter through individually controlled ports distributed above the molten glass on both side walls. The combustion gases and other gaseous effluents are exhausted through a ceiling port at the opposite end of the chamber and directed to the off-gas treatment system.

The horizontal melter can be designed to accommodate a variety of process rates as well. Penberthy[12] states that a furnace 1.2 m wide by 7 m long would be capable of treating wastes at 220 kg/h. Larger furnaces may be constructed to accommodate higher processing rates.

8.9.4 APPLICATIONS AND LIMITATIONS

Molten-glass processes can be applied to a variety of waste types. Many of the waste constituents are destroyed in the process, while the remaining constituents are immobilized into a compatible glass product. The destruction of wastes applies mainly to solid and liquid organics and nitrate-bearing chemicals associated

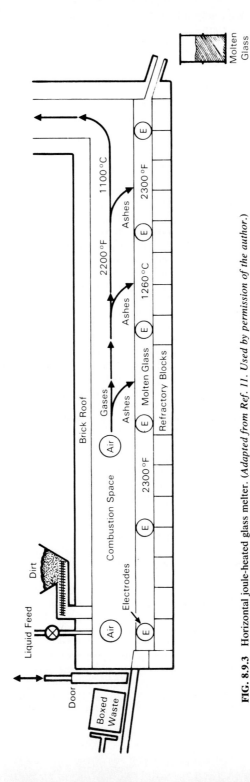

FIG. 8.9.3 Horizontal joule-heated glass melter. (*Adapted from Ref. 11. Used by permission of the author.*)

with the waste. Immobilized constituents consist of heavy metals, radionuclides, combustion ash, and other nonvolatile and semivolatile constituents such as silica, alumina, and alkaline oxides.

Solid and Liquid Organics

Organics are destroyed by pyrolysis and subsequent combustion when they come in contact with combustion air supplied to the molten-glass process. An extensive data base on the destruction of solid and liquid organics has been established for all three types of molten-glass processes.

Although low-molecular-weight liquid organics have not been tested extensively with the ISV process, destruction data are available for higher-molecular-weight liquid organics and solid combustibles. The ISV process has demonstrated a single-pass destruction and removal efficiency (DRE) of 99.9% with polychlorinated biphenyls (PCBs) exclusive of the off-gas system.[13] The overall DRE accounting for the off-gas treatment system was demonstrated in excess of 99.9999%. Experience with this high-boiling-point organic showed a negligible tendency for migration in the soil away from the vitrification region during processing. Beginning with an initial PCB contamination level in the soil of 500 ppm, the soil was converted to glass, thereby destroying the PCBs to nondetectable levels. Outlying concentrations in the soil ranged from 0 to 0.7 ppm, which is lower than the 40 *CFR* 761 U.S. Environmental Protection Agency (EPA) criteria for cleanup. Analysis for secondary decomposition products in the off gas showed relatively minute concentrations of furan and dioxin in the off-gas system, indicating complete destruction of the PCBs.

Oma, Farnsworth, and Timmerman[14] report good combustion efficiency with the ISV process for solid combustibles without the use of a secondary combustion chamber. Carbon monoxide concentrations in the off gas exiting the hood covering the area being vitrified are almost always less than 0.1% (v/v).

Light-hydrocarbon pyrolysis products have been detected in very low concentrations during a test with solid combustibles. Concentrations of CH_4 and C_2H_6 did not exceed $5.3 \times 10^{-3}\%$ (v/v) in the samples analyzed. Despite the high combustible loading within the soil, the excess air provided to the hood of the ISV process resulted in near complete combustion of the pyrolytic gases. As a result, buildup of condensible organics or other combustible gases does not occur within the off-gas treatment system, and the destruction of hydrocarbon pyrolysis products is essentially complete.

Oma et al.[15] have established a set of limiting criteria for the amount of organics that can be vitrified in situ in a variety of geometrical configurations. The large-scale off-gas system with a capacity of 104 standard m^3/min is capable of containing and combusting gaseous effluents from a variety of waste configurations. Pyrolyzed gases generated from the vitrification of combustible volumes as large as 0.9 m^3 can be contained without losing hood vacuum while continually supplying a 20% excess of combustion air. The same off-gas system is capable of containing the gases generated from a void volume of 4.3 m^3 with a design factor of 2. For homogeneous mixtures of combustibles and soil, the large-scale off-gas system is capable of providing a 20% excess of combustion air and containing combustion gas from 3,200 kg of combustibles for every meter of depth being vitrified per setting. The predicted performance of the process under these conditions is also based on a design factor of 2.

Likewise Buelt[16] has shown zero pyrolyzed hydrocarbon or carbon monoxide content when vitrifying ion-exchange resins (solid combustibles) in a vertical joule-heated glass melter. Although most of the combustion takes place in the drop tube and below the surface of the glass, occasionally a bright flash of flame can be seen at the glass surface. Fluctuations in the melter vacuum, which is generally operated at 25 cm H_2O, are minimal. The oxygen concentration in the off-gas stream is controlled arbitrarily to a minimum of 15% to keep combustion efficiency high and particulate entrainment low; however, this limit has not been optimized. No carbon monoxide has been detected in any of the off-gas samples analyzed by gas chromatography or mass spectrometry.

Combustion air is controlled to a minimum of 12% oxygen in the off gas for complete combustion in horizontal melters.[17] The horizontal melter has undergone a DRE test for principal organic hazardous constituents (POHCs).[18] The Resource Conservation and Recovery Act (RCRA) under Federal Regulation 40 *CFR* 264 (*Code of Federal Regulations*) states that the minimum allowable DRE for POHCs is 99.99%. This means that the rate at which an organic exits the stack must be less than 0.01% of the rate at which it is being fed to the melter. Peters[18] reports that in tests conducted by the Mound Laboratory the horizontal melter achieved overall DREs ranging from 99.97 to 99.99999%. Only one of the 54 measurements was below the 99.99% standard, and this result was related to a product of incomplete combustion. These data demonstrate the high destruction potential of molten-glass processes.

Nitrate Destruction

The ISV process has proven to be an excellent destructor of nitrates associated with contaminated soils and sludges.[19] Destruction efficiencies of greater than 99.6% have been demonstrated with the large-scale process. Numerous technical studies have been performed on the behavior of nitrates in thermal treatment processes to explain the NO_x destruction characteristic of ISV. In a study by Mori and Ohtake[20] the focus of NO decomposition was in the temperature region of 650°C to 950°C. Measurements were made to determine the effects of the reducing species, carbon monoxide and hydrogen, in combustion gas.

Mori and Ohtake concluded that NO decomposition is obtained by the decomposition reactions of the reducing species. In an NO–CO–Ar mixture the NO decomposition rate is in linear proportion to the CO concentration. The overall reaction is expressed as follows:

$$NO + CO \rightarrow \tfrac{1}{2}N_2 + CO_2$$

In an NO–H_2–Ar mixture the NO decomposition rate increased almost in proportion to the H_2 concentration. The overall reaction is given as

$$NO + H_2 \rightarrow \tfrac{1}{2}N_2 + H_2O$$

The decomposition rates in the reactions between NO and CO, and NO and H_2 depend on the CO or H_2 concentration when the NO concentration is higher than about 400 ppm but do not depend on the NO concentration itself. However, for NO concentrations lower than 250 ppm the decomposition rate decreases with the NO concentration. Therefore, this mechanism of NO decomposition may play an important role in ISV in the molten soil where temperatures are high and reducing components exist.

Armstrong and Klingler[10] have demonstrated the destruction of nitrates and NO_X in a horizontal melter. Their paper for the U.S. Department of Energy (DOE) describes the successful test results and mechanisms to achieve a high NO_X destruction efficiency in the joule-heated glass melter. A combination of advantageous characteristics that the joule-heated glass melter possesses led to their first consideration of this process for the destruction of nitrate wastes. These characteristics include high operating temperature (up to 1315°C), an electric heat supply to eliminate the need for oxygen to support fuel combustion, flexibility in control parameters such as temperature and vessel atmosphere, and a glass bed to immobilize inorganic constituents into a glass matrix. As with ISV, the presence of reducing agents at high temperature will permit complete destruction of the nitrate wastes to N_2 and O_2.

Armstrong and Klingler explain, however, that ammonia performs selective reduction of NO_X in air (i.e., reduction of NO_X takes precedence over that of oxygen). Faucett et al.[21] show the relevant reactions as follows:

$$4NH_3 + 4NO + O_2 \rightarrow 4N_2 + 6H_2O$$

$$4NH_3 + 2NO_2 + O_2 \rightarrow 3N_2 + 6H_2O$$

However temperature is an extremely important control parameter when using NH_3 as a reducing agent, as can be seen by the following reaction at 1100°C reported by Hurst:[22]

$$NH_3 + O_2 \rightarrow NO + H_2O$$

Hurst also explains that the reaction below 850°C needs to be supplemented with small amounts of H_2 to maintain NO_X destruction efficiencies. Because the use of H_2 as a reducing agent is not recommended in joule-heated glass melters, operating temperatures of 850 to 1100°C are desirable.

Armstrong and Klingler confirmed this theory in their screening tests and extended operation studies. They were able to achieve good destruction efficiencies (to below 500 ppm) exhausting from the melter when temperatures in the plenum above the melt were kept above 850°C. However, at temperatures of 750°C, destruction efficiencies began to fall off rapidly.

Although not demonstrated empirically, this same performance is conceivably achievable with vertical melters. In the case of vertical melters, NH_4OH or another suitable reducing agent could be added directly to the waste stream to achieve NO_X destruction. Reaction temperatures in this furnace are easily controlled within the NO_X destruction range by feed rate and power input. Successful demonstration of this principle would make all three types of molten-glass processes excellent destructors of nitrates and solid and liquid organics.

Immobilization of Residual Solids

After the combustible and nitrate portions of the hazardous waste are destroyed by the process, the residual solids are fluxed into the molten glass. The residual solids (such as ash and heavy-metal oxides) and nontoxic solid components (such as silica, alumina, and lime) are incorporated into the glass and become part of its matrix. The final product is generally reduced in volume by factors of from 2 to 100, depending on the waste characteristics and product-quality requirements. When cooled, the potentially toxic elements remain fixed and immobilized in a glass waste form that does not dissolve in water, has high leach resistance, and exhibits strength properties better than those of concrete. The ISV waste form possesses hydration properties similar to those of obsidian, which hydrates at rates of less than 1 mm/10,000 yr.[15] At these rates the life of the vitrified waste can be expected to exceed 1 million years.

The degree to which the residual solids are retained in the molten glass during processing is dependent upon the type of molten-glass process, the processing conditions, and the chemical element itself.

Retention Data on ISV

A summary of soil-to-off-gas decontamination factors (DFs) for the pilot-scale ISV unit is presented in Table 8.9.1. The DFs are ratios of the amount of waste components originally in the waste to the amount being evolved from the soil. If required, an off-gas system can provide an additional DF of 10^4 to 10^5 for each element. These data illustrate that most species are retained quite well (i.e., have high DFs) by the molten soil, exclusive of the off-gas system. The DFs for Co, Mo, Sr, Ce, La, and Nd are very high, ranging from 10^2 to more than 10^4. Retention of Cs is also quite good, with DFs ranging from 3×10^1 to 2×10^3, depending on burial depth and operating condition. Cadmium, however, consistently exhibits a relatively high loss (DF of 3 to 4) because of its volatile nature at high temperatures ($\hbar 1400°C$).

Element releases to the off gas are highly dependent on burial depth. More than 99.9% of the nonvolatile elements are retained at burial depths of 1.35 m or greater. The retention of semivolatiles—Cs, Sb, and Te—also increases to greater than 99% at a 1.35-m depth. In large-scale operations, it is expected that the vast majority of hazardous species will be buried deeper than 1.35 m, resulting in high retention values for nonvolatile and semivolatile components.

Large-scale ISV data confirm conclusions derived from engineering- and pilot-scale data that burial depth has a tremendous effect on retention. The DF for Sr,

TABLE 8.9.1 Soil-to-Off-Gas Decontamination Factors for Pilot-Scale in Situ Vitrification

Trace element	Decontamination factors
Co	80–640
Mo	200–>100,000
Sr	800–50,000
Cs	30–2,000
Sb	30–800
Te	2–100
Ce	90–700
La	90–5,000
Nd	100–5,000
Cd	3–4
Pb	7–30

for instance, was measured at 3×10^4 at a burial depth of 1.6 m, which is equivalent to a retention value of 99.997%.[19] This is consistent with the upper range of values reported for the pilot-scale unit, indicating that increased process scale and/or burial depth helps retain a greater percentage of potentially toxic particulates.

Large-scale ISV performance data show that 98.7% of fluorides are retained when vitrifying contaminated soil. The high retention figure demonstrated is important not only in keeping fluorides from reaching groundwater, but also in reducing the corrosive effects of the gaseous effluents on the off-gas treatment system. Sulfates, on the other hand, are not retained well and evolve with the off gas as SO_x. They are removed by a wet scrubbing system that is pH-neutralized to minimize corrosion of the off-gas system.

Retention Data on Melters

Typical DF data generated with the vertical melter during a test of a slurry-fed melter are summarized in Table 8.9.2.[23] In most cases, the loss of retention in the glass is due to entrainment of particulates in aerosol form. The only significant gas-phase losses are with Cl, S, and B, which form volatile acid gases. This is not to say that melter-induced volatilization has no influence upon melter losses of other feed component elements. On the contrary, the lower DFs associated with the semivolatile elements Cd, Cs, and Te clearly underscore the importance of this volatilization process. Thus, apart from the volatilization mechanism responsible for producing airborne effluents, particulate transport through the off-gas system is the predominant loss mechanism associated with slurry-fed melter operation.

The elements lost are recoverable; they can be removed by a recirculating scrub solution in the off-gas treatment system, centrifuged or sorbed to ion-exchange columns, and recycled to the melter. This technique would minimize or eliminate generation of secondary waste streams.

Buelt[16] reports the single-pass DFs of Cs and Sr particulates associated with combustible wastes are 6.3 and 20, respectively, in the vertical melter. These values are equivalent to retention values of 84% and 95%, respectively. These DFs

TABLE 8.9.2 Component Decontamination Factors for a Slurry-Fed Vertical Melter

Element	Average DF total
Al	22,000
B	100
Cd	9.9
Cl	2.9
Cs	14
Fe	1,800
La	2,100
Mn	1,800
Na	300
S	5.5
Sr	1,800
Te	3.0
Zr	22,800

are less than those measured for noncombustible wastes in the same melter because in a combustible operation the particulates must be absorbed into the glass phase from the combustion gas by mass transfer. However, the drop-tube feeding concept forces contact of the combustion gases with the molten glass to ensure the maximum achievable retention in the glass on a single pass. As with the slurry-fed system, particulates escaping the melter can be scrubbed, centrifuged or sorbed onto ion-exchange resins, and recycled to the melter to achieve zero secondary waste discharge and increase overall retention and DF values.[8]

Peters[18] reports that CdO was introduced with combustible feeds to determine the behavior of metal oxides in the horizontal glass melter. This technique relies on the deentrainment chamber of the melter to allow particulates to fall to the surface of the glass and react with the glass formers to form a glass matrix. Experimental results showed retention figures of 39 to 62% for Cd on a single pass. Previous tests with the horizontal melter indicated retention figures of 75% for Pu, 40 to 80% for Cs, and 70 to 88% for Co.

8.9.5 PROCESS COST

Because of the diversity of waste types, molten-glass processes, and scale of equipment, it is impossible to project reliable cost estimates for the molten-glass processes. Cost evaluations must be performed on a case by case basis. However, for purposes of this Handbook, it is prudent to provide examples of estimated costs for specific cases for each of the molten-glass processes.

The cost of using the large-scale ISV process as an in-place stabilization technique, for example, has been estimated for three general categories of contaminated soil[15,24] and escalated to 1985 dollars. The types of waste considered are wet and dry (25 and 5 wt % moisture) soil contaminated with hazardous waste, and wet industrial sludge (55 to 70 wt % moisture). The cost estimate includes the following categories: (1) site preparation, (2) amortized equipment costs, (3) labor and operating costs, and (4) electrodes, energy, and other consumable costs. Labor costs are based on the demonstrated operating times of actual large-scale operations. Electrode cost estimates assume a single-use set of four molybdenum electrodes per vitrification setting. Energy costs are provided for various electrical rates.

The cost of electrical power can affect the economics of the process significantly as seen in Fig. 8.9.4. Costs range from $70 to $320 per cubic meter ($53 to $240 per cubic yard) or $0.06 to $0.20 per kilogram depending on the waste type and cost of electrical power. In some cases, the amount of soil moisture in the area being vitrified can also have a significant effect on process economics. Soil moisture increases the operating cost of the process because more energy and operating time are required to vitrify a given volume of contaminated soil. This does not hold true when the contaminated soil or sludge is fully saturated, or when the soil or sludge begins to swell. Soil moisture concentrations beyond this point displace the solids, resulting in costs per unit volume that are independent of moisture.

The principal reason for differences in the costs of processing hazardous soils and industrial sludges is the complexity of equipment and number of electrodes. Each type of waste requires different types of off-gas systems, ranging from a train of wet and dry treatment systems inside for hazardous contaminated soils to simple filtration systems for industrial sludges. Reduced complexity of equipment

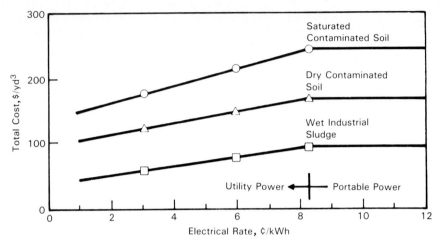

FIG. 8.9.4   Total cost of large-scale in situ vitrification.

results in lower annualized equipment charges and lower labor costs for operation and maintenance. Industrial sludges, which generally exhibit high volume reductions when vitrified, require fewer electrodes because more waste can be processed in a single setting. This drastically reduces the number of vitrification settings and correspondingly reduces costs for electrode materials.

An estimate of costs for the vertical melter adapted to combustible and noncombustible low-level nuclear-waste slurries has been provided by Buelt.[8] These costs are based on a completely self-contained system, which includes an extensive off-gas treatment system and electrical, air, and cooling services. Estimated processing costs include the energy from diesel generators, cost of glass-forming materials and receiving containers, operation, transportation from site to site, radiation monitoring, and a high allowance for nonroutine maintenance (13% of capital per year). The costs ranged from $1.80 per kilogram for noncombustible, concentrated liquids to $3.90 per kilogram for combustible slurries. Costs which include only energy, glass formers, and operational labor range from $0.63 to $1.92 per kilogram. These estimated costs for nuclear wastes (generally more expensive to treat), have not been optimized and are expected to be drastically reduced for hazardous wastes.

Armstrong and Klingler[10] have published a preliminary cost estimate for treating nitrate wastes in a horizontal melter. Again, the associated costs have not been optimized and represent an extension of process conditions employed during developmental studies. The cost of $1.25 per kilogram for a nitrated liquid waste includes the costs for energy, glass components and other chemical compounds, and operational labor. Optimization of the operational parameters will undoubtedly reduce the cost of operation for this melter as well.

8.9.6 PUBLIC-PERCEPTION CONSIDERATIONS

Aside from the attainable volume reduction by factors between 2 and 100, the primary attribute of molten-glass processes is the quality of the waste form they produce. Extensive leach studies with waste glasses have shown the waste form to be highly stable, possessing leach characteristics better than granite, marble, and common bottle glass.[15] Because of its high durability it is expected to pass EPA toxicity test procedures for delisting. These characteristics should help molten-glass processes gain rapid acceptability.

8.9.7 REFERENCES

1. J. Stanek, *Electric Melting of Glass,* Elsevier Scientific Publishing, New York, 1977.

2. R. A. Brouns, J. L. Buelt, and W. F. Bonner, *In Situ Vitrification of Soil,* U.S. Patent 4,376,598, March 15, 1983.

3. J. L. Buelt and J. G. Carter, *Description and Capabilities of the Large-Scale In Situ Vitrification Process,* PNL-5738, Pacific Northwest Laboratory, Richland, Washington, 1986.

4. R. W. Palmquist, "Corning's Moly-Lined Melter," *Glass Industry,* **65:**(7)14 (July 10, 1984).

5. R. A. Brouns and M. S. Hanson, *The Nuclear Waste Glass Melter—An Update of Technical Progress,* PNL-SA-12220, Pacific Northwest Laboratory, Richland, Washington, presented to the American Nuclear Society International Meeting on Fuel Reprocessing and Waste Management, Jackson, Wyoming, August 26–29, 1984.

6. R. A. Brouns, G. B. Mellinger, T. A. Nelson, and K. H. Oma, *Immobilization of High-Level Defense Waste in a Slurry-Fed Electric Glass Melter,* PNL-3372, Pacific Northwest Laboratory, Richland, Washington, 1980.

7. J. M. Perez, Jr., and R. R. Nakaoka, *Vitrification Testing of Simulated High-Level Radioactive Waste at Hanford,* PNL-SA-13360, Pacific Northwest Laboratory, Richland, Washington, presented to the Waste Management '86 Symposium, Tucson, Arizona, March 2–6, 1986.

8. J. L. Buelt, *A Mobile Encapsulation and Volume Reduction System for Wet Low-Level Waste,* PNL-5533, Pacific Northwest Laboratory, Richland, Washington, 1985.

9. *Penberthy Pyro-Converter Liquid Thermal Redox Reactor Producing Hydrochloric Acid,* PC-4, Penberthy Electromelt International, Inc., Seattle, Washington, 1984.

10. K. M. Armstrong and L. M. Klingler, *Nitrate Waste Processing by Means of the Joule-Heated Glass Furnace,* MLM-3304, Mound Laboratory, Miamisburg, Ohio, 1985.

11. H. L. Penberthy, *Method and Apparatus for Converting Hazardous Material to a Relatively Harmless Condition,* U.S. Patent 4,299,611, November 10, 1981.

12. H. L. Penberthy, *Use of Electric Glass Melting Furnaces for Destruction of Hazardous Wastes,* Penberthy Electromelt International, Inc., Seattle, Washington, 1986.

13. C. L. Timmerman, "In Situ Vitrification of PCB-Contaminated Soils," Pacific Northwest Laboratory, Richland, Washington, in *Proceedings of the 1985 EPRI PCB Seminar,* Seattle, Washington, March 1986.

14. K. H. Oma, R. K. Farnsworth, and C. L. Timmerman, "Characterization and Treatment of Gaseous Effluents from In Situ Vitrification," in *Radioactive Waste Management and the Nuclear Fuel Cycle,* vol. 4(4), Hardwood Academic Publishers, 1984, pp. 319–341.

15. K. H. Oma, D. R. Brown, J. L. Buelt, V. F. FitzPatrick, K. A. Hawley, G. B. Mellinger, B. A. Napier, D. J. Silviera, S. L. Stein, and C. L. Timmerman, *In Situ Vitrification of Transuranic Wastes: Systems Evaluation and Applications Assessment,* PNL-4800, Pacific Northwest Laboratory, Richland, Washington, 1983.

16. J. L. Buelt, *A Vitrification Process for the Volume Reduction and Stabilization of Organic Resins,* Pacific Northwest Laboratory, Richland, Washington, 1982.

17. H. L. Penberthy, "Penberthy Pyro-Converter Detoxifies Hazardous Materials," *Hazardous Materials & Waste Management,* **4:**(1)14 (January–February 1986).

18. J. A. Peters, *RCRA Demonstration Test Results of a Pilot-Scale Glass Melt Furnace Incineration System at Monsanto Research Corporation,* Mound Laboratory, Miamisburg, Ohio, 1985.

19. J. L. Buelt and J. G. Carter, *In Situ Vitrification Large-Scale Operational Acceptance Test Analysis,* PNL-5828, Pacific Northwest Laboratory, Richland, Washington, 1986.

20. V. Mori and K. Ohtake, "Decomposition Rate of Nitric Oxide on Alumina Surface in the Temperature Range 920–1220K," *Combustion Science and Technology,* **16**:11–20 (1977).

21. H. L. Faucett et al., *Technical Assessment of NO_x Removal Process for Utility Application,* EPA-600/7-77-127, U.S. EPA, 1977.

22. B. E. Hurst, "The Noncatalytic Denitrification Process for Glass Melting Furnaces," *Glass Technology,* **24**:(2) (April 1983).

23. R. W. Goles and G. J. Sevigny, *Off-Gas Characteristics of Defense Waste Vitrification Using Liquid-Fed Joule-Heated Ceramic Melters,* PNL-4819, Pacific Northwest Laboratory, Richland, Washington, 1983.

24. J. L. Buelt and S. T. Freim, *Demonstration of In Situ Vitrification for Volume Reduction of Zirconia/Lime Sludges,* Battelle, Pacific Northwest Laboratories, Richland, Washington, 1986.

SECTION 8.10
DEEP-SHAFT WET-AIR OXIDATION

John M. Smith
J. M. Smith & Associates
Cincinnati, Ohio

8.10.1 CURRENT STATUS OF THE TECHNOLOGY: SUBCRITICAL AND SUPERCRITICAL PROCESSES

The concept and practice of wet oxidation of organic waste materials is well-established in the United States. The Zimpro process, based on the early work of Zimmerman, has been practiced in the United States and Europe since the early 1960s for the conditioning and oxidation of municipal sludge.

The introduction of a new process configuration, that for deep-shaft wet-air oxidation, as a more satisfactory and economical engineering approach to the application of wet-oxidation concepts for municipal wastewater-treatment wastes came in 1975 with the U.S. Environmental Protection Agency's (EPA's) support of the Vertical Tube Reactor Pilot Plant Project in Colorado.[1]

Configuration of the deep-shaft wet-air oxidation reactor is unique to its purpose. Wastewater flows down the center tube of two concentric vertical tubes and returns in the annular space. Use of this vertical-tube configuration has multiple purposes. Tube diameter and length are designed so that sufficient reaction time and desired pressure can be attained during fluid-waste oxidation. Pressure is developed naturally by the hydrostatic liquid head above the waste flowing down the tube. Heat resulting from the exothermic combustion reaction maintains the downhole temperature required to sustain the reaction. Excess heat can be recovered at the surface for use as an additional energy source.

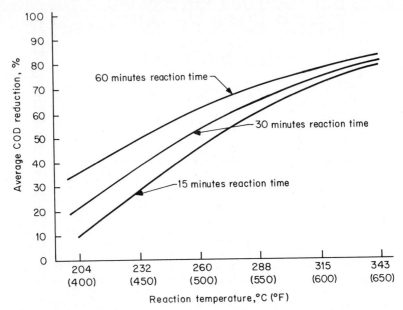

FIG. 8.10.1 Typiçal chemical oxygen-demand (COD) reduction vs. reaction temperature and time using laboratory reactor data.[1]

Deep-shaft wet-air oxidation can be classified as subcritical or supercritical. *Subcritical oxidation* refers to maximum operating reactor temperatures below the critical point of water, 374°C (705.5°F) and 22,120 kN/m² (3,206 psia). The supercritical process for deep-shaft wet-air oxidation attains maximum reactor temperatures and pressures above the critical point of water.[2]

Subcritical Process Status

In 1973, the Vertical Tube Reactor (VTR) Corporation of Denver, Colorado, began a laboratory research program to determine the applicability of a process for deep-well subcritical wet-air oxidation to the treatment of sewage sludge. Figure 8.10.1 shows typical laboratory test results.

A pilot plant was constructed and operated intermittently from 1977 to 1981 by VTR Corporation at a site near Denver. The chemical oxygen demand (COD) removals obtained from this reactor were comparable to laboratory runs made at similar conditions of temperature and detention time. The pilot plant demonstrated the overall technical feasibility of the deep-well chemical oxidation process and verified the original heat-loss and hydrodynamic models used to describe the process.[3]

A demonstration VTR system was constructed in Longmont, Colorado, over a 12-month period and became operational in June 1983. The project was constructed by the VTR Corporation*, with research being supported by the EPA's

---

*VTR Corporation is now VerTech Treatment Systems, Inc. The system installed at Longmont, Colorado, used deep-well reaction technology developed, owned, and patented by VerTech Treatment Systems, Inc., of Denver, Colorado. VerTech is the sole source of that particular system.

Office of Research and Development. The system was designed to process 5,682 kg (6.25 tons) of COD per day of 10,000 mg/L sludge produced by the 0.35 m³/d (8 million gpd) municipal wastewater-treatment plant. The reactor consisted of a pressure vessel 1,585 m (5,200 ft) in length and 25.4 cm (10 in) in diameter that enclosed a concentric tube 17.78 cm (7.0 in) in diameter that acted as the downcomer. Air was used as the oxidant.

In September 1984, addition of an oxygen-enrichment system allowed the use of pure oxygen and the system was able to process a higher-strength waste (30,000 mg/L COD), and was able to destroy over 18,200 kg (20 tons) per day of municipal-sludge solids.

Examination of the downhole pipe after 15 months of operation showed no evidence of pitting or localized corrosion. General corrosion rates experienced are compatible with a 20-year operating life.[4,5]

Supercritical Process Status

The supercritical deep-shaft wet-oxidation process is at an earlier stage of development than the subcritical technology and has not yet been piloted at full scale. The Vertox Corporation[†] has developed this technology based on two recent patents for the application of the technology in deep-well reactors. The process is being developed for the destruction of hazardous wastes to the 99.9999% destruction efficiency level. To date, all testing of supercritical wet-air oxidation has been conducted in continuous-flow and batch laboratory-testing units.

MODAR, Inc., of Natick, Massachusetts, has completed a series of laboratory- and bench-scale studies on the applicability of supercritical-water (SCW) oxidation to the destruction of various hazardous and toxic wastes. Among the classes of compounds which have been tested are aliphatic hydrocarbons (cyclohexane), halogenated aliphatics (1,1,1-trichloroethane), aromatic hydrocarbons (o-xylene), halogenated aromatics (PCB 1254), ketones (methyl ethyl ketone), and organic nitrogen compounds (2,4-dinitrotoluene). In all cases, extremely high destruction efficiencies were observed.[6,7,8]

8.10.2   FUNDAMENTAL PROCESS DESCRIPTION FOR DEEP-SHAFT WET-AIR OXIDATION: SUBCRITICAL AND SUPERCRITICAL PROCESSES

Fundamental Relationships for Subcritical and Supercritical Reactors

Wet-air oxidation is the destruction of organic matter in aqueous solutions under elevated temperatures and pressures. It differs from conventional oxidation in that the reaction occurs entirely in the liquid phase with the heat of combustion being released to the pressurized liquid in the reaction zone. The basic heat-release relationships for the oxidation process are shown as follows:

[†]Vertox, Inc., changed its name to Oxidyne Corporation subsequent to preparation of this manuscript. Vertox, Inc. (now Oxidyne Corporation), was formed in 1984 and contracted to install its first deep-well reaction system in Houston, Texas, in 1988–89. VerTech and Oxidyne did not pioneer the development of deep-well reaction systems. Oxidyne systems are substantially different from the VerTech system and Oxidyne is not associated in any way with VerTech or prior developments of VerTech.

$$C + O_2 = CO_2 + 14,100 \text{ Btu/lb C}$$

$$2H_2 + O_2 = 2H_2O + 61,000 \text{ Btu/lb } H_2$$

The application of this process for oxidation of waste organic materials requires moderate- to high-pressure reactor vessels, efficient heat exchange, the ability to dissolve stoichiometric oxygen into the reaction vessel, and sufficient heating value of the organic waste stream to support the heat losses of the process. The subcritical-water-temperature process is normally operated at temperatures of 190 to 315°C (375 to 600°F) and pressures of 10,342 to 13,790 kN/m^3 (1,500 to 2,000 psig), with organic-waste strengths of 5,000 to 60,000 mg/L of COD. Air or pure oxygen is used as the oxidant depending on the strength of the waste and the degree of oxidation required.

The deep-shaft wet-air oxidation process provides an alternative to aboveground processes by the use of the natural hydrostatic head of deep-well reactors and efficient countercurrent heat exchange. A vertical-well reactor consists of a pressurized vessel that is enclosed within a specially drilled and cased well with insulation placed between the reactor vessel and the well casing. Concentric tubes are placed within the vertically hung pressure vessel for separation of downflowing and upflowing streams. The weight of the liquid provides the pressure necessary for the reactions to take place, thereby eliminating the need for a high-pressure pump. The only pumping losses are frictional losses of flow through the downcomer and upcomer portions of the reactor.

Oxygen is injected into the downflowing liquid to provide the oxygen necessary for the oxidation reaction. The oxygen may be introduced near the surface, at the bottom of the well, or at intervals as the liquid is flowing downward. Mathematical models are used to determine the characteristics and behavior of the reactor operating under this complex, two-phase flow regime.

Because the process is a "plug-flow" process, the oxygen and liquid move together through the reactor, thereby bringing the reactants together in a more efficient manner than conventional wet-air oxidation. The advantage of the process is that it accomplishes the heating, pressurization, reaction, cooling, and depressurization processes in a single, vertical, subsurface unit. Because the high-pressure reactions take place beneath the surface of the earth the process is inherently safe. Figure 8.10.2 is a profile of a typical deep-shaft wet-air oxidation unit.[8]

Properties of Supercritical Water

One of the limiting factors of conventional wet-air oxidation is the limited solubility of oxygen in water. The solubility of oxygen can be increased by raising the operating temperature and pressure, but the limited solubility still presents an inherent limitation to the liquid-phase oxidation reaction. At 315°C (600°F) and 13,790 kN/m^2 (2,000 psi) the solubility of oxygen is still only 0.56% by weight.[9] This gas-to-liquid mass-transfer limitation is eliminated at supercritical water temperatures as described later. The terms *liquid* and *gas* are not applicable at these conditions and the term *supercritical fluid* is used instead.

A *supercritical fluid* may be defined as a single-phase form of matter with properties intermediate between those of the gaseous state and the liquid state. In the supercritical region, water exhibits a density of 0.1 to 0.5 g/cm^3 and a dielectric constant of 3 to 10, with close to 100% solubility of organics. When temperatures reach 500°C (932°F), near total insolubility of inorganics takes place. Fig-

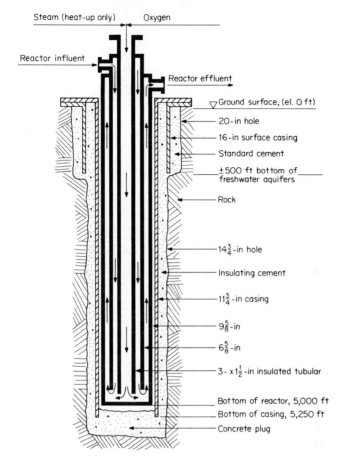

FIG. 8.10.2 Typical reactor and well cross section (wellhead detail not shown).[2]

ure 8.10.3 illustrates the property changes in water as the temperature and pressure approach and surpass the critical point.[10]

Supercritical water has been shown to be a viable processing medium for the thermal oxidation of hazardous wastes.[6,7,8] In this state, the thermal energy contained in the water molecules reduces hydrogen bonding with the result that SCW becomes an excellent solvent for nonpolar organic materials. Normally insoluble organic compounds such as oils and greases dissolve completely in SCW. Most important, oxygen mixes completely with the SCW and organic materials, resulting in a completely miscible mixture. The surface tension of the water becomes zero at SCW conditions, allowing the oxygen to penetrate even the smallest pores and oxidize any organic materials present. Under these conditions the organic materials are rapidly oxidized to carbon dioxide and water in less than 1 min. This very rapid and complete oxidation reaction destroys virtually 100% of the organic species, leaving only carbon dioxide, water, and the heat of reaction.

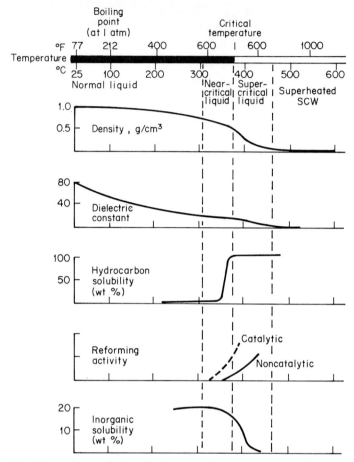

FIG. 8.10.3 Properties of water in the supercritical region for pressure range of 218 to 300 atm. Reforming activity = the ability to reform or break down organic molecules. (*Michael Modell, MODAR, Inc.*)

The Supercritical-Water Deep-Shaft Wet-Oxidation Reactor

The flow configuration for this unit is similar to that for the subcritical reactor; wastes flow down an inner annulus and up through the outer annulus. In this reactor, however, oxygen is injected near the supercritical zone—the bottom region of the reactor. Table 8.10.1 compares the design criteria for the subcritical and supercritical reactors.

A major difference between the subcritical and the SCW reactor is that in the supercritical reactor, the reaction time is much shorter (1 to 5 min compared with 30 min for the subcritical reactor) and hence the reaction zone is much shorter. Another major difference is that velocities in the supercritical reactor are greater than in the subcritical reactor, and are controlled by hydrodynamic considerations rather than oxidation kinetics for most wastes.

TABLE 8.10.1   Comparison of Supercritical and Subcritical Design Criteria

Parameter	Subcritical	Supercritical
Well depth, m (ft)	1,200–1,700 (4,000–5,500)	2,400–3,058 (8,000–12,000)
Temperature, °C (°F)	260–320 (500–600)	400–510 (750–950)
Pressure, kN/m² (psia)	10,000–14,000 (1,500–2,300)	22,000–31,000 (3,200–4,500)
Detention time, min	30–40	0.1–2.0
COD removal, %	80–85	100
Oxidant	Oxygen gas	Oxygen gas
Injection point	Lower ½	Lower 1/20
Velocity, m/s (ft/s)	1.2–2.7 (4–9)	1.5–4.6 (5–15)
Applicability	Sludges, organic wastes, some hazardous wastes	All organic wastes; all hazardous wastes
DRE, %	Depends on organic species range: 80–99.9	99.99 +

Source: Ref. 2.

The fundamental pressure-enthalpy relationship for a supercritical reactor is presented in Fig. 8.10.4. In this figure, the approximate temperature, pressure, and enthalpy changes that take place in the reactor are shown by the pathway at the upper left.[2] The supercritical conditions are reached at the uppermost portion of this pathway, which is above the vapor-liquid envelope at the supercritical point.

A supercritical reactor system includes provision for storage of hazardous wastes, SCW downhole processing, and storage of the treated material. Provision would also be made for recycling if necessary to achieve any desired level of destruction or to conserve water. Normally, a single pass is sufficient to provide greater than 99.99% removal for most organics. A hydrocarbon concentration of 0.5 is sufficient to provide autothermal operation. The design for a 63 L/s (1,000 gpm) supercritical reactor would require a 3.94-cm (10-in) reactor installed in a conventionally cased 5.51-cm (14-in) diameter well that is 3,048 m (10,000 ft) deep.

The SCW process is best-suited for large-volume [3.154 to 63 L/s (50 to 1,000 gpm)], dilute (1.0 to 100,000 mg/L COD) aqueous wastes. Where the waste does not have sufficient heating value to sustain the operating temperatures, nonhazardous wastes with high heating value can be mixed with the low-heating-value hazardous wastes to provide the heat energy needed to make the process self-sustaining. A proposed demonstration project featuring a supercritical oxidative reactor is shown in Fig. 8.10.5.

Several proprietary computer simulation models which calculate heat balances and model the hydrodynamic parameters within the reactor are used to design the

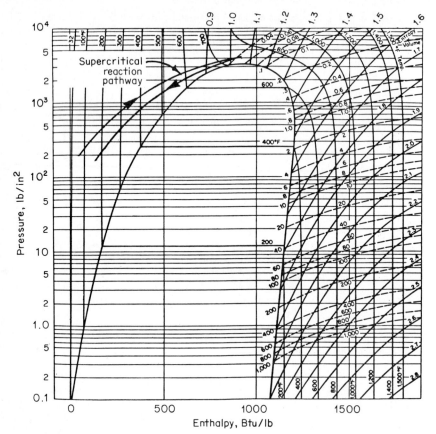

FIG. 8.10.4 Pressure-enthalpy diagram for water showing flow pathway for the supercritical oxidation process. (*From Perry and Chilton,* Chemical Engineer's Handbook, *5th ed., McGraw-Hill, New York, 1973*)

system. These sophisticated digital simulation models enable the designer to investigate the operation of both subcritical and supercritical reactors under a wide variety of operating conditions. One such program predicts the heat loss to various types of rock formations during operation. This program can calculate the effect of operating temperature, flow rates, reactor length, casing thickness, and length of operation on total energy losses, and can take into account rock formations with varying conductive properties. A differential heat-balance model which solves the differential equations which govern the heating, reaction, and cooling operations within the reactor is also used. A third model, the hydrodynamic program, calculates downhole pressures, velocities, pressure losses, two-phase flow dynamics, and residence times for the reactions. The hydrodynamic model uses the data from the differential heat-balance model, providing a detailed simulation of actual energy and flow characteristics for a wide variety of possible reactor configurations.[8]

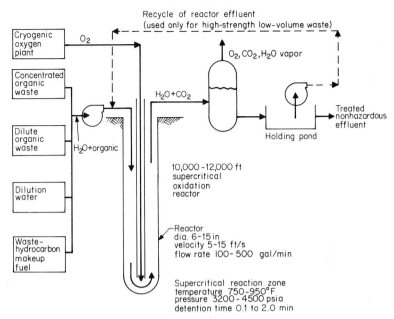

FIG. 8.10.5 Schematic flow diagram for a Vertox 10,000- to 12,000-ft supercritical-water oxidation reactor.[2]

8.10.3 PERFORMANCE DATA ON DEEP-SHAFT WET-AIR OXIDATION: SUBCRITICAL AND SUPERCRITICAL PROCESSES

Performance data for a full-scale plant is available from the subcritical-temperature EPA-sponsored facility at Longmont, Colorado. The data released from that study shows that 75 to 85% of the COD of the sludge is destroyed with remaining COD mainly as acetic acid; 90 to 96% of the volatile suspended solids (VSS) were destroyed. These results were obtained with a residence time of 45 to 50 min in the oxidation zone. According to the process owner, the observed demonstration-plant COD reductions exceeded laboratory bench-test results by 3 to 5%.[4,5] Figure 8.10.6 shows the correlation observed at Longmont between maximum downhole temperature (DHT) and COD reduction.

Ash removed from the process has been tested and found to easily pass the EP toxicity test for heavy metals as shown on Table 8.10.2. It is expected that ash from the SCW processing of industrial wastes containing heavy metals will also pass the EP toxicity test.

Data on supercritical-water wet-air oxidation is available from laboratory- and bench-scale tests. Table 8.10.3 is a partial list of compounds which have been tested in SCW oxidation at temperatures ranging from 404°C (759°F) to 574°C (1065°F). The treatability basis is either "B," indicating bench-scale testing, or "T," indicating that the class of compounds is theoretically oxidizable in SCW conditions.[6,7,8]

The laboratory- and bench-scale units used to produce the data in Table 8.10.3 are flowthrough units where the waste tested is injected into a flowing stream of

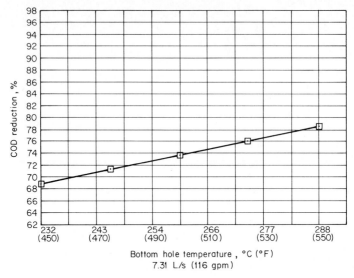

FIG. 8.10.6 Effect of DHT on COD reduction.[4]

supercritical water and oxygen as the mixture enters a reaction chamber. All the data presented in the table are for water and pure compounds.

A recent study conducted by Vertox Corporation of supercritical wet-air oxidation of sludge from an industrial wastewater-treatment process produced the results listed in Table 8.10.4 from stirred- and unstirred-autoclave tests. The sludge tested was from an industrial wastewater-treatment plant that treats wastewater from several industries, with a paper plant contributing the bulk of the flow.[12]

It is important to note the differences in the detention time required by the subcritical and supercritical processes. This is due to the much higher oxidation kinetics, total miscibility of reactants, and absence of phase boundary limitations at supercritical conditions. A major advantage of the SCW oxidation process is the high level of destruction of a very wide variety of hazardous organic compounds. The complete destruction of the organic material in the supercritical pro-

TABLE 8.10.2 Typical EP Toxicity Test Values for Composited Ash and Cleaning Residue

Test metal	Limit, mg/L	Composite, mg/L
As	5.0	0.002
Hg	0.2	0.001
Se	1.0	0.006
Cr	5.0	0.2
Cd	1.0	0.01
Ag	5.0	0.01
Pb	5.0	0.02
Ba	100.0	1.0

Source: Ref. 5.

TABLE 8.10.3 Applicability Summary for Some Organic Compounds in SCW Oxidation.

Compound	Treatability basis	Time, min	Temperature °C	°F	Removal, %
Aliphatic hydrocarbons					
Cyclohexane	B	<1	>500	>932	99.99+
Halogenated aliphatics					
1,1,1-Trichloroethane	B	<1	>500	>932	99.99+
1,2-Ethylene dichloride	B	<1	>500	>932	99.99+
Hexachlorocyclohexane	B	<1	>500	>932	99.99+
1,2-Dichloroethane	B	<1	>500	>932	99.99+
Aromatic hydrocarbons					
Toluene	B	<1	>500	>932	99.99+
Benzene	B	<1	>500	>932	99.99+
Biphenyl	B	<1	>500	>932	99.99+
o-Xylene	B	<1	>500	>932	99.99+
Halogenated aromatics					
o-Chlorotoluene	B	<1	>500	>932	99.99+
Hexachlorocyclopentadiene	B	<1	>500	>932	99.99+
1,2,4-Trichlorobenzene	B	<1	>500	>932	99.99+
4,4-Dichlorobiphenyl	B	<1	>500	>932	99.99+
PCB 1234	B	<1	>500	>932	99.99+
PCB 1254	B	<1	>500	>932	99.99+
Aliphatic alcohols	T	<1	>500	>932	99.99+
Aromatic alcohols	T	<1	>500	>932	99.99+
Aldehydes, ketones, esters,					
methyl ethyl ketone	B	<1	>500	>932	99.99+
Organic nitrogen compounds					
2,4-Dinitrotoluene	B	<1	404	759	92.5
2,4-Dinitrotoluene	B	<1	457	854	99.8
2,4-Dinitrotoluene	B	<1	513	955	99.9
2,4-Dinitrotoluene	B	<1	574	1065	99.998
Urea	B	<1	>500	>932	99.99+

Source: Refs. 6 and 7.

cess eliminates the need for additional treatment of this effluent except for ammonia nitrogen removal where required.

Off gases can be directly vented to the atmosphere and the liquid meets the requirement for discharge to surface waters since destruction of organics is complete.

8.10.4 CONSTRUCTION PROCEDURES, MATERIALS OF CONSTRUCTION, AND SYSTEM COSTS

Construction Procedures

The reactor for deep-shaft wet-air oxidation is constructed utilizing conventional oil-field equipment and installation techniques. The 1,524- to 3,048-m (5,000- to

TABLE 8.10.4 Laboratory Autoclave Testing of Oxidation of Industrial Wastewater at Supercritical Water Temperature

Parameter	Test 1*	Test 2†	Test 3†
Max. temp., (°F) °C	(906) 486	(921) 494	(921) 494
Max. press., (psia) MN/m^2	(4,930) 34.0	(4,601) 31.7	(4,825) 33.3
Time over 480°C, min	15	5	5
Liquid analysis			
Start COD, mg/L	14,750	29,500	29,500
End COD, mg/L (soluble)	300	890	370
COD reduction, %‡	97.5	97.0	98.7
End TOC, mg/L (soluble)	52	490	260
Start TSS, mg/L	31,625	27,250	27,250
End TSS, mg/L	§	14,910	18,095
Start VSS, mg/L	8,175	16,350	16,350
End VSS, mg/L¶	§	3,683	1,882
VSS reduction, %	§	77.4	88.5
Off gas			
O$_2$, % (v/v)	78.7	91.1	95.7
CO$_2$, % (v/v)	32.6	8.76	8.01
CO, % (v/v)	0.17	0.20	0.12
CH$_4$, % (v/v)	0.26	0.035	0.034

*Test 1 run in 300-mL stirred autoclave with 25-mL sample and 775-psig pure oxygen in head space. Total time to reach 375°C (707°F), 10 min from start-up temperature of 21°C (70°F).

†Tests 2 and 3 run in 300-mL stirred autoclave with 25-mL sample and 760-psig pure oxygen in head space. Total time to reach 482°C (900°F), 14 min.

‡COD removals from continuous-flow studies are expected to be much higher because of the inherent limitation of supercritical autoclave test procedures.

§It was necessary to filter the product from this run because of contamination from graphite stirrer packings. Therefore, end TSS and VSS are not available.

¶End VSS material believed to be the result of contamination of graphite stirrer packings.

Source: 1986 Vertox Corporation study.

10,000-ft) deep well is drilled using a large, rotary oil-well drilling rig. The carbon-steel casing is cemented in place from surface to well bottom to provide a leaktight containment vessel to protect the surrounding formations. Drilling and installation of the well casing can be completed in 30 to 60 days. The diameter of the well and reactor pipes is determined by the volume of waste to be treated.

Installation of the reactor, which consists of threaded sections of pipe 9.14 to 15.24 m (30 to 50 ft) in length, can be accomplished in one to three weeks. The remainder of the topside equipment and appurtenances take 6 to 18 months for complete installation. The item with the longest lead time is the cryogenic pure-oxygen plant used to supply the process oxygen. Multiple reactors are used for treatment of large quantities of waste and for process redundancy.

Construction Materials

Topside equipment and appurtenances such as pumps, valves, and tanks are off-the-shelf components selected for ease of maintenance and reliable operation.

The cryogenic oxygen plant is custom-engineered to provide the pure oxygen at the desired oxygen-injection pressure of 1,000 to 2,000 psi for the subcritical units and 3,000 to 4,000 psi for the supercritical unit.

Downhole reactor materials must be able to resist corrosion at high temperatures [177 to 510°C (350 to 850°F)], both with and without oxygen being present. High-temperature portions of the reactor are constructed of a high-nickel alloy material while lower-temperature portions of the reactor are constructed of a Duplex stainless steel.[2] Autoclave testing of the specific piping materials to be used with the specific waste to be treated are conducted during design to ensure that the materials used can withstand design operating conditions. Laboratory bench-test units for surface supercritical-water processes have demonstrated the suitability of the high-nickel alloys for the process.

Costs

A brief estimate of the total costs of supercritical deep-shaft wet-air oxidation as well as the costs of existing commonly used hazardous-waste disposal technologies is presented here. Table 8.10.5 compares the costs of several hazardous-waste treatment and disposal alternatives.

Figure 8.10.7 shows the distribution of the construction and operating costs for the 2,134-m (7,000-ft) supercritical reactor.

Total costs have been projected for supercritical reactors processing 100 gpm to 750 gpm of wastes at detention times of 1 to 5 min. Table 8.10.6 provides a breakdown of the total costs of such systems.[14] The costs for the process range from $0.07 to $0.15 per gallon of throughput. The waste stream used in this analysis has a heating value of approximately 581.5 kJ/kg (250 Btu/lb) of reactor feed. Estimated treatment costs of higher-heating-value waste streams are arrived at by diluting the waste stream to 581.5 kJ/kg (250 Btu/lb) and then using the appropriate dollar per gallon cost for the flow rate with the diluted waste stream.

The cost of the supercritical reactor and the well amounts to approximately one-half of the total system costs. Operating and maintenance costs include all costs for production of cryogenic oxygen, power, labor, and chemicals and all monitoring requirements.

TABLE 8.10.5 Comparison of Quoted Prices for Off-Site Disposal of Hazardous Waste by Nine Hazardous-Waste Firms in 1981

Type of waste management	Type or form of waste	Price 1981 $/gal
Landfill	Drummed	0.64–0.91
	Bulk	0.19–0.28
Land treatment	All	0.02–0.09
Incineration	Clean liquids, high Btu value	0.05–0.20
	Liquids	0.20–0.90
	Solids, highly toxic liquids	1.50–3.00
Deep-well injection	Oily wastewater	0.06–0.15
	Toxic rinsewater	0.50–1.00

Source: Ref. 13.

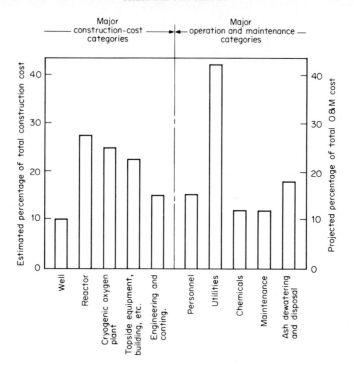

FIG. 8.10.7 Cost breakdown for proposed 2,134-m (7,000-ft), 15.77 L/s (250 gpm) supercritical Vertox reactor (*From 1986 internal Vertox Corporation study*)

TABLE 8.10.6 Projected Treatment Costs for the Vertox Supercritical Reactor*

Plant size range, gpm	Total treatment-system costs, $/gal†
50–150 (100)‡	0.13–0.15
150–400 (250)	0.09–0.12
400–1,000 (750)	0.07–0.10

*Based on 15-year life with annualized capital costs developed using 9% interest. (CRF = 0.1241)

These costs are total treatment costs including amortization of capital, O&M, site development, and residuals disposal. The costs have been developed from detailed process designs.

Total costs vary as function of maximum reactor temperature, detention time, reactor hydrodynamic design, supercritical-fluid design density, geology, influent waste characteristics, and oxygen-supply systems.

†The cost shown is for the volumetric flow rate through the reactor.

‡Figures shown in parentheses are the volumetric flow rate used in developing costs (figures in gpm).

To maintain reaction temperatures, reactor feed strength must be above 34.9 kJ/kg (15 Btu/lb). Wastes with lower heating values may require the addition of a combustible material such as municipal sludge, waste oil, or other organics to provide the needed heat energy.

Practical hydraulic limitations for the supercritical process appear to be close to the 3.154 L/s (50 gpm) flow rate; however, special applications may be feasible for wastewater flow rates even lower than this.

8.10.5 REFERENCES

1. J. J. McCarthy, *Technology Assessment of the Vertical Well Chemical Reactor,* U.S. EPA, EPA-600/2-82-005, February 1982, pp. 6, 7.

2. J. M. Smith and T. J. Raptis, *Supercritical Deep Well Wet Oxidation of Liquid Organic Wastes,* paper presented at the International Symposium on Subsurface Injection of Liquid Wastes, March 3–5, 1986, p. 2.

3. J. L. McGrew, G. L. Hartmann, C. B. Cassetti, J. E. Barnes, and W. G. Purdy, *Wet Oxidation of Municipal Sludge by the Vertical Tube Reactor,* U.S. EPA EPA-600/52-86/043, June 1986, pg 4.

4. G. B. Morrill, *Municipal Sludge Oxidation with the Vertical Tube Reactor,* paper presented at the 57th Annual Water Pollution Control Federation Conference, September 27–October 2, 1985.

5. G. C. Rappe, W. L. Schwoyer, and P. E. Dempsey, *Operating and Materials Performance of a Below-Ground Waste Destruction System for Municipal and Industrial Wastes,* presented at the 58th Annual Meeting, NY WPCA, January 13, 1986.

6. M. Modell, G. Gaudet, M. Simson, G. T. Hong and K. Biemann, "Supercritical Water: Testing Reveals New Process Holds Promise," *Solid Waste Management,* August 1982.

7. T. B. Thompson and M. Modell, "Supercritical Water Destruction of Aqueous Wastes," *Hazardous Waste,* **1**:(4)(1984).

8. *Vertox Technical Bulletin,* M-851, Vertox Corporation, Dallas, Texas.

9. L. M. Zoss, S. N. Suciu, and W. L. Sibitt, "The Solubility of Oxygen in Water," *Journal of the American Society of Mechanical Engineers,* January 1954, p. 69.

10. J. Josephson, "Supercritical Fluids," *Environmental Science and Technology,* **16**:(10).

11. R. A. Olexsey, *Issue Paper on Supercritical Fluids Processing for Hazardous Wastes,* IERL, Cincinnati, Ohio, June, 1980.

12. Vertox Corporation, *Study of Supercritical Wet Air Oxidation of Industrial Sludge,* report on testing for Gulf Coast Waste Disposal Agency, prepared by Vertox Corporation, Dallas, Texas, August 1986.

13. *Technologies and Management Strategies for Hazardous Waste Control,* report from the Office of Technology Assessment, March 1983.

14. J. M. Smith, *Vertox Corporation Cost Estimate for Supercritical Process,* Vertox Corporation, Dallas, Texas, August 1985.

SECTION 8.11

SUPERCRITICAL-WATER OXIDATION

Michael Modell
Modell Development Corporation
Framingham, Massachusetts

The process discussed in this section has been referred to as *supercritical-water oxidation (SCWO), supercritical wet oxidation,* and *supercritical wet-air oxidation*. It is the oxidation of organics, with air or oxygen, in the presence of a high concentration of water under temperatures and pressures above the critical-point values of water: 374°C and 22 MPa (705°F and 218 atm). In practice, the oxidation is usually conducted at 400 to 650°C (752 to 1200°F) under 253×10^5 Pa (250 atm).

The process underwent extensive development and testing from 1980 to 1985. It has been scaled from the bench to the pilot level. A mobile demonstration unit was successfully tested in 1985. No commercial installations have yet been built.

As an oxidation process, the efficiency can be measured as the degree of destruction of organic feed materials (e.g., DRE) or as the degree of conversion of organic carbon to carbon monoxide and carbon dioxide (e.g., DE), the latter being a more stringent measure. The DEs of SCWO can range from 99.9% at 400 to 450°C (752 to 842°F) with a residence time of 5 min, to 99.9999% at 600 to 650°C (1112 to 1200°F) with less than 1 min. The higher range of efficiencies are comparable to or greater than those of hazardous-waste incinerators.

A broad spectrum of organic compounds have been shown to be oxidized effectively in SCWO. At the higher end of the temperature range, 600 to 650°C (1112 to 1200°F), reported DEs are usually limited by the level of detectability of organics in the effluent stream. However, tests to date have involved only organic liquids and solids which are either directly soluble in supercritical water (SCW) or reform to products which are soluble in SCW. Organic solids which do not dissolve in SCW (e.g., char) will probably oxidize more slowly.

For aqueous wastes with low to moderate heat content [e.g., 8.14×10^5 to 4.652×10^6 J/kg (350 to 2000 Btu/lb)], SCWO is less costly than controlled incineration or activated-carbon treatment and more efficient than wet oxidation. Unlike incineration, the SCWO process can be rapidly bottled up by emergency shutdown procedures so as to avoid discharge of contaminated effluents during an upset or off-spec operation.

8.11.1 APPLICABILITY

In theory, SCWO is applicable to any waste stream containing organics: concentrated organic liquids, vapors, and solids; aqueous solutions containing dissolved or suspended organics; or inorganic solids with adsorbed organics. In the oxidation step, organics are converted to CO_2 and H_2O; heteroatoms form the corresponding oxyacids or salts if cations are present in the waste or added to the feed. Under supercritical conditions, salts may remain dissolved in the SCW medium [374 to 450°C (700 to 842°F)] or condense as a concentrated brine solution or as a solid particulate [$\hbar$450°C (>—½¼°F)]. Heavy metals may form oxides or carbonates, which may or may not precipitate, depending on their volatility; inert solids will largely be unaffected by the medium and remain as solids.

From an economic viewpoint, SCWO is far more attractive than incineration for treating aqueous wastes. SCWO can sustain oxidation without the addition of fuel for wastes containing 4.07×10^6 J/kg (1750 Btu/lb) (ca. 10 to 15 wt % organic), whereas incineration requires at least 30 wt % organic for autogenic oxidation. With aqueous wastes that can be preheated effectively (e.g., without char formation or corrosion), as little as 8.14×10^5 J/kg (350 Btu/lb) is required for autogenic SCWO.

The process is not applicable to wastes which are primarily inorganic (e.g., metal sludges). In its present form, SCWO has major drawbacks in treating aqueous wastes with high salt content or corrosive aqueous wastes with heating values of less than 2.09×10^6 J/kg (900 Btu/lb).

In practice, the applicability of the process may be limited by waste pressurization, char formation, or solids removal, as described here.

Waste Pressurization

Pumping is a consideration when the waste is a solid, slurry, or suspension. Solids can usually be slurried in water. Presently, high-pressure pumps are available for slurries with particulates smaller than 100 μm. Thus, the applicability of SCWO to wastes containing solids depends upon the ability to comminute the feed solids to an appropriate particle size.

Char Formation

Supercritical water has been shown to react with organics by what are conventionally referred to as *reforming reactions*. If organics are rapidly brought to supercritical-water conditions, they will reform to lower-molecular-weight species which readily dissolve in the SCW medium. On the other hand, if they are heated slowly, charring may occur. Char appears not to be solubilized rapidly by supercritical water.[1,2,3] Char should be oxidizable if given sufficient residence time, but conditions for effective oxidation have yet to be determined. Depending upon the materials of construction of the reactor, residence times beyond 1 to 5 min may be impractical.

Solids Removal

Formation of solids in the reactor may occur when inorganic salts are present. Such solids have a tendency to be "sticky" and may adhere to reactor walls or lines downstream of the reactor, leading to plugging. This scale can be removed by periodically cooling and washing the reactor and lines, but the down time may significantly impair the overall economics.

Beyond the potential plugging problems, several approaches have been suggested for removing solids which exit the reactor.[4] In some cases it is more desirable to remove inorganics as dry solids, but this approach entails the practical problems associated with depressurizing a bed of solids from 25.3 MPa (250 atm) to ambient. It is far simpler to remove the solids as a concentrated brine or slurry (e.g., by collecting the solids in a cooled vessel where a portion of the SCW condenses to liquid water), but the equipment associated with the intermediate temperature regime will have to be protected from a highly corrosive fluid. Alternatively, the solids can be carried out and cooled with the main stream of reactor effluent, resulting in a dilute brine or slurry, which may have to be posttreated for removal of toxic metals prior to discharge.

The problems described above are primarily related to mechanical equipment or material considerations; they are not inherent limitations of the SCWO process. Given the potential advantages of the process, it is not unreasonable to expect that adequate solutions will be developed in the not too distant future.

8.11.2 DESCRIPTION OF THE TECHNOLOGY

Supercritical-water oxidation of organics is significantly superior to wet oxidation in destruction efficiency. The major advantages of operating supercritically are (1) enhanced solubility of oxygen and air in water, eliminating two-phase flow; (2) rapid oxidation of organics in short residence times; (3) complete oxidation of organics, eliminating the need for auxiliary off-gas processing; (4) potential removal of inorganic constituents as solids and brines; and (5) recovery of the heat of combustion as high-temperature process heat or power. These advantages arise primarily from the unusual properties exhibited by water under supercritical conditions.

The Properties of Supercritical Water

Above the critical temperature and pressure, the properties of water are quite different from those of the normal liquid or atmospheric steam. For example, organic substances are completely soluble (i.e., miscible in all proportions) in water under some supercritical conditions, while salts are almost insoluble under other supercritical conditions. These solubility characteristics are strongly dependent upon density.

The critical point (CP), which lies on the vapor-liquid saturation dome, occurs at 374°C (705°F) and 0.3 g/cm³. The supercritical region lies above 374°C (705°F). Properties of water as a function of temperature at a supercritical pressure of 25.3 MPa (250 atm) are shown in Fig. 8.11.1. Near the critical point, the density ρ varies very rapidly with relatively small changes in temperature.

Insight into the structure of the aqueous fluid in this region has been obtained from measurements of the static dielectric constant ϵ, which is also shown in Fig. 8.11.1.[5,6] Normal liquid water has an ϵ of 80, largely as a result of strong hydrogen bonding. The dielectric constant of the saturated liquid decreases rapidly with increasing temperature, even though the density falls slowly. Since hydrogen-bonding forces are strong only when molecules are in close proximity, small increases in density parallel relatively large decreases in short-range order, resulting in a rapid decline in ϵ. At 130°C (ρ = 0.9 g/cm³), the dielectric constant is about 50, which is near that of formic acid; at 260°C (500°F) (ρ = 0.8 g/cm³), the ϵ of 25 is similar to that of ethanol. At the critical point, the dielectric constant is 5. Raman spectra of hydrogen deuterium oxide (HDO) in this region indicate little, if any, residual hydrogen bonding.[7] The major contribution to ϵ is now mo-

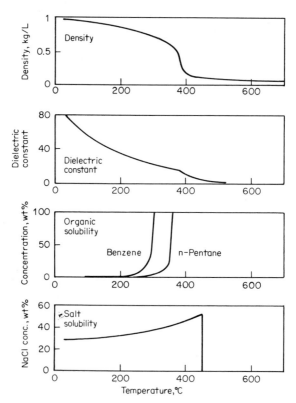

FIG. 8.11.1 Properties of water at 25.3 MPa (250 atm).

lecular association due to dipole-dipole interactions, which gradually decrease with increasing void volume.

While dielectric constant is not the sole determinant of solubility behavior, the solvent power of water for organics is consistent with the variation of ϵ, as shown in Fig. 8.11.1. Benzene solubility in water is a good example.[8,9] At 25°C (77°F), benzene is sparingly soluble in water (0.07 wt %). Between 260 and 300°C (500 and 572°F), benzene solubility increases from a few percent to completely miscible. Other hydrocarbons exhibit similar solubility behavior.

For densities less than 0.7 g/cm^3, the solubility of inorganic salts in water is as unusual as that of organics. At 25.3 MPa (250 atm), the solubilities of salts reach a maximum at 300 to 450°C (572 to 842°F). Beyond the maximum, the solubilities drop very rapidly with increasing temperature (see Fig. 8.11.1). For example, NaCl solubility is about 40 wt % at 300°C (572°F) and about 100 ppm at 450°C (842°F); $CaCl_2$ has a maximum solubility of 70 wt % at subcritical temperatures, which drops to 10 ppm at 500°C (932°F).[10] Given the fact that the dielectric constant of water is about 2 at 490°C (914°F) and 25.3 MPa (250 atm), it is not surprising that inorganics are practically insoluble.

Coincident with the loss of solvating power for inorganic salts, supercritical water also loses the ability to dissociate salts. For example, the dissociation constant of NaCl at 400 to 500°C (752 to 932°F) and densities in the range of 0.35 g/cm^3 is of the order of 10^{-4}.[11] Thus, strong electrolytes become weak electrolytes in supercritical water.

Process Description

Supercritical-water oxidation can be applied to wastes with a wide range of organic concentration. A schematic flowsheet for a process for treating an aqueous waste containing 10 wt % organic is given in Fig. 8.11.2. This process consists of the following steps:

1. The waste, as either an aqueous solution or a slurry, is pressurized and delivered to the oxidizer inlet. It is heated to supercritical conditions by direct mixing with recycled reactor effluent.

2. Oxygen is supplied in the form of compressed air, which is used as the motive fluid in an eductor to provide recycling of a portion of the reactor effluent. This inlet mixture is then a homogeneous phase of air, organics, and supercritical water.

3. The organics are oxidized in a controlled but rapid reaction. The heat released by combustion of readily oxidized components is sufficient to raise the fluid phase to temperatures at which all organics are oxidized rapidly.

4. The effluent from the oxidizer is fed to a cyclone. Inorganic salts that are originally present in the feed or which form in the combustion reactions precipitate out of the fluid phase in the oxidizer and are separated here.

5. The fluid effluent of the solid separator is a mixture of H_2O, N_2, and CO_2. A portion of this is recycled through the eductor to provide supercritical conditions at the oxidizer inlet.

6. The remainder of the effluent is available as a high-temperature, high-pressure fluid for energy recovery. This stream is cooled to a subcritical temperature in a heat exchanger which serves to generate low-pressure or high-pressure steam.

7. Now at a subcritical temperature, the mixture has formed two phases and en-

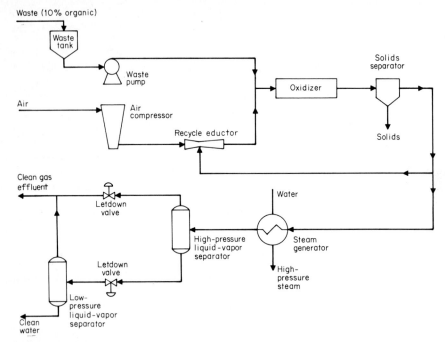

FIG. 8.11.2 Process schematic for oxidation of an aqueous waste with a heating value of 4.07×10^6 J/kg (1750 Btu/lb).

ters a high-pressure liquid-vapor separator. Practically all the N_2 and most of the CO_2 leave with the gas stream. The liquid consists of water with an appreciable amount of dissolved CO_2.

8. The gas stream can then be expanded through a turbine to extract the available energy as power. A portion of the power is used for compression of the inlet air.

9. The liquid from the high-pressure separator is depressurized and fed to a low-pressure separator. The vapor stream is primarily CO_2, which is vented with the gas-turbine effluent. The liquid stream is clean water.

The DE is a function of reactor temperature and residence time. It has been found that a reactor effluent temperature in the range of 600 to 650°C (1112 to 1200°F) and residence time of 5 s are sufficient for DEs of more than 99.999%.[18] Higher temperatures could be used to reduce the residence time. However, at 5 s, the reactor cost is a small fraction of total capital cost and, therefore, there is not much incentive to try to reduce reactor volume by operating above 650°C (1200°F).

For the process configuration illustrated in Fig. 8.11.2, the reactor exit temperature is a direct function of the heating value of the feed. To attain a temperature of 600 to 650°C (1112 to 1200°F), the waste should contain about 4.07×10^6 J/kg (1750 Btu/lb), which is the heating value of an aqueous solution of about 10 wt % benzene [heat of combustion, 4.07×10^7 J/kg (17,500 Btu/lb)] or 14 wt % ethanol [2.944×10^7 J/kg (12,800 Btu/lb)]. The energy released by combustion is contained within the reactor effluent as thermal energy. As shown in Fig. 8.11.2, it could be recovered as heat in the steam generator and/or power from an expansion turbine. The energetics of the SCW oxidation process are such that the amount of power available for recovery is substantially more than that required to compress the air and waste. The overall process is somewhat analogous to a gas-turbine power cycle. However, many applications require relatively small systems, which are capital-intensive. In those cases, power recovery cannot be justified on economic grounds and, thus, the heat of combustion and the energy input for air compression are simply recovered as steam.

For wastes with heating values below 4.07×10^6 J/kg (1750 Btu/lb), auxiliary fuel could be added to make up the required heating value. The fuel cost can be appreciable when treating a very dilute waste. In such cases, it is more economical to omit the steam generator and eductor and use instead a regenerative heat exchanger, as shown in Fig. 8.11.3. In this manner, the minimum heating value of the feed for autogenic operation is 8.14×10^5 J/kg (350 Btu/lb), which is a concentration of 2 wt % benzene-equivalent.

Advantages of Supercritical-Water Oxidation

As a waste-destruction process, SCW oxidation has several advantages over conventional processes. The chemical reactions that occur are carried out in a closed

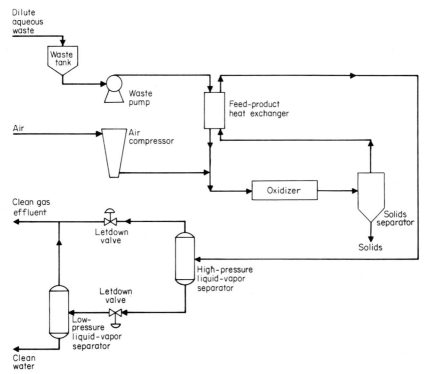

FIG. 8.11.3 Process schematic for oxidation of an aqueous waste with a heating value of 8.14×10^5 J/kg (350 Btu/lb).

system, making it possible to maintain total physical control of waste materials from storage, through the oxidation process, to the eventual discharge of the products of combustion. This feature provides positive assurance of environmental protection. In addition, bench-scale results have demonstrated essentially complete combustion of chemically stable hazardous and toxic materials. The process can be adapted to a wide range of feed mixtures and scales of operation. Systems can be designed as skid-mounted, transportable units and as larger-scale stationary units. The process is capable of generating all the power required for air compression and feed pumping and, thus, can have no net energy requirement for system operation.

8.11.3 DESIGN CONSIDERATIONS

The reactor and heat exchanger are the key components in the SCWO system and their design involves considerations of residence time, temperature, and materials of construction. The choice of reactor material influences both the maximum temperature permissible (tensile strength of alloys drops off rapidly above some temperature limit) and the cost of the reactor (volume depends upon residence time).

Materials of Construction

To date, reactors have been constructed of Hastelloy C-276 and Inconel 625.[13,14] These high-nickel alloys exhibit excellent corrosion resistance in aggressive environments.[15] In addition, these materials retain adequate tensile strength at high temperature to be used for pressure vessels [550°C (1022°F) for Hastelloy C-276 and 650 to 700°C (1200 to 1292°F) for Inconel 625].

Corrosion is known to be one of the major problems in wet-air oxidation systems. Nelson and Van Kirk[16] conducted accelerated-corrosion tests for a number of alloys and found that Hastelloy C-276 and Inconel 625 showed no signs of stress corrosion cracking, slight pitting tendencies, and minimal corrosion rates. Their results were in agreement with those of Jagow,[17] who selected Hastelloy C-276 for the reactor of a wet-oxidation system for a spacecraft.

MODAR, Inc.,[14] conducted accelerated-corrosion tests of Hastelloy C-276 and Cabot Alloy 625 under SCWO conditions. Prestressed coupons were exposed to chloride levels of 5,000 and 10,000 ppm for several hundred hours at 550°C (1022°F) and 25.3 MPa (250 atm). The chloride was presumably in the form of HCl, formed by oxidation of trichloroethylene. The results indicated that corrosion at reactor conditions is not a problem for the two materials studied.

From these results, it appears that SCWO conditions are significantly less corrosive than the wet-oxidation environment. Although wet oxidation involves lower temperatures [e.g., 150 to 300°C (300 to 572°F)], the lower density and dielectric constant of water under SCWO conditions result in a less ionic environment. The fact that sodium chloride does not dissociate appreciably above 400°C (752°F) is confirmation that water has lost its ability to support electrochemical phenomena such as corrosion.

Reactor Design

As illustrated in Figs. 8.11.2 and 8.11.3, the SCWO reactor is usually designed to operate above 400°C (752°F) and below 650°C (1200°F). The lower limit is dictated by residence-time considerations. Below 400°C (752°F), residence times in excess of 5 min are required to reach DEs of 99.9%. Reactor size and cost can become prohibitive. The upper temperature is limited by the strength of the materials of construction. At the present time, 650°C (1200°F) appears to be the practical limit.

In order to maintain reactor temperatures within this range, there is an upper limit to the heating value of the feed. If the reactor inlet were at 400°C (752°F), then a feed mixture entering the reactor having a heat of combustion of 8.14×10^5 J/kg (350 Btu/lb) will liberate sufficient energy to bring the reactor effluent to 650°C. If the waste has heat content in excess of this value, it must be mixed with sufficient water (or reactor effluent, as in Fig. 8.11.2) to dilute it accordingly.

Heat-Exchanger Design

Regenerative heat exchange from reactor effluent to pressurized feed is the major feature that improves the overall thermal efficiency of SCWO over incineration for treatment of dilute aqueous wastes (see Fig. 8.11.3). The driving force for regenerative heat exchange is the temperature differential created by oxidation of waste in the reactor. To minimize the heat-transfer area required, one would prefer to operate with the maximum temperature rise, which was given above as 400 to 650°C (752 to 1200°F), corresponding to a heating value of 8.14×10^5 J/kg (350 Btu/lb). Wastes with lower heating values can be treated at a higher cost, dictated by a trade-off of increased heat-exchanger area versus addition of fuel.

Heat-exchanger fouling or scale buildup can be a major factor limiting the kind and concentration of salts that can be processed. Salts exhibiting inverse solubilities (i.e., decreasing solubility at higher temperature) tend to build up as scale on heat-transfer surfaces. Calcium salts are well-known examples; many sodium salts also exhibit this phenomenon.[11] The problem can be dealt with by periodically flushing the exchanger with warm water or dilute nitric acid, but at a cost of down time.

Oxidant

Both oxygen and air have been used for SCWO with equal success.[14] The choice is dictated by economics. Oxygen compression is considerably less expensive than air (on an equivalent-oxygen basis), but represents an additional raw-material cost. Where oxygen can be purchased or manufactured for $30 to 50 per ton, it is the oxidant of choice. Otherwise, it is usually less expensive to compress ambient air.

8.11.4 ENVIRONMENTAL IMPACT

SCWO produces aqueous and sometimes solid effluents. When solids are removed as dry particulates, they represent an ultimate disposal issue similar to that posed by incinerator ash. The liquid effluent is water which is essentially free of organics [e.g., total organic carbon (TOC)<1 ppm] but may contain dissolved salts or acids and inert particulates (depending upon the mode of solids removal).

The gaseous effluent is typically a mixture of nitrogen (when using air as the oxidant), carbon dioxide, oxygen, and small amounts of carbon monoxide (1 to 100 ppm). Stack-gas scrubbing of SCWO effluent is not required because scrubbing is effected in situ during cooling of the reactor effluent.

As with other oxidation processes, the key to success is avoiding upsets or off-spec conditions, which can lead to incomplete oxidation and contaminated effluents. Like incineration, SCWO has a built-in early warning system in that increasing carbon monoxide concentration in the off gas is the first sign of off-spec conditions. Unlike incineration, the SCWO process can be rapidly bottled up by emergency shutdown procedures so as to avoid discharge of contaminated effluents into the environment. Contaminated aqueous effluent can be contained in a holding tank and later recycled by mixing with aqueous waste. Thus, by giving adequate attention to process control hardware and software, SCWO treatment can avoid the pitfalls (and consequent difficulties in permitting) that have befallen conventional incineration systems.

8.11.5 RESULTS OF TESTS

Most of the destruction tests reported to date have been conducted at the bench scale using synthetic mixtures of pure chemicals. Bench-scale units have been described by Timberlake[13] and MODAR, Inc;[14] both were of similar design. They consisted of a continuous-flow system with a throughput of 0.06 m^3/d (20 gal/d) of water and 0.003 m^3/d (1 gal/d) of organics. Operating temperatures ranged from 400 to 700°C at 250 atm. Pure water was pumped to operating pressure and heated to supercritical temperature. Organic liquid and oxidant were pressurized, mixed with the SCW, and fed to a tubular reactor. Residence time was controlled by reactor length and SCW flow rate. After exiting the reactor, the effluent was cooled to room temperature, depressurized to 1 atm, and fed to a gas-liquid separator. The two effluent streams were analyzed to determine DEs.

A compilation of results of bench-scale tests is given in Table 8.11.1. In general, these data indicate that DEs of the order of 99.9% can be obtained at 400 to 500°C (752° to 932°F) with 1- to 5-min residence time; 99.99% at 500 to 550°C (932 to 1022°F) with about 1 min; and 99.999% at 550 to 600°C (1022 to 1112°F) with less than 1 min. It has also been stated that 99.9999% is achievable at 600 to 650°C (1112 to 1200°F) with a residence time of seconds,[18] but such data have not been published.

Bench-scale tests are now in progress for destruction of urine and feces.[19] A slurry-pumping module has been constructed and successfully demonstrated, but results have not been published as of this writing.

A bench-scale test has been conducted on biopharmaceutical materials.[20] The synthetic mixture contained *Bacillus stearothermophilus* spores, trichloroacetic acid, fluorescein, ethylene glycol, sodium chloride, and sodium hydrogen phosphate. No viable spores or intact cells survived the treatment. Effluent TOC was below 1 ppm.

MODAR, Inc., has constructed a mobile pilot plant for field demonstrations of SCWO using actual wastes. The unit has a nominal capacity of 0.15 m^3/d (50 gal/d) of organic material. The first evaluation tests were concluded in 1985.[28] Two types of wastes were processed. The first was an aqueous waste of isopropyl alcohol and included supplemental components of 2-chlorophenol,

Table 8.11.1   Results of Bench-Scale Evaluation Tests

Compound	Temp, °C	Time, min	DE, %	Reference no.
Aliphatic hydrocarbons				
Cyclohexane	445	7.	99.97	14
Aromatic hydrocarbons				
Biphenyl	450	7.	99.97	14
o-Xylene	495	3.6	99.93	4
Halogenated aliphatics				
1,1,1-Trichloroethane	495	3.6	99.99	4
1,2-Ethylene dichloride	495	3.6	99.99	4
1,2-Dichloroethane				
1,1,2,2-Tetrachloroethylene	495	3.6	99.99	4
Hexachlorocyclohexane				
Halogenated aromatics				
o-Chlorotoluene	495	3.6	99.99	4
Hexachlorocyclopentadiene	488	3.5	99.99	4
1,2,4-Trichlorobenzene	495	3.6	99.99	4
4,4-Dichlorobiphenyl	500	4.4	99.993	4
DDT	505	3.7	99.997	4
PCB 1234	510	3.7	99.99	4
PCB 1254	510	3.7	99.99	4
Oxygenated compounds				
Methyl ethyl ketone	460	3.2	99.96	14
Methyl ethyl ketone	505	3.7	99.993	4
Dextrose	440	7.	99.6	14
Organic nitrogen compounds				
2,4-Dinitrotoluene	457	0.5	99.7	12
2,4-Dinitrotoluene	513	0.5	99.992	12
2,4-Dinitrotoluene	574	0.5	99.9998	12

nitrobenzene, and 1,1,2-trichloroethane. Approximately 3.40 m^3 (900 gal) was processed and destruction efficiencies of up to 99.9999% were reported. In a second test using a transformer fluid containing PCBs, the effluent PCB was below the detection limit, which equated to a destruction efficiency of over 99.99%.

8.11.6   CURRENT RESEARCH

At 600 to 650°C (1112 to 1200°F), the destruction efficiency of SCWO is comparable to or greater than that of hazardous-waste incineration even though the temperature of SCWO is considerably lower than that of the flame [2000 to 2500°C (3632 to 4532°F)] or thermal oxidation zone [900 to 1200°C (1652 to 2192°F)] of incinerators. Although the mechanism of SCWO is not yet clearly understood, the longer residence time of SCWO may more than compensate for the lower temperature. It may also be that the shorter mean free path in the supercritical-water medium prevents loss of free radicals (e.g., by collision with walls) and is therefore more efficient than thermal oxidation at ambient pressure.

The fundamental mechanism of SCWO has only recently been the subject of research. Helling[22] has studied the kinetics of oxidation of CO in supercritical water in the range of 400 to 540°C (752 to 1000°F), 24.3 MPa (240 atm), and 6- to 13-s residence time. Empirical Arrhenius parameters and reaction order were 28.7 kcal/(g · mol) activation energy; $10^{7.25}$ preexponential factor; first order in carbon monoxide and zero order in oxygen. At the lower end of the temperature range, the water gas shift reaction contributed significantly to carbon monoxide disappearance and hydrogen production. Helling extrapolated existing models for high-temperature, free-radical CO oxidation to the temperature range under study. Although he concluded that the fit was not good, it was probably better than one would expect given the large temperature range encompassed by the extrapolation.

Helling also attempted to study the kinetics of ammonia oxidation, but found the rate too slow under his experimental conditions.

8.11.7 ECONOMIC ESTIMATES

The only costs for SCWO published to date are those of MODAR, Inc.[12,14] Since no commercial installations have yet been built, the published costs should be considered only as estimates.

Costs for SCWO may vary with throughput and heating value of the waste. Two cases have been treated in the literature: heating values of 4.07×10^6 J/kg and 8.14×10^5 J/kg (1750 and 350 Btu/lb).[12] The former is in reference to a 10% organic waste because a 4.07×10^6 J/kg (1750 Btu/lb) heating value is equivalent to 10 wt % benzene in water. The latter is equivalent to 2 wt % benzene, but is simply referred to as *dilute waste*.

10% Organic

For wastes with a heating value of 4.07×10^6 J/kg (1750 Btu/lb), a schematic proposed by MODAR, Inc., was given in Fig. 8.11.2. The waste is introduced, without preheating, directly into an oxidizing reactor. A portion of the effluent is recycled through an eductor so as to heat the feed. This approach provides an effluent at 600 to 650°C (1112 to 1200°F), which is cooled in a steam-generating heat exchanger.

Table 8.11.2 contains cost estimates recently published by MODAR, Inc., for a unit with a capacity of 9,070 kg/d (10 tons/d) of a waste with 10 wt % organic content. Since these estimates provide both capital and operating costs, one can extrapolate from these to other throughputs. Table 8.11.3 was generated from the estimates of Table 8.11.2 on the assumption that the capital cost follows a 0.6 power with throughput. Two sets of costs are shown in Table 8.11.3, for capital-recovery factors of 25% and 12%.

These estimated costs compare very favorably to those for incineration of aqueous wastes with comparable heating value.[23] For throughputs higher than 9,070 to 18,140 kg/d (10 to 20 tons/d) organic, the total estimated costs of SCWO treatment are less than the fuel cost alone of liquid-injection incineration with effluent scrubbers.

Also shown in Table 8.11.3 is a projected cost for a 90,700 kg/d (100 tons/d) organic capacity for an advanced system with a turboexpander for power recovery (instead of the steam generator shown in Fig. 8.11.2). Since the expander in-

Table 8.11.2   Projected Costs for 10 tons/d Throughput

Annual capacity	10 tons/d × 330 d/yr = 3,300 tons/yr
Capital-cost estimate (CCE)	
Tanks, pumps, eductor	
Heat exchanger, reactor, and instrumentation	
Separators, letdown units, start-up furnace	
Air compressor	
Contingencies	
Total	$5,320,000
Processing costs	
Chemicals and utilities (elec @ 4¢/kWh;	
H_2O @ 5.5¢/10^6 gal	
Operating labor and plant overhead	
Maintenance (5% CCE)	
By-product steam credit ($6.00/$10^6$ lb)	
Cost of capital (25% CCE)	
Total operating cost	$2,143,000
Unit cost/gal organic	$2.38
Unit cost/gal wastewater at 10%	
organic content	$0.26

volves development of a new piece of turbomachinery, one would not expect such a system to be available in the near future.[14] However, the projected costs are extremely attractive because the system would provide an alternative for concentrated wastes which is comparable in cost to current costs of deep-well injection.

Dilute Wastes

MODAR, Inc., has proposed to treat dilute aqueous wastes by the process displayed in Fig. 8.11.3. In this case, a regenerative heat exchanger is used to pre-

Table 8.11.3   Extrapolated Costs for SCWO Treatment of an Aqueous Waste with 10 wt % Organic Content

Throughput		Capital-cost estimate, $ million	Operating Costs, $/gal	
Organic, tons/d	Waste, gal/min		25% capital recovery	12% capital recovery
2	3	2.0	0.43	0.27
4	7	3.1	0.35	0.22
8	14	4.7	0.28	0.19
10	17	5.3	0.26	0.18
12	21	5.9	0.25	0.17
20	35	8.1	0.21	0.15
50	87	14.	0.17	0.12
100	174	21.	0.14	0.11
100	174	15. *	0.04*	0.02*

*Projected costs for advanced systems with turboexpander for power recovery.

heat the feed and cool the reactor effluent. The heating value of the feed provides for heat losses as well as for differences across the heat exchanger. The lower the heating value of the feed, the lower the temperature differences across the heat exchanger and the larger the heat exchanger.

In research reported in early publications, MODAR, Inc., used 2 wt % organic as the minimum concentration that could be oxidized without the addition of auxiliary fuel. More recent publications suggest that the lower limit might be 1 wt % organic if oxygen is used instead of air.

The projected costs for treating dilute wastes are thought to be comparable, per unit volume of waste, to those of more concentrated wastes for the same throughput of waste.[12] Consider a unit processing 9,700 kg/d (10 tons/d) organics at 10 wt % organic, which is 17 gpm of waste, versus one processing 9,700 kg/d (10 tons/d) organics at 2 wt % organic, which is 87 gpm of waste. The two processes would require the same air-compression system because the oxygen requirement is directly proportional to the organic flow rate. Although the dilute waste would require a reactor with 5 times the volume of the 10 wt % case, the reactor cost is not a dominant factor when operating at 600 to 650°C where the residence time is less than 1 min; excluding heat exchangers, the major cost is that of oxidant compression.[18] Thus, for the two cases being compared, the systems differ primarily in that the dilute case requires the addition of the regenerative heat exchanger. According to Thomason and Modell,[12] the 10 wt % case would have a processing cost of $0.26 per gallon, while the 2 wt % case would have a cost of $0.17 per gallon (Table 8.11.3, 25% capital recovery for 17 gpm and 87 gpm, respectively). The dilute case processes 5 times the waste at about two-thirds the unit cost.

Other Waste Concentrations

As discussed previously, the highest heating value that can be treated by an SCWO system is 4.07×10^6 J/kg (1750 Btu/lb) (see Subsec. 8.11.3). More concentrated wastes could always be treated by diluting the feed with lower-heating-value aqueous wastes, recycled aqueous effluent, or water. The processing costs for more concentrated wastes would then be the entry in the row in Table 8.11.3 for the actual organic throughput, multiplied by the ratio of waste throughput in column 2 to the actual waste throughput. For example, a 20 wt % organic waste at 10 tons/d organics would have a processing cost (at 25% capital recovery) of $0.26 \times (17/8.5)$, or $0.52 per gallon of waste.

For wastes with heating values between 8.14×10^5 J/kg (350 Btu/lb) and 4.07×10^6 J/kg (1750 Btu/lb), one can bracket the cost by using the procedures above for estimating the costs of 10 wt % organic and dilute wastes.

For wastes with heating values less than 8.14×10^5 J/kg (350 Btu/lb), fuel can be added to bring the heating value up to 8.14×10^5 J/kg (350 Btu/lb). At the extreme of a waste with no heating value, the fuel requirement is about 0.02 gallon of fuel per gallon of waste, which corresponds to a fuel cost of $0.02 per gallon at a unit cost of fuel of $1 per gallon.

8.11.8 VENDORS

Supercritical-water oxidation systems, as described in this section, are offered for commercial applications by MODAR, Inc., 3200 Wilcrest, Suite 220, Houston,

Texas 77042, (713) 785-5615. Special applications, laboratory testing, and development services are available through Modell Development Corp., 39 Loring Drive, Framingham, Massachusetts 01701, (617) 820-9213. Supercritical-water oxidation in downhole applications is available through Oxidyne, Inc., 14881 Quorum Drive, 9th Floor, Dallas, Texas 75240, (214) 701-8200.

8.11.9 REFERENCES

1. M. Modell, R. C. Reid, and S. I. Amin, *Gasification Process,* U.S. Patent 4,113,446, September 9, 1978.

2. G. V. Deshpande, G. D. Holder, A. A. Bishop, and J. Gopal, "Extraction of Coal Using Supercritical Water," *Fuel,* **63**:956–960 (1984).

3. M. Modell, "Gasification and Liquefaction of Forest Products in Supercritical Water," chap. 6 in *Fundamentals of Thermochemical Biomass Conversion,* R. P. Overend, T. A. Milne, and L. K. Mudge (eds.), Elsevier Scientific, 1985.

4. M. Modell, *Processing Methods for the Oxidation of Organics in Supercritical Water,* U.S. Patent 4,543,190, September 24, 1985.

5. A. S. Quist and W. L. Marshall, "Estimation of the Dielectric Constant of Water to 800°," *J. Phys. Chem.,* **69**:3,165 (1965).

6. M. Uematsu and E. U. Franck, *J. Phys. Chem. Reference Data,* **9**:(4)1,291–1,306 (1980).

7. E. U. Franck, "Properties of Water," in *High Temperature, High Pressure Electrochemistry in Aqueous Solutions,* (NACE-4), 1976, p. 109.

8. J. F. Connelly, "Solubility of Hydrocarbons in Water Near the Critical Solution Temperature," *J. Chem. Eng. Data,* **11**:13 (1966).

9. C. J. Rebert and W. B. Kay, "The Phase Behavior and Solubility Relations of the Benzene-Water System," *AIChE Journal,* **5**:285 (1959).

10. O. I. Martynova, "Solubility of Inorganic Compounds in Subcritical and Supercritical Water," in *High Temperature, High Pressure Electrochemistry in Aqueous Solutions,* (NACE-4), 1976, p. 131.

11. W. L. Marshall, "Predicting Conductance and Equilibrium Behavior of Aqueous Electrolytes at High Temperatures and Pressures," in *High Temperature, High Pressure Electrochemistry in Aqueous Solutions,* (NACE-4), 1976, p. 117.

12. T. B. Thomason and M. Modell, "Supercritical Water Destruction of Aqueous Wastes," *Hazardous Waste,* **1**:(4)453 (1984).

13. S. H. Timberlake, G. T. Hong, M. Simson, and M. Modell, "Supercritical Water Oxidation for Wastewater Treatment: Preliminary Study of Urea Destruction," *SAE Tech. Paper Ser.* No. 820872 (1982).

14. "Detoxification and Disposal of Hazardous Organic Chemicals by Processing in Supercritical Water," Final Report, U.S. Army Contract No. DAMD 17-80-C-0078, prepared by MODAR, Inc., Houston, Texas, 1987.

15. A. J. Asphahani, "Overview of Advanced Materials Technology," *Chem. Eng. Prog.,* **82**:(6)33 (1986).

16. J. A. Nelson and J. W. Van Kirk, "Stress Corrosion Cracking Evaluation of Alloys in the Accelerated Wet Ox Environment," Information Corrosion Forum, paper no. 43, Atlanta, March 12, 1979.

17. R. B. Jagow, "Development of a Spacecraft Wet Oxidation Waste Processing System," ASME paper no. 72-ENAs-3, 1972.

18. M. Modell, "Supercritical Water Treatment of Organic Wastes," HAZMACON 85 Conference, Oakland, April 25, 1985.

19. "Supercritical Water Oxidation of Urine and Feces," Progress Report, NASA Contract No. NAS 2-12176, prepared by MODAR, Inc., Houston, Texas, April 1986.

20. V. L. Cunningham, P. L. Burk, J. B. Johnston, and R. E. Hannah, "The MODAR Process; An Effective Oxidation Process for Destruction of Biopharmaceutical By-products," AIChE National Meeting, Boston, August 26, 1986.

21. "MODAR Toxic Waste Destruction Process Demonstration Program," Final Report to the U.S. EPA and NYSDEC, from CECOS International and MODAR, Inc., 1986.

22. R. K. Helling, "Oxidation Kinetics of Simple Compounds in Supercritical Water: Carbon Monoxide, Ammonia and Ethanol," Sc.D. thesis, M.I.T., Cambridge, Massachusetts, 1986.

23. M. Modell, "Incineration of Low Heat Content Waste," AIChE National Meeting, Boston, August 27, 1986.

24. W. M. Flarsheim, Y-M. Tsou, I. Trachtenberg, K. P. Johnston, and A. J. Bard, "Electrochemistry in Nearcritical and Supercritical Fluids. 3. Studies of Br − , I − , and Hydroquinone in Aqueous Solutions," *J. Phys. Chem.*, **90**:3,857 (1986).

25. M. Modell, *Processing Methods for the Oxidation of Organics in Supercritical Water*, U.S. Patent 4,338,199, July 6, 1982.

26. M. Modell, G. G. Gaudet, M. Simson, G. T. Hong, and K. Biemann, "Supercritical Water," *Solid Waste Management*, August 1982.

27. A. R. Wilhelmi and P. V. Knopp, "Wet Air Oxidation—An Alternative to Incineration," *Chem. Eng. Prog.*, **75**:(8)46 (1979).

28. C. N. Staszak, K. C. Malinowski, and W. R. Killilea, "The Pilot-Scale Demonstration of the MODAR Oxidation Process for the Destruction of Hazardous Organic Waste Materials," *Environmental Progress*, **6**(1):39–43 (February 1987).

8.11.10 *FURTHER READING*

There are a number of related fields of activity wherein research findings may provide insight into SCWO theory or applications: supercritical-water extraction of fossil fuels (see, e.g., Ref. 2); supercritical-water re-forming of organics (see, e.g., Ref. 3); and electrochemical reactions in the near and supercritical regions (see, e.g., Ref. 24). Helling (Ref. 22) contains a review of much of the literature prior to 1986 and is recommended as a good starting point for those who wish to delve deeper into these subjects.

SECTION 8.12

PLASMA SYSTEMS

C. C. Lee, Ph.D.

U.S. Environmental Protection Agency
Hazardous Waste Engineering Research Laboratory
Cincinnati, Ohio

Plasma systems technology uses a *plasma-arc* device to create extremely high temperatures (temperatures approach 10,000°C) for destruction of highly toxic wastes. Gaseous emissions (mostly H_2, CO), acid gases in the scrubber, and ash components in scrubber water are the residuals. The system's advantages are that it can destroy refractory compounds, the equipment can be made portable, and typically the process has a very short on-off cycle. There are two plasma-arc systems currently under development. One is plasma arc for liquid-waste destruction and the other is for solid-waste destruction.

8.12.1 APPLICABILITY OF THE TECHNOLOGY

Although plasma arc itself has been successfully applied in the metals industry for many years, application of the plasma arc to hazardous-waste destruction is still in a pilot-scale research stage. Initial test results have shown that the plasma arc is a promising alternative for destruction of difficult to treat wastes such as dioxin-contaminated sludges. If the cost of this technology is competitive with conventional incinerators, then the applicability of the plasma arc to waste destruction could be significant. It could be used to destroy not only part or all of the liquid organic hazardous wastes generated in the United States, but also capacitors used in the utility industry.

8.12.2 DESCRIPTION OF THE TECHNOLOGY

The most common method of plasma generation is electrical discharge through a gas. The gas used is relatively unimportant in creating the discharge, but will ultimately affect the products formed. In passing through the gas, electrical energy is converted to thermal energy and is absorbed by gas molecules, which are activated into ionized atomic states, losing electrons in the process. Arc temperatures up to 10,000°C may be achieved along the centerline of the recirculation vortex. Radiation is emitted when molecules or atoms relax from the highly activated states to lower energy levels.

The plasma, when applied to waste disposal, can best be understood by thinking of it as an energy-conversion and energy-transfer device. As the activated components of the plasma decay, their energy is transferred to waste materials exposed to the plasma. The wastes are then atomized, ionized, pyrolyzed, and finally destroyed as they interact with the decaying plasma species. Theoretically speaking, the destruction of wastes should result in simple molecules or atoms such as hydrogen, carbon monoxide, carbon, and hydrochloric acid. The off gases from the plasma system are scrubbed to remove hydrochloric acid and are then flared.

One of the key factors in waste destruction by plasma arc is radiative heat transfer. The following is an example to demonstrate how rapidly a carbon particle can increase its temperature from an atmospheric temperature to 2200°C under the radiative heat-transfer environment.

The radiative heat-transfer equation can be expressed as[2,3]

$$P = QGT^4$$

Where P = blackbody radiative power radiated per unit area, W/cm^{-2}
  Q = Stefan-Boltzmann constant, $5.3. \times 10^{-13}$ $W/(cm^{-2} \cdot K^{-4})$
  G = emissivity of the substance, $0<G<1$
  T = absolute temperature of the substance, K

This function is plotted in Fig. 8.12.1. Inspection of Fig. 8.12.1 shows that radiation as a heat-transfer method does not become significant until temperatures on the order of 1650°C are reached.

The spectrum of the blackbody radiator changes, of course, with temperatures. Fig. 8.12.2 shows the relative spectral distribution of the blackbody.

In the near-infrared, many materials exhibit very strong absorption coefficients: coal, for example, has an absorption coefficient on the order of 10^5 cm^{-1} which means (to a first approximation) that most of the incident energy is absorbed in a surface layer on the order of 10^{-5} cm thickness.

The rate of temperature rise on the surface of a coal particle can be estimated by the equation

$$\frac{T}{t} = \frac{P}{MC} \quad \text{and} \quad M = \frac{d}{a}$$

Where T = temperature rise
  t = unit time
  P = radiative power

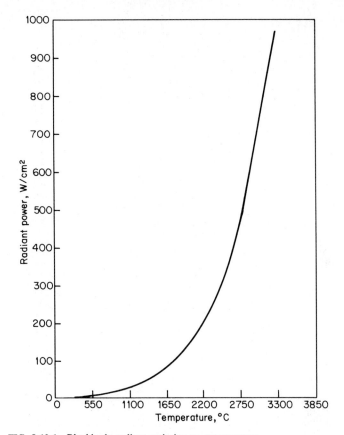

FIG. 8.12.1 Blackbody radiant emission vs. temperature.

C = specific heat
M = mass
d = density
a = absorption coefficient

At 2200°C, the blackbody radiative power density is on the order of 200 W/cm^2. The specific heat of coal is 0.9 J/(g · °F) and its density is 1.4 g/cm^3; therefore the rate of temperature rise on the surface of a coal particle is

$$\frac{T}{t} = \frac{200 \times 10^5}{1.4 \times 0.9}$$

$$= 1.58 \times 10^7 \text{ °C/s}$$

where 1 W = 1 J/s.

Thus, the time needed to raise the temperature to 2200°C (4000°F) on the surface of an absorbing coal particle is

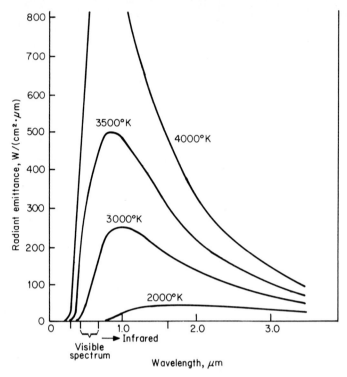

FIG. 8.12.2 Spectral output of blackbody radiators at various temperatures.

$$t = \frac{2,200}{1.58 \times 10^7}$$
$$= 1.4 \times 10^{-4} \text{ s}$$

Plasma Arc for Liquid-Waste Destruction

A schematic of the plasma-arc system for destruction of liquid wastes is shown in Fig. 8.12.3. Its major components include a liquid-waste feed system, plasma torch, reactor, caustic scrubber, on-line analytical equipment, and flare.[1] The system is rated at 4 kg/min or approximately 55 gal/h of waste feed and is housed in a 45-ft trailer. Its product-gas rate before and after flare is about 5 to 6 m^3/min and 30 to 40 m^3/min, respectively. For the purpose of sampling, a flare-containment chamber and a 10-m stack were constructed to facilitate testing.

The system was constructed in 1982 under a joint research program of the U.S. Environmental Protection Agency (EPA) and the New York State Department of Environmental Conservation (NYSDEC). After having been successfully tested in Kingston, Canada, the system is currently in Love Canal, New York, for further testing.

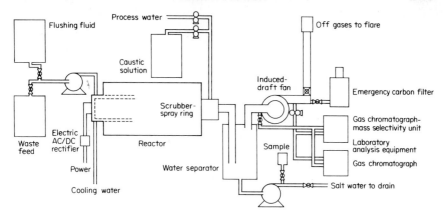

FIG. 8.12.3 Process schematic of plasma pyrolysis system

Plasma Arc for Destruction of Unopened Capacitors

A schematic of the plasma-arc system for destruction of unopened, PCB-containing capacitors is shown in Fig. 8.12.4. This system is sponsored by the Electric Power Research Institute (EPRI) and Arc Technologies Company (ATC), which is a joint venture of Chemical Waste Management, Inc. and Electro-Pyrolysis, Inc. The purpose is to destroy whole, unopened PCB capacitors at the rate of 666 to 1,111 kg/h (3,000 to 5,000 lb/h). The sponsors are currently applying for permits under the Resource Conservation and Recovery Act (RCRA) for the construction of the system in Model City, New York.[4]

Basically the process involves four steps as follows:

Step 1: A whole, unopened PCB capacitor is fed into a bath of molten metal (iron).

Step 2: The capacitor shells melt and the internal organic components (PCBs) are subjected to intense radiation from the plasma arc and heat from the molten metal. As a result, the PCB fluids evaporate and decompose.

Step 3: The PCB and/or decomposition products are further directed through the plasma arc in the vicinity of the high-current (DC) arc for complete destruction.

Step 4: The products are channeled to a scrubber system, where any inorganic materials from the capacitors are converted to inert solid residues. Metals may be recovered if desired from both the metal bath and scrubber ash.

8.12.3 ADVANTAGES AND DISADVANTAGES

Advantages.

1. Since radiative heat transfer proceeds as a function of the fourth power of a temperature, a plasma system has very intense radiative power and therefore is capable of transferring its heat much faster than a conventional flame.

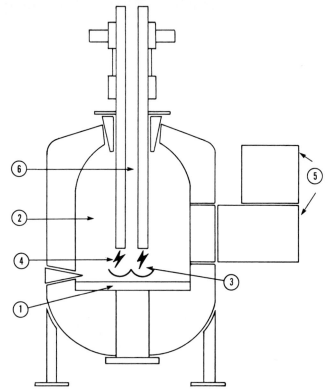

FIG. 8.12.4   Plasma-arc process for PCB destruction.[4] (1) Zone 1: molten metal, approx. 3000°F. (2) Zone 2: furnace chamber, approx. 3000°F. (3) Zone 3: plasma zone, approx. 11,000°F. (4) Zone 4: plasma arc >11,000°F. (5) Sealed loading system. (6) Gas exit.

2. Organic chlorides are known to dehydrogenate when excited by ultraviolet radiation, which is abundant from thermal plasmas.

3. Because the plasma arc for waste destruction is a pyrolysis process, it requires virtually no oxygen. Compared with conventional incinerators, which normally require about 150% excess air to ensure proper combustion, the plasma arc will save the energy required to heat the excess air to the combustion temperatures and will also produce significantly less gaseous by-product to be treated downstream.

4. The process has a very short on-off cycle.

5. Because of its compactness, a plasma-arc system has potential for use as a mobile treatment system; the system would be housed in a trailer and moved from site to site.

Disadvantages.

1. Because temperature is so high (about 10,000°C at arc centerline), the durability of the arc and the refractory materials could be a potential problem.

2. Because the arc is very sensitive to many factors (e.g., sudden drops in voltage, energy and mass balance of the system) execution of the system requires highly trained professionals.

8.12.4 DESIGN CONSIDERATIONS

In general, a plasma-arc system includes the following major components:

- Waste feed
- Plasma arc
- Recombination zone
- Scrubber
- Flare
- Monitoring

The key component is the electrodes to produce the necessary arc. Because plasma-arc systems are still in the developmental stages, detailed design information is mostly confidential and is not available to the public.

8.12.5 ENVIRONMENTAL EFFECTS

A plasma-arc system generally can destroy hazardous wastes to more than "six nines" destruction and removal efficiency (DRE). Only limited environmental data are currently available, and considerable efforts are still needed to characterize the environmental effects from plasma-arc application and to compare environmental impacts of the plasma arc and conventional incinerators.

8.12.6 CASE STUDIES

To date, the only case study results available are from the plasma-arc system for liquid-waste destruction. The system has been tested during three 1-h burns of carbon tetrachloride (CCl_4), and during three 1-h and three 6-h burns of polychlorinated biphenyls (PCBs). The results of the three 1-h carbon tetrachloride burns and the three 1-h PCB burns are currently available and are provided in Tables 8.12.1 and 8.12.2, respectively.[1]

Carbon tetrachloride is regulated under the Resource Conservation and Recovery Act and PCB under the Toxic Substances Control Act (TSCA). Tables 8.12.1 and 8.12.2 show that the test results for CCl_4 and PCBs meet the RCRA and TSCA DRE standards, respectively.

8.12.7 ENGINEERING ECONOMIC EVALUATION

Again, because the plasma-arc system for hazardous-waste destruction is still in its developmental stage, cost information is virtually nonexistent. Nevertheless,

TABLE 8.12.1 Carbon Tetrachloride Test Results

	Run 1	Run 2	Run 3
Date	2/18/85	2/26/85	2/26/85
Sampling time, min	60	60	60
Feed rate, L/min			
CCl_4	0.63	0.63	0.63
MEK mixture	2	1.6	2
Total mass fed, CCl_4, kg	60.0	60.6	60.6
Chlorine loading, mass %	35	40	35
Reactor operating temperature, °C	974	1008	1025
Plasma torch power, kW	280	298	−300
Stack-gas parameters			
Average flow rate, dscmm	38.13	29.69	29.81
(dscfm)	(1,346.3)	(1,048.2)	(1,052.7)
Average temperature, °C	893.3	807.1	677.3
NO_X conc., ppm (v/v)	106	92	81
Emission rate, kg/h	0.46	0.31	0.28
CO conc., ppm (v/v)	48	57	81
Emission rate, kg/h	0.13	0.12	0.17
O_2, %	12.7	14.4	15.1
CO_2, %	6.0	5.7	4.9
HCl, mg/dscm	*	137.7	247.2
Emission rate, kg/h	*	0.25	0.44
CCl_4 conc., ppb	\2†	\2†	\2†
Emission rate, mg/h	\29.3	\22.8	\22.9
Scrubber-effluent parameters			
Scrubber-effluent flow rate, L/min	30.0	30.0	30.0
CCl_4 conc., ppb	1.3	5.5	3.3
Discharge rate, mg/h	2.3	9.9	5.9
Destruction and removal efficiency‡			
CCl_4, % DRE	99.99995	99.99996	99.9999

*Invalid data.
†The detection limit of 2 ppb CCl_4 in the stack gas was used to calculate the CCl_4 mass emission rate for each run.
‡The DRE is based on stack emissions and excludes scrubber effluent.

from a practical viewpoint, the following factors may be worthy of consideration in evaluating costs for plasma-arc systems:

1. A plasma-arc system is a pyrolysis process. It does not need the energy to heat excess air required by conventional incinerators.

2. A plasma-arc system needs a smaller capacity of downstream cleanup systems, because no excess air is involved.

3. A plasma-arc system uses electricity as an energy source. Electricity is more expensive, compared with the oil used to fire incinerators.

TABLE 8.12.2 PCB 1-h Test Results

	Run 1		Run 2		Run 3	
Date	12/5/85		12/17/85		1/16/86	
Sampling time, min	50		60		60	
Stack-gas parameters						
Flow rate, dscmm	37.9		45.0		38.1	
Temperature, °C	836		678		962	
NO_x, ppm	117		N/A		139	
HCl, mg/dscm	N/A		43		68	
O_2, %	14		14.5		16.5	
CO_2, %	5.5		5.0		3.0	
CO, %	0.01		0.01		0.01	
Total PCB, µg/dscm	\0.013	*	0.46	*	3.0	*
	\0.013	†	0.32	†	\0.011	†
Total dioxins, µg/dscm	0.076	‡	\0.43		\0.13	
Total furans, µg/dscm	0.26		1.66		\0.30	
Total B[*a*]P, µg/dscm	0.18		0.45		2.8	
Destruction and removal efficiency						
PCB, % DRE	99.99999	*	99.99994	*	99.9999	*
	99.999999	†	99.99997	†	99.999999	†

*These values are based upon monodecachlorobiphenyl.
†These values are based upon tridecachlorobiphenyl.
‡No tetra- or pentadioxins were detected at 0.05 ng on a GC column, except for run 1 where 0.06 ng tetra dioxin was reported.

8.12.8 REFERENCES

1. N. P. Kolak et al., U.S. EPA 12th Symposium, Cincinnati, Ohio, April 21–23, 1986.
2. E. Matovich, AIChE Southern California Section, April 21, 1981.
3. C. Lee, *Proceedings of the First Annual Hazardous Materials Conference*, Philadelphia, July 12–14, 1983, pp. 278–303.
4. Arc Pyrolysis Project, SCA Chemical Services, Inc., Model City, New York, 1986.

SECTION 8.13

MOBILE THERMAL TREATMENT SYSTEMS

Steven G. DeCicco
IT Corporation
Knoxville, Tennessee

William L. Troxler
IT Corporation
Knoxville, Tennessee

8.13.1 APPLICABILITY TO SPECIFIC WASTES

The Office of Technology Assessment of the U.S. Congress estimates that approximately 50 years and $300 billion (in 1986 dollars) will be required to clean up wastes that have been improperly disposed of in the United States.[1] As of May 20, 1986, a total of 888 hazardous-waste sites were either on or proposed to be on the National Priorities List of the U.S. Environmental Protection Agency (EPA). In addition to these sites, several hundred federal government installations, primarily within the Department of Defense and the Department of Energy, will probably require remedial action. Many states also have cleanup lists, and hundreds of active industrial sites that are not on either the EPA or state lists will require remedial action.

A combination of factors has generated interest in utilizing mobile thermal treatment systems for site-remediation activities, for example,

- High cost, risk of accidents, and public opposition to the transportation of hazardous wastes
- Perpetual generator liability if wastes are stored rather than destroyed
- Decreasing availability and rising disposal costs at permitted landfill facilities

- Lack of available capacity, limited capability, and the high cost of using commercial incineration facilities to handle large quantities of bulk wastes (especially contaminated soil)
- 1984 Hazardous and Solid Waste Amendments to the Resource Conservation and Recovery Act (RCRA) requiring the EPA to consider restricting certain types of wastes from land disposal
- Ban enacted in the Toxic Substances Control Act (TSCA) on land disposal of liquids contaminated with PCBs at concentration levels greater than 500 ppm

On-site treatment by a mobile thermal treatment system may be considered for a site which contains several thousand tons of materials which are considered toxic, reactive, or not readily amenable to treatment by other technologies. Examples of such wastes include materials contaminated with PCBs, dioxins, chlorinated phenols, pesticides, herbicides, and explosives and propellants. Thermal treatment can destroy organic substances but does not destroy inorganic substances including heavy metals.

The following chemical characteristics are key design criteria in waste treatment by thermal technologies: (1) the presence of high concentrations of organic chlorine, sulfur, or phosphorus (because of acid-gas removal capabilities and corrosion considerations); (2) the presence of alkali metal salts (leading to formation of submicron particulates and problems with the formation of low-melting slag eutectics and metal-corrosion); and (3) the presence of heavy metals (preventing delisting of ash and of residues from the gas-cleaning system).

8.13.2 DESIGN CONSIDERATIONS

Transportation Considerations

The term *mobile thermal treatment system* as defined in this section of Chap. 8 applies both to systems which are truly *mobile* and to systems which may be defined as *transportable*. Truly mobile systems are generally truck-mounted, have minimal field-erection and installation requirements (often not requiring foundations), and can be relatively easily mobilized and demobilized. Throughput capacity for truly mobile systems is less than the capacity for transportable units because of over-the-road weight and size restrictions for fully assembled equipment, as shown in Table 8.13.1. The thermal capacities of most truly mobile incineration systems are 2.5 to 7.5 Gcal/h (10×10^6 to 30×10^6 Btu/h).

Transportable thermal treatment systems are large by comparison and constructed as preassembled, skid-mounted, process modules. Mobilization, erection, and demobilization require more effort than for truly mobile systems because some of the high-capacity unit operations may require the interconnection of multiple skids and the construction of proper foundations. Refractory, if used, may have to be removed each time the unit is moved to bring the weight of the unit within transportation limits. The thermal capacities of most transportable thermal treatment systems are 7.5 to 15.0 Gcal/h (30×10^6 to 60×10^6 Btu/h).

Mobile systems have an economic advantage at small sites (e.g., 5,000 t) because of their lower capital, mobilization, and demobilization costs. However, the unit treatment cost ($/t) becomes less sensitive to capital, mobilization, and demobilization costs at medium (10,000 to 25,000 t) and large-size sites (25,000 to

TABLE 8.13.1 Highway Transportation Size and Weight Limits

Parameter	Without special permits		With special permits*	
Weight, kg (lb)†	36,364	(80,000)	45,455	(100,000)
Width, m (ft)	2.6	(8.5)	3.7	(12)
Height, m (ft)	4.1	(13.5)	4.6	(15)
Trailer length, m (ft)	14.5	(48.0)	19.8	(65)

*Typical maximum values. Actual weight and size limits vary from state to state.
†Includes tractor, trailer, and load. Typical weight for tractor and lowboy trailer is 14,545 to 15,455 kg (32,000 to 34,000 lb).

100,000 t). At such sites, the higher processing rate of transportable systems can produce a lower unit treatment cost.

Performance Requirements

Performance requirements for mobile thermal treatment systems may be imposed by federal (RCRA, TSCA, Clean Air Act, Clean Water Act), state, and local regulations. The RCRA requires (1) 99.99% destruction and removal efficiency (DRE) for principal organic hazardous constituents (POHCs) (99.9999% for dioxins), (2) a particulate emission rate of less than 180 mg/m^3 corrected to 7% O_2, and (3) 99% HCl removal or an HCl emission rate of less than 1.8 kg/h (4.0 lb/h). Carbon monoxide, hydrocarbon, and nitrogen oxide limitations in the stack and treatment criteria for delisting of ash and gas-cleaning system residuals are negotiated on a case by case basis.

The TSCA requires a 99.9999% DRE for PCBs and imposes minimum requirements for operating temperature, residence time, and excess air for processing of liquids containing PCBs (1200°C, 3% O_2 for 2 s; or 1600°C, 2% O_2 for 1.5 s).

Even though policy under the Comprehensive Environmental Respose Compensation and Liability Act (CERCLA) requires no federal permits for remediation of CERCLA (Superfund) sites, it is likely that testing will be required to verify compliance with TSCA and RCRA performance criteria.

Factors Affecting Processing Capacity

The primary factors determining the throughput or processing capacity of a thermal treatment system are the waste characteristics (i.e., water and heat content, amount and type of organic and ash), the operating temperature, excess-air requirements (for oxidative units), and heat-transfer considerations. Upper processing limits are typically set by either gas-handling or heat-transfer considerations, with the controlling criteria dependent upon the type of equipment and the type of waste being processed.

Figures 8.13.1 and 8.13.2 illustrate examples of relationships between processing capacity and excess air at the exit of the primary chamber, operating temperatures in the primary and secondary combustion chambers, and soil moisture content. These relationships were developed for a 7.5 Gcal/h (30 × 10^6 Btu/h) rotary-kiln incinerator with secondary combustion chamber, processing a soil with 5% organic content and 275 cal/kg (500 Btu/lb) heat content. All points on

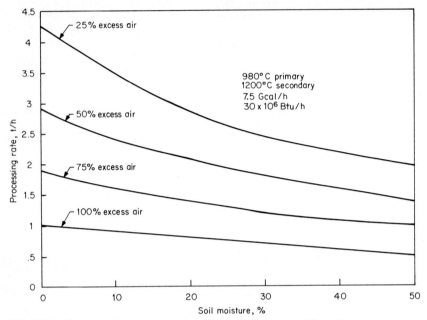

FIG. 8.13.1   Soil-processing rate vs. soil mositure and excess air at kiln exit.

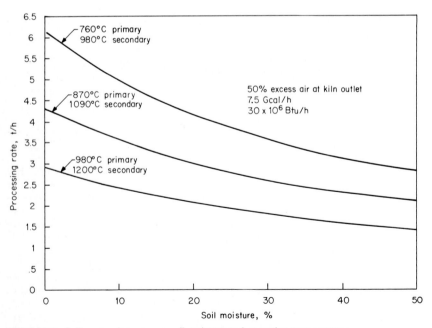

FIG. 8.13.2   Soil-processing rate vs. soil moisture and operating temperature.

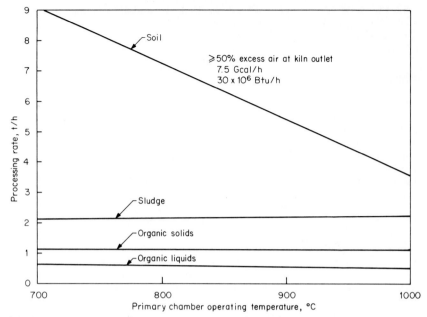

FIG. 8.13.3 Waste-processing rate vs. waste type and operating temperature.

the curves represent operating conditions generating the same thermal duty and approximately the same quantity of flue gas at the exit of the secondary combustion chamber. The processing rate falls as the amount of moisture requiring evaporation rises. The processing rate rises as the amount of excess combustion air is reduced. Increasing the operating temperature decreases the processing rate.

Figure 8.13.3 illustrates the relationship between waste type and primary chamber processing temperature for the same rotary-kiln system. (Characteristics of the waste streams are shown in Table 8.13.2.) These curves are developed for a primary chamber operating at a minimum of 50% excess air and a secondary chamber operating 220°C (428°F) hotter than the primary chamber. The system capacity decreases as the heat content of the waste increases. The relationship between capacity and operating temperature depends on the heat content; for materials with low heat content (e.g., soil), they are inversely related. For mate-

TABLE 8.13.2 Example Waste Characteristics

Waste type	Organics, %	Water, %	Ash, %	Heat content cal/kg	Heat content Btu/lb
Contaminated soil	5	20	75	386	700
Sludge	25	50	25	1,933	3,500
Organic solids	50	10	40	3,865	7,000
Organic liquid	100	0	0	7,730	14,000

rials with moderate to high heat content (e.g., organic liquids, sludges, and solids), capacity and operating temperature become more independent as the heat content of the waste approaches autogenous conditions.

Process Components

The major process components of a mobile thermal treatment system include (1) the materials-handling and feed system with controls for fugitive particulate and gaseous emissions and spills, (2) thermal treatment system with a safe emergency vent system, (3) ash-handling and disposal system with fugitive emission controls, (4) gas-cleaning system with efficient collection of submicron particles and acid gases, (5) wastewater-treatment and disposal system, (6) process-control system with an uninterruptable power supply, (7) electric power distribution and grounding system, and (8) continuous emission-monitoring system.

Ancillary systems are required and must be either provided as a component of the mobile treatment system or obtained from available on-site services: emergency diesel generator with automatic switch gear to enable a controlled shutdown during power failures; fuel supply; caustic, lime, or limestone storage and supply; process and potable water supply; process and instrument air supply; spare parts storage and maintenance facilities; on-site analytical laboratory; personnel decontamination, shower, and change facilities; equipment decontamination facilities; offices; and site lighting and security.

In addition to these major and ancillary systems, other components become important in specific applications. Examples include an emergency combustion system to oxidize hazardous organic vapors generated by the waste inventory in the primary thermal treatment device during a system malfunction or shutdown; heavy-metals removal capability in the wastewater-treatment system; automatic data logging on the process-control system; the capability of supplying power with diesel generators at sites where commercial electric power is unavailable; and automatic data logging and report generation on the continuous emission-monitoring system.

Materials of construction should be carefully selected to withstand difficult service conditions such as high temperatures in the presence of alkali metal salts, abrasive conditions, high and low pH, and hot brine solutions. Instruments and control devices should be prewired for easy setup and installed for easy maintenance with minimum down time. Equipment should be laid out in a way that minimizes worker exposure to hazardous wastes and raw materials and hazardous physical conditions during routine operation and maintenance. An on-site inventory of critical spare parts should be maintained to minimize down time.

Process Functions

Waste forms could include bulk soil and debris, drummed wastes, tars and heavy sludges, organic chemical liquids, and contaminated water. Wastes are brought into the waste preparation area where they are sampled, segregated, physically sized, chemically stabilized, blended, and then stored. All equipment and storage facilities may require fume and spill controls.

Solid, liquid, and sludge wastes are then pumped or conveyed to the thermal treatment devices under monitored and controlled conditions. These waste-handling operations are often the most challenging part of plant operations. Each site may require modifications to existing equipment, or possibly an entire system suited to a unique waste material.

Thermal processing is usually done in two stages. The primary stage treats solids along with other wastes. The secondary stage treats the gases produced in the primary stage, along with other liquid wastes. The primary stage is operated within a portion of a broad range of operating conditions. In the oxidative mode, a burner is fired under excess air conditions to heat the wastes, combust the organics, and volatilize the inorganics at 760 to 980°C (1400 to 1800°F). In the pyrolytic mode, a burner is fired with carefully controlled sources of air. The wastes are heated, the volatile contaminants are evaporated and partially combusted, and the residuals are thermally decomposed at 500 to 760°C (1000 to 1400°F). In the desorption mode, heat is supplied to the chamber from an external source rather than from an internal burner. The wastes are heated to 250 to 450°C (500 to 850°F) and volatile constituents are evaporated and separated from the nonvolatile substrate. No air is admitted to the chamber, so the wastes do not combust.

Gases generated in the primary stage are combusted in a secondary chamber equipped with a burner. If the gases are laden with organic vapors from operation of the primary device in desorption and pyrolytic modes, the addition of air may be all that is needed to sustain complete combustion. Secondary treatment for desorption gases could include condensation of organic vapors to the liquid form where they can be treated in a chemical rather than thermal manner. A critical consideration in designing secondary-combustion and gas-cleaning systems and for soil process is the entrainment of soil particles in the gas leaving the primary chamber. These particles, which tend to settle and accumulate in the secondary combustion system, place a significant burden on the gas-cleaning system.

Mobile thermal treatment systems utilize ash-handling systems to cool the ash by exchanging heat with either air or water and add water to moisturize the ash to minimize dust emissions. In small systems the hot ash can be stored until it can slowly cool by natural radiation and convection. In most on-site thermal treatment applications, the cooled, moisturized ash will be stored under controlled conditions, analyzed and certified, then backfilled at the waste excavation site. If heavy metals are present in the ash, the ash may have to be subjected to stablization and fixation processes prior to final fisposal.

Because of cost, complexity, and fouling considerations, very few mobile thermal treatment systems incorporate any type of energy-recovery equipment; however, exceptions do exist. Energy is being recovered as steam by one vendor's system and used in an ejector which provides the prime moving force for the combustion gases. Energy can also be recovered from the flue gas by preheating combustion air.

Alternative gas-cleaning systems include (1) wet scrubbing systems, (2) in situ acid-gas neutralization by lime or limestone injection in the primary combustion device followed by collection of dry particulates in cyclones and baghouses, and (3) dry scrubbing systems utilizing lime-slurry injection followed by spray drying and particulate collection in baghouses. Wet scrubbing systems are the most common types of gas-cleaning systems used for mobile thermal treatment systems.

Acid gases are produced when halogenated wastes are combusted. The gases are neutralized and converted into salts in the gas-cleaning system. These salts must be continuously purged from the system at the rate they are produced. This is a major consideration in on-site thermal treatment. Wet scrubbing systems produce an aqueous blowdown stream that may require pH adjustment, heavy-metal removal, and/or suspended-solids removal before disposal. In some cases, this stream can be disposed of by mixing it with the hot ash from the incinerator. The "dry" systems produce a solid salt residue during the acid-gas neutralization step that must be disposed of.

8.13.3   COMMERCIAL MOBILE THERMAL TREATMENT SYSTEMS

Short descriptions of the major features of the various thermal treatment systems are given in this subsection. More detailed descriptions of several of the thermal treatment unit operations are presented in other sections of Chap. 8.

Rotary Kiln with Secondary Combustion Chamber

A number of companies are developing mobile thermal treatment systems utilizing *rotary kilns with secondary combustion chambers*. Combustion gases from the refractory-lined kiln flow to a refractory-lined secondary combustion chamber which may be configured either horizontally or vertically. Auxiliary fuel is burned in the secondary to raise the temperature of the flue gas to between 93 and 315°C (200 and 600°F) above the temperature of the flue gas at the kiln exit. The flow of waste-feed materials and combustion gases may be either cocurrent or countercurrent. Operating temperatures in the kiln are normally in the range of 760 to 980°C (1400 to 1800°F) at excess-air levels ranging from 25 to 150%.

Rotary kilns are the most common type of equipment used for portable thermal treatment systems because they represent a commercially proven technology, offer flexibility in their ability to handle a wide variety of physical forms of waste materials with minimal feed pretreatment, and provide good mixing and long residence time for solids. However, they generally operate at higher excess-air levels than other types of thermal equipment, resulting in relatively high fuel consumption, the potential for particulate carry-over from the kiln, and the requirement for large downstream gas-processing equipment. The weight of refractory also limits the maximum size of mobile systems or may require that the refractory be removed for over-the-road transportation of larger transportable systems.

Rotary kilns used in the ore- and mineral-processing industries differ from conventional rotary-kiln incinerators in that the mode of operation is countercurrent, with tight control on the supply of conbustion air. IT Corporation has commercialized and patented a mobile version of this technology for site-remediation activities involving thermal treatment of contaminated soils and hazardous wastes. The combustion characteristics of this type of rotary kiln achieve the following advantages over conventional cocurrent rotary kilns: (1) higher soil-processing capacity, (2) lower off-gas volume to treat, and (3) the potential for more consistent and higher-quality ash residue.

Indirectly Heated Thermal Desorber with Secondary Combustion Chamber

Two European companies have each developed *indirectly heated thermal desorbers* for the treatment of lightly contaminated soils. These large systems are on the upper end of the range of what is considered transportable. Each system consists of a rotating, unlined, horizontal primary chamber with a steel shell and a refractory-lined secondary chamber. The primary chamber is heated indirectly by transferring heat to the inside of the chamber from hot gases circulating through an external jacket. The heat may be supplied by recirculating gas from the secondary combustion chamber or by firing a burner in the jacketed space.

Water and volatile compounds are vaporized in the primary chamber at 250 to 450°C (500 to 850°F). The volatiles are then combusted in a secondary combustion chamber at a temperature of 800 to 1200°C (1500 to 2200°F). By keeping the burner, its flue gases, and excess air out of the chamber, the gas flow consists only of the vaporized water and organics. Relatively small equipment can thus be used for the secondary combustion chamber and the air-pollution control equipment. For cases where low operating temperatures produce adequate decontamination, high soil-processing capacities can be achieved for a given thermal input to the unit.

Indirectly heated thermal desorbers have been primarily used to treat soils that are lightly contaminated with highly volatile compounds, such as materials from oil and gasoline spills and soils contaminated by residues from coal-gasification plants. The unlined shell imposes a maximum operating temperature that is below that of a refractory-lined kiln, limits the heat-transfer rate, and severely limits the maximum concentration of chlorine and sulfur in the feed material (because of corrosion considerations).

Fluidized Bed

Fluidized-bed incinerators may be of either the circulating-bed or the bubbling-bed type. Both types consist of a single refractory-lined combustion vessel using high-velocity air to either fluidize the bed (bubbling bed) or entrain the bed (circulating bed). In the circulating-bed design, air velocities are higher and the solids are blown overhead, separated in a cyclone, and returned to the combustion chamber. Operating temperatures are normally maintained in the 760 to 870°C (1400 to 1600°F) range and excess-air requirements range from 20 to 40%.

Fluid-bed incinerators are primarily used for sludges or shredded solid materials. To allow for good distribution of solids within the bed and removal of solids from the bed, all solids generally require prescreening or crushing to less than 5 to 7.5 cm (2 to 3 in) in diameter. Fluid-bed incinerators offer high gas-to-solids heat-transfer efficiencies, high turbulence in both gas and solid phases, uniform temperatures throughout the bed, and the potential for in situ acid-gas neutralization by lime addition. However, fluid beds may have the potential for solids agglomeration in the bed if salts are present in waste feeds and may have a low residence time for fine particulates.

Infrared Furnace

The *infrared-furnace system* consists of horizontal, rectangular primary and secondary steel chambers lined with a ceramic-fiber insulation material. Presized solid wastes are fed to the primary chamber and are conveyed through the chamber on an alloy wire-mesh belt. Heat is supplied to the two chambers by either silicon carbide resistance heating elements or a gas-fired burner. Design operating temperature in the primary chamber ranges from 260 to 1010°C (500 to 1850°F). Oxidizing, reducing, or neutral atmospheres can be provided in the primary. The design temperature for the secondary combustion chamber is 1260°C (2300°F).

The lightweight ceramic insulation allows the unit to be transported with the insulation intact. Supplying heat to the primary via infrared heating elements results in low flue-gas flows and low particulate carry-over. Control over the residence time and temperature profile of solids can be achieved by controlling the belt speed and controlling the input of electrical energy to the primary chamber. Limitations of the system include the allowable concentration of inorganic chlorides in feed materials (because of corrosion considerations) and the inability to process materials with free liquids.

Cascading-Bed Reactor

The *cascading-bed reactor* consists of an unlined, rotating, horizontal steel shell with internal lifters. As the cylinder rotates, solids are mechanically lifted and then cascade through the flue gas, promoting good heat transfer between the solid and gas phases. An internal recycling system that works on the Archimedes' screw principle picks up heated solids from the middle of the reactor and transports them to the feed end. Feed material is mixed with the hot recycled materials to dry the feed and condition it so that wet solids will not stick to the walls of the reactor. The maximum operating temperature for the reactor is approximately 760°C (1400°F).

The system offers a high throughput capacity per unit volume. However, the system is limited to operation at low primary temperatures because of materials of construction considerations, and the process does not include a secondary combustion chamber.

Plasma Arc

The *plasma-arc system* consists of a plasma torch that uses electricity to generate temperatures in excess of 5000°C (9000°F). A colinear electrode arrangement is used to create an electric arc. Pressurized air is passed through the arc, producing a plasma. Waste liquids are injected into the plasma and pyrolyzed. Destruction of waste liquids by pyrolysis rather than by oxidation minimizes the production of flue gases and results in a high throughput per volume of equipment. The system cannot treat solids. For a given application, the cost of electricity must be weighed against the cost of fossil fuels for other types of thermal treatment systems.

8.13.4 CASE STUDIES

The use of mobile thermal-treatment systems for on-site cleanups began slowly in the early 1980s and attained a commercial pace by the late 1980s. The experience base with technology and waste handling methods is broadening rapidly. Table 8.13.3 presents the initial field demonstrations. These demonstrations include both systems that were already commercialized and systems that were in the developmental stage. A summary of the results of these field demonstrations is presented in Table 8.13.3. Several other major on-site cleanup projects were under contract at the time of this publication.

8.13.5 ECONOMIC EVALUATION OF THE TECHNOLOGY

Cost Components in a Site-Remediation Project

Table 8.13.4 presents the major cost components for remediating a site using mobile thermal treatment technology. The table is structured for a mobile, state of the art, self-sufficient hazardous-waste management facility, operated 24 hours per day, 7 days per week. Unlike fixed facilities, mobile technologies require repetition of the planning, site preparation, equipment mobilization, commissioning, demobilization, site closure, and possibly the trial-burn and permitting activities at each site.

Major Factors Affecting Treatment Cost

Figure 8.13.4 presents a comparison of unit treatment cost ($/ton) as a function of waste type, waste quantity, processing rate, and capital cost. Treatment costs were developed using the cost components presented in Table 8.13.4 by aggregating actual site data with technical and cost information from several manufacturers and users. This figure is intended to illustrate the relative effects of the variables and not to indicate the actual unit treatment price offered by a specific technology at a specific site. The following general trends may be observed:

- Unit treatment cost ($/ton of waste) decreases as the quantity of waste increases. Unit cost is higher at small sites because fixed costs (i.e., planning, permitting, site preparation, mobilization, commissioning, trial burns, demobilization, site closure) are recovered over a smaller number of tons.

- Unit treatment cost decreases as the processing rate of wasted in tons/h increases (see Subsec. 8.13.2 for a discussion on processing rate). High capacity and high reliability mean fewer operating days. Each day of operation brings additional costs, so reducing the number of operating days reduces the overall treatment cost. Several of the major daily operating-cost components (e.g., labor, analytical) are independent of technology or equipment size.

- Unit treatment cost is more sensitive to processing rate than to capital cost. Doubling the capital cost to achieve twice the processing rate results in a lower unit cost, except at small sites where the higher fixed costs for a large system are recovered over a smaller number of tons.

- Treatment costs are typically higher for sites containing significant quantities of organics than for sites with lightly contaminated soil. The processing rate typically decreases as the waste's organic content increases because of limitations on heat-release and flue-gas-handling capacity with mobile equipment. Consequently, the number of operating days for a given quantity of organic waste is larger than for the same quantity of lightly contaminated soils.

TABLE 8.13.3 Field Demonstrations of Mobile Thermal Treatment Systems

Company	Treatment system type	Demonstration location	Waste type	Waste feed rate, kg/hr	Compound(s)	Destruction and removal efficiency, %	Reference
IT Corporation	Counter-current, controlled-air rotary kiln	Cornhusker Army Ammunition Plant, Grand Island, Nebr.	Soil contaminated with explosives	13,000	TNT, RDX	99.9999	18,19
IT Corporation	Countercurrent, controlled-air rotary kiln	Louisiana Army Ammunition Plant, Shreveport, La.	Soil and lagoon sediments contaminated with explosives	13,000	TNT, RDX	—	18,19
IT Corporation/ U.S. Air Force	Thermal desorber	U.S. NavalConst. Battalion Command, Gulfport, Miss.	Soil contaminated with dioxin	14–44	Dioxin	—	14
IT Corporation/ U.S. Air Force	Thermal desorber	Defense Nuclear Agency, Johnston Island	Soil contaminated with dioxin	23–91	Dioxin	—	—
IT Corporation/ U.S. EPA	Rotary kiln	Edison, N.J.	Fuel oil (spiked)	94–146	PCBs o-Dichlorobenzene Carbon tetrachloride	99.99991 99.99998 99.99996	12
IT Corporation/ U.S. EPA	Rotary kiln	Denney farm site, Verona, Mo.	Soil, still bottoms contam. with dioxin	907	Dioxin	99.999973– 99.999995	11
Ecotechniek	Thermal desorber	Utrecht, Netherlands	Soil contaminated with VOCs and cyanide	27,200– 45,300	Phenol, xylene, toluene, benzene naphthalene, cyanide	—	3
Ensco	Rotary kiln	Sydney Mines, Brandon, Fla.	Oil-contaminated sludge and soil	2,040	Petroleum oils	—	4,5
Ensco	Rotary kiln	Lenz Oil, Chicago, Ill.	Soil and sludge contaminated with oil	6,400	Petroleum oils, PCBs	99.9999	—
Ensco	Rotary kiln	U.S. Naval Const. Battalion Command, Gulfport, Miss.	Soil contaminated with dioxins	6,400	Dioxins	99.9999	—

TABLE 8.13.3 Field Demonstrations of Mobile Thermal Treatment Systems (*Continued*)

Company	Treatment system type	Demonstration location	Waste type	Waste feed rate, kg/hr	Compound(s)	Destruction and removal efficiency, %	Reference
J. M. Huber Corp.	Advanced elec. reactor	Times Beach, Mo.	Soil contaminated with dioxin	—	Dioxin	>99.99999	8
J. M. Huber Corp.	Advanced elec. reactor	U.S. Naval Const. Battalion Command, Gulfport, Miss.	Soil contaminated with dioxins	6,400	Dioxins	99.9999	—
Shirco	Infrared furnace	Times Beach, Missouri	Soil contaminated with dioxin	22	Dioxin	99.99997– 99.999994	8
Shirco	Infrared furnace	—	Simulated creosote waste		Pentachlorophenol*	99.99614– 99.99986	8
O.H. Materials	Infrared furnace	Florida Steel Site, Indiantown, Fla.	Soil contaminated with PCBs	7,300	PCBs	99.9999	—
Westinghouse/ Haztech	Infrared furnace	Peak Oil, Brandon, Fla.	Soil contaminated with PCBs	5,600	PCBs, oil	99.9999	—
Roy F. Weston	Rotary kiln	Beardstown, Ill.	Soil contamianted with PCBs	4,000	PCBs	99.9999	—
U.S. Army/Roy F. Weston	Rotary kiln	Savannah Army Depot, Savannah, Ill.	Sediment contaminated with explosives	180	TNT,RDX,HMX	ND†	15
U.S. Army/Roy F. Weston	Thermal screw	—	Soil contaminated with VOCs		Dichloroethylene Trichloroethylene Tetrachloroethylene Xylene	ND† ND ND ND	16
U.S. EPA/Haz. Waste Tech. Service	Asphalt drier	Fla.	Soil contaminated with VOCs		Trichloroethane Trichloroethene Toluene, ethyl benzene Xylene	‡ ‡ ‡ ‡	17

*Results reported for 15 compounds.
†Not detected in stack gas; detection levels not reported.
‡Objective was to remove contaminates from soil. No unit operation applied for ultimate destruction of contaminants.

TABLE 8.13.4 Generic Cost Components in Site Remediation Using Mobile Thermal Treatment Technologies

Major project phases	Specific activities
1. Planning and procurement	Survey the site and develop layout drawings, design foundations, design utility & waste-disposal systems, plan transportation and mobilization, plan health and safety program and QA/QC program, implement public relations program, develop site-security plan, develop operations plan and procedures, develop environmental monitoring plan.
2. Permitting	Identify permits and specific information requirements, prepare draft permit applications and trial burn plans, client and agency review, finalize permit applications, conduct public hearings, negotiate final operating permits.
3. Site preparation	Mobilize site-preparation equipment; set up site containment and security; grade, grub, and fill site; pour foundations and pads; construct access roads and parking; connect utilities; install environmental monitoring system; set up support facilities; prepare waste- and residuals-handling facilities.
4. Equipment mobilization	Transport the process and utility equipment and personnel facilities, unload equipment, erect all equipment modules, interconnect instruments and control systems, interconnect electrical distribution system, connect emission-monitoring system, interconnect all utility systems.
5. Commissioning	Conduct site personnel training, check out electrical and instrumentation systems, conduct hydrostatic testing, align rotating equipment, check containment systems, check winterization systems, check fire protection systems, check emergency procedures, start up the plant and bring the process into equilibrium.
6. Trial burns	Check out monitoring systems; deploy sampling teams; prepare waste feeds; excavate and execute trial burns; conduct laboratory analyses of feeds, treated ashes and wastewater, gaseous emissions; analyze results and prepare report to agency; conditionally operate or mothball system during agency review.
7. Operation	Excavate waste; analyze waste; pretreat and blend wastes; thermally treat wastes; store, analyze, and delist residuals; dispose of treated ashes, treated wastewater, and residuals from the gas-cleaning and wastewater-treatment systems; sample and analyze groundwater well samples.
8. Equipment demobilization	Clean and decontaminate equipment; dispose of wastes generated during decontamination; conduct required equipment maintenance; disconnect power, electrical, utility, and stack-monitoring systems; disassemble process modules; load and transport equipment to next site.
9. Site disassembly and closure	Disconnect and remove site utilities, remove personnel support facilities, remove waste-handling facilities, demolish and remove foundations, remove access roads and parking, grade and vegetate the site.

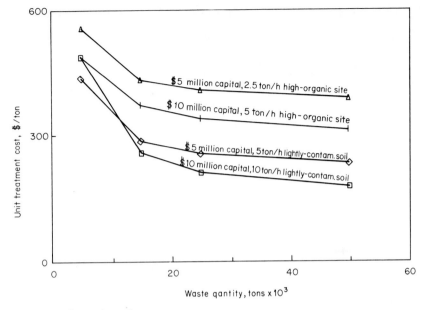

FIG. 8.13.4 Comparison of treatment costs.

Operating costs are also sensitive to the technology's inherent fuel consumption rate per unit of waste and the amount of labor required to remediate the site and operate the treatment system.

8.13.5 REFERENCES

1. J. Main, "Who Will Clean Up?" *Fortune,* March 17, 1984, pp. 96–102.

2. J. F. Frank and R. Kuykendall, "Illinois Plan for On-Site Incineration of Hazardous Wastes," in *Proceedings of the National Conference on Hazardous Wastes and Hazardous Materials,* Atlanta, Georgia, March 4–6, 1986, pp. 229–232.

3. D. Hoogendoorn, "Review of the Development of Remedial Action Techniques for Soil Contamination in the Netherlands," in *Proceedings of the 5th National Conference on Management of Uncontrolled Hazardous Waste Sites,* Washington, D.C., November 7–9, 1984, pp. 569–575.

4. G. N. Dietrich and G. F. Martini, "Today's Technology Offers On-Site Incineration of Hazardous Waste," *Hazardous Materials and Waste Management,* January–February 1986, pp. 27–28.

5. N. N. Hatch and E. Hayes, "State-of-the-Art Remedial Action Technology Used for The Sydney Mine Waste Disposal Site Cleanup," in *Proceedings of the 6th National Conference on Management of Uncontrolled Hazardous Waste Sites,* Washington, D.C., November 4–6, 1985, pp. 285–290.

6. D. L. Vrable and D. R. Engler, "Transportable Circulating Bed Combustor for the Incineration of Hazardous Waste," in *Proceedings of the 6th National Conference on Management of Uncontrolled Hazardous Waste Sites,* Washington, D.C., November 4–6, 1985, pp. 285–290.

7. W. R. Schofield, J. Boyd, D. Derrington, and D. S. Lewis, "Electric Reactor for the Detoxification of Hazardous Chemical Wastes," in *Proceedings of the 5th National Conference on Management of Uncontrolled Hazardous Waste Sites,* Washington, D.C., November 7–9, 1984, pp. 382–385.

8. P. L. Daily, "Performance Assessment of Portable Infrared Incinerator," in *Proceedings of the 6th National Conference on Management of Uncontrolled Hazardous Waste Sites,* Washington, D.C., November 4–6, 1985, pp. 383–386.

9. J. N. McFee, G. P. Rasmussen, and C. M. Young, "Design and Demonstration of a Fluidized Bed Incinerator for the Destruction of Hazardous Organic Materials in Soils," *Journal of Hazardous Materials,* **12**:(2)129–142 (November 1985).

10. T. G. Barton and J. A. G. Mordy, "Mobile Plasma Pyrolysis," paper presented at EPRI 1983 PCB Seminar, Atlanta, Georgia, December 6–8, 1983.

11. "EPA Mobile Incinerator Safely Destroys Dioxin," *Environmental News,* U.S. EPA Region 7, Kansas City, Kansas, May 30, 1985.

12. J. Yezzi, J. E. Brugger, I. Wilder, F. Freestone, R. A. Miller, C. Pfrommer, and R. Lovell, "Results of the Initial Trial Burn of the EPA-ORD Mobile Incineration System," in *Proceedings of the ASME 1984 National Waste Processing Conference, Engineering: The Solution,* Orlando, Florida, June 3–6, 1984, pp. 514–534.

13. H. M. Freeman and R. A. Olexsey, "A Review of Treatment Alternatives for Dioxin Wastes," *Journal of the Air Pollution Control Association,* **36**:(1)67 (January 1986).

14. R. Helsel, E. Alperin, T. Geisler, A. Groen, and R. Fox, "Technology Demonstration of a Thermal Desorption/UV Photolysis Process for Decontaminating Soils Containing Herbicide Orange," paper presented at ACS Meeting, New York, April 1986.

15. J. Noland and W. E. Sisk, "Incineration of Explosives Contaminated Soils," in *Proceedings of the 5th National Conference on Management of Uncontrolled Hazardous Waste Sites,* Washington, D.C., November 7–9, 1984.

16. J. W. Noland, N. P. McDevitt, and D. L. Koltuniak, "Low Temperature Thermal Stripping of Volatile Compounds," in *Proceedings of the National Conference on Hazardous Wastes and Hazardous Materials,* Atlanta, Georgia, March 4–6, 1986, pp. 229–232.

17. D. Hazaga, S. Fields, and G. P. Clemmons, "Thermal Treatment of Solvent Contaminated Soils," in *Proceedings of the 5th National Conference on Management of Uncontrolled Hazardous Waste Sites,* Washington, D.C., November 7–9, 1984, pp. 404–406.

18. S. G. DeCicco and M. L. Aident, "Transportable Hybrid Thermal Treatment System," in *Proceedings of the 24th AIChE/ASME National Heat Transfer Conference,* Pittsburgh, Pa., August 9–12, 1987.

19. M. L. Aident, D. C. Burton, and R. J. Lovell, "Cornhusker Army Ammunition Plant Remediation Project Case History," in *Proceedings of the International Conference on Incineration of Hazardous/Radioactive Wastes,* San Francisco, Calif., May 3–6, 1988.

SECTION 8.14
CATALYTIC INCINERATION

Yen-Hsiung Kiang, Ph.D.
Four Nines, Inc.
Conshohocken, Pennsylvania

Catalytic incineration is an incineration process which uses catalysts to increase the oxidation rate of wastes at lower temperature than conventional thermal incineration processes. A catalytic incineration process always requires a lower oxidation temperature [less than 537°C (1000°F)] than thermal incineration [greater than 760°C (1400°F)]. The products of combustion of a catalytic incineration reaction are the same as those of thermal incineration. The heat of incineration liberated by catalytic incineration is the same as that of a thermal incineration. Catalysts are used to increase the oxidation rate at lower temperatures.

The advantages of catalytic incineration over thermal incineration are the lower fuel requirement, and less severe service requirement for materials of construction. The disadvantages are limited applications for liquid wastes, not applicable for solid wastes, and fouling, suppressing, and aging of catalysts.

8.14.1   APPLICABILITY OF THE TECHNOLOGY

The primary application of catalytic incineration systems is the oxidation of organic compounds in gaseous streams. Liquid wastes can also be treated in a catalytic incineration system. However, application of catalytic incineration to liquid-waste disposal is rather limited. Catalytic incineration systems cannot be used for solid-waste disposal.

The poison and masking of catalysts are a major concern in the application of catalytic incineration and must be studied before a catalytic system is selected.

8.14.2   DESCRIPTION OF THE TECHNOLOGY

Principle

The thermal oxidation of a combustible waste normally follows the following hydrocarbon oxidation reaction:

$$a\text{HC} + b\text{O}_2 \rightarrow c\text{CO}_2 + d\text{H}_2\text{O}$$

The reaction follows the rate equation

$$-\frac{dC}{dt} = kC$$

where C = contaminant concentration. This differential equation can be solved to determine the combustion efficiency of wastes, as follows:

$$\text{Eff} = 100 \times \left(1 - \frac{C}{C_0}\right) = 100 \times (1 - e^{-kt})$$

where Eff = conversion efficiency (in percent) and C_0 = initial contaminant concentration. The rate constant k is an empirical number that has to be determined experimentally for the burning of various hydrocarbons. However, the variation of values of k with temperature is predictable from kinetic theory, and follows the Arrhenius equation:

$$k = A \cdot e^{-(E/RT)}$$

Where A = collision coefficient
            E = energy of activation
            R = universal gas constant
            T = absolute temperature

    The rate of reaction can be increased by adsorbing gas molecules on catalytically active sites of the catalyst. At room temperature, there is a film of adsorbed gases on the catalyst surface. This film is quite stable at low temperature. However, at elevated temperatures, this film becomes unstable. Gases are then desorbed from the catalyst surface leaving a partial film. It is in this film that the catalytic oxidation reactions take place.

    In summary, the gas-phase catalytic reaction consists of the following steps (refer to Fig. 8.14.1):

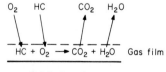

FIG. 8.14.1   Mechanism of catalytic reaction.

- Gas-phase heat and mass transfer of reactants to the surface film on the catalyst
- Diffusion of adsorbed reactants within the surface film
- Oxidation reaction between reactants
- Diffusion and desorption of reaction products
- Gas-phase heat and mass transfer of products leaving the catalyst surface

There are different mechanisms by which reaction rates can be increased. The catalyst may function as a concentrator, bringing about a higher concentration of reactive materials at the surface than is presented in the bulk gas phase. More concentrated reactive materials will increase the collision constant A in the Arrhenius equation. Or, it is possible that the catalyst may modify a molecule of adsorbed gas by adding or removing an electron or by physically opening a bond. This has the effect of decreasing the activation energy E in the Arrhenius equation. In either case, it is necessary for the reactive materials to reach the active surface of the catalyst by diffusion through the gas phase, and for the reaction products to leave the surface.

Most oxidation reactions are extremely rapid and take place principally on the outer surfaces of the catalyst. Catalysts with a very great area of internal microsurface do not show much advantage over catalysts with lower internal surface area.

The chemical and physical processes involved in the catalytic reaction have just been discussed. The slowest of these processes controls the overall rate of reaction. In catalytic incineration applications, heat transfer is necessary to maintain a constant incineration temperature.

Process Description

The basic catalytic incineration process is presented in Fig. 8.14.2. The incinerator can be either a fixed-bed (Fig. 8.14.2*a*) or a fluidized-bed (Fig. 8.14.2*b*) unit. Waste, together with combustion air, is introduced into the catalytic bed where the oxidation reaction takes place. The overall system may include heat-recovery and air-pollution control systems. A fluidized-bed unit can be designed with integrated boiler. An external heat-recovery system can be used in either type of system. An air-pollution control system may include particulate and acid-gas removal equipment.

8.14.3 DESIGN CONSIDERATIONS

The important design parameters for a catalytic incinerator include the superficial-gas velocity, the catalysts used, the geometric support of catalysts (i.e., the form of the catalyst element), the depth of the catalyst bed, waste type and composition, and the incineration temperature.

Process Design

The superficial-gas velocity for a catalytic incinerator is usually 3 to 6 m/s (10 to 20 ft/s) for a normal heat-release system, and higher for high-release systems. The

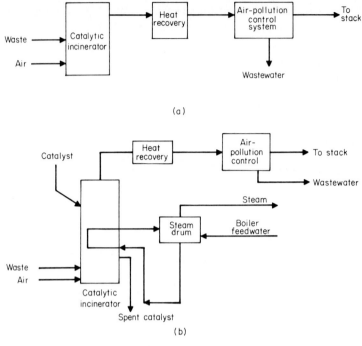

FIG. 8.14.2 Catalytic incineration systems. (*a*) Fixed-bed incinerator. (*b*) Fluidized-bed incinerator.

selection of the catalyst and the support design is normally based on the waste type and composition, and for optimal equipment design. The depth of the catalyst bed is dependent on the maximum allowable pressure drop, and the incineration temperature is usually fixed by the destruction-efficiency requirement.

To maintain satisfactory incineration temperatures, temperature indicators must be used to indicate the temperature of the catalyst bed at one or more points, and of the combustion-product gases leaving the catalyst bed. In most cases, controllers are used to maintain one or more of these temperatures. The temperature measurements of the bed and product gases can be used to control an auxiliary fuel burner so that when the indicated temperature falls, auxiliary fuel is used to maintain the required incineration temperature. The same measured temperatures can also be used to control the flow of cooling medium (air or water). When temperature rises in the catalyst bed, the control will reduce the auxiliary-fuel requirement and then introduce cooling medium to maintain the incineration temperature. An oxygen analyzer can be used to maintain required oxygen content in the product gases by controlling either the air flow rate or the waste flow rate. These are typical control systems; depending on the individual requirement, the control system can be designed to suit different applications.

Incinerator Equipment

There are two types of equipment which can be used in catalytic incineration: fixed bed and fluidized bed. For a fixed-bed incinerator, catalysts are manufactured on ceramic or metal supports, which are installed in the incinerator. The reactor design for this type of catalytic incinerator is simple; however, the system design to compensate for catalyst fouling is rather complicated. The fixed-bed system has been applied successfully for the incineration of gases with no or little masking and poison agents. High concentration of the masking and poison agents will lead to installations that either do not perform the incineration function for which they are designed or that require frequent and costly replacement of the catalyst elements and interruption of the process operation. Fixed-bed design is not suitable to liquid-waste applications.

The problem of catalyst deactivation cannot always be entirely eliminated. Thus, continuous replacement of a portion of the catalyst bed during normal operation is required to maintain continuing operation at high efficiency even in the presence of poisoning agents. This requirement leads to the development of the fluidized-bed incinerator. A fluidized-bed incinerator has the advantage that catalysts can be added and withdrawn during operation.

Furthermore, the use of a fluidized bed will provide added protection against fouling because of the self-abrading action of the catalyst particles in the fluidized bed can keep the surfaces clean. However, the abrasion of the catalyst also becomes a problem in the fluidized-bed incinerator. The minute amounts of wear would result in significant rates of catalyst loss and, potentially, in a particulate-emission problem. Therefore the catalyst used for a fluidized-bed unit must be very hard and abrasion-resistant. Also, air-pollution control systems must be able to control the fine particles generated.

Characteristics of Catalysts

A catalyst bed consists of three elements: the catalyst, the carriers, and the promoters. The catalysts are normally made of metals or metal compounds. The destruction efficiency of a catalytic incineration system depends heavily on the catalyst selected. In most applications, platinum is the primary choice. Other precious metals used for catalytic incineration are rhodium, palladium, iridium, and gold. Precious-metal alloys such as rhodium–platinum alloy can also be used. The disadvantage of precious metals and metal compounds is their high cost. For catalytic incineration applications, it is necessary to develop a practical nonprecious-metal catalyst of sufficient activity to compete favorably with the precious-metal catalyst. The price of a nonprecious-metal catalyst may be low enough that poisoned or deactivated catalyst can be discarded without implementing costly and cumbersome procedures for recovery of precious metals.

A catalyst system usually consists of small, metal catalyst particles distributed uniformly on the surface of a support, or *carrier*. The carriers are inert metal oxides with large surface area. Typical materials used as carriers are alumina, magnesia, asbestos, china clay, activated carbon, porcelain rods, and metal wire or ribbon.

The activity of catalysts may be increased by the addition of one or more components. These types of compounds are called *promoters*.

The reaction kinetics of catalytic incineration differ with the selection of catalysts and the waste materials. The reaction data are usually available through the manufacturer. For new or different applications, pilot-plant testing is required to obtain the necessary data.

Catalyst Deactivation

During normal operation, most catalysts deteriorate or gradually lose their activity. This phenomenon is referred to as *deactivation*. The factors which will cause catalyst deactivation are

1. *Overheating:* Overheating will change the catalyst surface and cause rapid loss of activity.
2. *Thermal aging:* *Thermal aging* is a normal thermal effect which will result in recrystallization of the catalyst, the carrier, or both. Recrystallization causes change in the catalyst's surface area and dispersion of the support material. It has an irreversible effect on catalyst activity.
3. *Poisoning:* Catalyst *poisoning* is the result of the chemical reaction of contaminants with the catalysts. Its effect usually is irreversible. Lead, antimony, cadmium, zinc, phosphorous, arsenic, and copper are usually considered to be poisonous to catalysts.
4. *Masking:* *Masking* results when contaminants accumulate on and mechanically coat the catalysts. This film prevents or slows down the diffusion of reactants and makes it more difficult for them to find unoccupied sites. Masking agents are high-molecular-weight organics (organic liquids or tars condensed at temperatures too low for catalytic oxidation and retained on the surface as the catalyst is heated), ordinary airborne dust and dirt, silicon dioxide ash remaining when silicone compounds are oxidized, and those materials listed as catalyst poisons where thermal conditions are not favorable for chemical reaction.

Halogens, sulfur dioxide, nitrogen dioxide, and numerous other materials act as catalyst-masking agents. These compounds tend to adsorb strongly on the catalytic surface. However, the masking agents will gradually be stripped off if there is no longer a concentration of suppressant materials in the reactant stream passing through the catalyst. For this reason, the catalyst appears to lose activity during periods when halogens, halogenated hydrocarbons, etc., are present in the reactant stream, but regains normal activity slowly after these types of masking agents are removed.

Catalyst Regeneration

Of the four catalyst-deactivating mechanisms, only masking is reversible. If masking is the only problem, catalysts can be regenerated to restore their activity. The fouling material on catalysts may be burned off by hot products of combustion of a fuel, or driven off by passing through a stream of superheated steam with or without the addition of air, or by washing or mechanical brushing. Such treatment causes oxidation of the catalyst, and the catalyst may require activation by passing through a stream of hydrogen-containing gas with or without the addition of hydrogen sulfide. Usually, before start-up of a catalytic incinerator, the catalyst is activated by passing the treating gas through the reactor at a required temperature until the catalyst becomes fully active.

Where fouling of the catalyst is known or likely to occur during operation, the following process-equipment arrangements can be used:

1. *Use two fixed-bed incinerators in series:* During normal operation, the gas flows through the reactors in series and the first reactor acts as a guard. The

catalyst in the first reactor absorbs the masking agents. The first reactor can be periodically isolated from the gas stream for removal of the fouled catalyst; during this period, the gas stream passes through only the second reactor.

2. *Use two or more fixed-bed incinerators in parallel:* During normal operation, one or more reactors are on line. If more than one incinerator is used, they are operated at different time intervals. One incinerator is the standby incinerator, and is isolated from the stream. When the catalyst in an incinerator loses activity, this incinerator is isolated from the gas stream and the standby incinerator is connected to the stream. The fouled catalyst in the incinerator that has been isolated may be regenerated in situ or may be withdrawn from the incinerator and replaced by new or regenerated catalyst.

3. *Use fluidized-bed incinerator:* When a fluidized-bed system is used, new or regenerated catalyst can be periodically fed and an equal quantity of catalyst removed from the bed while the reactor is in continuous operation.

8.14.4 ENVIRONMENTAL EFFECTS

- *Air:* Air-pollution control systems can be used so that air emissions from a well-run catalytic incinerator will not affect the environment.

- *Water:* The wastewater discharge is normally from the air-pollution systems; it may require further treatment.

- *Residue:* The residues discharged from a catalytic incineration system are spent catalysts. They can be landfilled or sold for precious-metal recovery.

8.14.5 FURTHER READING

Cheremisinoff, P. N., and A. C. Moores, *Benzene, Basic and Hazardous Properties,* Marcel Dekker, New York, 1979.

Hardison, L. C., and E. J. Dowd, "Emission Control via Fluidized Bed Oxidation," *Chemical Engineering Progress,* August, 1977, pp. 31–35.

Kiang, Y. H., and A. Metry, *Hazardous Waste Processing Technology,* Ann Arbor Science, Ann Arbor, Michigan, 1982.

SECTION 8.15

HAZARDOUS WASTE AS FUEL IN INDUSTRIAL PROCESSES

Fred D. Hall, P.E.

PEI Associates, Inc.
Cincinnati, Ohio

Robert E. Mournighan

U.S. Environmental Protection Agency
Hazardous Waste Engineering Research Laboratory
Cincinnati, Ohio

Large quantities of liquid hazardous wastes can be destroyed by incineration. Unfortunately, however, the cost of building and operating a hazardous-waste incinerator is high. In an effort to find a more cost-effective answer to this disposal approach, the use of various high-temperature industrial processes has been investigated for destruction of these wastes. The first processes to be investigated were industrial boilers and cement kilns, which were chosen partially because of their ubiquity and partially because of the interest many boiler and kiln operators expressed in recovering the heat value of wastes. Other processes, the focus of this section, are also potential users of hazardous wastes as fuels. These include industrial kilns, metallurgical furnaces, and glass furnaces as shown in Table 8.15.1 along with key combustion parameters.

8.15.1 APPLICABILITY OF INDUSTRIAL PROCESSES

Not every incinerable hazardous waste is suitable for use as a fuel in industrial processes. The most suitable waste fuel is a liquid with a high energy value, a low water content, and a low concentration of metals.[1] Contaminated soils and

TABLE 8.15.1 Industrial Processes That Are Candidates for Hazardous-Waste Incineration

Process	Process temperature °C	Process temperature °F	Typical gas residence time, s
Lime kiln	1093	2000	8
Aggregate kiln	1093	2000	2
Asphalt-plant dryer	427	800	2
Iron and steel blast furnace	1870	3400	1
Brick tunnel kiln	1200	2200	4
Glass furnace	1260	2700	4
Copper reverberatory furnace	1426	2600	2
Lead blast furnace	1200	2200	6

drummed wastes are not good candidates. At several plants, the minimum heat content of acceptable materials has been established at 2.326×10^6 J/kg (10,000 Btu/lb) to ensure that only fuel-quality materials will be used.[2,3] As a general guideline, separable water should be limited to 1% by volume, and solids content must be sufficiently low to prevent pluggage of the burner and allow the fuel to be pumped. Generally, hazardous-waste fuel (HWF) should contain less than 20% solids, and it must be fine enough to pass through a ⅛-in screen. These limitations, which may vary from plant to plant and process to process, are established primarily because of handling constraints. A maximum viscosity of approximately 1,400 SSU ensures that the supplemental fuel is readily pumpable at normal ambient temperatures.[3]

Heavy-metals content (e.g., lead, arsenic, chromium, nickel, and cadmium) is also a concern as these metals may affect product quality, or become concentrated in the waste dust, which can change the character of the air emissions. The latter may necessitate a modification or addition to the emissions-control system. The presence of certain heavy metals in the waste dust may also cause the dust to be classified as hazardous. Should this occur, the plant would have to treat and dispose of the dust as a hazardous waste.

Generally, the following types of wastes are considered undesirable for incineration in industrial processes because of institutional factors:

Organic cyanides

PCBs

Insecticides

Pesticides

Radioactive materials

These institutional factors include public relations, worker relations, and safety.

When a waste material is being considered, the compatibility of the constituents of this material with other wastes being considered or used must be confirmed. This compatibility must be verified by laboratory tests. The waste must also be compatible with the containment (storage) facility and its components; i.e., it must not accelerate corrosion or deterioration of the facility. Existing storage tanks may have to be provided with additional protection, such as painting, coating, or lining with an impervious material. Compatibility of wastes with other equipment (e.g., pumps, valves, and pipe joints) must also be considered.

8.15.2 TECHNOLOGY DESCRIPTIONS

Lime Kilns

The term *lime* is a general term that includes the various chemical and physical forms of quicklime and hydrated lime, the two types generally produced. Lime producers at 137 plants in 38 states sold or used 14.6 million t (16.1 million tons) of lime in 1984. The lime-manufacturing process is similar to that of cement in that the raw material (usually limestone or dolomite) is quarried, crushed and sized, and calcined in a kiln at about 1200°C (2000°F). Although a variety of kiln types can be used, about 85% of the U.S. producers use the rotary kiln. The largest lime kiln is 164 m long and 5.6 m in diameter (500 ft long and 17 ft in diameter) and is capable of producing more than 1090 metric t (1,200 tons) of quicklime per day.

About 7×10^9 J (6.7 million Btu) of energy is required for each ton of quicklime produced.[4] The cost of this high energy requirement has led to increased energy efficiency in the industry and to the use of more readily available and lower-cost fuels, especially coal. Hazardous-waste fuel can be used as a supplementary fuel by feeding the HWF into the kiln via a separate burner near the primary-fuel burner.

Figure 8.15.1 shows a schematic diagram of the lime-kiln process. The calcining drives off nearly half the limestone's weight as carbon dioxide (CO_2) and leaves a soft, porous, highly reactive lime known as quicklime (CaO). Heating beyond this stage can result in lumps of inert, semivitrified material (known as *overburned* or *dead-burned* lime), which is often used in the manufacture of refractory materials. The quicklime is discharged at the lower end of the kiln into the cooling system, where it is air-cooled, and then stored in silos. A portion of the quicklime is hydrated before storage. Hydrated lime is produced by reacting quicklime with sufficient water to cause formation of a dry, white powder.

The measured destruction and removal efficiencies (DREs) during a lime-kiln test burn were:[5]

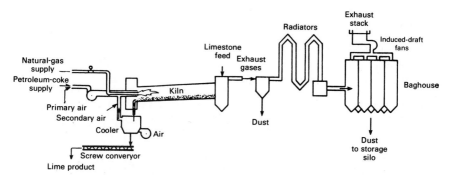

FIG. 8.15.1 Schematic diagram of lime-kiln process.

POHC or waste component	Destruction efficiency, %
Methylene chloride	99.9947–99.9995
Methyl ethyl ketone	99.9992–99.9997
1,1,1-Trichloroethane	99.9955–99.9982
Tetrachloroethylene	99.997–99.9999
Toluene	99.995–99.998

These high DREs were obtained during the tests even though there were fluctuations in CO and poor fuel mixing during combustion.[5] The CO fluctuations may have been due, in part, to the inability to "fine tune" the kiln to minimize operational fluctuations when cofiring waste fuel.[4] The waste fuel was burned only 8 hours per day, whereas at least 24 hours of operation is generally required to make appropriate adjustments.[5] Wide CO fluctuations were attributed not only to firing waste fuel but also to a wet supply of primary fuel (petroleum coke), which resulted in clumps of coke being fed into the kiln, and therefore, to excess-fuel conditions, and to normal variations in the fuel feed rate. The waste-fuel feed and burner system (a fuel pipe laid on top of the main burner) did not allow adequate mixing of the fuels.[5] At low waste-fuel feed rates, this caused puffing of the flame.

Aggregate Kilns

Lightweight aggregates encompass a variety of raw materials that provide the bulk in concrete products. Some occur naturally (e.g., pumice and volcanic cinders), some are by-products (e.g., slag and cinders), while others can be manufactured by heating expandable clay, shale, or slate in a rotary kiln to temperatures of about 1200°C (2000°F). In most cases, the lightweight-aggregate plant is composed of a quarry, a raw-material preparation area, a kiln, a cooler, and a product-storage area.

Such kiln operating parameters as flame temperature, excess air, feed size, material flow, and speed of rotation vary from plant to plant and are determined by the general characteristics of the raw material, including moisture content. Approximately 80 to 100% excess air is forced into the kiln to aid in expanding the raw material. The design of an appropriate burner system is perhaps the most critical feature in the lightweight-aggregate rotary kiln. Maintaining the correct relationship between the retention time of the feed and the temperature along the axis of the kiln is essential to achieving the desired product quality. An HWF system for an aggregate kiln would be similar to those for cement and lime kilns (i.e., an atomizing HWF burner alongside the primary-fuel burner).

The test results for the principal organic components from two aggregate kilns burning HWF are shown in Table 8.15.2. To the extent possible, normal operating conditions with respect to temperatures, total fuel input [J/kg (Btu/h)], feed and production rates, and combustion air were maintained during each test. The lower destruction values may have been caused by the low concentrations of the chemical components in the waste feed (many less than 1,000 ppm) and by formation of products of incomplete combustion (PICs).

Asphalt Plants

There are approximately 4,500 asphalt plants in the United States. Many of them are transportable, which allows them to be set up near paving operations. Firing

TABLE 8.15.2   Test Results for the Principal Organic Components from Two Aggregate Kilns Burning Hazardous-Waste Fuels

Site	POHC* or waste component	DRE, %
A	Methylene chloride	ħ99.99996–>99.99998
	1,2-Dichloroethane	99.91–>99.9993†
	1,1,1-Trichloroethane	99.9998–99.9999†
	Carbon tetrachloride	99.8–99.995†
	Trichloroethylene	99.996–99.993†
	Benzene	99.75–99.993†
	Tetrachloroethylene	99.998–99.9998
	Toluene	99.997–99.9998
	Chlorobenzene	99.92–99.97†
	Methyl ethyl ketone	99.996–>99.999992
	Freon 113	99.99991–99.99998
B	Methyl ethyl ketone	99.992–99.999
	Methyl isobutyl ketone	99.995–99.999
	Tetrachloroethylene	99.995–99.999
	Toluene	99.998–99.999

*POHC = principal organic hazardous constituent.
†Waste-component concentration less than 1,000 ppm. Testing and analytical error as well as component contribution from PICs caused by either primary fuel and/or waste combustion may have resulted in "lower than actual" DRE.
Source: Refs. 6 and 7.

waste oil and hazardous wastes in the asphalt-plant dryer is economically beneficial to the asphalt manufacturers since it is a relatively cheap source of fuel. Stacks are generally short and the units are often set up in populated areas, which makes them an environmental concern especially when firing waste oil or hazardous wastes.

There are two basic plant designs:

1. *Batch-mix plants:*   The heating, screening, and blending production steps are separate. One batch is mixed at a time; rotary dryer is started and stopped many times during the day.

2. *Drum-mix plants:*   The aggregate is sized first, then the heating and blending steps are combined. Plant usually runs continuously for longer periods than a batch plant. Most new asphalt plants are drum-mix types.

Figure 8.15.2 shows the schematic of a typical batch-mix plant. The rotary dryers are designed to quickly heat the rock and drive off most of the surface and pore moisture that would interfere with the proper bonding with the asphaltic cement. The fuel being combusted (gas, oil, waste oil) in the burner is converted into heat and exhaust gases. Outside the actual flame, the gases contact the cold stone and quickly drop in temperature to about 177°C (350°F) or less.[8] The desired flame pattern is short and wide but without impingement on the sides of the dryer.[8]

Burning HWF in asphalt-plant dryers can be considered normal operation and, as such, does not appear to adversely affect plant capacities, materials of construction, or other plant parameters. Emissions testing on asphalt plants indicates that both types of plants, drum mix and batch type, are capable of meeting the

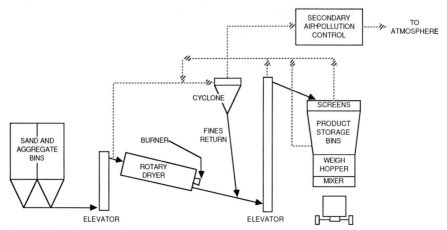

FIG. 8.15.2 Typical batch-mix asphalt-concrete plant.

Resource Conservation and Recovery Act (RCRA) hazardous-waste incinerator standards for particulates (less than 180 mg/m^3 (0.08 gr/dscf) at 7% O_2) and hydrogen chloride emissions (less than 4 lb/h). The results of the tests also showed that drum-mix asphalt plants can meet 99.99% DREs for hazardous organic constituents but were inconclusive about DRE in batch-type plants.[8]

Blast Furnaces

Blast furnaces are currently operating at about 30 locations throughout the country. A typical blast furnace configuration is shown in Fig. 8.15.3.[9] Iron ore, pellets, sinter, limestone, and coke are charged into the blast furnace from the top. Hot air [780 to 1220°C (1400 to 2200°F)] is introduced at the bottom of the furnace through tuyeres. The coke burns at the tuyere level, and the hot gases rise through the descending mass and transfer the sensible heat to the descending raw materials. A temperature of around 1930°C (3500°F) is attained at the tuyere level, and the top gases leave the furnace at around 150°C (300°F). The exit gases pass through a dust catcher (cyclone) and then through a scrubber system. Part of this cleaned gas is fired in stoves used for preheating the combustion air for the blast furnace. The molten iron accumulates in the hearth, and the lighter slag forms a layer on top of the molten iron. Iron and slag are periodically drained from the surface. In most blast furnaces in the United States some kind of fuel is injected through the tuyeres to reduce coke consumption and increase productivity. The fuels used for this purpose include oil, tar, pitch, natural gas, coke-oven gas, and pulverized coal, or combinations thereof.[10] An HWF could also be injected at this location.

A 3630-t/d (4,000-ton/d) blast furnace was tested while burning waste oil as a supplemental fuel.[9] Burning the HWF did not affect blast-furnace operation and for the major organic constituents in the HWF (toluene and o-xylene) the process was found to have a destruction and removal efficiency of greater than 99.99%.

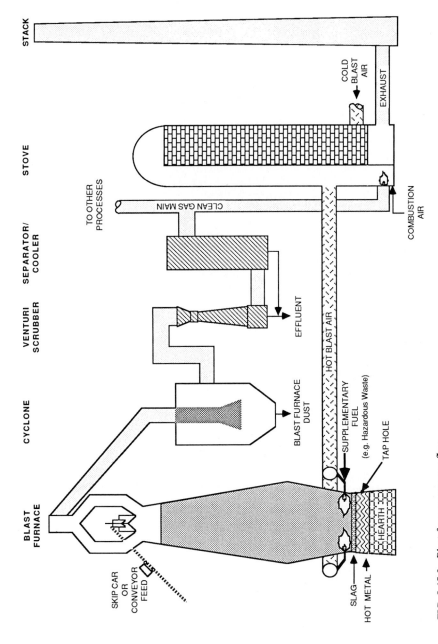

FIG. 8.15.3 Blast-furnace process flow.

8.209

Brick Tunnel Kilns

Tunnel kilns are used in most modern brick and tile plants. The product is loaded onto a flatcar, which enters one end of the kiln, a refractory-lined tunnel. As the car progresses along the length of the kiln, it passes through zones of increasing temperature. In most cases, the kiln has two sets of burners. The main (or upper) fuel-fired burners are located along the middle section of the kiln and provide the bulk of heat necessary for firing. Secondary (or lower) burners are located along the length of the kiln between the preheating zone and the firing zones. Air is passed through the kiln countercurrent to the movement of the ware (i.e., bricks). Processing is continuous, and the average tunnel kiln produces 100,000 bricks per day. The inside dimensions of a typical kiln are 98 m (321 ft) long by 3.4 m (11.2 ft) wide by 1.5 m (4.9 ft) high.[10] Historically, tunnel kilns are fired with natural gas.

Newer kilns are being designed to accept more than one fuel type; for example, either sawdust and/or oil. These new kilns are also significantly more fuel-efficient; they consume about 1860 kJ/kg (800 Btu/lb) of brick compared with the current average of 2790 kJ/kg (1200 Btu/lb). Hazardous-waste fuels can potentially be used to replace existing fuels in these brick tunnel kilns. HWF quality, however, would need to be carefully monitored because changes in the kiln's atmosphere resulting from the incineration of organic wastes could discolor the brick. The pink color, for example, often associated with brick used for residential construction comes from iron impurities. Brick kilns are also more susceptible to abrasion and thermal shock than cement and lime kilns because no particulate coatings are built up on the refractory surface during operation. The addition of abrasive HWFs could, therefore, increase normal wear and tear on the refractory.

No test data are available that permit determination of the effectiveness of brick tunnel kilns for destroying hazardous components of HWF.

8.15.3 ENGINEERING ECONOMIC EVALUATIONS

If HWF is to be burned in an industrial process, provisions must be made for the receiving, storage, and handling of large quantities of this material. For example, at 1.2 kg/L (10 lb/gal) and 2.326×10^6 J/kg (10,000 Btu/lb), a typical lime kiln would require about 113 L (30 gal) of HWF per ton of quicklime produced if operated at a 50:50 HWF/base fuel ratio. A 1360 t/d (1,500 ton/d) or 450,000 t/yr (500,000 ton/yr) kiln would consume about 7,200 L (1,900 gal) of HWF per hour. Requirements for an HWF system would include a receiving facility, blending tanks, a laboratory, a working tank, and a fuel-delivery system (from the working tank to the kiln burner). The estimated capital cost of such a system is more than $1 million, as shown in Table 8.15.3.

The annual costs of owning and operating the HWF system just discussed would run approximately $430,000, including capital charges of $220,000.[4] The capital charges are based on a capital-recovery factor that assumes a 15% interest rate and a 10-year equipment life. These costs are also shown in Table 8.15.3.

In addition to the basic costs described above, consideration must be given to institutional costs such as those associated with obtaining and maintaining the necessary permits, community and employee education programs, and employee training. Figure 8.15.4 presents the total cost data (including the institutional costs) for plants over a range of sizes.

TABLE 8.15.3   Estimated Costs of Hazardous-Waste Supplemental Fuel System for a 454,000 t/yr (500,000 ton/yr) Lime Kiln*

(Thousands of 1985 dollars)

Capital costs	
Four 95,000-L (25,000-gal) storage and blending tanks	240
One 570,000-L (150,000-gal) working tank	150
Pumps, motor, and auxiliary equipment and instrumentation	180
Containment system, sumps, and paved areas	70
Laboratory building	100
Laboratory equipment and safety equipment	120
Fuel-delivery system	50
Contingencies, 20%	180
	1,090

Annualized costs	
Materials and supplies	30
Maintenance	20
Operating labor and general and administration expenses	160
Capital recovery	220
	430

*Does not include costs related to obtaining permits and community and employee relations.
Source: Ref. 4.

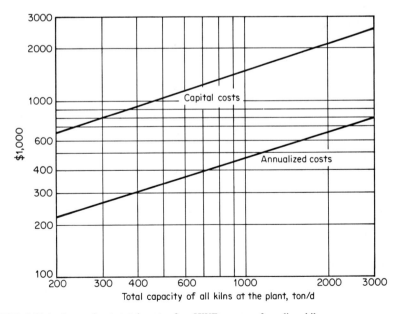

FIG. 8.15.4   Approximate total costs of an HWF program for a lime kiln.

8.15.4 REFERENCES

1. State of California, *The Use of Waste Chemicals as Fuel Supplements for Cement Kilns in California: Briefing Document,* prepared by the Cement Kiln Task Force, Sacramento, Calif., 1982.

2. Research Triangle Institute and Engineering Science, *Evaluation of Waste Combustion in Cement Kilns at General Portland, Inc., Paulding, Ohio,* prepared for the U.S. EPA Industrial Environmental Research Laboratory under Contract No. 68-02-3149, Cincinnati, Ohio, March 1984.

3. J. F. Chadbourne, *The General Portland Supplemental Fuel Program, Fredonia, Kansas Cement Plant,* Review of Major Issues, General Portland, Inc., Paulding, Ohio, July 1983.

4. *Guidance Manual for Co-Firing Hazardous Wastes in Cement and Lime Kilns,* prepared for the U.S. EPA under Contract No. 68-02-3995, by PEI Associates, Inc., Cincinnati, Ohio, July 1985.

5. D. R. Day and L. A. Cox, *Evaluation of Hazardous Waste Incineration in a Lime Kiln: Rockwell Lime Company,* prepared by Monsanto Research Corporation for the U.S. EPA, Industrial Environmental Research Laboratory, under Contract No. 68-03-3025, Cincinnati, Ohio, 1984.

6. D. R. Day and L. A. Cox, *Evaluation of Hazardous Waste Incineration in an Aggregate Kiln: Florida Solite Corporation,* prepared by Monsanto Research Corporation for the U.S. EPA, Industrial Environmental Research Laboratory, under Contract No. 68-03-3025, Cincinnati, Ohio, 1984.

7. A. W. Wyss, C. Castaldini, and M. M. Murray, *Field Evaluation of Resource Recovery of Hazardous Wastes,* prepared by Acurex Corporation for the U.S. EPA, Industrial Environmental Research Laboratory, under Contract No. 68-02-3176, Cincinnati, Ohio, August 1984.

8. R. A. Baker et al., *Summary Test Report: Sampling and Analysis of Hazardous Waste and Waste Oil Burned in Three Asphalt Plants,* prepared by Engineering-Science for the U.S. EPA, Hazardous Waste Engineering Research Laboratory, under Contract No. 68-03-3149, Cincinnati, Ohio, May 1986.

9. *Destruction and Removal of POHC's in Iron Making Blast Furnaces,* prepared by Radian Corporation for the U.S. EPA, Hazardous Waste Engineering Research Laboratory, under Contract No. 68-03-3148, Cincinnati, Ohio, December 1985.

10. *Evaluation of the Feasibility of Incinerating Hazardous Waste in High-Temperature Industrial Processes,* prepared for the U.S. EPA under Contract No. 68-03-3036, by PEI Associates, Inc., Cincinnati, Ohio, April 1983.

SECTION 8.16
INFECTIOUS-WASTE INCINERATION

Frank L. Cross, Jr., P.E.
President
Cross/Tessitore & Associates, P.A.
Orlando, Florida

Rosemary Robinson
Project Engineer
Cross/Tessitore & Associates, P.A.
Orlando, Florida

The disposal of infectious wastes has become an issue of growing concern. Currently, most infectious wastes are incinerated; however, infectious wastes are often inadvertently integrated with general wastes by hospital personnel and are landfilled. For this reason, many counties in Florida are restricting hospital wastes from their landfills.

Another problem associated with infectious wastes is the improper design and/or operation of infectious-waste incinerators. Testing of several types of incinerators shows that improper design and/or operation of these incinerators can result in inefficient destruction of pathogenic organisms.[1] Some factors which govern the sterilization process include load size (with respect to incinerator capacity), moisture content of refuse, and retention time. These factors are directly attributable to design and operation. Therefore, it is of the utmost importance that the design process include a detailed, site-specific waste study (quantities and composition), careful equipment selection, and thorough operator training.

8.16.1 WASTE QUANTITIES AND COMPOSITION

Infectious waste is generated from a number of sources including

- Hospitals
- Research centers (e.g., National Institutes of Health)
- Veterinary clinics
- Universities
- Schools of veterinary medicine

The U.S. Environmental Protection Agency (EPA)[2] has recommended that the following types of waste be considered infectious waste:

- Isolation wastes
- Cultures and stocks of infectious agents and associated biologicals
- Human blood and blood products
- Pathological wastes
- Other wastes from surgery and autopsy
- Contaminated laboratory wastes
- Contaminated sharps (hypodermic needles, etc.)
- Dialysis-unit wastes
- Contaminated animal carcasses, body parts, and bedding
- Discarded biologicals
- Contaminated food and other products
- Contaminated equipment

As the greatest quantities of these wastes are generated at hospitals, composition and quantities of hospital waste will be focused on in this section. Estimates of hospital waste quantities vary from 10 to 20 lb/d per bed with 17 lb/d per bed being a reasonably representative sample. Recently, however, the quantity of disposables has increased and a recent study of South Florida hospitals indicates a generation rate as high as 23 lb/d per bed.[3] Representative hospital-waste characteristics are presented in Table 8.16.1. Generally, the composition of hospital

TABLE 8.16.1 Hospital-Waste Characteristics

Waste type	Wt %*	Genera-tion rate, lb/d/bed	Heating value, Btu/lb	Weighted value, Btu/lb
Trash (Class 0—mostly paper and cardboard)	70	11.9	8500	5950
Plastic	15	2.55	19,500	2925
Garbage (Class 3—food wastes)	10	1.70	4,500	450
Pathological (Class 4)	5	0.85	1,000	50
Total	100	17	N/A	9375

*Courtesy of Simonds Manufacturing Corporation

waste is quite uniform. (A more extensive discussion of infectious waste can be found in Sec. 4.4 of the Handbook.)

The average heating value of hospital wastes varies depending upon the amount of disposables, especially plastics, in the waste stream. For purposes of incinerator design, a value from 8500 to 9500 Btu/lb should be used if all hospital wastes are to be incinerated. (See Table 8.16.1.)

8.16.2 INCINERATION OF HOSPITAL WASTES

Before selecting an incineration system, a detailed solid-waste study should be conducted. The study should determine waste types, quantities, and composition and evaluate current handling practices and capacity for heat recovery. This information is necessary for incinerator design and equipment selection. Figure 8.16.1 outlines a systematic approach to establishing a solid-waste management program.

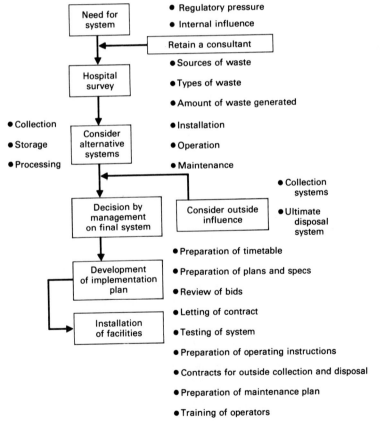

FIG. 8.16.1 Systematic approach to decision making for hospital solid-waste management.

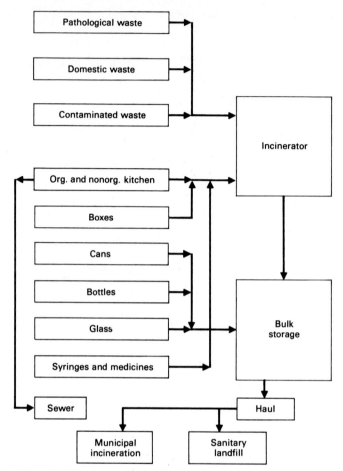

FIG. 8.16.2   Incineration of hospital wastes.

Following the transport of waste to the incinerator site (by carts, by chutes, or pneumatically), the waste, contained in color-coded plastic bags, is fed into the incinerator (either manually or by a feed system). (See Fig. 8.16.2.) The majority of the incinerators used by hospitals are controlled-air units, although some rotary-kiln systems are being used where the economics are favorable (i.e., depending on size and heat-recovery considerations). Table 8.16.2 shows a comparison between rotary-kiln and controlled-air incinerators.

Since controlled-air incinerators are those most often used for destruction of infectious waste, these units will be discussed further. The controlled-air incinerator is a two-chamber, hearth-burning, pyrolytic unit. Waste is fed into the primary chamber, where it starts to burn with less than stoichiometric air. Volatiles are liberated and are burned in the secondary combustion chamber (Fig. 8.16.3). Controlled-air units are efficient incineration systems with a high degree of burn-out and low fly-ash generation. These units were batch-operated, but the newer

TABLE 8.16.2 Comparison of Alternative Incinerators

Advantages	Disadvantages
Controlled-air incinerator	
Small space requirements for incinerator as secondary chamber is above primary combustion chamber.	May have poor burnout of carbonaceous material in ash.
Low emissions without control device (0.03 gr/ft³).	Tends to operate on batch mode; if continuous, may be problems with clinkers.
Relatively low capital cost.	Push plates for ash removal tend to buckle.
Rotary-kiln incinerator	
Extremely good burnout of carbonaceous material in ash.	May require additional space because secondary combustion chamber separate from kiln.
Good for continuous operation.	Emissions tend to be higher than with controlled-air units; characterized by high turbulence and air exposure.
Will incinerate materials passing through a melt phase.	Problems of maintaining seals at ends of drum.
Residence time can be adjusted by drum speed.	High capital cost.
Up to 50% turndown ratio.	May have undesirable noise and vibration.

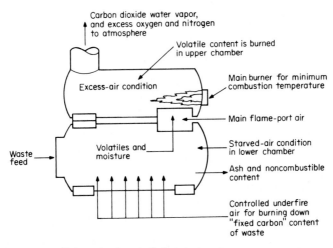

Carbon dioxide water vapor, and excess oxygen and nitrogen to atmosphere

Volatile content is burned in upper chamber

Main burner for minimum combustion temperature

Excess-air condition

Main flame-port air

Waste feed

Volatiles and moisture

Starved-air condition in lower chamber

Ash and noncombustible content

Controlled underfire air for burning down "fixed carbon" content of waste

FIG. 8.16.3 Schematic of controlled-air incinerator.

units have continuous ash removal (pusher plates are used). Some (e.g., Ecolaire) use steam injection in the primary chamber to reduce slag formation. The production of slag sometimes occurs with 24-h operation of controlled-air units.

The big advantage of controlled-air incinerators is that proper operation produces relatively low particulate emissions. Typical rates are 0.08 to 0.10 gr per standard ft^3, with rates as low as 0.04 gr per standard ft^3 in some cases.[4] Thus, air-pollution control equipment is not needed on controlled-air units for particulate emissions, but only if acid-gas emissions are high.

Excessive emission rates generally result from one or more of the following causes:

1. High set point for secondary burner too low.
2. Excessive infiltration air.
3. Excessive negative draft in primary chamber.
4. Excessive primary air.
5. Excessive secondary combustion air.
6. Waste characteristics prevent operation at design conditions.

Most controlled-air incinerators operate in the range of −0.10 to −0.25 in w.c. in the primary chamber. This negative draft should be sufficient to prevent discharge of smoke and odors when the charging door is open. A greater negative draft in the primary chamber tends to lift particulates from the refuse bed into the discharge gas stream, resulting in visible emissions from the stack.

8.16.3   WASTE-HEAT RECOVERY

From the secondary chamber, hot exhaust gases resulting from the combustion process approach temperatures of 900 to 1000°F. With this available energy, heat recovery is often very economically attractive. That is, these hot exhaust gases can be directed from the secondary combustion chamber either to a "dump" stack or to heat-recovery equipment. Waste-heat boilers, the usual heat-recovery unit, are identical to firetube boilers with the usual variations (single-pass, double-pass, etc.), but without a burner. (See Figs. 8.16.4 and 8.16.5.) Space must be provided for easy access to the boiler tubes, as operators report that it is necessary to clean the firetubes weekly.

If desired, a standard gas or oil burner can be added to the boiler to provide steam if the incinerator is out of operation or if there is inadequate fuel.

In addition to the waste-heat boiler, another type of heat-recovery unit is the air-to-air heat exchanger. This exchanger is available from several manufacturers. As a general rule, however, heat-recovery efficiency of air-to-air heat exchangers is substantially lower than that of waste-heat boilers, so heat exchangers are usually used just for space heating. Because of this low efficiency, most owners of heat-recovery equipment either do not utilize or express dissatisfaction with air-to-air heat exchangers.

When heat-recovery equipment is used, the flue gases are drawn through the heat-recovery equipment by an induced-draft fan. Specifications on this fan should include capability of working with gases up to 600°F.

The economics of heat recovery using controlled-air incinerators at hospitals are dependent upon the demand for steam generation and the amount of waste

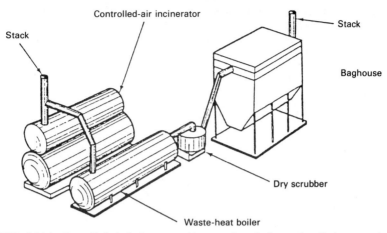

FIG. 8.16.4 Controlled-air incinerator with waste-heat boiler and pollution-control equipment.

FIG. 8.16.5 Controlled-air incinerator with heat-recovery equipment.

available for incineration. Generally, these two factors depend on hospital size. With this consideration, it is economically feasible to install heat-recovery equipment only in hospitals larger than 800 beds (assuming 90% occupancy). Heat recovery also looks favorable for central incinerator facilities which incinerate wastes for several hospitals.

8.16.4 POLLUTION-CONTROL SYSTEMS FOR CONTROLLED-AIR INCINERATORS

With most incinerators, the primary concern in air-pollution control is with particulate emissions. One of the advantages of controlled-air incinerators is the low particulate emissions (approximately 1.4 lb/ton). Most hospital incinerators currently are not equipped with pollution-control devices. However, with stringent standards and concerns about air toxics and acid-gas emissions, future units should be equipped with control systems. The advantages of control systems for hospital incinerators are that they

- Reduce particulate emissions (0.01 to 0.03 gr/ft^3)
- Reduce acid emissions (HCl)
- Minimize visible emissions during start-up, shutdown, and malfunction
- Remove air toxics
- Reduce pathogens

The two most suitable devices for use in the future on hospital incinerators are wet and dry scrubbers (see Fig. 8.16.4). Table 8.16.3 shows a comparison of

TABLE 8.16.3 Comparison of Air-Pollution Control Systems

Collection Principle Involved	Advantages	Disadvantages
	Dry scrubber	
Reacting particles and gases in spray-dryer section & collection of particles in a baghouse collector	Good particulate removal (down to 0.01 gr/standard ft^3) Good acid-gas removal (up to 90%) Small space req'd compared to baghouse Solids removed dry	Requires skilled operator High capital costs & operating costs Have not been used extensively on hospital incinerators
	Wet scrubber	
Impaction of particles & reaction of gases, with countercurrent-flow reactor	Good particulate removal Very small space required Excellent acid-gas removal Moderate capital costs Modest operating costs	Requires skilled operator Solids removed wet Needs quench section

TABLE 8.16.4 Laboratory Analysis of Ash

Parameter	EP toxicity: acceptable limit, mg/L	Value from incinerator ash analysis, mg/L
Arsenic	5.0	0.002
Barium	100.0	0.223
Mercury	0.2	0.0005
Cadmium	1.0	0.108
Lead	5.0	2.14
Chromium	5.0	0.01
Silver	5.0	0.005
Selenium	1.0	0.003

these air-pollution control devices.

Once pollution-control equipment is added, the residual from the control device (e.g., scrubber water) must be disposed of. This residual, as well as the ash from the incinerator, must be analyzed for EP toxicity prior to disposal. Based on tests in Florida,[5] hospital ash appears to be acceptable for disposal without further treatment as shown in Table 8.16.4.

8.16.5 SPECIAL PROBLEMS

As pollution-control regulations become more restrictive, not only will additional control be needed on infectious-waste incinerators, but also attention will be

---

Basic Engineering
24 W. 161 Hill
Glen Ellyn, Ill. 60137

ECP-Environmental Control Products
Southern Corporation
10 Southern Place
Clover, S.C. 29710

Industronics, Inc.
489 Sullivan Avenue
South Windsor, Conn. 06074

Kelly Company, Inc.
6720 N. Teutonia Avenue
Milwaukee, Wis. 53209

Simonds Manufacturing Corporation
304 Progress Road
Auburndale, Fla. 33823

FIG. 8.16.6 Equipment vendors for controlled-air incinerators.

directed to the operation of these units. Presently, many of the violations of regulations are directly attributable to the operation of the system. This problem arises from actions of untrained or improperly trained operators. Where tests have been conducted on several types of incinerators (i.e., in-line industrial-refuse incinerator, vertical in-line incinerator, and others), evidence shows that only when the unit was operated properly did complete destruction of pathogens occur.[6] However, no testing of microorganism emissions from controlled-air incinerators has been conducted as yet.

8.16.6 REFERENCES

1. M. S. Barbeito, and G. G. Gremillion, "Microbiological Safety Evaluation of an Industrial Refuse Incinerator," *Applied Microbiology,* **16**:291 (February 1968).

2. U.S. Environmental Protection Agency, Office of Solid Waste, *Draft Manual for Infectious Waste Management,* Washington, D.C., September 1982, sec. 2.5.

3. *Centralized Incinerator Study for the South Florida Hospital Association,* Cross/Tessitore & Associates, P.A., Orlando, Florida, December 1985, p. 14.

4. F. L. Cross and H. E. Hesketh, *Controlled Air Incineration,* Technomic Publishing, Lancaster, Pennsylvania, 1985, p. 69.

5. *Solid Waste Management Study, Phase I Report,* prepared for the Sarasota Memorial Hospital by Cross/Tessitore & Associates, P.A., Orlando, Florida, July 1985, p. 40.

6. M. L. Peterson and F. J. Stutzenberger, "Microbiological Evaluation of Incinerator Operations," *Applied Microbiology,* **18**:8 (July 1969).

SECTION 8.17

GUIDE FOR INCINERATOR TRIAL BURNS

Paul G. Gorman
Principal Chemical Engineer
Midwest Research Institute
Kansas City, Missouri

All hazardous-waste incinerators are required by the U.S. Environmental Protection Agency (EPA) to submit a Part B Permit Application under the authority of the Resource Conservation and Recovery Act (RCRA). One section of that application is a trial-burn plan. After the trial-burn plan has been approved by the EPA and appropriate state agencies, the actual trial burn must be conducted and results reported, before the operating permit for the incinerator is issued. Important aspects in preparing the trial-burn plan and in conducting the actual trial burn are presented in this section of Chap. 8.

8.17.1 Trial-Burn Plan

What Is a Trial Burn?

The *trial burn* is a test of the incinerator to demonstrate that it meets the RCRA [or Toxic Substances Control Act (TSCA)] requirements when operating under specific test conditions. These test conditions are chosen, for the most part, by the incinerator operators with the objective of providing them with the widest possible operating flexibility. Usually, the trial-burn test conditions are specified as the operating limits for the incinerator in the operating permit issued subsequently by the EPA. The performance requirements that must be met are:

RCRA Performance Requirements

- Destruction and removal efficiency (DRE) of 99.99% or greater for each specified principal organic hazardous constituent (POHC) in the waste feed

- Particulate emission concentration of 180 mg/dscm or less (corrected to 7% O_2)
- Hydrogen chloride (HCl) removal efficiency of 99% or greater, or HCl emissions not exceeding 1.8 kg/h

TSCA Performance Requirements (for PCB Wastes)

- DRE of 99.9999% or greater for total PCBs in the waste feed
- Combustion efficiency (CE) >99.9%

$$CE = \frac{\%CO_2}{\%CO_2 + \%CO} \times 100$$

The major point about the trial burn is that the plant's objective of wide operating flexibility means that the test should be conducted under the "worst-case" conditions, but operating under these conditions will maximize the chances of failing to achieve the performance requirements. To repeat, the trial-burn test conditions will likely be specified as the permit operating limits for the incinerator, thus restricting all subsequent operations.

The worst-case operating parameters usually include those listed in Table 8.17.1. However, it is often not possible to achieve all these worst-case conditions at the same time. In fact, some are mutually exclusive (e.g., maximum air feed rate and minimum O_2 in stack gas). Consequently, the trial burn may need to include testing at more than one set of operating conditions. The EPA's practice is to require three replicate runs (tests) at each set of operating conditions, so the number of test conditions needs to be kept to a minimum (usually no more than two test conditions with six runs required).

Whatever the test conditions, the objective of the trial burn is to demonstrate compliance with the performance requirements. This basically involves sampling and analysis of waste feeds to determine the input rate of each POHC, followed by sampling and analysis of stack effluent to determine the POHC emission rate. From the input and emission rates, the DRE can be calculated. At the same time, the stack effluent is sampled and analyzed to determine particulate concentrations and the HCl emission rate. Each test is done over a 2- to 3-h period, and three replicate tests are performed. This is the essence of a trial burn, although supplemental sampling and analysis is often included, as discussed later.

TABLE 8.17.1   Trial-Burn Conditions

Waste Characteristics

Maximum concentration of selected POHCs
Maximum Cl content
Maximum ash content
Minimum heating value (HHV) of waste feed

Operating Conditions

Maximum heat input rate
Minimum combustion temperature
Maximum waste feed rate
Minimum O_2 concentration in stack gas
Maximum air input rate (maximum gas flow rate
  to yield minimum residence time)
Maximum CO content in stack gas

When Is a Trial-Burn Plan Necessary?

A trial-burn plan is unnecessary only under two circumstances:

- If the waste to be incinerated is hazardous (ignitable, corrosive, etc.) but does not contain any Appendix VIII constituents (see 40 *CFR* 261)
- If previous test data can be submitted in lieu of the trial burn

Some wastes that are incinerated may be hazardous but not contain any Appendix VIII compounds. The EPA regulations do not necessarily require analysis of wastes for all Appendix VIII compounds, since such analysis is almost impossible to do for the 300+ compounds listed. Rather, the incinerator must be able to provide general analysis data for what is present, and must be able to support an argument that no Appendix VIII compounds are reasonably expected to be present at any significant concentration ($\hbar$100 ppm). If *any* of the Appendix VIII compounds are expected or known to be present, then a trial burn is probably required, unless data in lieu of a trial burn can be submitted.

Data in lieu of a trial burn can be submitted in some cases. Practically speaking, this means other trial-burn data for a similar incinerator burning a similar waste at similar operating conditions. Guidance from the EPA on the meaning of the term *similar* is provided in Ref. 1. But in practice, the term almost means *identical*. That is, other trial-burn data must be available for an incinerator that is essentially identical and is operated at the same or less severe (lower-temperature) conditions, and burns an essentially identical waste. Such situations do occur, but they are rare.

If neither of the above two circumstances apply, then a trial-burn plan and trial burn will likely be necessary.

Contents of a Trial-Burn Plan

The trial-burn plan is a part of the RCRA Part B Permit Application. It is Section D, headed "Process Information." (See Ref. 2 for a model permit application for an existing incinerator.) An outline of the contents of Section D is given here as Table 8.17.2. Much of Section D is relatively simple and straightforward. The selection of operating conditions and specification of the sampling and analysis methods and the associated quality assurance and quality control (QA/QC) for

TABLE 8.17.2   Outline of Contents of a Trial-Burn Plan

D-5b (2) Trial-Burn Plan
a. Detailed engineering descriptions of incinerator
b. Detailed description of sampling and monitoring procedures
c. Test schedule
d. Test protocol for each test
e. Descriptions and planned operating conditions for the emission-control equipment
f. Shutdown procedures
g. Not used
h. Quality-assurance and quality-control procedures

Source: Ref. 2.

the trial burn is difficult, however, and usually requires assistance from sampling and analysis contractors experienced in those aspects of trial burns. Further discussion of some of the more difficult aspects is presented below, relative to

- Selection of operating conditions
- Selection of POHCs
- Sampling and analysis methods
- Quality assurance and quality control (QA/QC)

Selection of Operating Conditions. The selection of operating conditions and waste-feed characteristics is one of the most important and difficult parts of the trial-burn plan because these conditions and characteristics probably will be the limiting conditions specified subsequently in the operating permit.

As shown in Table 8.17.1, the worst-case operating conditions and waste characteristics need to be selected to provide the necessary operating flexibility. The conditions to be used during the trial burn must be set out in the trial-burn plan. If more than one set of operating conditions appears to be necessary, then each set of operating conditions must be set out.

One problem that has arisen more than once is that an incinerator operator may desire to test at a maximum feed rate with an associated operating temperature (e.g., 2000°F or 1100°C). But the incinerator may not be able to operate at as high a temperature when firing at a lower feed rate. Operating at a lower feed rate might be allowable under the operating permit, but not at a lower temperature. Consequently, the trial burn would have to include a second set of operating conditions (in this case, lower temperature).

Achieving all of the worst-case waste-feed characteristics can also be a problem, especially maximum ash content. In some cases, the feed must be spiked with fly ash or diatomaceous earth for the trial burn to provide maximum ash content. But such spiking can cause other difficulties, so close coordination with the incinerator's operating staff is essential for planning the trial-burn test requirements.

Another more common problem, even for existing incinerators that have been operating for several years, is the operation of the incinerator at all the worst-case conditions simultaneously for relatively long periods (8 h per test). This often leads to considerable difficulty—and delays—while conducting the tests.

Selection of POHCs. The incinerator operator does not select the POHC compounds. Rather, the operator recommends selection of POHCs in the trial-burn plan, and the permit reviewer has final authority for their selection. Most commonly, the trial-burn plan presents information on the concentration of Appendix VIII compounds expected to be present in the wastes, and from that list, two to six compounds are selected as POHCs. At least two are usually recommended, as follows:

- The one compound expected to be present at the highest concentration
- The one compound having the lowest higher heating value (HHV) (because it is the most difficult to destroy). (See Ref. 1 for HHVs of all Appendix VIII compounds.)

Again, the waste characteristics include specification of the selected POHCs and their concentrations. Quite often, incinerator operators must prepare synthetic wastes or spike wastes to ensure that the specified POHCs are present at the proper concentrations. Further, it should be noted that the selected POHC should be present at concentrations of at least 5%, or higher if possible. The reason for this is that higher concentrations usually yield higher DREs.[3]

Sampling and Analysis Methods. All the sampling and analysis (S&A) methods to be used in the trial burn must be specified in the trial-burn plan. This is one of the most difficult parts of the plan, and it is one of the major reasons that operators often utilize experienced contractors for help in preparing the trial burn plan and in conducting the actual tests. It is also the part of the plan that generates the most questions from regulatory agencies.

Examples of sampling and analysis methods that would be specified in a trial-burn plan are shown in Tables 8.17.3 and 8.17.4.[4] The two primary sources for the recommended methods are Ref. 5 and Ref. 6. Table 8.17.5 (from Ref. 1) is often useful for waste-analysis methods other than the POHCs.

Most of the recommended sampling and analysis methods, and especially

TABLE 8.17.3   Example Sampling Methods and Analysis Parameters

Sample	Sampling frequency for each run	Sampling method*	Analysis parameter†
1. Liquid-waste feed	Grab sample every 15 min	S004	VPOHCs SVPOHCs, Cl⁻, ash, ult. anal., viscosity, HHV
2. Solid-waste feed	Grab sample of each drum	S006, S007	VPOHCs SVPOHCs Cl⁻, ash, HHV
3. Chamber ash	Grab one sample after all three runs are completed	S006	VPOHCs, SVPOHCs, EP toxicity
4. Stack gas	Composite	MM5 (3 h)	SVPOHCs, particulate, H_2O, HCl
	Three pairs of traps, 40 min each pair	VOST (2 h)	VPOHCs
	Composite in Tedlar gas bag	S011	VPOHCs‡
	Composite in Mylar gas bag	M3 (1–2 h)	CO_2 and O_2 by Orsat
	Continuous (3 h)	Continuous monitor	CO (by plant's monitor)

*VOST = volatile-organic sampling train.
 MM5 = EPA Modified Method 5.
 M3 = EPA Method 3.
 SXXX denotes sampling methods found in Ref. 5.
†VPOHCs = volatile principal organic hazardous constituents.
 SVPOHCs = semivolatile POHCs.
 HHV = higher heating value.
‡Gas-bag samples may be analyzed for VPOHCs only if VOST samples are saturated and not quantifiable.

TABLE 8.17.4   Example Analytical Methods

Sample	Analysis parameter	Sample preparation method	Sample analysis method
Liquid-waste feed	VPOHCs	8240	8240
	SVPOHCs	8270	8270
	Cl⁻	——	E442-74
	Ash	——	D482
	HHV	——	D240
	Viscosity	——	A005
Solid-waste feed	VPOHCs	8240	8240
	SVPOHCs	8270	8270
	Cl⁻	——	D-2361-66 (1978)
	Ash	——	D-3174-73 (1979)
	HHV	——	D-2015-77 (1978)
Ash	VPOHCs	8240	A101
	SVPOHCs	P024b, P031	A121
	Toxicity	——	C004
Stack gas			
MM5 train			
Filter and	Particulate	M5	M5
probe rinse	SVPOHCs	P024b, P031	A121
Condensate	Cl⁻	——	325.2
	SVPOHCs	P021a	A121
XAD resin	SVPOHCs	P021a	A121
Caustic impinger	Cl⁻	——	325.2
VOST	VPOHCs	——	A101
Tedlar gas bag	VPOHCs	——	A101*
Gas bag	CO_2, O_2	——	M3 (Orsat)
Continuous monitor	CO	——	Continuous monitor

Note: Four-digit numbers denote methods found in Ref. 6. Numbers with prefixes A, C, and P denote methods found in Ref. 5. Method No. 325.2 (for Cl⁻) is from *Methods for Chemical Analysis of Water and Wastes*, EPA-600/4-79-020, March 1979. Numbers with prefixes D and E denote methods established by the American Society for Testing and Materials Standards (ASTM). M3, M5 refer to EPA testing methods found in the *Federal Register*, vol. 42, no. 160, August 18, 1977.

those for POHCs, are discussed in Ref. 5. Both Ref. 5 and Ref. 6 take considerable time to understand, however; and experience helps a great deal in selecting the most appropriate methods. Neither reference alone provides a true appreciation of the complexity of the sampling and analysis methods associated with a trial burn.

Very briefly, the trial-burn sampling and analysis procedure usually consists of the following, as a minimum:

Waste feed

Sampling: Obtain representative samples of all waste feeds (e.g., sample liquids every 15 min to form composites)

TABLE 8.17.5 Acceptable Analytical Methods for Waste Analysis

Parameter	Method(s)*	Comments
Heating value	SA A006	Methods D2015 and D3826 are applicable to solid wastes and D240 is applicable to liquid wastes.
Chlorine (organically bound)	SA A004 ASTM D2361, E442	Combustion method, may be combined with determination of carbon, hydrogen, and sulfur.
Hazardous metals Mercury	SA A021 SW 7470, 7471	Summary of atomic-absorption and ICAP methods. These methods are based on detection of mercury vapor by atomic-absorption spectrophotometer, and are subject to interferences. Spiked samples should be analyzed to establish recovery. Methods involving strong oxidation, such as ASTM D3223, should be avoided because of the possibility of explosions. Alternatively, atomic absorption may be used with a graphite furnace.
Arsenic Selenium	SW 7060, 7061 SW 7740, 7741	Gaseous hydride generation coupled with atomic-absorption detection is recommended. This method is subject to interferences, so spiked samples should be analyzed to establish recovery. Colorimetric methods, such as EPA 206.4 or ASTM D3081, should not be used because of interferences. Alternatively, atomic absorption may be used with a graphite furnace.
Barium Beryllium Cadmium Chromium	SW 7080, 7081, EPA 208.1 SW 7090, 7091, EPA 210.1 SW 7130, 7131, EPA 213.1 SW 7190, 7191, 7195, 7196, 7197, EPA 218.1	These methods are for direct-aspiration, flame, atomic-absorption spectroscopy. Sample preparation should be performed in accordance with Section 200.1 of the EPA manual. Generally, the sensitivity achieved with the graphite-furnace techniques is not required with hazardous-waste samples, and the furnace methods are subject to interferences.
Nickel Thallium Lead Silver Antimony	SW 7520, 7521, EPA 249.1 SW 7840, 7841, EPA 279.1 SW 7420, 7421, EPA 239.1 SW 7760, 7761, EPA 272.1 SW 7040, 7041, EPA 204.1	
Hazardous constituents, including PCBs	Sampling and Analysis Manual	Hazardous constituents listed in Appendix VIII of Part 261 and those in Table 1 of 261.24 may be analyzed by methods in SW-846.

TABLE 8.17.5 Acceptable Analytical Methods for Waste Analysis (*Cont.*)

Parameter	Method(s)*	Comments
Kinematic viscosity	SA A005 ASTM D445 or D88	A variety of methods may be employed using various types of instruments, including rational, piston, float, vibrating-probe or capillary types.
Percent solids	ASTM D1888	A distinction should be noted between water-insoluble solids and solids not soluble in organic solvents. Any of a variety of separation techniques may be employed, e.g.; vacuum filtration, centrifugation, pressure filtration.
Sulfur	ASTM D3177, E443	Combustion methods.
Ash	SA A001-A002 ASTM D3174 or D482	D3174 is for solid wastes and D482 is for liquid wastes.
Flashpoint	ASTM D93, D3278, or D1310	Methods D93 and D3278 are pursuant to the definition of ignitable wastes in Section 261.21 of the regulations. D1310 provides comparable results.
Carbon and hydrogen	ASTM D3178	Combustion method.
Moisture	SA A001-A002 ASTM D85, D3173	D95 is a xylene codistillation and is recommended for most wastes. D3173 and A001-A002 are intended for solid wastes, but the oven heating will drive off volatile compounds as well as water. D1796 is a centrifuge method intended for use with liquids.

*SA refers to *Sampling and Analysis Manual for Hazardous Waste Incineration*, 1st ed., Dec. 1983.
SW refers to *Test Methods for Evaluating Solid Waste*, SW-846, 2d ed., July 1982.
ASTM refers to American Society for Testing and Materials Standards.
EPA refers to *Chemical Analysis of Water and Wastes*, EPA 600/4-79-020, March 1979.
Source: Ref. 1.

Analysis: Analyze samples for POHC concentrations, HHV, ash, and Cl
Stack effluent

Sampling:

1. Sample by EPA Method 5 (M5) for determination of gas flow rate and to determine particulate and HCl concentration
2. Sample with Modified Method 5 (MM5) for semivolatile POHCs
3. Sample with volatile-organic sampling train (VOST) for volatile POHCs

Analysis:

1. Extract MM5 samples with methylene chloride and analyze by gas chromatograph and mass spectrometry (GC/MS) for semivolatile POHCs
2. Heat-desorb VOST traps with analysis by purge-and-trap GC/MS method for volatile POHCs[7]

Sampling of the waste feed is relatively straightforward but can get complicated if solid feeds are involved. Sampling of the stack effluent by M5 and MM5 is the most complex part of the sampling in a trial burn. It is also the most difficult, as can be realized from the three accompanying photographs (Figs. 8.17.1, 8.17.2, and 8.17.3). Figure 8.17.3 shows an MM5 train in place on the rail system, extending beyond the edge of the platform.

Diagrams of an M5 train and an MM5 train are shown in Figs. 8.17.4 and 8.17.5. The basic difference between the two is that the MM5 train includes a

FIG. 8.17.1   (*Photo by Tom Walker*)

FIG. 8.17.2 *(Photo by Tom Walker)*

FIG. 8.17.3 *(Photo by Tom Walker)*

8.232

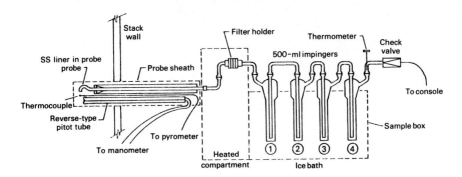

1 Modified Greenburg-Smith, reversed, empty
2 Greenburg-Smith, 200 ml of distilled H_2O, 200 ml capacity
3 Greenburg-Smith, 100 ml of 0.1N KOH
4 Modified Greenburg-Smith, SiO_2

FIG. 8.17.4   Method 5 sampling train (M5).

condenser and an XAD resin trap. However, recovery of samples from the MM5 train for POHC analysis is much different than for the M5 train. Also, no silicone rubber gaskets or joint-sealing compound should be used in the MM5 train since these can interfere with the analysis.

The VOST train is used for sampling volatile POHCs (see Fig. 8.17.6). It is relatively simple to use but does require very specific methods for its use, as de-

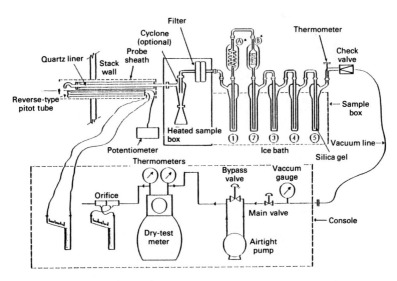

1 Modified Greenburg-Smith, reversed, empty
2 Greenburg-Smith, 50 ml of double distilled in glass H_2O
3 Greenburg-Smith, 100 ml of 0.1 N KOH
4 Modified Greenburg-Smith, empty
5 Modified Greenburg-Smith, SiO_2

A Condenser
B XAD resin cartridge

* Icewater jacket

FIG. 8.17.5   Modified Method 5 (MM5) sampling train.

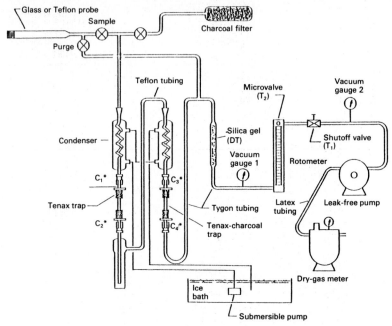

FIG. 8.17.6 Volatile-organic sampling train (VOST).

tailed in Ref. 7. Also, preparation of the VOST traps must follow rigorous procedures to avoid contamination.

From the above discussion, one notes that the POHCs fall into three groups relative to the sampling and analysis requirements:

- *Volatiles:* Organic compounds with boiling point less than 100°C, usually sampled with VOST
- *Semivolatiles:* Organic compounds with boiling point greater than 100°C, usually sampled with MM5
- *Other:* Organic compounds with unique characteristics (e.g., highly reactive or highly water-soluble), requiring other sampling and analysis methods

In terms of sampling POHCs in the stack effluent for determination of DRE, one important question that often arises is the detection limit of the sampling and analysis method. The concentration of a POHC that may be expected in the stack effluent can be estimated, in most cases, by the following rule of thumb:

$$100 \text{ ppm in waste feed} = 1 \ \mu g/m^3 \text{ in stack gas, at } 99.99\% \text{ DRE}$$

This rule of thumb equation can be used to estimate stack-gas concentration for any waste-feed concentration (e.g., 2,000 ppm in waste = 20 $\mu g/m^3$ in stack gas, at 99.99% DRE). Table 8.17.6 shows the procedure for carrying out selection of sampling methods and estimating the concentration of POHC to be detected in the stack effluent.

TABLE 8.17.6 Procedure for Identifying Necessary Stack-Sampling Methods

Step no.	Action	Comment
1	Determine whether each POHC is a volatile or semivolatile compound.	Volatile compounds are generally those that have boiling points below 100°C. Most can be sampled with VOST. The best way of determining the sampling method needed is to refer to Appendix B of Ref. 5. If the POHC shows sampling by "particulate/sorbent" then *MM5* is required. If it shows "sorbent" or "gas bulb" then *VOST or gas bags* will be the sampling method. Regardless of whether or not a POHC is a volatile or semivolatile, some POHCs require special sampling methods as indicated in Appendix B of Ref. 5 (e.g., formaldehyde).
2	Estimate concentration of each POHC in the stack gas, assuming a DRE of 99.99%.	Estimation of the concentration of each POHC requires some knowledge or approximation of POHC concentration in waste feeds, waste feed rates, and stack-gas flow rates. Using that information, concentrations of each POHC in the stack gas can be estimated, for an assumed DRE of 99.99% (see example calculation in Table 8.17.7). For semivolatile POHCs, MM5 is suitable for any stack-gas concentration above 1 $\mu g/m^3$. For volatile compounds, VOST should be used when stack-gas concentrations fall within the range of 1 to 100 ng/L. However, if the estimated stack-gas concentration exceeds 100 ng/L, then gas bags should also be used and analyzed in the event that the VOST sample concentrations saturate the GC/MS system.

TABLE 8.17.7   Example Calculation for Estimating POHC Concentration in Stack Gas, at DRE of 99.99%

Basis	
Waste-feed flow rate, gpm	#4
Density of waste, lb/gal	#9
POHC concentration in waste feed, ppm	Near minimum significant level of 200*
Stack-gas flow rate	Unknown but total heat input to incinerator is approximately 30×10^6 Btu/h with 100% excess air

Calculation	
Rule of thumb (applies in most, but not all cases)	Each 100 Btu of heat input produces about 1 dscf of flue gas at 0% excess air, or 2 dscf of dry flue gas at 100% excess air
1. Flue-gas flow rate:	$\left(30 \times 10^6 \dfrac{\text{Btu}}{\text{h}}\right) \left(\dfrac{\text{h}}{60 \text{ min}}\right) \left(\dfrac{2 \text{ dscf}}{100 \text{ Btu}}\right) = 10{,}000 \text{ dscf/min}$
	$\left(10{,}000 \dfrac{\text{dscf}}{\text{min}}\right) \left(\dfrac{1 \text{ m}^3}{35.3 \text{ ft}^3}\right) = 283 \text{ m}^3/\text{min}$
2. Waste feed rate:	$\left(4 \dfrac{\text{gal}}{\text{min}}\right) \left(\dfrac{9 \text{ lb}}{\text{gal}}\right) \left(\dfrac{454 \text{ g}}{\text{lb}}\right) = 16{,}300 \text{ g/min}$
3. POHC input rate:	$\left(16{,}300 \dfrac{\text{g feed}}{\text{min}}\right) \left(0.000200 \dfrac{\text{g POHC}}{\text{g feed}}\right) = 3.26 \text{ g POHC/min}$
4. POHC stack output rate, at 99.99% DRE:	$(3.26 \text{ g/min}) (1.0 - 0.9999) = 0.000326 \text{ g/min}$
5. POHC concentration in stack gas (at 99.99% DRE):	$\dfrac{0.000326 \text{ g/min}}{283 \text{ m}^3/\text{min}} = 0.0000012 \dfrac{\text{g}}{\text{m}^3} = 1.2 \text{ } \mu\text{g/m}^3 = 1.2 \text{ ng/L}$

*200 ppm = 0.000200 g POHC/g feed

An example calculation using a different method is shown in Table 8.17.7, which relies on another useful rule of thumb:

Each 100 Btu of heat input produces 2 dscf of flue gas, at 100% excess air.

Most incinerators operate at excess-air levels in the neighborhood of 100%, so this rule of thumb equation can be used to estimate stack flow rate based on the more commonly known firing rate of the incinerator (e.g., 30×10^6 Btu/h). Such estimates are very helpful in estimating POHC concentrations and in assessing the detection-limit requirements for sampling and analysis methods.

Quality Assurance and Quality Control (QA/QC).   Quality-assurance and quality-control requirements are another aspect of the sampling and analysis methods that must be included in the trial-burn plan, and are often questioned by regulatory agencies. It is important in planning the trial burn to stipulate exactly what

TABLE 8.17.8   Example QA Activities for a Trial Burn

- There should be written calibration procedures for all equipment used in sampling and analysis activities. Copies of procedures and documentation of the most recent calibration should be available.
- Traceability procedures (not necessarily chain of custody) should be established.
- A GC/MS performance-check sample should be analyzed each day prior to sample analysis. If results are outside acceptable limits, samples should not be run.
- All samples from at least one run should be analyzed in triplicate to assess precision.
- Wherever possible, all samples from at least one run should be spiked with POHCs or surrogates, to assess recovery efficiency for those compounds.
- A minimum frequency of check standards (5% is suggested) should be used with each sample batch. Analysis of actual samples should be suspended if check standards are outside the desired range.
- Blank samples should be analyzed to assess possible contamination and corrective measures should be taken as necessary. Blank samples include field blanks, method blanks, and reagent blanks.
- Field audits and laboratory performance and systems audits may be included in some cases. Cylinders of audit gases for volatile POHCs are available from EPA.
- A minimal level of calculation checks (e.g., 10%) should be established.

QA activities will be done and also to know why each is needed. Some QA activities may be desirable in specific cases but not essential. Blanket statements that "full QA" will be employed in the trial burn are not definitive enough, while excessive QA can drastically increase costs. An example list of basic QA needs for a trial burn is given in Table 8.17.8.

   One example of a QA activity that may be specified without adequate foresight is "chain of custody." The number of samples collected in a trial burn normally ranges from 100 to 300. Adherence to chain of custody procedures for all of these samples requires considerable time and effort, with associated cost impacts. Unless there is reason to believe that sample results will be a part of some judicial proceedings, chain of custody procedures on all samples may be an unnecessary added cost, whereas traceability procedures would suffice.

Schedules Related to the Trial Burn

The time factors involved in the trial-burn plan and the trial burn itself are outlined in Table 8.17.9. One of the factors that can vary the most is the time involvement after the initial submission of the Part B Permit Application and the trial-burn plan contained therein. Regulatory agencies almost always request additional information and details. Preparing responses to the agencies, awaiting their responses, and holding meetings to discuss the plan can consume as little as 1 month or more than 6 months.

8.17.2   PRACTICAL ASPECTS OF THE TRIAL BURN

Many practical aspects of the actual trial burn should be considered both in preparing the trial-burn plan and in preparing for and conducting the trial burn itself.

TABLE 8.17.9   Time Factors Involved in a Trial Burn

---

1. Notification to submit Part B application
2. Evaluate all conditions at which plant desires to be permitted (1 month)
3. Prepare trial-burn plan and submit to EPA (required 6 months after notification)
4. Prepare responses to EPA on any questions or deficiencies in the trial-burn plan (1 month)
5. Make any additions or modifications to plant that may be necessary (1 to 3 months)
6. Prepare for trial burn
    a. Prepare for all S&A, or select S&A contractor (2 to 3 months)
    b. Select date for trial burn, in concert with S&A staff or contractor (completed 1 month prior to test)
    c. Notify all appropriate regulatory agencies (1 month)
    d. Obtain required quantities of waste having specified characteristics
    e. Calibrate all critical incinerator instrumentation (2 weeks)
7. Conduct trial-burn sampling (1 week)
8. Sample analysis (1 to 1½ months)
9. Calculate trial burn results (½ month)
10. Prepare results and requested permit operating conditions for submittal to EPA (½ to 1 month)
11. Obtain operating permit

This subsection describes the most important aspects and those that commonly cause problems. The following will be discussed:

- Sampling and analysis contractor—cost and scheduling
- Preparation of wastes
- Number of runs
- Sampling and analysis methods
- QA/QC requirements
- Process data
- Operating problems

Sampling and Analysis Contractor—Cost and Scheduling

A sampling and analysis contractor is most often engaged to prepare or help prepare the trial-burn plan, and to conduct most or all of the sampling and analysis for the trial burn. As stated earlier, many incinerator operators do not have sufficient knowledge about the sampling and analysis portion of a trial-burn plan, and they often need the input of experienced contractors. In addition, a contractor can be very helpful in responding to questions and requests from regulatory agencies as the trial-burn plan is being reviewed.

For the trial burn itself, most incinerator operators do not have the necessary equipment or expertise to perform the complex sampling and analysis required. Also, an outside contractor is often perceived as being unbiased, which lends greater credibility to the results.

Activity directed to selecting a sampling and analysis contractor must start at least three months before the trial burn. The steps that may be involved are:

- Prepare request for proposal (RFP)
- Submit RFP to potential contractors
- Evaluate proposals and select contractor

Then the contractor

- Conducts pretest survey
- Conducts test

A request for proposal is usually based on the sampling and analysis contained in the trial-burn plan. In any case, the requirements must be clearly defined, including all associated QA/QC requirements. The RFP should allow one month for contractors to prepare and submit their proposals.

Trial-burn procedures are relatively new and are much more sophisticated than a normal EPA Method 5 test for particulate emissions. About 10 to 20 organizations in the United States have trial-burn sampling and analysis experience, and there are probably several more who have the capability. If a facility would like to know whom to contact for such services, they should make inquiries at state and federal regulatory agencies or contact other incinerator facilities who may have already conducted trial burns.

Final selection of a contractor should be made at least 1 month before the scheduled trial burn, and two to three months ahead if at all possible. Longer periods increase the probability that the contractor does not have a scheduling conflict on the desired test date. Also, at least one month is necessary for the contractor to conduct a pretest survey and for the one to two weeks needed for preparation of all test equipment.

Costs associated with the trial-burn contractor are often somewhat of a shock to the facility. In general, these costs may be:

- Assistance in preparation of trial-burn plan—$10,000 to $20,000
- Trial-burn sampling and analysis—$50,000 to $150,000

Depending on the number and extent of requests received from regulatory agencies for additional information, the costs for preparation of the trial-burn plan may be higher than those given above.

The cost of the trial burn itself is highly site-dependent. In general, the trial-burn costs will depend on the following factors:

- Number of runs
- Number and type of waste-feed samples
- Number of effluent samples
- Number of different analyses performed on each sample
- Complexity of the QA/QC plan
- Modifications required to prepare facility for trial burn

As stated above, the normal range for a trial burn sampling and analysis program conducted by an outside contractor is $50,000 to $150,000+. This range does *not* include plant modifications, or other preparations on site for the test (e.g., preparation of waste feeds). Costs for PCB trial burns are usually near the high end of the range because of the complexity of PCB analysis and the fact that these tests almost always include analysis of the stack samples for dioxins and furans.

The breakdown of costs for a trial burn is roughly one-third for the field sampling program; one-third for sample analysis and QA/QC; and one-third for preparation, engineering calculations, and reporting. These are rough estimates only. Frequently, the analysis portion of the program involves as much as half of the total cost.

In summary, the sampling and analysis part of a trial burn is costly; and each time another run, another sample, or another analysis is added to the test plan, the cost will rise. Each such addition needs to be carefully considered in order to hold costs down.

Preparation of Wastes

The preparation of wastes for the trial burn in the quantities required is very important but often causes some problems. The first step is to determine the planned waste feed rates and then to calculate the total quantity of each waste required for each run, assuming that the incinerator will need to operate at that feed rate for an 8-h period in each run for at least three runs. To take an example:

Liquid waste at 5 gpm = 2,400 gal per run

Solid feed of 1 drum every 5 min = 95 drums per run

Accumulating all the waste required is a problem in itself, but this problem is further complicated by the fact that the wastes must possess all the required characteristics (e.g., POHC content, high ash, high Cl, low HHV) specified in the trial-burn plan.

Three methods can be used to prepare the required quantities of wastes possessing the correct characteristics. These three methods pertain mainly to POHC characteristics, but they may be used to achieve any of the necessary characteristics. The three methods are:

1. Use actual wastes
2. Use synthetically prepared wastes
3. Continuously spike POHCs into the waste during the trial burn

Usually, method 1 is desirable if it is possible to acquire actual wastes that have the necessary characteristics or to achieve those characteristics by the blending of actual wastes. Method 2 involves using actual wastes mixed with purchased chemical compounds (i.e., POHCs). Method 3 is similar to method 2 except that it applies mainly to continuous liquid feeds and the purchased chemical compounds are pumped continuously into the feed line.

Although all three of the above methods require that the waste feeds be well-mixed, mixing is especially important for methods 2 and 3. Lack of good mixing for any waste feed can cause (and has caused) problems in trial burns. For method 3 the trial burn may involve continuous spiking of POHCs (pure components or mixtures thereof) into a liquid-waste feed line. When this is to be done, an in-line mixer may be needed to help ensure that the spiked components are well mixed with the waste and that the samples collected are representative of the resulting mixture.

Another important factor in waste preparation is time. The quantities of waste involved can be rather large, and it may take several weeks for a facility to acquire sufficient quantities of wastes to prepare a homogeneous batch with the proper characteristics. Storage space for these "special" wastes over a period of time can affect normal plant operations. And finally, additional time may be needed to sample and analyze the wastes to be sure they possess the necessary characteristics.

Adequate time must be allowed for numbering, weighing, and sampling of drummed solids before the trial burn. One should remember that, since the number of drums may exceed 300, the problem of weighing and sampling may be more complex than initially thought. Also, samples of drummed solids must be representative of those drums used in each run. Representative samples may be obtained by sampling each drum during the run, or by "staging" the drums to be used in each run and then sampling them prior to the trial burn.

Number of Runs

As noted earlier, EPA practice is to require three runs at each set of operating conditions. Thus, most trial burns consist of three runs at one condition, or six runs at two conditions. Each run will require at least 2 to 4 h. In addition, the sampling crew will need 2 h of preparation before each run and 2 to 4 h after each run for sample recovery and related activities. Therefore, it is advisable to plan only one run per day.

Sampling and Analysis Methods

Selection of sampling and analysis methods was discussed previously in terms of its having to be done well in advance of the actual tests. However, that discussion did not illuminate many of the practical aspects of performing the trial-burn sampling and analysis. Some of these aspects are covered here.

A brief list of the capabilities necessary for trial-burn sampling and analysis is shown in Table 8.17.10. The main factors involved in preparation for sampling are listed in Table 8.17.11. A typical example of the number of sampling personnel required is given in Table 8.17.12. An example list of the data forms used for sampling is given in Table 8.17.13.

Another very important aspect of sampling in a trial burn is preparation of sample labels. Careful thought must be given to identifying each and every specific sample that will be taken during each run, including all replicates and blanks. Label preparation also helps identify all the sizes and types of sample containers that will be needed, how they must be prepared, and the number of each required (including spares). This activity often reveals that 50 to 100 individual samples (and labels) will be involved in each run. Given the large number of samples, preprinted labels with specific sample names and a consistent numbering system should be prepared before the trial burn to help avoid confusion and errors that can occur if labels are prepared later in the field.

The main activities involved in the actual sampling are summarized below.

Equipment Setup

- Set up sampling train
- Set up waste-feed sampling

TABLE 8.17.10 Capabilities Necessary for Trial-Burn Sampling and Analysis

1. Sampling equipment for solid waste feeds (especially drummed wastes)
2. Stack sampling equipment, usually including the following:
   a. EPA Method 5 equipment, and all associated test equipment (e.g., EPA Methods 1, 2, and 3)
   b. Method 5 equipment adaptable to Modified Method 5 (greaseless) and associated XAD-resin preparation, extraction, and analysis facilities
   c. Volatile-organic sampling train (VOST) equipment with at least 18 pairs of VOST traps; also, all facilities needed for preparing, checking, and analyzing the traps
   d. Gas bags and associated sampling equipment (See Fig. 7, Ref. 3.)
   e. Field laboratory equipment for sample recovery
3. Facilities for analyzing all samples, including:
   a. Laboratories containing relevant safety equipment such as hoods and equipped with sample-preparation equipment including Soxhlet extractors; separatory funnels; continuous extractors; blenders, Sanifiers, or other equipment to homogenize waste-feed samples; sodium-sulfate drying tubes, Kuderna-Danish glassware, etc.
   b. Equipment for preparing VOST traps to allow simultaneous heating and purging of the traps. Ideally the traps should be prepared and stored in an organics-free laboratory.
   c. All required compounds to prepare calibration standards and surrogate-recovery spiking solutions.
   d. Computerized GC/MS instrumentation.
   e. Established QA procedures for assessing precision and accuracy of analytical methods.
4. Knowledge and preferably experience in all of the sampling and analysis methods and also in calculation and reporting of results.
5. Process-monitoring experience, especially in quantification of waste feed rates and documentation of plant operating conditions.

TABLE 8.17.11 Major Activities Preparatory to Sampling

- Calibrate all MM5 consoles
- Condition and check all VOST traps
- Clean and pack all MM5 resin traps (XAD)
- Preweigh all MM5 filters
- Clean all glassware and sample bottles
- Purchase special reagents and solvents
- Modify equipment for special sampling situations
- Reserve time for use of analytical instrumentation
- Collect and pack all necessary field-sampling and laboratory equipment

TABLE 8.17.12 A Typical Example of Sampling Personnel Required

Job	Number of personnel	Experience required
Sample liquid feed (once every 15 min)	1	Technician with sampling experience and safety training
Drum-solids sampling and recording (once every 5–10 min)	1	Technician with sampling experience and safety training
Sample ash and scrubber waters (every ½–1 h)	1	Technician with safety training
Stack sampling, MM5	2	Experienced console operator and technician for probe pushing
VOST	1	Experience with VOST operation
Process monitor to record operating data every ¼–½ h and determine waste feed rates	1	Engineer or other person experienced in plant operations and trial-burn requirements
Field laboratory	1	Experienced chemist for check-in and recovery of all samples, and preparation of sampling equipment for each run
Crew chief	1	Person experienced in all aspects of trial-burn sampling to direct all activity and solve problems that may occur
Total	9	

TABLE 8.17.13 Example List of Data Forms

Traverse Point Locations
Preliminary Velocity Traverse
Method 5 Data Sheets
Isokinetic Performance Work Sheet
M5 Sample Recovery Data
Integrated Gas Sampling Data (Bag)
Orsat Data Sheet
VOST Sampling Data

Drum Weighing Record
Drum Sampling Record
Liquid Waste Feed Sampling Record
Fuel Oil Sampling Record
Drum Feed Record

Process Data (Control Room)
Miscellaneous Process Data (In-Plant)
Tank Level Readings

Log of Activities

Ash Sampling Record
Scrubber Waters Sampling Record
Sample Traceability Sheets

Note: Units of measure must be shown for each item on each data sheet.

Preliminary Testing

- Velocity traverse
- Cyclonic flow check
- Moisture measurements

Actual Testing

- Waste-feed sampling
- Process monitoring and determination of waste feed rates
- Sampling of ash and scrubber waters, etc.
- Stack sampling
- Sample recovery
- Labeling and sample packaging and storage
- Preparation of equipment for next run

Equipment Dismantling and Packing.   The most common problems that occur in the actual sampling are

- Sampling-equipment setup problems
- Plant operating problems
- Weather
- Sampling-equipment operating problems

Problems, or at least difficulties, can also be created by the fact that trial-burn periods may extend into the night, as shown in Fig. 8.17.7.

After all the trial-burn sampling has been completed, the samples are returned to the laboratory for sample check-in and analysis. At that point it is very important to prepare a sample-analysis memo that clearly specifies the following information for *each* and *all* samples:

- Sample number and type of sample (e.g., "141: waste feed")
- Notation of any safety or hazard considerations in handling or analyzing samples
- Analysis parameters and analytes (POHCs, HHV, Cl, ash, etc.)
- Analysis method for each parameter or analyte
- Analysis sequence (where necessary)
- Indication of whether sample is to be analyzed in duplicate (or triplicate)
- Indication of samples to be spiked with analytes and surrogate recovery compounds for determination of percent recovery efficiency

The memo should also indicate expected concentrations, analysis priorities, and the schedule for reporting of results of the analysis. Preparation of the analysis memo is complex, it takes time, but it is very important. The memo documents the analysis needs and will be very helpful to those who must perform the analyses. Most importantly, it helps ensure that all the analyses necessary to satisfy the trial-burn requirements will be done.

Analysis of samples can involve several problems, but three of the most common are sample inhomogeneity, analytical interferences, and time requirements. These three are discussed briefly under the headings that follow.

FIG. 8.17.7   (*Photo by Tom Walker*)

Sample Inhomogeneity.   When the analysis of samples is initiated, a problem often encountered is nonhomogeneity of samples, especially waste-feed samples. Liquid-waste feeds often separate during shipment and storage. Steps must be taken to homogenize samples before portions are removed for analysis. Solid-waste samples can present even more difficult problems, which should be discussed with the analysis task leader.

Analytical Interferences.   Waste-feed samples are organically complex and usually contain many constituents other than POHCs. These other constituents may interfere with the POHC analysis, especially if the POHC concentrations are low in comparison to concentrations of the other constituents. Some type of sample cleanup may be required.

Time.   Trial burns produce numerous samples, and numerous analyses are performed on each sample, all of which require considerable time. A common problem associated with this is that the incinerator operators often seem to think that all analyses will be completed one to two weeks after testing. In most cases, however, the analyses, associated calculations, and report preparation will require about two months.

QA/QC Requirements

Some aspects of QA/QC requirements were discussed earlier, but this section describes problems that can occur in this regard, including

- High blanks
- Poor precision
- Low or high recovery efficiency
- Differences between acutal and expected results

High Blanks.   "High blanks" may occur for any parameter or analyte, but probably they occur more frequently for POHCs. Blank levels can be used to "correct" sample levels, but this is not possible if blank levels are near or exceed sample levels. If both the blanks and the sample levels are high, causing the uncorrected sample value to yield a DRE below 99.99%, no useful information is obtained for that sample. Every precaution must be taken in the laboratory and in the field to prevent contamination of samples.

Poor Precision.   The QA protocol usually requires triplicate analysis of critical samples from at least one run, with a precision goal of ±30% for POHCs. Wide variability in the results of this triplicate analysis may occur if samples are not homogeneous, if the samples contain some interfering component, or if other problems occur with the analytical technique. In any case, the precision obtained provides an indication of the possible variability in results reported for each sample and analyte. For POHCs, knowledge of this variability may be quite important if the calculated DREs are close to 99.99% (e.g., 99.989%).

Poor Recovery Efficiency.   Another normal part of the QA/QC is spiking of samples with known amounts of POHCs and/or surrogates to determine recovery efficiency for the analysis method. The desired recovery range is normally 70 to 130%, but results may in some cases actually be much lower or much higher, for any of several reasons. Sample results usually are not corrected based on recovery-efficiency results, but knowledge of the recovery efficiency is important for the same reason as knowledge of the precision. Also, poor precision or poor recovery efficiency may indicate a need to reanalyze the samples. It is not possible to reanalyze some types of samples, however (e.g., VOST traps).

Actual versus Expected Results.   In certain cases the sampling and analysis task leader has some idea of what the analytical results should be. For example, the amount of POHC added to a waste-feed tank might be known, so there is some concentration of the POHC expected in those samples. Usually the analytical results are in good agreement with the expected results, but in some instances they may disagree. The analyses and calculations then must be rechecked and the QA/QC data scrutinized more closely for a clue. Mathematical errors are the most common cause of such disagreements, but other possible causes (e.g., a poorly mixed feed tank) cannot be overlooked.

Recording the Process Data

Essentially all process data should be manually recorded during each run, and data on all critical process parameters should be recorded once every 15 min (see Fig. 8.17.8). These critical process parameters usually include

- Auxiliary-fuel feed rate
- Waste feed rates

FIG. 8.17.8 *(Photo by Tom Walker)*

- Combustion temperature
- Stack CO concentration
- Air flow rate (or combustion-gas velocity)
- Water flow rate through scrubber

All other available process data should also be recorded so that records will be as complete as possible; complete records will help in evaluating problems that may come up later (e.g., failure to achieve particulate limit). Strip charts for the test periods should not only be conserved, but also carefully identified with instrument number, units of measure, and so forth. Strip charts are often difficult to interpret later, and this is one reason that manual logging of data during the trial burn is highly recommended. Manual logging should be done separately from the routine record keeping (i.e., the operator's usual log sheets), with care taken to identify each instrument, the units of measure used (e.g., gpm, lb/h, °F, psig), and the scale factors (e.g., Rdg × 20 = gpm).

Plant Operating Problems

Many problems can occur with all aspects of a trial burn, but the most common one is plant operating problems. It is important to realize that the trial burn is intended to be carried out under a specific set of operating conditions, most or all of which represent worst-case conditions (e.g., maximum feed rate, minimum temperature, maximum Cl and ash in waste feed). This situation almost always represents abnormal operating conditions that can result in operating problems and cause test delays or interruptions in sampling. This is one reason that a pretest should be considered. It is much better, and cheaper, to identify problems during a pretest than during the official trial burn. In conclusion, two important recommendations are:

- Conduct a pretest at the trial-burn operating conditions, with trial-burn waste characteristics, to help ensure that the incinerator can sustain those conditions for extended periods (8 h).
- Include trial-burn sampling and analysis in the pretest to determine whether the EPA limits are being achieved.

8.17.3 REPORTING TRIAL-BURN RESULTS

Trial burns involve taking many samples, performing multiple analyses, recording other associated test data, and making numerous calculations. Converting all of this information into a complete and clearly understandable report is difficult and problems can arise in the process. It would be desirable to include an example report in this chapter but that is not possible given the space limitation. The best that can be done is to provide references to actual test data that are available and to discuss some of the problems that are commonly encountered in preparing trial-burn test reports. The EPA previously sponsored a project involving trial-burn-type testing at eight different incinerators. Results reported for those facilities are contained in Ref. 8. Earlier tests at another facility are reported in Ref. 9.

The usual elements of the trial-burn report are listed in Table 8.17.14. One should not infer from this table that the report must include step by step calculations. Rather, it must contain data and results formatted in such a way that the reader can easily see, in every table, what data values were used to compute the results shown in that table (see example Table 8.17.15). In some cases the data values shown in one table will be average values whose calculations were shown in a previous table.

Determinations of DREs for the report involve calculations that are rather simple but that sometimes raise questions. For that reason, the data necessary to calculate DRE are listed in Table 8.17.16 and an example calculation is shown in Table 8.17.17. Since most analytical results are reported in metric units, the calculations are usually done in metric units.

More specifically, when preparing trial-burn test reports, problems are commonly encountered in the following areas:

- Calculation based on dry standard conditions
- Significant figures and DRE
- Rounding off DRE results
- Blank corrections

Calculation Based on Dry Standard Conditions

Calculation of DRE requires use of M5 data on stack-gas flow rates. One need not be confused by the fact that such flow rates will be expressed in different

TABLE 8.17.14   Elements of a Trial-Burn Report

Summary

Introduction and process description

Description of sampling and analysis methods

Tabulation of process data
Tabulation of continuous-monitoring data (CO, O_2, etc.)

Waste-feed characteristics and feed rate
POHC concentrations in waste feed
POHC input rates

Summary of M5 test data (stack flow rate)
VOST analysis results for POHCs
MM5 analysis results for POHCs

POHC stack-emission rates

DRE results

Other POHC analysis results (scrubber waters, ash)

M5 particulate-concentration results
M5 HCl-emission rate results

Appendix
  A Detailed sample-analysis results with QA/QC analysis results
  B Detailed M5 and MM5 stack-sampling data

TABLE 8.17.15   Example Table of VOST Results
(Concentrations in ng/L—not Blank-Corrected)

	1st pair	2nd pair	3rd pair	Average
Run 1				
Carbon tetrachloride	2.3	0.47	0.57	1.1
Trichloroethylene	20	1.8	1.6	7.8
Benzene	2.2	2.3	2.2	2.2
Toluene	6.2	0.99	2.1	3.1
Run 2				
Carbon tetrachloride	2.3	1.7	1.7	1.9
Trichloroethylene	17	1.8	1.0	6.6
Benzene	2.0	7.4	2.6	4.0
Toluene	21	7.5	4.3	11
Run 3				
Carbon tetrachloride	3.1	0.58	0.45	1.4
Trichloroethylene	4.8	0.95	0.66	2.1
Benzene	6.0	7.1	6.2	6.4
Toluene	15	9.7	5.7	10

TABLE 8.17.16 Data Necessary for Calculating DRE

Measured value	Example units	How value is obtained
Mass-flow rate of feed	g/min	Measured during test or calculated from flow and density
Volumetric flow of feed	L/min	Measured during test
Density of feed	g/mL	Density analysis from lab
Concentration of POHC in feed	g/g	Analysis of waste feed samples
Total quantity of POHC in stack sample	μg	Reported by lab for each sample taken during test
Volume sampled of stack gas	Nm^3	For VOST this is found in the sample-train data
		For M5 this is found with the M5 train data
		For gas bags this is reported as the volume analyzed by the laboratory
Total quantity of POHC in blank samples	μg	Analysis of "blanks"
Volumetric stack flow rate	Nm^3/min	Reported as result of pitot traverse with M5 train

Note: $1 \ \mu g = 10^{-6} \ g$
$1 \ ng = 10^{-9} \ g$
$1 \ \mu g/m^3 = 1 \ ng/L$
Nm^3 = normal cubic meters = dry standard cubic meters

ways, for example, actual cfm, actual cfm dry basis, standard cfm, and dry standard cfm.

The best method for calculation of all results is *always to use dry standard conditions for expressing gas-sample flow rates, gas-sample volumes, and stack flow rates.* This will eliminate confusion about actual flow rates, or actual sample volumes, or sample concentrations. Moreover, most of the sampling volumes for gases are measured by dry-test meters, and sample volumes are easily converted to dry standard basis. The important thing to be realized is that multiplication of stack flow rate (e.g., dscfm) times a POHC concentration value (e.g., μg/dscfm) yields the correct POHC emission rate (μg/min). There is no need to be concerned with actual stack flow rates or to be confused by moisture content of stack gas in making the calculations, so long as all values are first correctly converted to dry standard basis. This point may seem obvious, but it has previously created some confusion in calculations and reporting results, or has made them unduly complex.

Significant Figures and DRE

One other point about DRE results that is often misinterpreted is the number of significant figures. For example, a reported DRE of 99.994% does not represent a value that is accurate to five significant figures. Actually, it represents only one significant figure. That is,

$$DRE = 100\%, \% \ penetration$$

TABLE 8.17.17 Example of DRE Calculation*

$$DRE = \frac{W_{in} - W_{out}}{W_{in}} \times 100$$

Determine input rate W_{in}

W_{in} = (organic waste flow rate × TCE concentration)
+ (aqueous waste flow rate × TCE concentration)
W_{in} = (4,010 g/min) (5,500 µg/g) + (5,380 g/min) (<1 µg/g)
= 22 × 10^6 µg/min = 22 g/min

Calculate output rate W_{out}

Stack flow rate = 76 Nm³/min (from M5 test results)

VOST concentration† avg = $\frac{20 + 1.8 + 1.6}{3}$
= 7.8 ng/L (not blank-corrected)

Blank correction

$VOST = \frac{< 1 \text{ ng/sample}}{18.5 \text{ L/sample}}$ = <0.05 ng/L

Blank-corrected value = 7.8 ng/L − <0.05 ng/L
= <7.8 ng/L = <7.8 µg/m³

POHC output rate

Mass flow‡ = (<7.8 µg/m³) (76 Nm³/min) (1 × 10^{-6} g/µg)
= <0.00059 g/min (corrected)

Calculate DRE

$$DRE = \frac{22 \text{ g/min} - <0.00059 \text{ g/min}}{22 \text{ g/min}} \times 100$$
= >99.9973%

---

*The above is a sample calculation showing the method used to convert the analytical results to DREs for trichloroethylene (TCE) using the VOST sample results for one run.
†Concentration values taken from Run 1 in Table 8.17.15.
‡Nm^3 = normal cubic meters = dry standard cubic meters.

What is actually being measured by the sampling is the amount of compound (POHC) which is *not* destroyed. Thus, a DRE of 99.994% represents a penetration of 0.006%, which has only one significant figure.

Rounding Off DRE Results

The rules on this are stated in the *Guidance Manual for Hazardous Waste Incineration Permits:* "...if the DRE was 99.9880 percent, it could not be rounded off to 99.99 percent."[1] In other words, the calculated value, after rounding to the proper number of significant figures, must equal or exceed 99.99% to meet the performance requirements.

Note. This same rule applies to rounding of HCl results to 99%.

Blank Corrections

Blank corrections on VOST samples and MM5 samples sometimes create problems or confusion. Since the number of blanks (e.g., blank traps) is usually limited, the practice generally applied by the author is

- To ignore any blank correction if blank values are less than 10% of the lowest sample value
- Not to blank-correct if blank values are greater than 10% of any sample value, but to express that sample value as a "less than" (<) value

The above practice almost always pertains only to stack POHC concentration, as measured by analyses of VOST or MM5 samples, and yields stack concentrations that are conservatively high. The only situation in which the above practice can create problems occurs when the results yield a DRE below 99.99%, but more rigorous application of blank corrections would yield a DRE above 99.99%. Fortunately, this does not happen very often. But if it does, the preparer of the trial-burn report will be forced to state what blank-correction values were used, how they were determined, and how they were applied.

One other convention in reporting results is that sample values are *not* corrected or changed based on QA/QC results for precision of analyses or for recovery efficiency determined for POHCs or surrogates spiked into the samples. It is not uncommon for recovery efficiencies to be above or below the goal of 70 to 130%, and it might seem that it would be valid to correct the analysis results based on the recovery-efficiency data. The conventional practice, however, is not to make such a correction. Rather, the QA/QC results are used to assess the precision and accuracy of the test results relative to their potential effect on the calculated DRE values: i.e., would DRE still be above 99.99% even if a correction for recovery efficiency were applied? However, if the QA/QC data indicate severe problems (e.g., recovery efficiency = 10%), then samples may have to be reanalyzed, or other more drastic action taken.

Numerous other problems or questions can arise in preparing trial-burn reports, but they are usually realized only by those who must convert the raw analysis numbers and other data to final reported values for emission rates, DREs, etc. The only advice that can be offered is to use best judgment, document what was done, and proceed. That same advice probably applies to many other aspects of an incinerator trial burn.

8.17.4 REFERENCES

1. U.S. Environmental Protection Agency, *Guidance Manual for Hazardous Waste Incinerator Permits,* SW-966, U.S. EPA, Washington, D.C., July 1983.
2. *RCRA Part B Model Permit Application for Existing Incinerator,* a report prepared for the U.S. EPA by A. T. Kearney, Inc., and Battelle Columbus Labs, Columbus, Ohio, under EPA Contract No. 68-01-6515, Work Assignment H00-013, February 11, 1983.

3. *Performance Evaluation of Full-Scale Hazardous Waste Incinerators,* vol. I, *Executive Summary,* a report prepared for the U.S. EPA by Midwest Research Institute, Kansas City, Missouri, under EPA Contract No. 68-02-3177, November 1984 (NTCS, PB85-129500).

4. *Practical Guide—Trial Burns for Hazardous Waste Incinerators,* a report prepared for the U.S. EPA by Midwest Research Institute, Kansas City, Missouri, under EPA Contract No. 68-03-3149, Work Assignment 12-7, June 1985.

5. *Sampling and Analysis Methods for Hazardous Waste Combustion,* a report prepared for the U.S. EPA by Arthur D. Little, Inc., Cambridge, Massassachusetts, under EPA Contract No. 68-02-3111, Technical Directive No. 124, December 1983.

6. U.S. Environmental Protection Agency, *Test Methods for Evaluating Solid Waste,* SW-846, U.S. EPA, Washington, D.C., July 1982.

7. U.S. Environmental Protection Agency, *Protocol for the Collection and Analysis of Volatile POHCs Using VOST,* EPA-60618-84-007, U.S. EPA, Research Triangle Park, N.C., March 1984.

8. *Performance Evaluation of Field Scale Hazardous Waste Incinerator,* report prepared for U.S. EPA by Midwest Research Institute, Kansas City, Missouri, under EPA Contract No. 68-02-3177, November 1984.
Vol. I: *Executive Summary,* PB85-129500
Vol. II: *Incinerator Performance Results,* PB85-129518
Vol. III: *Appendices A and B,* PB85-129526
Vol. IV: *Appendices C and J,* PB85-129534
Vol. V: *Appendices K and L,* PB85-129542

9. U.S. Environmental Protection Agency, *Trial Burn Protocol Verification at a Hazardous Waste Incinerator,* EPA 600/2-84/048, U.S. EPA, Cincinnati, Ohio, February 1984, PB84-159193.

CHAPTER 9

BIOLOGICAL PROCESSES

SECTION 9.1
AEROBIC PROCESSES

Robert L. Irvine
Professor
University of Notre Dame
Notre Dame, Indiana

Peter A. Wilderer
Professor
Technische Univesitat Hamburg-Harburg
Hamburg, West Germany

Hazardous materials are present in solids, concentrated liquids (e.g., industrial solvents stored in drums), concentrated and dilute industrial wastewaters, and domestic discharges. About 68% of the hazardous wastes generated each year comes from the chemical industry.[1] In 1984, 81 member companies of the Chemical Manufacturers Association (CMA), or about 50% of the total number of CMA member companies, reported that 253 million t of hazardous wastes were treated and disposed.[2] Of this, 99.4% was wastewater and 0.6% was solid waste. A report prepared by the American Institute of Chemical Engineers indicated that of the wastes that are disposed, 51% goes to landfills, land treatment, landfarm, or impoundments and the remaining 49% to deep underground injection wells.[1] About 67% of the wastes were categorized as being treated: 1% by incineration and 15% by physical, chemical, or biological means with the remaining 84% being placed in tanks or impoundments.[1] An historic view of the problem can be simply summarized by noting that the Office of Technology Assessment has estimated that there are 10,000 nonfederal sites that may require cleanup by Superfund.[3]

Virtually all domestic wastewaters and many dilute industrial discharges (e.g., less than 0.1% hazardous materials) are treated biologically, while heavily contaminated leachates from Superfund sites and from other landfills, land "treatment," landfarm, or impoundments (whether categorized as for treatment or disposal) are not. It is the hazardous organics in these leachates and in the many heavily contaminated wastewaters that are presently disposed of by underground injection to deep wells that will be the focus of the aerobic biological processes discussed in this section. Because of the intense recent interest in the biological detoxification of contaminated soils, the discussion will be extended to include this topic as well.

9.1.1 APPLICABILITY OF AEROBIC BIOLOGICAL PROCESSES

A Brief Word About Microbial Growth

Some microbes grow best when attached as a biofilm to a surface; others, when unattached (i.e., in suspension as dispersed organisms or in clusters called *flocs*). *Strict aerobes* can only grow in the presence of oxygen; *strict anaerobes* are destroyed by oxygen; and *facultative anaerobes* grow in the presence or absence of oxygen.

All organisms require both an energy and a carbon source for growth, with both having a major impact on which organism or organisms will grow and flourish. The energy source may be either sunlight or some reduced inorganic (such as ammonia nitrogen or ferrous ion) or an organic compound. When sunlight or an inorganic supplies energy, carbon dioxide must be provided as the carbon source for the production of cellular materials (e.g., proteins, carbohydrates, fats, nucleic acids).

Collectively the cellular materials contain a wide variety of elements including carbon, hydrogen, oxygen, nitrogen, phosphorus, and sulfur. Many proteins, when used as enzymes, require trace concentrations of metals such as iron, magnesium, and calcium. As a result, all of these elements must be present as nutrients or micronutrients for organisms to grow. Other organisms, because they are not able to synthesize some of the required intracellular organics (e.g., amino acids or vitamins), must be supplied both the inorganic nutrients and the needed preformed organic intermediates for growth. In this case, the microbial consortium that develops will not include these organisms unless the needed preformed intermediates either are present in the raw waste supply (an unlikely condition for most hazardous wastes) or are released [e.g., through death (lysis)] by other members of the microbial community.

In summary, microorganisms will grow if provided the proper conditions. They need a carbon and energy source. Many hazardous organics satisfy either or both of these requirements. They need nutrients such as nitrogen, phosphorus, and trace metals. Aerobic organisms need a source of oxygen. Some organisms can use oxidized inorganics (e.g., nitrate) as a substitute for oxygen. The design engineer must ensure that all of the necessary ingredients are supplied in sufficient quantities. The temperature and pH must be controlled as needed and substances that are toxic to the organism (e.g., heavy metals) must be removed. In general, it is not difficult for the designer to engineer a system which will meet all of these requirements for the biological treatment of contaminated liquids. The in situ biological treatment of contaminated soil may, however, be considerably more complex to handle.

Range of Capabilities

Many individuals who deal with the treatment of hazardous wastes mistakenly assume that biological treatment is not possible because they assume that hazardous materials are toxic to living organisms. They conclude that, since micro-organisms are living organisms, biological treatment is not a viable alternative. As a result, a typical flowsheet depicting the treatment of hazardous materials includes an array of physical, chemical and thermal operations and processes. Biological treatment, if any, is present in the form of a final holding pond which may or may not include an aeration device. Clearly a more thoughtful and aggressive approach to biological treatment is required if its full potential, in terms of removal efficiency and cost, is to be exploited.

Historically there have been two basic systems used in the aerobic biological treatment of wastewaters. The most commonly used of the two is generically referred to as *activated sludge*. This is a suspended-growth, mixed-culture system. The other, the *trickling filter,* is a fixed-film, attached-growth system. The objective of either system is to develop a microbial community which will convert a wide variety of organic compounds to new cell material which can then be removed by conventional separation techniques, and to inorganic, nonhazardous substances such as carbon dioxide and water. Obviously some organics are recalcitrant and may not be degraded in a particular environment, while others are only partially degraded and are converted to other secondary organics. This common sense understanding of biodegradability of organics is applicable to all organics whether or not they are classified as hazardous compounds. The interesting characteristic of hazardous organics is that any one of them *may* be toxic to some or all members of a given microbial consortium when that compound is present above some critical concentration. The existence of such a critical toxic concentration is, of course, dependent upon the compound in question, its concentration, its exposure route, and the makeup of the microbial consortium being asked to degrade the compound.

The basic message of this chapter is that most "hazardous organics" can be treated biologically provided that the proper organism distribution can be established. A substance that is hazardous for one group of organisms may be a valuable food source for another group. It is only by viewing biological treatment systems as a dynamic population of microbes whose composition is subject to various selection pressures that proper systems can be designed, maintained, and corrected when problems arise. As a result, this chapter emphasizes how fundamental principles can be used to develop common sense approaches to establishing the desired microbial consortium either in biological reactors designed for the treatment of heavily contaminated leachates, or in soils, using in situ decontamination techniques.

As a final note, it should be remembered that it is tough to remain nonbiodegradable! It seems that for every report that describes a severe problem with the biodegradability of a specific compound, there is a second report, somewhat later in time, that discusses the rate of biological destruction of that compound. In many cases, chemicals persist in the environment not because they are inherently toxic to all organisms, but because they are inaccessible to potential degraders.

Limitations of Biological Treatment

Hazardous materials are present in low to high concentrations in wastewaters, leachates, and soils. The treatment of highly contaminated liquids and soils is emphasized in this chapter. These wastes are characterized by high organic content (e.g., up to 40,000 mg/L total organic carbon), low and high pH (2 to 12), elevated salt levels (sometimes over 5%), presence of heavy metals, and, of course, presence of hazardous organics. The limitations of the aerobic biological process used to treat such liquids and soils can be summarized by considering those factors that must be taken into account in the overall system design. These factors are

- Extent of biodegradability
- Applicability of conventional technology
- Effects of toxic and nontoxic inorganics
- Effects of toxic and nontoxic organics
- Load variations
- Nutrient requirements
- pH
- Oxygen supply
- Temperature

Each of these is discussed separately in this subsection.

Extent of Biodegradability.   While many hazardous organics are quite biodegradable, some are not. For example, 2,3,7,8-tetrachlorodibenzo-*p*-dioxin (2,3,7,8-TCDD) has maintained a somewhat elitist status by remaining at or near the top of the nonbiodegradability list for a rather extended period of time. In addition, the question of biodegradability is very system-specific. Unless the proper microbial consortium can be both developed and maintained in a given system, a compound "known" to be biodegradable will not be degraded in that system. Even if good biological removals are achieved, the aerobic biological process may have to be used in conjunction with other physical and chemical destruction technologies in order to achieve the levels of treatment that are required for a particular cleanup operation. In any event, biological systems can reduce the cost of downstream processes by reducing organic load.

Applicability of Conventional Technology.   Leachates and highly contaminated wastewaters containing hazardous materials can be treated in systems that utilize conventional reactor technology. In such systems adequate mixing, oxygen supply, nutrient addition, and temperature and pH control can be provided. Although this level of control has not yet been demonstrated for contaminated soils, especially in situ, some rather interesting and innovative designs have been developed which allow reasonably high levels of control. Whether dealing with contaminated soils or waters, however, there are very few full-scale aerobic biological systems in operation today that were specifically designed to remove hazardous organics. As a result, the full extent of the problems that can develop with such systems is not well understood. This has contributed to the reluctance of many potential users to adopt biological processes for the treatment of hazardous materials.

Toxic and Nontoxic Inorganics.   While a microbial consortium can be developed which is resistant to moderately high levels of heavy metals, the deposition and accumulation of precipitates of these metals on the biomass may severely inhibit the activity of the community developed. As a result, a pretreatment system designed to remove these contaminants will reduce the likelihood of system failure. This is easily accomplished for systems treating leachates and wastewaters but may be more difficult to achieve for contaminated soils. As far as nontoxic inorganics are concerned, organism selection and overall treatment efficiency can be adversely affected, especially for those cases in which the salt content can range from 2 to 5% or more. It has been the authors' experience that the salt matrix within which the organism community must work can have a strong impact on system performance. Short-term variations in the salt concentration are of special concern. Proper selection and control procedures can be developed, however, which will allow for the development of a reasonably efficient microbial consortium.[4,5]

Toxic and Nontoxic Organics.   Selective removal of a particular organic toxin by chemical or physicochemical means is difficult to achieve. Pretreatment with activated carbon or with some "reasonably" selective resin or with an immobilized enzyme[6] may be useful but is likely to be inefficient and costly. If the organic toxin is biodegradable and if the organisms possessing the specific capabilities needed can be enriched and maintained in the microbial community, then the system can be designed so that the toxin is maintained below toxic levels.

Note.   The question of toxicity can be somewhat confusing. It must be remembered that we are dealing with several microbes within the total microbial community. As a result, a toxin may impair the activity of one or more members of the community that are essential for good overall removals while serving as an excellent carbon and energy source for other members of the consortium.

If the toxin is nonbiodegradable, more serious, and perhaps insurmountable, problems can develop. In this case, microbial strains that are "resistant" to the toxin must be enriched; otherwise only limited removals may be possible biologically.

Load Variations.   The concept of load variations has limited applicability to soil-decontamination systems. It is quite relevant, however, to systems treating leachates and highly contaminated wastewaters. Because of rainfall or because of a number of other reasons, the types of wastes received at a hazardous-waste disposal site and the mass flow rate of the total load of organics and of the individual hazardous compounds in the waste stream can vary notably with time. This means that the concentration of certain organics may fluctuate above the toxic limit during selected periods, causing the performance of the biological system to deteriorate. Installation of an equalization and storage facility prior to the biological system would reduce the possibility of upset. Because biological processes are known to be sensitive to load variations even when treating nonhazardous wastewaters, care must be taken to incorporate all possible information on load variation into the design of a biological system.

Nutrient Requirements.   Nitrogen and phosphorus are the two most common nutrients that may have to be added to a biological system. If either or both are not present in sufficient quantities in a usable form, biological treatment will be severely retarded. Both nitrogen and phosphorus may be covalently bound to some of the organics in the hazardous materials. Unless hydrolyzed to the inorganic form (e.g., ammonia nitrogen and phosphate), a supplemental supply will be required. In the case of phosphorus, some problems may result because the phosphate may precipitate with the cations present in the hazardous materials. This problem can be mitigated somewhat by adding phosphate with a chelating agent.

pH.   Most of the organisms capable of treating hazardous substances grow best in the 6 to 8 pH range. Adjustment of the pH of contaminated soils and pH control of biological reactors treating wastewaters may be required if either the biological activity or the hazardous-material supply causes a marked change in the pH. Conventional technology is readily available to deal with this problem.

Oxygen Supply.   This is one of the more difficult challenges facing in situ soil detoxification. Many hazardous organics are readily degraded aerobically but, because the depth of oxygen penetration in soils is limited, the rate and extent of biological detoxification is also limited. Some of the innovative approaches which have addressed this problem will be discussed later in this chapter. As far as biological treatment of leachates and other heavily contaminated wastewaters is concerned, conventional aeration equipment and design approaches are directly applicable. It has been noted, however, that the stripping of volatile organics is greater in mechanical aeration systems than in diffused-air systems.[7]

Temperature.   The classical view of the impact of temperature on treatment performance is that the rate of removal of organics increases with increasing temperature (until some maximum temperature is reached) and decreases with decreasing temperature. This view is certainly appropriate for short-term (e.g., daily) fluctuations in temperature. The impact of long-term or seasonal changes in temperature on removals is less easy to define because the microbial consortium that is established at one temperature may be different than that consortium which is established at another temperature. As a result, the rate and extent of removal of any individual hazardous organic can vary with temperature much more than would be expected from short-term considerations only. A third, and perhaps more important, impact of temperature on the removal of *xenobiotics* (manufactured rather than biologically produced compounds that may or may not be on some list of hazardous materials) has been observed by the authors. Some microbial communities experience a dramatic drop in activity at about 10 to 15°C. In one case, the activity was restored by slowly bringing the temperature down below 15°C (at about 1°C decrease per week) and remaining at each new temperature until "reasonable" activity was observed.[8] The initial drop in activity was too abrupt to be explained by the need for a shift in the organism community. One explanation for the success of gradual low-temperature acclimation may lie in the observation that the fluidity (and, therefore, the transport properties) of the cell membranes is strongly influenced by temperature. Many, if not all, microorganisms can restructure the lipid content of the membranes if they are to function properly at the reduced temperatures.[9,10]

The question of temperature reduction is important in colder climates. While insulated reactors being supplied oxygen with compressed air are able to maintain a temperature some 5 to 8°C above the temperature of influent wastewaters,[11] a heat-exchanger system may be necessary to keep the temperature above critical levels.

Temperature control for systems used to detoxify contaminated soils, especially in situ, is obviously a much more difficult problem.

A final note: Shifts in microbial populations at elevated temperatures, either in insulated reactors in colder climates or in noninsulated reactors in warmer climates, can reduce performance and should be considered in the final system design.[8]

9.1.2 DESCRIPTION OF SELECTED AEROBIC BIOLOGICAL PROCESSES

Leachates and Highly Contaminated Wastewaters

Biological Principles. A thorough examination of the basic microbiological, biochemical, and genetic fundamentals that are applicable to the biological treatment of hazardous organics can be found in the text by Bailey and Ollis (Ref. 12). While any abbreviated description of these fundamentals will suffer from serious shortcomings, there are some basic principles that must be understood.

First, a microorganism will grow if it can compete effectively for the available carbon and energy sources. The simplest level of competition requires that the microbe contain the genetic information necessary to produce catabolic enzymes which extract energy from the available organics and convert the organics to intermediate products. Enzymes for anabolic pathways use the extracted energy and intermediates to synthesize the compounds necessary for the microbe to reproduce.

Second, genetic information for the bulk of the catabolic enzymes and all of the anabolic enzymes is present on the chromosome. Additional genetic information, especially for the production of enzymes that catalyze the key breakdown steps for xenobiotics, is typically found on plasmids. As a result, the plasmid genes allow a microbe to compete for substrates (i.e., carbon and energy sources) that would otherwise not be available to that microbe.

Third, those microbes that are able to convert the largest fraction of the available substrates to cellular material will be the predominant organisms in the mixed culture. The importance of kinetics can be seen by noting that if two different organisms have the necessary genetic information but one is more efficient than the other (i.e., converts the substrates at a faster rate), the more efficient organism will be present in higher numbers.

Fourth, if the required genetic information is not present either in the seed (i.e., start-up culture) or in the microbes present in the raw waste flow, it is not likely to be developed from a spontaneous mutation. As a result, a compound that is nonbiodegradable in that system will probably remain nonbiodegradable independent of the length of the acclimation period.

Fifth, genetic information, especially that present on plasmids, can be transferred from one organism to another. As a result, if the desired genetic information is present in either the feed or the start-up culture, an acclimation period will be productive provided that the organisms that produce the necessary genes can compete or the genes can be transferred to other organisms that can compete.

Sixth, plasmids contain genes that produce enzymes that protect organisms from toxins. The enzymes may simply alter the organic to a nontoxic form. Either the plasmid-bearing organisms or other organisms in the mixed culture may be able to utilize the altered compound as a carbon or energy source. Clearly, one organism's ability to compete can strongly influence the competitive position of one or more other organisms by either supplying substrate (e.g., an altered toxin) or eliminating a toxin.

Seventh, the microbial consortium (and, therefore, the pool of genetic information) that develops can be strongly influenced by the specific treatment process selected and the operating strategies used to start up and then to run the process. For example, a start-up strategy that allows a biodegradable toxin to accumulate in the system may eliminate an essential member of the microbial consortium from the seed culture. If this organism is not present in the feed stream to the bioreactor or if the essential genetic information is lost, general poor performance is virtually assured.

Obviously, the aerobic biological process selected must allow the proper environmental conditions to be established so that the required organisms will have a competitive edge over those which may also grow but take up only components of minor concern. Temperature, pH, and oxygen supply are factors of importance in this context. From practical experience it appears that making periodic changes in major environmental factors is of particular benefit. For instance, periodic changes in oxygen availability and oxygen deficiency are required to maintain facultative anaerobes in the biocommunity. These organisms may be required to allow breakdown of target compounds under anaerobic conditions and biodegradation and assimilation under the subsequent aerobic conditions.

General Flow Diagram.   The actual flowsheet for the biological treatment of leachates and wastewaters containing hazardous materials will, of course, depend upon the specific case in question. In general, however, the flowsheet (see Fig. 9.1.1) would be expected to include equalization and storage, pH adjustment, chemical precipitation and clarification, biotreatment, final polishing (e.g., with activated carbon), and sludge handling. In some instances, treatment of the exhausted gas must be considered. *Sizing of the individual unit operations and processes will require input from bench and pilot-scale treatability studies.* All sludges produced (e.g., inorganic from precipitations and organic from biotreatment) will have to be handled as hazardous materials.

Bioreactor Flow Diagrams.   Because of the paucity of full-scale aerobic biological processes treating hazardous materials, it is not possible to identify a typical flow diagram. Two full-scale systems that have been utilized will be described here. Both are suspended-growth activated-sludge systems. One is a conventional continuous-flow system (CFS); the other, a time-oriented periodic process called the sequencing batch reactor (SBR).

The flow diagram for one possible CFS is depicted in Fig. 9.1.2. As can be seen, the pretreated water enters a completely mixed bioreactor which provides suitable conditions for organism growth and concomitant substrate removal. The organism suspension passes from the bioreactor to the clarifier where the biomass is separated from the treated effluent. The clarified effluent then moves on for any polishing that may be required. The biomass, removed from the bottom of the clarifier, is returned to the bioreactor to maintain the desired biomass concentration in the reactor independent of the actual rate of biomass production. Biomass in excess of that required for proper operation of the bioreactor is sent to the sludge-handling facility.

The SBR is shown in Fig. 9.1.3. This is a time-oriented process that accomplishes in a single tank the same functions that a conventional CFS carries out in a series of separate tanks.[13] In order to provide for the continuous treatment of wastewater, the SBR must either be preceded by an equalization basin (a must for the treatment of leachates and highly contaminated wastewaters) or be accompanied by additional identical reactors.

A typical SBR system would involve the use of two reactors. Wastewater flow is first directed to one reactor and then to the other. When the flow is directed back to the first tank, that tank is said to have completed one cycle. Each cycle is divided into five discrete periods. They are FILL, REACT, SETTLE, DRAW, and IDLE. As one tank is filling, the other is completing the remaining periods of the cycle.

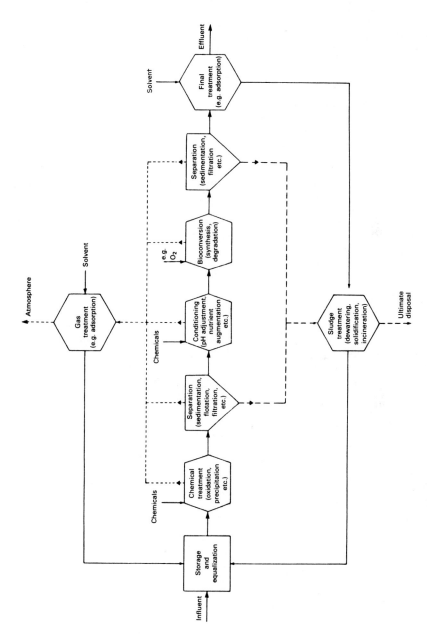

FIG. 9.1.1 General flowsheet for the biological treatment of liquid wastes.

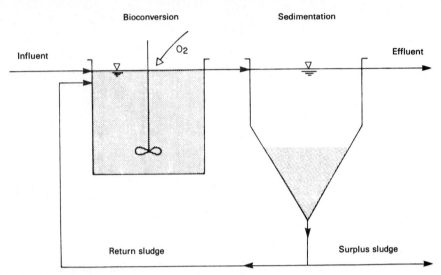

FIG. 9.1.2   Flow diagram for a continuous-flow system.

During FILL, wastewater enters a partially full tank containing acclimated biomass. Aerobic or anaerobic reactions can be initiated during FILL by providing either aeration or only mixing. After the tank reaches its predetermined maximum liquid level, the flow is diverted to the other tank by means of a control system. The reactions initiated during FILL are continued during REACT. The time set aside for REACT must be sufficient to allow the desired effluent requirements to be met. After REACT, the biomass is allowed to SETTLE quiescently for a predetermined period of time by shutting down the mixing and aeration equipment. The treated, clarified effluent is then removed during DRAW. This tank now waits during IDLE for the liquid in the other tank to reach its maximum level so that it can begin a new cycle.

As can be seen from the above description, the SBR mimics a conventional CFS by providing for piecewise equalization (during FILL), biotreatment, and sedimentation in the same tank but at different times. In fact, the time set aside for each period in the SBR can be directly related to some corresponding detention time in a CFS. As a result, it is both appropriate and useful to equate the fraction of the time dedicated to a particular period with the fraction of the total SBR system volume that has been set aside for that period. Clearly, then, if the time (i.e., volume) allowed for one period (say SETTLE or IDLE) is required for another function (say treatment), a simple adjustment to the control system is all that is needed.

The flexibility of the SBR time continuum is not easily matched by the discrete spatial arrangement offered by the CFS. In other words, tanks cannot be interchanged, enlarged, or decreased in size easily in a CFS but can be (in principle) in the SBR by changing the time allocated to a particular function. This, of course, is not a problem for a CFS if all the tanks are sized exactly. Because treatability studies are imperfect and because leachates and highly contaminated wastewaters change character with time, the SBR offers an excellent treatment alternative to the conventional CFS.

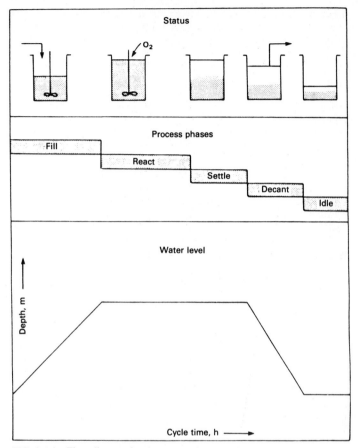

FIG. 9.1.3 Flow diagram for the sequencing batch reactor.

Contaminated Soils

Soils can be excavated and placed in bioreactors and then mixed and aerated. In this sense, the biological principles discussed above for liquids are directly applicable to soils. As a result, strategies for organism selection such as alternating mixing (anaerobic) and aeration (aerobic) during FILL and REACT in the SBR or in a series of tanks in a conventional CFS can be tested and applied. Unfortunately, the mixing and aeration of soils in a bioreactor can be an expensive proposition. A typical bioreactor for the treatment of wastewater is operated with a biomass concentration (on a dry weight basis) of about 2 to 4 kg/m^3. If soils were added to a reactor, the energy required for mixing and aeration would be quite high if as little as 15 kg of soils was added per cubic meter of reactor volume. To put this in perspective, 57,000 kg (125,000 lb) of soil would have to be added to a 3,785 m^3 (1 million gal) reactor to achieve a soil concentration of 15 kg/m^3. At a soil density of about 1.8 g/cm^3, this 57,000 kg represents about 32 m^3 (41 yd^3) of soil. Either a very high treatment rate or a very large reactor would be necessary to detoxify even small quantities of contaminated soils using this approach.

The alternative to excavation is in situ treatment. In this case it is important to ensure that the soils are kept moist, nutrients are provided, the pH adjusted, and oxygen supplied so that suitable conditions for microbial growth are maintained. Two problems that have received the greatest attention are (1) how to supply oxygen in large enough quantities to satisfy the needs of the microbes and (2) how can the existing soil microbial consortium be best enriched and/or augmented so that maximum possible rates are achieved. Oxygen-supply solutions have included the use of sparger wells or hydrogen peroxide.[14] Alternatively, nitrate instead of oxygen has been successfully applied to allow soil microbes to gain the energy required to rapidly grow and detoxify contaminated soils.[15] "Adapted" laboratory-produced microbes have been added to contaminated soils to augment the naturally present bacteria.[14]

The simplest of all the systems devised involves a recirculation loop where waters enriched with nutrients and an oxygen or nitrate supply are added to the subsurface with injection wells and then pumped to the surface via recovery wells for replenishment of nutrients and oxygen. Organic contaminants present in the recirculated waters can be removed at the surface by air stripping or adsorption or biological degradation.[14] Alternative approaches which include the application of landfarming techniques are discussed in Subsec. 9.1.4.

9.1.3 RESULTS OF CASE STUDIES

Two case studies are discussed. Both deal with the treatment of liquids rather than soils because soil decontamination systems are in an embryonic stage of development. One is a conventional CFS, augmented with powdered activated carbon, that was designed to treat industrial wastewaters generated at Du Pont's Chambers Works site at Deepwater, New Jersey. In 1980 this facility opened its doors to non-Chambers Works wastes and thus became a candidate for inclusion in this discussion. The other system is an SBR which treats leachates and other hazardous materials that are processed at CECOS International's hazardous-waste disposal facility in Niagara Falls, New York.

Du Pont's Chambers Works PACT Processes

In 1975 Du Pont put a 151,400 m^3/d (40 $\times$ 10^6 gal/d) industrial wastewater-treatment plant into operation at its Chambers Works site.[16,17] By 1981 only 50% of the organics-removal capacity of the facility was being utilized. As a result, Du Pont advertised the facility as a treatment and disposal system for non-Chambers Works wastes. At the present time over 30% of the waste load handled by the plant is from outside sources, including landfill leachates and chemical-process wastes.

The system employs the patented powdered activated-carbon treatment (PACT) process. This involves the addition of powdered activated carbon directly to the bioreactor so that nonbiodegradable organic carbon can be removed by adsorption.

With the exception of the addition of powdered activated carbon, the PACT process is a conventional CFS. Wastewater is first neutralized with lime in three 757 m^3 (0.2 million gal) neutralization tanks operated in parallel. The overflow from these tanks then passes through four parallel 3,785 m^3 (1 million gal) clarifiers before entering (in parallel) four 15,140 m^3 (4 million gal) aerated bioreactors. The effluent is then clarified in two 9,463 m³ (2.5 million gal) clarifiers before being discharged to a river. Powdered activated carbon is dosed to the bioreactor at a rate of 120 mg/L.

Because of the high flow and relatively low waste levels generated at the Chambers Works plant, the addition of outside wastes has had no significant effect on the quality of the treated effluent. [Note that the influent five-day biological oxygen demand (BOD₅) for the raw wastewater for the three years preceding the acceptance of outside wastes averaged 175 mg/L. This is quite low, being similar in strength to many domestic wastewaters. No data was provided on the feed strength after the acceptance of outside wastes.] The prime message here is that conventional technology can be used to treat hazardous wastes but there seems to be a need to work in conjunction with some adsorbent (in this case, powdered activated carbon). While this CFS works with a dilute wastewater, there is no reason to believe that higher-strength wastes could not be handled as well.

CECOS' SBR

Bench-scale treatability studies were conducted in 1981 and 1982 to test the treatment potential of the SBR on Hooker Chemical Company's [now Occidental Chemical Corporation (OCC)] leachate from their Hyde Park Landfill site.[4] The studies were initiated because OCC was using granular activated-carbon technology to remove the hazardous organics from the leachates. The cost forecast for purchase of activated carbon alone was $21 million for the next 10 years. The treatability studies demonstrated that roughly 90% of the total organic carbon compounds and about 50% of the chlorinated organics could be degraded biologically in the SBR. As a result, a system which included the SBR followed by activated-carbon columns was designed.

The benefits gained from the biological removal of organics are clear. First, the quantity of activated carbon required is markedly reduced. Second, the adsorption sites on the activated carbon are made available for the removal of nonbiodegradable hazardous organics rather than for both hazardous and nonhazardous biodegradable compounds.

The problems faced by CECOS International were similar to those of OCC. They were using granular activated carbon to remove the organics at their hazardous-waste disposal site in Niagara Falls. Treatability studies demonstrated that about 75% of the organics could be removed biologically in the SBR. An 1,892 m³ (0.5 million gal) SBR designed to handle up to 757 m³/d (0.2 million gal/d) was put into operation in June of 1984.[5,8] The effluent from the SBR was directed through the existing activated-carbon columns.

The desire to reduce costs was the prime motivating factor behind CECOS' integration of biological treatment into the overall waste-management system. Costs were saved in three basic areas: less regeneration of activated carbon; lowered electrical power usage because of the reduction in both pre- and postcarbon aeration; and reduced surcharge fees paid to the Niagara Falls publicly owned treatment works because of an overall improvement in effluent quality. Costs were incurred through the capital depreciation of the SBR and its associated operating costs. A cash-flow analysis showed a return on capital investment of 12.1% with a payback period (excluding interest) of four years.[8]

The SBR at CECOS was basically the first controlled, full-scale biological treatment facility used in the United States to remove organics from a hazardous-waste disposal site. The experience gained from the operation of this system served as a basis for several of the design factors discussed in Subsec. 9.1.1. Awareness of two major considerations resulted from the operation of this system, however. They are: A program of quality control and chemical equalization on the entering streams must be established so that potentially harmful or persistent substances are excluded; and temperature variations, especially during the transition to cold weather conditions, must be controlled. Both of these factors must be addressed during treatability studies so that the most effective system can be designed.

9.1.4 CURRENT RESEARCH AND ADVANCES

Research Triangle Institute was commissioned by the U.S. Environmental Protection Agency (EPA) to conduct a workshop and prepare a report which would provide recommendations for research projects in the general area of biotechnology for hazardous-waste destruction.[18] Fourteen short-term projects (i.e., to be conducted during the next five years), ranging from the development of a testing protocol to the determination of those factors which affect the ability of organisms to survive and function, were proposed. In West Germany, the Minister of Research and Technology granted a project to study the applicability of biological processes to treatment of hazardous leachates emitted from a landfill in Hamburg.[19] All of the project proposals recognized that considerable research must be conducted before the full potential of biological methods for the destruction of hazardous organics can be realized.

As can be seen from Subsecs. 9.1.2 and 9.1.3, a considerable amount of research, development, and application has already been conducted. It is clear, however, that further advances will require input from additional full-scale treatment systems. Results from the operation of these facilities will then be used both to provide research studies with better direction (we don't need any more studies that simply confirm that many xenobiotics are biodegradable in the laboratory under controlled conditions!) and to develop more realistic design criteria.

Additional full-scale experience is needed with the biological degradation of hazardous organics in both liquids and soils. Only soil-detoxification systems will be discussed in this subsection and this discussion will be limited to current research and applications being developed by the authors of this section.

The first system to be described integrates the treatment of leachates and contaminated soils. Irvine, Ketchum, and Kulpa have developed an egg-shaped SBR (to minimize mixing requirements, to capture stripped organics easily, and to separate biomass from inorganics) to generate specialized bacteria in the field which can be added directly to the soil.[20] The SBR would be fed leachates and low levels of soils (e.g., 1 g/L) supplemented with nutrients and be operated in such a manner that microbes acclimated to the leachates and soils can be enriched. The biomass thus produced (i.e., the specialized bacteria) would then be applied to the soil and disked in to a level of about 0.3 m (1 ft). The treated effluent from the SBR would be applied to the surface to maintain moisture levels in the soil. Natural aeration and organism development would be maintained by repeated additions of specialized bacteria and disking. Depending upon the depth of contamination, the soils can be arranged in windrows to enhance atmospheric oxygen transfer.

Wilderer and Sekoulov have developed a system that uses silicone-rubber membrane tubing to transfer oxygen to the soil.[21] Two designs have been proposed. One is similar to the landfarming approach described above. In this case mats made up of silicone-rubber tubing (silicone is highly permeable to oxygen) are sandwiched between layers of soil and microorganism sludges grown in a nearby fermenter or in a wastewater-treatment plant. They have also proposed supplying nitrogen to the soil organisms by adding ammonia to the gas and other gases, such as hydrogen and methane, as a source of substrate for the organisms should that be necessary for cometabolism (the simultaneous removal of the organic compound being used as the energy source and some other organic that is not utilized otherwise). The temperature of the gas can be adjusted to a level which allows both enhanced permeability and bioactivity. Any volatile organics that may have permeated into the silicone tubing can then be removed by an adsorber.

The second system involves the placement of loops of silicone tubing in small wells (inoculated with acclimated organisms) driven into the soil. The two ends of the tubing associated with each well are connected to a gas-supply system as described before. A second series of wells is used to recirculate the groundwater from the subsurface to the surface and back by means of spraying through the soil past the loops of silicone tubing.

9.1.5 LIST OF EQUIPMENT VENDORS

Aquifer Remediation Systems
FMC Corporation
P.O. Box 8
Princeton, N.J. 08540

Groundwater Technology Division
Oil Recovery Systems, Inc.
1420 Providence Highway
Norwood, Mass. 02062

Jet Tech, Inc.
P.O. Box 8
100 Prairie Village Drive
Industrial Airport, Kan. 66031

O. H. Materials Corp.
16406 U.S. Route 224 East
Findlay, Ohio 45840

9.1.6 REFERENCES

1. *Technology Needs for Improved Management of Hazardous Wastes,* prepared by the Hazardous Waste Task Force, Government Programs Steering Committee, American Institute of Chemical Engineers, Washington, D.C., 1986.

2. *Results of the 1984 CMA Hazardous Waste Survey,* prepared by the Chemical Manufactures Association and Engineering-Science, Austin, Texas, 1986.

3. U.S. Congress, Office of Technology Assessment, *Superfund Strategy,* OTA-ITE-252, Washington, D.C., 1985.

4. R. L. Irvine, S. A. Sojka, and J. F. Colaruotolo, "Enhanced Biological Treatment of Leachates from Industrial Landfills," *Hazardous Waste,* **1**:123 (1984).

5. P. A. Herzbrun, R. L. Irvine, and K. C. Malinowski, "Biological Treatment of Hazardous Waste in Sequencing Batch Reactors," *J. Water Pollut. Control Fed.,* **57**:1,163 (1985).

6. M. D. Aitken and R. L. Irvine, "Use of Enzymes to Treat Xenobiotics in Industrial Wastewaters," presented at the Summer National Meeting, American Institute of Chemical Engineers, Boston, Massachusetts, 1986.

7. P. V. Roberts, M. Christoph, and P. Dandliker, "Modeling Volatile Organic Solute Removal by Surface and Bubble Aeration," *J. Water Pollut. Control Fed.,* **56**:157 (1984).

8. C. N. Staszak, *Full-Scale Demonstration of a Sequencing Batch Reactor for a Hazardous Waste Disposal Site,* NYSERDA Report 85-21, Albany, New York, 1985.

9. R. N. McElhaney, "Modifications of Membrane Lipid Structure and Their Influence on Cell Growth, Passive Permeability, and Enzymatic and Transport Activities in *Acholeplasma laidlawii* B[1]," *Biochem. and Cell Biol.,* **64**:58 (1986).

10. J. L. Ingraham, O. Maaloe, and F. C. Neidhardt, *Growth of the Bacterial Cell,* Sinauer Associates, Inc., Sunderland, Massachusetts, 1983.

11. K. L. Norcross, "A Full-Scale Evaluation of Thermal Efficiency for a Mechanical Surface and a Submerged Aeration System," presented at the 58th Annual Conference of the Water Pollution Control Federation, Kansas City, Missouri, 1985.

12. J. E. Bailey and D. F. Ollis, *Biochemical Engineering Fundamentals,* 2d ed., McGraw-Hill, New York, 1986.

13. R. L. Irvine and A. W. Busch, "Sequencing Batch Reactors—An Overview," *J. Water Pollut. Control Fed.,* **51**:235 (1979).

14. "CPI Go Below To Remove Groundwater Pollutants," *Chem. Eng.,* June 9, 1986, p. 14.

15. G. Battermann and P. Werner, "Beseitigung einer Untergrundkontamination mit Kohlenwasserstoffen durch mikrobiellen Abbau," *gwf-wasser/abwasser,* **125**:336 (1984).

16. H. W. Heath, Jr., "Combined Powdered Carbon/Biological ("PACT") Treatment to Destroy Organics in Industrial Wastewater," in *Proceedings of International Conference on New Frontiers for Hazardous Waste Management,* U.S. EPA, EPA 600/9-85/025, 1985.

17. ———, "Bugs and Carbon Make a PACT," *Civil Engineering,* April, 1986, p. 81.

18. *Biotechnology for Hazardous Waste Destruction—Workshop Summary,* Research Triangle Institute, Research Triangle Park, North Carolina, 1986.

19. P. A. Wilderer and I. Sekoulov, *Behandlung der Stauflusigkeiten aus der Deponie Georgswerder,* German BMFT research project, Grant No. 1440 386 B/7, 1985.

20. R. L. Irvine, L. H. Ketchum, Jr., and C. P. Kulpa, Jr., *Phase I Laboratory Test Program for On-Site Biological Remediation of Soils at the Harbor Point Site,* University of Notre Dame, Notre Dame, Indiana, 1986.

21. P. A. Wilderer and I. Sekoulov, *Verfahren und Vorrichtung zur biologischen Behandlung kontaminierter Bodenkorper,* West German Patent (submitted 1986).

SECTION 9.2
ANAEROBIC DIGESTION

Michael F. Torpy, Ph.D.
General Manager, Laboratory Services
Retec
Kent, Washington

9.2.1   DESCRIPTION OF THE TECHNOLOGY

Anaerobic digestion is a sequential, biologically destructive process in which hydrocarbons are converted, in the absence of free oxygen, from complex to simpler molecules, and ultimately to carbon dioxide and methane. The stepwise destruction of the molecules is conducted by uniquely capable groups of microorganisms that thrive during the catabolic process, and capture the energy of the hydrocarbons to conduct their respiration, growth, and reproduction. The process is mediated through enzyme catalysis and depends on maintaining a balance of populations within a specific set of environmental conditions. Understanding how to monitor and control these conditions is the path to success for engineers and scientists who wish to exploit these valuable organisms through applications of biotechnology.

Hazardous waste streams often consist of hydrocarbons in high concentrations of chemical oxygen demand (COD). Depending on the nature of the waste, the organic constituents may be derived from a single process stream, or from a mixture of streams. The treatability of the waste depends on the susceptibility of the hydrocarbon content to anaerobic biological degradation, and also on the ability of the organisms to resist detrimental effects of biologically recalcitrant, and toxic organic and inorganic chemicals. The practice of sophisticated waste management, an understanding of the microbiology, chemistry, of anaerobic digesters, and development of reactor-control techniques for the digesters, are important developments in waste treatment that enable anaerobic digestion to perform at high and stable rates of organic removal.

Anaerobic digestion has several attributes that are considered favorable in biological treatment.[18] In comparison to anaerobic digestion's conventional counterpart, aerated activated sludge, several benefits accrue. The power requirements for aeration are avoided; solids generation from growth of biological cells is about 20 times less, and therefore solids disposal is minimized; extremely high organic destruction rates can be achieved; the methane product can become a financial credit when it is sold, or burned for its heat value; significantly less land is required with the commercially available high-rate reactors; and certain organic constituents can be degraded with anaerobic digestion. Advanced reactor designs and enlightened operating practices have become essential to promoting stable and efficient operating conditions. With these attributes, a reactor can resist upsets and resume operations even when a toxic waste inadvertently contaminates the reactor. Modern waste-management practices of segregating streams and testing the treatability of individual streams increase the likelihood of success in anaerobic treatment of hazardous organic wastes. By understanding useful cotreatment and pretreatment practices for otherwise troublesome wastes, treatment by anaerobic digestion for a wide array of process streams can be accomplished at a particular facility. Anaerobic treatment technology suffers from some inferior characteristics that are often considered significant by those who would otherwise install this technology. Some types of reactors take months before the populations of anaerobic organisms are sufficient for operation; experience in scale-up from pilot scale or bench scale is inadequate, and imparting an acceptable degree of certainty in planning for reactor design and size is sometimes an elusive and frustrating task to the planning engineers. With a knowledge of the types of available reactors and the general characteristics of the waste, these concerns can be minimized. The following presents a brief, somewhat consolidated but comprehensive overview of anaerobic biotechnology (from references including 1, 7, 14, 15, 16, 31, 34).

The microbiology of anaerobic digestion consists of microorganisms present at several trophic levels, depending on the complexity of the organic wastes. At the higher levels, organisms attack molecules through hydrolysis and fermentation, reducing the molecules to simpler hydrocarbons. They are finally transformed to end products (methane and carbon dioxide) in the last two steps of the destructive process, where a delicate balance must be maintained between two distinct and essential microbial populations: the *acetogens* and the *methanogens*. In the acetogenic phase, simple carbohydrates can be converted to organic acids including acetate, propionate, butyrate, lactate, valerate, succinate, and ethanol, as well as hydrogen and carbon dioxide gas. When the partial pressure of H_2 is kept at very low levels (below 10^{-3} atm) by the methanogens,[13] the acetogens convert most of the carbohydrate to acetate, CO_2, and H_2, and a minimal amount to the other organic acids. In the methanogenic phase, organic acids and CO_2 are converted by the methanogens to CH_4 in the presence of the available H_2. When the H_2-utilizing methanogens are unable to convert the available H_2 at appropriate rates, the acetogens produce the organic acids, depleting the alkalinity and causing the pH to fall with the accumulation of organic acids, resulting in a "stuck" or "sour" reactor. The methanogens are generally the slower growers of the two populations, and careful maintenance of their environment and the balance of metabolic activities between the two populations is essential to the reactor's stability.

9.2.2 OPERATING CONDITIONS

The metabolic interactions between the various groups of organisms are essential to the successful and complete mineralization of the organic molecules. Various parameters are monitored to maintain efficient operating conditions within the reactor. They relate to the influent quality, the biological activity of the reactor, and the quality of the reactor environment.

The concentration of the organic constituents can be measured as the biochemical oxygen demand (BOD), but the COD parameter is usually more useful. Generally, higher ratios of BOD and COD ensure more highly degradable materials, but significance of a BOD measurement is rather dubious for hazardous wastes. The COD of the influent should remain relatively consistent from day to day, and equalization basins can become essential for providing this requirement. Changes in the organic loading rates (OLR) to reactors operating on the threshold of their capacity promote unstable metabolic conditions. The methanogens are the rate-limiting organisms and when their populations are not adequate for higher organic loads, the acetogens will outproduce the methanogenic populations. Reactors are often operated at less than threshold conditions, however, and it is not uncommon with some reactors to be able to handle occasional shock loads.

Oxygenated molecules, such as sulfates and nitrates, become the first electron acceptors, and the formation of methane is restricted until these molecules are reduced with the available hydrogen. Streams with concentrations of ammonia as high as 2,000 mg/L and cyanide at 5 mg/L were treated. Speece[25] presented a useful nomograph for determining the required, allowable, and toxic levels of sulfide concentration in a nomograph. Anaerobic metabolism relies on sulfides[26,30] and they are also useful for precipitating heavy metals from solution in the reactor, thus preventing metal toxicity by some wastes,[10,11,22] but may also become inhibitory and prevent methane formation.[5]

The temperature of the influent is critical to the metabolism of the anaerobes. Usually, the reactors are operated with mesophyllic organisms in the range of about 30 to 40°C, whereas thermophilic reactors are operated in the range of 50 to 75°C.[35] Changes of more than 2°C per day should not occur in the reactor.[23]

The pH of the reactor is maintained approximately in the 6 to 8 range, and the optimal pH for most methanogens is near 7.0. The alkalinity of the reactor is usually maintained at a balance with volatile acids (VA). A sudden shift in the VA concentration is a certain sign of a metabolic imbalance, and can result in depressed pH values in the liquor. Unless the cause of the imbalance is corrected, the reactor will be destined for failure, with suppression of methanogenic activity. Changes in the pH can be corrected with the addition of buffers such as calcium hydroxide. Sodium buffers should be used when sodium concentrations can be maintained below 10,000 mg/L with acclimated cultures, and less than 6,000 mg/L for unacclimated cultures. Reactors are generally operated at VA concentrations below about 400 mg/L, and a ratio of total VA (as acetic acid) to total alkalinity (as calcium carbonate) of less than 0.1 is desirable.[23]

The quality of the biogas depends on the type of substrate being degraded, as well as other factors, such as the source of alkalinity. The amount of biogas produced from a kilogram of COD destroyed is about 0.35 m³ (Ref. 15).

9.2.3 TREATABILITY TESTING

The nature of the materials introduced to the digester affects the ability of the organisms to mineralize the organic constituents. The organisms react differently to various types of wastes, and testing new materials with anaerobic cultures to evaluate the waste's treatability is good management practice. The biochemical methane potential (BMP) assay[19,24,27] is a conventional test used to evaluate the anaerobic treatability of the waste. A sample of the waste is mixed with an anaerobic culture in a defined medium of essential nutrients (except carbon sources) and held in an anaerobic environment (such as a sealed serum bottle). During the period of incubation, the volume of gas produced from the sample is intermittently measured and the bottle's pressure reduced. The amount of gas produced during the testing period with respect to the amount of carbon associated with the sample (measured as COD) is an indication of the extent to which the material can be degraded with the anaerobic consortium. A variation of this test is the anaerobic toxicity assay (ATA) that measures the relative rate at which a simple organic substrate (such as sucrose) is degraded in the presence of the waste sample, in comparison to the rate at which the substrate is degraded in the absence of the sample. These assays are useful in determining the treatability of the sample, and are efficient tests for evaluating the effects of variations in the treatment conditions. A large number of assays can be conducted within a relatively limited period of time and in a relatively small space, and can be applied to evaluate conditions such as the effects of pretreatment and cotreatment strategies. The effects of various microbial cultures on the degradation of the sample

TABLE 9.2.1   Organic Chemicals Degradable through Anaerobic Digestion

Acetylsalicylic acid	Di-*n*-butylphthalate	3-Methylbutanol
Acetaldehyde	Diacetone gulusonic acid	Methyl acetate
Acetic anhydride	Diethylphthalate	Methyl acrylate
Acetone	Dimethoxy benzoic acid	Methyl ethyl ketone
Acrylic acid	Dimethylphthalate	Methyl formate
Adipic acid	Ethanol	Nitrobenzene
1-Amino-2-propanol	Ethyl acetate	*o*-Nitrophenol
4-Amino-butyric acid	Ethyl acrylate	*p*-Nitrophenol
Aniline	Ferulic acid	1-Octanol
p-Anisic acid	Formaldehyde	Pentaerythritol
o-Anthranilic acid	Formic acid	Pentanol
Benzoic acid	Fumaric acid	Phenol
Benzyl alcohol	Glutamic acid	Phloroglucinol
Butanol	Glutaric acid	Phthalic acid
sec-Butanol	Glycerol	Polyethylene glycol
tert-Butanol	Hexanoic acid	Propanal
sec-Butylamine	2-Hexanone	Propanol
2,3-Butanediol	Hydroquinone	2-Propanol
Butylbenzylphthalate	*o*-Hydroxybenzoic acid	Propionic acid
Butyraldehyde	*p*-Hydroxybenzoic acid	Propylene glycol
Butylene glycerol	3-Hydroxybutanone	Protocatechuic acid
Butyric acid	Isobutyric acid	Pyrogallol
Catechol	Isopropanol	Resorcinol
m-Chlorobenzoic acid	Isopropyl alcohol	Sorbic acid
o-Chlorophenol	Lactic acid	Syringaldehyde
Cresol	Maleic acid	Syringic acid
m-Cresol	Methanol	Succinic acid
p-Cresol	*o*-Methoxyphenol	Valeric acid
Crotonaldehyde	*m*-Methoxyphenol	Vanillic acid
Crotonic acid	*p*-Methoxyphenol	Vinyl acetate

Source: Refs. 2, 8, 12, 20, 25, and 33.

can also be tested in a relatively limited space and time. The assay offers a convenient method for promoting and evaluating the acclimation of the anaerobic consortium over time. Normally, the BMP assays are monitored for about 30 days, but easily degraded materials can be detected within periods of a week or less with efficient biogas formation. It is evident in some cases, where the purpose of the testing is to elicit acclimation of the culture to the "new" material, that periods of months may be required before methanogenesis can be detected, resulting in relatively sudden and high biogas production rates.

Bench-scale testing with continuous-feed operations is conducted after successful BMP testing, to identify the maximum OLRs that can be achieved, to identify the effects of variations in sample quality, and to assess the stability of the reactor, the need for nutrient supplementation, and other factors that may be important to operations at full scale.

Treatability testing has been conducted by many investigators for many compounds and for many different types of wastes. Table 9.2.1 is a summary of some of the organic constituents that have been treated with anaerobic digestion. In addition, Table 9.2.2 lists several other organic materials that have been shown to be at least partially degraded by anaerobic digestion. It can be seen from these tables that applications of anaerobic digestion are quite appropriate for the control of diverse types of wastes including the aromatics, phthalates, esters, organic acids, alcohols, halogenated materials, and other complex molecules.

TABLE 9.2.2 Description of Organic Compounds Transformed through Anaerobic Digestion

	Source: Ref. 32
2,4-Dinitrophenol bis(2-Ethylhexyl)phthalate 2,4-Dimethyl phenol 4-Chloro-*m*-cresol 4,6 Dinitro-*o*-cresol	Yielded 10% of theoretical methane volume at concentrations of 20 mg/L.
	Source: Ref. 3
Tetrachloroethylene 1,1,1-Trichloroethane Chloroform Carbon tetrachloride	Degraded slowly but only in the presence of another metabolite which can support the growth of the organisms as a carbon source.
	Source: Ref. 21
Aldrin	Transformed to Dieldrin.
a-Hexachlorocyclohexane	Dechlorinated to pentachlorocyclohexane & HCl.
g-Hexachlorocyclohexane	Transformed to 1,3,4,5,6-pentachlorocyclohexane.
DDT	Dechlorinated to DDE.
Toxaphene	Reduced and dechlorinated.
1,1-Dichloroethane	Biodegradation.
1,2-Dichloroethane	Biodegradation.
Carbamates (*N*-methyl)[9]	Degradation depends on destruction of monomethylamine.-

TABLE 9.2.2   Description of Organic Compounds Transformed through Anaerobic Digestion (*Continued*)

	Source: Ref. 28
Halogenated Benzoates	Degraded after dechlorination to CO_2, CH_4.
Iodo- and Bromo-	2-,3-, or 4-substituted positions.
Chloro-	3-substituted position only.
	Source: Ref. 17
Benzene	Degraded to CO_2, CH_4.
Isopropylbenzene	*Note:* Anaerobes also catalyze aryl
Ethylbenzene	dehydroxylation, decarboxylation,
Toluene	demethoxylation, and demethylation (Ref. 29).
	Source: Ref. 29
3,4-, 2,5-, and 3,5-Dichlorobenzoate 2,4,5-Trichlorophenoxyacetate 5-Bromo-2-chlorobenzoate 4-Amino-3,5-dichlorobenzoate	Dehalogenation was elicited from a cell moiety derived from anaerobic organisms.

9.2.4   REACTOR CONFIGURATIONS

While anaerobic digestion is not a novel concept in waste treatment, recent and important innovations in the technology promote it as an important one for effective destruction of organic chemicals. Design improvements of "modern" reactors are related to promoting effective methods for maintaining the microbiological consortium within the reactor, and efficiently exposing the organisms to the organic substrate. Several reactor types with these features are commercially available and suitable for hazardous-waste treatment.

Anaerobic Filter

The bacteria of this type of reactor attach as a thin biofilm to a surface within the reactor.[18,23,25]

1. *Fixed-bed reactors* contain biological solids that grow on inert, high-surface-area material, and the waste stream is circulated along the vertical length of the reactor. It can be operated either in an upflow or downflow direction, the choice being dependent, in part, on the possible effects of insoluble solids entering or forming within the reactor. High organic loading rates (OLR) of 10 to 20 kg/($m^3 \cdot$ d) [0.6 to 1.2 lb/($ft^3 \cdot$ d)] can be achieved with COD-removal efficiencies of 80 to 95%. Start-up can be exceptionally long, but the reactor is stable once the organisms are established. The reactor is relatively stable to toxic and shock loading.

2. *Fluidized-bed reactors* have inert materials such as sand that have extremely high surface areas covered with biological growth. These reactors are operated with hydraulic velocities sufficient to maintain an expanded bed, but low enough to avoid solids washout. Recycling of some of the effluent stream is

usually required to maintain the design velocities, and the reactor can be operated with high efficiency at OLRs greater than 15 kg/(m³ · d) [0.9 lb/(ft³ · d)]. Particulate solids and solids buildup on the media surfaces are not a concern, nor are problems of hydraulic channeling and plugging. The reactor is relatively stable to toxic and shock loading. The cost of pumping power for this type of reactor can be a significant part of the operating costs.

Upflow Anaerobic-Sludge Blanket (UASB)

This type of reactor contains highly settleable sludge particles.[18,23,25] The waste stream is fed from the bottom of the reactor and passes through a blanket of sludge granules as it moves toward the top of the reactor. The granules consist of the active anaerobic biomass and tend to entrain biogas and become buoyant. An overhead baffle system is located at the top of the reactor to capture and promote degassing of the rising granules. The granules subsequently settle to the sludge blanket and in this way the biomass of the reactor is efficiently controlled. Maintaining the organisms while treating high COD concentrations—sometimes as high as 40 kg/(m³ · d) [2.5 lb/(ft³ · d)]—is not uncommon. Start-up can be accomplished by importing the granules from other UASB reactors. While the effects of introducing insoluble solids to the reactor must be monitored, problems of toxicity and maintaining stable operating conditions during intermittent and large shifts in OLRs can be managed.

Other Reactors

Various designs and reactor configurations may be appropriate for particular wastes and are commercially available or their operating experiences are reported in the literature.

1. *Two-stage reactor:* The special attributes of this type of reactor may become exceptionally useful in hazardous-waste treatment as anaerobic digestion becomes more commonly used. In this type of reactor, organisms that hydrolyze insoluble organics, and ferment other organic constituents are maintained in the first stage, and their catabolic products become the nourishment for the organisms associated with efficient methanogenesis in a second reactor. The particular usefulness of this type of reactor lies in at least two important areas of interest. When wastes contain high concentrations of organic materials that are either insoluble, or slowly degraded during acetogenesis, the first reactor would be operated to degrade these types of compounds to other, more simple forms. This type of reactor may be appropriate when particular compounds, toxic to methanogens, are degraded in the first stage, and rendered amenable to methanogenesis in the second stage. No vendor of this technology is known, although ample operating and design information is found in the literature.[4,6]

2. *Lagoon with full-length biomass contact:* These reactors are constructed with a sealed liner and cover and operate by contacting the waste stream with the biomass along the length of the lagoon. Generally, the reactor is considered slow-rate and requires relatively long hydraulic retention times. This type of reactor is especially appropriate in applications where relatively small vol-

umes of high-COD waste are generated, and where land costs are minor factors in the choice of treatment. Relatively high solids content can be accommodated, and maintenance costs are minimal.

3. *Hybrid:* Recent engineering and design incorporate the features of at least two of the reactors to form a hybrid reactor. For example, a reactor may consist of a UASB on the bottom, and a fixed-film reactor on the top to promote higher treatment efficiencies. Other combinations and configurations may be found to be useful as the number of applications increases and as various types of wastes are degraded in anaerobic digesters.

9.2.5 MARKET SUPPLIERS AND PROPRIETARY SYSTEMS

The equipment available for anaerobic digestion is becoming more specialized with the development of unique applications. Various manufacturers are able to supply media for fixed-film reactors, others provide proprietary equipment for gas cleanup, and others may provide, for example, special tanks. Several consulting engineering firms have associated themselves with owners of proprietary and licensed systems in order to have wider access to potential clients. Also consulting engineers with various degrees of experience offer their services. Table 9.2.3 summarizes the various types of proprietary systems known to be available.

TABLE 9.2.3 Proprietary Anaerobic-Digestion Systems

Company name	Process name	Reactor type
A. C. Biotechnics	ANAMET	Contact
ADI	——	Contact
Bacardi	Bacardi	Fixed film
Badger	Celrobic	Packed bed
Biomechanics	Bioenergy	Contact
Biothane	Biothane	UASB
Dorr-Oliver	ANITRON	Fluidized bed
Dorr-Oliver	MARS	Contact
Ecolotrol	HY-FLO	Fluidized bed
Gore & Storrie	HYAN	Hybrid
Paque-Lavalin	BIOPAQ	UASB

9.2.6 REFERENCES

1. C. Anthony, "Methanogens and Methanogenesis," chap. 11 in *The Biochemistry of Methylotrophs,* Academic Press, 1983.

2. E. J. Bouwer and P. L. McCarty, "Removal of Trace Chlorinated Organic Compounds by Activated Carbon and Fixed-Film Bacteria," *Environ. Sci. Technol.,* **16**:(12) (1982).

3. E. J. Bouwer and P. L. McCarty, "Transformations of 1- and 2-Carbon Halogenated

Aliphatic Organic Compounds Under Methanogenic Conditions," *Applied and Environ. Microbiol.*, **45**:1,286–1,294 (1983).

4. A. Cohen, "Two-phase Digestion of Liquid and Solid Wastes," in *Proceedings from the Third International Symposium on Anaerobic Digestion*, Boston, August 1983, pp. 123–138.

5. P. Corbo and R. C. Ahlert, "Anaerobic Treatment of Concentrated Wastewater," *Env. Prog.*, **4**:(1)22–26 (1985).

6. S. Ghosh, J. P. Ombregt, and P. Pipyn, "Methane Production from Industrial Wastes by Two-phase Anaerobic Digestion," *Water Research*, **19**:(9)1,083–1,088 (1985).

7. C. P. L. Grady, Jr., and H. C. Lim, *Biological Wastewater Treatment: Theory and Applications*, Marcel Dekker, 1980, pp. 833–881.

8. J. B. Healy, Jr., and L. Y. Young, "Catechol and Phenol Degradation by a Methanogenic Population of Bacteria," *Applied and Environ. Microbiol.*, **35**:(1)216–218 (1978).

9. R. P. Kiene and D. C. Capone, "Stimulation of Methanogenesis by Aldicarb and Several Other *n*-Methyl Carbamate Pesticides," *Applied and Environ. Microbiol.*, **51**:(4)1,247–1,251 (1986).

10. I. J. Kugelman and K. K. Chin, "Toxicity, Synergism, and Antagonism in Anaerobic Waste Treatment Processes," chap. 5 in *Anaerobic Biological Treatment Processes*, Advances in Chemistry Series 105, American Chemical Society, 1971, pp. 55–90.

11. A. W. Lawrence and P. L. McCarty, "The Role of Sulfide in Preventing Heavy Metal Toxicity in Anaerobic Treatment," *J. Water Pollut. Control Fed.*, **37**:(3)392–406 (1965).

12. G. Lettinga, W. de Zeeuw, and E. Ouborg, "Anaerobic Treatment of Wastes Containing Methanol and Higher Alcohols," *Water Research*, **15**:171–182 (1981).

13. P. L. McCarty and D. Smith, "Effect of Hydrogen Concentration on Population Distribution and Kinetics in Methanogenesis of Propionate," chap. 5 in *Proceedings from Biotechnological Advances in Processing Municipal Wastes for Fuels and Chemicals*, A. A. Antonopoulos (ed.), ANL/CNSV-TM-167 Argonne National Laboratory, Argonne, Illinois, 1985.

14. M. J. McInerney and M. P. Bryant, "Basic Principles of Bioconversions in Anaerobic Digestion and Methanogenesis," chap. 15 in *Biomass Conversion Processes for Energy and Fuels*, S. S. Sofer and O. R. Zaborsky (eds.), Plenum Press, 1981.

15. *Wastewater Engineering Treatment/Disposal/Reuse*, Metcalf & Eddy, Inc., McGraw-Hill, New York, 1979, pp. 455–461.

16. D. P. Nagle, Jr., and R. S. Wolfe, "Methanogenesis," chap. 22 in *Comprehensive Biotechnology—The Principles, Applications and Regulations of Biotechnology in Industry and Medicine*, vol. 1, *The Principles of Biotechnology: Scientific Fundamentals*, A. T. Bull and H. Dalton (eds.), Pergamon Press.

17. A. S. Ng, C. M. Rose, and M. F. Torpy, "Anaerobic Treatment of Industrial Wastes," in *Proceedings from the Second National Conference on Anaerobic Digestion of Industrial Wastes*, M. F. Torpy (ed.), in preparation, presented 1986.

18. M. Olthof and J. Oleszkiewicz, "Anaerobic Treatment of Industrial Wastewaters," *Chem. Eng.*, November 15, 1982, pp. 121–126.

19. W. F. Owen et al., "Bioassay for Monitoring Biochemical Methane Potential and Anaerobic Toxicity," *Water Research*, **13**:485–492 (1979).

20. G. F. Parkin and S. W. Miller, "Response of Methane Fermentation to Continuous Addition of Selected Industrial Toxicants," in *Proceedings of the 37th Industrial Waste Conference*, Ann Arbor Press, Ann Arbor, Michigan, 1982.

21. D. J. Richards and W. K. Shieh, "Biological Fate of Organic Priority Pollutants in the Aquatic Environment—Review Paper," *Water Research*, **20**:(9)1,077–1,090 (1986).

22. A. L. Rivera, "Heavy Metal Removal in a Packed-bed, Anaerobic Upflow (ANFLOW) Bioreactor," *J. Water Pollut. Control Fed.*, **55**:(12) (1983).

23. H. Sahm, "Anaerobic Wastewater Treatment," in *Advances in Biochemical Engineering/Biotechnology,* A. Fiechter (ed.), Springer, 1984.

24. D. R. Shelton and J. M. Tiedje, "General Method for Determining Anaerobic Biodegradation Potential," *Applied and Environ. Microbiol.,* **47**:(4)850–857 (1984).

25. R. E. Speece, "Anaerobic Biotechnology for Industrial Wastewater Treatment," *Environ. Sci. and Technol.,* **17**:(9)416A–427A (1983).

26. R. E. Speece and G. F. Parkin, "Nutrient Requirements for Anaerobic Digestion," chap. 15 in *Proceedings from Biotechnological Advances in Processing Municipal Wastes for Fuels and Chemicals,* A. A. Antonopoulos (ed.), ANL/CNSV-TM-167, Argonne National Laboratory, Argonne, Illinois, 1985.

27. D. C. Stuckey et al., "Anaerobic Toxicity Evaluation by Batch and Semicontinuous Assays," *J. Water Pollut. Control Fed.,* **52**:(4) (1980).

28. J. M. Suflita et al., "Dehalogenation: A Novel Pathway for the Anerobic Biodegradation of Haloaromatic Compounds," *Science,* **218**:(10)1,115–1,117 (1982).

29. J. M. Suflita, University of Oklahoma, Norman, Oklahoma, unpublished information, 1985.

30. L. Van Den Berg et al., "Effects of Sulfate, Iron, and Hydrogen on the Microbiological Conversion of Acetic Acid to Methane," *J. Applied Bact.,* **48**:437–447 (1980).

31. R. S. Wolfe, "Microbial Biochemistry of Methane—A Study in Contrasts: Part 1-Methanogenesis," in *Microbial Biochemistry, International Review of Biochemistry,* vol. 21, J. R. Quale (ed.), University Park Press, 1979, pp. 267–299.

32. L. Y. Young, New York University Medical Center, New York, unpublished information, 1985.

33. L. Y. Young and M. D. Rivera, "Methanogenic Degradation of Four Phenolic Compounds," *Water Research,* **19**:(10)1,325–1,332 (1985).

34. J. G. Zeikus, "The Biology of Methanogenic Bacteria," *Microbiological Reviews,* **41**:(2)514–541 (1977).

35. S. H. Zinder, T. Anguish, and S. C. Cardwell, "Effects of Temperature on Methanogenesis in a Thermophilic (58°C) Anaerobic Digestor," *Applied and Environ. Microbiol.,* **47**:(4)808–813 (1984).

SECTION 9.3
COMPOSTING OF INDUSTRIAL WASTES

George M. Savage
Vice President
Cal Recovery Systems, Inc.
Richmond, California

Luis F. Diaz
President
Cal Recovery Systems, Inc.
Richmond, California

Clarence G. Golueke
Director of Research and Development
Cal Recovery Systems, Inc.
Richmond, California

The purpose of this section is to provide useful information regarding: (1) the rationale in the use of composting for the treatment of toxic and hazardous organic industrial wastes; (2) the principles involved in so doing, including definitions, microbiological aspects, important factors, and major parameters; (3) applicable compost technology, including a critical evaluation of available systems; and (4) the utility of composting in hazardous-waste treatment, including pertinent constraints and precautions.

Before proceeding with the discussion of the four subjects, attention is called to the fact that in this section, the term *hazardous waste* refers *only* to organic hazardous wastes.

9.3.1 RATIONALE FOR COMPOSTING

Successes experienced in land treatment[1,2,3] and favorable findings made in microbiological investigations involving the degradation of recalcitrant organic materials[4-9] give reason for cautious optimism regarding the composting of certain hazardous residues. Examples of recalcitrant organic compounds and the microorganisms capable of metabolizing them are listed in Table 9.3.1.

TABLE 9.3.1 Metabolizable Hazardous Organic Compounds and Associated Microorganisms

Compound	Microorganism
Phenylmercuric acetate	*Pseudomonas, Arthrobacter, Citrobacter, Vibrio*
Raw rubber, hevea latex	*Actinomycetes,* bacteria
Detergents	*Nocardia, Pseudomonas*
PCBs	Not identified
Malathion	*Trichoderma, Pseudomonas*
Endrin	*Arthrobacter, Bacillus*
Lindane	*Clostridium*
DDT	*Escherichia, Hydrogenomonas, Saccharomyces*
Dieldrin	*Mucor*

Source: Data were obtained from Ref. 7.

Despite the large number and broad range of isolated organisms, the qualification "cautious optimism" remains applicable, because most of the reported studies deal with monocultures and idealized carbonaceous substrates observed under carefully controlled laboratory conditions. Nevertheless, the existing information can serve as a starting point in a search for and selection of suitable microbial populations to serve as "seed" either for starting or for accelerating the composting of biodegradable hazardous wastes.

9.3.2 PRINCIPLES

The principles involved in composting of industrial wastes are the same as those in the composting of all organic materials. Moreover, except for moderate modifications, most are common to all types of biological waste-treatment systems. This commonality is not surprising inasmuch as composting is essentially a biological process.

A glossary in which are given definitions and meanings of terms used in this chapter appears as Table 9.3.2.

Microbiology

The microbiology of hazardous-waste composting differs from that of composting in general in that the use of an inoculum is indicated—at least initially. The rea-

TABLE 9.3.2 Glossary of Composting Terms

Term	Definition or meaning
Composting	A system of waste treatment that involves the engineered biological decomposition of solid or solidified biodegradable residues under controlled conditions and independently of the soil medium
Engineered	Distinguishes composting from the uncontrolled decomposition in nature
Solid	Distinguishes composting from biological processes designed for liquefied wastes
Solidified	Describes a liquid waste that has been dewatered in preparation for composting
Independently of soil medium	Distinguishes composting from landfarming
Factors	Environmental conditions that influence or determine microbial activities
Parameters	Standards for evaluating or gauging performance and operational responses and for evaluating products
Bulking material	Dry material added to impart or improve porosity of compost substrate (e.g., sawdust, straw)

son is that certain types of hazardous-waste molecules can be degraded by only one or a very few microbial species, which may not be widely distributed or abundant in nature.

In the absence of an inoculum, recourse initially must be had to "enrichment" techniques.[9] Enrichment techniques involve the establishment of conditions that foster the proliferation of the few or more representatives of suitable species which may happen to be in the substrate. An alternative would be to introduce material that would be likely to contain the desired organisms. In view of the favorable experience had with landfarming, garden loam could be a good example of such material. Other possibilities are composted sewage sludge, crop residues, and manures.

In an ongoing operation, recirculated compost product would be an ideal inoculum. The flow of the process would be designed such that a portion (10 to 20%) of the end product would be diverted to the incoming feed.

Factors

Factors of importance in composting are those that govern all biological reactions.

Physical. The principal physical factors are the shapes and dimensions of the particles of the material to be composted. Theoretically, the smaller the size, the better in terms of composting efficiency and rate;[10,11] however, other factors exert a decided influence in practice. The most important of the other factors is porosity. Porosity is particularly a function of the structural integrity and size of the individual particles. For example, the permissible dimensions of sawdust particles used as a bulking agent may be as small as a few millimeters, whereas paper particles should be on the order of 2.5 to 10 cm (1 to 4 in). If municipal refuse is used as a bulking agent, particle-size distribution should be on the order of 2.5 to 5 cm (1 to 2 in).[12]

Environmental. The environmental factors of interest in a practical operation are temperature, pH, available oxygen, moisture, and nutrient availability.

Temperature. Temperature has a decisive influence on rate of composting. In practice, little is to be done about temperature, because heat generated by microbial metabolism under proper composting conditions is sufficient to raise the temperature of the composting mass to thermophilic levels. Inasmuch as microbial activity is inhibited at temperatures higher than 50 to 55°C (122 to 131°F), the composting mass should be manipulated such that its temperature does not exceed that level.[13,14]

Hydrogen Ion Level. The pH level most conducive to microbial activity and efficiency is within the range of 6.0 to 7.5. However, in practice an initial pH level of 5.5 or 8.0 would not have a serious effect on composting time and overall composting progress. A low pH can be adjusted through the addition of $Ca(OH)_2$ or any other biologically innocuous alkaline material to the mixture to be composted. Nitrogen loss from the mass accelerates with rise of the pH level of the composting mass above 7.5. An excessively high pH level could be brought down with the use of an innocuous acidic material [e.g., $(NH_4)_2SO_4$].

Availability of Oxygen. Although composting can be done either anaerobically or aerobically, the aerobic approach is followed almost exclusively in modern practice. (However, an anaerobic phase may be essential to the successful composting of certain halogenated organic pesticides and hydrocarbons.[15,16,17]

The primary and usually the sole source of the oxygen consumed in microbial respiration is the atmospheric O_2 trapped in the interstices of the composting mass. That being the case, the physical characteristics of the composting mass should be adjusted such that it becomes and remains porous throughout the composting period. If the hazardous waste is lacking in the characteristics needed to make it porous, a bulking agent should be added.

Moisture Content. In practical composting, moisture content does not become limiting until it drops to 40 to 45%. Thereafter, activity declines with drop in moisture content until a level of between 5 and 12% is reached. At this level, all biological activity ceases. If paper or municipal refuse constitutes the principal bulking agent, the maximum permissible moisture content would be on the order of 45 to 55%. On the other hand, if the bulking agent is either straw or sawdust, the limit could be as high as 85 to 90%, although prudence would dictate 70 to 80%.

Nutrition. The essential elements are carbon, nitrogen, phosphorus, and an assortment of other elements. If the waste is deficient in an element or elements, the deficiency must be removed by adding the required element either in a chemical form (commercial fertilizer) or as a waste that contains the required nutrient element. Inasmuch as most toxic organic wastes are hydrocarbons, nitrogen probably would be deficient. Examples of the wastes rich in nitrogen are listed in Table 9.3.3.

Parameters

The more important of the parameters are temperature level, O_2-to-CO_2 ratio, change in appearance of the composting mass, and disappearance (reduction in

TABLE 9.3.3 Nitrogen Content of Several Wastes

Material	Nitrogen, %	C/N ratio
Night soil	5.5–6.5	6–10
Urine	15–18	0.8
Blood	10–14	5.0
Animal tankage	——	4.1
Cow manure	1.7	18
Poultry manure	6.3	15
Sheep manure	3.8	
Pig manure	3.8	
Horse manure	2.3	25
Raw sewage sludge	4–7	11
Digested sludge	2–4	
Activated sludge	5	6
Grass clippings	3–6	12–15
Wheat straw	0.3–0.5	128–150
Oat straw	1.1	48
Sawdust	0.1	200–500

Source: Adapted from Ref. 10.

concentration) of the hazardous component. Generally, the interpretation of and response to parameters is based upon the common practice of dividing the compost process into an *active* (high-temperature) stage and a *maturation* stage.

Temperature Level.   Change in temperature level not only is the easiest of the parameters to determine and evaluate, it also is the most significant. Although its utility as an indicator depends upon the existence of conditions such that rate of heat loss from the composting mass does not exceed that of heat generation within it, conventional compost practice is such that the composting mass does retain sufficient heat to make temperature level and its change a useful parameter. Our experience indicates that, if sufficiently sheltered from wind and rain, an actively composting mass at least 1 m^3 (35 ft^3) in volume would be large enough to meet this condition.

A decline in microbial activity and the accompanying drop in temperature can be due either to the development of an inhibitory condition or conditions (e.g., desiccation, anaerobiosis, nutritional imbalance or deficiency) or simply to the aging of the microbial populations. With respect to the inhibitory conditions, the obvious recourse is to make the necessary adjustments. If aging were the cause, the temperature drop would be irreversible and would be indicative of the onset of the maturation stage. When with the final drop the composting mass approaches the ambient temperature level, the composting material may be considered to be sufficiently stabilized. If a hazardous compound were the principal source of carbon for the microbial population, the final drop could indicate that the compound has been eliminated or has been transformed into an intermediate.

The sequence and extent of the changes in temperature level typically encountered in composting are indicated by the curve in Fig. 9.3.1.

Oxygen/Carbon Dioxide Ratio.   Reported rates of O_2 uptake range from as low as 1.0 to as high as 13.6 mg O_2/(g volatile matter · h) [0.001 to 0.013 lb O_2/(lb volatile matter · h)]. Other examples of interest pertain to approximations of O_2 concentrations. Thus, an O_2 pressure greater than 14% of the total indicates that more

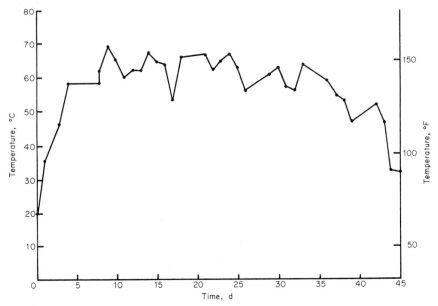

FIG. 9.3.1   Average temperature of compost pile.

than two-thirds of the oxygen in the air remains to be consumed. The optimum level of oxygen is within the range of 14 to 17%. Supposedly, aerobic composting comes to an end when the oxygen concentration drops to 10% by volume. The CO_2 in the exhaust gas should be between 3 and 6% by volume.

Change in Physical Characteristics.   Changes most readily detected by the human sensory apparatus are the visual (e.g., color, overall texture of the mass, moisture) and the olfactory (e.g., foul odors, earthy odors). With the exception of color, the visual changes are simultaneously tactual—e.g., texture, dryness.

   Visual and Tactual.   Improvement in texture is a reliable indicator, whereas the taking on of a dark color is not. An overall condition of moisture deficiency is indicated by a "dusty" appearance. In routine composting, a rule of thumb with respect to satisfactory moisture content is a glistening sheen on the composting particles.

   Olfactory.   A foul odor indicates anaerobiosis. Consequently, increasing the rate of aeration causes the malodor to disappear. If aeration had been inadequate because of excessive moisture, amplifying aeration would dry the composting to the extent required. The acquisition of an earthy odor does not necessarily indicate sufficient stability or maturity.

   If the hazardous substance had a particular odor (e.g., chlordane, gasoline), the loss of that odor would indicate the degradation of the compound.

Disappearance of the Hazardous Compound.   If a microbial population capable of degrading the hazardous component is present, rate and extent of disappearance of the component are indicators of the efficiency of the compost operation. However, the complete loss of the hazardous substance and the attainment of stability of other components of the composting mass are not necessarily coincidental.

9.3.3 TECHNOLOGY

Compost technology can be divided into two broad classes, namely *windrow* (*open pile*) and *in vessel* (*enclosed*). The windrow class may be further subdivided into *turned* and *forced aeration* (*static pile*). The term *open pile* does not preclude the erection of a structure or structures to shelter the composting piles from the elements. The number of subdivisions for the class *in vessel* matches that of the systems on the market.

In the description and discussion that follows, it is assumed that the material to be composted has been suitably prepared for composting.

Windrow

In the application of the windrow system, the material is stacked to a height theoretically determined by the structural and physical characteristics of the individual particles of the composting mass, i.e., a height at which porosity remains undiminished throughout the mass to be composted. In practice, however, height is further determined by the capacity of the materials-handling equipment. For example, although the composting mass would remain sufficiently porous at a windrow height of 2.5 m (8.0 ft), the equipment used in manipulating the mass might be able to stack the material only to a height of 1.0 to 2.0 m (3.3 to 6.6 ft). On the other hand, the optimum height might be on the order of 1.5 to 2.0 m (5.0 to 6.6 ft) with manual stacking.[11] The proper width of the windrow would be the one most suitable to the method used in stacking the material. In practice, widths usually range from 2.0 to 3.0 m (6.6 to 10 ft). Length is a function of pile arrangement and manipulation and of site characteristics.

The configuration most suitable for a windrow depends upon a number of conditions, among which are climatic conditions, capacity of the materials-handling equipment, and method of aeration. Because they are conducive to shedding water, conical or triangular cross sections are the best configurations for a rainy climate. If heat retention is essential and rainfall is very moderate, a flattened top would be suitable. The conical and triangular configurations are best suited for aeration by turning, whereas the flattened or loaf-shaped configurations are best for forced aeration. Ultimately, the choice of which of the configurations to use depends upon the equipment used in constructing the piles and in turning them.

Windrows should always be underlain by a hard surface, and arrangements should be made for collecting and disposing of runoff and leachate from the materials. These strictures are especially applicable when toxic organics are concerned. Regardless of the nature of the raw material, leachate from a composting mass is very concentrated in terms of objectionable components. Obviously, leachate from composting toxic organics would be much more objectionable than that from nonhazardous wastes. When hazardous compounds are involved, enclosure of the windrows in a shelter during the active stage would be mandatory because of the need to prevent the release of dangerous emissions. The shelter should be designed such that all gaseous emissions from the enclosure could be monitored and dangerous substances removed.

Crop residues contaminated with toxic organics (e.g., pesticides) can be satisfactorily and safely treated by windrow composting. Concentrations of contaminants on crop residues usually are relatively low and the contaminants usually are of short duration because of vulnerability to physical conditions that develop during composting (e.g., temperature rise, pH variation) and especially because of susceptibility to microbial action. Safety is assured by careful monitoring and by exercising careful control of incoming material and of the operational environment. The situation may be different when highly toxic or resistant contaminants are concerned. Unless extraordinary measures are taken, the requisite tight control cannot be supplied in windrow composting.

Turned Windrow. The term *turned* in the expression *turned windrow* is taken from the method of aeration followed in this type of composting. In essence, "turning" consists in tearing down and immediately reconstructing the windrow or pile of composting material. Aeration results from the displacement of oxygen-depleted air in the interstices in the original pile and replacement with the fresh air contained in the new interstitial complex formed by the reconstruction of the pile. It follows that satisfactory aeration depends upon a thorough agitation of the material during the reconstruction of the windrow. At this point it should be mentioned that some aeration results from the limited diffusion of ambient air into the outer layer of the windrow.

Inasmuch as the O_2 supply for the microbial population is largely from the air contained in the interstices, it follows that required frequency of turning depends in part on the volume of the interstitial air. Therefore, if the interstices were partially filled with moisture, the turning frequency would have to be correspondingly increased. Consequently, moisture content becomes a factor in frequency along with O_2 requirement.

Several types of automatic turning equipment are available.[10,11,12] Most depend upon a revolving drum equipped with protruding spikes for agitating the composting mass and a shaping blade for reconstructing the windrow. One of the machines available depends upon a front-end revolving auger designed to tear down, agitate the composting mass, and reconstitute the windrow in a single pass.

Forced-Aeration (Static) Windrow. In forced-aeration systems, turning is replaced by mechanical pulling (negative pressure) or pushing (positive pressure) of air through the windrow. However, an occasional turning may be applied so as to ensure uniform decomposition.

One of the various designs for forced-aeration composting calls for constructing the pile upon an elevated, perforated base through which air is forced into the pile.[18] Another approach, the *Beltsville method,*[19] involves embedding perforated ducts in a bottom layer of wood chips or other comparable material, upon which the material to be composted is stacked. Changes in the original Beltsville system include the introduction of (1) a sequence of positive-pressure aeration followed by negative-air-pressure aeration, (2) deodorization by passing the exhaust air through a secondary small pile, and (3) addition of an outer layer of fully composted material or wood chips.

According to one report,[12] an air flow of about 5 m^3/min (175 ft^3/min) at the blower inlet was enough to provide an oxygen concentration of 5 to 15% in a 40-t (44-ton) pile of combined sludge cake and wood chips. Dimensions of the pile were $12 \times 6 \times 2.5$ m high ($40 \times 20 \times 8.4$ ft high). In operations in which forced aeration is used for moisture removal and temperature regulation as well as aeration, amounts and rates of air input will depend upon oxygen demand, moisture content, temperature, and their interrelationship. Recent studies[13,14] have shown that drying and destruction of volatile solids are greatest at high aeration rates, e.g., 22 m^3/(h · t) [880 ft^3/(h · ton)] initial wet weight, and low process temperatures, i.e., lower than 55°C (131°F).

For a forced-aeration windrow operation to be successful:

1. The composting material should be granular.
2. Particle size should be relatively uniform.
3. Particles should be resistant to compaction.
4. The composting mass should not be compacted.
5. The material must not be excessively moist.

Our experience indicates that the first of the preceding five requirements probably would rule out oily waste sludges as suitable candidates for forced-aeration composting.

In-Vessel (Enclosed) Composting

The major emphasis in the designs of the various in-vessel systems seems to be on aeration. Basically, all in-vessel aeration systems are based upon agitation of the particles, or upon forced aeration, or upon a combination of the two. Agitation may be accomplished by one or more of the following: tumbling, dropping from one level to another, and stirring.

In one classification,[20] in-vessel systems are divided into four classes, namely, *agitated-bed systems, silo-type systems, tunnel-type systems,* and the *enclosed static pile.* The agitated-solids bed systems are further subdivided into types that use rectangular reactors and those that use circular reactors. In another classification,[11] in-vessel systems are divided into three categories, namely, *vertical-, inclined-,* and *horizontal-flow* units. The systems are further divided on the basis of the condition of the mass in the reactor and the method of agitating the composting mass.

Aside from its relatively high capital and maintenance costs, a serious drawback of in-vessel composting is the lack of operating experience and data for evaluating the various systems.

9.3.4   UTILITY OF COMPOSTING

The most likely use for composting is in the disposal of biodegradable pesticides and solid wastes that have been contaminated with organic hazardous materials. Liquid hazardous wastes cannot be composted unless they are converted into a solid form through concentration or through absorption by a bulking agent.

The use of a bulking agent has made composting a very promising means of disposing oily wastes.[21,22,23] The utility of composting in this respect is indicated by its advantages over incineration and landfarming. Within the limits imposed by the constraints named in Subsec. 9.3.5, the advantages are these:

1. Operational and capital costs of composting are lower than those for incineration, and the environmental impact is not as adverse.
2. Composting is safer than landfarming in terms of impact upon aqueous, land, and air resources; retention times are shorter; and land requirements are far less.
3. A combination of physical and chemical factors generated in composting (pH and temperature) hastens the destruction of less persistent pesticides such as Malathion and carbaryls.[24]

9.3.5 CONSTRAINTS AND PRECAUTIONS

Composting is by no means a panacea for the hazardous-waste problem. On the contrary, the technology for composting of hazardous wastes is burdened with many important constraints in addition to those named and discussed in the preceding subsections. For example, a serious constraint could be the fact that the nature of hazardous wastes is such that costs of and work involved in preparing the waste for composting may be so great as to render the undertaking economically infeasible. Another consideration is the possibility that the cost of the precautions (enclosure, etc.) that must be taken with composting may be such as to make composting noncompetitive with other means of hazardous-waste disposal. Despite these constraints, composting may be competitive in the disposal of crop residues, contaminated municipal solid waste, and oily wastes.

When considering the future of hazardous-waste composting, heed must be paid to the advantages and disadvantages inherent in composting as compared with those inherent in physical, chemical, and thermal methods of waste treatment. The economic advantage that accounts for the increase in land spreading of sewage sludge at the expense of incineration, also holds true with the composting of hazardous agricultural wastes and petroleum sludges. With respect to agricultural residues, the economic advantage over incineration follows from the sizable volumes of the agricultural residues involved. With petroleum sludges, it follows from the difficulties and expenses attending the incineration of the sludges. With respect to the latter, an extensive record of successful landfarming has paved the way for a switch to composting.

In closing, it must be remembered that the full potential of composting in the treatment of hazardous wastes can be realized only after an intensive and extensive program of research and study has been completed. Past experience has been sketchy, and much remains to be learned.

9.3.6 REFERENCES

1. D. J. Norris, "Landspreading of Oily and Biological Sludges in Canada," in *Proceedings of the 35th Industrial Waste Conference,* Purdue University, West Lafayette, Indiana, 1980, p. 14.
2. *Land Treatment Practices in the Petroleum Industry,* Environmental Research and Technology, Inc., prepared for the American Petroleum Institute, 1983.
3. *Hazardous Waste Land Treatment,* K. W. Brown, G. B. Evans, Jr., and B. D. Frentrup (eds.), Butterworth Publishers, Boston, 1983.
4. *Biodegradation Techniques for Industrial Organic Wastes,* D. J. De Renzo, (ed.), Noyes Data Corp., Park Ridge, New Jersey, 1980, pp. 10–103.
5. A. E. Prince, "Microbiological Sludge in Jet Aircraft Fuel," *Dev. Ind. Microbiology,* **2**:197 (1961).
6. E. J. Nyns, J. P. Auguiere, and A. L. Wiaux, "Taxonomic Value of the Properties of Fungi to Assimilate Hydrocarbons," *Antonie von Leeuwenhoek J. Microbiol. Serol.,* **34**:441–457 (1986).
7. J. D. Nelson, W. Blair, F. E. Bringkman, R. R. Colwell, and W. P. Iverson, "Biodegradation of Phenylmercuric Acetate by Mercury-Resistant Bacteria," *Applied Microbiology,* **26**:321–326 (1973).
8. C. E. Zobell, "Assimilation of Hydrocarbons by Microorganisms," *Adv. Enzymol.,* **10**:443–486 (1950).

9. R. Mitchell, *Introduction to Environmental Microbiology,* Prentice-Hall, Englewood Cliffs, New Jersey, 1974.

10. C. G. Golueke, *Biological Reclamation of Solid Wastes,* The J. G. Press, 1977.

11. C. G. Golueke and L. F. Diaz, *Organic Wastes for Fuel and Fertilizer in Developing Countries,* report prepared for UNIDO by Cal Recovery Systems, Inc., Richmond, California, 1980.

12. L. F. Diaz, G. M. Savage, and C. G. Golueke, *Resource Recovery from Municipal Solid Wastes,* vol. II, *Final Processing,* CRC Press, Boca Raton, Florida, 1982, pp. 95–120.

13. M. S. Finstein, "Composting Ecosystem Management for Waste Treatment," *Biotechnol.,* **1**:347 (1983).

14. G. A. Kuter, H. A. J. Hoitink, and L. A. Rossman, "Effects of Aeration and Temperature on Composting of Municipal Sludge in a Full-Scale Vessel System," *J. Water Pollut. Control Fed.,* **57**:(4)309–315 (1985).

15. D. Ghosal, L. S. You, D. K. Chatterjee, and A. M. Chakrabarty, "Microbial Degradation of Halogenated Compounds," *Science,* **228**:(4,696)135–142 (1985).

16. W. W. Rose and W. A. Mercer, *Fate of Insecticides in Composted Agricultural Wastes,* Progress Report, Part I, National Canners Association (presently National Food Processors Association), 1968.

17. E. Epstein and J. E. Alpert, "Composting of Industrial Wastes," in *Toxic and Hazardous Waste Disposal,* R. B. Pojasek (ed.), Ann Arbor Science, The Butterworth Group, Ann Arbor, Michigan, 1980, pp. 243–252.

18. U.S. Environmental Protection Agency, *Process Design Manual for Sludge Treatment and Disposal,* EPA 625/1-70-011, U.S. EPA, Center for Environmental Research Information Transfer, 1979, pp. 12-5 to 12-49.

19. G. B. Wilson, J. F. Parr, E. Epstein, P. B. Marsh, R. L. Chaney, D. Colacicco, W. D. Burge, L. J. Sikora, C. F. Tester, and S. Hornick, *Manual for Composting Sewage Sludge by the Beltsville Aerated-Pile Method,* EPA 600/8-80-022, U.S. EPA, Office of Research and Development, Cincinnati, Ohio, 1980.

20. J. Anderson, M. Ponte, S. Biuso, D. Brailey, J. Kantorek, and T. Schink, "Case Study of a Selection Process," in *The BioCycle Guide to In-Vessel Composting,* The J. G. Press, Emmaus, Pennsylvania, 1986, pp. 19–47.

21. G. M. Savage, L. F. Diaz, and C. G. Golueke, "Disposing of Organic Hazardous Wastes by Composting," *BioCycle,* **26**:(1)31–34 (1985).

22. ———, "Biological Treatment of Organic Toxic Wastes," *BioCycle,* **26**:(7)30–34 (1985).

23. *Composting as a Waste Management Alternative for Organic Chemical Wastes,* Phase I, Final Report, unpublished report prepared by Cal Recovery Systems, Inc., Richmond, California, for the U.S. EPA Office of Research and Development, Washington, D.C., May 1985.

24. M. E. Mount and F. W. Oehme, "Carbaryl: A Literature Review," *Residue Reviews,* **80**:1–64 (Springer-Verlad New York) (1981).

SECTION 9.4

LAND TREATMENT OF HAZARDOUS WASTES

Tan Phung
Senior Project Scientist
Woodward-Clyde Consultants
Santa Ana, California

Land treatment is a waste treatment and disposal process whereby a waste is mixed with or incorporated into the surface soil and is degraded, transformed, or immobilized through proper management. Synonyms include *land cultivation, landfarming, land application,* and *sludge spreading.* Compared to other land disposal options (landfills and surface impoundments), land treatment carries lower long-term monitoring, maintenance, and potential cleanup liabilities. Because of the potentially lower liabilities and relatively low initial and operating costs, land treatment has received considerable attention as an ultimate disposal alternative. Unlike landfills, where waste is buried in subsurface cells, land treatment uses the surface soil as a treatment medium and is based primarily on the principle of aerobic decomposition for organic hazardous-waste constituents. Thus, proper management of the treatment zone and monitoring of the unsaturated zone are keys to the effectiveness of land treatment and protection of air and water resources.

In a 1980 survey, 197 land treatment facilities treating hazardous wastes were identified in the United States.[3] Over half of these facilities were at petroleum refineries, and about 15% were associated with chemical production. Geographically, land treatment facilities were concentrated in the southeast (from Texas to the Carolinas), with a few scattered in the Great Plains and far west regions.

The key issue in the current standards for land treatment[10] is the treatment program, which covers (1) the wastes to be disposed, (2) the design and operating measures, and (3) the monitoring program for the unsaturated zone (as a land disposal facility, a land treatment site is also required to have a groundwater-monitoring program). In addition, a treatment demonstration is required to establish that the hazardous constituents of the waste can be completely degraded, transformed, or immobilized within the treatment zone. Site monitoring, proper closure and postclosure care, and record keeping are also required. In certain states such as California that require stringent groundwater protection measures (e.g., double liners and a leachate-collection system) and air-quality monitoring, land treatment becomes a less attractive alternative.

Industrial-waste engineers have questioned the acceptability of land treatment as a disposal alternative for hazardous wastes because of the lack of firm design criteria.[6] There has been a lack of research funding from the U.S. Environmental Protection Agency (EPA) and private industry to further explore the land treatment option for many industrial hazardous wastes other than refinery wastes. Even so, many waste streams that are currently disposed of by incineration or in landfills may be candidates for land treatment. Examples are some spent halogenated solvents and pesticides that would require pretreatments to enhance their biodegradability. Large quantities of soil contaminated with petroleum hydrocarbons (gasoline, diesel fuel, jet fuel) from leaky underground storage tanks can also be land-treated with considerable cost savings when compared with disposal in landfills.

This section describes current land treatment technology; discusses site and waste evaluation; summarizes design and operating criteria, and advantages and disadvantages of land treatment; and assesses the potential environmental effects of land treatment. Much of the section derives from previously published documents.[1,5,6,7,11]

9.4.1 DESCRIPTION OF THE TECHNOLOGY

If properly managed, soil may serve as an effective treatment and disposal sink for many hazardous wastes. Land treatment is a dynamic, management-intensive process involving waste, site, soil, climate, and biological activity as a system to degrade or immobilize waste constituents. In land treatment, the organic fraction must be biodegradable at reasonable rates so as to minimize environmental problems associated with migration of hazardous-waste constituents. Some principal factors affecting the operation of the system are discussed in this subsection.

Waste Characteristics

Biodegradable wastes are suitable for land treatment. Radioactive waste; highly volatile, reactive, or flammable liquids; and inorganic wastes such as heavy metals, acids and bases, cyanide, and ammonia are not considered candidates for land treatment.

Laboratory procedures are available for measuring the treatability of a waste. Laboratory studies using pure cultures are useful; however, the data can be misleading when improperly extrapolated to predict leachability or biodegradability of organic chemicals under field conditions. Land treatability of organic constituents often follows a predictable pattern for similar types of compounds. Chemical structure, molecular weight, water solubility, and vapor pressure are a few characteristics that determine the ease of biodegradation.

Soil Characteristics

The rate of biodegradation and leaching of the waste applied, the availability of nutrients and toxicants to microorganisms, and the fate of hazardous-waste constituents are determined to a large extent by application rate as well as by the soil's chemical and physical characteristics or reactions. Principal soil characteristics affecting land treatment processes are pH, salinity, cation-exchange capacity (adsorption), redox behavior, texture, aeration, moisture-holding capacity, internal drainage, and soil temperature. Some of these characteristics can be improved through application of soil amendments (e.g., nutrients, lime), tillage, and irrigation, or through adjustments of loading rate, frequency, and time of waste application.

Microorganisms

Soil normally contains large numbers of diverse microorganisms, consisting of several groups, that are predominantly aerobic in well-drained soils. The types and populations of microorganisms present in a waste-amended soil depend on the soil's moisture content, available oxygen, nutrient composition, and other characteristics.

The principal groups of microorganisms present in surface soils are bacteria, actinomycetes, fungi, algae, and protozoa.[1] In addition to these groups, other micro- and macrofauna, such as nematodes and insects, are often present.

The bacteria are the most numerous and biochemically active group of organisms, especially at low oxygen levels. The genera of bacteria most frequently isolated from soil include *Arthrobacteri, Bacillus, Pseudomonas, Agrobacterium, Alcaligenes,* and *Flavobacterium.*[1]

Actinomycetes usually predominate in soil of low moisture content and limited nutrient supply. They are heterotropic organisms that use organic acids, lipids, proteins, and aliphatic hydrocarbons.

Fungi often grow vigorously after initial decomposition. They have extensive mycelial branching and compete effectively with bacteria for simple carbohydrates and bioproducts. One of the major activities of fungi in the mycelial state is the degradation of complex molecules. Fungi can withstand a wide range of pH and temperatures.

Soil animals make up a large proportion of soil microorganisms. The more prominent ones with respect to waste degradation are the protozoa, nematodes, and earthworms. All soil animals may assist during waste decomposition. Earthworms are important in mixing organic waste with soil. This mixing improves soil structure, aeration, and fertility. However, some nematodes may impede waste degradation, by feeding on bacteria, fungi, algae, protozoa, and other nematodes.

Waste Degradation

Generally speaking, conditions favorable for plant growth are also favorable for the activity of soil microorganisms. The factors affecting waste degradation may be adjusted in the design and operation of a land treatment facility, as described here.

Soil pH.   The optimum pH for bacterial growth is near 7. This pH is generally maintained by liming to promote microbial activity and to immobilize heavy metals in the waste.

Soil Moisture Content. Decomposition is typically not restricted by moisture if the soil moisture content is maintained above a certain minimum (usually between 30 and 90% of the water-holding capacity of the soil). However, excess water reduces the available oxygen levels and can retard microbial decomposition of the waste applied. Soil moisture content can be adjusted through irrigation or timing of waste application.

Soil Temperature. Soil microbial activity has been shown to decrease greatly at 10°C and essentially to cease at 5°C.[4] Thus, waste application rates require seasonal adjustments in cold regions. Decomposition of oily waste may be expected in cold regions, but at a rate perhaps 15% below that possible in the southern United States.[9] Soil temperature may be raised by such methods as covering the surface with heat-absorbing material and linking the process with heat-generating sources.

Nutrients. Soil microorganisms require basically all nutrient elements required for the growth of higher plants. When a highly carbonaceous waste is applied to the soil, inorganic nitrogen is rapidly consumed by the microorganisms, eventually resulting in slowdown of the degradation. Additional nitrogen must therefore be supplied to attain optimum waste decomposition rates. Other nutrients such as phosphorus, potassium, sulfur, and trace elements are usually present in adequate quantities in most organic wastes or soil to satisfy the needs of the majority of soil microorganisms.

Other Factors. There are a number of design and operating parameters that can be modified to enhance microbial degradation. Modification techniques include (1) repeated application of small amounts of waste, (2) spiking with laboratory-prepared bacterial culture, (3) addition of limited amounts of organic matter (e.g., sewage sludge), and (4) pretreatment of recalcitrant waste constituents to render a waste more amenable to degradation.

9.4.2 EVALUATION OF SITES AND WASTES

Potential candidate sites and waste streams intended for land treatment must be assessed so as to minimize potential adverse environmental effects and risks to public health and safety. The assessment results are used in planning for site design, operation, and closure. During operation, management practices can be adjusted to increase treatment effectiveness and to ensure that human health and the environment are adequately protected.

Site Characterization

As with other land disposal facilities, assessments of candidate sites for land treatment should include regional and site geology, hydrology, topography, soil, climate, and land use.

Land treatment facilities should not be located in an aquifer recharge zone or within 200 ft of a fault which has had displacement since the Holocene Epoch. Current EPA standards[10] require the treatment zone to be less than 4.5 ft deep and at least 3 ft above the seasonal water table. Although topography can be modified to some extent by grading, the site should not be so flat as to avoid ponding nor so steep as to cause excessive erosion and runoff. The site should not be subject to flooding or washout.

As a treatment medium, the soil should be evaluated in terms of its assimilative capacity (retention and degradation), and the potential for erosion and leaching of hazardous-waste constituents. In general, land treatment facilities should not be established on deep, sandy soil because of the potential for groundwater contamination. Silty soils with crusting problems are undesirable because of the potential for excessive runoff.[11] Generally, soils suitable for land treatment are loam, sandy clay loam (ML, SC), silty clay loam, clay loam (CL), and silty clay (CH).[11] Soil properties that should be evaluated include soil depth, texture, drainage, pH, organic matter, soluble salts, cation-exchange capacity, moisture-holding capacity, and microbial counts.

Wind, temperature, and rainfall are three variables of climate that are generally considered in site selection. Careful design and well-planned management can overcome most climatic constraints.

Existing and future land use for the site of the land treatment facility should be evaluated in the planning stages. Zoning restrictions, potential effects on environmentally sensitive areas, proximity to existing or planned developments, and effects on the local economy should all be addressed.

Waste Characterization

Initially, the waste stream considered for land treatment can be evaluated using available in-house data, such as percent moisture or oil, waste constituents present, raw materials used, or products and by-products generated. However, the waste stream should be analyzed in greater detail if possible. The waste analysis may include density, percentage solids (or oil), pH, salinity, known hazardous constituents, heavy metals, EPA organic priority pollutants, EP toxicity, and acute toxicity (microbial toxicity and phytotoxicity). If more than one waste stream is considered, compatibility of the waste streams must be evaluated. To determine the acceptability of a waste for land treatment, it is necessary to evaluate the results of the waste analysis in conjunction with laboratory experiments and/or field plot studies on waste degradability, sorption, and mobility in site soil, and incorporate them into the final design and monitoring program.

The operator may choose to modify in-plant processes or pretreat the waste stream for volume reduction, odor reduction, or elimination of certain hazardous constituents to improve treatability.

9.4.3 ADVANTAGES AND DISADVANTAGES OF LAND TREATMENT

The advantages of land treatment are summarized as follows:

- The waste is degraded, transformed, or immobilized; thus, the long-term liability is lower than for other land disposal options.
- The treatment area is continually monitored; thus, remedial action can be taken immediately when there are signs of waste migration from the treatment zone.
- The cost for land treatment is lower than for landfills and incineration.

- The closed land treatment site can be converted to beneficial uses, such as parks and playgrounds.

The disadvantages of land treatment are:

- Waste storage will be required periodically for reasons such as inclement weather and equipment breakdown.
- Land treatment is land-intensive and management-intensive.
- Air emissions and odors are a source of nuisance and probably a health hazard.
- There is potential for adverse environmental impacts if the facility is not properly designed and managed.
- Site selection and permitting can be time-consuming and costly; delays and high costs result from public opposition and lengthy negotiations with and approval processes of various regulatory agencies.
- Land treatment is suitable only for selected wastes.

9.4.4 DESIGN AND OPERATING CONSIDERATIONS

Land treatment is a management-intensive endeavor; to be effective and environmentally sound, the land treatment facility must be properly designed and operated. Figure 9.4.1 shows some design and operating considerations. A few of the more salient features are discussed below.

Area and Facility Layout

Land treatment generally requires the commitment of a large parcel of land to accommodate the treatment area, buffer zone, roads, waste storage, equipment storage, retention pond, and on-site structures. For this reason, the acreage available plays a significant role in the layout and in many design and operating features of the facility. The land treatment area can have either a single or multiple

• Land requirements	• pH adjustment and nutrient supply
• Facility layout	• Frequency and rate of waste application
• Access road	• Application and mixing methods
• Equipment selection	• Contingency plan
• Run-on and runoff control	• Vegetation
• Erosion control	• Site security and inspection
• Odor and air-emission control	• Site monitoring
• Waste storage	• Record keeping
• Land preparation	• Closure and postclosure care

FIG. 9.4.1 Design and operational considerations.

plots, depending upon the amount of waste to be disposed, waste and soil characteristics, disposal schedule, and climatic conditions. In the multiple-plot system, the plots are treated sequentially, tilled, and then revegetated. This sequence is planned to maximize the efficiency of the erosion- and runoff-control systems.

Equipment

A wide variety of equipment can be used for land treatment of waste. Equipment selection depends primarily on the characteristics of the waste applied, site conditions, and the constraints imposed by regulations. For example, tank wagons may be adequate for surface application of sludge at some facilities, whereas subsurface-injection equipment may be required by local regulations (e.g., for odor control) at other facilities. Many types of agricultural equipment, such as the manure spreader and rototiller, are suitable for waste application and disking; other equipment, such as sludge injectors, is designed specifically for subsurface injection of sludge. Care must be exercised to obtain compatible implements (adaptors).

Water Management

Systems for control of run-on and runoff are crucial to the protection of natural waters and the effectiveness of land treatment as these systems are designed to avoid saturation of the treatment zone or washout of the hazardous constituents. Typical designs for run-on control are earthen berms or ditches that intercept and redirect the flow of surface water. Runoff can be controlled by using diversion structures to channel water to a retention pond or ponds at one end of the site. Runoff is either stored, or treated and released under a National Pollutant Discharge Elimination System (NPDES) permit. The retention pond must be designed to contain runoff from the 25-year, 24-h return period storm. It should also have an emergency or flood spillway.

Soil Erosion

To prevent hazardous-waste contaminants from moving off site, control of soil erosion caused by wind or water is generally implemented at land treatment sites. Wind erosion can be minimized by maintaining a vegetative cover and moist soil. Soil erosion caused by water flow can be minimized by proper construction terracing and by maintaining vegetated waterways. Air emissions can be controlled by subsurface injection, use of odor-control chemicals, watering, or planting a windbreak.

Waste Application

Waste-loading specifications, including application rate, frequency, and schedule, as well as methods of waste application and incorporation, are generally a part of the overall management plan. Although these specifications may change in a given year as a result of inclement weather, they will be developed through integration of the volume and characteristics of waste generated, site and climatic conditions, feasibility studies, and regulatory constraints. Application rates and annual loading rates are likely to be determined by the most limiting constituent or constituents (e.g., water content, degradation rate of an organic constituent, metal loading). Optimum degradation rates are often achieved when small waste applications are made at frequent intervals. Some hazardous wastes are "foreign" to soil microorganisms, and heavy doses may cause microbial toxicity or genetic damage. A waste should be applied as uniformly as possible at a design rate over the intended treatment area. Costs, effectiveness, service life, and compatibility of implements are a few considerations when choosing application and incorporation methods.

Facility Inspection and Record Keeping

The land treatment facility should be inspected weekly for diversion structures and regularly for equipment. In addition, the treated plot should be inspected daily for uniformity of waste spreading and incorporation and for adequate freeboard in the retention pond. Careful records should be maintained of waste disposed in each plot, monitoring results, climatic data, maintenance schedule, and accidents. A contingency plan should be developed to address preparedness for operational and environmental emergencies.

Site Monitoring

Current federal standards on land treatment require the monitoring of groundwater, of the unsaturated zone (soil cores and soil-pore liquid), and of waste to be applied.[9] A summary of site-monitoring requirements is presented in Table 9.4.1. Monitoring of air quality, vegetation, runoff water, and soil in the treatment zone is not regulated by the EPA, but is important to ensure an environmentally sound practice and to reduce liability risks.

Closure and Postclosure

When the waste-assimilating capacity of the site soil is reached, the land treatment facility must be properly closed. Site management and monitoring continue during site closure until the organic constituents are sufficiently degraded. In some cases, the treatment zone is removed and contaminated soil is hauled to an off-site disposal facility. The site should be graded and revegetated. Management and monitoring activities are reduced during the postclosure period, which can last as long as 30 years. Postclosure care is intended to complete waste treatment and stabilization of the remaining soil and waste residuals while checking for any unforeseen long-term changes in the system.

9.4.5 POTENTIAL ENVIRONMENTAL IMPACTS

Soil, a natural acceptor of wastes, has been viewed as a physical, chemical, and biological filter that can effectively deactivate, decompose, or assimilate a wide

TABLE 9.4.1 Monitoring Program at Land Treatment Facilities

Medium to be monitored*	Purpose	Sampling frequency	Parameters to be analyzed
Waste	Quality change	Quarterly composites if continuous stream; each bath if intermittent generation	At least rate- and capacity-limiting constituents, plus those within 25% of being limiting, principal hazardous constituents, pH, and EC
Soil cores	Determine slow movement of hazardous constituents	Quarterly	All hazardous constituents in the waste or the principal hazardous constituents, metabolites of hazardous constituents, and nonhazardous constituents of concern
Soil-pore liquid	Determine highly mobile constituents	Quarterly, preferably following leachate-generating precipitation or snowmelt	All hazardous constituents in the waste or the principal hazardous constituents, mobile metabolites of hazardous constituents, and important mobile nonhazardous constituents
Groundwater	Determine mobile constituents	Semiannually	Hazardous constituents and metabolites or select indicators
Vegetation (if grown for food-chain use)	Phytotoxic and hazardous transmitted constituents (food-chain hazards)	Annually or at harvests	Hazardous metals and organics and their metabolites
Runoff water	Soluble or suspended constituents	As required for NPDES permit	Discharge permit and background parameters plus hazardous organics
Soil in the treatment zone	Determine degradation, pH, nutrients, and rate- and capacity-limiting constituents	Quarterly	
Air	Personnel and population health hazards	Quarterly	Particulates (adsorbed), hazardous constituents, and hazardous volatiles

*Current EPA standards require monitoring for only the first four media (Ref. 9).
Source: Ref. 10.

range of waste materials. Factors affecting this assimilative capacity must be understood and considered in order to develop sound management systems for land treatment. It is desirable to predict the behavior of the proposed waste within the soil system at the outset of a land treatment operation to predetermine the impact of the practice on the receiving environment. However, predicting soil-waste interactions is difficult because of the inherent variability of the waste and of the treatment medium (soil), and the unpredictable influence of climatic variability. Without a detailed analysis of land treatment components and a sound design and management plan, adverse environmental effects can result from land treatment practices.

Water Quality

Operators of land treatment facilities must adequately protect the surface and subsurface waters from contamination with hazardous-waste constituents or by-products from waste degradation. Waste constituents that are not degraded by microorganisms, transformed, or immobilized in soil may leach to the groundwater. Runoff resulting from excess waste application or heavy rains may carry the constituents and contaminated sediments to nearby streams and lakes.

Incorporation of wastes into the surface soil may result in soluble constituents moving downward by leaching. Thus, it would be a race between mobility and degradability of organic waste constituents. Under current federal regulations, the seasonal groundwater table beneath a land treatment facility can be as shallow as 8 ft below grade, or about 3 ft below the treatment zone. In some cases, an impermeable layer with a leachate-collection system placed a few feet below the treatment zone can prevent waste constituents or their metabolites from contaminating the groundwater.

Surface-water contamination resulting from improper water management or damaged diversion structures and retention ponds is probably a major environmental concern. In a land treatment operation, wastes are concentrated in the soil surface; the concentration of waste constituents in the runoff water may be sufficiently high to have deleterious effects on certain trophic levels in the aquatic ecosystem. Waste constituents may be transported as particulate, dissolved, or bound on eroded soil particles. Surface runoff water should be monitored to meet NPDES permit requirements.

Air Quality

Malodorous emissions from organic wastes can have detrimental aesthetic and economic effects on a community, as well as negative effects on the mental and physical health of nearby residents. Air quality at a land treatment site can be impaired by odors, dust, volatile substances, and aerosols emitted from waste-spreading and incorporating processes. Many industrial hazardous wastes contain heavy metals and organic substances that are highly toxic in particulate or volatile form. The level of volatile compounds present in the air is not necessarily related to the severity of the odor problems. There are insufficient data to fully assess the potential adverse effects of land treatment practices on air quality.

Human Health and Safety

Harmful effects associated with land treatment of hazardous wastes can be of a biological, chemical, physical, mechanical, or psychological nature. Hazards associated with field operations include those linked to equipment operations, fire and explosion, dust and aerosols, and odor problems. Vegetation grown on the treatment area may absorb hazardous-waste constituents (e.g., cadmium, pesticides) through its root system. These constituents may pose a health hazard if they enter the food chain.

9.4.6 REFERENCES

1. M. Alexander, *Introduction to Soil Microbiology,* 2d ed., Wiley, New York, 1977.
2. *A Survey of Existing Hazardous Waste Land Treatment Facilities in the United States,* prepared for the U.S. EPA by K. W. Brown and Associates, Inc., College Station, Texas, under EPA Contract No. 68-03-2943, 1981.
3. M. Bluestone et al., "Microbes to the Rescue–Hazardous Waste Biotreatment Fights for Recognition," *Chemical Week,* October 29, 1986, pp. 34–40.
4. J. T. Dibble and R. Bartha, "Effect of Environmental Parameters on Biodegradation of Oil Sludge," *Appl. Environ. Microbiol.,* **37**:729–738 (1979).
5. W. H. Fuller and A. W. Warrick, *Soil in Waste Treatment and Utilization,* vol. 1, *Land Treatment,* CRC Press, Boca Raton, Florida, 1984.
6. M. R. Overcash and D. Pal, *Design of Land Treatment Systems for Industrial Wastes—Theory and Practice,* Ann Arbor Science, Ann Arbor, Michigan, 1979.
7. J. F. Parr, P. B. Marsh, and J. M. Kla, *Land Treatment of Hazardous Wastes,* Noyes Data Corp., Park Ridge, New Jersey, 1983.
8. T. Phung et al., *Cultivation of Industrial Wastes and Municipal Solid Wastes: State-of-the-Art Study,* EPA-600/2-78-140a, U.S. EPA Office of Research and Development, Hazardous Waste Engineering Research Laboratory, Cincinnati, Ohio, August 1978.
9. M. J. Rowell, "Land Cultivation in Cold Regions," in *Disposal of Industrial and Oily Sludges by Land Cultivation,* Resource Systems and Management Association, Northfield, New Jersey, 1980, pp. 249–264.
10. U.S. Environmental Protection Agency, "Hazardous Waste Management System; Permitting Requirements for Land Disposal Facilities," *Federal Register,* vol. 47, no. 143, pp., 32,274–32,388, July 26, 1982.
11. U.S. Environmental Protection Agency, *Hazardous Waste Land Treatment,* rev. ed. EPA SW-874, U.S. EPA Office of Solid Waste and Emergency Response, Washington, D.C., April 1983.

SECTION 9.5

BIODEGRADATION OF ENVIRONMENTAL POLLUTANTS

P. R. Sferra, Ph.D.
Biologist
U.S. Environmental Protection Agency
Hazardous Waste Engineering Research Laboratory
Cincinnati, Ohio

The synthesis of molecules useful for subsistence has resulted in a lifestyle of convenience, protection, health, and amusement heretofore unequalled but at the same time burdened with the danger of harm to people and the environment. Some of this danger is due to pollution from chemical syntheses and other manufacturing processes. Only recently has there been a directed effort toward controlling rates of pollution and eliminating the pollution already in place.

Among the methods considered for the destruction of hazardous and toxic wastes is the use of living organisms or their products to detoxify or destroy the polluting chemicals by means of a process known as biodegradation. In biodegradation there is the potential for low-cost, highly effective treatment with little or no consequential harm to the environment. These methods are part of what is known today as *biotechnology,* a term of increasing popularity meaning "the application of organisms, biological systems or biological processes."[10]

Concurrent with the gain in notoriety of hazardous wastes to the point where a crisis atmosphere now exists, there has been a series of major discoveries in the biological sciences leading to a part of biotechnology now known as *gene engineering.* Techniques in microbiology, biochemistry, molecular genetics, and other disciplines have been combined in gene engineering to rearrange or recombine genetic material. Varied potential benefits can be derived from the application of genetically engineered organisms. These include control of frost damage to crops, increased herbicide and disease resistance, and crops with improved nutritional qualities. Biological processes responsible for mineral extraction, oil recovery, and conversion of biomass into energy can be improved through gene-engineering applications.

Biological processes are also capable of degrading toxic or hazardous chemicals in waste sites and in aqueous waste streams. Success in the development of biodegradation technology for contaminated soils and industrial waste streams is expected to bring cost-effective and environmentally safe cleanup processes to a world beset with hazardous chemicals.

9.5.1 HAZARDOUS-WASTE APPLICATIONS

The different biological processes or biotechnologies can be classified as being either conventional or emerging. Among the conventional systems are (1) aerobic treatments such as activated sludge, aerated lagoons, trickling filters, and rotating biological contactors and (2) anaerobic treatments, of which there are three main types: anaerobic digestion, anaerobic contact, and anaerobic fixed-film.[26] While the active agents in these systems are microorganisms producing enzymes that are catalyzing the reactions going on in the systems, characterization of the systems by identification of the species and strains is usually not practiced.

The search for useful biological agents to destroy hazardous wastes consists of an assortment of methodologies ranging from attempts to isolate pure strains from real-life contaminated situations to sophisticated genetic-engineering applications to produce a tailor-made organism capable of degrading a specific compound. Some of these methodologies are innovative or emerging, e.g., the construction of gene-engineered organisms, and others employ conventional methodologies, e.g., the discovery that a naturally occurring organism, white rot fungus, has the capability to degrade an assortment of chlorinated aromatic compounds. A major difference between the aforementioned conventional methodologies and this type of conventional methodology is the degree of identification and characterization of the organism or organisms involved.

9.5.2 TECHNICAL DESCRIPTION

Biodegradation has been defined as "the molecular degradation of an organic substance resulting from the complex action of living organisms" by the Biodegradation Task Force, Safety of Chemicals Committee, in Brussels in 1978.[24] Several recent general references covering biodegradation of hazardous-waste compounds are available for more detailed information.[1,8,20,21,24,25,27] The hazardous-waste problem is mainly one of organic chemicals polluting the environment and in association with this environmental problem a few relevant terms have shown increased usage. Three which have become popular are *xenobiotic, anthropogenic,* and *recalcitrant.* Hutzinger et al.[12] have carefully explained usage and meaning of the terminology of microbial degradation.

Each of these terms can be applied to the polluting chemicals we are concerned with. Because of their *recalcitrance,* or persistence, there is no simple or inexpensive way to get rid of them. The use of the term *anthropogenic* needs some explanation in that while in usage it refers to chemicals produced by human beings it really means "man-making." In purely etymological terms, an anthropogenic chemical would be one that "makes" a human being. The most important term of the three in the context of the subject of biodegradation is *xenobiotic,* which translates as "stranger to life." A *xenobiotic compound* would be one that is either a chemical substance which is not produced by any natural enzymatic process or a compound having unnatural structural features. It is a "chemical out of place in the biotic world." Natural products occurring in concentrations increased as a result of human activity are also considered to be xenobiotic.

We have in the environment then two kinds of chemicals, xenobiotic and natural. The products of the industrial chemical revolution include many chemical structures which can be considered foreign to living organisms. What matters is that living organisms have become exposed to chemicals for which there has been no evolutionary experience. There could be several possible consequences. When xenobiotic compounds are released into the environment they are subjected to the processes of adsorption, sedimentation, and chemical reactions including biotransformation and photolysis. Biodegradation is a *biotransformation* process and biotransformation is whatever the living organism can do to the compound. The organism may not get the chance to do anything if the compound is toxic and interferes with some step in the organism's life-sustaining processes.

Any given population of a species, however, will exhibit among its members variations in species characteristics. This phenomenon is the basis of the theory of natural selection. These variations range from the obvious and easily measurable, such as size, shape, weight, color, and speed, to variations in metabolic capability. Consider an organism a chemical machine that has inherited the instructions for producing all the enzymes, or catalysts, for all the chemical reactions it will need to sustain life and pass these instructions on to the next generation. These chemical reactions collectively are referred to as *metabolism.*

The organism then is an energy-processing machine tearing down molecules and converting their bond energy into biologically useful energy to drive the reactions required for growth, maintenance, and reproduction. The building processes of metabolism are referred to as *anabolism* and the tearing down processes are referred to as *catabolism.* The catabolic reactions are the source of energy and building blocks, e.g., carbon skeletons, for the molecules the organism needs to make for growth and all its other functions. The raw materials for the catabolic reactions generally are from an outside source. Each species has its own particular ecological niche and this niche consists of everything the organism requires to maintain life—actually to maintain the species. The species has its own temperature, humidity, space, pH, light, and other requirements. These make up its niche. Included also would be a food or nutrient source for energy, building blocks, and certain compounds that are essential but cannot be synthesized by the organism. The chemical processes are enzyme-catalyzed and the enzymes are coded for in the DNA the organism inherited. So life for the species will proceed without difficulty as long as its niche is intact or complete or undisturbed.

The appearance of a xenobiotic compound could be considered a perturbation of the niche. If the compound is toxic and in high enough concentration then the population conceivably could be wiped out. If it enters in lesser concentrations, only some of the population might lose out and the survivors make it because they are "resistant" or less "susceptible." What is likely is that the survivors are variants with the capacity to process the xenobiotics; i.e., they have inherited the metabolic machinery to neutralize, dechlorinate, detoxify, or even mineralize the xenobiotic. They may get nothing out of the process other than survival or they may be opportunists and be capable of getting energy or building blocks out of the xenobiotic by tearing it down. In other words, in microbial metabolism of xenobiotic compounds the chemical that is metabolized may support growth and serve as a source of carbon or other elements and energy. Alternatively the chemical may be metabolized but not serve as a source of nutrients; the process by which this takes place is known as *cometabolism.*

Microorganisms adapt to degrade new synthetic compounds either by utilizing catabolic enzymes they already possess or by acquiring new metabolic pathways. In pollution control the need is for the organism that can process the xenobiotic. The promise is very attractive especially if the organism can be released on a hazardous-waste site, destroy all the unwanted pollutants, cause no environmental damage, and be so reliant upon the pollutant that by destroying the pollutant it is eliminating its sole means of survival and will eventually leave no trace of itself. It should also be cheap to produce and safe to handle.

The number of factors upon which the success of such applications of biodegradation depends seemingly makes the task insurmountable. These factors include environmental conditions, such as nutrient availability, light, temperature conditions, pH, and the presence or absence of oxygen. The microorganisms being developed must have specific available or inducible enzymes to effectively degrade the xenobiotic compound and they must exist with specific characteristics and in sufficient quantity to survive and be able to operate in the environment being treated. Of importance also are the characteristics of the structure of the polluting compound upon which the biodegradation processes must work. What is needed, then, is the right organism to work on the right compound under the right environmental conditions.

Among the innovative or emerging methods are techniques to develop microorganisms specifically suited to degrade hazardous wastes such as chlorinated aromatic compounds. One of these methods, called *plasmid-assisted molecular breeding,* is used to accelerate selection for strains of bacteria with the desired biodegradation capability.[17] It has already resulted in a strain able to exist with 2,4,5-T as its sole carbon source. This pure culture is the bacterium *Pseudomonas cepacia* with the strain designation AC1100. This effort commenced with the inoculation into a chemostat system of soil samples from dump sites and selected strains of bacteria harboring a variety of plasmids. By manipulating the substrate concentrations and other conditions in the chemostat, in time the mixed culture and subsequently the pure culture developed with the ability to utilize 2,4,5-T as its sole carbon source. This technique does not involve gene isolation, cutting, and splicing of the type done in the so-called gene-engineering procedures. It is, however, a way of recombining DNA through the action of the plasmid vectors. It is a way of accelerating the evolution of an organism to a form with desirable traits. The ingredients for the process were combined in the chemostat, conditions under which the chemostat operated were carefully controlled and adjusted, and the process of adaptation happened. This was, in reality, a means of applying strong selection pressure on the microorganism in the chemostat and what resulted was the AC1100 strain that had behaved as an opportunist in adopting useful genetic instructions available in the plasmid assortment combined with some of its own instructions to enable it to metabolize the 2,4,5-T molecules.[7,8,9,15,16,17,18,19]

An alternative to the natural recombination that might occur in plasmid-assisted molecular-breeding process would be the use of gene-engineering techniques where the researcher directly affects the genotype of the organism. An example would be the work being done on yeast by Loper et al.[13,14,22,23] Here the group is subjecting yeast to recombinant-DNA procedures to produce a strain of *Saccharomyces cerevisiae* or some other yeast species with the degradation mechanism found in another type of organism. This work is based upon the knowledge that the problem compounds, once they have entered the mammalian body, are capable of inducing changes in gene expression which result in changes in metabolic enzymes in the organism. Some types of these induced enzymes catalyze metabolic activities which detoxicate the problem compounds. The enzymes of interest in this work are complex oxidative enzymes known as the cytochrome P-450 monooxygenases. The research approach is to isolate and identify any gene and its enzyme product possessing the ability to inactivate specific compounds, clone this gene, and then establish it in a microorganism, such as yeast, which could be cultivated for field treatment of a hazardous-waste site or for use in an industrial waste-treatment process.

The organism of choice is the yeast *Saccharomyces cerevisiae* because the formal genetics and recombinant-DNA techniques are quite advanced for this species. Yeasts, rather than bacteria, are chosen because the yeasts are eucaryotic organisms as are mammals and yeasts contain particulate P-450 systems.

In order to accomplish the objectives, the team must be able to manipulate genes in yeast, specifically the genes coding for the P-450 enzymes. They must also know the what and where of the regulatory genes that govern the genes coding for the P-450 enzymes and the signals affecting these regulatory genes. Another phase of this project is aimed at obtaining mammalian gene sequences for P-450 expression in a form useful for testing in yeast. In addition, classic mapping will determine the inheritance of the P-450 stability-lability trait. This will help in gaining an understanding of the DNA signals for modifying expression of desired traits.

The overall strategy then of this project is first to establish the gene-engineered control of a yeast P-450 gene in yeast and then to establish any available P-450 gene sequence from higher organisms for expression in yeast.

Other research is under way to develop microorganisms capable of destroying hazardous organic compounds under anaerobic conditions. One procedure is to isolate from aqueous environments microorganisms which dehalogenate organochlorine compounds under anaerobic conditions. Promising isolates would be characterized and from these isolates plasmids containing genes responsible for anaerobic dehalogenation would be isolated and characterized. Then, using genetic manipulative techniques, strains will be constructed which exhibit improved anaerobic dehalogenation activities against single and multicomponent waste systems.

There remains also the option to investigate the ability of naturally occurring organisms to degrade toxic chemical pollutants. One technique involves a process known as *enhancement of indigenous microorganisms,* whereby a contaminated site possibly can detoxify itself after human intervention to add whatever is needed to optimize the biodegradation process. This technique depends upon the biodegrading organisms' presence in situ; the human intervenor attempts to accelerate the process the surviving (successful) indigenous organisms utilize in adapting to a potentially threatening life situation.

A very promising naturally occurring organism with high degradation potential is the white rot fungus, *Phanerochaete chrysosporium.*[3,4,4a,5] The degradation potential of a product of white rot fungi is being investigated by several research teams. This product is a ligninolytic enzyme system, a powerful oxidant, which may prove effective as a degrader of xenobiotic compounds.

A feasibility study to determine the effectiveness of white rot fungi to degrade chlorinated aromatic pollutants has shown that in liquid culture this type of organism will degrade DDT, DDE, Lindane, polychlorinated biphenyls, Chlordane, chlorodioxins, benz[a]pyrene Mirex, and PCP to CO_2.[4,4a] This activity is accomplished by means of the lignin-degrading mechanism the organism already possesses. Factors affecting the rate of degradation were found to be the atmospheric oxygen concentration, the concentration of nutrient nitrogen, and the source of carbon. The pollutants studied were tagged with carbon-14 and any radioactive CO_2 was measured by liquid-scintillation spectrometry.

Data from this study indicate that this fungus oxidizes, by means of its lignin-degradation system, a broad spectrum of recalcitrant organopollutants. Thus, treatment with this organism, *P. chrysosporium,* fortified with a suitable carbohydrate source, under nitrogen-limiting conditions, may prove to be an effective and economical means for the detoxification or destruction of hazardous chemical wastes.

Another study has confirmed the mineralization of polychlorinated biphenyls by *P. chrysosporium*.[6] In this work four other species of white rot fungi and four other species of microorganisms were tested. A radioactive Aroclor 1254 at a concentration of 300 ppb was used as the pollutant substrate and after five weeks 7 times more CO_2 was released in the *P. chrysosporium* experiment than from any of the others.

9.5.3 ENVIRONMENTAL EFFECTS

In February of 1984 the House Subcommittee on Investigations and Oversight transmitted to the Committee on Science and Technology a staff report on the environmental implications of genetic engineering.[2] This report is based on hearings during which testimony was received from experts representing academia, industry, and government. The issues covered included the benefits and risks associated with the release of organisms with new genotypes into the environment, the potential for predicting the impact of the introduction of a particular organism into a given environment, and the adequacy of the existing regulatory framework for permitting purposes to assess the associated risks and benefits to the extent that the benefits of biotechnology can be realized with minimum risk.

In the finding of this inquiry the benefits were easy to enumerate and included the degradation of toxic and hazardous materials. Risks, however, were less easy to define and depend on the nature of the organism released. Although no detrimental effects of any genetically engineered organisms on an ecosystem have been documented, severe negative and beneficial impacts from newly introduced naturally occurring organisms are well known. The potential risks are best described as "low probability of high-consequence risks"; that is, while there is only a small possibility that damage could occur, that damage that could occur is great.

The testimony also indicated that predicting the specific type, magnitude, or probability of environmental effects associated with deliberate release of genetically engineered organisms will be extremely difficult, if not impossible, at the present time. This is mainly because no historical or scientific data base exists for these phenomena. It is believed, however,, that it would be possible to devise procedures to produce generalized estimates of the probability of environmental damage by, and survival and growth of, a genetically engineered organism, although specific risk assessment may not be achievable.

Finally, the report states that the current regulatory framework does not guarantee that adequate consideration will be given to the potential environmental effects of a deliberate release. No single agency or entity presently has both the expertise and the authority to properly evaluate the environmental implications of releases from all sources. The issue is still under consideration and other meetings have been held. In June, 1985, a cross-disciplinary symposium was sponsored by the American Society for Microbiology to consider the issues of engineered organisms in the environment.[11] Major items addressed included the definition of genetically engineered organisms and how they can be designed for maximum benefits and minimum risk, how the benefits and risks can be assessed, and how the release of genetically engineered organisms can be regulated.

9.5.4   REFERENCES

1. M. Alexander, "Biodegradation of Chemicals of Environmental Concern," *Science,* **211**:132–138 (9 January 1981).

2. *The Environmental Implications of Genetic Engineering,* staff report prepared by the Subcommittee on Investigations and Oversight transmitted to the Committee on Science and Technology, U.S. House of Representatives, U.S. Government Printing Office, Washington, D.C., 1984.

3. J. A. Bumpus and S. D. Aust, "Studies on the Biodegradation of Organopollutants by a White Rot Fungus," presented at the International Conference for New Frontiers for Hazardous Waste Management, 1985.

4. J. A. Bumpus, M. Tien, D. Wright, and S. D. Aust, "Oxidation of Persistent Environmental Pollutants by a White Rot Fungus," *Science,* **228**:1,434–1,436 (21 June 1985).

4a. J.A. Bumpus and S. D. Aust, "Biodegradation of Chlorinated Organic Compounds by *Phanerochaete chrysosporium,* a Wood-Rotting Fungus," in *Solving Hazardous Waste Problems: Learning from Dioxins,* J. H. Exner (ed.), ACS Symposium Series 338, 1987, pp. 340–349.

5. ———, "Biodegradation of Environmental Pollutants by the White Rot Fungus *Phanerochaete chrysosporium,*" in *Proceedings of the Eleventh Annual Research Symposium. Incineration and Treatment of Hazardous Waste,* EPA/600/9-85/028, pp. 120–126.

6. D. C. Eaton, "Mineralization of Polychlorinated Biphenyls by *Phanerochaete chrysosporium:* a Ligninolytic Fungus," *Enzyme Microb. Technol.,* **7**:194–196 (1985).

7. K. Furukawa and A. M. Chakrabarty, "Involvement of Plasmids in Total Degradation of Chlorinated Biphenyls," *Applied and Environ. Microbiol.,* **44**:619–626.

8. D. Ghosal, I. S. You, D. K. Chatterjee, and A. M. Chakrabarty, "Microbial Degradation of Halogenated Compounds," *Science,* **228**(4696):135–142 (1985).

9. ———, "Plasmids in the Degradation of Chlorinated Aromatic Compounds," in *Plasmids in Bacteria,* D. R. Helsinki et al. (eds.), Plenum Press, New York and London, 1985.

10. J. M. Grainger and J. M. Lynch, "Introduction: Towards an Understanding of Environmental Biotechnology," in J. M. Grainger and J. M. Lynch (eds.), *Microbiological Methods for Environmental Biotechnology,* Academic Press, London, 1984, p. 2.

11. H. O. Halvorson, D. Pramer, and M. Rogul (eds.), *Engineered Organisms in the Environment: Scientific Issues,* American Society for Microbiology, Washington, D.C., 1985.

12. O. Hutzinger and W. Veerkamp, "Xenobiotic Chemicals with Pollution Potential," in *Microbial Degradation of Xenobiotics and Recalcitrant Compounds,* T. Leisinger et al. (eds.), FEMS Symposium No. 12, Academic Press, London, 1981, pp. 3–45.

13. V. F. Kalb, J. C. Loper, C. R. Dey, C. Woods, and T. R. Sutter, "Isolation of a Cytochrome P-450 Structural Gene from *Saccharomyces cerevisiae,*" *Gene.*, **45**(3):237–245 (1986).

14. V. F. Kalb, C. W. Woods, T. G. Turi, C. R. Dey, T. R. Sutter, and J. C. Loper, "Primary Structure of the P-450 Lanosterol Demethylase Gene from *Saccharomyces cerevisiae,*" *DNA*, **6**(6):529–537 (1987).

15. J. S. Karns, S. Duttagupta, and A. M. Chakrabarty, "Regulation of 2,4,5-Trichlorophenoxyacetic Acid and Chlorophenol Metabolism in *Pseudomonas cepacia* AC1100," *Applied and Environ. Microbiol.*, **46**:1182–1186 (1983).

16. J. S. Karns, J. J. Kilbane, S. Duttagupta, and A. M. Chakrabarty, "Metabolism of Halophenols by 2,4,5-Trichlorophenoxyacetic Acid-Degrading *Pseudomonas cepacia,*" *Applied and Environ. Microbiol.*, **46**:1,176–1,181 (1983).

17. S. T. Kellogg, D. K. Chatterjee, and A. M. Chakrabarty, "Plasmid-Assisted Molecular Breeding: New Technique for Enhanced Biodegradation of Persistent Toxic Chemicals," *Science,* **214**:1,133–1,135 (1981).

18. J. J. Kilbane, D. K. Chatterjee, et al., "Biodegradation of 2,4,5-Trichlorophenoxyacetic Acid by a Pure Culture of *Pseudomonas cepacia,*" *Applied and Environ. Microbiol.*, **44**:72–78 (1982).

19. J. J. Kilbane, D. K. Chatterjee, and A. M. Chakrabarty, "Detoxification of 2,4,5-Trichlorophenoxyacetic Acid from Contaminated Soil by *Pseudomonas cepacia,*" *Applied and Environ. Microbiol.*, **45**(5):1,697–1,700 (1983).

20. H. Kobayashi and B. E. Rittmann, "Microbial Removal of Hazardous Organic Compounds," *Environ. Sci. Technol.*, **16**:170A–181A (1982).

21. T. Leisinger et al. (eds.), *"Microbial Degradation of Xenobiotics and Recalcitrant Compounds,"* FEMS Symposium No. 12, Academic Press, London, 1981.

22. J. C. Loper, C. Chen, and C. R. Dey, "Gene Engineering of Yeasts for the Biodegradation of Hazardous Waste," *Hazardous Waste and Hazardous Materials,* **2**:131–141.

23. J. C. Loper, J. B. Lingrel, V. F. Kalb, and T. R. Sutter, "Gene Engineering of Yeast for the Degradation of Hazardous Waste," abstract published in *Genetic Control of Environmental Pollutants,* G. S. Omenn and A. Hollaender (eds.), Plenum Press, New York, 1984, p. 375.

24. F. Matsumura and C. R. Krishna Murti in F. Matsumura and C. R. Krishna Murti (eds.), *Biodegradation of Pesticides,* Plenum Press, New York, 1982, preface.

25. G. S. Omenn and A. Hollaender (eds.), *Genetic Control of Environmental Pollutants,* Basic Life Sciences, vol. 28, Plenum Press, New York, 1984.

26. T. O. Peyton, "Biological Disposal of Hazardous Waste," *Enzyme Microb. Technol.,* **6**:146–154 (1984).

27. M. L. Rochkind, J. W. Blackburn, and G. S. Sayler, *"Microbial Decomposition of Chlorinated Aromatic Compounds,"* EPA 600/2-86/090, Cincinnati, Ohio, September 1986.

SECTION 9.6
ENZYME SYSTEMS AND RELATED REACTIVE SPECIES

John A. Glaser, Ph.D

U.S. Environmental Protection Agency
Hazardous Waste Engineering Research Laboratory
Cincinnati, Ohio

Enzymes capable of transforming hazardous chemicals to nontoxic products can be harvested from microorganisms grown in mass culture. The resulting cell-free enzymes can be applied to contaminated matrices—e.g., water, soil—for detoxification treatment. Such crude enzyme extracts derived from microorganisms have been shown to convert pesticide chemicals into less toxic and less persistent products. The parallel industrial use of crude or purified enzyme extracts, in solution or immobilized in reactors, to catalyze a wide range of synthetic reactions, continues to grow at an exceptional pace.[1]

The use of enzymes to detoxify chemical pollutants may appear to be somewhat distant. Such technology is often dismissed as "high tech" and consequently fraught with a myriad of attendant control problems that would prevent its use in the "real world." This viewpoint is based on an inadequate understanding of enzymes and their potentials. As with each control technology there are limitations but with enzymes the assets are overwhelming when compared with strictly chemical means of detoxification.

Enzymes are complex proteins ubiquitous in nature. (See Table 9.6.1 for a summary of the major classes of enzymes and their functions.) The proteins are composed of the naturally occurring amino acids linked together via peptide bonds and in the case of enzymes there are additional chemical bonds which stabilize fixed macromolecular geometries that are implicated in enzyme action. Every living organism has countless enzyme systems performing dedicated tasks to maintain the well-being of the organism. The importance of enzymes to living or-

TABLE 9.6.1   International Enzyme Classification

Enzyme type	Function
1. Oxidoreductase	Enzymes of this group catalyze oxidation-reduction reactions, i.e., the insertion of oxygen into carbon-hydrogen bonds (oxidation) or the addition of hydrogen to a variety of chemical bonds (reduction).
2. Transferase	The transfer of one-carbon groups—i.e., the movement of aldehyde, ketone, phosphoryl, acyl, sugar, and sulfur functionalities from one molecule to another—is mediated by this group of enzymes.
3. Hydrolase	This group of enzymes are responsible for the hydrolytic dissociation of functional groups such as esters, anhydrides, amides, and other acyl derivatives.
4. Lyase	Additions to or the formation of double bonds is catalyzed by this group of enzymes.
5. Isomerase	Structural modifications of the substrate without any loss of the atoms forming the substrate are catalyzed by this group.
6. Ligase	These enzymes catalyze the formation of carbon-oxygen, carbon-sulfur, carbon-nitrogen, and carbon-carbon bonds with assistance from ATP.

ganisms is obvious but the scope of reactions encountered at temperatures and reaction speeds in a living system is truly amazing. The alacrity with which organisms dispose of cellular chemistry is solely attributable to the functioning of a myriad of enzyme systems. This apparent ease of reaction in aqueous media displayed by enzyme systems offers considerable power to applications set apart from the cell or unprotected from the surroundings. (See Table 9.6.2 for a list of

TABLE 9.6.2   Enzyme-Selection Properties and Factors

Property or Factor	Range
Molecular weight, Daltons	12,000–1,000,000
pH	1.5–11.0 (usually neutral conditions)
Temperature, °C	20–60 (activity observed above and below this range)
Specificity	Wide vs. narrow substrate capability
Catalytic power	Enzyme-catalyzed reaction rates are 10^9 to 10^{12} times faster than the corresponding nonenzymatic reaction rate
Activators and inhibitors	Control the facility with which the catalyzed reaction ensues
Cost	Unit conversion for process

Source: Ref. 1.

TABLE 9.6.3 Selected Soil Enzyme Reactions

Enzyme	Catalyzed reaction
1. Oxidoreductases	
Dehydrogenases	$XH_2 + A \rightarrow X + AH_2$
Catalase	$2H_2O_2 \rightarrow 2H_2O + O_2$
Peroxidases and polyphenol oxidases	$A + H_2O_2 \rightarrow$ oxidized $A + H_2O$
Catechol oxidase (phenolase)	1,2-dihydroxy aromatic $+ \frac{1}{2}O_2$ $\rightarrow$ 1,2-quinone $+ H_2O$
Diphenol oxidase	1,4-dihydroxy aromatic $+ \frac{1}{2}O_2$ $\rightarrow$ 1,4-quinone $+ H_2O$
2. Hydrolases	
Phosphatases	Phosphate ester $+ H_2O$ $\rightarrow$ R—OH $+ PO_4^{3-}$

enzyme properties and other factors that affect the activity of enzymes both in-side and outside living systems and are therefore relevant to their selection for use in cell-free systems.)

Enzymes produced by microorganisms capable of detoxifying hazardous-waste compounds may be separated from cells grown in bulk culture and subsequently applied to a contaminated matrix, i.e., soil. The natural regulatory mechanisms of the cell can be altered either by natural selection of specific mutants of a microorganism or through genetic engineering to cause the overproduction of certain cellular constituents, including enzymes.[2,3] The cell-free extract is usually made with the least amount of preparation possible in order to minimize cost.

The use of enzymes as a means to detoxify hazardous-waste constituents is a new and growing research area. The reactions of detoxifying enzymes are not limited to intracellular conditions but have been demonstrated through the use of immobilized enzyme extracts on several liquid waste streams.[4] Even the application of cell-free extracts to soil has met with success.[5]

The soil itself exhibits a wide range of enzymatic activity (see Table 9.6.3), some of which has been attributed to extracellular enzymes secreted by a variety of soil microorganisms.[6] Each soil may have its own characteristic array of specific enzymes.[7] Soil enzymatic activity is not static but fluctuates with changes in biotic and/or abiotic conditions. The obvious factors of moisture, temperature, aeration, soil structure, organic-matter content, seasonal variations, and the availability of soil nutrients influence the presence and abundance of enzymes.

9.6.1 BASIC CONSIDERATIONS[8]

In order to evaluate the ability of enzyme systems to accomplish the detoxification transformations on the substrate molecule it is necessary to determine the rate of reaction associated with the transformation and any other rates that may be limiting.[9] Kinetic studies generally involve the study of the rate of reaction as a function of the concentration of reacting molecules. Analysis of this data will establish a rate expression that relates the rate of reaction to the initial concentration of substrate and reactant molecules participating in the rate-limiting step. This analysis when used with enzyme systems allows the discrimination between similar enzyme systems by normalizing the reaction behavior to a fixed set of constraints assuming that the enzyme systems follow the same rate expressions. Knowledge of the characteristic kinetic constants for enzymatic reactions (especially Michaelis-Menten and inhibition constants and any attendant pH dependencies) is very useful although not required for the extension of this data to treatment applications. Relatively, a small amount of time is required to obtain the kinetic parameters for an enzyme-catalyzed treatment when compared with competing detoxification technology. These parameters enable the investigator to rationally control the activity of an enzyme system for treatment applications.

In the simplest case, where substrate S is converted into a product P by enzyme E, for a process following Michaelis-Menten kinetics without inhibition by products or other materials present in the reaction mixture, the rate of reaction is given by the familiar equation

$$S + E \underset{k_{-1}}{\overset{k_1}{\rightleftharpoons}} S*E \overset{K_{1/4}}{\rightarrow} P + E$$

$$\frac{d(P)}{dt} = \frac{k_2\,(E)}{1 + \dfrac{K_m}{[S]}} = \frac{V_{max}}{1 + \dfrac{K_m}{[S]}}$$

The rate of reaction will increase as the concentration of substrate is increased until all the available active sites on the enzyme are occupied. Subsequent increases in the substrate concentration do not affect the rate of reaction. Enzyme deactivation does eventually occur if the substrate concentration is sufficiently increased. Maximum enzyme efficiency is achieved in the region of substrate concentration in which the enzyme concentration limits the rate (V_{max} conditions). Should the substrate concentration be too low, a certain fraction of the enzyme's active sites will be unoccupied at equilibrium; however, when the substrate concentration is too high, the enzyme is deactivated. The *Michaelis constant* K_m is an invaluable measure to judge the minimum substrate concentration necessary to achieve efficient use of the enzyme; V_{max} indicates the most rapid rate of product formation that can be achieved for a specified quantity of enzyme under ideal conditions. Both K_m and V_{max} are measured under laboratory assay conditions. Values of K_m are often roughly the same under assay conditions as those obtained under treatment conditions. Where immobilized enzymes are employed, higher apparent values for K_{max} may be observed if mass-transport limitations are important. V_{max} under assay conditions is often larger than the maximum rate observed under such nonideal situations as treatment conditions since in these situations inhibition of enzymatic activity is common.

9.6.2 LIMITATIONS

The action of enzymes is limited by two basic considerations. The physical constraints of the active site within the enzyme molecule and the attendant kinetic expression relating the rate of disappearance of substrate to time result in a series of kinetic limitations referred to as *enzyme inhibition* that are integral to an understanding of enzyme use. A subtle limitation is the resistance of the enzyme to destruction by outside agents.

Kinetic Limitations

The discrepancy between the catalytic activity of an enzyme under ideal (assay or V_{max}) conditions and lower values for the maximum rate encountered in nonideal applications can be a major source of confusion. This discrepancy may have several origins: nonspecific deactivation of the enzyme by high or inappropriate concentrations of substrates, buffers, or ions; loss of activity or immobilization because of chemical modification or mass-transport limitations; or reversible kinetic inhibition by product.

The different classes of inhibition are important to recognize since optimal performance of an enzyme system is often controlled by one or more types of inhibition (see Fig. 9.6.1). The types of inhibition can be characterized as follows:

- *Competitive inhibition,* in which the product or inhibitor and substrate bind competitively to the active site.

- *Noncompetitive inhibition,* in which the product or inhibitor binds tightly to the active site; this effectively curtails catalytic activity by the enzyme.

- *Mixed inhibition,* in which more than a single type of inhibition is governing the operation of the system.

- *Uncompetitive inhibition,* in which the product or inhibitor is irreversibly bound to the active site of the enzyme.

These last two forms of inhibition influence V_{max} and K_m simultaneously.

To optimize a treatment, it is important to derive, from kinetic analysis of the enzyme, values for K_m and V_{max} since they are critical in choosing starting values for the substrate and product concentrations. Product inhibition must be identified if present and approximate values for k_1 should be determined in order to select the correct strategy of either increasing the substrate concentration and/or

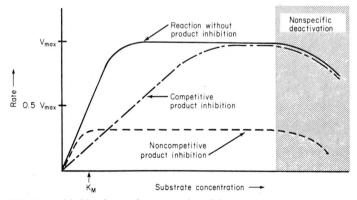

FIG. 9.6.1 Limiting factors for enzymatic activity.

decreasing the product concentration to limit its unfavorable influence on the rate of reaction.

Established applications have not necessarily been optimized. Operating conditions can be found to be sufficient for some treatment applications but others may not follow the same optimization criteria due to the wide differences in environmental setting for the control technology.

Biological Limitations

The use of cell-free extracts in an environmental setting requires the additional inquiry as to the stability of the "naked" enzyme. Proteases (peptide bond-breaking enzymes), in microorganisms, and any free in the polluted matrix, will degrade enzymes and are potential agents which will reduce the efficacy of the enzyme preparation.

9.6.3   SYNZYMES AND ARTIFICIAL ENZYMES[11]

A number of research groups have been investigating small molecular constructs to evaluate enzymatic activity at a more accessible model of the enzyme's active site.[12] This chemical simplification of enzymatic behavior has increased the general knowledge of enzymatic activity[13] and has made possible design of chemical analogs that may have some utility for applications such as detoxification. There is a basic debate concerning the level of correlation between these simplified constructs and the corresponding enzyme.[14] The obvious macromolecular complexity of enzymes is currently deemed necessary for the desired rate of reaction and attendant specificity.

9.6.4   APPLICATIONS

The detoxification of hazardous-waste chemicals by enzymes implies the ability on the part of the enzyme to conduct some structural alteration on the substrate.[15] *Detoxification* refers to the conversion of a toxicant to innocuous metabolites; it does not necessarily mean that the substrate has been mineralized. *Mineralization* refers to the conversion of an organic substrate to inorganic products. The ability of microorganisms to convert a toxic substrate into less complex metabolites plays an important role in the eventual mineralization of the substrate. Once a single degradation step has occurred the toxicity of the substrate may be reduced substantially. A less toxic metabolite will generally lead to easier degradation of the starting toxic substrate.

Oxidoreductases

Ligninases[16,17].   The decomposition of dead wood and other forest detritus has been accomplished in the natural setting through the activity of fungi and other microorganisms that can utilize such materials as a food source. Recent work with a white rot fungus, *Phanerochaete chrysosporium,* has shown that it has a very powerful oxidizing enzyme system capable of degrading hazardous-waste chemicals. This organism secretes a very complicated series of enzymes with broad specificity that act outside the fungal cell to break down lignin.[18]

These *ligninases* are responsible for the breakdown of the naturally occurring polymer lignin. Structurally, lignin is chemically complex and serves as the "glue" that holds the living wood structure together. Since these enzymes are both extracellular and exhibit the ability to detoxify hazardous-waste chemicals they are a prime candidate for use as a cell-free extract. Research leading to a better understanding of this enzyme system is underway. Future applications will clearly depend on the outcome of these investigations but the future looks exceptional based on current knowledge.

Horseradish Peroxidase.   Researchers have recently elaborated the use of peroxidase derived from horseradish to remove organic compounds from drinking water[19] and wastewater[20] derived from several processes. The treatment process involves the addition of horseradish peroxidase and hydrogen peroxide to solutions containing pollutants such as phenols and aromatic amines. The treatment is based on the enzymatic cross-linking of the substrates (Fig. 9.6.2) to form

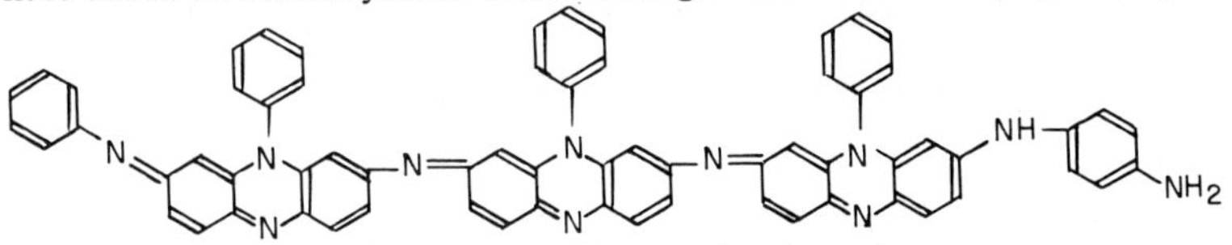

FIG. 9.6.2   Selective horseradish peroxidase-catalyzed reactions.

insoluble polymers (Fig. 9.6.3) which precipitate from solution. Over 40 different chemical pollutants have been removed from water by this precipitation. The removal efficiencies (see Table 9.6.4) based on precipitation are 99% or better, thereby reducing the concentrations to 1 ppm or less.[21] The more easily removed phenols and anilines were found to assist the removal of less reactive substrates. The oxidative process has been depicted as the formation of intermediate radicals

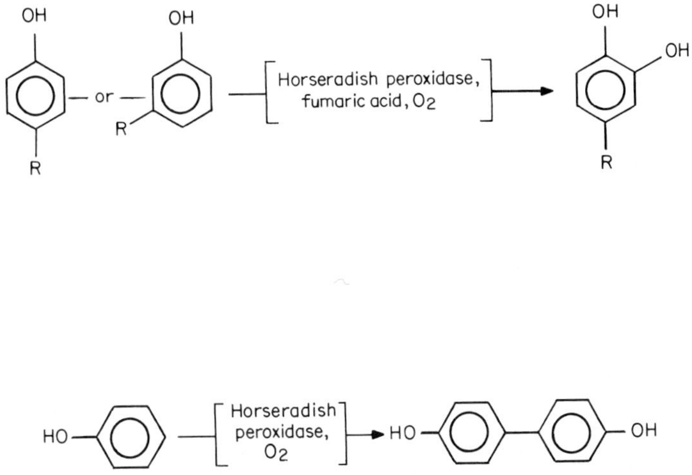

FIG. 9.6.3   Structure of peroxidase-catalyzed oxidation of aniline.

TABLE 9.6.4 Treatment of Aromatic Amines and
Phenols by Horseradish Peroxidase and H_2O_2*

Compound	Removal efficiency, %
Benzidine	99.94
3,3'-Dimethoxybenzidine	99.9
3,3'-Diaminobenzidine	99.6
3,3'-Dichlorobenzidine	99.9
3,3'-Dimethylbenzidine	99.6
1-Naphthylamine	99.7
2-Naphthylamine	98.3
5-Nitro-1-naphthylamine	99.6
N,N'-Dimethylnaphthylamine	93.2
Phenol	85.3
2-Methoxyphenol	98.0
3-Methoxyphenol	98.6
4-Methoxyphenol	89.1
2-Methylphenol	86.2
3-Methylphenol	95.3
4-Methylphenol	85.0
2-Chlorophenol	99.8
3-Chlorophenol	66.9
4-Chlorophenol	98.7
2,3-Dimethylphenol	99.7
2,6-Dimethylphenol	82.3
Aniline	72.9
4-Chloroaniline	62.5
4-Bromoaniline	84.5
4-Fluoroaniline	86.4
1,3-Diaminophenol	98.6
Diphenylamine	80.5
1-Naphthol	99.6
2-Nitroso-1-naphthol	98.9
4-Phenylphenol	99.9
8-Hydroxyquinoline	99.8

*Conditions: 100 ppm of a phenol or aromatic amine, 3-h
treatment at room temperature, pH 5.5, 100 units/L peroxidase, 1
mM H_2O_2. Some compounds required greater excesses of
peroxidase and hydrogen peroxide—namely, the naphthylamines
and several phenols (phenol, 2-methylphenol, 3-methylphenol, 1-
naphthol, 2-nitroso-1-naphthol and 4-phenylphenol).

which diffuse from the active site of the enzyme into bulk solution, where polymerization occurs. The products are polyaromatic compounds which are nearly
insoluble in the water phase. Total removal could then be effected by simple filtration and sedimentation.

Hydrolases[22]

This group of enzymes generally performs a simple reaction involving a chemical
functionality that is susceptible to hydrolysis. Due to the reaction's simplicity,

the enzymatic reaction does not require additional components such as cofactors for the reaction to ensue and this simplicity is a prime reason for the development of applications of these enzymes to environmental detoxification. If enzymes can be found that greatly reduce the toxicity of a substrate chemical through simple steps such as hydrolysis, then subsequent steps leading to complete mineralization may not be prohibitive.

The chemical conversion of nitriles to the corresponding amides or carboxylic acids is an energy-consuming process and is complicated by the disposal of side products. A group of Japanese investigators have explored this conversion using enzyme systems derived from microorganisms for the purpose of full-scale production of the chemical intermediates.[23,24] This same technology can be used for the control of nitriles in process waters (Fig. 9.6.4).

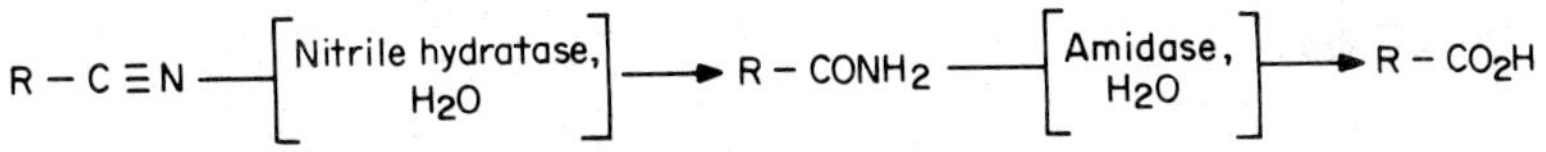

$$R-C\equiv N \longrightarrow \left[\begin{array}{c}\text{Nitrile hydratase,}\\ H_2O\end{array}\right] \longrightarrow R-CONH_2 \longrightarrow \left[\begin{array}{c}\text{Amidase,}\\ H_2O\end{array}\right] \longrightarrow R-CO_2H$$

FIG. 9.6.4 Enzymatic detoxification of nitriles.

The products of enzymatic detoxification of a series of pesticide chemicals are listed in Table 9.6.5.[22,25] The entries in this table are organized according to the type of hydrolytic reaction that is accomplished through the use of the enzyme. The detoxification factor is the simple ratio of the substrate's toxicity to that of the metabolic product. Any number less than 1 indicates an accretion of toxicity. Enzymatic detoxification of substrates such as phenylureas, acylanilides, and phenoxyacetates is not conducive to environmental application because of the

TABLE 9.6.5 Toxicity Reduction of Selected Pesticides through Enzymatic Treatment

Substrate	Metabolic product	Toxicity*	Detoxification factor†
Organophosphates			
Parathion		6	58
	p-Nitrophenol	350	
Paraoxon		1.8	195
	p-Nitrophenol	350	
Carbamates			
Carbaryl		500	7
	1-Napthol	2,590	
Phenylureas			
Monuron		1,480	0.2
	4-Chloroaniline	300	
Diuron		430	1.3
	3,4-Dichlorophenol	580	
Linuron		1,500	0.07
	3,4-Dichloroaniline	100	

TABLE 9.6.5 Toxicity Reduction of Selected Pesticides through Enzymatic Treatment (*Continued*)

Substrate	Metabolic product	Toxicity*	Detoxification factor†
Acylanilides			
Alachlor		1,800	1.5
	2,6-Diethylaniline	2,690	
Propanil		560	0.18
	3,4-Dichloroaniline	100	
Phenoxyacetates			
2,4-D		370	1.6
	2,6-Dichlorophenol	580	
Silvex		650	0.5
	2,4,5-Trichlorophenol	320	
Dithioates			
Azinphos-methyl		13	824
	Anthranilic acid	4,650	
Guthion-E		9	12–40
	3-Mercaptomethyl-1,2,3-benzotriazin-4(3*H*)-one	?	
Disulfoton		6	50–100
	2-(Ethylthio)ethanol	300–600	

*Toxicity, LD_{50} oral, rst, mg/kg.
†Detoxification factor calculated by dividing the metabolic product toxicity by the substrate toxicity.

small differences in toxicity between the substrates and their metabolites. Since toxicity is a function of chemical structure, concentration, and contact time, enzymatic hydrolysis may be displaced as the detoxification methodology of choice in some instances where the products are too toxic and less biodegradable.

The information in Table 9.6.6[4] is offered for comparison purposes and is organized to correlate with Table 9.6.5. We see that the specific activity of the hydrolytic enzymes assayed so far spans several orders of magnitude with the same substrate. There are additional factors such as pH dependence that serve to complicate the basic comparison offered in Tables 9.6.5 and 9.6.6.

9.6.5 REACTOR TECHNOLOGY[26]

Enzymes are of their own nature delicate and require sufficient protection to operate productively. Common applications of cell-free enzyme extracts have generally involved some sort of immobilization.[27] The enzyme can be chemically bound to a support and is then assembled into a bed through which a solution containing the substrate is passed. Additional techniques such as entrapment can be used for immobilization. Direct application of the cell-free enzyme preparation to the contaminated matrix (e.g., soil) can be and has been used for detoxification. Preparation of the cell-free extract is optimized for the least

TABLE 9.6.6   Specific Activity of Cell-Free Enzymes for Selected Hydrolase Reactions

Substrate	Enzyme	Specific activity*
Organophosphates		
Parathion	Esterase	3,000, 7,000
Paraoxon	Esterase	3,600, 13.8
Phenylureas		
Monuron	Amidase	4
Diuron	Amidase	
Linuron	Amidase	0.2, 18, 20, 130
Acylanilides		
Propanil	Amidase	2, 3, 555, 249
Phenoxyacetates		
2,4-D	Hydrolase	16.5
Dithioates		
Azinphos-methyl	Esterase	87

*Specific activity is expressed in terms of nanomoles of substrate transformed per minute per milligram of protein.

amount of handling and work.[28,29] Higher degrees of purification are only undertaken where necessary. Immobilization of a hydrolytic enzyme has been studied for the detoxification of pesticide-production wastewaters.[30] The enzyme preparation could detoxify parathion and eight other organophosphate pesticides. Using the immobilized enzyme, hydrolysis kinetics were studied for parathion at flow rates up to 96 L/h and the influent substrate concentration of 10 to 250 mg/L. More than 95% of the parathion added to industrial wastewaters was hydrolyzed.

9.6.6   COST

Although there is information[31] detailing the expenses of industrial enzyme scale-up, very little cost-comparison information is available for this type of enzymatic application. Klibanov offered a cost analysis for the application of horseradish peroxidase to wastewater treatment of a series of phenols. For the treatment of 1,000 L (264 gal) contaminated with a total phenol concentration of 105 ppm, the raw-material costs (crude enzyme preparation) were $0.69.[20]

9.6.7   REFERENCES

1. *Enzyme Nomenclature,* Academic Press, New York, 1979.
2. K. Toda, "Induction and Repression of Enzymes in Microbial Culture," *J. Chem. Tech. Biotechnol.,* **31**:775–790 (1981).

3. P. H. Clarke, "Experiments in Microbial Evolution: New Enzymes, New Metabolic Activities," *Proc. R. Soc. London,* **B207**:385 (1980).

4. D. M. Munnecke, L. M. Johnson, H. W. Talbot, and S. Barik, in *Biodegradation and Detoxification of Environmental Pollutants,* A. M. Chakrabarty (ed.), chap. 1, "Microbial Metabolism and Enzymology of Selected Pesticides," p. 1.

5. S. Barik and D. M. Munnecke, "Enzymatic Hydrolysis of Concentrated Diazinon in Soil," *Bull. Environm. Contam. Toxicol.,* **29**:235 (1982).

6. J. Skujins, "Extracellular Enzymes in Soil," *CRC Critical Reviews in Microbiology,* **4**:(4)383 (1976).

7. J. Ladd, in *Soil Enzymes,* R. G. Burns (ed.), Academic Press, New York, 1978, p. 51.

8. A. Fersht, *Enzyme Structure and Mechanism,* 2d ed., W. H. Freeman, New York, 1985.

9. K. M. Plowman, *Enzyme Kinetics,* McGraw-Hill, New York, 1972.

10. C. Walsh, *Enzymatic Reaction Mechanisms,* W. H. Freeman, San Francisco, 1979.

11. G. P. Royer, "Enzyme-like Synthetic Catalysts (Synzymes)," *Adv. Catal.,* **29**:197 (1980).

12. E. T. Kaiser and D. S. Lawrence, "Chemical Mutation of Enzyme Active Sites," *Science,* **226**:87 (1984).

13. I. Tabushi, "Design and Synthesis of Artifical Enzymes," *Tetrahedron,* **40**:269 (1984).

14. M. Mutter, "The Construction of New Proteins and Enzymes—A Prospect for the Future," *Angew. Chem. Int. Ed. Engl.,* **24**:639 (1985).

15. D. M. Munnecke, "The Use of Microbial Enzymes for Pesticide Detoxification," in *Microbial Degradation of Xenobiotics and Recalcitrant Compounds,* T. Leisinger, R. Hutter, A. M. Cook, and J. Neusch (eds.), Academic Press, New York, 1981, p. 251.

16. D. C. Eaton, "Mineralization of Polychlorinated Biphenyls by *Phanerochaete Chrysosporium*: a Ligninolytic Fungus," *Enzyme Microb. Technol.,* **7**:194 (1985).

17. J. A. Bumpus, M. Tien, D. Wright, and S. D. Aust, "Oxidation of Persistent Environmental Pollutants by a White Rot Fungus," *Science,* **228**:1434 (1985).

18. T. K. Kirk, S. Croan, M. Tien, K. E. Murtagh, and R. L. Farrell, "Production of Multiple Ligninases by *Phanerochaete Chrysosporium*: Effect of Selected Growth Conditions and Use of Mutant Strain," *Enzyme Microb. Technol.,* **8**:27 (1986).

19. S. W. Maloney, J. Manem, J. Mallevialle, and F. Fiessinger, "Transformation of Trace Organic Compounds in Drinking Water by Enzymatic Oxidative Coupling," *Environ. Sci. Technol.,* **20**:249 (1986).

20. A. N. Alberti and A. M. Klibanov, in *Detoxication of Hazardous Waste,* J. H. Exner (ed.), Ann Arbor Science, Ann Arbor, Michigan, 1982, p. 349; A. M. Klibanov, B. N. Alberti, E. D. Morris, and L. M. Felshin, "Enzymatic Removal of Toxic Phenols and Anilines from Waste Waters," *J. Appl. Biochem.,* **2**:414 (1980); A. M. Klibanov and E. D. Morris, "Horseradish Peroxidase for the Removal of Carcinogenic Aromatic Amines from Water," *Enzyme Microb. Technol.,* **3**:(4)119 (1981); and A. M. Klibanov, T.-M. Tu, and K. P. Scott, "Peroxidase-Catalyzed Removal of Phenols from Coal-Conversion Waste Waters," *Science,* **221**:259 (1983).

21. A. M. Klibanov, "Enzymatic Removal of Dissolved Aromatics from Industrial Aqueous Effluents," in *Environmental Engineering and Pollution Control Processes,* EPA/600/X-85/106, June 1985.

22. D. M. Munnecke, "Detoxification of Pesticides Using Soluble or Immobilized Enzymes," *Process Biochemistry,* **13**:(2)14 (1978).

23. Y. Asano, Y. Tani, and H. Yamon, "A New Enzyme 'Nitrile Hydratase' which Degrades Acetonitrile in Combination with Amidase," *Agric. Biol. Chem.,* **44**:(9)2,251 (1980).

24. Y. Asano, K. Fujishiro, Y. Tani, and H. Yamada, "Aliphatic Nitrile Hydratase from *Arthrobacter* sp. J-1, Purification and Characterization, *Agric. Biol. Chem.,* **46**:(5)1,165 (1982).

25. L. M. Johnson and H. W. Talbot, Jr., "Detoxification of pesticides by microbial enzymes," *Experientia,* **39**:1,236 (1983).

26. O. R. Zaborsky, *Immobilized Enzymes,* CRC Press, Cleveland, Ohio, 1973.

27. A. M. Klibanov, "Enzyme Stabilization by Immobilization," *Anal. Biochem.,* **93**:1(1979); A. M. Klibanov, "Immobilized Enzymes and Cells as Practical Catalysts," *Science,* **219**:722 (1983); A. Rosevear, "Immobilized Biocatalysts—A Critical Review," *J. Chem. Tech. Biotechnol.,* **34B**:127 (1984).

28. D. M. Munnecke, "Hydrolysis of Organophosphate Insecticides by an Immobilized-Enzyme System," *Biotechnol. Bioengng.,* **21**:2,247 (1979).

29. B. N. Volesky and J. H. T. Luong, "Microbial Enzymes: Production, Purification and Isolation," *CRC Critical Reviews in Biotechnology,* **2**:(2)119 (1985).

30. H. H. Weethall and W. H. Pitcher, Jr., "Scaling Up an Immobilized Enzyme System," *Science,* **232**:1,396 (1986).

31. G. Street, "Large Scale Industrial Enzyme Production," *CRC Critical Reviews in Biotechnology* **1**:(1)59 (1983).

CHAPTER 10

LAND STORAGE AND DISPOSAL

SECTION 10.1

HAZARDOUS-WASTE LANDFILL CONSTRUCTION: THE STATE OF THE ART

Thomas D. Wright, P.E.
SCS Engineers
Long Beach, California

David E. Ross, P.E.
SCS Engineers
Long Beach, California

Lori Tagawa
Hydrogeologist
SCS Engineers
Long Beach, California

10.1.1   OBJECTIVES OF A HAZARDOUS-WASTE LANDFILL

A hazardous-waste landfill contains and isolates hazardous wastes that are not presently recoverable to ensure present and long-term environmental protection. To accomplish these objectives, the landfill must be planned, designed, constructed, operated, and maintained in accordance with federal, state, and local regulations. These regulations are addressed in Handbook Chap. 1.

10.1.2 CHARACTERISTICS OF HAZARDOUS WASTES SUITABLE FOR LANDFILLS

Many but not all hazardous wastes can be disposed of on land in properly designed landfills. To minimize potentially adverse environmental effects from wastes deposited at hazardous-waste landfill sites, the U.S. Environmental Protection Agency (EPA) has developed specific regulations regarding the characteristics of wastes suitable for landfilling. These regulations (40 *CFR* 265) include a prohibition on the placement of

- Noncontainerized hazardous wastes containing free liquids, whether or not absorbents have been added.
- Containers holding free liquids unless all freestanding liquid has been removed by decanting or other methods or has been mixed with absorbent or solidified so that freestanding liquid is no longer observed. Such containers must be at least 90% full or, if empty, reduced in size as much as possible via crushing or shredding prior to disposal.

The following containers are exempt from the above regulations:

- Very small containers, such as ampules, and containers holding liquids for use other than storage, such as batteries, which may be disposed directly in a hazardous-waste landfill.
- Small lab-pack containers of hazardous waste if they are first placed in nonleaking, larger containers. These containers must be filled to capacity and surrounded by enough absorbent material to contain the liquid contents of the lab pack. The resultant container must then be placed in a larger container packed with absorbent material which will not react with, become decomposed by, or be ignited by the contents of the inside containers. Incompatible wastes may not be packed and disposed of together in this manner.

10.1.3 HAZARDOUS-WASTE LANDFILLING METHODOLOGIES

In most cases, hazardous wastes are initially deposited below grade in an existing natural or excavated depression (e.g., a quarry) or in a pit excavated for that purpose. Generally, landfilling then proceeds above grade, forming a mound which increases landfill capacity and causes drainage of incident precipitation from the landfill surface.

Hazardous wastes delivered to a landfill site are placed in a manner such that only compatible wastes are disposed of together. To accomplish this objective, wastes are placed either in separate areas or in individual control cells within a single, larger landfill.

Control Cells

Control-cell sizes vary depending on the daily rate of delivery and characteristics of hazardous wastes received. Cell heights generally range from 4.9 to 6.1 m (16 to 20 ft), although this range can be exceeded on either end, depending on the type of waste disposed and the rate of delivery. For instance, cell dimensions would differ for regular deliveries of 208-L (55-gal) drums versus occasional deliveries of contaminated soils. The optimal cell height is set to conserve available cover soil while adequately accommodating as much waste as possible.

Control cells receiving bulk hazardous wastes are constructed by placing and spreading the waste in 0.61 to 0.91-m (2- to 3-ft) layers and then compacting it using standard earth-moving equipment. A compacted 0.30-m thick (1-ft) layer of cover soil is placed over the wastes at the end of each working day.

Control cells receiving containers of solidified hazardous waste are constructed by placing the containers upright in the cell using a forklift or barrel snatcher, leaving sufficient space around the containers for the placement and compaction of compatible bulk hazardous wastes or soil. A compacted 0.3-m (1-ft) layer of cover soil is also placed over all containers at the end of each working day.

Cover soils must be able to impede rainfall infiltration and to control odors and airborne emissions from deposited wastes.

Control cells are generally laid out in accordance with an engineered grid system prepared for the site. Corner monuments for the grid are established in the field by a survey team. The two-dimensional grid system along with a lift designation (a lift is the distance from the bottom of one cell to the bottom of the cell above) provides a three-dimensional record by source, type, and date of the location of all wastes placed at the site.

Figures 10.1.1 and 10.1.2 illustrate a conceptual design for a typical control-cell grid system for a hazardous-waste landfill.

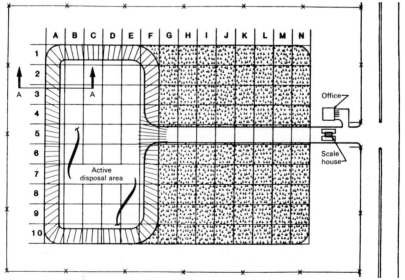

FIG. 10.1.1 Conceptual design for a typical grid system for hazardous-waste landfill (plan view).

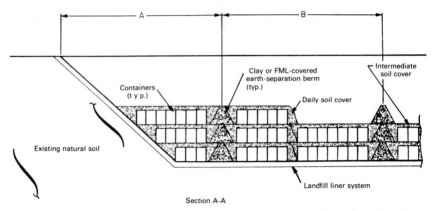

FIG. 10.1.2 Conceptual design for control cells for hazardous-waste disposal (section view). FML = flexible-membrane liner.

10.1.4 ENVIRONMENTAL PROTECTION FEATURES

Environmental protection, including control of leachate emissions, gas emissions, and surface-water erosion and waste saturation, is an important aspect of site selection and construction for the hazardous-waste landfill.

Control of Leachate Emissions

In general, hazardous-waste landfills must be provided with a system to control and collect leachate. Typically, at a minimum a double-liner system is placed at the bottom and side slopes of the facility prior to initial placement of hazardous wastes. This liner system will contain leachate that may later be generated in the deposited wastes and then flow downward by gravity. Leachate that accumulates in the liner is directed to one or more central collection sumps through a series of

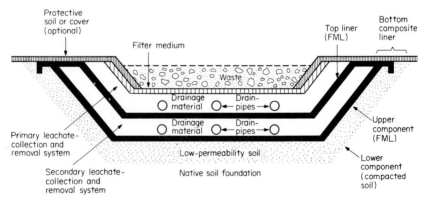

FIG. 10.1.3 Schematic of an FML plus compacted-soil double-liner system for a landfill. (Drawing not to scale.) (*U.S. EPA, EPA/530/SW-85-012, Washington, D.C., 1985*)

perforated plastic collection pipes; from these it can be pumped out for treatment and ultimate disposal.

According to EPA regulations, the bottom liner may be either a layer of recompacted clay or other natural material with a minimum thickness of 0.91 m (3 ft) and a permeability of no more than 1×10^{-7} cm/s, or a flexible-membrane liner (FML). The upper liner must be an FML, since EPA regulations dictate that there can be no migration of leachate *into* the top liner. Characteristics of acceptable leachate-control liners for these landfills are discussed in subsequent subsections of Sec. 10.1.

To promote removal of leachate from the liner, the landfill's bottom is sloped, and a sufficient number of drainage pipes are provided to assure that the leachate depth over the liner does not exceed 0.3 m (1 ft).

Figures 10.1.3 through 10.1.5 illustrate a conceptual design for a bottom-liner and leachate-collection system configuration.

Leachate generation is also reduced through the placement of a low-permeability cap over completed portions of the landfill. This cap must have a

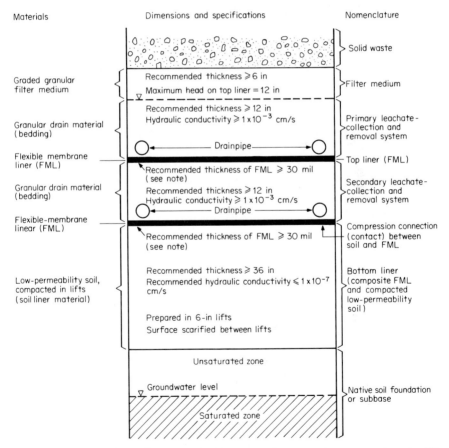

FIG. 10.1.4 Schematic profile of an FML plus composite double-liner system for a landfill. NOTE: FML thickness >45 mil is recommended if liner is not covered within three months.

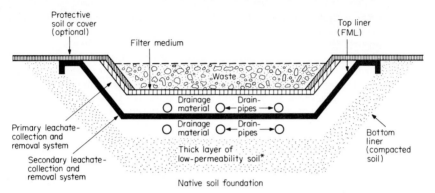

*Thickness to be determined by breakthrough time.

FIG. 10.1.5 Schematic of an FML plus composite double-liner system for a landfill. (Drawing not to scale.) (*U.S. EPA, EPA/530/SW-85-012, Washington, D.C., 1985*)

permeability less than or equal to the permeability of the bottom-liner system. Further information regarding landfill caps is presented in subsec. 10.1.7.

In some cases, a hazardous-waste landfill can be exempted from double-liner requirements for leachate emission control if, among other things, the hydrogeologic setting would preclude leachate migration to ground or surface water. For example, some hazardous wastes may be minimally prone to leaching or migration. Also, hydrogeologic factors at the site could provide a degree of protection; these include on-site soils with a significant attenuative capacity (e.g., certain clays) and/or thickness located between the landfill and ground and surface water.

Landfill-Gas Control

If organic hazardous or nonhazardous waste is deposited at the landfill, landfill gas (LFG) production can be expected. Landfill gas is produced by anaerobic decomposition of organic material, and consists primarily of methane and carbon dioxide, but could also contain small concentrations of other volatile organic gases such as vinyl chloride.

Methane gas is odorless and nontoxic; however, the LFG mixture can be odiferous in the presence of other gases. The methane component is explosive when present in air at concentrations between 5 and 15%, and flammable at higher concentrations in air. Landfill gas can move vertically and laterally through soils from a landfill under a pressure gradient or a concentration gradient via diffusion.

Landfill gas from hazardous-waste sites is currently either vented into the atmosphere or collected and flared or incinerated. Atmospheric vent systems usually consist of a series of horizontal, perforated collection pipes located atop the landfilled material and under the final cap; the LFG is vented to the atmosphere via vertical riser pipes. Alternatively, the LFG can be collected via an extraction blower and flared or incinerated if it is too toxic or odiferous for direct ventilation. Figures 10.1.6 and 10.1.7 show a typical LFG vent system.

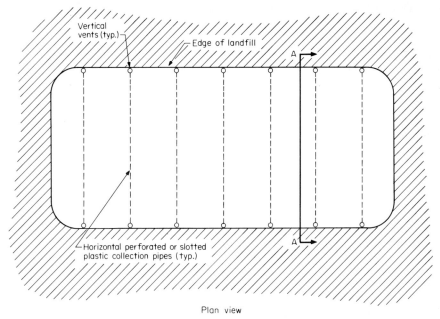

FIG. 10.1.6 Conceptual layout for landfill-gas vent system. (plan view).

Surface-Water Control

Surface water originating off site is often directed around or away from the landfill via half-round corrugated-metal pipe (CMP) or ditches lined with asphaltic concrete (AC). These ditches are designed such that off-site peak run-on from at least a 25-year storm can be diverted.

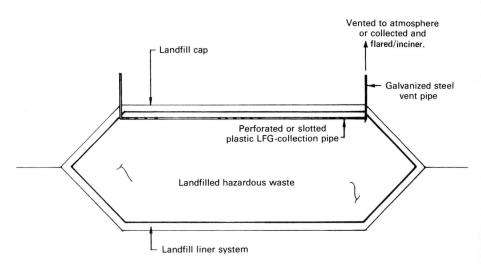

FIG. 10.1.7 Conceptual landfill-gas vent system (section view).

On-site surface water is generally handled as follows:

- Runoff from nonlandfilled areas is diverted directly off site.
- Runoff from exposed excavation areas is directed to siltation basins and then discharged off site.
- Runoff from the active landfill area is directed to localized holding sumps where it is sampled for contamination. Uncontaminated water is directed off site, and contaminated water is treated and then discharged off site.
- Runoff from completed landfill surfaces is directed to localized holding sumps (different sumps than used for runoff from active landfill areas since the runoff volume will be greater and the runoff less likely to be contaminated), where it is sampled for contamination. Uncontaminated water is directed off site and contaminated water is treated and then discharged off site.

Drainage ditches located on nonlandfilled areas are generally earth ditches, half-round CMP, or AC- or gunite-lined earth ditches.

Temporary earth berms and drainage ditches are generally used to divert on-site runoff away from active landfilling areas. Permanent drainage ditches located on the landfill must be able to accommodate some differential settlement, and are thus frequently constructed of half-round CMP with overlapping flexible joints; AC-lined earth ditches are also commonly used.

All on-site runoff-collection and holding facilities (e.g., sumps and basins) must be able to accommodate on-site runoff volumes resulting from a 24-h, 25-year storm.

10.1.5   OTHER PERTINENT LANDFILL FEATURES

Access Roads

A paved on-site road is normally provided to connect the disposal areas to the public road system.

Permanent on-site roadways are usually 6.1 to 7.3 m (20 to 24 ft) wide to allow two-way traffic flow. For low-volume facilities, a road 4.6 m wide (15 ft) is acceptable. On-site roads are often paved; however, roadways with gravel surfaces can also be used.

Access-road grades are generally less than 7% for uphill grades, and less than 10% for downhill grades.

Temporary roads are constructed for delivery of waste to the working area from the paved on-site road system. Temporary roads may be constructed by compacting the local soils and by controlling drainage onto and off of the roadway, or by topping the road with a supporting surface such as a layer of gravel, crushed stone or concrete, or asphalt binders to make the roads more serviceable during wet weather.

All access roads constructed over FML material are provided with a sufficient earth foundation—at least 0.61 m (2 ft) deep—to prevent damage from vehicular traffic.

Permanent access roads located over completed landfill areas are often constructed to accommodate some differential settlement by first placing a geotextile on the landfill surface and then placing the aggregate drainage base and final AC surface.

Cover Soil

Cover-soil handling is an important part of hazardous-waste landfilling. Consequently, this activity is well-planned as discussed here:

- Soil that is determined to be readily excavatable and suitable for use as cover is removed and stockpiled for later use before waste filling buries it.
- Accelerated excavation programs are sometimes implemented during warm weather to avoid the need to excavate frozen soil during cold weather.
- Soil stockpiles are generally located away from waste-delivery vehicles, but convenient for daily placement of cover. This minimizes costly soil-handling activities.
- Soil stockpiles are generally laid out and maintained to minimize erosion from rainfall runoff.

Special Working Areas

Special working areas are often designated on the site plan for use during inclement weather or other contingency situations. Access roads to these areas are generally of all-weather construction, and the areas are kept grubbed and graded. Arrangements for special working areas often include locating such areas as close as possible to the landfill's entrance gate.

In addition to being readily accessible, areas for use during inclement weather are often constructed to allow unhindered waste disposal. Runoff is diverted around or away from these areas to prevent vehicles from becoming mired in a muddy surface. They are either constructed on natural ground or over waste material which has been in place for preferably over a year, and which is covered with a compacted soil layer at least 0.61 m (2 ft) deep. In either case, a final, low-permeability, hardened surface (such as crushed AC) is generally placed atop the inclement-weather area.

Buildings and Structures

Buildings and structures typically provided at a hazardous-waste landfill site include

- An office building.
- Employee facilities (e.g., lunchroom, showers, sanitary facilities).
- A laboratory for analysis of incoming wastes and monitoring samples. The lab can be housed in the office building or in its own structure.
- An equipment storage and maintenance building.
- A weather station.
- Emergency wash and first aid areas.

Buildings on sites that will be used for less than 10 years are often temporary, mobile structures. Where possible, all structures, whether mobile or more permanent, are located on natural ground a sufficient distance from the landfill to preclude hazard from subsurface LFG migration. A good rule of thumb is a *minimum* distance of 305 m (1,000 ft) from the landfilling area. If this is impractical, gas movement must be considered in the design and location of all structures; also, if structures are to be located over hazardous-waste-containing organic material, the differential settlement caused by decomposition must be considered.

Wash Rack

To assure that vehicles leaving the facility carry no contaminants and mud onto surrounding roadways, hazardous-waste sites usually have a truck wash rack. This is particularly true at larger facilities, in wet climates where the soil is heavy, and where trucks drive through or near residential or commercial areas.

Landscaping

Erosion-control landscaping for a landfill facility is generally selected based on the following criteria:

- Proposed interim and end uses
- Degree of tolerance to adverse soil conditions (e.g., concentrations of LFG in the soil interstices)
- Depth of root zone
- Amount of water required to sustain growth
- Final landfill configuration
- Ability to control erosion

   To control erosion, the final grade of the landfill surface must be kept relatively flat while maintaining good drainage. Slopes ranging from 3% minimum to 3:1 maximum are typical. Usually, a mixture of grasses and legumes adapted to local soil and climatic conditions, planted in combinations of two or three species, is used. Consultation with horticulturists familiar with local conditions is advisable. The site is generally irrigated only to establish vegetation, except where there is insufficient precipitation to maintain vegetation once it is established.

   Deep-rooted plants are not normally planted over landfilled areas unless sufficient topsoil is provided to assure that the low-permeability cap is not destroyed by the root system.

Monitoring Facilities

Environmental monitoring is an integral part of both ongoing and postclosure operations at the hazardous-waste landfill. Monitoring of the vadose zone, groundwater, and surface water for leachate emissions, and the perimeter soils and atmosphere for LFG emissions is generally required by regulations and permit conditions.

Vadose-Zone Monitoring.   Monitoring of the *vadose zone,* i.e., the unsaturated zone, of soil beneath a hazardous-waste landfill is not required by current federal regulations. However, vadose-zone monitoring is required by some states.

   The most commonly used monitoring tool in the vadose is the suction

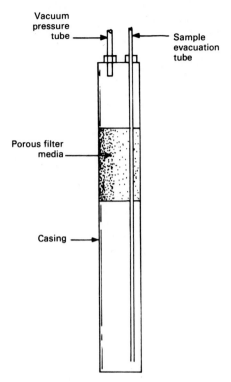

FIG. 10.1.8 An example of a casing lysimeter.

lysimeter (see Fig. 10.1.8). Suction lysimeters are installed in the vadose zone beneath the landfill site prior to installation of the bottom, low-permeability liner. Generally, lysimeters are located under leachate-collection sumps or other low points in the landfill bottom since these are the most likely places for potential leachate leakage after landfilling commences. Backup lysimeters are often installed beneath the primary lysimeters to assure that leakage is detected. Figure 10.1.9 shows a conceptual layout of a vadose-zone monitoring system.

Groundwater Monitoring. Site-specific characteristics that influence the placement of monitoring wells include the nature of the aquifer (e.g., artesian); characteristics of potential leachate; and groundwater depth, flow rates, and direction of flow. In general, monitoring wells should be installed to a depth of at least 3.0 m (10 ft) below the maximum historic groundwater depth. Based on assumptions and data about the characteristics of leachate to be generated, approximate permeability of soils in the zone of aeration, and directions and velocities of groundwater flow, the maximum probable areal extent of contaminant migration can be estimated as a basis for establishing the position of monitoring wells.

Proper location and installation of monitoring wells are essential to a monitoring program. A minimum of four groundwater-monitoring wells are typically installed at a disposal facility: one up-gradient well and three down-gradient wells. However, site hydrogeology is often too complex for only four wells to provide adequate detection of groundwater contamination.

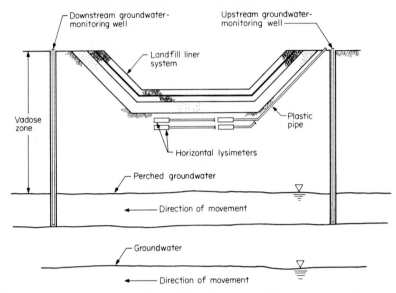

FIG. 10.1.9   Conceptual layout of system for vadose-zone and groundwater-monitoring.

Up-gradient wells are placed beyond the up-gradient extent of contamination expected from the landfill so that "natural" background groundwater quality will be reflected. These wells are screened at the same stratigraphic horizon(s) as the down-gradient wells to ensure comparability of data. A sufficient number of up-gradient wells are provided to obtain representative background water-quality information.

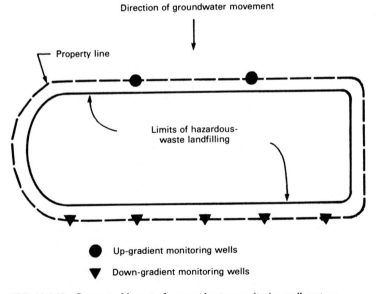

FIG. 10.1.10   Conceptual layout of a groundwater-monitoring well system.

At least three down-gradient wells are located to ensure that releases of hazardous waste or hazardous-waste constituents from the hazardous-waste management unit(s) to the uppermost aquifer will be quickly detected. The specific number of wells to be included in a detection system depends on the horizontal spacing between well locations and the vertical sampling interval of individual wells.

Down-gradient monitoring wells are screened to reflect the following:

- Hydrogeologic factors that influence potential contaminant pathways of migration to the uppermost aquifer
- Waste characteristics that affect movement and distribution of chemicals in the aquifer
- Hydrogeologic factors likely to affect contaminant movement

Figure 10.1.10 depicts a conceptual layout of a groundwater-monitoring well system.

Casing used for groundwater-monitoring well construction should be inert or at least chemically compatible with the potential contaminants to be monitored. Stainless steel 316 and polytetrafluoroethylene (PTFE, also called Teflon) are the generally recommended materials for casing groundwater-monitoring wells. These materials are more highly resistant to corrosion from the wide variety of chemical species likely to be encountered than are other materials such as polyvinyl chloride (PVC) and stainless steel 304. Figure 10.1.11 shows a conceptual monitoring-well cross section.

Surface-Water Monitoring. Surface-water monitoring is implemented as a routine component of a total monitoring network. Surface-water samples will generally be taken from stormwater-detention sumps and basins. Indicator parameters and analytical methods used for surface-water samples are consistent with selected procedures for groundwater sample testing. The effects of surface-water mixing and interference with contaminants should be considered.

LFG Monitoring. An LFG-monitoring system is generally installed to determine whether a potential hazard exists from subsurface migration of LFG from the landfill. [The Resource Conservation and Recovery Act (RCRA) requires that methane levels be maintained at 5% or less at the property boundary of the landfill, and less than 1.25% in any on-site structures.] A typical monitoring system consists of a series of monitoring wells constructed between the LFG-producing area of the landfill and the area of concern (e.g., an inhabited structure). The wells are generally equipped with one to three monitoring probes for detecting LFG concentrations. Figure 10.1.12 shows a typical monitoring-well configuration.

10.1.6 EQUIPMENT TYPES AND CHARACTERISTICS

Heavy equipment used for earth moving, waste handling, and compaction at hazardous-waste landfills includes rubber-tired loaders, vibratory compactors, forklifts and/or barrel snatchers, scrapers, graders, and water trucks. The functions of each equipment type are briefly described below.

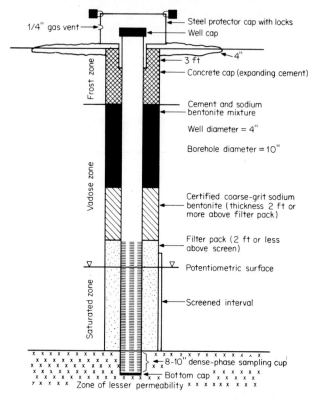

FIG. 10.1.11 Conceptual monitoring well (cross section). (*U.S. EPA, EPA/530/SW-85-012, Washington, D.C., 1985*)

Track and rubber-tired loaders.   Track and rubber-tired loaders can serve multiple functions, including the spreading and compacting of bulk hazardous wastes, the transport of rolls of FML and soil-liner materials, and the transport and spreading of cover soils in confined areas. Rubber-tired loaders are also used to transport and spread rolls of flexible-membrane liners.

Vibratory compactors.   Vibratory compactors are used to compact soils used for liner and waste cover. They compact best on flat and moderate slopes, but lack traction for operation on steep slopes (greater than 4:1).

Forklifts and/or barrel snatchers.   This equipment is used to unload incoming containers and to place them in the disposal area. Forklifts can also be used to transport rolls of FML and other materials.

Scrapers, graders, and water trucks.   Scrapers are used to excavate on-site soils and for spreading cover soils over exposed liners and/or hazardous waste. Graders are used for road and drainage-structure construction. Water trucks are used to control dust emissions from on-site roadways and sometimes to irrigate newly planted landscape.

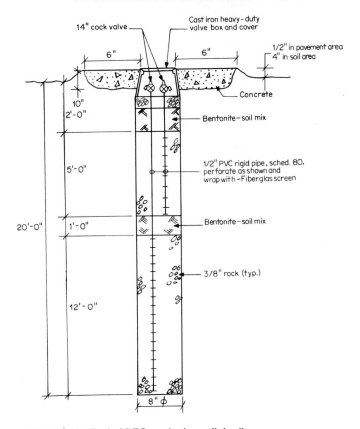

FIG. 10.1.12 Typical LFG-monitoring well detail.

Any equipment that is used on top of the protective cover soil layer of an FML system should be limited to the lighter, rubber-tired, front-end loaders and forklifts and barrel snatchers. The heavier equipment, such as scrapers or water trucks, should not be used until there is a minimum of 0.61 m (2 ft) of protective soil in place over the liner system.

10.1.7 CLOSURE AND POSTCLOSURE

Closure

Upon final closure of the landfill or upon completion of a cell, a final cover is applied which is constructed to

- Have a permeability less than or equal to the permeability of any bottom-liner system or natural subsoils present in order to provide long-term minimization of migration of liquids through the closed landfill's surface
- Function with minimum maintenance

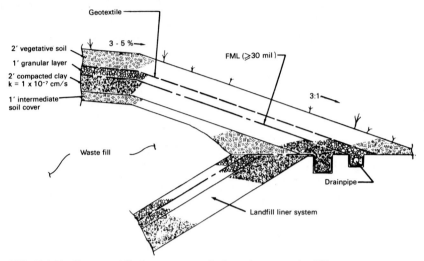

FIG. 10.1.13   Conceptual final-cover system for hazardous-waste landfill.

- Promote drainage and minimize erosion or abrasion of the cover
- Accommodate settling and subsidence so that the cover's integrity is maintained

For landfills with all or a portion of the wastes in above-grade cells, final side slopes are generally up to 3:1 horizontal to vertical. A 4.6- to 6.1-m wide (15 to 20 ft) slope bench is placed every 7.6 to 9.1 m (25 to 30 ft) of vertical rise to minimize slope erosion. Top surface slopes generally range from 3 to 5% to promote runoff and to accommodate some settling of the landfill. Figure 10.1.13 illustrates a conceptual final-cover configuration for a hazardous-waste landfill.

Postclosure

After final closure of a disposal facility, the following postclosure maintenance and monitoring activities are conducted throughout the postclosure care period:

- Perform routine final-cover maintenance. The integrity and effectiveness of the final cover is regularly maintained by filling and regrading depressions in the cover and surface as necessary to correct adverse effects such as settling, subsidence, erosion, or animal burrowing. Also, drainage and flood-control channels should be kept clear of accumulated debris. Dead vegetation should also be replaced.
- Prevent run-on and runoff of precipitation from eroding or otherwise damaging the final cover.
- Protect and maintain surveyed benchmarks.
- Maintain the environmental-monitoring facilities.
- Obtain samples from monitoring facilities and analyze and report on results in accordance with applicable permit requirements.

Postclosure monitoring during the first 12 to 18 months will generally involve relatively frequent site inspections and sampling (e.g., once per month). Monitoring frequencies can generally be reduced after this if no problems are encountered during that time.

10.1.8   DESIGN CONSIDERATIONS

Facility Life and Size

Design of the facility's useful life for disposal is determined by several factors including the area of the site, cover-soil availability, quantity and characteristics of the waste to be delivered, and the landfilling methods used. The usable area of the site is determined by excluding from the gross area land for (1) buffer between adjacent property at the site boundary and the filling area, (2) access roads, (3) soil stockpile area outside the fill surface, and (4) on-site structures and equipment-storage areas. Typically, the usable fill area ranges from 50 to 80% of the site's gross area.

Topography

Landfills are generally located in existing natural or excavated depressions or pits excavated for the purpose in areas with relatively flat terrain. Sites located in depressions (e.g., abandoned coal strip mines, clay or gravel pits, and natural depressions or canyons) generally have little suitable cover soil available on site. Therefore, the site's capacity for waste depends on the availability of soil from off-site sources for cover.

When a flat site is used for landfilling, the final site surface is designed to form a mound above the surrounding grade. This is done to optimize filling capacity of the site and to promote surface-water runoff. The final height of the mound is usually based on esthetic considerations and availability of cover soil. Cover soil is ideally obtained from excavation of on-site soils. The depth and aereal extent of excavation is dependent on several factors including the ease of soil excavation, the depth to groundwater, the ultimate height of the fill surface, and the projected waste input.

Unless there is a ready demand for soil at another location the designer aims to balance the volume of cover soil excavated and the soil required for landfill cover or other on-site uses (e.g., construction of a noise-control berm). Off-site drainage is diverted around this type of site, especially during the time that filling operations are underway in the excavated area.

Surface-Water Drainage

Design efforts involve mapping of existing bodies of surface water and water courses on or near proposed sites. Also, current and planned future use of this water are determined. Certain areas, such as 100-year floodplains, are avoided. Where it is necessary to construct a landfill in a 100-year floodplain, the EPA requires that the proponent be prepared to demonstrate that no adverse effects on human health or the environment will result if a washout occurs.

A knowledge of local rainfall intensities and off-site topography is required to determine run-on and runoff volumes to be handled by the site's drainage facilities. The drainage facilities at hazardous-waste landfills must be capable of handling off-site peak run-on from at least a 25-year storm, and all on-site collection and holding facilities must be able to accommodate on-site runoff volume from a 24-h, 25-year storm.

Soils

A knowledge of on-site soil characteristics is an important design consideration for a hazardous-waste landfill. Soils are used as construction materials at landfills to:

- *Provide cover:*   Cover soil serves several functions including (1) control of water infiltration, which in turn affects leachate production; (2) support of vegetative growth; (3) encapsulation of the deposited waste to isolate it from the local environment, and to reduce odor emissions and esthetic impacts.

- *Attenuate potential contaminants:*   Soil characteristics such as pH and cation-exchange capacity (CEC) influence the ability of a soil to attenuate cations in leachate that may form. Cation-exchange capacity and pH are in part determined by clay content, free iron oxide content, organic matter, the lime concentration of a soil, and soil permeability. In general, as CEC and pH increase, heavy metals are more readily retained. Similarly, as the clay content, free iron oxide content, and lime concentration of a soil increase, its pH and CEC (i.e., attenuation capacity) generally increase.

Soils to be used at a hazardous-waste landfill and site geology are thoroughly characterized. The volume of on-site soil of suitable properties is determined and compared against the needs for soil for liner construction and waste covering (daily, intermediate, and final) throughout the life of the landfill. Alternatively, soil can be imported from off-site locations, but importation can be expensive.

The ease with which the on-site soil can be excavated is a major design consideration. For example, excavation of bedrock or hardpan is costly, and such material is not usually well suited for cover. Seasonal variations in workability are considered; soils with a high proportion of fines may be easy to excavate when moist, but may behave like hardpan when dry or frozen. Soil types can also influence vertical and lateral LFG migration from the site.

Soils can be modified to enhance their attenuation capacity. One approach is to apply lime to the soil surface to increase pH. The clay and free iron oxide content can also be modified, but with greater difficulty and questionable success. Recent studies show that metal concentrations in leachate decrease when flow rates (or flux rates) decrease; however, further research is needed to determine the mechanisms associated with this phenomenon.

Geology

Knowledge of a site's geology is a necessity since the geology is an important design consideration. Formations that have faults, major fractures, joint systems, and other discontinuities or are soluble are avoided or provisions made in the landfill design to protect against groundwater contamination. In general, limestone, dolomite, and heavily fractured crystalline rock are less desirable than siliceous sandstone, siltstone, and other consolidated alluvial bedrock.

Groundwater

Hazardous-waste landfill designers consider several groundwater features of a site including

- Depth to groundwater from the bottom of the proposed waste fill (including historical highs and lows), and the physical and chemical properties of subgrade soils.
- Direction of groundwater movement. Based on the direction of groundwater movement, potential impacted areas can be identified and locations for on-site groundwater-monitoring wells can be determined. Hydraulic gradients can be computed from data obtained during the predesign subsurface investigations.
- Quality and availability of groundwater, its current and projected use, and the location of primary recharge and discharge zones. For example, a good landfill location may be a site overlying poor-quality and/or low-yielding groundwater where the groundwater basin does not discharge to a nearby watercourse.

Hazardous waste cannot be placed where there is a potential for direct contact with the groundwater table. Also, major recharge zones are eliminated from consideration, particularly in areas overlying EPA-designated sole-source aquifers which currently provide or could provide significant quantities of drinking water. The separation between the bottom of the fill and the highest known level of groundwater is generally maximized. State and local regulations may stipulate a minimum acceptable separation. A 1.5-m (5-ft) separation is a common regulatory stipulation, but this distance may vary from state to state.

Sources of data on groundwater quality and movement include the U.S. Geological Survey Groundwater Data Network, local well drillers, U.S. Department of Agriculture soil surveys, state geological surveys, state health departments, other state environmental and regulatory agencies, and samplings from nearby wells. The USGS also publishes an annual report entitled *Groundwater Levels in the United States* in its Water-Supply Paper Series. The data for these reports are derived from approximately 3,500 observation wells located throughout the United States.

10.1.9   ENGINEERING ECONOMIC EVALUATIONS

Hazardous-waste landfill costs are generally categorized into (1) capital costs, and (2) operating costs.

Capital Costs

Capital costs include all initial expenses required prior to the start-up of landfilling operations. Capital costs generally include

- Land acquisition.

- Planning and design.
- Permitting, including preparation of environmental impact assessments and special site investigations.
- Site preparation, including clearing and grubbing, road construction, surface-water and leachate controls, excavation of disposal areas, preparation of soil stockpiles, and installation of monitoring wells.
- Construction of facilities (i.e., offices, on-site laboratory, personnel shelters, garages, utilities, etc.). These types of facilities are generally assumed to have a 15-year useful life.
- Equipment purchase. The cost of equipment is often a significant portion of initial expenditures. The landfill-equipment market is competitive, but rough approximations of costs can be obtained from local equipment suppliers. As a rule of thumb, a piece of landfill equipment used for excavating, spreading, and compacting has a useful life of five years or 10,000 operating hours, whichever is less. Because of the high cost of equipment, however, it may be prudent to overhaul the machine at the end of five years or 10,000 operating hours, and then replace the machine after another two to three years, or 5,000 to 7,000 operating hours.

Closure costs can be significant. A closure fund is often established to raise monies during the operating life of the landfill to cover capital expenses to be incurred in the future.

Operating Costs

It is common to compute annual operating costs based on a cost per unit of waste received (e.g., dollars per ton). Operating costs for a landfill will vary significantly since local conditions and labor rates are site-specific.

A range of from $5 to $50 per ton is a guideline for design-level projections. Operating costs generally cover

- Labor, including operations, administration, and maintenance.
- Equipment fuel.
- Equipment maintenance and parts. Equipment maintenance and repair costs vary widely. Assuming a useful life of 10,000 h, maintenance costs for heavy earth-moving equipment can be expected to total approximately one-half of the initial cost of the machine. To make these costs more predictable, most equipment dealers offer lease agreements and maintenance contracts. Long down times usually associated with major repairs can be reduced by taking advantage of leasing and other programs offered by equipment dealers.
- Expendable supplies and materials.
- Utilities (e.g., electricity, heating oil, water, sewer, gas, and telephone).
- Ongoing inspection and engineering.
- Laboratory analyses (for samples of water, leachate, and LFG obtained through the environmental-monitoring program).
- Postclosure maintenance. These costs will not be incurred until after the facility is closed to receipt of additional waste deliveries.

10.1.10   FURTHER READING

Chemical Waste Land Disposal Facility Demonstration Grant Application. Interim Report, SW-87d.i, prepared by Barr Engineering Company for the Minnesota Pollution Control Agency, Minneapolis, 1976.

Civil Engineering: Solid Waste Disposal; Final Design Manual 5.10., prepared by SCS Engineers for the Pacific Division, Naval Facilities Engineering Command, Pearl Harbor, Hawaii, October 1985.

Cost Estimating Handbook for Transfer, Shredding and Sanitary Landfilling of Solid Waste. Final Report, SW-124c, prepared by Booz, Allen and Hamilton for the Office of Solid Waste Management Programs, U.S. Environmental Protection Agency, Washington, D.C., August 1976.

Everett, L. G., E. W. Hoylman, L. G. McMillion, and L. G. Wilson, *Constraints and Categories of Zone Monitoring Devices,* Kaman Tempo, Santa Barbara, California, August 1983.

Everett, L. G., L. G. Wilson, and E. W. Hoylman, *Vadose Zone Monitoring for Hazardous Waste Sites,* Noyes Data Corp., Park Ridge, New Jersey, 1984.

Everett, L. G., L. G. Wilson, and L. G. McMillion, "Vadose zone monitoring concepts for hazardous waste sites," *Ground Water* **20**:312–324 (1982).

Hansen, W. G., and R. A. Carr, *Cost Estimate for Closure and Post-Closure at the Pollution Control Center, Arlington, Oregon. Project Report,* SCS Engineers, Bellevue, Washington, for Chem-Security Systems, May 1981.

McGahan, J. F., "The secure landfill disposal of hazardous wastes," in *Toxic and Hazardous Waste Disposal,* R. B. Pojasek (ed.).

Minimum Technology Guidance on Double Liner Systems for Landfills and Surface Impoundments—Design, Construction, and Operation, second version, May 24, 1985.

Rishel, H. L., T. M. Boston, and C. J. Schmidt, *Costs of Remedial Response Actions at Hazardous Waste Sites,* SCS Engineers, Long Beach, California, for the Municipal Environmental Research Laboratory, Cincinnati, Ohio, October 1981.

Solid Waste Landfill Design and Operation Practices, prepared by SCS Engineers under EPA Contract #68-01-3915, Long Beach, California, April 1981.

Summary Report: Remedial Response at Hazardous Waste Sites, EPA-540/2-84-002a, prepared by JRB Associates for the Municipal Environmental Research Laboratory, Cincinnati, Ohio, March 1984.

U.S. Environmental Protection Agency, *RCRA Ground-Water Monitoring Technical Enforcement Guidance Document, Draft,* U.S. EPA Office of Water Programs Enforcement, Washington, D.C., August 1985.

U.S. Environmental Protection Agency, *Standards for owners and operators of hazardous waste treatment, storage, and disposal facilities, 40 CFR Part 264, amended to November 27, 1985, effective March 31, 1986.*

SECTION 10.2
SYNTHETIC LININGS

Louis R. Hovater
Hovater Engineers
Laguna Hills, California

According to Webster, a *lining* is defined as the material used or suitable for the covering of the inner surface of something. For hydraulic purposes, a *lining* is defined as the material used to line a structure to control or prevent seepage out of or into the structure. Such structures include surface impoundments (ponds, lagoons, reservoirs, lakes), landfills, underground tanks, and portions of buildings which are below ground surface. Because purposes and conditions of use vary with most structures, the materials used for linings must also vary.

Nature has provided us with the oldest, most widely accepted, used, and proven lining material—earth. However, on many projects impervious earth is not available, and a substitute must be used. Synthetic linings have been used successfully for over 30 years, and most linings used today are of the synthetic type. Synthetic linings are generally known as *flexible-membrane linings* (FMLs), and are classified as varieties of plastic, rubber, and asphalt. Today plastic and rubber in thin sheets dominate the lining field for disposal of hazardous wastes on land.

The state of the art relative to FMLs has become very complicated with almost constant changes occurring in lining materials and governmental regulations. Multiple linings are required for disposal of hazardous wastes, and triple linings are being used to control leakage. Regardless of the number of linings making up the lining system of a structure, the best results are obtained by relying on proper design, the best materials, adequate inspection, and top-quality construction techniques. The environment is priceless and its protection is a must. FMLs play a vital role in providing a part of the necessary protection.

10.2.1 APPLICATION

Once the decision has been made to use a synthetic lining (FML) for a particular project, selection of the proper FML is of paramount importance. The success or failure of the entire project could depend on the selection. Following selection, proper design, installation, and inspection assure that application of the selected FML is the best solution to the containment problem.

Definition

A *synthetic lining* [*flexible-membrane lining* (*FML*)] is by definition a continuous sheet of thin, flexible material such as plastic, rubber, or asphalt used to line a hydraulic structure (landfill, surface impoundment). Only plastic and rubber FMLs will be discussed in this chapter.

Characteristics

The most important characteristics of an FML are that it must be

- Flexible
- Proper for intended use
- Chemically compatible with the contained material
- Impermeable
- Nondecaying
- Durable
- Installed without difficulty
- Cost-effective

All lining characteristics must be evaluated to their fullest extent along with the design criteria to determine the best FML for the particular structure.

Types

Today plastic and rubber FMLs are used mostly for the containment of hazardous waste. Plastic FMLs include

- Polyvinyl chloride (PVC)
- High-density polyethylene (HDPE)
- Chlorinated polyethylene (CPE)

Rubber FMLs include

- Chlorosulfonated polyethylene (CSPE)
- Ethylene propylene diene monomer (EPDM)

Many other FMLs are available, but their use has been limited, primarily because they are not cost-effective compared with those listed. Except for HDPE, the FMLs listed above are available either unsupported or supported with a reinforcing fabric.

Uses

FMLs are used to line containments for hazardous wastes, such as landfills and surface impoundments. They are also used in other types of containments and structures for many purposes, but primarily to control or prevent seepage out of or into a facility (pond, reservoir, tank, building). Selection of an FML which is proper for the intended use is of paramount importance.

Chemical Compatibility

FMLs are susceptible to damage caused by chemical incompatibility with the contained hazardous waste. Evaluating and testing of the hazardous waste with potential FMLs are necessary to eliminate FMLs that are incompatible. A widely accepted industry compatibility test is known as Test Method 9090 and was developed by the U.S. Environmental Protection Agency (EPA). It involves testing of FMLs with the actual (or simulated) hazardous waste and takes about 120 days to complete. Most independent laboratories operating in the FML industry can perform EPA Test Method 9090.

10.2.2 DESCRIPTION

FMLs are generally described by (1) their industry-accepted chemical name, which is usually indicative of their composition—e.g., polyvinyl chloride (PVC), which is produced by any of several polymerization processes from vinyl chloride monomer—and (2) their physical properties. Other descriptive terms include available thicknesses, sizes and shapes, manufacturing processes, and factory fabrication.

Properties

Physical properties of FMLs are generally listed in tabular form and include properties with recommended test methods (ASTM, FTMS) such as

- Thickness
- Specific gravity
- Tensile strength
- Elongation
- Tear resistance
- Resistance to low temperature
- Dimensional stability

- Bonded-seam strength
- Peel adhesion

Thickness

FMLs are available in thicknesses ranging from 0.25 mm (10 mils) to over 2.5 mm (100 mils). Normal usage is as follows:

- *PVC:*   0.5 to 0.75 mm (20 to 30 mils)
- *HDPE:*   1 to 2.5 mm (40 to 100 mils)
- *CPE:*   0.75 mm (30 mils) unsupported
  0.9 to 1.15 mm (36 to 45 mils) supported
- *CSPE:*   0.75 mm (30 mils) unsupported
  0.9 to 1.15 mm (36 to 45 mils) supported
- *EPDM:*   1.15 to 1.5 mm (45 to 60 mils)

Sizes and Shapes

Most FMLs are manufactured by the calendering process into roll stock and factory-fabricated into various sizes and shapes; the exception is HDPE, which is manufactured by various extrusion processes, and, generally, is not factory-fabricated. Weight and configuration determine sizes and shapes. Most FMLs are rectangular, flat sheets weighing a maximum of 1,800 kg (4,000 lb), but some rolls of HDPE have weighed as much as 3.6 to 4.6 t (4 to 5 tons).

Manufacturing

Unsupported FMLs are manufactured by the calendering or extrusion processes. Supported FMLs are manufactured by laminating calendered roll stock onto both sides of an open-weave reinforcing fabric or by coating both sides of a closed-weave reinforcing fabric with the polymer in paste or semiliquid form. The latter is sometimes called a coated fabric and is a common type for PVC FMLs.

Factory Fabrication

FMLs are seamed into large sheets of various sizes and shapes in a factory using a number of techniques and seams such as

- Dielectric seams
- Solvent seams
- Thermal seams
- Vulcanized seams

After fabrication, the FMLs are accordion-folded in both directions and packaged for shipment to the field. Often FMLs are custom-fabricated to fit a particular structure, such as a tank or a pit, and are shipped from the factory as a one-piece lining to ease installation in the field.

Advantages

Many advantages are offered by FMLs, such as

- Impermeability
- Flexibility
- Toughness
- Durability
- Large sheets
- Chemical resistance
- Large temperature range
- Cost effectiveness

Of course, these advantages are most applicable when the selected FML is proper for the intended use and properly installed.

Disadvantages

Assuming that the decision to use an FML is correct, and that the installation of the FML is proper, about the only disadvantage of using an FML is that it can be easily damaged by weather, human activity, and animals. It is vulnerable to damage from these sources and generally must be protected by some means such as an earth cover, riprap, soil cement, or conventional roof. Depending on the particular facility, wind and wave action can destroy an FML in a short period of time unless protected, or unless the FML consists of a heavy-duty material with adequate stiffness and toughness to offset the effect of wind and waves.

10.2.3 DESIGN CRITERIA

Many factors contribute to the design criteria for a facility using an FML. These factors must be established before a suitable FML can be selected.

Use

The intended use of an FML is a criterion of utmost importance. Whether it will be used as a primary or secondary lining in a landfill or surface impoundment is a matter of vital concern; and whether it will contain hazardous waste or potable water is critical to the design criteria. After use has been determined, the design criteria should be established based on other factors governing design.

Size and Shape

The size and shape of a landfill or surface impoundment greatly affects the selection of an FML. If the facility is small, such as a few thousand square meters (square feet), the cost-effective answer may be different than for a facility involving millions of square meters (square feet). A small pond to contain oily waste may demand a more costly lining than a large pond used for evaporation of wastewater. Further, choice of shape for a facility can minimize waste of the selected FML. Irregularly shaped, free-form structures are more costly to line than square and rectangular ones, because most FMLs are available in rectangular sheet form. However, the real estate on which the structure is located often determines size and shape.

Depth

Theoretically, there is no limit to depth relative to an FML. If the FML were to be installed on an adequate supporting surface and protected from puncture by the stored material, very thin FMLs could be used. Practically, however, the risk is too great to depend on thin FMLs for many applications even of shallow depth: so as a result, thicker FMLs than theoretically necessary to contain the stored liquids are used. There used to be a rule which said "1 mil (0.025 mm) of thickness of FML for each foot (0.3 m) of hydraulic head." Thus, for a 6-m (20-ft) depth of liquid, use a 0.5-mm (20-mil) thick FML as a minimum. Today, with de minimis leakage requirements, and the availability of thicker FMLs (such as HDPE), the state of the art has negated this rule in favor of utilizing thicker FMLs to minimize the risk. Thickness is not always best, however, since some FMLs have failed because they were thicker than necessary and not as flexible as required.

Steepness of Slope

For most facilities, 3:1 slopes are used because of ease of construction. Steeper slopes create problems concerning cost of earthwork and cost of FML installation, since both become more difficult to do. Generally, flatter slopes are not cost-effective because of the additional real estate and quantity of FML required for the same volume. However, FMLs are used on all slopes, including vertical. Figure 10.2.1 shows a typical section of an FML on a slope.

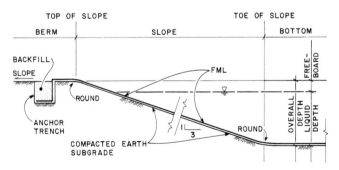

FIG. 10.2.1 FML on a slope. (*Hovater Engineers*)

Supporting Surface

The supporting surface for most FMLs is earth. Sandy silts and silty sands are best, because when properly prepared, they present a desirable smooth surface to support the FML. Earth with gravel and pieces of fractured rock must be prepared—by removal or embedment into the earth—to eliminate their ability to puncture, or covered with a protective material such as a geotextile or a layer of sand. Other types of supporting surface include concrete, gunite, asphalt concrete, metal, and wood. These surfaces must also be prepared to receive the FML, by eliminating all voids and projections, and to present a smooth clean surface to support the FML.

Open vs. Closed Facilities

Installation of an FML in an open facility is generally less costly versus installation in a closed one with roof columns. About the only time a closed facility has any advantage is during inclement weather, and with hot or cold ambient temperatures. Closed facilities require lights and ventilation. Accessibility can be a problem, but most open facilities offer ease of access with many places of entrance.

Accessibility

The difficulty of installation of an FML depends on accessibility to the work. If located on top of a mountain reachable only by a steep, winding road and surrounded by a narrow berm that will not accommodate a pickup truck, installation could be difficult versus the facility in flat country with wide berms and an area to work. Many times the bottom will provide a space to use as a staging area for installation of the FML, but an area that can be used adjacent to the work for storage of materials and equipment will help minimize installation costs. Whether the facility is open to the sky or enclosed beneath a roof, ease of access for installation of the FML greatly affects cost.

Life Expectancy

Many facilities will have a short life (less than 10 years), whereas others are almost "forever" since they will be abandoned in place. An average life expectancy is about 30 years, though some FMLs have a longer one. Some are guaranteed by the manufacturer against normal weather aging for 20 years on a prorated basis. Regardless of anticipated life expectancy at time of design, many FMLs remain in service long after their design life has expired.

Miscellaneous

There are numerous other factors to be considered at time of design such as

- Geographic location
- Climatic conditions

- Operation and maintenance
- Type of ownership

All factors important to the design criteria must be evaluated to the fullest extent. Generally, there is not just one FML that will do the job: two or three may be applicable. Selection of the proper one is like programming a computer—one collects all the available data and information—establishes design criteria—puts it all together as related to the problems—and, in most cases, the proper FML answer will be obvious.

10.2.4 INSTALLATION

Best results of installation are obtained by relying on state of the art design, materials, construction, and inspection. If one of these is missing, chances for successful performance of the FML are nil.

Supporting Surface

If the supporting surface is earth, it should be compacted and rolled to provide a smooth and even surface without abrupt breaks. All rocks, clods of earth, and debris must be removed or eliminated. If the supporting surface is concrete, asphalt, or some other material, all voids and projections must be removed or eliminated, and the surface cleaned of all foreign matter prior to installation of the FML. Figure 10.2.2 shows equipment preparing an earth subgrade.

FIG. 10.2.2 Preparation of earth subgrade. (*Hovater Engineers*)

Layout

Layout of the sheets comprising the completed FML must be given serious consideration by the installation contractor. Usually the project specifications require shop drawings indicating sheet size and layout to minimize field seams. Generally, horizontal seams on slopes are not a good practice, even though fabricated in a factory; and transverse seams at the toe of slope should be avoided. Most seams on slopes should be parallel with the slope and transverse seams on the bottom should be 1 to 1.5 m (3 to 5 ft) away from the toe of slope. Figure 10.2.3 shows a field crew installing an FML sheet.

Field Seams

Seams fabricated in the field to join sheets of an FML are generally made by lapping the edges of the sheets 5 to 15 cm (2 to 6 in) (depending on the type of FML and thickness), and using one of several techniques to provide effective field seams, such as adhesives, solvents, heat, and extrusions. With the increased use of FMLs, demand for improved techniques for making field seams has caused manufacturers and contractors to seek better equipment and materials for this purpose. Since this scenario changes almost constantly, the best practice is to obtain state of the art seaming techniques and materials from the FML manufacturer, or from experienced FML installation contractors. In some cases, requirements for handling hazardous waste are dictating a "belt and suspenders" solution for field seams, and a combination of more than one type of field seam is being used. For instance, a field seam in an HDPE lining may require both the lap weld and fillet weld using extrudate and/or heat. Regardless of the techniques

FIG. 10.2.3 Installation of an FML sheet

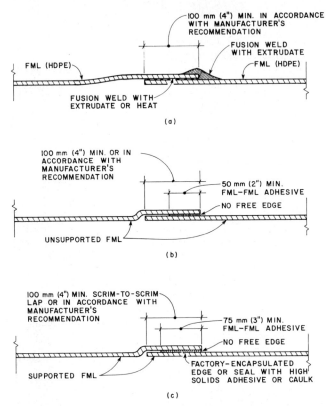

FIG. 10.2.4 Various types of field seams. (*a*) HDPE flexible-membrane lining. (*b*) Unsupported flexible-membrane lining (not applicable to HDPE). (*c*) Supported flexible-membrane lining (not applicable to HDPE). HDPE = high-density polyethylene. (*Hovater Engineers.*)

used, it is imperative that all field seams have no free edges on the exposed surface, and that all edges of the FML be tightly sealed. Edges that are tightly sealed do not assure seam strength, but they are a first line of defense against leakage. Figure 10.2.4 shows various types of field seams.

Anchors

Anchors of FMLs to earth, concrete structures, metal, and wood all vary with the FML. Anchors to earth are almost always located at the perimeter of an FML and involve insertion of the edge of an FML into an anchor trench. The trench is later backfilled with the excavated earth or sometimes with selected earth, concrete, or soil cement. Anchors to concrete structures are made by using stainless-steel anchor bolts and battens, or by plastic reglets cast in the concrete structures. The bolts and battens serve to clamp the FML to the structure, and a hard rubber rod performs the same function with the reglet. Attachments to metal and

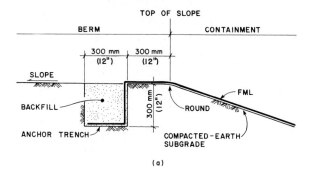

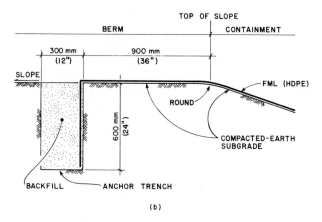

FIG. 10.2.5 Anchor trenches. (*a*) Flexible-membrane lining anchor trench (not applicable to HDPE). (*b*) HDPE flexible-membrane anchor trench. HDPE = high-density polyethylene. (*Hovater Engineers.*)

wood may be by stainless-steel bands to pipe, and wooden battens and galvanized nails to wood. Figure 10.2.5 shows typical anchor trenches for FMLs.

Seals

Seals are made in conjunction with anchors, and vary with the FML. Adhesives, sealants, caulks, tapes, and rubber pads (part or all) make up a typical seal. Generally, the best practice is to follow the manufacturer's recommendations concerning the use of these materials. Figure 10.2.6 shows typical anchor and seal of FML to concrete.

Protective Cover

It has been said many times that all FMLs, regardless of type or use, should be protected from damage by weather, human activity, and animals. Earth, riprap,

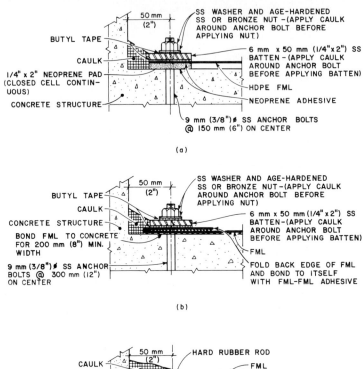

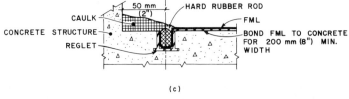

FIG. 10.2.6 Anchor and seal of FML to concrete. (*a*) HDPE flexible-membrane lining concrete structure. (*b*), (*c*) Flexible-membrane lining concrete structure (not applicable to HDPE. HDPE = high-density polyethylene. (*Hovater Engineers*)

concrete, or geotextiles are examples of proper protective covers. However, if these materials are improperly placed, as is often the case, damage can be done to the underlying FML. Many FMLs have failed because improper materials were improperly applied on top of the FML. For instance, an earth cover containing fractured rock applied in thin lifts by heavy equipment can easily damage even the thicker FMLs. Extreme care must be taken in specifying proper materials for a protective cover and in specifying acceptable procedures for installation. This work should only be done under the watchful eye of the engineer. Figure 10.2.7 shows a typical section of an FML with a protective cover. Figure 10.2.8 shows equipment installing an earth protective cover.

Geotextiles and Drainage Nets

These two types of geosynthetics have greatly assisted in the increased use of FMLs. Geotextiles are primarily used to provide support and protection for

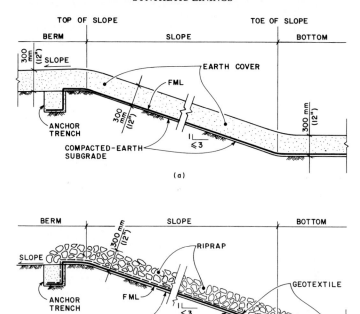

FIG. 10.2.7 FML with protective cover. (*a*) With earth cover or flatter. (*b*) With riprap cover or flatter. (*Hovater Engineers*)

FMLs, with some secondary use as drainage media. Drainage nets are primarily used to provide a drainage medium for FMLs, particularly between two FMLs, and, secondarily, to provide protection. One major use of a geotextile on a slope beneath an FML is to prevent the migration of the earth subgrade downslope due to flapping action of the FML against the earth subgrade caused by wind and wave action. The geotextile takes the abuse from the flap-

FIG. 10.2.8 Installing protective earth cover. (*Hovater Engineers*)

ping action and holds the soil particles in place. Further, using a thin drainage net instead of 300 mm (12 in) of gravel is a very cost-effective answer to a very tough question of construction procedure: how to place 300 mm (12 in) of gravel on an FML without damage.

Testing

If possible, it is imperative that all FMLs be tested before, during, and after installation. Resins, batching, calendering and extrusion processes, and factory fabrication should be tested according to industry standards prior to shipment of FML sheets to the field. During installation all seaming and sealing operations must be tested, and a watchful eye must be kept by all on the job to find imperfections. Before and during installation, there are two types of sampling for testing and two types of testing—destructive and nondestructive. Destructive sampling is the removal of samples for testing from the installed FML. Nondestructive sampling is the fabrication (during the work) of samples for testing using the same materials and seaming techniques as those being used in the actual work. Destructive testing is testing of the samples until they fail in tension, shear, or peel. Nondestructive testing is generally visual, and involves testing all field seams with a probe, vacuum box, spark tester, pressure gauge, and/or ultrasonic tester. Selection of one or all of the above methods depends on the type of FML and its thickness.

After installation, the final and best test of the FML is the hydraulic test. This test is generally performed only on facilities of small volume—say less than 40 million L (10 million gal)—because it requires the containment to be filled with water to some level (usually overflow or operating), and measurement of any leakage taken at a point such as a monitoring well, or by observing drop in water level over a period of time. All imperfections discovered during testing must be noted, repaired, and retested. In most cases, it is recommended that all testing be handled by a third party, except where installation contractors need to test for their own quality-control (QC) program as a guide to their own performance during installation.

Miscellaneous

There are many other things to be concerned about during installation of the FML, and some are

- Climatic conditions
- Contractor's QC program
- Contractor's personnel (including level of experience)
- Contractor's equipment
- Contractor's schedule
- Time of completion
- Test results

10.2.5   COST

There is no general rule of thumb pertaining to cost. Years ago before the FML field became crowded, with so many types and thicknesses available, the rule of thumb for the average FML installed was \$0.10 per square meter per 0.025 millimeter (\$0.01 per square foot per mil) of thickness—e.g., a 0.5-mm (20-mil) FML cost about \$2 per square meter (\$0.20 per square foot) installed. This cost did not include any other costs such as earthwork or protective cover. Cost today depends on the type of containment, the selected FML, the quantity of FML, and the difficulty of installation.

Material

Most FMLs used today are based on sheeting of polymeric materials. Since raw materials come from similar sources and manufacturing processes are comparable, finished FML materials of the various types are very competitive in cost. Material costs generally make up the bulk of the cost of an installed FML, unless, of course, the particular containment is labor-intensive, in which case labor may be the dominant cost. Such a containment could be a closed facility with many roof columns and pipe penetrations. For the average installation, material costs account for about two-thirds of the installed cost.

Installation

If on the average material costs are two-thirds, then one-third of the installed cost is for labor and equipment, assuming overhead and profit are spread evenly between materials and installation costs. Some FMLs require very little equipment for seaming purposes, whereas other FMLs require expensive equipment. A good example is the seaming of PVC with a plastic squeeze bottle and/or paint brush versus seaming of HDPE with a self-propelled lap welder. Also some FMLs are available in very large sheets requiring fewer field seams. Currently, PVC is probably the least expensive to install on a square meter (square foot) basis and EPDM the most expensive.

Capital

Capital costs are always of paramount importance, and it is the primary task of the design engineer to find the most cost-effective FML for the owner. This is why establishment of the design criteria is so critical: because a few cents per square meter (square foot) difference in installed cost of an FML can make or break the project economically. Time is money and spending more time in selecting the proper FML may result in large savings in capital costs.

Operating and Maintenance

Information and data relative to operation and maintenance costs of FMLs are not readily available because of their relatively short track records and because, in general, not many owners have bothered to keep such records. The EPA is attempting to gather available information and data, and to encourage future record keeping, but in the meantime, such costs are a guess.

10.2.6 ENVIRONMENTAL AND OTHER GOVERNMENT REGULATIONS

Preservation of the environment has no cost limitation according to most federal, state, and local regulations. As a result, cost-effectiveness of an FML is the ultimate determinator of its use. For a discussion of governmental regulations, please refer to Handbook Chap. 1.

10.2.7 CASE HISTORIES

As mentioned previously, FMLs have a very short track record, or history—only about 30 years. And their increased use to contain hazardous waste is most recent—only about 10 years. Case histories, as a result, are hard to find, particularly for waste piles and closures. To date, only a few new hazardous-waste containments of any sort have been constructed that conform to present minimum governmental requirements, and many existing ones need to be retrofitted.

10.2.8 VENDORS

For the purposes of this chapter, *vendors* are described as organizations that supply materials and services pertaining to FMLs for profit, such as manufacturers, fabricators and suppliers, and contractors. Generally, a vendor does not play all three roles of manufacturer, fabricator and supplier, and contractor, and selects one role only for profitable operation. This industry is very competitive, and there is a constant changeover among all three types of vendors.

10.2.9 FURTHER READING

U.S. Environmental Protection Agency, *Method 9090, Liner Compatibility Test,* proposed rule promulgated by the U.S. EPA Office of Solid Waste, *Federal Register,* vol. 49, no. 191, October 1, 1984.

SECTION 10.3

SUBSURFACE INJECTION OF LIQUID HAZARDOUS WASTES

Don L. Warner

School of Mines and Metallurgy
University of Missouri-Rolla
Rolla, Missouri

Until the mid-1960s, the subject of this section of Chap. 10 was described as *deep-well disposal*. Some still use this terminology. However, the majority now seem to prefer the terminology *subsurface* or *underground injection of wastewater* or *waste liquid*. In any case, what is being discussed here is the introduction of liquid hazardous wastes into the subsurface through drilled deep wells.

When used in this context, the word *deep* cannot be given any specific value, but refers to the depth required to reach a porous, permeable, saline-water-bearing rock stratum that is vertically confined by relatively impermeable beds. The minimum depth of burial, the necessary thickness of confining strata, and other site-specific factors must be determined in each individual case; these will be discussed later.

The section contains a brief overview of the extent of use of wastewater injection wells, and discussions of injection-well siting, of the evaluation of wastewater characteristics and preinjection treatment, of injection-well design and construction, of injection-system operation, of monitoring, and of regulation. Much of the chapter is summarized from Warner and Lehr.[1] The U.S. Environmental Protection Agency's (EPA's) 1985 Report to Congress on injection of hazardous waste is another important source.[2] The proceedings of the International Symposium on Subsurface Injection of Liquid Wastes that was held March 3 to 5, 1985, in New Orleans, Louisiana, under the sponsorship of the Underground Injection Practice Council and the National Water Well Association is an invaluable source of current hazardous-waste injection-well technology.

10.3.1 *HISTORY AND CURRENT USE*

It is not certain where controlled wastewater injection was first practiced outside the oil field, but Harlow, in an article published in 1939[3] described problems encountered by Dow Chemical Company in disposing of waste brines from chemical manufacturing by subsurface injection. Inventories by various individuals and groups have succeeded in locating no more than four such wells constructed prior to 1950. A 1963 inventory by Donaldson[4] listed only 30 wells. Subsequent inventories published in 1967,[5] 1968,[6] 1972,[7] and 1974[8] listed 110, 118, 246, and 278 wells, respectively. The most recent comprehensive inventory[9] showed that a total of 322 industrial and municipal injection wells had been drilled as of January 1975, and 209 of those were reportedly operating at that time. In an inventory limited to wells injecting hazardous wastes, the U.S. EPA[2] identified 112 facilities which inject hazardous wastes through 252 wells. Of those facilities, 90 were active and injected hazardous waste into 195 wells during 1984. The other 57 wells were inactive. Of the 195 active wells, 152 operated continuously and 43 intermittently. Of the 57 inactive wells, 41 were abandoned, 3 were shut-in or in the process of changing type of operation, and 13 had a permit pending or were under construction.

Active hazardous-waste injection wells are presently located in 15 states. The vast majority of the wells are located along the Gulf Coast and near the Great Lakes. Louisiana and Texas alone account for 66% of the wells. Other states with sizable numbers of hazardous-waste wells are Michigan, Indiana, Ohio, Illinois, and Oklahoma. Historically, these states have had experience in underground injection mainly because of oil- and gas-related activities, which have provided abundant data on deep formations. Geologically, formations in these states are amenable to efficient and environmentally safe injection. Another common characteristic, though not exclusive to these two regions, is that both are highly industrialized.

According to the EPA data,[2] most of the wells were drilled between the mid-1960s and the mid-1970s. There has been no significant increase in the rate of construction of new wells since 1980. The EPA's 1985 report[2] shows that the biggest user of hazardous-waste injection wells is the chemical industry. Manufacturers of organic chemicals account for about 44% of the wells and about 50% of the volume. The petroleum-refining industry accounts for 20% of the wells and 25% of the volume. Other chemical manufacturers (agricultural, inorganic, and miscellaneous) account for 17.5% of the wells and 12.6% of the volume. The metals and minerals industry accounts for 8.2% of the wells and 5.8% of the volume. The aerospace industry accounts for 1% of the wells and 1.5% of the volume.

In 1985, only 4.4% of the total injected volume was handled by commercial waste disposers with 9.2% of the wells (18 wells at 13 facilities).[2] These are classified as *off-site wells* because they inject hazardous waste which has been generated at other locations.

10.3.2 *INJECTION-WELL SITING*

The general technical requirements of a suitable site for a hazardous-waste injection well include

1. A saline-water-bearing injection interval that is sufficiently thick, extensive, porous, and permeable to accept wastewater at the necessary rate at safe in-

jection pressures and that is not economically more valuable for its contained mineral resources or other uses.

2. Overlying and underlying strata (confining beds) sufficiently thick, extensive, and impermeable to confine waste to the injection interval.

3. The absence of solution-collapse features, faults, extensive joints, or unplugged or improperly plugged abandoned wells that would permit escape of injected wastewater from the injection interval into adjacent aquifers.

Knowledge of the geologic and hydrologic characteristics of the subsurface environment at an injection-well site and in the surrounding region is fundamental to the evaluation of the suitability of the site for wastewater injection and to the design, construction, operation, and monitoring of injection wells. In defining the geologic environment, the subsurface rock units that are present are described in terms of their lithology, thickness, aereal distribution, structural configuration, and engineering properties.

Examination of a site for a hazardous-waste injection well begins at the regional level, then is narrowed to the vicinity of the site, and finally focuses upon the immediate well location. Table 10.3.1 lists the factors important to regional and local site evaluation.

TABLE 10.3.1 Factors to Be Considered for Geologic and Hydrologic Evaluation of a Site for Subsurface Waste Injection

Regional Geologic and Hydrologic Framework

Physiography and general geology

Structure

Stratigraphy

Groundwater

Mineral resources

Seismicity

Hydrodynamics

Local Geology and Geohydrology

Structural geology

Geologic description of subsurface rock units

General rock types and characteristics

Description of injection horizons and confining beds: lithology, thickness and vertical and lateral distribution, porosity (type and distribution as well as amount), permeability (same as for porosity), reservoir temperature and pressure, chemical characteristics of reservoir fluids, formation breakdown or fracture pressure, hydrodynamics

Freshwater aquifers at the site and in the vicinity: depth, thickness, general character, quality of contained water, amount of use and potential for use

Mineral resources and their occurrence at the well site and in the immediate area: oil and gas (including past, present, and possible future development), coal, brines, other

Source: Ref. 1.

10.3.3 WASTEWATER CHARACTERISTICS

Table 10.3.2 lists the characteristics of the untreated wastewater that must be considered in evaluating its suitability for disposal by subsurface injection. Preliminary examination of these factors will show, in general, whether the effluent is such that more detailed appraisal is warranted. During a detailed study, an attempt is made to define all design and operational problems related to the wastewater and to provide for these by plant process control, wastewater pretreatment, design modification, and operational procedure.

It is not usually necessary for injection purposes to know the exact chemical composition of a wastewater, because empirical tests can be run to determine the reactivity and stability of a waste. However, this information should be obtained so that the environmental effect of the injected fluid can be assessed.

Volume

One of the most constraining limitations in managing a wastewater by subsurface injection is the volume that can be safely injected for the desired length of time. This is because the intake rate or life of an individual well is limited by the properties of the injection interval, which cannot be changed much. The variable limiting the injection rate or well life can be the injection pressure required to dispose of the produced waste. Injection pressure is a limiting factor because excessive pressure causes hydraulic fracturing and possible consequent damage

TABLE 10.3.2 Factors to Be Considered in Evaluating the Suitability of Untreated Wastes for Deep-Well Disposal

A. Volume

B. Physical characteristics
  1. Density
  2. Viscosity
  3. Temperature
  4. Suspended-solids content
  5. Gas content

C. Chemical characteristics
  1. Dissolved constituents
  2. pH
  3. Chemical stability
  4. Reactivity
    (a) With system components
    (b) With formation waters
    (c) With formation minerals
  5. Toxicity

D. Biological characteristics

Source: Ref. 1.

to confining strata. In addition, the pressure capacity of injection-well pumps, tubing, and casing is limited.

The initial pressure required to inject waste at a specified rate and the rate at which injection pressure increases with time can be calculated if the physical properties of the aquifer and waste are known. The average intake rate* of hazardous waste-injection wells now in use ranges from less than 6.3×10^{-5} m^3/s to more than 4.4×10^{-2} m^3/s (less than 1 gpm to more than 700 gpm).[2] The average rate for all active hazardous-waste injection wells is about 7.4×10^{-3} m^3/s (120 gpm). The EPA estimated that a total of 43.7×10^6 m^3 (11.5 billion gal) of wastewater was injected through the 151 hazardous-waste injection wells in operation in 1983.

Waste Types

The U.S. EPA[2] reported data compiled in 1983 that showed that, based on analysis of 108 injection wells, the hazardous wastes being injected were acids (41.3%), organic chemicals (36.3%), heavy metals (1.4%), inorganic chemicals (0.08%) and other wastes (21%). More than half of the injected wastes were nonhazardous but, because of the presence of the hazardous wastes, the entire injected volume is considered hazardous.

Compatibility

In the siting, design, and operation of an injection well, the compatibility of the injected wastes with the mechanical components of the injection-well system, with the injection and confining intervals, and with the natural formation water is an important consideration. For example, a corrosive acid stream, when injected without neutralization, can react with all of the above, sometimes with very unfortunate results.

Corrosion is, of course, the most common undesirable reaction between a waste and injection-well components. Reactions with injection and confining intervals include dissolution of carbonate minerals and a variety of possible interactions with clay minerals. The formation of precipitates or polymerization of injected chemicals, both of which can lead to plugging of the injection interval, are examples of undesirable wastewater–formation water reactions.

Suspended solids may pose no problem when injection is into a limestone or dolomite with large pore spaces but may need to be removed to the greatest possible extent when injecting into a fine-grained sandstone.

The potential for the various undesirable types of reactions is anticipated on the basis of theory, experience, and laboratory testing. If an incompatibility problem is identified, it can often be addressed either by materials selection or by pretreatment of the wastewater prior to injection.

Preinjection Treatment

To ensure success of a subsurface waste-disposal operation, surface preinjection treatment of the wastewater is generally required. Treatment can be quite expensive, but it can make the difference between a successful, environmentally cog-

---

*The average intake rate in gpm is the annual injected volume divided by the minutes per year.

TABLE 10.3.3   Operational Problems Related to Wastewater Character

Problem of concern	Means of evaluating	Means of controlling undesirable effects
Reaction		
Wastes and formation minerals	Laboratory tests and observation of system	Preinjection waste treatment
Wastes and formation water	Laboratory tests	Preinjection waste treatment or a buffer zone
Autoreaction of waste at formation temperature and pressure	Laboratory tests	Preinjection waste treatment
Wastes and system components	Laboratory tests and observation of system	Preinjection waste treatment, addition of corrosion inhibitors to waste, and use of corrosion-resistant materials
Microorganisms	Laboratory tests and observation of system	Preinjection waste treatment and addition of biocides
Suspended solids and oils	Laboratory tests and observation of system	Preinjection waste treatment or formation treatment
Entrained or dissolved gases	Laboratory tests	Chemical or mechanical degasification

nizant operation and one subject to repeated difficulties and even failure. The principal objective of preinjection treatment is to modify the physical and chemical character of a waste so that it is compatible with the surface and subsurface equipment that is used and with the injection interval and its contained water. Table 10.3.3 is a brief summary of operational problems related to wastewater character along with suggested means for evaluating and controlling them.

As suggested in Table 10.3.3, pretreatment processes are not the only means of solving some of the problems related to wastewater character. For example, corrosion control can be achieved by use of corrosion-resistant materials and a buffer zone can be established to prevent reaction between injected fluids and aquifer water. The choice of control method will be dictated by economics, engineering feasibility, and regulatory controls.

The first step in development of a preinjection scheme is characterization of the wastewater in terms of source, volume, and physical, chemical, and biological characteristics. Then, if the waste is classified as suitable for injection, the necessary treatment processes can be selected.

10.3.4 INJECTION-WELL DESIGN AND CONSTRUCTION

After the suitability of a possible injection-well site has been confirmed regionally and locally, and the suitability of the wastewater to be injected has been determined, it is appropriate to develop a well-design and construction program. Since the well is such an important part of the injection system, its proper design and construction are obviously fundamental to the success of the injection program. It has been found by experience that minor mistakes in design or construction can result in damage to the well and subsequent economic loss and possible degradation of the environment. The most common deficiency in well design has been inadequate provision for corrosion protection. In construction, an oversight as small as improper selection of the drilling fluid used while drilling the injection interval can result in irreversible loss of permeability and possibly even in well abandonment.

Planning a Well

The first step in planning a well is the acquisition of geologic and engineering information for the well site and for any previously constructed wells in the area. Determination of the geologic and engineering characteristics of the site will have been accomplished during the regional and local site evaluation. These characteristics must now be evaluated in view of their significance to well design and construction. Local experience in well design and construction is an invaluable aid in planning a well and the importance of obtaining information concerning local design and construction practices cannot be overemphasized.

A major decision that is made early in the planning of a well is the choice of the bottom-hole completion method. A wide variety of bottom-hole completion methods are used, but in general they can be categorized as methods applied to wells completed in competent formations and methods applied to wells completed in incompetent formations. *Competent formations* are limestones, dolomites, and consolidated sandstones that will stand unsupported in a borehole. The most commonly encountered *incompetent formations* are unconsolidated sands and gravels. They will cave readily into the borehole if not artificially supported. There are, of course, intermediate cases of formations that are generally competent but require some support for such purposes as preventing sloughing of occasional incompetent intervals or preventing fractured blocks from falling into the borehole.

Examination of the records of wastewater injection wells that have been constructed in the United States shows that almost all the wells constructed thus far have been completed by one of three methods or close variations of them. The methods are

1. Open-hole completion in competent formations
2. Screened, or screened and gravel-packed in incompetent sands and gravels
3. Fully cased and cemented with the casing perforated in either competent or somewhat incompetent formations

Most wastewater injection wells will be constructed with injection tubing inside the long casing string, and with a packer set between the tubing and the casing near the bottom of the casing (Fig. 10.3.1). This design is not entirely free of problems, particularly with the packer, but experience has proved it generally superior to other designs. Some wells are completed with an annulus open at the bottom. The annulus is filled with a lighter-than-waste liquid which "floats" on the aqueous waste. This type of well completion has been referred to as a *fluid-seal completion*.

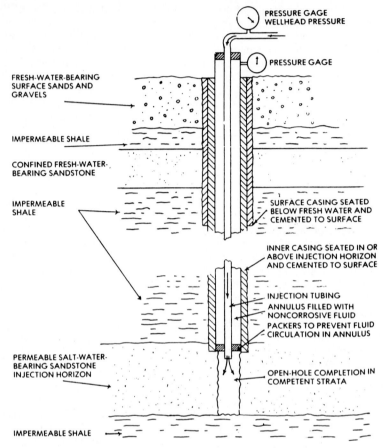

FIG. 10.3.1 Schematic diagram of an industrial-waste injection well completed in competent sandstone (*After Ref. 10*)

Injection-System Operation

The period between the completion of injection-well construction and initiation of full-scale operation affords an excellent opportunity to undertake several very important tasks. The well operator may run several types of logs and final tests, including mechanical integrity tests, to obtain the data needed to establish the baseline characteristics of the well. The operator can also use this time to instruct supervisory and operating personnel in all aspects of injection-well operation, as well as in proper procedures for dealing with both common problems and emergencies.

The operating program for an injection system should be adapted to the geological and engineering properties of the injection interval and to the volume and chemistry of the waste fluids. The geological and engineering properties of the injection interval are defined during construction and testing, and the final operating program is then developed.

Injection rates and pressures must be considered jointly, since the pressure will usually depend on the volume being injected. Pressures are limited to values that will prevent damage to well facilities or to the confining formations. The maximum bottom-hole injection pressure is commonly specified on the basis of well depth. It ranges upward from about 0.5 psi per foot of depth but must be based on local hydraulic fracturing gradients and will seldom exceed about 0.8 psi per foot.

Operating Procedures

Wastewater injection-well systems are normally operated by technicians, with supervision by professional engineers or scientists. Because such systems can be complex, and because the operating personnel will usually be inexperienced in injection-well operation, all details of the system's operating procedures should be documented as thoroughly as possible in an operator's manual. Since the operating procedures will be unique to each system, no attempt will be made here to specify the contents in detail. However, some of the essential subject matter would include instructions relating to the sources and composition of the wastewater streams to be handled, to the nature and operation of preinjection wastewater-treatment systems, to monitoring procedures, to regulatory requirements, and to any emergency procedures. It is common to provide training sessions to educate operating personnel when a new system is to be put into operation.

Contingency Plan

A plan should be developed by the well operator and approved by the state regulatory agency for an alternative waste-management procedure if the injection well should become inoperative or if it needs to be shut down. Such a plan requires the availability of a standby facility, which could be a standby well, holding tanks or ponds, or a waste-treatment plant.

Full-Scale Operations

The transition from the last stage of development to full-scale operation should be gradual. The well operator must recognize that, as the pressure front and the leading edge of the effluent radiate outward, exploration and development are a continuous process.

When waste injection into the subsurface is begun, the operator immediately incurs an obligation. The requirement that the operator know "what it is," "where it is," and "what it is doing," is not relieved until the project is both chemically and hydrodynamically inactive. This may be a period of months, years, or decades after the cessation of injection.

10.3.5 MONITORING

The principal means of surveillance of wastewater injection that is presently practiced is monitoring at the injection well of the volume, flow rate, chemistry, and biology of the injected wastewater and of the injection and annulus pressures (Fig. 10.3.1). To some this apparently seems inadequate. However, if all the necessary evaluations have been made during the planning, construction, and testing of the well, then this may be a satisfactory program when combined with periodic inspection of surface and subsurface facilities. This is because the greatest risk of escape of injected fluids is normally through or around the outside of the injection well itself, rather than from leakage through permeable confining beds, fractures, or unplugged wells.

The purpose of monitoring the volume and chemistry of injected wastewater is to allow for estimates of the distance of wastewater travel, to allow for interpretation of pressure data, and to provide a permanent record. This record is needed as evidence of compliance with restrictions, as an aid in interpretation of well behavior, as an aid in well maintenance, and as a precaution in the event that a chemical parameter should deviate from design specifications. Some characteristics that have been monitored continuously are flow, suspended solids, pH, conductance, temperature, density, dissolved oxygen, and chlorine residual. Complete chemical analyses are frequently made on a periodic basis on composite or grab samples. Because bacteria may have a damaging effect on reservoir permeability, periodic biological analysis of some wastewaters may be desirable to ensure that organisms are not being introduced.

Injection pressure is monitored to provide a record of reservoir performance and as evidence of compliance with regulatory restrictions. Injection pressures are limited to prevent hydraulic fracturing of the injection reservoir and confining beds, or damage to well facilities. As with flow data, injection pressure should be continuously recorded.

Other methods of injection-well monitoring also deserve mention. The corrosion rate of well tubing and casing may be monitored by use of corrosion coupons inserted in the well. *Corrosion coupons* are weight-loss specimens of the metal of which the tubing or casing is made. They are carefully cleaned and weighed before exposure, exposed for a carefully measured period of time, and then removed, cleaned, and reweighed. The weight is divided by the specimen area and time of exposure to give the corrosion rate, usually in mils per year.

A conductivity probe may be used to detect a change in the chemistry of the fluid in the casing-tubing annulus. In wells with packers, the conductivity probe can be used to detect tubing leaks, and in wells without packers to detect shifts in the interface between the injected fluid and the casing-tubing fluid. Another technique that has been used to monitor the casing-tubing annulus is continuous cycling of the annulus fluid and analysis of the return flow for evidence of contamination by wastewater.

Monitoring Wells

The subject of monitoring wells has been a controversial one in regulation of wastewater injection. Such wells are routinely used in shallow groundwater studies but are less frequently used in conjunction with wastewater injection, for reasons that will be examined.

At least three hydrogeologically different types of monitor wells can be and have been constructed, each with different objectives as shown in Table 10.3.4.

Monitor wells constructed in the receiving aquifer are normally nondischarging because a discharging well would defeat most of the purposes of this type of monitor well. Also, the produced brines would have to be disposed of. Although it is not normally necessary to monitor pressure in the receiving

TABLE 10.3.4 Objectives of Various Kinds of Monitoring Wells

Well type	Objective
Constructed in receiving aquifer—nondischarging	Obtain geologic data
	Monitor pressure in receiving aquifer
	Determine rate and direction of wastewater movement
	Detect geochemical changes in injected wastewater
	Detect shifts in freshwater–saline water interfaces
Constructed in or just above confining unit—nondischarging	Obtain geologic data
	Detect leakage through confining unit
Constructed in a freshwater aquifer above receiving aquifer	Obtain geologic data
	Detect evidence of freshwater contamination

aquifer except at the injection well, special monitor wells may be desired where pressure at a distance from the injection well is of concern because of the presence of known or suspected faults or abandoned wells that may be inadequately plugged. The pressure response in a monitor well at such locations would indicate the extent of danger of flow through such breaches in the confining beds and possibly also indicate whether leakage was occurring.

Constructing a monitor well or wells in the receiving aquifer is the only direct means of verifying the rate and direction of wastewater movement. However, more than one well will frequently be necessary to meet this objective, because monitor wells of this type only sample wastewater plumes that pass directly through the well bore; and nonuniformity in aquifer porosity and permeability can cause wastewater to arrive very rapidly or perhaps not at all at a particular well.

Negative factors should be considered in any case where deep monitor wells are contemplated: monitor wells in the receiving aquifer may be of limited usefulness, and they provide an additional means by which injected wastewater might escape from the receiving aquifer.

For detection of leakage, the use of nondischarging monitor wells completed in the confining beds or in a confined aquifer immediately above the confining beds has been widely discussed in principle but has been little practiced. This type of well has the potential for acting as a very sensitive indicator of leakage by allowing measurement of small changes in pressure (or water level) that accompany leakage. A well of this type is best-suited for use where the confining unit is relatively thin and well-defined and where the engineering properties of the two aquifers are within a range such that pressure response in the monitored aquifer will be rapid if leakage occurs.

The type of monitor well most commonly in use is that which is completed in a freshwater aquifer above the injection horizon; the purpose of these wells is detection of freshwater contamination. In a number of locations, this type of monitoring is provided by wells that are a part of the plant's water-supply system. In other cases, the wells have been constructed particularly for monitoring and are not used for water supply. Wells for detection of freshwater contamination should be discharging wells because they then sample an area of aquifer within their cone of depression. As previously mentioned, nondischarging wells are of limited value for detection of contamination because they sample only that water that passes through the well bore. Wells for monitoring freshwater contamination should be located close to the anticipated sources or possible routes of contamination, which are

1. The injection well itself
2. Other nearby deep wells, active or abandoned
3. Nearby faults or fracture zones

Since the objectives for each of the types of monitoring wells discussed are worthwhile ones, why are monitor wells not more widely used? The answer to this question is that the potential benefits are often judged to be small in comparison with the costs and negative aspects. Therefore, such wells may not be voluntarily constructed by the operating companies nor required by the regulatory agencies. In particular, monitor wells constructed in the receiving aquifer are often difficult to justify because such wells are the most expensive form of surveillance and may yield very little information that is important for regulation. It can reasonably be concluded that monitor wells should not be arbitrarily required, but should be used where local circumstances justify them.

10.3.6 REGULATIONS

The principal means of federal regulation of hazardous-waste injection wells is through the Underground Injection Control (UIC) program, which was mandated in the Safe Drinking Water Act (SDWA) of 1974. Some other regulation of such wells is through the Clean Water Act and the Resource Conservation and Recovery Act (RCRA).

Under the SDWA, the national regulations, which were promulgated in 1980, define minimum standards for effective state UIC programs. Requirements become applicable to owners and operators of injection wells in a particular jurisdiction when the administrator approves a state's UIC program or promulgates a federally implemented program for a state, except that injectors of hazardous waste are subject to the interim standards under RCRA.

Under UIC regulations, injection wells are classified according to their use and ownership. Wells for injection of hazardous industrial wastes are Class I.

Because of the environmental concern regarding hazardous wastes, Class I hazardous-waste injection wells must meet very strict construction and operating requirements. These technical requirements are set forth in 40 *CFR* Part 146, Subparts A and B. Subpart A contains general specifications used for permitting and repermitting all Class I wells. Subpart B provides for specific construction, operation, monitoring, and reporting requirements that take into account the site characteristics for a well. These characteristics include the geology, hydrology, types of waste, and construction techniques.*

A stated purpose of the Safe Drinking Water Act is the delegation of the UIC program to the states. The EPA has delegated the UIC program in those states that have most of the hazardous-waste injection wells, and provides technical and financial assistance to these states for implementation of the program.

---

*The EPA has recently considered amendments in "Proposed Rules Underground Injection Control Program, Class I Wells," *Federal Register*, vol. 52, no. 166, August 27, 1987, pp. 32446–32476.

In the event that a state fails to submit an application, or if a state application is disapproved, the EPA must promulgate the UIC program for that state and assume primary enforcement responsibility.

10.3.7 REFERENCES

1. D. L. Warner and J. H. Lehr, *An Introduction to the Technology of Subsurface Wastewater Injection,* U.S. EPA-600/2-77-240, 1977; reprinted as D. L. Warner and J. H. Lehr, *Subsurface Wastewater Injection,* Premier Press, Berkeley, California.

2. U.S. Environmental Protection Agency, *Report to Congress on Injection of Hazardous Waste,* EPA 570/9-95-003, U.S. EPA Office of Drinking Water, Washington, D.C., 1985.

3. I. F. Harlow, "Waste Problem of a Chemical Company," *Ind. Eng. Chem.,* **31**:1,345–1,349 (1939).

4. E. C. Donaldson, *Subsurface Disposal of Industrial Wastes in the United States,* U.S. Bureau of Mines Information Circular 8212, 1964.

5. D. L. Warner, *Deep-Wells for Industrial Waste Injection in the United States—Summary of Data,* Federal Water Pollution Control Research Series Publication No. WP-20-10, 1967.

6. R. E. Ives and G. E. Eddy, *Subsurface Disposal of Industrial Wastes,* Interstate Oil Compact Commission, Oklahoma City, Oklahoma, 1968.

7. D. L. Warner, *Survey of Industrial Waste Injection Wells* (3 vols), Final Report, U.S. Geological Survey Contract No. 14-080001-12280, University of Missouri, Rolla, Missouri, 1972.

8. U.S. Environmental Protection Agency, *Compilation of Industrial and Municipal Injection Wells in the United States* (2 vols.), EPA-520/9-74-020, U.S. EPA Office of Water Program Operations, Washington, D.C., October 1974.

9. L. Reeder et al., *Review and Assessment of Deep-Well Injection of Hazardous Waste,* Final Report for EPA Contract No. 68-03-2013, Program Element No. 1DB063, U.S. EPA Solid and Hazardous Waste Research Laboratory, Cincinnati, Ohio.

10. D. L. Warner, *Deep Well Injection of Liquid Waste,* U.S. Department of Health, Education and Welfare, Public Health Service Publication No. 99-WP-21, 1965.

11. D. L. Warner and D. H. Orcutt, "Industrial Wastewater-Injection Wells in the United States—Status of Use and Regulation, 1973," in *Underground Waste Management and Artificial Recharge,* Jules Braunstein (ed.), American Association of Petroleum Geologists, Tulsa, Oklahoma, 1973, pp. 552–564.

SECTION 10.4

SURFACE IMPOUNDMENTS

David C. Anderson
Vice President
K. W. Brown and Associates, Inc.
College Station, Texas

A *surface impoundment* (*SI*) is an excavation or diked area designed to contain an accumulation of liquid wastes or wastes which contain free liquids. There are two basic types of SIs used to manage hazardous waste: those designed for the ultimate disposal and those designed for the temporary storage of hazardous wastes. *Disposal SIs* are designed to allow closure as landfills, while *storage SIs* are temporary structures which eventually require the removal of contaminated components and subsoil at closure. Nearly all SI facilities can be considered *treatment SIs* since some treatment occurs in all impoundments. Several recent studies have been conducted to gather background information on SI's.[1,2] Findings from each of these studies are summarized below, followed by a discussion of current regulations for hazardous-waste SIs.

10.4.1 RECENT SURVEYS

The Surface Impoundment Assessment conducted by the U.S. Environmental Protection Agency (EPA)[1] was a survey to account for as many of the SIs in the United States as possible. State and federal agencies, universities, and consultants combined efforts to locate and characterize over 180,000 SIs. A majority of these SIs were used either as pits for oil- and gas-production fluids or as municipal sewage and wastewater-treatment facilities. A substantial number of other SIs were associated with agricultural and mining activities. The survey also revealed that there were nearly 30,000 SIs used by industry.

Another recent survey, conducted by the Chemical Manufacturers Association, found that 99.4% of all treated and disposed hazardous waste was wastewater, while only 0.6% was solid waste.[3] Approximately 90% of the wastewater is treated, stored, or disposed of in SIs. Over a quarter of the solid waste was also placed in SIs.[3]

The largest amounts of hazardous waste placed in SIs were generated by the organic-chemical plants. Two types of industrial plants [represented by Standard Industrial Classification (SIC) Codes 2869 (Miscellaneous Organic Chemicals) and 2821 (Plastics)] accounted for 53% and 71% of the solid waste and hazardous wastewater placed in SIs. Other SIC codes which place large volumes of hazardous waste in SIs include the following:[3]

1. 2819 (Miscellaneous Inorganic Chemicals)

2. 2816 (Inorganic Pigments)

3. 2865 (Cyclic Crudes and Intermediates)

The largest amounts of general waste placed in industrial SIs were generated by food processors, organic- and inorganic-chemical plants, oil refineries, primary- and fabricated-metal industries, pulp and paper plants, and commercial waste-management facilities.[1]

Most industrial SIs are at least partially used for treatment of waste.[1,3] The treatment processes most widely used are neutralization, settling, anaerobic or aerobic digestion, pH adjustment, and polishing. Less than one-third of all industrial SIs were used for the ultimate disposal of wastes.[1]

Industrial SIs range in size from less than 400 m² (0.1 acre) (29%) to greater than 400,000 m² (100 acres) (1%).[1] Approximately 40% of all industrial SIs were between 400 and 400,000 m² (0.1 and 1.0 acre), with another 20% falling between 4,000 and 20,000 m² (1 and 5 acres) in area. Only 11% of industrial SIs were larger than 20,000 m² (5 acres). Other findings of the EPA survey included the following:[1]

1. Over 50% of industrial SIs are located over either very thin or very permeable unsaturated soils. These conditions would allow for little attenuation of hazardous constituents prior to the interception of groundwater by any leakage that may occur.

2. Nearly 80% of industrial SIs are located over thick and very permeable aquifers that would allow relatively rapid movement of any contaminant plume that may develop.

3. Only about 7% of all SI sites are located in areas with hydrogeologic settings that offer the maximum protection from groundwater contamination. Less that 2% of all sites are located more than 1 mi from a source of high-quality drinking water (\10,000 ppm total dissolved solids).

10.4.2 RECENT PRACTICES

A recent report prepared for the EPA detailed findings from nine case studies and several interviews with technical experts on hazardous-waste SIs.[2] Major findings from the case studies included the following:

1. Land availability appears to be the primary criterion used to select a particular SI site. Any siting studies performed are usually conducted to justify the selected site.

2. Nearly all small, on-site SIs are designed by the owner and constructed using

contract labor. Outside consultants and technical experts are used primarily with larger on-site and commercial SIs.

3. Selection of a particular flexible-membrane liner (FML) for a site is generally based on generic liner-waste compatibility data provided by the suppliers.

4. While there are several exceptions, experiences with FML systems have generally not been very satisfactory. Common failure modes include cracking, brittleness, and seam separations.

5. Quality assurance and quality control (QA/QC) incorporated into design and construction of SIs varies from a simple inspection by the owner to comprehensive specifications and guidelines. Common deviations from construction quality-assurance (CQA) programs include failure to adhere to design specifications and lack of adequate inspections to identify and document these failures. Cases are documented where such inadequacies in the CQA program were responsible for SI failures.

6. Inspection and maintenance programs generally include at least periodic visual inspections and SI dredging. Maintenance activities included liner repairs and replacement, construction of dikes, filling of sinkholes, and surface grading to improve drainage.

7. Performance of leak-detection systems varies widely. Case studies revealed systems that are plugged and ineffective as a result of deviations from design specifications during construction. One system failed to detect large-volume leakage at locations not directly over subdrains. In another case, adequacy of the leak-detection system was verified by in-place testing.

Interviews with experts on SIs provided the following interesting opinions:

1. A majority preferred soil liners to FMLs. This was due to the belief that current installation procedures cannot guarantee the integrity of installed FMLs or covers and that there are many documented FML failures. Advantages given for soil liners include a considerable experience base to draw on, relative thickness of soil liners, attenuative capacity for waste constituents, availability, and less susceptibility to damage during installation.

2. It was recognized that there has been considerable recent improvement in the design and construction of leak-collection systems. Clogging was believed to remain as a major problem area. Damage to the system during construction was cited as a major cause for failures. It was believed that potential problems could be substantially reduced by proper design, source control, and QA/QC.

3. Leak-detection systems which allow direct observation of leakage (such as leak-collection systems) were said to have great merit. Advantages were early detection of failures, the ability to monitor large areas under the liners, and more reliable results than obtained from indirect methods (e.g., thermistor or resistivity devices).

4. Factors to consider in site selection, design, construction, and monitoring were said to be similar in most cases for landfills and SIs. Important SI-specific factors included fluctuations of liquid level as a result of drawdown, discharges, and wave action; freeze-thaw action on soil liners; continuous contact between the waste and liners; differences in the stresses between the bottom and sidewalls; stresses under wastewater discharge points; and desiccation cracking of soil-liner sidewalls. Surface impoundments were considered to be less permanent, more amenable to corrective action, and to have a greater design flexibility than landfills.

5. Regardless of type or number of liners in an SI, the best defense against groundwater contamination was thought to be siting in suitable geological formations. Underlying this philosophy was a consensus opinion that all liners will eventually leak.

6. Even with the soundest of designs, adequacy of a completed SI cannot be guaranteed without an effective CQA plan. Key elements of such a program were said to be a thorough program for construction inspections, a competent inspector, and detailed documentation and record keeping.

7. Major areas of uncertainty that were identified for research and development included evaluation of new design concepts, waste-liner compatibility, and monitoring systems.

Surface impoundments represent one of the nation's most serious threats to groundwater quality. Prior to 1980, only 28% of industrial SIs were lined and less than 10% had monitoring programs.[1] Today, SIs are the principal source of environmental contamination at over 27% of the priority Superfund sites.[4] Recommendations given in the Citizens For a Better Environment (CBE) 1983 report[4] included the following:

1. Ban the use of unlined SIs for treatment, storage, or disposal of hazardous waste.

2. Initiate effective enforcement programs to ensure compliance with groundwater-monitoring requirements.

3. Establish air-emission standards and monitoring requirements for SIs.

4. Develop and use alternative waste-management technologies that reduce, recycle, treat, or destroy hazardous waste.

10.4.3 REGULATIONS

Subpart K of 40 *CFR* 264 gives the regulations applicable to owners and operators of new SIs used to manage hazardous waste. Regulations covering the design, operation, monitoring, and closure of SIs are discussed in this subsection.

Design and Operation

Final rules for design and operation of liner systems for SIs were issued by the EPA in November 1986. These rules represent the codification of the Hazardous and Solid Waste Amendments of 1984. Basic provisions in the final rule are as follows:

1. Each new SI or expansion of an existing SI must be equipped with two or more liners and a leachate-collection system between such liners (Figs. 10.4.1 and 10.4.2). At a minimum, the liners and leachate-collection system must meet the following requirements:
   a. Have a top liner designed, operated, and constructed of materials such as to prevent the migration of any hazardous constituent into the liner during the active life and postclosure care period

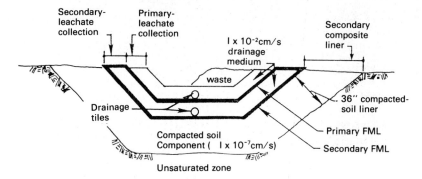

FIG. 10.4.1 Landfill double-liner options: FML plus composite.

b. Have a bottom liner composed of one of the following:
   i. A layer of compacted soil thick enough to prevent the release of any waste constituents throughout the active life and postclosure period (Fig. 10.4.1)
   ii. A composite liner consisting of an upper component with characteristics similar to the top liner (see above) and a lower component of at least 90 cm of compacted soil with a permeability no greater than 1×10^{-7} cm/s (Fig. 10.4.2)
2. Other requirements for the liners include the following:
   a. Construction with materials that have appropriate chemical properties and sufficient strength and thickness to prevent failure
   b. Placement on materials capable of providing support to the liners
   c. Coverage of all surrounding earth likely to be in contact with the waste or leachate

The leachate-collection system between the liners must be designed, constructed, maintained, and operated to detect, collect, and remove liquids that leak through any area of the top liner during the active life and postclosure care period. As with the liners, the materials used to construct the leachate-collection system must have the appropriate chemical properties and sufficient strength to prevent failure. In addition, the leachate-collection system must be designed and

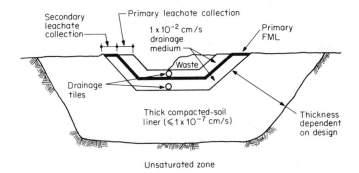

FIG. 10.4.2 Landfill double-liner option: FML and compacted soil.

operated so as to function without clogging during the active life and postclosure care period.

Detailed guidance on the design, operation, and construction of FMLs, compacted-soil liners, and leachate-collection systems has been issued by the EPA's Office of Solid Waste.[5,6] The EPA has also issued detailed guidance concerning the quality-assurance procedures that should be used for the construction of liners, leachate-collection systems, dikes, and other components of SIs.[7]

Other design and operating requirements are given in Subpart K concerning prevention of overtopping and dike stability. A surface impoundment must be designed, constructed, maintained, and operated to prevent overtopping resulting from any of the following:

1. Normal or abnormal operations
2. Overfilling
3. Wind and wave action
4. Rainfall
5. Run-on
6. Malfunctions of level controllers, alarms, and other equipment
7. Human error

Dikes used in an SI must be designed, constructed, and maintained with sufficient structural integrity to prevent massive failure of the dikes. Dike-stability analysis should be conducted under the presumption that the liners will leak during the active life of the SI.

Monitoring

During construction and installation, liners and cover systems must be monitored to ensure tight seams and joints and the absence of tears, punctures, or blisters. Soil liners must be monitored for imperfections including lenses, cracks, channels, root holes, and other structural nonuniformities that may cause an increase in the permeability of the liner or cover.

During operations, an SI must be monitored weekly and after storms to detect evidence of any of the following:

1. Deterioration, malfunctions, or improper operation of overtopping-control systems;
2. Sudden drops in the liquid level in the SI
3. Presence of liquids in the secondary leachate-collection system
4. Erosion or other signs of deterioration in dikes or other containment devices

Prior to obtaining a permit and after extended periods when the SI is not in service, the owner or operator must have the dikes certified by a qualified engineer. The certification must establish that the dike

1. Will withstand the stress of the pressure exerted by the types and amounts of wastes to be placed in the SI
2. Will not fail due to scouring or piping even if the liner systems fail

Closure

Two options are given in the regulations for closure of SIs. One option is to remove or decontaminate all waste residues, SI components, and contaminated subsoil and all associated structures or equipment contaminated with waste or leachate. All these materials must be managed as hazardous waste unless they have been demonstrated not to be hazardous through the delisting process of the EPA.

Another option is in-place closure of the SI. For in-place closure the following specific steps must be taken:

1. Eliminate free liquids by removing liquid wastes and solidifying the remaining wastes and waste residues
2. Stabilize all remaining waste to a bearing capacity sufficient to support the final cover system
3. Place a final cover

Final covers must be designed and constructed to

1. Minimize long-term migration of liquids through the closed SI
2. Function with minimum maintenance
3. Promote drainage and minimize erosion and abrasion of the final cover
4. Accommodate settling and subsidence so as to maintain integrity of the final cover
5. Incorporate a hydraulic barrier system with a permeability less than or equal to that of either the liner system or the subsoils underlying the facility, whichever is lower

If the in-place closure option is selected, the following postclosure requirements are given in the regulations:

1. Maintain the integrity and effectiveness of the final cover, including but not limited to making repairs necessary to correct the effects of settling, subsidence, and erosion
2. Maintain and monitor the groundwater-monitoring system
3. Prevent run-on and runoff from eroding or otherwise damaging the final cover

If the waste is to be removed at closure, the owner or operator must develop both a contingency plan for in-place closure and a contingency postclosure care plan, in case it is not practical to remove all contaminated subsoils at closure.

10.4.4 REFERENCES

1. U.S. Environmental Protection Agency, *Surface Impoundment Assessment National Report*, EPA-570/9-84-002, U.S. EPA, 1983.
2. U.S. Environmental Protection Agency, *Assessment of Hazardous Waste Surface Impoundment Technology, Case Studies and Perspectives of Experts*, EPA-600/2-84-173, U.S. EPA, 1984.
3. Chemical Manufacturers Association, *Results of the 1984 CMA Hazardous Waste Survey*, Chemical Manufacturers Association, 1986.
4. Citizens for a Better Environment, *Hazardous Waste Surface Impoundments: The Na-*

tions Most Serious and Neglected Threat to Groundwater, Citizens for a Better Environment, 1983.

5. U.S. Environmental Protection Agency, *Minimum Technology Guidance on Double Liner Systems for Landfills and Surface Impoundments: Design, Construction, and Operation,* EPA/530-SW-35-012, U.S. EPA, 1985.

6. U.S. Environmental Protection Agency, *Minimum Technology Guidance on Single Liner Systems for Landfills and Surface Impoundments: Design, Construction, and Operation,* EPA/530-SW-35-013, U.S. EPA, 1985.

7. U.S. Environmental Protection Agency, *Construction Quality Assurance for Hazardous Waste Land Disposal Facilities,* EPA/530-SW-85-021, U.S. EPA, 1986.

SECTION 10.5
SOIL LINERS

David C. Anderson

Vice President
K. W. Brown and Associates, Inc.
College Station, Texas

Soil liners are a required component of double-liner systems for hazardous-waste disposal facilities. This requirement was established in the Hazardous and Solid Waste Amendments of 1984; it was further defined by the U.S. Environmental Protection Agency (EPA)[1] and later supported by regulations under Title 40 Parts 264 and 265 of the *Code of Federal Regulations*. The EPA[1] provided two basic options for the design of soil liners (Fig. 10.5.1). The two basic options for use of compacted soils in liner systems for hazardous-waste facilities include a compacted soil as a component of a composite liner (Fig. 10.5.1*a*) and a secondary liner consisting only of compacted soil (Fig. 10.5.1*b*). While there are important differences in the characteristics and requirements for these two soil-liner configurations,[2] the most critical performance characteristic of any soil liner is hydraulic conductivity. Consequently, this chapter focuses on laboratory and field testing of hydraulic conductivity and the factors that influence the hydraulic conductivity of soil liners.

10.5.1 LABORATORY VS. FIELD HYDRAULIC-CONDUCTIVITY TESTING OF SOIL LINERS

For decades, engineers and scientists have assumed that soil liners constructed with compacted clayey materials would form practically impermeable hydraulic barriers. These assumptions were based largely on laboratory hydraulic-conductivity tests using small, undisturbed or remolded samples of the liner. However, no case histories or full-scale field tests have been published which support the critical assumption that laboratory hydraulic-conductivity tests give an accurate assessment of actual field hydraulic conductivity. In addition, several

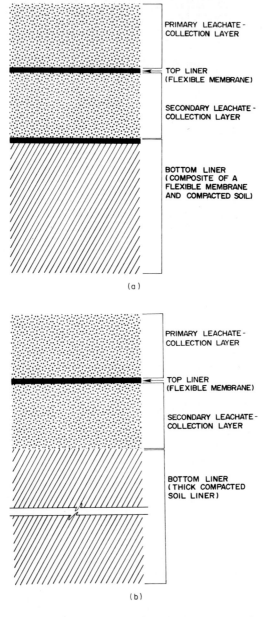

FIG. 10.5.1 Two basic options for liner systems for hazardous-waste disposal facilities. (*a*) Composite of a flexible membrane and compacted soil. (*b*) Thick compacted soil.

recent studies have indicated that laboratory tests tend to underestimate the actual hydraulic conductivity obtained in the field.

Laboratory Hydraulic-Conductivity Testing of Soil Liners

Hydraulic conductivity of a soil liner can be evaluated in the laboratory by either fixed-wall or flexible-wall permeameters. Both methods have shortcomings and advantages but it is not practical to use either method with samples large enough to be representative of field conditions. Both methods are, however, valuable tools for comparative analysis of various soils, soil conditions, and compactive efforts.

Fixed-wall permeameters (Fig. 10.5.2) have rigid sidewalls, which have been

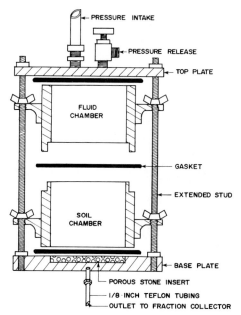

FIG. 10.5.2   Fixed-wall permeameter.

found to occasionally exhibit excess flow near the sidewall.[3] There are, however, several means to detect excess sidewall flow in samples. First, fixed-wall tests are sufficiently inexpensive to allow the running of several concurrent tests. If one or more samples have excess sidewall flow, this will be evident in the form of a significantly higher hydraulic conductivity. Double-ring permeameters (Fig. 10.5.3) are equipped with special base plates (Fig. 10.5.4), which separately collect flow through the matrix of the soil sample and flow near the sidewall.[4]

Flexible-wall permeameters (Fig. 10.5.5) have soil samples encased in flexible membranes. Confining pressure is applied to the outside of the membrane, thereby pressing the membrane against the soil and thus preventing excess flow along the sides of the soil sample. A major disadvantage of this permeameter con-

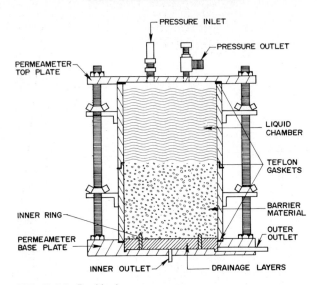

FIG. 10.5.3 Double-ring permeameter.

figuration is that the confining pressure may press together any cracks that exist in the matrix of the sample, resulting in substantial reductions in the apparent hydraulic conductivity.

Beyond the shortcomings of individual laboratory methods is the overriding concern that small-scale tests cannot simulate field conditions. Several other reasons cited for large discrepancies between laboratory and field-derived hydraulic-conductivity values include the following:

1. Soil samples prepared in the laboratory are not subjected to the climatic conditions (such as desiccation and freezing) which can cause cracking and hydraulic-conductivity increases in field-prepared soil liners.[5]

2. There may be a tendency to run laboratory tests on samples of selected, finer-textured soil samples.[6]

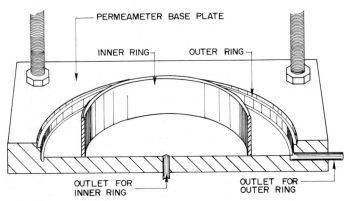

FIG. 10.5.4 Base plate of the double-ring permeameter.

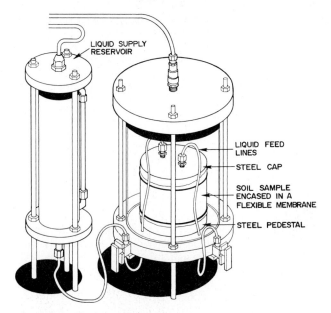

FIG. 10.5.5   Flexible-wall permeameter.

3. Laboratory samples generally receive a much greater degree of homogenization (such as grinding and removal of rocks, roots, and rubbish) than is practical under field conditions.

Field Hydraulic-Conductivity Testing of Soil Liners

The EPA has for years stated a requirement that compacted-soil liners have hydraulic conductivities of 1×10^{-7} cm/s or less. It has only recently become practical to verify that this performance standard has been obtained in the field. While
there have long been a number of field methods for hydraulic-conductivity testing, it was not possible with these methods either to measure very low hydraulic
conductivities or to evaluate a large enough area to account for a representative
distribution of macrofeatures. In the past several years, however, several field
infiltrometers have been developed and field-tested which can test large areas (1
to 5 m²) and low hydraulic conductivities (less than 1×10^{-7} cm/s).

It is difficult to exactly quantify field hydraulic-conductivity values which are
substantially less than 1×10^{-7} cm/s.[4] It is much easier, however, to simply verify that the field hydraulic conductivity is 1×10^{-7} cm/s or less.[7] Such a verification would indicate that a soil liner meets or exceeds the EPA performance
standard.

Field hydraulic-conductivity tests are necessary to verify the performance of a
soil liner. If conducted on the actual liner, however, these tests can cause substantial delays in construction and result in other problems arising from prolonged exposure of the liner to the effects of weather. In addition, it would be
very costly to rip out and replace a whole soil liner if it did not meet the performance standard. Field tests could be conveniently conducted on the scaled-down

soil liner of a test fill. Such field tests would be valid only if the test fill and full-scale facility were constructed according to the same strictly followed construction quality-assurance plan.[1]

Comparison of Field and Laboratory Hydraulic-Conductivity Test Results

Griffin et al.[8] conducted both laboratory and field hydraulic-conductivity tests on clayey soil used as a liner for a hazardous-waste landfill. Laboratory-derived hydraulic-conductivity values were all 2×10^{-7} cm/s or less. In contrast, hydraulic-conductivity values obtained in the field were as much as three orders of magnitude greater than the laboratory values. Routine monitoring of the actual hazardous-waste landfill site revealed that organic contaminants were migrating two to three orders of magnitude faster than had been predicted.[9]

Daniel[10] compared laboratory and field hydraulic-conductivity values with the actual leakage rates of four clay liners. In all four cases, the laboratory values were much lower than those measured in the field. In the two cases where field hydraulic-conductivity tests were performed on the actual liner, values obtained from these field tests accurately reflected the leakage rate of the entire liner.

Other studies have also found large differences between field-measured and laboratory-measured hydraulic conductivities.[11,12,13] One of the reasons cited for the higher values obtained in field tests is that laboratory samples can be more readily prepared without the defects that can greatly increase hydraulic conductivity. Methods used to prepare soil liners in the field are difficult to simulate in the laboratory. One example is method of compaction. Soil liners are often compacted in the field with a kneading action using sheepsfoot rollers. In contrast, impact compaction is usually used to prepare soil-liner samples in the laboratory. Even though identical densities and water contents may be obtained with these different compaction methods, the resulting samples may have very different hydraulic conductivities.[14]

Several reasons have been cited for the large discrepancies between laboratory- and field-derived hydraulic-conductivity values. Probably the most widely cited and most important cause of the observed differences, however, is that field tests can evaluate much larger and, hence, more representative samples than is practical in laboratory tests.

10.5.2 MACROFEATURES IN COMPACTED SOIL AND IN SITU CLAYEY DEPOSITS

Macrofeatures are common in poorly compacted soil liners[10] and in situ deposits.[15,8,16,17] In the past, acceptably low hydraulic-conductivity values obtained from laboratory tests created a false sense of security. As stated earlier, however, laboratory values have since been found to often underestimate the actual hydraulic conductivity of clayey materials.

There are many causes for macrofeatures within the in situ clay deposits used as soil liners. For instance, the removal of overburden (i.e., the excavation of a site for a landfill) can cause clay soils to develop deep fissures.[16] Shrinkage cracks may form as the result of dehydration[18] or synaeresis.[19,15] *Synaeresis,* or chemical "drying," has been cited as causing permanent shrinkage cracks in saturated clay deposits below the top of the water table.[15]

Fissures and joints are common in clay soils and can cause order of magnitude decreases in strength[17] and large increases in hydraulic conductivity.[9] Sand lenses can be eliminated in compacted-soil liners by thoroughly mixing the soil prior to compaction. In contrast, there are no adequate methods for detecting every silt or sand lens hidden within an in situ soil deposit. Lenses of permeable soil have been cited as the probable cause for high hydraulic-conductivity values found for in situ clayey soil deposits used to line landfills.[9]

Other macrofeatures that can occur within an in situ deposit include fillings,[15] faults, bedding planes, sheeting,[17] root holes, burrows,[20] slickensides, subsurface erosion pipes,[8] blocky peds, and prismatic structure.[21] Nearly all in situ deposits, have some macrofeatures. Additional structural features may be added to an in situ deposit as a result of changes in overburden pressure,[16] water content, or solution chemistry within the soil.[9]

Macrofeatures tend to be the dominant factor in determining the effective hydraulic conductivity of in situ deposits and compacted soil liners.[20] While a construction quality-assurance program can minimize the number of macrofeatures in compacted soil, there is little that can be done to detect every macrofeature within undisturbed in situ deposits.

10.5.3 CLOD STRUCTURE IN SOIL LINERS

One of the most prevalent types of macrofeatures found in compacted-soil liners is clod remnants. These remnants are the result of the incomplete remolding of the clods that are an inevitable characteristic of uncompacted clayey borrow material. When a clayey soil is excavated from a borrow site, it is typically broken into clods varying in diameter from 3 to 20 cm. If the compaction process does not completely remold the borrow material into a homogeneous mass, there can be clod remnants throughout the thickness of a compacted soil. This leaves the large interclod pores intact, which results in high field hydraulic conductivities.

Olsen[22] showed that the hydraulic conductivity of a compacted soil or a structured, in situ soil deposit is controlled more by interclod voids than by the smaller interparticle voids within a clod. Daniel[10] found that when soil was compacted with clods that were not remolded in the compaction process, the hydraulic conductivity of the sample increased with the maximum clod size (Table 10.5.1).

It is impractical to depend on soil processing to remove clod remnants. Even power tillers can only reduce the maximum clod size of clayey soil to approximately 1 cm. It is practical, however, to destroy clod remnants and remove interclod pores by the use of compaction equipment that completely remolds the soil. The compactor rollers should have feet that are at least as long as the loose-lift thickness. This allows the weight of the compactor to be borne by the compactor feet rather than the roller drum. In addition, the compactor should be suf-

TABLE 10.5.1   Hydraulic Conductivity of Compacted Clay Soil with Different Maximum Clod Sizes*

Maximum clod size, cm	Hydraulic Conductivity cm/s
1.0	2.5×10^{-7}
0.5	1.7×10^{-8}
0.2	8.5×10^{-9}

*Soil prepared at same water content and compacted to similar densities.
Source: Modified from Ref. 10.

ficiently massive to exert a greater shear stress in the soil than the undrained shear strength of the clods.[23]

It is important to note that the large interclod pores may represent a small fraction of total porosity. Consequently, even though a given compactor produces a high density, the liner may still have a high hydraulic conductivity if clod remnants have not been removed. There is, therefore, need to supplement density tests with field hydraulic-conductivity tests in any construction quality-assurance program for soil liners.

10.5.4 REFERENCES

1. U.S. Environmental Protection Agency, *Minimum Technology Guidance Document on Double Liner Systems for Landfills and Surface Impoundments,* EPA/530-SW-85-014, U.S. EPA, Washington, D.C., 1985.

2. D. C. Anderson, "Soil Liners for Hazardous Waste Disposal Facilities," in *Proceedings of the National Conference on Hazardous Waste and Hazardous Materials,* Atlanta, Georgia, March 4–6, 1986, pp. 206–209.

3. D. E. Daniel, D. C. Anderson, and S. S. Boynton, "Fixed-Wall Versus Flexible-Wall Permeameters," In *Hydraulic Barriers in Soil and Rock,* ASTM STP 874, American Society for Testing and Materials, Philadelphia, Pennsylvania, 1985, pp. 107–126.

4. D. C. Anderson, J. O. Sai, and A. Gill, *Surface Impoundment Soil Liners: Permeability and Morphology of a Soil Liner Permeated by Acid and Field Permeability Testing for Soil Liners,* report prepared for the U.S. EPA, Washington, D.C., by K. W. Brown and Associates, Inc., College Station, Texas, under EPA Contract #68-03-2943, 1984.

5. U.S. Environmental Protection Agency, *Soil Properties Classification and Hydraulic Conductivity Testing,* SW-925, U.S. EPA Washington, D.C., 1984.

6. R. E. Olson and D. E. Daniel, "Field and Laboratory Measurement of the Permeability of Saturated and Partially Saturated Fine-Grained Soils," in *Permeability and Groundwater Contaminant Transport,* ASTM STP 746, American Society for Testing and Materials, Philadelphia, Pennsylvania, 1981, pp. 18–64.

7. S. R. Day and D. E. Daniel, "Field Permeability Test for Clay Liners," in *Hydraulic Barriers in Soil and Rock,* ASTM STP 874, American Society for Testing and Materials, Philadelphia, Pennsylvania, 1985, pp. 276–288.

8. R. A. Griffin, R. E. Hughes, L. R. Foller, C. J. Stohr, W. J. Morse, T. M. Johnson, J. K. Bartz, J. D. Steels, K. Cartwright, M. M. Killey, and P. B. Du Montelle, "Migration of Industrial Chemicals and Soil-Waste Interactions at Wilsonville, Illinois," in *Proceedings of the Tenth Annual Research Symposium on Land Disposal of Hazardous Waste,* EPA/600/9-84-00, U.S. EPA, Washington, D.C., 1984.

9. R. A. Griffin, B. L. Herzog, T. M. Johnson, W. J. Morse, R. E. Hughes, S. F. J. Chow, and L. R. Follmer, "Mechanisms of Contaminant Migration Through a Clay Barrier—Case Study, Wilsonville, Illinois," in *Proceedings of the Eleventh Annual Research Symposium on Land Disposal of Hazardous Waste,* EPA/600/9-85-013, U.S. EPA, Washington, D.C., 1985.

10. D. E. Daniel, "Protecting Hydraulic Conductivity of Clay Liners," *J. Geotech. Eng.,* **110**:(2)285–300 (1984).

11. B. L. Herzog and W. J. Morse, "A Comparison of Laboratory and Field Determined Values of Hydraulic Conductivity at a Disposal Site," in *Proceedings of the Seventh Annual Madison Waste Conference,* University of Wisconsin-Extension, Madison, Wisconsin, 1984, pp. 30–52.

12. G. C. Boutell and V. R. Donald, "Compacted Clay Liners for Industrial Waste Disposal," presented at the ASCE National Meeting, Las Vegas, Nevada, 1982.

13. M. E. Gordon and P. M. Huebner, "An Evaluation of the Performance of Zone of Saturation Landfills in Wisconsin," presented at the Sixth Annual Madison Waste Conference, University of Wisconsin, Madison, Wisconsin, 1983.

14. J. K. Mitchell, *Fundamentals of Soil Behavior,* Wiley, New York, 1976.

15. W. A. White, "Origin of Fissure Fillings in a Pennsylvanian Shale in Vermilion County, Illinois," *Transactions: Illinois Academy of Science,* **57**:208–215 (1964).

16. B. E. Kulkarni, N. K. Phadke, and B. S. Kapre, "Genesis of Formation of Fissures in Fissured Clay of Maharashtra, India," in *Proceedings of Fourth Southwest Asian Conference on Soil Engineers,* Institute of Engineering Malaysia, Kuala Lumpur, Malaysia, 1975, pp. 1.27 to 1.35.

17. A. W. Skempton, R. L. Schuster, and D. J. Petley, "Joints and Fissures in the London Clay at Wraysbury and Edgeware," *Geotechnique,* **19**:205, 217 (1969).

18. S. S. Boynton, *An Investigation of Selected Factors Affecting the Hydraulic Conductivity of Compacted Clay,* M.S. thesis (Thesis GT 83-4), University of Texas, Austin, 1983.

19. K. Kallstenius, *Studies on Clay Samples Taken With Standard Piston Sampler,* Swedish Geotechnical Inst., Proc. 21, 1963.

20. D. J. Folkes, "Fifth Canadian Geotechnical Colloquium: Control of Contaminant Migration by the Use of Liners," *Canadian Geotechnical Journal,* **19**:320–344 (1982).

21. L. D. Baver, W. H. Gardner, and W. R. Gardner, *Soil Physics,* Wiley, New York, 1972.

22. H. W. Olsen, *Hydraulic Flow Through Saturated Clays,* vol. II, *Clays and Clay Minerals,* 1962, pp. 131–161.

23. B. R. Elsbury, J. M. Norstrom, D. C. Anderson, J. A. Rehage, J. O. Sai, R. L. Shiver, and D. E. Daniel, *Optimizing Construction Criteria for a Hazardous Waste Soil Liner,* Phase I Interim Report, U.S. EPA, Cincinnati, Ohio, EPA Contract No. 68-08-3250, 1985.

SECTION 10.6
DISPOSAL IN MINES AND SALT DOMES

Ronald B. Stone
GEOMIN, Inc.
Tulsa, Oklahoma

The use of mined space for hazardous-waste storage and disposal, whether the mined space was created for extraction of minerals or specifically for hazardous-waste storage or disposal, is an economically viable alternative to the methods of storage or disposal presently favored and utilized. *Mined space* can be conventional mines converted to this purpose, new mined space, or space created by solution-mining methods in salt beds or domes.

10.6.1 APPLICABILITY OF THE TECHNOLOGY

Historical Precedent: Mines and Solution Mines

The use of mined underground space for storage and disposal of hazardous materials has a long history of successful application in our industrial society. The petroleum industry has utilized underground space, both conventionally mined and solution-mined in salt, for storage of crude oil and products for over 30 years. The hazardous-waste industry has been using mined space for storage since 1972 in Germany. The chemical industry has been using solution-mined caverns in salt for many years for disposal of hazardous chemicals in several locations in the world.[1]

10.6.2 DESCRIPTION OF THE TECHNOLOGY

The current technology for storing hazardous waste in underground openings allows the disposal or storage of a wide range of wastes. Conventional mines can

10.73

store packaged, dry, nonexplosive, nonflammable, zero-vapor-pressure wastes.[2] Solution mines can store nonreactive liquids and slurries in bulk form; no packages are required. In general, it can be stated that if the hazardous material can be transported to the facility it can be stored in the underground.

Conventional Mines

Conventional mines have been created around industrial centers to provide basic raw materials for our industrial society. Some of these mines have been excavated using conservative methods; these are candidates for hazardous-waste storage and disposal. The ones that are dry and structurally stable offer large volumes of economical, available space which can be converted to this use. If no existing mines are available, new, special-purpose mined openings can be constructed at reasonable cost. The prime use for mined space would be to serve as a storage warehouse until economical waste-treatment methods are available for disposal.

Solution Mines

Solution mines in salt or potash offer a great potential for economical storage and disposal of hazardous waste primarily because of the wide geographic distribution of these deposits in the United States (Fig. 10.6.1) and the low cost of creating solution-mined cavern space. Many of the hazardous-waste generators are located near or on top of these deposits; thus transportation of the hazardous waste could be minimized.[3]

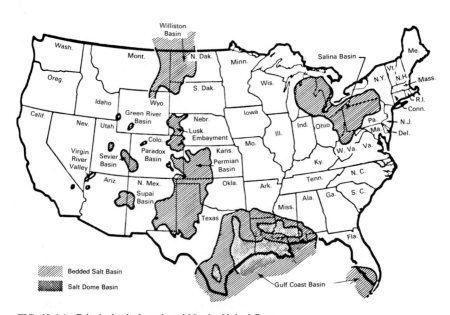

FIG. 10.6.1 Principal salt deposits within the United States.

10.6.3 DESIGN CONSIDERATIONS

Conventional Mines

Types of Mines (Rock Types). There are several types of mines which would be suitable for use as repositories. These include salt, potash, gypsum, limestone, and underground granite mines. The primary requirement for any mine to be used for this purpose is that it be dry, without water leakage from the surface. Because of the nature and low cost of the minerals produced from these mines, they have been excavated using bulk mining techniques, thus producing a large volume of space at a reasonable cost.

Depth. Conventional mines can range from 50 ft to 3,000 ft in depth depending on the deposit being mined. The depth can be much greater but stability and access problems would make deeper mines undesirable for hazardous-waste storage or disposal at this time. Near-surface mines often have fractures in their roofs that allow surface water to drain into the mine, which is detrimental to the operation of a hazardous-waste facility.

Rock Mechanics; Mine Stability and Integrity. The integrity of the mine must be such that it will remain stable and not cave in or close due to plastic flow of the rock for a very long period of time. This is achieved by conservative design and careful attention to the structural properties of the mineral being mined. The mines may be deep mines which require a vertical shaft for access or shallow mines which may be entered by tunnels in a hillside. A shallow mine would be preferred for ease of access, while a deep mine would be preferred for isolation of the waste.

Layout and Access (Room-and-Pillar Configuration). The design of the mine openings for stability usually results in a pattern known as a room-and-pillar or room-and-tunnel configuration and can be readily adapted to the storage or disposal of hazardous waste. There are a wide range of options available in opening size and mine layout: some of these are shown in Fig. 10.6.2. Room width can range from 15 ft to 80 ft, and room height can range from six ft to 100 ft; lengths are limited only by the extent of the deposit or property boundaries.

Safety: Spill-Cleanup and Emergency-Exit Provisions. Because underground mines are confined space they present a somewhat more hazardous working environment than surface facilities. Many mines have excellent safety records and a mine either converted to or newly mined for hazardous-waste placement can be safe. The introduction of hazardous waste into the mine environment imposes an additional burden of safety requirements. The design must include provisions for worker protection in the event of a spill of the hazardous material. The mine must be designed so that it is filled from the back or farthest mined area first, retreating toward the opening or shaft. Spill contingency plans must be considered as a part of the design. Access to the entrance should be maintained clear at all times for emergency traffic. It is desirable that more than one entrance be maintained for safety. The route of the hazardous waste into the mine should be dedicated to that purpose and equipped with suitable safety monitors.

Ventilation: Fresh Air and Exhaust. Ventilation of the mine is a very important consideration in developing a hazardous-waste storage or disposal facility. The

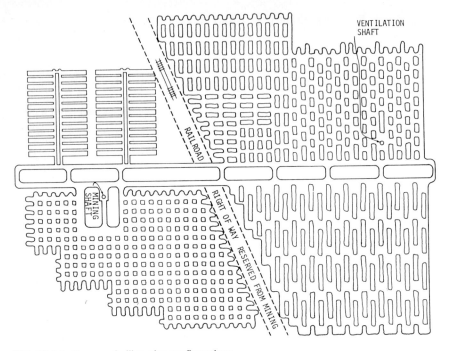

FIG. 10.6.2 Room-and-pillar mine configurations.

workers in the facility must always work in the fresh-air stream to avoid contamination in the event of an inadvertent spill or other mishap. Exhaust air must always be assumed to be contaminated.

Surface Facilities: Offices, Process, Cleanup, Packaging. The extent of any surface facilities will depend upon the specific wastes to be handled. A repackaging facility would probably be a minimum requirement, with dehydration, solidification, and chemical treatment being added as the wastes become more complex. Decontamination facilities for the transporters will be required. Offices and shops will be necessary even for a small facility.

Solution Mines: Bedded Salt Deposits and Salt Domes

Salt deposits occur either as bedded deposits or as dome deposits. The bedded deposits cover the greatest geographic area of the United States, while the dome deposits are located on the Gulf Coast of the United States.

Depth Range. Salt deposits extend to great depths beneath the surface at various locations in the United States and the world. Given current technology, stability considerations dictate that the depth for a solution-mined cavern for hazardous-waste storage not exceed 3,000 ft. The top of a cavern should be at least 300 ft below the top of the salt formation in dome salt. The top of a solution-mined cavern in bedded salt should be at least 10 ft below the top of the beds.

Rock Mechanics; Mine Stability and Integrity. The integrity and stability of solution-mined caverns has been proven over the last 35 years by the petroleum industry, which uses specially constructed caverns for storage of many millions of dollars worth of petroleum products. During this time much study has been done on how to predict structural stability and comprehensive computer programs are now available for design of stable caverns. The petroleum-storage caverns are liquid-filled and rely upon a balancing internal pressure created by the liquid or gas product to help achieve a stable cavern. This pressure acts to counterbalance the earth forces trying to force the cavern closed. Early studies of proposed hazardous-waste caverns indicate that, because of contamination problems caused by the hazardous waste, they may have to be operated as atmospheric caverns. This means that hazardous-waste caverns may have to be built smaller to achieve the required stability. The required technology is available to design and build stable salt caverns for hazardous-waste storage and disposal.

Layout and Access; Cavern Spacing. Typically, solution-mined caverns are spaced from 300 ft to 800 ft apart depending upon the proposed finished diameter of the cavern. The finished diameter of a cavern will range from 50 ft to 300 ft and the height of a cavern can range from 50 ft to 2,000 ft. The size selected for a particular facility depends upon the local geological conditions and the proposed use of the cavern. Some states now regulate the spacing between caverns. A typical solution-cavern facility plan and section are shown in Fig. 10.6.3.

Safety: Spill-Cleanup Provisions. Solution-mined caverns are most suited to storage and disposal of hazardous-waste slurries and liquids; thus the design of a surface facility must include holding tanks, pipelines, and pumps. Appropriate design codes for these types of facilities must be followed. Spill basins and ditches

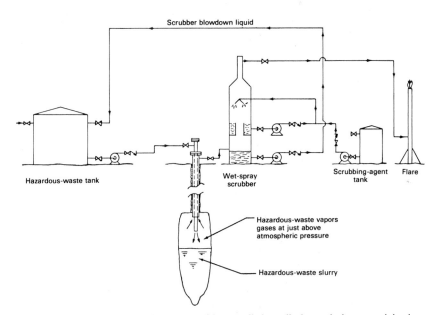

FIG. 10.6.3 Atmospheric solution cavern with controlled gas discharge during waste injection.

under critical unloading and transfer points must be included. A spill-cleanup plan should be written very early in the design so that all cleanup provisions can be included in the design.

Cavern Discharges: Liquid and Vapor.   The present concept for safe storage and disposal of hazardous waste in solution caverns is to pump out the solution-mining liquid and allow the cavern to assume atmospheric pressure. This eliminates the problem of contaminated-liquid handling and disposal when the cavern is being filled with waste. The cavern will have a gas or vapor discharge when being filled if pressure buildup is to be avoided. This will require a scrubber and a flare stack as a minimum to reduce the gases or vapors to permissible levels.

Surface Facilities: Offices, Process, Cleanup, Blending.   The extent of any surface facilities will depend upon the specific wastes to be handled. A blending facility could possibly be needed to neutralize the liquids prior to injection into the cavern. A mixing circuit may be needed to optimize the slurry consistency for pumpability. Decontamination facilities for the transporters will be required. Offices and shops will be necessary even for a small facility.

10.6.4 ENVIRONMENTAL EFFECTS OF THE PROCESS

Air and Water Exposure: Conventional and Solution Mines

- Conventional underground mines would have a very low exposure for air contamination because of the nature of the waste to be stored. The only exposure would occur during container-transfer or repackaging operations.
- Solution mines would have a higher possibility of air contamination because of the more volatile nature of the waste to be handled. Air contamination could occur during transfer operations and during filling of the cavern when the off gas is released to allow the cavern pressure to equalize.
- In shallow-mined space the waste would be contained above the groundwater table.
- In deep-mined space the waste would be located below the potable aquifers.
- If required, waste can be isolated from the hydrological environment by encapsulation or containerization.

Residuals: Conventional and Solution Mines

- In conventional mines the requirement that all waste be containerized will preclude contamination of the mine, except for inadvertent spills, which must be cleaned up to acceptable standards.
- Solution mines are presently proposed as bulk-storage containers with the cavern walls serving as the containment vessel; also, the waste would be stabilized with a cementing agent. This condition would result in the cavern being permanently contaminated.

Recycling Possibilities

- If retrievability for recycling were desired, a conventional mine would be used as a long-term underground warehouse.
- Because it would be solidified, the material in the solution-mined cavern could not be recycled.

10.6.5   CASE STUDIES AND EVALUATIONS

Mines

Germany.   The Herfa-Neurode facility has operated for 15 years since its start-up in 1972. Approximately 270,000 t of hazardous waste have been placed in the mine. The annual volume presently is running between 35,000 and 40,000 t. Approximately 25% of this tonnage originates outside Germany. The re-use of stored waste is possible and over 1,000 t of waste has been retrieved and sent back to the generator.[4]

Minnesota.   The Minnesota Waste Management Board financed a study entitled *Subsurface Isolation of Hazardous Wastes,* which was completed in 1982. This study was conducted by the University of Minnesota and was oriented to deep geologic disposal in crystalline bedrock within the state. Eighteen mine study sites were identified and the plan was to reduce this number to five sites for further investigation. Further investigation was to consist of air and water surveys and core drilling to determine the rock quality. When the five sites were announced, public opposition was immediate and vocal. During a meeting held on February 24, 1984, the Minnesota Waste Management Board dropped the crystalline-rock concept from further consideration because of its high cost.[5]

Missouri.   The discovery of dioxin-contaminated soil at Times Beach, Missouri, and the subsequent detailed investigation of the state for other contaminated sites led to the identification of 42 contaminated sites throughout the state with others being suspected to exist. Of these sites, 27 are located in the St. Louis area, with another sizable contaminated area near Springfield. The EPA's Hazardous Waste Engineering Research Laboratory in Cincinnati, Ohio, conducted a study of the use of mined space for permanent storage of the dioxin-contaminated soil from each site. The EPA was assisted by PEI Associates, Inc., and the Missouri Department of Natural Resources. The state of Missouri is well-known for its utilization of mined space for warehouses, offices, manufacturing facilities, and recreational facilities; thus, the use of mined space for this purpose is a natural extension of the existing technology and use.

In general, this concept is a viable alternative to other treatment and disposal options and may be as cost-effective or even more cost-effective than some of the other technologies. The packaging and transportation of the contaminated material will constitute the largest part of the cost of operating a facility. Because this proposed facility would have a single use (dioxin storage), the design can be simplified to a near-surface room-and-pillar mine, located in a massive sandstone or limestone formation above the water table. A new mine facility, possibly on state-owned land, may be required because of the long-term liability problems posed by the dioxin-contaminated soil.[5]

Ohio. The conversion of the PPG Industries, Inc. (Barberton, Ohio), limestone mine to a waste-storage facility proposed by that company in 1981 and 1982 was dropped. This facility was proposed as a general hazardous-waste repository with all types of waste being acceptable. Waste was to arrive at the facility by rail and truck and be transported underground. A public-information program was started and the local residents were invited to tour the mine to see for themselves the integrity and containment potential. Subsequently, public opposition was organized and pledged to fight the establishment of a hazardous-waste facility at this site.

The project did not reach the permit application stage. A public notice discontinuing the project was issued by PPG Industries, Inc., on July 12, 1982. The company cited U.S. economic conditions as the reason for discontinuing plans for this facility.[6]

Canada. Giant Yellowknife Mines Limited at Yellowknife (Yukon Territory) uses cemented backfill to replace reinforced concrete in the construction of pressure bulkheads for underground chambers. These chambers are designed to contain arsenic dust produced through the roasting of arsenopyrite. The disposal of arsenic dust is into seven separate underground chambers, located in the permafrost zone at about 75 m below ground level. Since 1951 when disposal was begun, 170,000 t of arsenic dust has accumulated in the chambers.[7]

Solution Mines

Canada. Mercury compounds have been deposited in three solution-mined salt caverns near a chloralkali plant in Saskatchewan. The mercury compounds take the form of clarifier sludge in one cavern and occur in suspension and in solution in the others.

A salt cavern in the Sarnia district of southwestern Ontario serves as a repository for heavy, chlorinated hydrocarbons which are injected down three disposal wells. Three other salt caverns contain waste purge gas from a gas-processing plant.

Holland. In 1938, Akzo, a chemical manufacturing firm in the Netherlands, started disposing of wastes from brine-purification units into existing salt caverns in the Henglo area. In 1965, Akzo began using salt caverns for the disposal of salty drilling muds. In 1978, it began disposing of magnesium chloride under the salt brine. Akzo has developed the original "string of pearls" concept, wherein a series of waste-disposal caverns are solution-mined, one above the other, from a single deep solution well.[8]

Texas. United Resource Recovery, Inc., of Houston, Texas, submitted an application to the Texas Department of Water Resources for a permit to store hazardous waste in the Boling Salt Dome. This application triggered a two-phase geological study to evaluate the acceptability of using salt domes in Texas for waste disposal and to recommend guidelines for waste storage in salt domes. This study was performed by the Texas Bureau of Economic Geology (BEG). Reports issued as a result of the study have defined the geological and technical issues involved in hazardous-waste storage in salt-dome caverns. The BEG concluded that the first year of study did not answer all critical questions, but that results did not disqualify salt domes as potential hosts for permanent isolation of toxic wastes.[9]

A second year of study (Phase III) was authorized and research covered the following subjects:

Subsidence and collapse

Structural patterns around Texas salt domes

Cap rock

Cap-rock hydrology

The contents of the report consisted of five individual research reports covering various factors for the Boling, Barbers Hill, and Damon Mound Domes. A summary of the results of the various studies as they applied to the overall problem was written by S. J. Seni. Conclusions reached were:

1. Salt domes may be suitable hosts for permanent isolation of some types of toxic waste in solution-mined caverns in salt.

2. Not all domes are appropriate sites for toxic-waste disposal owing to uncertainties about dome size, shape, depth, salt heterogenities, cap-rock lost-circulation zones, hydrologic and structural stability, growth history, and effects of resource exploration and development.

3. The large number of negative aspects associated with the Boling Dome override the positive aspects and the conclusion was reached that the use of the Boling Dome as the first site for toxic-chemical waste disposal should be discouraged.

4. Methods for slurry transport and for disposal of solidified toxic waste in solution-mined caverns are needed. No specific studies of the waste or in situ solidification within solution-mined caverns are available.[10]

Louisiana. In 1983, Empak, Inc., of Houston, Texas, submitted an application to the Louisiana Department of Conservation for a permit to build a hazardous-waste facility at the Vinton Salt Dome in Southwest Louisiana. This project envisioned using solution-mined caverns in the salt dome as final storage for hazardous waste. After this project was announced, a state law was passed (in 1983) forbidding emplacement of hazardous waste in salt domes for a period of two years. This was intended to allow the state time to evaluate the proposed use prior to issuing a permit. The option for exclusive use of the Vinton Salt Dome was dropped and the application withdrawn due to regulatory and legislative delays.[11]

New York. The International Salt Company at Watkins Glen, New York, began placing the residual natural wastes from their salt-production operation into three interconnected solution caverns in 1971. These wastes consist primarily of calcium sulfate solids and of trace minerals such as iron, copper, cobalt, and sulphur that are used in the manufacture of cattle feeds and salt blocks. The heavy metals were allowed to settle out in a solution cavern, displacing the lighter brine to the surface for injection into a subsurface contaminated aquifer called the "black water" horizon and also known as the Cherry Valley formation. This formation contains hydrogen sulfide and other organic contaminants.[12]

10.6.6 CURRENT RESEARCH

The EPA has sponsored studies and development projects in recent years covering the fields of encapsulation, storage, and fixation technologies. The developments from these and related projects have provided valuable background information on containment which did not previously exist and would apply directly to the use of mined space for storage of hazardous waste.[13]

In addition to the EPA projects, the U.S. Department of Energy (DOE) and its predecessor agencies have sponsored research and testing for siting, designing, construction, and maintenance of facilities for storing radioactive wastes in mined spaces. This research, administered by the Office of Nuclear Waste Isolation (ONWI) via various national laboratories and their contractors, was initially directed at storage of such wastes in salt deposits. Research has since been broadened to include other types of geological settings. Present studies, tests, and demonstrations for nuclear-waste storage are continuing in salt, basalt, tuff, and crystalline-rock geologic formations.[14]

10.6.7 REFERENCES

1. R. Stone, "Update: Storage of Hazardous Waste in Mined Space," in *Proceedings of the Eleventh Annual Research Symposium, Land Disposal of Hazardous Waste,* sponsored by the U.S. EPA Office of Research and Development, Hazardous Waste Engineering Research Laboratory, Cincinnati, Ohio, 1985, pp. 341–342, 344–350.

2. R. Stone, P. Aamodt, M. Engler, P. Madden, *Evaluation of Hazardous Wastes Emplacement in Mined Openings,* Contract No. 68-03-0470, EPA-600/2-75-040, prepared by Fenix and Scisson, Inc., Tulsa, Oklahoma, for the U.S. EPA Solid and Hazardous Waste Research Division, Municipal Environmental Research Laboratory, Cincinnati, Ohio, December 1975.

3. R. Stone, K. Covell, T. Moran, C. Sparkman, L. Weyand, *Using Mined Space For Long-Term Retention of Nonradioactive Hazardous Waste* (2 vol.), prepared by Fenix & Scisson, Inc., Tulsa, Oklahoma, under EPA Contract No. 68-03-3191, for the U.S. EPA Land Pollution Control Division, Hazardous Waste Engineering Research Laboratory, Office of Research and Development, Cincinnati, Ohio, 1984.

4. A. Finkenwirth and G. Johnsson, "Subsurface Waste Disposal Facility, Herfa-Neurode Near Heringen on the Werra," in *Proceedings of the 5th Symposium on Salt,* vol. 1, Northern Ohio Geological Society, Cleveland, Ohio, 1979, p. 239.

5. M. Pat Esposito, W. E. Thompson, J. S. Gerber, and J. C. Ponder, of PEI Associates, Inc., Cincinnati, Ohio, "Mined Space—A Viable Alternative to Landfilling for the Long-Term Retention of Contaminated Soils and Other Hazardous Wastes in Missouri," in *Proceedings of the 6th National Conference on Management of Uncontrolled Hazardous Waste Sites,* November 4–6, 1985, Washington, D.C.

6. *Disposal of Hazardous Wastes in Mines; A Case Study at Norton, Ohio, Hearing before the Subcommittee on Commerce, Transportation and Tourism of the Committee on Energy and Commerce,* House of Representatives, 97th Cong., 2d Sess., Washington, D.C., Government Printing Office, 1982.

7. F. Simpson, "Potential for Deep, Underground Storage in Canada," in M. Bergman (ed.), *Proceedings of the First International Symposium on "Storage In Excavated Rock Caverns"—ROCKSTORE 77,* Stockholm, Sweden, June 1977, Pergamon Press, New York, pp. 37–41.

8. H. Wassmann, *Cavity Utilization in The Nederlands,* Akzo Zout, Chemic Nederland, Postbox 25, 7550 G.C. Henglo, The Netherlands, 1983.

9. S. Seni, H. Hamlin, and W. Mullican III, *Technical Issues for Chemical Waste Isolation in Solution-Mined Caverns in Salt Domes,* prepared under Contract No. IAC (84-85)-1019 for the Texas Department of Water Resources, by the Bureau of Economic Geology, The University of Texas at Austin, 1984.

10. S. Seni, E. Collins, H. Hamlin, W. Mullican III, and D. Smith, *Phase III: Examination of Texas Salt Domes as Potential Sites for Permanent Storage of Toxic Chemical Waste,* prepared under Contract No. IAC (84-85)-2203 for the Texas Department of Water Resources, by the Bureau of Economic Geology, The University of Texas at Austin, November 1985.

11. M. Hooper, J. Geiselman, T. Noel, J. Piskura, and F. Gentry, *Vinton Salt Dome Project,* Empak, Inc., Houston, Texas, 1983.

12. C. Jacoby, "Underground Mining, Mine Storage, and Other Attendant Uses," in *Proceedings of the English Foundation Conference on Need for National Policy for the Use of Underground Space,* Proceedings, Papers, and Summaries, American Society of Civil Engineers, New York, 1983.

13. J. Jones, M. Bricka, T. E. Hayes, D. Thompson, Of The Waterways Experiment Station, Vicksburg, Mississippi, "Stabilization/Solidification Of Hazardous Waste," in *Proceedings of the 11th Annual Research Symposium,* EPA/600/9-85/013, sponsored by the U.S. EPA Office of Research and Development, Hazardous Waste Engineering Research Laboratory, Cincinnati, Ohio, April 29–30, May 1, 1985.

14. Office of Nuclear Waste Isolation (ONWI), Battelle Memorial Institute, Columbus, Ohio. A list of publications may be obtained from the above office.

SECTION 10.7
ABOVEGROUND DISPOSAL

K. W. Brown

Professor
Soil and Crop Sciences Department
Texas A&M University
College Station, Texas

David C. Anderson

Vice President
K. W. Brown & Associates, Inc.
College Station, Texas

10.7.1 BACKGROUND

Even after application of all available waste-reduction and treatment technologies, there will still be a need for long-term disposal or storage of the residual solids. These residues, including ashes, sludges, salts, contaminated soils, and solidified liquids are now being placed in belowground lined landfills. Unfortunately, even with the best technology now available, water leaks into these landfills, resulting in the accumulation of leachates contaminated with hazardous-waste constituents. Much of the accumulated leachate can be removed via leachate-collection systems during the 30 years postclosure as required by law. Operation of such collection systems minimizes the potential for the leachate to breach the liner immediately after closure. However, leachates will continue to accumulate following the postclosure period, eventually penetrate the liners, and endanger the underlying groundwater. Cartwright et al.[1] suggest that some migration of leachate from wastes buried in the ground in humid areas will always occur.

To prevent the hazardous constituents of waste from contaminating the adjacent environment, the waste must be kept dry. This can best be done by confining the waste aboveground under a waterproof cover, rather than belowground where it is difficult, if not impossible, to keep the waste dry. If the stored waste can be kept dry, then little leachate can be generated and the possibility of contaminating groundwater resources is greatly diminished.

There are two aboveground options which may be used; these are (1) warehouse storage and (2) underdrained, aboveground storage mounds. In a sense, anything we do with our solid-waste residue is only storage, since it will likely remain hazardous for at least hundreds of years. Thus, the warehouse option may be viewed as short-term storage, while the aboveground storage mound may be viewed as long-term storage. Either option will require some degree of maintenance for as long as the waste remains hazardous.

While the warehouse option has been considered for storage periods of 10 or 20 years until better options are available, the current regulatory environment, as well as the high cost of maintenance, makes this option undesirable. Thus, the only remaining aboveground option is the use of storage mounds.

10.7.2 ABOVEGROUND STORAGE MOUNDS

The essential features of an aboveground storage mound were described by Brown and Anderson[2] and are shown schematically in Fig. 10.7.1. They consist of a sloping base which can support the weight of the waste and the cover. The base must be overlain by a low-permeability liner system to retard vertical percolation of pollutants from the thin film of leachate which may be present on the liner prior to placement of the cover and during periods when the cover develops leaks. The liner system should consist of an upper flexible-membrane liner and a lower composite liner (a sandwich of a compacted clay overlain by a flexible membrane). A drain system consisting of stable aggregates and an appropriate system of collection pipes should be installed over the liner. This layer could act as both a primary leachate-collection and a leakage-detection system. In addition, there should be a similar layer between the upper and lower liners to serve as both a secondary leachate-collection system and a leakage-detection system for the liner (Fig. 10.7.2). The leachate-collection systems must be designed so that they drain freely by gravity and so that the drainage exits the mound aboveground on the

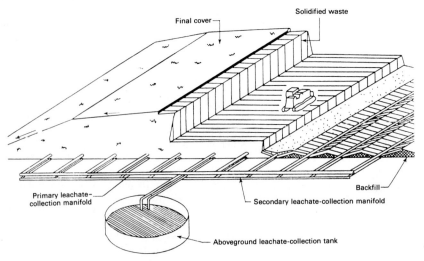

FIG. 10.7.1 Schematic diagram of an aboveground storage mound.

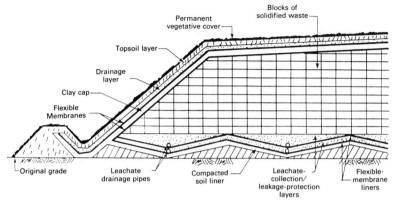

FIG. 10.7.2 Primary and secondary leachate-collection layers.

downhill side. As shown in Fig. 10.7.1, an ideal setup is to construct the base in a sawtooth fashion so that the drainage from different segments of a large facility can be isolated. This facilitates construction, minimizes the size of the working face, and thereby minimizes the amount of precipitation which contributes to the contaminated leachate. This also allows the installation of flexible-membrane liners with elevated overlapping strips, which decreases the reliance on seaming to prevent leakage. An additional advantage of isolating segments of the drainage systems is that in the event that the cover leaks at some time in the future, the drain system will serve to indicate which cover segment is leaking.

Provisions must also be made for collecting and disposing of the leachate. However, with proper design and operation, it should be possible to keep the amount of leachate generated during filling to a minimum, and if the cover does not leak, leachate production will cease soon after closure.

Depending on the nature of the waste and, in particular, the presence of toxic and mobile constituents, it may be desirable to install adsorption layers above the leachate-collection system to remove undesirable constituents leaking from the waste solids. A layer of calcareous gravel could be used to precipitate heavy metals, while a lignious or activated-carbon layer could serve to adsorb mobile organic species.

Ideally the waste to be placed in the facility would be compacted in lifts during placement to achieve the strength needed to provide the required physical support for the cover. The primary function of the cover is to prevent water from entering the waste. An ideal cover would consist of the following: (1) a gas-collection layer, (2) a composite hydraulic barrier layer combining a compacted-clay and a flexible-membrane liner, (3) a biotic barrier and drainage layer, (4) a topsoil layer, and (5) a permanent vegetative cover. Components of such a system are shown in Fig. 10.7.3. The sequence of layers and the thickness will depend on the hydrologic load which must be deflected. Much care must go into the design and construction of the cover, since this is the most critical component of the facility for limiting the long-term production of leachate.

Advantages of Aboveground Storage Mounds

There are many advantages to utilizing aboveground storage mounds for the containment of hazardous solid waste. Perhaps the greatest advantage over conven-

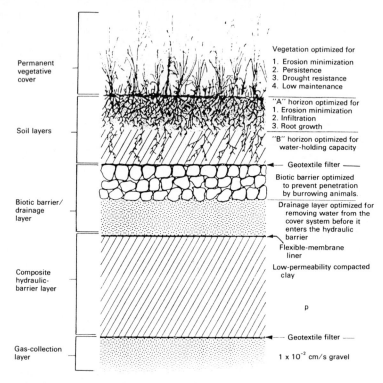

FIG. 10.7.3 Components of a landfill cover system.

tional belowground landfills is that the leachate is removed immediately by gravitational drainage, thus eliminating the primary driving force behind the migration of contaminants and thereby reducing the possibility of groundwater contamination. If, after closure, the cover is breached as a result of inadequate facility design or construction or by burrowing animals, the appearance of leachate will signal the need to repair the cover. Thus, the primary leachate-collection system acts as a complete and continuously operating leakage-detection system for the cover. Its continuous operation will collect and remove leachate, minimizing the potential for movement of contaminants into the groundwater, until repairs can be made to the cover. Since the primary leachate-collection system serves also as a cover leakage-detection system, the generation of leachate will be immediately evident to even the most casual observer. Thus, if leakage does become a problem because of a faulty cover, it will become apparent soon after the fault develops. Leaking covers over belowground facilities may only become apparent as the result of groundwater contamination.

Although gas generation is likely to be small compared to the large volumes which emanate from municipal-waste landfills, aboveground storage mounds are ideally suited to contain the gases and assist in their collection. Consequently, the possibility of gas migrating underground to adjacent structures is minimized.

There are several good reasons to isolate different wastes in separate cells in these facilities. First, by doing so it becomes easier to select an appropriate adsorption layer to retain any mobile constituents. Another reason is that the waste may become a resource in the future or the technology may become available to detoxify the waste, and in either case, it is much more efficient to recover the waste from a pile than to excavate it from a conventional belowground landfill.

Another advantage of aboveground storage mounds is that much of the technology which has been developed in support of belowground hazardous-waste landfills is directly applicable to aboveground facilities. Furthermore, aboveground facilities have been permitted under the current federal regulating guidelines. Although visual appearance may at first be considered a disadvantage, appropriately designed facilities would not necessarily need to be eyesores, and their appearance may be a distinct advantage in that we would be less likely at any point in the future to lose track of where our hazardous wastes are stored.

Site selection for aboveground storage mounds is somewhat less critical than that for a conventional belowground landfill. Aboveground facilities are particularly well suited to areas with high water tables, and there is less need to locate a thick deposit of clay. While it is desirable to double-line any storage landfill used for the storage of hazardous waste, the leachate will be continuously removed, thus decreasing the potential for deterioration of the liner material and greatly decreasing the head which would otherwise drive the accumulated leachate through flaws in the liner system.

Disadvantages of Aboveground Storage Mounds

Aboveground storage mounds are not without their disadvantages. Perhaps the greatest disadvantage is the potential for erosion of the topsoil layer in the cover. This problem is not, however, unique to aboveground storage mounds: all modern landfills are closed with caps which are mounded to increase runoff and are thus also subject to erosion. The slopes on aboveground storage mounds may, however, be somewhat greater than those on the caps of most conventional landfills. In either case, a certain amount of maintenance will be required to maintain the integrity of the cover. As pointed out above, in the case of the mound, the appearance of leachate in the drain system will signal the need to repair the cover, even if visual evidence of deterioration is not immediately apparent.

Embankments along highway road cuts and other earth-covered structures are successfully maintained at rather steep grades in a wide variety of climatic settings, thus indicating that we have the technology to deal with erosion. Furthermore, a study of prehistoric earthen burial mounds[3] gives compelling archeological evidence that large, aboveground earthen structures can be constructed so as to remain sound and intact for periods comparable to those required to retain radioactive-tailing piles, if the mounds are properly constructed. Many of the ancient mounds have survived for thousands of years, with some in the United States still in existence after 3,000 years.

10.7.3 COSTS

Cost has been suggested as an obstacle to the use of aboveground storage mounds. Soil must be excavated and brought on site for construction. With proper site selection however, the volumes of soil and the distances the soil may need to be transported do not differ greatly from those required for conventional landfills. A comprehensive cost-comparison study was conducted,[4] the results of which are given in Table 10.7.1. The comparisons for different size facilities in-

TABLE 10.7.1   Comparison of Cumulative 20-Year and per Ton Costs

	Waste capacity, tons/yr (kg/yr)		
	5,000 (4,535,000)	30,000 (27,210,000)	80,000 (72,560,000)
Total cumulative costs, $ millions			
Conventional landfill			
Double liner	19.0	39.3	71.1
Single liner	18.7	38.9	67.8
Aboveground storage mound			
Double liner	19.4	40.2	68.8
Single liner	18.8	37.9	64.0
Per ton costs, $/ton			
Conventional landfill			
Double liner	190.0	67.0	44.0
Single liner	187.0	64.0	42.0
Aboveground storage mound			
Double liner	194.0	67.0	43.0
Single liner	188.0	63.0	40.0

Source: Adapted from Ref 4.

dicate that for all practical purposes, there is no cost difference between the two options. If the potential for groundwater contamination resulting from belowground landfills and the cost of cleanup is figured in, the long-term liability results in much greater costs for conventional belowground landfills.

Many landfills have been extended above the ground surface simply as a matter of convenience. Once the landfill excavation has been filled, waste has continued to be mounded on top. It is not uncommon for municipal-waste landfills to extend 30 to 40 m above the original surface, while hazardous-waste landfills have more commonly been constructed to heights 10 to 20 m above the original surface. While these structures indicate that it is possible to mound waste to such heights and keep it in place, they have not been constructed with gravity-operated underdrains and thus are not true aboveground storage mounds.

Recent interest in aboveground mounds has resulted in the design and construction of such facilities along the Gulf Coast, and in the states of Kansas, Wisconsin, Alabama, and Pennsylvania. The cleanup plans for the Rocky Mountain arsenal in Colorado include the use of aboveground storage mounds.[5] As conventional belowground landfills continue to be closed, the aboveground storage mound remains as one of the few economical options for the long-term storage of solid hazardous-waste residue.

10.7.4 REFERENCES

1. K. Cartwright, R. H. Gilkeson, and T. M. Johnson, "Geological Considerations in Hazardous Wastes Disposal," *J. Hydrol.*, **54**:357–369 (1981).

2. K. W. Brown and D. C. Anderson, "The Case for Aboveground Landfills," *Pollut. Engr.*, **15**(11):28–29, (1983).

3. C. G. Lindsey, J. Mishima, S. E. King, and W. H. Walters, "Survivability of Ancient Man-Made Earthen Mounds: Implications for Uranium Mill Tailings Impoundments," NUREC/CR-3061, a report prepared for the U.S. Nuclear Regulatory Commission by Pacific Northwest Laboratory, Richland, Washington, 1983, pp. 1–33.

4. J. B. Hallowell, E. P. DeNiro, and J. S. Larsen, Jr., "Comparative Assessment of Alternatives for Waste Disposal and Storage," in *Proceedings of the Hazardous Waste and Environmental Emergencies,* March 12–14, 1984, Houston, Texas, pp. 269–274.

5. D. L. Campbell and W. N. Quintrell, "Cleanup Strategy for Rocky Mountain Arsenal," in *Proceedings of the 6th National Conference on Management of Uncontrolled Hazardous Waste Sites,* November 4–6, 1985, Washington, D.C., p. 36.

SECTION 10.8

AIR POLLUTION FROM LAND DISPOSAL FACILITIES

Paul R. de Percin

U.S. Environmental Protection Agency
Hazardous Waste Engineering Research Laboratory
Cincinnati, Ohio

Air pollution from hazardous-waste land disposal facilities has not been a great concern in the past and, until recently, has not been studied to any great degree. Unlike groundwater problems that take some time to occur and appear, air pollution has a more immediate and direct impact. Also, evaluation and control of air pollution from land disposal air emissions is difficult because air emissions are greatly affected by climate, weather, and the great variety of facility designs and operations. Examples of air-emission sources at land disposal facilities are surface impoundments, landfills, land treatment operations, and drum and tank storage. It is not unusual for the groundwater- and leachate-treatment systems to be air-emission sources themselves. These treatment, storage, and disposal operations can be sources of both particulates and volatile organic compounds (VOCs).

The Resource Conservation and Recovery Act of 1976 (RCRA), Public Law 94-580, and the disposal regulations based on this law (*Federal Register,* vol. 45, no. 98, May 19, 1980) did little more than identify the concern and need for regulation, and place some restrictions on blowing dust. This was changed, however, in the Hazardous and Solid Waste Amendments of 1984 (HSWA), which specifically stated that within 30 months regulations would be promulgated for the monitoring and control of air emissions from land disposal facilities. For a detailed discussion of the hazardous-waste laws refer to Chap. 1.

The air-emission regulations for disposal facilities must do two things: (1) protect the disposal facility's workers and neighbors, and (2) prevent the deterioration of the regional air quality. These requirements address the two impacts, acute and chronic, that the air emissions can have on the environment. Acute or toxic effects are a local concern and dependent on the specific chemicals being emitted. Chronic effects are usually attributable to the total sum of the chemical concentrations—e.g., total particulate or total VOCs—and are a regional concern.

As indicated earlier, air-emission sources at land disposal facilities vary widely in purpose, design, and operation. A general rule of thumb is: The greater the exposed area of waste, the greater the air emissions will be. This is logical because both VOCs and particulate emissions depend on wind to carry the pollutant away from the site, i.e., to create and maintain the dispersive force. Based on this rule of thumb, the major emission sources at land disposal facilities are

Surface impoundments

Landfills

Land treatment

Drum and tank storage

Treatment systems

Surface impoundments, open landfills, and land treatment sites can have large emission rates because these operations have large exposed waste areas. Closed landfills, drum and tank storage, and treatment systems, e.g., air strippers, have low emission rates because waste is generally not exposed to the air. These disposal operations are discussed in the subsequent subsections.

10.8.1 SURFACE IMPOUNDMENTS

Surface impoundments are used in land disposal facilities to treat, store, and dispose of liquid hazardous waste. In this discussion of air emissions, we define *surface impoundments* to include pits, ponds, lagoons, open-top tanks, and wastewater-treatment systems—any unit that can hold an accumulation of liquid waste and has a liquid surface exposed to the atmosphere. These units are generally operated in one of two modes: quiescent or aerated. *Quiescent* impoundments are used for equalization, clarification and settling, storage, biological treatment, and disposal. *Aerated* impoundments can be used for solar evaporation, but are primarily used for biological treatment. For a detailed discussion of design and operation of surface impoundments refer to Sec. 10.4.

Surface impoundments are sources of volatile organic compounds (VOCs) and particulates during several phases of their operation. During the operating life, only VOC emissions are expected from the impoundment. When an impoundment is decommissioned, however, both particulates and VOCs can be emitted. This is because either a surface impoundment must be entirely removed (liner and all) or the remaining hazardous sludge must be stabilized. Both closure procedures require a great deal of solids handling and exposure of waste. This discussion will examine the emissions from an operating impoundment.

The volatilization mechanism for VOCs from surface impoundments is well understood, but the emissions are not easy to accurately estimate. The VOC emissions are dependent on which of the competing mechanisms dominates the fate of hazardous components in surface impoundments. Volatilization, biological or chemical degradation, and adsorption are the three possible fates for VOCs that are not discharged out of the surface impoundment. In biological treatment systems this competition is particularly important. For surface impoundments with sterile systems, volatilization is the key fate mechanism. Adsorption onto the settled solids is not thought to be a large factor. This concept can be described by a mass-balance equation:

$$M_i - M_o = M_o + M_a + E \qquad (1)$$

Where M_i = mass into the impoundment
M_o = mass out of the impoundment
M_d = mass destroyed by biological or chemical degradation
M_a = mass adsorbed by the solids
E = mass volatilized to air

Models have been proposed for both the biological and volatilization mechanisms. The volatilization emission model, i.e., equation, is based on chemical mass-transfer phenomena. Mass transfer from water into the air is expressed as

$$E = KA\,(C_l - C_g) \qquad (2)$$

Where E = air emissions from the liquid surface, g/s
K = overall mass-transfer coefficient, m/s
A = liquid surface area, m^2
C_l = concentration of solute in the liquid phase, g/m^3
C_g = concentration of solute in the gas phase, g/m^3

The overall mass-transfer coefficient K is estimated from a two-phase (gas and liquid) resistance model. The resistance model is fully explained in Ref. 1.

The adsorption and especially the biological degradation rate models are now being developed and tested.

An effort has been made to check the accuracy of the emission models discussed above by measuring surface-impoundment VOC emissions. Impoundments are fugitive-area sources, i.e., not contained or emitting from a single point, making sampling very difficult. Three procedures have been used to sample impoundment emissions: (1) isolation flux chamber, (2) concentration profile, and (3) transect technique (modified upwind-downwind).[2] The isolation flux chamber uses an enclosure device (flux chamber) to sample gaseous emissions from a defined surface area (see Fig. 10.8.1). The vertical flux (emissions) is calculated using the concentration-profile method to measure the wind velocity, VOC concentration, and temperature profiles in the boundary layer above the liquid waste. The transect technique has a horizontal array of samplers to measure VOC concentrations within the effective cross section of the fugitive-emission plume. The VOC emission rate is calculated by spatial integration of the measured concentrations over the assumed plume area.

Each of these three sampling procedures has measurement strengths and weaknesses. While the isolation flux chamber (IFC) is the preferred procedure because of its ease of use and lower variability, it is not appropriate for large or highly aerated liquid surfaces. Unlike the isolation flux chamber, the concentration-profile (C-P) and transect procedures can sample aerated impoundments, but must have uniform weather conditions. The concentration-profile method was designed to measure emissions from large impoundments, with or

FIG. 10.8.1 Surface impoundment being sampled by the isolation flux chamber method.

without aeration. The transect technique is appropriate for the in-between cases, e.g., smaller, highly agitated or aerated impoundments or spray impoundments.

The difficulty in sampling surface-impoundment emissions is reflected in the variability in the sampling results. Table 10.8.1 presents examples of field measurements for four impoundments.[3] These data are the average of more than two measurements and are not presented as typical or representative of any other surface impoundments. Despite these results being averages, the variation between different sampling results (i.e., for impoundment 6, indicates that further development of the sampling methods is needed.

Four approaches have been identified for controlling or limiting the VOC emissions from surface impoundments: (1) pretreatment controls, (2) design and operating practices, (3) in situ controls, and (4) posttreatment techniques.[5] Pretreatment controls include administrative measures—e.g., waste banning—and technical measures—e.g., remove the VOCs from the waste before the waste is put in the surface impoundment (air stripping and distillation). While it is unusual to design and operate a surface impoundment for air-emissions considerations, it is possible to reduce air emissions by minimizing liquid surface area, by reducing influent waste temperature, by installing submerged fill pipes, and by less frequent dredging and cleaning. In situ controls are covers and vapor barriers used to prevent volatilization of the wastes. These controls can interfere with the impoundment's operation, especially with biological treatment. Posttreatment

TABLE 10.8.1 Measured Emission Rates for Four Surface Impoundments

Surface impoundments	Sampling approach	Carbon emission rates	
		kg/(ha · d)	(kg/d
1	IFC	10	1.4
2	IFC	49	7.1
3	Transect*	54	2.7
6	IFC	2.7	1.4
6	C-P	0.8	0.4

*Field data questionable.
Source: Ref. 3.

FIG. 10.8.2 An activated wastewater-treatment system with a dome and gas-phase carbon-adsorption unit as a posttreatment control.

controls involve collection of the impoundment's emissions by means of a cover, followed by treatment of the collected emissions. Figure 10.8.2 is an example of posttreatment control of surface-impoundment emissions: an air-supported dome with activated treatment of the exhaust gases.

10.8.2 LANDFILLS

Landfills are containment systems used to fully enclose hazardous waste that has been ultimately disposed. The landfill is broken up into separate areas called *cells,* which are used to isolate classes of wastes that are incompatible. Wastes are placed in a cell in either drum or bulk form. After a cell has been filled with waste to a height of about 6 ft (1.8 m), the waste or drums are covered with a soil layer about 1 ft (0.3 m) thick. This layer of waste is called a *lift,* and another lift is built on top of the lift just completed. Figure 10.8.3 is an example of an open landfill cell. When all the cells are full and no more lifts are possible, the landfill is closed by the construction of a final cap. For a detailed discussion of landfill design and operation refer to Sec. 10.1.

Landfills are sources of VOC and particulate emissions during their active, i.e., operating, phases and a source of VOCs after the landfill has been capped. Under both these conditions, the landfill emissions will be dependent on the following variables:

1. *Waste-related factors:* Composition, volatility, solubility, diffusion coefficient, biodegradability

2. *Landfill construction:* Soil type, compaction, cap design, vents

3. *Operating procedures:* Intermediate soil covers (lifts), form of waste, frequency of soil-cover application

4. *Environmental factors:* Wind, rain, temperature, barometric-pressure variations

From this list it can be seen that active landfills can be changed to reduce present

FIG. 10.8.3   An open cell of an active landfill.

and future emissions more easily than closed landfills. More of the variables that affect emissions can be modified.

The following is a list of potential emission pathways; which ones are of concern depends on the landfill's design and construction.

1. Volatilization and erosion of exposed waste
2. Diffusion through the landfill cap
3. Cracks in the cap
4. Gas vents
5. Access manways
6. Leachate-collection systems

The pathway that has received the most study is diffusion through soil caps, which is similar for both landfill caps and intermediate soil covers. At present the available emission models are incomplete because the other emission sources are not taken into account. The emission model that describes the soil-cap diffusion mechanism is

$$Q = \frac{A \cdot D_{\text{eff}} \, (C_i - C_o)}{T_c} \tag{3}$$

Where Q = volatilization rate, lb/h (kg/h)
A = area of soil cover, ft^2 (m^2)
D_{eff} = effective diffusivity, ft^2/h (m^2/h)
C_i, C_o = concentration of vapor on underside and top of cover, respectively, lb/ft^3 (kg/m^3)
T_c = thickness of cover, ft (m)

The calculation of the effective diffusivity D_{eff} is very involved and can be found in Ref. 1.

Measuring landfill emissions is as difficult as identifying the emission pathways and estimating the emissions. Generally, the isolation flux chamber (IFC) can be used to sample the VOCs from landfill caps, intermediate soil covers, and cracks in the cap. The transect technique can be used to measure the VOC emissions from exposed waste and open cells. Methods to measure emissions from access manways and leachate-collection systems have not been addressed. One of the main problems is that without biodegradation there is little or no measurable gas flow. Biodegradation generates gas that carries VOCs out of the landfill, substantially increasing the emissions and the control requirements. Table 10.8.2

TABLE 10.8.2 VOC Emissions from Landfill Caps and Vents

Landfill	Sampling approach	Carbon emission rate	
		kg/(ha · d)	kg/d
10	Transect	3.8	9.5
	IFC	4.5	1.1
D, cell	IFC	4.1	1.6
Q, cell	IFC	0.8	0.015
7, cell	IFC	<0.1	—
	Vent	—	<0.01
C, cell	IFC	<0.1	—
	Vent	—	<0.001

shows examples of VOC emissions from landfill caps and vents. Again these results are not necessarily typical or representative of all landfills. No biological activity is thought to occur at these landfills.

Considering the complexity of landfills, the emissions are easily controlled. Municipal-waste landfills have been studied and the problem of controlling methane migration out of landfills has been solved. Most of the design, operating, and cost information for landfill-gas control at municipal landfills is applicable to hazardous-waste landfill gas.[4] Both active and passive control systems will work. Active systems use gas vents and wells with vacuum pumps to collect the landfill gas for treatment. A passive system uses vapor barriers and collection systems that operate on diffusion and natural pressure.

10.8.3 LAND TREATMENT

Land treatment, or *landfarming,* is the spreading of hazardous waste on soil, with the natural processes then allowed to degrade and immobilize the hazardous components. Typically, the waste is either surface-applied or subsurface-injected. The soil and waste mixture is then tilled and watered at regular intervals to promote biological activity. The waste-application rates and intervals vary depending on the soil condition and weather. Because biodegradation is the objective of land treatment, mostly organic wastes are disposed of in this manner. For a detailed description of the design and operation of land treatment facilities refer to Sec. 9.4.

FIG. 10.8.4   Sampling of a land treatment operation with the isolation flux chamber method.

Both particulate and gaseous emissions are possible from land treatment operations. Particulate emissions can be caused by wind erosion of the contaminated soil and gaseous emissions by the volatilization of the VOCs. No measurement data could be found on the fugitive particulate emissions from land treatment, but procedures have been proposed to estimate these emissions based on fugitive particulate-emission models for agricultural fields and on the organic concentration in the soil.[5]

VOC emissions from land treatment operations have been measured using the isolation flux chamber and the concentration-profile sampling approaches. Figure 10.8.4 shows a land treatment operation being sampled by the isolation flux chamber method. The results of this sampling are presented in Table 10.8.3. The

TABLE 10.8.3   VOC Emission Rates from a Land Treatment Operation as Sampled by Two Methods

	Carbon emission rate	
Sampling approach	kg/(ha · d)	kg/d
Isolation flux chamber	53–626	3–35
Concentration profile	831–1080	46.5–60.5

range of emission rates is due to the samples being taken at different times after spreading occurred. The time period represented by these results is from just after spreading to two days after spreading.[3] The sharp drop-off of the emissions is easy to understand. The diffusion of VOCs through the soil to the soil-air interface is much slower than the VOC volatilization from the soil surface. Information from sampling reports indicates that tilling causes a temporary increase in emissions by bringing waste-rich soil to the surface, where volatilization can take place.

Emission models have been proposed for VOCs from land treatment systems,[3] but these models have assumed the VOCs are not adsorbed by the soil and are not biologically degraded. Both these assumptions are at odds with the way land treatment is to operate.

Air emissions from land treatment can be limited, primarily by proper operational practices.[6] Particulates can be controlled by regular application of water and tilling to maintain rich soil consistency. Wind fences and other basic fugitive-particulate controls will work here as long as the surface is not covered and can be tilled. Subsurface injection may reduce VOC emissions by preventing waste from pooling on the soil surface, and by allowing good mixing with the soil. The better the soil and waste are mixed, the greater the possibility that biological degradation will take place before the waste component can be volatilized.

10.8.4 DRUM AND TANK STORAGE

Drums and stationary storage tanks are used extensively to store hazardous waste generated in the United States. Drums can contain almost any waste, and may be used for storage, disposal, or both. Stationary tanks are usually limited to liquid storage. Both these storage vessels are essentially closed systems but gaseous emissions are possible depending on the nature of the facility's operation.

Drum-storage areas are sources of VOC emissions only when the drums are not sealed. This occurs during drum loading and unloading, and from leaking drums. The VOC emissions in these three cases can be reduced or limited by using the appropriate operating procedures. Leaking drums should not be allowed and can be detected with regular visual checks. Drums opened for filling and emptying should not be left open if not being handled immediately.

At many storage facilities, the drums are kept on curbed, concrete pads exposed to the climate. This makes sampling drum VOC emissions very difficult. There are a few drum-storage areas, however, inside buildings (for ease of handling). In these cases it is possible to measure the VOC emissions from the drums by sampling the building vents. The vent of a storage building containing about 1,500 drums was sampled and determined to have a carbon emission rate of 0.2 kg/d. No leaking drums or drum handling was observed in the building during this sampling.[3]

Storage tanks hold liquid wastes until the waste is disposed, treated, or recycled. Hazardous-waste tanks are similar to those types used for the storage of petroleum liquids, e.g., fixed roof, floating roof, pressure, with the same construction materials and vapor controls. The emissions models developed for petroleum storage tanks are applicable to hazardous-waste storage tanks. The vapor pressure of the waste must be substituted for the petroleum product's vapor pressure.[7,8]

10.8.5 TREATMENT SYSTEMS

Hazardous-waste treatment systems are processes used to modify the chemical and/or physical characteristics of the waste to prevent the release of the hazardous components to air and water. These treatment systems are mostly standard chemical process and unit operations that can remove, destroy or contain the hazardous components of the waste. During the waste processing, however, the treatment system can be an air-pollution emission source.[9,6] These emissions will be primarily VOCs because solids handling and processing mostly occurs in a wet form, a low-dust situation.

Removal processes separate VOCs from the waste before the waste receives further treatment or is disposed.[10] Examples of VOC-removal processes are air and steam stripping, solar evaporation, distillation, and solvent extraction. Without further processing or control, the VOCs removed in these treatment systems will cause pollution problems. Controls that have been used and found effective for these emissions are carbon adsorption and incineration.

Destruction processes convert the VOCs in the waste to carbon dioxide and water using chemical and biological reactions, thereby eliminating any potential emissions. Examples of VOC-destruction processes are chemical oxidation, ozonation, radiolysis, wet oxidation, and biological treatment. These processes involve competing mechanisms, i.e., destruction versus volatilization, that determine the air emissions (recall the previous discussion of surface impoundments). For those destruction processes in which there are exposed liquid surfaces, the surface-impoundment emission controls would be appropriate. The removal processes are used as pretreatment controls for surface impoundments.

Containment processes use stabilization, encapsulation, and solidification techniques to modify the waste form. This modified waste form is designed to prevent the VOCs from leaching and diffusing from the waste into the air and water. Air emissions are expected during the mixing of the waste and additive, i.e., stabilizer, and during the curing of the modified waste. Emissions of VOCs from the disposed modified waste will probably be low because the modified waste must pass the EPA's liquid and leaching tests.

Many of these treatment systems are discussed in detail in Chaps. 6, 7, and 9.

10.8.6 SUMMARY

The major treatment, storage, and disposal operations at hazardous-waste facilities have been identified as air-pollution sources. Initial air measurements have confirmed this observation, but further evaluations are necessary to determine whether these air emissions are significant, i.e., a health hazard.

10.8.7 REFERENCES

1. C. Springer, P. D. Lunney, and K. T. Valsaraj, *Emission of Hazardous Chemicals from Surface and Near Surface Impoundments to Air,* U.S. EPA Hazardous Waste Engineering Research Laboratory, Cincinnati, Ohio, December 1984.

2. W. D. Balfour, and C. E. Schmidt, "Sampling Approaches for Measuring Emission Rates for Hazardous Waste Disposal Facilities," *Proceedings of 77th Annual Meeting of the Air Pollution Control Association,* San Francisco, California, June 1984.

3. W. D. Balfour, R. G. Wetherold, and D. L. Lewis, *Evaluation of Air Emissions for Hazardous Waste Treatment, Storage, and Disposal Facilities,* EPA 600/2-85/057, U.S. EPA Hazardous Waste Engineering Research Laboratory, Cincinnati, Ohio, June 1984.

4. R. A. Shafer, A. Renta-Bobb, J. T. Bandy, E. D. Smith, and P. Malone, *Landfill Gas Control at Military Installations,* U.S. Army Corps of Engineers, Construction Engineering Research Laboratory, Technical Report N-173, Vicksburg, Miss., January 1984.

5. J. H. Turner, M. Bascome, et al., *Fugitive Particulate Emissions from Hazardous Waste Sites,* EPA 600/2-87/066, U.S. EPA Hazardous Waste Engineering Research Laboratory, Cincinnati, Ohio, September 1984.

6. U.S. Environmental Protection Agency, *Evaluation of Emission Controls for Hazard-*

ous Waste Treatment, Storage, and Disposal Facilities, EPA 450/3-84-017, U.S. EPA Office of Air Quality Planning and Standards, Durham, North Carolina, November 1984.

7. U.S. Environmental Protection Agency, *Evaluation of Methods for Measuring and Controlling Hydrocarbon Emissions from Petroleum Storage Tanks,* EPA 450/3-76-036, U.S. EPA, Research Triangle Park, North Carolina, 1976.

8. U.S. Environmental Protection Agency, *Compilation of Air Pollution Emission Factors,* 3d ed., U.S. EPA, Research Triangle Park, North Carolina, 1977.

9. U.S. Environmental Protection Agency, *Control Techniques for Volatile Organic Emissions from Stationary Sources,* EPA 450/2-78-022, U.S. EPA, Research Triangle Park, North Carolina, 1978.

10. J. J. Spivey, C. C. Allen, D. A. Green, J. P. Wood, and R. L. Stallings, *Preliminary Assessment of Hazardous Waste Pretreatment as an Air Pollution Control Technique,* EPA 600/2-86/028, U.S. EPA Hazardous Waste Engineering Research Laboratory, Cincinnati, Ohio, October 1984.

CHAPTER 11

COMPREHENSIVE HAZARDOUS-WASTE TREATMENT FACILITIES

CHAPTER 11
COMPREHENSIVE HAZARDOUS-WASTE TREATMENT FACILITIES

Peter Daley
Director, Research and Development
Chemical Waste Management
Riverdale, Illinois

The hazardous-waste management industry has evolved simultaneously in several directions. This is the result of a generator and service-industry tendency to focus on single solutions to single problems and to search for ways to resolve each problem through one of the traditional disposal methods: burning, burying, or discharging. Resource recovery, augmented by both traditional and new separation technologies, has extended the scope of hazardous-waste operations, but the overall trend has nevertheless been to extend lines of practice with only occasional regard for how systematic management of a variety of wastes can improve overall effectiveness of waste management.

Indeed, there are notable exceptions to this trend. A number of businesses exist in the United States and elsewhere that owe their existence to clever use of different waste materials to mutually neutralize their hazardous characteristics. Waste acids are used to neutralize waste alkalis, and waste oxidizers are used to oxidize cyanides and organic materials. In Japan, Europe, and the United States a few comprehensive waste-treatment operations exist which attempt to optimize these synergistic reactions, but even these offer considerable room for improvement. A comprehensive hazardous-waste management facility, designed from the ground up to systematically optimize the benefits of innovative and traditional treatment processes, offers substantial advantages over the "add-a-process" approach. A state of the art comprehensive hazardous-waste management facility focuses on enhancing efficiency and effectiveness by using common processes to treat or recover a variety of materials. Equally important, it also offers considerable opportunity to develop improved materials-handling, environmental-monitoring, and management systems that would be impossible or uneconomical for less broadly based facilities.

Obstacles to the development of comprehensive facilities in the United States have been generally sociopolitical rather than technical. The ability of local interest and political groups to successfully oppose siting of new hazardous-waste management facilities, as well as new processes at existing facilities, will continue to make the development of comprehensive waste-management facilities difficult. Nevertheless, the many advantages of optimized comprehensive hazardous-waste management facilities dictate that they take a role in resolving hazardous-waste problems. The purpose of this chapter is to review these benefits and to study the basic structure of such a facility. Since it is impossible to develop a complete engineering design in these pages, our goal is to provide planners, engineers, and scientists with enough detail to demonstrate the strengths of the concept so that they may develop facility designs, given an opportunity to do so.

11.1 WHY COMPREHENSIVE FACILITIES?

There are three principal reasons why comprehensive waste-management facilities are a highly efficient means for managing hazardous wastes: policeability, process synergism, economy of scale.

Policeability: Key to Compliance

We must demand and ensure the highest practicable standards for managing our wastes. Without control, we have learned through hard lessons that the fouling of our global nest is the likely result. We have also learned that both the proper management of wastes and the regulatory systems required to ensure proper management can be costly. With tens of thousands of generators and waste-management operations to oversee, the task can only be economically accomplished through voluntary compliance, supplemented by focused enforcement activity. Any step that increases the ratio of regulatory resources to industrial activity will necessarily improve the quality of our waste-management system.

The U.S. Environmental Protection Agency (EPA) has issued permits for thousands of hazardous-waste generators and disposers. Each generator must store, treat, or dispose of wastes. Improper disposal has historically produced the most costly and environmentally damaging waste-related problems. Improper treatment and storage have been a less visible problem. This does not mean that they are less important; poor treatment often leads to poor disposal, and poor storage can result in seriously damaging leaks and contamination.

Treatment of discharges to air, land, and water is increasingly more common; policing of these treatment operators is a growing burden. For example, the May 8, 1985, ban on land disposal of bulk liquids created a massive new treatment requirement. Additional requirements for treatment, and consequently the number of treatment operations, will grow as a result of the steady tightening of all disposal standards. While tightened standards and increased cost induce waste-reduction and waste-elimination activity, more and better treatment is an inevitable corollary.

Contrary to treatment, disposal operations will trend toward lower numbers of operators for many years, although the numbers are likely to remain large. Better treatment eliminates wastes from the hazardous-waste category. Unsuitable disposal sites and methods are steadily being eliminated. These trends, however, are tempered by the desire of many generators to maintain close control of their own wastes by using on-site disposal facilities.

On balance, the task of policing hazardous-waste management operations will grow, primarily driven by the demand for better treatment and resource recovery in place of disposal. The policing task is greatly simplified by consolidation of treatment facilities in treatment centers. A single large treatment center can easily process the wastes of several thousand generators. The nation's largest treatment and disposal operation in 1985 handled over 40,000 waste streams. A few dozen treatment centers could be constructed to process a large part of the nation's hazardous waste.

Regarding land disposal, it is not technically impractical for all currently off-site land disposal operations to be concentrated at a few large facilities with appropriate treatment capability. In 1984, the five largest land disposal facilities in the United States handled over half of all off-site land-disposed hazardous waste. Effective comprehensive waste-treatment systems and volume-reduction programs, as well as increasing use of on-site treatment and disposal for remedial actions will, nevertheless, prevent this resource from being exhausted, even in the face of closure of other land disposal facilities. The point is not that land disposal is or is not a desirable endpoint for hazardous waste, but that a relatively small number of facilities can, in fact, manage the nation's hazardous wastes.

In conclusion, there is a great regulatory economy associated with a hazardous-waste program including a system of large, comprehensive treatment and disposal facilities. Given the relatively small number of such treatment facilities that would be required nationally, it would be reasonable to have full-time regulatory inspectors at each one. In fact, several large facilities currently have such full-time surveillance practices in place. This is a powerful argument for developing the comprehensive treatment facility approach.

Waste-Treatment Process Synergisms

In the introduction I mentioned some of the obvious synergisms that can be realized by managing a variety of wastes in a single operation, e.g., neutralization of waste acids with waste alkalis. When a wide variety of sophisticated treatment processes is available on a single site, many opportunities become available that can greatly increase operational efficiency. Table 11.1 lists but a few of these.

Only when a wide variety of process options exist at a single site do these synergistic opportunities begin to be significant. They are only exploited if it is clear to the operator that the processes and streams they depend upon can be reliably obtained in quantities large enough to warrant installation of specialized treatment equipment. This often means a large, even national, supply must be drawn upon to minimize short-term supply fluctuations and maintain an economic production rate. Because of the large capital investment often involved, a time frame of several years for this supply must be reasonably assured.

The most successful hazardous-waste management facilities are those that minimize cost by carefully managing the opportunities to get effective use of

TABLE 11.1   Some Synergistic Opportunities in Comprehensive Hazardous-Waste Management Operations

Waste acids and alkalis to neutralize one another.

Waste oxidants to treat cyanides and organic contaminants in water.

Organic materials separated from water and sludges used as fuel for site's incinerator and steam boiler.

Metal ions to form hydroxide flocs in precipitation processes.

Acids to break emulsions.

Salts and acids to "salt out" organic compounds from wastewater.

On-site incinerators to dispose of organic vapors and waste activated carbon generated by other on-site processes.

Ash and calcium and magnesium oxides to aid in stabilization processes.

Production of blended liquid fuels from combustible solids and liquids.

Use of advanced separation processes—e.g., supercritical extraction, solvent extraction, and freezing—that are not cost-effective in small scale, to recover and concentrate wastes.

Use of advanced treatment processes not effective on small scale.

wastes as process chemicals, and multiple uses of process equipment. The design features of some of these facilities discussed later exemplify how synergisms are identified and exploited.

The concept of comprehensively planning waste-treatment processes to use wastes to replace purchased materials and use single process units for multiple applications is not new. However, it is only since about 1980 that engineers have seriously thought about and begun to build these facilities in significant numbers. National facilities in Denmark and Finland, commercial facilities in the United States and France, and provincially chartered facilities in Alberta and Ontario have led the way. Plans to build several new facilities are on the drawing board.

Even the newest facilities have not yet taken full advantage of the potential to optimize treatment by applying modern data-management techniques. Computer models of the physical and chemical processes at a treatment facility, including storage systems, will lead to better and cheaper treatment. They will also augment site safety and environmental programs by precluding dangerous situations and providing warnings of dangerous conditions. More sophisticated models could be developed to coordinate delivery of specific wastes to treatment facilities and coordinate exchange of materials between various facilities that might be linked together through common data bases.

Large Scale Fosters Advanced Technology, Safety, and Economy

On a scale frequently encountered in hazardous-waste treatment, costs for traditional treatment equipment, e.g. filters, precipitators, liquid incinerators, afterburners, typically range from $25,000 to $250,000 per unit. Advanced processes, e.g. two-stage solids incinerators, wet oxidizers, plasma arcs, can easily cost 10 times these amounts. In addition, the costs of many of the environmental and safety programs needed to support hazardous-waste operations tend to be nonlinear, increasing at a steadily decreasing rate compared to the capacity of a treatment facility. For example, the cost of a test burn for an incinerator is

$50,000 to $150,000 and depends primarily on the complexity of the wastes burned, not the size of the system.

Given the relatively high capital and operating costs of advanced processes and the nonlinearity of environmental- and safety-program costs, it is understandable that only large facilities capable of distributing these costs over large volumes can operate them economically. "Niche" applications for specific wastes-treatment technologies undoubtedly are, and will be, important for on-site treatment facilities, but this only makes economic sense when size of the operation justifies the investment. For smaller operations to have access to costly advanced processes one must take the waste to the system or the system to the waste. Thus, centralized facilities or mobile systems must be used.

Both avenues, centralized and mobile, have special advantages (see Table 11.2).

The key benefits of mobile systems are that they eliminate risks associated with transportation of waste material and that there is less public objection to siting temporary (versus fixed) systems. On the other hand there is added risk associated with over-the-road transport of process equipment, fresh start-ups, and making and breaking connections at each site. In addition, a fixed site, especially a large fixed site with a broad permanent staff of environmental, safety, scientific, and engineering personnel can respond more effectively to any accident that may occur than can the limited staff of a mobile treatment system. The largest hazardous-waste management facility in the U.S. today employs over 300 people, including approximately 40 professional engineers and scientists.

Mobile processes are most desirable when scale is large enough to justify taking a unit to the site and too small to justify permanent facilities. They are especially desirable when treatment products can be re-used or ultimately disposed of on site. This avoids another transportation task. Mobile processes are least desirable when additional product treatment is required or when creation of a costly, secure land disposal facility is required.

Mobile systems are, therefore, useful for many situations but are not a universal answer.

In short, there are many scale factors in addition to other tangible and intan-

TABLE 11.2 Relative Merits of Mobile and Centralized Hazardous-Waste Treatment Processes

Centralized	Easily monitored.
	Receiving, storage, treatment, and disposal systems are safe from the "perils of the road" and errors due to making and breaking connections.
	Overhead costs of field crews avoided.
	No size and weight limits on equipment.
	Permits for generator-site operations not required.
	Immediately available support staff of specialists (e.g., industrial hygiene, safety, analytical chemists) more easily maintained.
Mobile	Reduced waste-transportation risk.
	Public objections to local fixed facilities are avoided.
	System transport, start-up, and making and breaking of connections increases risk of malfunctions and processing accidents.
	Cost of mobile systems is generally higher than fixed.
	No permitting system for mobile processes (except PCB treatment) exists so permits must be tied to each operational site.

gible factors that favor centralized, off-site treatment over on-site treatment by fixed or mobile systems. The many advantages of centralized facilities demonstrate that they can play an important role in hazardous-waste treatment, especially when treatment systems require large amounts of capital and costly-to-duplicate environmental and safety systems.

11.2 THE OBJECTIONS TO COMPREHENSIVE FACILITIES

Given the many merits of comprehensive facilities and the decisions of many governments to foster and even subsidize their development, why are they opposed? There are three principal reasons: first and foremost, people don't want them nearby; second, generators and communities perceive a high risk associated with transportation of wastes; and third, generators have seen that poorly run central treatment facilities can lead to large cleanup costs for which they may be liable.

These objections, justified or unjustified, have greatly curtailed growth of centralized treatment operations and will affect the types and locations of facilities that will be built. The objections have major components which are creatures of local laws and governmental practices. Local influence is very important in the United States. The strong U.S. tradition of public participation in decision making is firmly established in the Resource Recovery and Conservation Act and its implementing regulations (the system that governs hazardous-waste management in the United States). The Act gives local individuals and organizations power to influence siting decisions for hazardous-waste treatment and disposal facilities. Only through extremely sensitive and well-planned (time-consuming and costly) development programs can siting and expansion programs succeed. These difficulties and their associated costs reduce the attractiveness of centralized facilities which are otherwise economically and environmentally sound. Even when well-planned and sensitive programs are conducted, failure may result. The general issue of establishing difficult to site facilities of all sorts (prisons, airports, highways, and a host of others) is a major topic worthy of a textbook and will not be discussed further here.

The transportation risk issue is also affected by public opinion and influence. There are clear lines of logic that support the opposing sides. Arguments favoring on-site treatment and disposal cite studies showing that transportation is often the biggest risk component in the disposal process. Opposing this are all the advantages of centralized facilities discussed in the foregoing sections, plus the argument that the transportation risk, as well as the other components of the risk equation, are individually and together so small that economic and policeability factors, factors not included in traditional risk assessment, dominate.

A general conclusion to this argument is unreachable. Decisions and judgments must be based on site-specific analysis. This discussion is presented here to alert planners to the many quantitative and subjective factors they must consider in selecting treatment and disposal options.

The potential financial liability stemming from poorly run commercial treatment and disposal facilities is a key factor affecting generators' decisions to develop captive treatment and disposal systems (both stream-specific and centralized). Today's perception of this liability is based on the mounting cleanup cost stemming from poor past practices. Poor past practices are also responsible for much of the public opposition to local siting of hazardous-waste treatment or disposal operations.

Our greatly increased awareness of the risks associated with uncontrolled emissions and effluents from all our activities, specifically from hazardous-waste-related activities, have greatly reduced environmental risks. A modern, comprehensive hazardous-waste facility includes a full-time environmental-management staff; close monitoring of effluents and waste products (solid, liquid, and gas); strict regulatory control (in many cases, full-time on-site regulatory personnel), and requirements for financial responsibility (both regulatory and as a matter of policy by generators). These are control systems that are easily policed and affordable only when associated with a manageable number of large facilities as discussed earlier. The combination of controls and awareness of risks (both environmental and financial) cannot assure that problems will not occur at comprehensive facilities, but it greatly reduces the risk. Those governments and corporations choosing to use and build comprehensive facilities base their decisions on the conclusion that comprehensive, centralized facilities are indeed the optimum solution to the problem of managing hazardous wastes which, if uncontrolled, threaten our environment and our health. Alternate solutions are more likely to be the result of the inability of governments and entrepeneurs to overcome local opposition to siting.

Generators selecting on-site options presume (knowingly or unknowingly) that they can identify risks and manage treatment and disposal operations better and cheaper than commercial operators. While this may be true in some cases, it may be a risky assumption. Generators are not in the waste-treatment and disposal business and their managers focus on the essential tasks of producing a quality product or service. This means that a crack in a waste-treatment vessel or other apparently "minor" problem may go unnoticed or ignored until serious damage has been done. On the other hand, a closely monitored comprehensive facility depends totally for its economic success on the maintenance of operational and fully permitted storage, treatment, and disposal systems. This is a powerful driving force. Combined with today's technologies, and with current permitting and regulatory systems, it greatly reduces the risk of environmentally damaging errors at closely monitored centralized facilities.

The many advantages of comprehensive, centralized facilities continue to lead to more decisions to pursue this option in spite of the difficulties and opposition. These facilities are not destined to manage all wastes for all industries, but they will play an essential role in the overall waste-management schemes of the United States and other countries. In fact, the Canadian provinces of Alberta and Ontario, several European countries (including France, Denmark, Finland, and Spain), and other countries are committed to central hazardous-waste processing. In the remaining sections of this chapter, we will discuss the general technical aspects of such facilities both generally and through specific examples.

11.3 THE DESIGN OF COMPREHENSIVE HAZARDOUS-WASTE FACILITIES

Design Goals and Objectives; The Christmas Tree Problem

The design goals and objectives for a comprehensive hazardous-waste facility must carefully balance the need for versatile equipment against the tendency to add something for everything. The tendency to the latter produces a costly

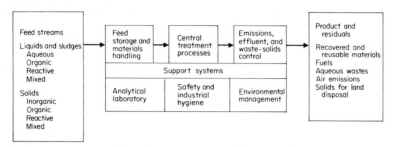

FIG. 11.1   The hazardous-waste treatment process and its support elements, and feed and product streams.

Christmas tree with so many processes that it cannot be effectively or efficiently operated.

Certain waste classes must inevitably be excluded from comprehensive plants and left to special systems, either on generator sites or off site. Current practice excludes explosives and many wastes containing some economically valuable components (e.g., precious metals) from comprehensive plant designs.

The optimum design is that which can effectively treat or recover the broadest range of waste feed streams economically. Looking at wastes as component classes and identifying process systems to treat each class is an essential step in process selection. Figure 11.1 identifies the basic waste-feed categories and schematically represents the entire treatment facility. Figure 11.2 expands the view of the process sections and provides an idea of the wide range of systems required.

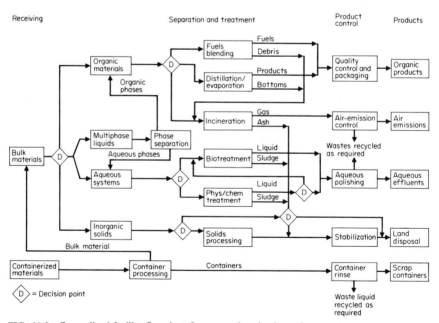

FIG. 11.2   Generalized facility flowchart for comprehensive hazardous-waste management.

The various feeds must be converted into a limited number of marketable products or environmentally innocuous residues, (air emissions, aqueous discharges, or solids for land disposal). Marketable products include recovered and re-usable materials as well as fuels blended from waste solids, sludges, and liquids. Between these feeds and products lie the plant's basic treatment and separation operations. Table 11.3 outlines the main functions of the subsystems required for plant operation. The three subsystems are described here as follows.

The Three Subsystems

The central processing equipment usually gets the most attention during the development period. Equal attention must be paid, however, to the "front-end" materials, receiving, handling, and storage systems and to the "tail-end" effluent-, emissions-, waste-solids-, and product-control systems. These ancillary systems absorb equivalent amounts of capital and operating effort and are as likely to cause plant failure or shutdown as the central process.

Materials handling at the front end typically presents significant safety prob-

---

TABLE 11.3   Facility Subsystems: Typical Functions and Processes

---

Receiving processes and functions (the "front-end")

Unloading (bulk and containers)
Inventory and documentation
Sampling
Composition verification
Weight and volume determinations
Container emptying
Reagent and waste storage
Distribution for processing
Spill control

Separation and treatment processes ("central processing")

Pretreatment
  Phase separation
  Size reduction
  Screening
  Dewatering

Advanced separation
  Liquid-liquid extraction
  Freezing
  Distillation and evaporation
  Drying
  Precipitation

Destruction and neutralization
  Incineration
  Pyrolysis
  Chemical oxidation and reduction
  Biooxidation
  Water reactions
  Acid-base neutralization

TABLE 11.3   Facility Subsystems: Typical Functions and Processes
(*Continued*)

Product-control processes and functions (the "tail-end")

Air processes
  Adsorption
  Acid scrubbing
  Electrostatic precipitation
  Condensation
  Filtration

Water processes
  Adsorption
  Precipitation
  Ion exchange
  Electrodialyses
  Reverse osmosis
  Filtration
  Dewatering

Solid processes
  Chemical stabilization
  Slagging

Organic-products processes
  Containerizing
  Quality control
  Tanker loading
  Product storage

Emissions, effluents, and disposal functions
  Product storage for quality control
  Product analysis
  Environmental monitoring

---

lems and requires costly, high-maintenance mechanical equipment to handle the wide variety of solids and liquids presented for treatment, as well as the associated containers. It is usual to expect rock, reinforcing rods, cans, rags, plastic sheeting, low-flashpoint liquids, oxidizers, and all sorts of other difficult-to-handle materials in the feed. Further, actual feeds may vary significantly from those described by generators. It must be presumed that they will contain volatile, low-flash, and reactive materials until it is demonstrated otherwise.

The product streams and the final treatment systems producing them require a high degree of quality control to assure the public and the environment are protected. Failure to meet predetermined product-quality standards not only risks adverse environmental and health impacts but also discredits the plant as a source of fuels and recovered materials. Even if quality failures are not regarded as significant for customers or the environment, failures to meet performance standards may discredit the site and its operators to the extent that future operations are jeopardized.

Material-Receiving Functions and Processes

Table 11.3 details the primary waste-receiving functions and processes. One overriding issue gives rise to the high costs and numerous problems encountered in this part of the processes: unplanned and wide variations in the physical and chemical quality of materials received.

Generators can and do throw an almost limitless variety of materials "in the garbage," and many of these eventually appear at the treatment facility; the problems they can cause are likewise limitless. What happens to a shredder that's fed a low-flashpoint material or a chunk of heavy-gauge aluminum? How do you get 400 pounds of sticky resin out of a drum? How do you avoid polymerization reactions that can turn entire tanks into a solid, plastic mass? How do you verify that a generator has not used illegal pretreatment processes before delivering the waste to you? How do you sample a tank truck or container that has 3, 4, or 5 physical phases of all physical states inside? How do you separate hammers and electric motors from potential fuel materials?

This is just a small sampling of the types of problems one may encounter over and above the range of problems visualized if one considers the bewildering range of materials listed on customer manifests. One U.S. hazardous-waste company handles over 100,000 different waste streams. An analysis of the compositions of these streams when compared to those listed on the shipping documents showed only a modest correlation. The great majority contained substantially less of the hazardous components than the manifest showed. In fact, many contained *no measurable hazardous-waste materials at all*. This was traced to the fact that many generators live in great fear of the liability associated with potentially "underclassifying" a waste, so they intentionally do the opposite. This practice is laudable from a safety standpoint, but makes operations and business planning based on data on the manifests (the data used by agencies in estimating the scope of the hazardous-waste problem) almost impossible.

This brief discussion should serve to alert engineers and managers to the great need for thoroughness and apparent overdesign in receiving systems.

The Central Processing System

The unit operations selected for the central processing system must convert wastes into a limited number of environmentally innocuous residuals (air emissions, aqueous discharges, or solids for land disposal) or marketable products. Marketable products include recovered and re-usable materials as well as fuels blended from waste solids, sludges, and liquids. Manufactured fuels are an essential product, for they provide a constructive use for many materials that should not be discharged into the environment (gases, liquids, or solids) in place of costly dedicated destruction processes. These fuels are best used in high-efficiency industrial combustion processes (e.g., boilers, blast furnaces, and kilns). Importantly, they can supply essential energy for on-site use.

A number of candidate processes for the central processing system are listed in Table 11.3. Figure 11.2 shows how these processes might be interconnected to optimize the performance of the system. Elements of this design are found in most of today's existing and proposed comprehensive plants. The principal ones are

1. *Incinerator: Essential but expensive.*   An incinerator or pyrolyzer capable of destroying organic solids, liquids, and gases is a fundamental and essential element. It is, nevertheless, an inefficient means of treating many materials, and much less costly processes should be sought out and developed. Candidate materials for alternative processing are containers, low-energy materials, and

materials that are chemically destructive to incinerators constructed of reasonably resistant materials.

2. *Biotreatment: Use should be maximized.*   Biotreatment is a cheap, effective method of treating many organic materials and should be considered seriously for most organics in dilute aqueous systems and many organic solids and sludges.

3. *Waste-derived fuels production: The principal means for medium- and high-energy organic-material disposal.*   Fuels blending for off-site and on-site use is the best way to dispose of materials with energy values exceeding approximately 8,000 Btu/lb. This includes solids that may be reduced in particle size so that with a small energy input they can be kept suspended in fuel.

4. *Separation process: Essential to assuring that the most economical approach for each material is used.*   Direct phase separations and separations induced by physical-chemical processes such as precipitation, solvent extraction, freezing, drying, and evaporation are necessary pretreatment steps to produce streams that can be directly disposed of or efficiently treated by low-cost treatment processes such as biodegradation. They are especially important in reducing the need for costly incineration of low-energy material.

5. *Recovery: Do it when possible.*   Product recovery is the best way to manage many materials, especially halogenated solvents as long as they are in relatively high concentrations. They are generally marketable and are expensive to treat in any process. For low concentrations, halocarbons can often be stripped with air, steam, or inert gas, or simply heat. The vapor stream is best handled by mixing it with incinerator combustion air.

6. *Stabilization of solid residuals: Land disposal should be minimized.*   Residuals requiring land disposal are an inevitable product, but they should be minimized and treated to reduce or eliminate their hazardous nature whenever required.

7. *Treatment of water-reactive material: Essential for safety.*   Water-reactive sulfides and cyanide solids must be reacted or stabilized prior to land disposal in order to avoid potential hydrogen cyanide or hydrogen sulfide releases upon wetting.

The flow chart in Fig. 11.2 implies that the plant is a continuous operation. In fact, it may be; however, many of the processes are better run in a batch mode. This is because waste is usually delivered in batches, and quality variations frequently necessitate customized treatment adjustments. In addition, chemical and biological batch processes almost always have higher average rates than continuous processes. This is especially important because the low final concentrations often required in waste treatment support only slow reactions (see Fig. 11.3).

An expansion of the logic of incinerator use is appropriate, since incinerators are, contrary to many expectations, best used sparingly. Incineration is one of the most costly processes to build and operate in the plant. Comprehensive-plant incinerators, because they must handle a wide variety of solid and liquid feeds, must include a number of design features not required in incinerators dedicated to specific waste streams. In addition, incineration requires introduction of large amounts of combustion air, which must be treated in the emission-control system and heated to incineration temperatures. In fact, pyrolysis systems which do not require addition of this excess air and oxygen burners which reduce gas volume offer many advantages over conventional incinerators and should be seriously

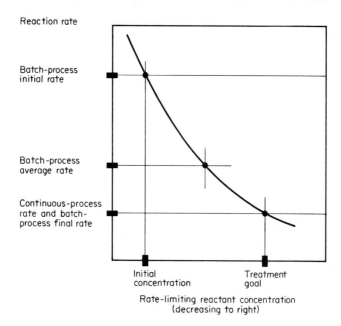

FIG. 11.3   Reaction rates for batch and continuous processes. Figure assumes second-order reaction.

considered to supplement or replace them. Likewise, wet-oxidation systems are usually more efficient for treating high-water, nonbiodegradable organic wastes.

The preceding guidelines are only a beginning in defining the treatment processes for a comprehensive treatment plant, but they demonstrate the type of thinking engineers must use in developing a successful plant. Briefly, pretreatment systems must be provided that can separate wastes into classes that minimize the use of high-cost processes, the number of processes must be minimized, and effectiveness and reliability must be prime process-selection considerations.

Product-Control Processes

The various feeds must be converted into a limited number of marketable products or environmentally innocuous residues, (air emissions, aqueous discharges, or solids for land disposal). Marketable products include recovered and re-usable materials as well as fuels blended from waste solids, sludges and liquids. Manufactured fuels are an essential product, for they provide an outlet for many combustible materials that should not be discharged into the environment (gases, liquids, or solids).

The quality-control standards established for the hazardous-waste business are very tough. Destruction to undetectable levels, which may mean less than one part in ten billion, is required for many materials. While virtually any material is likely to arrive at the plant, regulators and the public expect the plant operator to demonstrate that *no* hazardous situations are created by plant operations. This often leads to conflicts, since zero risk is unattainable.

From a practical standpoint, this demand means different things for marketable products, air emissions, aqueous effluents, and waste solids. For incinerator emissions it means complying with the demonstrated "test-burn" performance through a wide range of continuous performance-monitoring systems. For miscellaneous air emissions it usually means a host of activated-carbon adsorption systems, condensers, and filters, all of which generate their own wastes and require continuous monitoring. In order to minimize air-emission measurement and quality-control problems, consolidation of streams and subsequent incineration by blending with incineration air is often economical.

For aqueous wastes, both batch and continuous discharges are useful. In order to minimize analytical costs and clearly demonstrate the quality of the effluent before discharge, batch analysis can be very effective. Unfortunately, this requires a large holding system which may be very costly to build and maintain. This is essentially true since extensive groundwater monitoring may be required. The alternative to batch discharge is continuous discharge with continuous or pe-

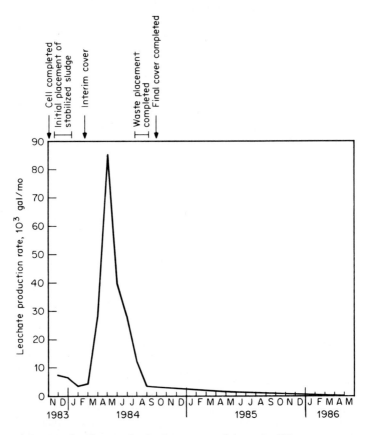

FIG. 11.4 Leachate production from a state of the art landfill.

riodic monitoring. The monitoring frequency must be short compared to the rate of change in the possible contaminant concentrations in the treatment process. For example, for a system discharging 100 gal/min, if the basic process volume is 5,000 gal, the process would turn over every 50 min, so effluent concentration measurements every 30 min may be required.

One of the most appealing quality-control systems for aqueous discharges is continuous, flowthrough bioassay. This typically is accomplished by discharging the waste or a continuous sample of the waste through an aquarium, tank, or lagoon containing fish. One problem with this system, as many hobbyists know, is that fish can die for many reasons totally unrelated to the chemical quality of their environment. Also, it is not unlikely that the discharge stream will be so clean that it lacks many of the basic minerals, nutrients, and electrolytes necessary to support fish life.

Solid residuals can usually be stored and assayed relatively easily. Difficulties, however, stem from the fact that existing regulations allow land disposal of these solids only with specific and years-long programs to demonstrate their innocuous nature. In many cases they must be handled as costly hazardous waste, even after treatment to environmentally safe products.

Efforts to use sludges and incinerator ash constructively in building materials, road beds, soil conditioning, landfill cover, and so forth have not been very successful to date. Many waste solids have high concentrations of highly soluble ions, particularly nitrates, that may make disposal in ordinary sanitary landfills undesirable because of their leachability. The most desirable disposal method for these materials today is probably in a dedicated secure landfill designed to minimize leaching. Current designs can do this very effectively, as shown in Fig. 11.4.

The most important message on quality control is that it must be absolutely first rate. Failures will lead to shutdowns, fines, bad publicity, and attendant business loss and, most importantly, to possibly unsafe or unhealthful environmental conditions. The costs associated with diligence are high but essential.

11.4 CONCLUSION

The world has entered an era in which we understand that it is indeed possible to foul our global nest. The result of this has been a rightful demand for the highest possible quality in our waste-management systems while productive use, total destruction, and discharge into the environment are the only alternatives available for hazardous-waste management.

Policing the systems necessary for these functions at tens of thousands of generator sites is virtually impossible, and many generators cannot afford the costly systems necessary for treating these materials in the relatively small volumes produced by any one generator.

Establishment of large, comprehensive waste-management facilities has proven to be a very effective solution to these problems in many locations. Policing of a few facilities is manageable, and large facilities can justify the wide variety of treatment systems necessary to service a locality or even a nation. In addition, many synergistic combinations of treatment systems and requirements develop in comprehensive facilities that greatly enhance their economics.

Specific designs of existing and developing comprehensive treatment facilities have many commonalities. All pay close attention to "front-end" receiving systems to assure that the endless variety of wastes received can be safely and effectively stored and introduced into the processing system. All pay especially close attention to product quality to protect the health and welfare of staff and public. All provide an array of unit processes, usually including incineration, biotreatment, and a range of physical and chemical separation systems, to produce the desired products.

These facilities are extremely costly and complex, rivaling the most complex industrial systems ever built. Nevertheless, they may be the best way available to us to assure a clean, healthful global nest for ourselves and our descendants.

CHAPTER 12

REMEDIAL-ACTION TECHNIQUES AND TECHNOLOGY

SECTION 12.1

SAMPLING AND MONITORING OF REMEDIAL-ACTION SITES

Thomas P. McVeigh

Project Manager
Roy F. Weston, Inc.
West Chester, Pennsylvania

12.1.1 REMEDIAL INVESTIGATIONS UNDER CERCLA AND INTERRELATIONSHIP OF REMEDIAL INVESTIGATION AND FEASIBILITY STUDY

Sampling and monitoring at remedial-action sites are part of the Remedial Investigation and Feasibility Study (RI/FS) process described in the National Oil and Hazardous Substance Contingency Plan (NCP) (47 *FR* 28,926, November 20, 1985; 40 *CFR* Part 300). An RI/FS [40 *CFR* Part 300.68(d)] is undertaken to determine the nature and extent of the threat presented by a release at a site and to evaluate proposed remedies. The remedial investigation and the feasibility study are interdependent and are usually conducted simultaneously. The integration of the two activities is illustrated in Fig. 12.1.1.

Data collection and site characterization are the primary elements of the remedial investigation. Additional activities include data management, bench and/or pilot studies, and site-investigation reports as shown in Fig. 12.1.2. The remedial-investigation data are also used in the feasibility phase to evaluate alternative design criteria as discussed in Subsec. 12.1.6.

The data evaluated and collected in the remedial-investigation process form the foundation of the choices between alternative treatment technologies that are made in the feasibility-study and site-remediation design phases. The National

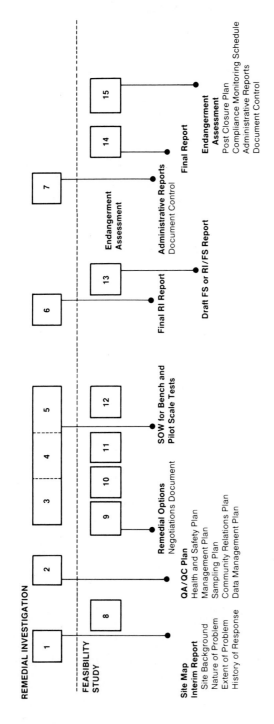

FIG. 12.1.1 RI/FS process.

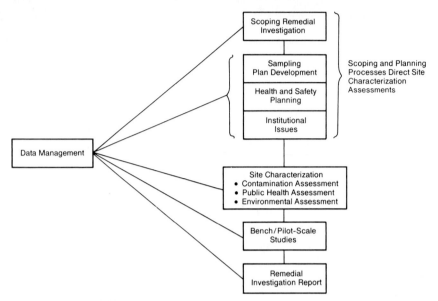

FIG. 12.1.2   Remedial-investigation process.

Contingency Plan outlines requirements for data collection and site characterization as part of the remedial-investigation process. This section will identify and discuss some of these key elements.

12.1.2   PRELIMINARY SITE ASSESSMENT

Prior to any sampling or monitoring activities at a remedial-action site, a preliminary assessment is performed. The assessment usually consists of the following four elements: collection of existing data, identification of site characteristics and pollutant pathways, evaluation of potential hazards, and identification of data requirements for site assessment and remedial action. The information from these activities will be used in the development of the sampling, health and safety, and quality-assurance plans for the site.

Collection of Existing Data

The objective of this activity is to evaluate the quantity and quality of the historical site information available and to avoid costly and time-consuming duplication of effort. The historical site investigation should concentrate on available site and waste data. Once these data have been collected, they must be evaluated for utility and validity. Evaluation criteria can be affected by legal considerations, technical competence and procedures, and assessment objectives. A good rule of thumb is to subject all existing data to the same quality-assurance and quality-control (QA/QC) procedures that are established for the site investigation.

TABLE 12.1.1   Data Collection Information Sources

Information source	Hazardous-waste sources	Migration pathways			
		Sub-surface	Sur-face	Air	Recep-tors
U.S. EPA Files	X	X	X	X	X
U.S. Geological Survey		X	X		
U.S. DA: Soil Conservation Service		X	X		
U.S. DA: Agricultural Stabilization and Conservation Service		X	X		
U.S. DA: Forest Service			X		X
U.S. DI: Fish and Wildlife Agencies					X
U.S. DI: Bureau of Reclamation	X	X	X		
U.S. Army Corps ofEngineers	X				
Federal Emergency Management Agency			X		
U.S. Census Bureau					X
National Oceanic and Atmospheric Administration				X	
State environmental protection or public-health agencies	X	X	X	X	X
State geological survey		X	X		
State fish and wildlife agencies					X
Local planning boards		X	X	X	X
County or city health departments	X	X	X	X	X
Town engineer or Town Hall	X				X
Local chamber of commerce	X	X			
Local airport				X	
Local library		X			X
Local well drillers		X			
Regional geologic and hydrologic publications		X	X		
Court records of legal action	X				
Department of Justice files	X				
State Attorney General files	X				
Facility records	X				
Facility owners and employees	X	X			X
Citizens residing near site	X	X	X	X	X
Waste haulers and generators	X				
Site visit reports	X		X	X	X
Photographs	X		X		X
Preliminary-assessment report	X	X	X	X	X
Field-investigation analytical data	X	X	X	X	
FIT/TAT reports	X	X	X	X	X
Site-inspection report	X	X	X	X	X

Sources of and access to site and hazardous-waste characteristics can vary widely. Table 12.1.1 lists some of these potential sources. Access to facility records usually depends on the type of investigation being conducted and the status of the legal disposition of the site.

Of particular importance in this step is a chronology of hazardous materials and waste practices at the site. Although facility records are usually the most comprehensive, other sources of materials-handling and waste information include permitting records, previous sampling activities, generator and/or transporter information, emergency response records, and enforcement documentation. Owner-operator and land-use information is also usually readily available from municipal records. From these sources, a preliminary list of potential waste characteristics can be determined.

Site Characteristics and Pollutant Pathways

A preliminary site assessment usually includes a site inspection to verify the information from the historical investigation as well as to update present site conditions. Physical site characteristics of particular importance include location and description, meteorology, geology, hydrogeology, and present land use. Site and Site Area maps to scale should be developed at this time, including all permanent features for use throughout the remedial investigation. A surveyor may be required for larger and more intricate sites. These maps can be used as base maps, with subsequent information, particularly sampling locations, portrayed on a series of overlays for visual presentations. The Site Area map should cover an area adequate to depict regional characteristics, including groundwater movement and topographic characteristics.

An inventory of hazardous materials and wastes currently stored on site should be recorded and located on the base map. This includes the physical state, quantity, and general characteristics of the materials. Storage vessels and their conditions should also be described. The data collected from the historical investigation and site inspection are used to identify potential pollutant pathways of the site. Figure 12.1.3 shows the potential dispersal pathways. The objective of the sampling and monitoring program at the site is to help define the extent of contamination of these pathways by the hazardous substances from the site.

Hazard Identification and Evaluation

Identification of the physical, toxicological, chemical, biological, and radiological properties of the known chemicals involved at the site is the first step in the evaluation of potential hazards to the environment, public health, and response personnel. Table 12.1.2 summarizes some of the hazard characteristics for each category.

Sources of information include basic reference manuals and on-line computer systems. Manuals such as *The Merck Index, Dangerous Properties of Industrial Materials,* and *The Condensed Chemical Dictionary* describe not only the properties of individual substances, but often their reactivity with other chemicals and the environment. On-line computer systems, such as Occupational Health Services (OHS) and the Chemical Information System (CIS) are also used frequently to identify properties of materials, as well as the associated regulatory standards.

Once the hazards associated with the known substances present at a site have

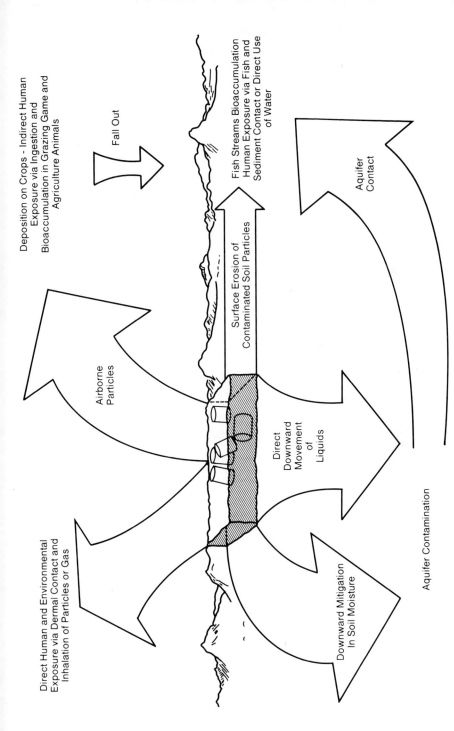

Deposition on Crops - Indirect Human Exposure via Ingestion and Bioaccumulation in Grazing Game and Agriculture Animals

Fall Out

Fish Streams Bioaccumulation Human Exposure via Fish and Sediment Contact or Direct Use of Water

Aquifer Contact

Surface Erosion of Contaminated Soil Particles

Airborne Particles

Direct Downward Movement of Liquids

Direct Human and Environmental Exposure via Dermal Contact and Inhalation of Particles or Gas

Aquifer Contamination

Downward Mitigation In Soil Moisture

FIG. 12.1.3 Dispersal pathways.

TABLE 12.1.2   Physical, Toxicological, Chemical, Biological, and Radiological Hazard Characteristics

Physical	Solubility
	Density
	Volatility
	Melting point and boiling point
Chemical	Combustibility
	Flammability
	Corrosivity
	Compatibility
	Reactivity
Toxicological	Dose response
	Routes of exposure
	Acute and chronic effects
	Teratogenic, mutagenic, carcino-genic effects
Biological	Viral
	Bacterial
	Fungal
	Parasitic
	Chlamydial
Radiological	Routes of exposure
	Ionization radiation

been identified, the threat to the public health and the environment needs to be evaluated. The criteria for this evaluation will depend upon the classification of the substance involved, the particular environmental media affected, and the detection limits of the analytical procedures employed.

Two groupings are used to classify the hazardous substances: the list of priority pollutants (PP) and the Hazardous Substances List (HSL). The groupings include volatiles, semivolatiles, pesticides, metals, cyanide, and phenolics. Compounds on the lists are reviewed periodically by the U.S. Environmental Protection Agency (EPA), and additions and deletions are made to the list periodically.

Different environmental media (e.g., soils, groundwater, surface water) are evaluated separately and sometimes according to different criteria. With the exception of drinking water, where acceptable levels of water quality have generally been defined by the federal Safe Drinking Water Act of 1975 (40 *CFR* 141), amended in 1977 (40 *CFR* 143), standards are still needed to determine what level of contamination in an environmental medium constitutes a requirement for remedial action. Two recent developments may help to define uniform action levels: guidelines published by the EPA for groundwater protection and the reauthorization of the Comprehensive Environmental Response Compensation and Liability Act (CERCLA) by Congress. The groundwater-protection plan is expected to set minimum quality standards for various types of geologic formations, while Congress is attempting to include specific cleanup levels for National Priority List (NPL) sites.

In evaluating existing data and assessing the potential hazard, the detection limit of the analytical methods used should be examined. *Detection limit* is the minimum or maximum concentration that an instrument or analytical method can accurately measure for a specific parameter. At the present time, the toxicity of the compounds at a remedial-investigation site is usually considered in establishing action levels. The detection limits must be sufficiently low to establish this dose-response relationship.

An important step in the evaluation process during the preliminary site assessment is the determination whether an immediate removal action is required as part of the remedial investigation. Section 300.65(b)(2) of the NCP lists some factors that should be considered:

- Exposure of nearby populations, animals, or food chain to the hazardous substances
- Contamination of drinking-water supplies
- Release of pollutants from bulk storage containers
- Migration of contaminants in the surface soils
- Extreme weather conditions that may cause migration or release
- Threat of fire or explosion
- Other factors that pose threats to the public health or environment

Removal-response actions are designed to stabilize the site and remove the release or threat of release of hazardous substances to the environment. They should be executed within the context of the remedial investigation.

Data Requirements for Site Assessment and Remedial Action

Preliminary site assessment concludes with identification of the data requirements for alternative remedial-action development in conjunction with the Feasibility Study. The data deficiencies in site assessment (hazardous-waste sources, pollutant pathways, and affected populations) are reviewed in light of the design requirements of the identified remedial actions. Environmental and hazardous-waste sampling provide an important means of completing the site assessment. Sampling activities include field investigation, plan development, identification of appropriate procedures, and data management and analysis.

12.1.3 REMEDIAL-ACTION SITE-INVESTIGATION PLANS

Sampling at remedial-action sites requires the development of three plans before the actual site activities begin: sampling, health and safety, and quality-assurance plans. These plans are interdependent and should be consistent with the evaluation of the results of the preliminary site assessment. They are intended to provide guidelines for field activities and are subject to frequent documented revisions during sampling activities.

Sampling Plans

Sampling plans are developed to maximize safety of sampling personnel, minimize sampling time and costs, reduce errors in sampling, and protect the integrity

TABLE 12.1.3   Sampling-Plan Elements

- Site background
- Goals and objectives
- Project organization and responsibilities
- Sample parameters and locations
- Sampling methodology
- Health and safety plan
- Quality-assurance plan
- Sample-handling procedures
- Sample-analysis request form
- Task scheduling
- Documentation requirements
- Report formats

of the samples after sampling. The purpose of sampling, in general, is the physical collection of a representative portion of an environment. Sampling plans are the "plan of action" to obtain this representation. Elements of the sampling plan are listed in Table 12.1.3, and discussed as follows. Additional elements may be required to meet site-specific needs.

Site Background.   Information from the preliminary assessment is included in this section. Site and Site Area maps, and all available information on composition and characteristics of the waste, storage and handling areas of waste on site, routes of potential migration of waste, and the site's operational history should also be included.

Goals and Objectives.   This section should provide guidelines to all personnel involved in the site investigation. It should list the data deficiencies identified in the site assessment for specific environmental media and contaminants. The objectives should relate to the development of remedial alternatives, and be of sufficient detail to allow modifications to the plan, caused by site conditions, without jeopardizing the goal of the investigation.

Project Organization and Responsibility.   Key individuals and their responsibilities are detailed in this section. Personnel should be assigned to the following tasks: project coordinator, sampling team leader, laboratory analysis, data processing, data quality, and overall project quality assurance. The number of personnel assigned depends on the complexity of the site, and individuals are often assigned dual roles on smaller sites.

Sampling Parameters and Locations.   This section includes the sampling-site locations, type of sample, sample matrix, parameters for analysis, sampling frequency, and number of samples. Sampling locations are usually shown on the Site and Site Area maps. Samples should be coded alphanumerically, with the first two code letters identifying the sample type [e.g., surface water (SW), soil (SO)] and numbers assigned progressively. A duplicate or triplicate sample receives an additional alphanumeric code, A or B. The type of sample describes whether the sample is environmental or hazardous, grab or composite, and the assessment purpose (source, pathway, population, or remedial design). The sample matrix describes the type of environmental and/or hazardous sample—e.g., drum, surface water, air. The parameters may vary by medium and should be described for each sample. Samples may also be taken from the same location at different time intervals in cases where variations in weather and seasons can affect the environmental media. Groundwater and surface water are examples. The number of samples to be collected depends on a number of variables: statistical validation of confidence levels, evaluation of existing site-characterization data, quality-assurance requirements, and regulatory guidelines. A discussion of the rationale for choosing the number of samples should be included in this section.

Sampling Methodology. This section describes the protocol that field samplers will employ for each environmental matrix. Standardized methods, or standard operating procedures (SOPs), are often referenced. In choosing a sampling method, the following criteria should be considered:

- Practicality
- Representativeness
- Economics
- Simplicity or ease of operation
- Compatibility with analytical considerations
- Versatility
- Safety

The methodology should describe the sampling devices, containers, preservation techniques, decontamination procedures, and holding times. A list of required field equipment should be compiled in this section. Calibration of monitoring equipment to be used in the field should also be stipulated here.

Health and Safety Plans. See under *Health and Safety Plans,* on p. 12.13.

Sample-Handling Procedures. This section refers to the chain of custody procedures to be followed in the investigation. For field sampling, chain of custody begins with the preparation of the sample containers and ends with the analysis of the samples. This section should describe the procedures for ensuring noncontaminating sampling devices, the laboratory procedure for assuring "clean" containers, the appropriate chemical preservation technique, the sample-labeling procedures, and the applicable DOT shipping, labeling, and packaging regulations that apply to the transport of the sample to the laboratory for analysis.

Sample Analytical Request Form. The laboratory analyzing the samples should be consulted throughout the development of the sampling plan to ensure compatibility of sampling procedures and analytical requirements. The sampling plan should specify the analytical method required for each sampling matrix. Turnaround times, number of samples, preservation methods, holding times, QA/QC samples, and required precision and accuracy should also be specified in the sampling as well as the quality-assurance plan. This section should also specify the data-reporting, reduction, and validating format required.

Task Scheduling.   A schedule for each phase of the project, from initiation to final report, should be established to ensure coordination of management, field, and laboratory personnel. The schedule should identify critical dates, such as project start-up, completion of analyses, and final report, but be flexible enough to ensure compliance with all investigation-plan requirements.

Documentation Requirements.   This section should specify the documentation procedures for photographs, field log books, field data books, sample tags, chain of custody, correspondence, and analytical records. For large sampling projects, a document-control system should be employed. The purpose of these documentation procedures is to centralize all pertinent information from the field investigation.

Report Format.   The sample plan should specify the number and format of the reports that are required throughout the length of the investigation. These reports include field-sampling reports, interim and final analytical reports, QA/QC review, audit findings, and final reports. These reports should address the goals and objectives of the field investigation and make some determination of the success of the operation. The reports should also identify additional data requirements. Finally, the reports should address alternative remedial actions for site mitigation.

Health and Safety Plans

Health and safety plans are designed to protect site workers, and the general public, from exposure to toxic substances during remedial-action investigations. Toxic substances can be inhaled, ingested, or absorbed through the skin. Generally, there are two types of potential exposure: acute and chronic. An *acute exposure* occurs when personnel are exposed for a short time to relatively high concentrations of a substance. A *chronic exposure* occurs from relatively low concentrations of contaminants over a long period of time. The health and safety plan also addresses other types of site hazards: fire or explosion, reactivity, radiation, and physical exposure.

Guidelines for worker safety are generally provided by the National Institute for Occupational Safety and Health (NIOSH) and the Occupational Safety and Health Administration (OSHA). The manual *Occupational Safety and Health for Superfund Activities* provides a framework for remedial-action site investigations. Specific regulations can be found in 29 *CFR* 1910, "Occupational Safety and Health Standards." In developing the health and safety plan for the site, specific references should be made for each contaminant to applicable standards of protection and sources noted.

The content, format, and execution of a site health and safety plan depend heavily on the health and safety program of the individuals involved. *All* personnel involved in a remedial site investigation should be *active* participants in their organization's program. Minimum elements of an organization's health and safety program should include:

- Designation of an overall safety coordinator to review site safety plans
- An established training program for hazardous-waste site personnel with minimum standards
- Standard operating safety procedures for accomplishing specific tasks in a safe manner

- A comprehensive inventory of personnel protection and monitoring equipment
- A medical surveillance and monitoring program for all personnel

The site-specific health and safety plan will vary greatly in detail depending upon the complexity of the site. However, it should be prepared by a qualified safety person and reviewed periodically to keep it current and technically correct. The object of the health and safety plan is to prevent injury and/or exposure to hazardous substances, to identify key personnel, and to specify procedures for normal work and emergency actions. At a minimum, the site health and safety plan must

- Evaluate the risks associated with the site and with each operation conducted
- Identify key personnel and alternates responsible for both site safety and response operations
- Address levels of protection to be worn by personnel during various site operations
- Designate work area (exclusion zone, contamination-reduction zone, and support zone), boundaries, size of zones, distance between zones, and access-control points into each zone (see Fig. 12.1.4)
- Determine the number of personnel and equipment needed in the work zones during the initial entries and/or subsequent operations
- Establish site emergency procedures, for example, escape routes, signals for evacuating work parties, emergency communications (internal and external), procedures for fire and/or explosions
- Determine location and make arrangements with the nearest medical facility (and emergency medical squad) for emergency medical care for routine-type injuries and toxicological problems
- Establish action levels for atmospheric and radiation conditions (see Table 12.1.4)

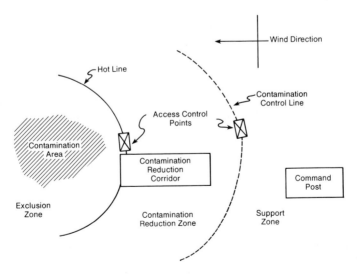

FIG. 12.1.4   Site work zones.

TABLE 12.1.4 Atmospheric and Radiation Guidelines

Monitoring equipment	Hazard	Ambient level	Action
Combustible-gas indicator	Explosive atmosphere	<10% LEL	Continue investigation.
		10–25%	Continue on-site monitoring with extreme caution as higher levels are encountered.
		<25% LEL	Explosion hazard; withdraw from area immediately.
Oxygen-concentration meter	Oxygen	<19.5%	Wear personal monitors. *Note:* Combustible-gas readings are not valid in atmosphere with <19.5% oxygen.
		19.5–25%	Continue investigation with caution. SCBA* not needed, based on oxygen content only.
		<25.0%	Discontinue inspection; fire-hazard potential. Consult specialist.
Radiation survey devices	Radiation	< 1 mR/h	Continue investigation. If radiation is detected above background levels, this signifies the presence of possible radiation sources; at this level, more thorough monitoring is advisable. Consult with a health physicist.
		>10 mR/h	Potential radiation hazard; evacuate site. Continue monitoring only upon the advice of a health physicist.
Colorimetric tubes	Organic and inorganic vapors/gases	Depends on species	Consult standard reference manual for air-concentrations/toxicity data.
Photoionizer	Organic vapors/gases	1. Depends on species	Consult standard reference manuals for air concentrations/toxicity data.
		2. Total-response mode	Consult EPA Standard Operating Procedures.
Organic-vapor analyzer	Organic vapors/gases	1. Depends on species	Consult standard reference manuals for air concentrations/toxicity data.
		2. Total-response mode	Consult EPA Standard Operating Procedures.

*SCBA = self-contained breathing apparatus.

- Implement a program for periodic air monitoring, personnel monitoring, and environmental sampling
- Train personnel for any nonroutine site activities
- Consider weather and other conditions which may affect the health and safety of personnel during site operations
- Implement control procedures to prevent access to the site by unauthorized personnel
- Specify decontamination procedures for personnel and equipment and disposal practices for these materials

Quality-Assurance Plans

The four basic factors which affect the quality of environmental data are sample collection, sample preservation, analysis, and recording. Field, laboratory, and data-management procedures are equally important elements of quality assurance. The quality of the field-sampling activities is dependent upon (1) collecting representative samples, (2) using appropriate sampling techniques, and (3) protecting the samples until they are analyzed (sample preservation). In the laboratory, quality assurance is maintained by monitoring the reliability, accuracy, and precision of the reported results, and by establishing methods to ensure program requirements for reliability (quality control).

A Quality-Assurance Project Plan is a written document which presents in specific terms the policies, organization, objectives, functional activities, and specific QA and QC activities designed to achieve the data-quality goals of a specific project.

A Quality-Assurance Project Plan is needed for sampling activities at a remedial-action site. Table 12.1.5 lists the essential elements of a QA plan. A description and direction for preparing each of these elements can be found in the EPA's *Interim Guidelines and Specifications for Preparing Quality Assurance Project Plans* (QAMS 005/80).

12.1.4 SAMPLING TECHNIQUES AT REMEDIAL-ACTION INVESTIGATION SITES

Sampling at remedial-action sites begins with the monitoring of ambient conditions during the preliminary site assessment and ends with the validation of analytical data by the laboratory. Field screening methods and mobile analytical devices are often used to assess the type and extent of contamination, in addition to sampling activities and standard analytical procedures. Calibration, maintenance, and performance standards are established for each type of equipment and procedure, and should be addressed in the quality-assurance and sampling plans.

Remedial-action site sampling usually includes hazardous-waste sources and the pollution migration pathways. In situ screening of samples is desirable to establish the type of sample (hazardous versus environmental). This screening provides valuable information in defining the safeguards required for the surrounding populations at the site, as well as the investigation and cleanup personnel, and the requirements of the U.S. Department of Transportation in 49 *CFR* for packaging, labeling, and marking sample shipments. The screening information also

TABLE 12.1.5 Quality-Assurance Plan Elements

- Title page with provision for approval signatures
- Table of contents
- Project description
- Project organization and responsibility
- QA objectives for measurement data in terms of precision, accuracy, completeness, representativeness, and comparability
- Sampling procedures
- Sample custody
- Calibration procedures and frequency
- Analytical procedures
- Data reduction, validation, and reporting
- Internal quality-control checks and frequency
- Performance and system audits and frequency
- Preventive-maintenance procedures and schedules
- Specific routine procedures to be used to assess data precision, accuracy, and completeness with regard to specific measurement parameters involved
- Corrective action
- Quality-assurance reports to management

---

provides the sampling team's partner, the analytical laboratory, with data to be used in sample preparation.

Sampling methods for remedial-action sites traditionally were adapted from standard operating procedures for environmental sampling. Through experience and technology development, they have been refined to render a more accurate picture of the targeted environment. Selection criteria have been established to assist in the proper selection of sampling methods and material. The following is a listing of these criteria.

- Practicality
- Representativeness
- Ease of operation
- Compatibility with analytical considerations
- Versatility
- Safety

Hazardous-Waste Sources

Hazardous-waste sources at a remedial-action site can be complex, multiphase mixtures of solids, sludges, sediments, and liquids. The chemical and physical properties of toxicity, flammability, corrosivity, incompatibility, explosivity, and so forth, are often present. For these reasons sampling devices should be made of stainless steel, or have their sampling surfaces coated with Teflon. Sampling devices also need to have tensile strength and durability, given the wide range of structural properties of hazardous-waste sources.

Containers of hazardous-waste on site include drums, barrels, storage tanks, vats, and bags. Liquid wastes, sediments, and sludges are often found in ponds, pits, and lagoons, while dry wastes may be contained in bags, drums, barrels, soils, and waste pits. Table 12.1.6 lists the sample devices recommended for various types of waste, and their limitations. The following is a discussion of the sampling procedures for each type of waste and how the sampling devices are employed.

Liquids, Sludges, and Slurries in Drums, Barrels, and Similar Containers. Visual inspection and ambient monitoring of contained liquid wastes are critical because

TABLE 12.1.6 Samplers Recommended for Various Types of Waste

Waste type	Recommended sampler	Limitations
Liquids, sludges, and slurries in drums, vacuum trucks, barrels, and similar containers	Coliwasa open-tube (thief); stratified-sample (thief)	Not for containers 1.5 m (5 ft) deep.
	(a) Plastic	Not for wastes containing ketones, nitrobenzene, dimethylformamide, mesityl oxide, or tetrahydrofuran.
	(b) Glass	Not for wastes containing hydrofluoric acid and concentrated alkali solutions.
	(c) PTFE	None
Liquids, sludges, and slurries in drums, vacuum trucks, barrels, and similar containers	Open tube	Not for containers 1.5 m (5 ft) deep.
	(a) Plastic	Not for wastes containing ketones, nitrobenzene, dimethylformamide, mesityl oxide, or tetrahydrofuran.
	(b) Glass	Not for wastes containing hydrofluoric acid and concentrated alkali solutions.
Liquids and sludges in ponds, pits, or lagoons	Pond sampler	Cannot be used to collect samples beyond 3.5 m (11.5 ft). Dip and retrieve sampler slowly to avoid bending the tubular aluminum handle.

TABLE 12.1.6 Samplers Recommended for Various Types of Waste (*Continued*)

Waste type	Recommended sampler	Limitations
Powdered or granular in bags, drums, barrels, and similar containers	(a) Grain sampler	Limited application for solids sampling; moist and sticky solids with a diameter 0.6 cm (¼ in).
	(b) Sampling trier	May incur difficulty in retaining core sample of very dry granular materials during sampling.
Dry wastes in shallow containers and surface soil	Trowel or scoop	Not applicable to sampling deeper than 8 cm (3 in). Difficult to obtain reproducible mass of samples.
Waste piles	Waste-pile sampler	Not applicable to sampling solid wastes with dimensions greater than half the diameter of the sampling tube.
Solid deeper than 8 cm (3 in)	(a) Soil auger	Does not collect undisturbed core sample.
	(b) Veihmeyer sampler	Difficult to use on stony, rocky, or very wet soil.
Wastes in storage tanks	(a) Weighted-bottle sampler	May be difficult to use on very viscous liquids.
	(b) Bacon Bomb	Volume restriction: 1-L maximum.
	(c) Kemmerer sampler	May need extra weight.

Source: Adapted from U.S. Environmental Protection Agency, *Samplers and Procedures for Hazardous Waste Streams,* EPA 600/2-80-018, Washington, D.C., 1980.

of the dangers of ignition, explosion, or chemical exposure when handling unknown container wastes. Full protective gear, including a splash apron, is recommended when opening a drum. Splashproof equipment is also recommended. A description of the container and any available identification should be entered in the field log book. If the drum is not under pressure, the container should be uprighted and restaged, with the bung upright and the container marked with its appropriate sampling number. If the container is under pressure, remote handling and drum opening are recommended.

The container should be opened with caution and a preliminary sampling of head-space gases should be performed. The head space needs to be monitored for organic vapors by portable detectors (organic-vapor analyzer or photoionization detector). Before sampling, the location of the solid-liquid interface phase should be determined. This is done by measuring the liquid phase and comparing it to the drum height. Bottom sludges, if present, should also be sampled. The liquid in the container can then be sampled by one of the following devices: Coliwasa, glass rods, stratified-sample thief, or vascom.

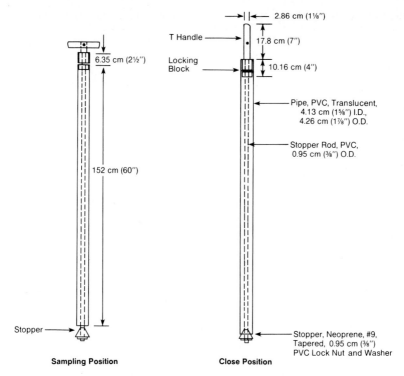

FIG. 12.1.5 Coliwasa sampler.

The *Coliwasa* (Fig. 12.1.5) is highly recommended for sampling containerized liquids. Its advantage is that it allows the representative sampling of multiphase wastes. Glass is the preferred material and can be used in all sampling efforts except strong alkali and hydrofluoric solutions. Because of its configuration, decontamination is difficult. The moving parts are disposable and easily replaced. The Coliwasa functions similarly to the glass rods and drum thieves, but allows less sample leakage. The Coliwasa is usually not used when sampling liquids of high viscosity.

Liquids and Sludges in Ponds, Pits, or Lagoons. Obtaining representative hazardous-waste samples from surface impoundments is a function of three factors: whether the samples can be taken from the shore of the impoundment; the depth and physical phases of the liquids; and the chemical properties and hazards of the impoundment's constituents. Sampling of an impoundment of shallow depth, small surface area, and low concentrations of chemical is relatively safe and easy. Sampling of an impoundment of significant depth, large surface area, and concentrated chemicals poses cost problems and physical hazards as well as toxicity problems. A preliminary survey of radiological, organic-vapor, and explosivity conditions should always be conducted prior to site operations, and continuous monitoring performed as needed.

Surface-water samples are most easily obtained using a pond sampler from the shore of an impoundment. A *pond sampler* consists of an adjustable clamp which holds beakers of various sizes, preferably glass or Teflon, and is attached to a telescoping aluminum tube that serves as a handle. The sampler is not commercially available, but is easily and inexpensively fabricated. A beaker with a spout should be used to facilitate transfer to a container. Depending upon the viscosity of the sampling material, grab samples can be obtained as far as 3.5 m from the edge of the pond. The location of the surface-water samples can be determined by sampling grids or visual observation of the surficial flow pattern. Aerial photography can assist in determining distribution patterns in the impoundments, and grab samples can be taken at discrete locations. Representative samples are obtained by statistical sampling of a grid, transect, or quadrant sampling configuration. Surface-sample methodology for impoundments that cannot be reached from the shore poses special problems. Boats, cranes, or even helicopters are used to allow sampling parties access to areas of an impoundment. All in some way limit the dexterity of the crew and affect the methodology to be used. The sampling plan should identify and address these logistical problems.

Sampling of shallow impoundments can be accomplished by using the pond sampler or a peristaltic pump and tubing. The range and depth of the pond sampler is limited by the length of the arm and the material's viscosity, while the range of the peristaltic pump is limited by the ability to lift the sample to the container. Maximum range for most peristaltic pumps is 8 m. Silicone tubing should be used to reduce contamination potential. Another consideration is the phase distribution of liquids and solids. The pond sampler can be used to obtain a semiliquid sample from shallow depths, while the silicone tubing on the peristaltic pump will most likely clog.

For depths greater than those obtainable by the pond sampler and peristaltic pump, the devices described in the section on sampling of waste-storage tanks should be used: Kemmerer depth-sampling device, Bacon Bomb, and weighted bottle. Personal safety must be considered, and sampling from anything other than a stable platform is *not* recommended with these devices.

Powdered or Granular Wastes in Bags, Drums, Barrels, and Similar Containers. Three devices are used in obtaining samples of dry materials from bags, drums, barrels, and similar containers: trowel, sampling trier, and grain sampler. In addition to the normal monitoring and personnel protection, the sampling crew must be protected from airborne particulates with proper respiratory protection. Dry materials exposed to the atmosphere may also experience chemical changes, and care should be taken in staging and opening these containers. Since most bulk materials are homogeneous, composite samples are usually taken from throughout the container.

For sampling dry wastes in shallow containers, a *laboratory scoop or trowel* is used. The scoop is usually made of materials less subject to corrosion or chemical reactions than the garden trowel and less likely to contaminate the sample.

Also known as the *grain thief,* the *grain sampler* is used for sampling powdered or granular waste materials in bags, fiber drums, sacks, or similar containers. The grain sampler consists of two slotted telescoping tubes, usually made of brass or stainless steel. The outer tube has a conical, pointed tip on one end that permits the sampler to penetrate the material being sampled. The sampler is opened and closed by rotating the inner tube.

The *sampling trier* is a long tube with a slot that extends almost its entire length. The tip and edges of the tube slot are sharpened to allow the trier to cut a core of the material to be sampled when rotated after insertion into the material. Sampling triers are usually made of stainless steel with wooden handles. This tool is preferred over the grain sampler when the powdered or granular material to be sampled is moist or sticky.

Waste Piles.   A *waste-pile sampler* is essentially a large sampling trier. It is used for larger waste piles and where the sampling trier is not long enough. The waste piles are usually divided into grids and a composite sample of various depths is taken. For representativeness, several composite samples are taken at various angles throughout the pile.

Solids Deeper Than 8 cm.   Sampling techniques are described under *Soils and Sediments,* below.

Wastes in Storage Tanks.   Three devices are available for sampling wastes at depth in storage tanks: weighted bottle, Bacon Bomb and Kemmerer depth-sampling device. Samples should be taken from the upper, middle, and lower sections of the tank. Field compositing of samples is not recommended. Tanks may also be subject to significant gas vapor pressure, and care should be taken in opening. Bottom sludge may be present and the phase distribution should be established. For safety purposes, the minimum number of people in a sample crew for storage-tank sampling is two.

The *weighted-bottle sampler* consists of a bottle, usually glass, a weight sinker, a bottle stopper, and a line that is used to open the bottle and to lower and raise the sampler during sampling. The sampler cannot be used to collect liquids that are incompatible, or that react chemically with the weight sinker and line.

The *Bacon Bomb* is a commercially available sampler designed for sampling petroleum products. It is useful for sampling large storage tanks because the internal collection mechanism is not exposed to product until the sampler is triggered. It is constructed of brass or stainless steel, and available in two sizes. Its disadvantage is that it tends to aerate the sample.

The *Kemmerer depth-sampling device* consists of an open tube with two sealing end pieces. The end pieces are maintained in the open position until a weighted messenger is sent down the line, releasing the end pieces and trapping the sample within the tube. Although the Kemmerer device can be used at any depth, the sampling tube is exposed to material while traveling down to sampling depth.

Soils and Sediments

Soil sampling can be divided into three categories: surface, shallow subsurface, and deep subsurface. Soils are composed of organic and mineral matter, and may vary in color, thickness, number of layers, and content of clay, salts, and organic matter. The soil strata are dynamic environs with constantly varying physical, chemical, and biological properties. The horizontal and vertical migration of contaminants in soils is dependent upon the attenuation and structural properties of soils. Sediments are deposited material underlying a water body. Streams, lakes, and impoundments will demonstrate significant variations in sediment composition with respect to distance from inflows, discharge, or other physical disturbances. Soil samples are usually environmental samples, while sediment samples can be environmental or hazardous.

Soil sampling is used to determine the existence of contaminants and their dispersion in the pollution pathways. Soil sampling also can

- Determine if the distribution of chemicals in the soils is threatening the groundwater

- Establish a relationship between depth of soil and its chemical and physical properties
- Discover if any physical or chemical characteristics of the material beneath a site can be used to predict the extent of contaminant dispersion

Sediment sampling is used to locate pollutants of low water solubility and high soil-binding affinity. Sediments are usually sampled in areas of low water velocity. On occasion they are exposed by evaporation, stream rerouting, or other means of water loss. In these instances they can be collected using soil-sampling techniques.

Two sampling methods are used to determine the number and location of soil samples: bias and statistical. The choice of method is dependent upon the objective of the sampling, the amount of site information available, the validity of the existing analytical data, and the general properties of soils and the suspected pollutant(s). The two methods are sometimes used in combination, and sometimes in phases. Background and/or control samples are also taken with both methods.

Bias, or *judgmental, sampling* is employed when the samplers are confident that the location, type of pollutant, time frame of contamination, and geologic characteristics are known. This information is often obtained from site records, aerial photographs, and visual observation of the soil staining or lack of vegetation. Bias sampling is also used to evaluate "worst-case" conditions when removal actions are being considered as part of the remedial-action investigation. Both grab and composite samples are usually taken when using the judgmental sampling techniques.

Statistical sampling is based on statistical variability and confidence levels of the data as established by sampling analysis. Phase sampling with an increasing confidence level is usually employed when few existing data are available for a site. In his report, *Preparation of Soil Sampling Protocol: Techniques and Strategies,* Benjamin Mason describes the following three types of statistical sampling methods.

- *Simple random sampling:* Although seldom used by itself, simple random sampling is the basis for all probability sampling techniques. By dividing the study area into grids and assigning coordinates, bias-free samples are obtained by sampling only those grids chosen by a random number table. Large statistical variations are usually produced, and the precision of samples and confidence levels is also usually low.

- *Stratified random sampling:* The precision of sampling effort is increased by using the stratified random-sampling method. The soil strata are selected according to their properties and characteristics to produce more homogeneous zones. Within each stratum, simple random sampling is performed. The method is based on consistent pollution behavior within a soil stratum.

- *Systematic sampling:* This method employs the statistical theory of "Kriging." Samples are collected in a regular pattern (triangular or grid) over the area under investigation. The starting point is established randomly, and the samples are collected at regular intervals along the transect lines. The interval spacing is determined by regionalized variable theory to establish the "range" of variance, and thus the confidence level.

Techniques of soil and sediment sampling vary from the use of a stainless-steel scoop to trench excavation. Surface soils are usually collected with the scoop or thin-wall tube sampler. Shallow subsurface samples are obtained using the auger or Veihmeyer sampler. Deep-subsurface soil samples are obtained by digging a trench and sampling along the exposed soil profile or by using a split-spoon sampler. Exposed sediments are obtained using soil-sampling methods, while the shallow sediment samples can be obtained with a gravity corer. The Ponar grabsampler is used to obtain sediment (or sludge) samples at depth. Four of the most commonly used soil and sediment samplers (thin-wall tube, gravity corer, Veihmeyer, Ponar grab) are briefly described here:

- *Thin-wall tube:* This system consists of an auger bit, a series of drill rods, a T-shaped handle, and a thin-wall tube corer. The auger bit is used to bore a hole to the desired sampling depth and then withdraw. The auger tip is then replaced with the tube corer, lowered down the borehole, and forced into the soil at the completion depth. The corer is then withdrawn and the sample collected. This system can be used in a wide variety of soil conditions. It can be used to sample both from the surface, and to depths in excess of 6 m.

    Interchangeable cutting tips on the corer reduce the disturbance to the soil during sampling and aid in maintaining the core in the device during removal from the borehole.

- *Gravity corer:* A *gravity corer* is a metal tube with a replacement tapered nosepiece on the bottom and a ball or other type of check valve on the top. The check valve allows water to pass through the corer on descent, but prevents a washout during recovery. The tapered nosepiece facilitates cutting and reduces core disturbance during penetration.

- *Veihmeyer:* The *Veihmeyer sampler* (Fig. 12.1.6) consists of four basic parts: a sampling tube, a drive head, a sampling tip, and a drop hammer. The sampling tube is constructed of chromium-molybdenum steel and comes in various standard lengths from 0.91 to 4.9 m (3 to 16 ft). This tube is calibrated every 30.5 cm (12 in). The drive head is attached to the upper end of the tube and acts to protect the top of the tube from deforming when the tube is driven into the ground. Interchangeable sampling tips, which provide the cutting action, can be attached to the bottom of the tube. A variety of tips are available for different types of soil and sampling situations.

- *Ponar grab:* The Ponar grab is a clamshell-type scoop activated by a counterlever system. The shell is opened and latched in place and slowly lowered to the bottom. When tension is released on the lowering cable, the latch releases and the lifting action of the cable on the lever system closes the clamshell.

Groundwater

Methods of collecting representative groundwater samples are difficult and expensive in this relatively inaccessible environment. The subsurface is an extremely complex system subject to extensive physical, chemical, and biological changes within small vertical and horizontal distance. The following elements should be considered when selecting a groundwater-sampling procedure:

- Objective of the sampling program
- Characteristic of the pollutants
- Nature of the pollutant source
- Hydrogeology of the area

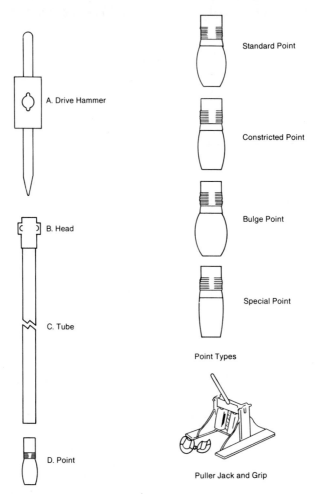

FIG. 12.1.6 Veihmeyer sampler.

- Construction technique for sampling wells
- Methods of purging and sampling monitor wells

The objective of groundwater sampling at a remedial-action investigation site is usually twofold: to determine the presence or absence of a pollutant in an aquifer, and to define the horizontal and vertical distribution of contamination and predict the eventual fate. Confirmation of the presence of a pollutant in the groundwater is achieved by sampling existing site wells or installing a few sampling wells. However, in order to determine pollutant fate and establish the extent of subsurface contamination, a monitoring-well system to characterize the subsurface conditions of an area is required.

The unstable nature of many chemical, physical, and microbial constituents in groundwater and the subsurface limits the sample-collection and analysis options. However, certain factors should be considered when collecting representative samples:

1. Groundwater moves slowly; therefore the rate of change of water-quality parameters is slow.

2. Temperatures are relatively constant in the subsurface; therefore the sample temperature may change significantly when the sample is brought to the surface. This change can alter chemical reaction rates, reverse cationic and anionic ion exchange on solids, and change microbial growth rates.

3. A change in pH can occur, resulting from carbon dioxide absorption and subsequent changes in alkalinity. Oxidation of some compounds may also occur.

4. Dissolved gases such as hydrogen sulfide may be lost at the surface.

5. Integrity of organic samples may be affected by problems associated with either absorption or contamination from sampling material, and with volatility.

6. Both soils and groundwaters may be so severely contaminated as to present a health or safety problem to the sampling crew.

The area of consideration, the time available for monitoring, and potential concentration levels of pollutants all influence the sampling procedure selected. A regional or large-area monitoring program may permit the use of existing wells, springs, or even the base flow of streams if these systems are compatible with the parameters of interest. If time is critical, existing sampling locations may be the only alternative. However, if the possible pollution source is relatively small, such as a landfill or lagoon, or if the pollutant concentrations are very low, such as with trace organics, special monitoring wells will almost surely be necessary. The number and location of additional wells needed depends on the purpose of the monitoring, aquifer characteristics, and the mobility of pollutants in the aquifer.

Geologic factors relate chiefly to geologic formations and their water-bearing properties, and hydrologic factors relate to the movement of water in the formations. Prior to initiating any field work, all existing geologic and hydrologic data should be collected, compiled, and interpreted. Data that may be available include geologic maps; cross sections; aerial photographs; and an array of well-well data including location, date drilled, depth, name of driller, water level and date, well-completion methods, use of well, electric or radioactivity logs, or other geophysical water-quality data. Gross groundwater flow patterns can be developed from water-level contours. However, the actual movement of a plume may be somewhat more complex. The hydrogeology is further complicated by different flow patterns of different pollutants, which move through the subsurface at different rates relative to the rate of water movement because of sorption, desorption, ion exchange, and biodegradation. Therefore, points of maximum orientation of the different pollutants along the groundwater flow path will probably vary considerably.

A necessary component of any groundwater-monitoring program is background sampling. One recommended monitoring method for detecting contamination at landfills is location of a background well upgradient from the landfill and a minimum of three wells downgradient and at an angle perpendicular to groundwater flow, penetrating the entire saturated thickness of the aquifer. Such an arrangement is illustrated in Fig. 12.1.7 and is applicable to most potential sources of contamination. The location, design, and construction of monitoring wells is usually the most costly and nonrepeatable factor in the success of a groundwater-monitoring program. Design factors include well diameter, well depth, intake por-

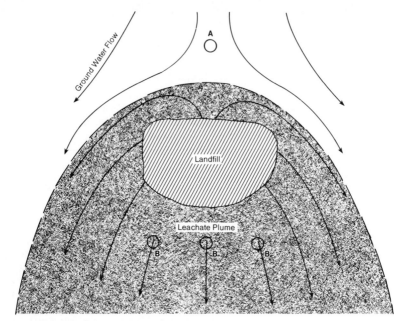

FIG. 12.1.7   Idealized monitoring network.

tion of monitoring wells, and well casings. The diameter of the casing for monitoring wells should be just sufficient to allow the sampling tool (bailer or pump) to be lowered into the well to the desired depth. The diameter of the hole into which the casing is placed must be at least sufficiently large for the casing to fit, and in many cases must be at least 2 in (5 cm) larger to permit placement of a grout seal around the outside of the casing. Casings and/or holes drilled much larger than the necessary minimum can have undesired effects on the data. The intake part of a monitoring well should be at the depth desired. The monitoring wells should be constructed to sample only from one specific layer without interconnection to other layers. Design criteria for the intake part of the well are:

1. The screen or intake part should have sufficient open area to permit the easy flow of water from the formation.
2. The slot openings should be just small enough to keep most of the natural formation out.
3. The well should be developed.

Well casings should be constructed of materials that have the least potential for affecting the quality parameters of the sample. Although PVC is less costly than Teflon, it may affect water-quality samples of lower concentration. The laboratory should be fully aware of the construction materials used.

Selection of the drilling method best suited for a particular job is based on the following factors in order of importance.

1. Hydrogeologic environment
   a. Types of formation(s)

b. Depth of drilling
c. Depth of desired screen setting below water table
2. Types of pollutants expected
3. Location of drilling site
4. Design of monitoring well desired
5. Availability of drilling equipment

In the *Manual of Ground-Water Quality Sampling Procedures* prepared by the Robert S. Kerr Environmental Research Laboratory, the principles of operation and the advantages and disadvantages of the more common drilling techniques suitable for constructing groundwater-monitoring wells are discussed. Four of the most common drilling methods at a remedial-action investigation site are mud rotary, cable tool, hollow stem, and air rotary.

Two recent methods that have helped reduce the number of monitoring wells required to characterize a site are gas surveys and geophysical techniques. Soil-gas analysis, using a field-operable gas chromatograph (GC), is used to estimate the extent and movements of volatile organic compounds (VOCs). The technique is based on the fact that VOCs volatilize from contaminated groundwater and move, by processes such as molecular diffusion, away from source areas toward lower concentrations in the overlying soil profile. A gaseous-phase concentration gradient from the water table to the ground surface is established. There are several geophysical techniques currently being used in the evaluation of hazardous-waste sites and groundwater contamination:

- Electromagnetic and resistivity methods to define groundwater-contaminant plumes
- Resistivity and seismic techniques
- Metal detectors and magnetometers to locate drums and other buried metal
- Ground-penetrating radar to define the boundaries of buried trenches and other subsurface installations

The collection of representative groundwater samples from a well consists of two steps: purging and sampling. The recommended length of time required to bail or pump a well before sampling is dependent on many factors including the characteristics of the well, the hydrogeological nature of the aquifer, the type of sounding equipment being used, and the parameters being sampled. The time required may range from the time needed to pump or bail one bore volume to the time needed to pump several bore volumes. A common procedure is to pump or bail the well until a minimum of 4 to 10 bore volumes have been replaced.

The type of sampling system used is a function of the type and size of well construction, pumping, level, type of pollutant, analytical procedures, and presence or absence of permanent pumping fixtures. Sampling methods include the use of bailers, suction-lift pumps, portable submersible pumps, air-lift samplers, gas-operated squeeze pump, and gas-driven piston pump. For a discussion of the operating procedures and the advantages and disadvantages of each sampling method, please refer to the *Manual of Ground-Water Quality Sampling Procedures*.

Surface Waters

Two types of water bodies are addressed in surface-water sampling: standing (lakes, ponds, impoundments) and flowing (streams, rivers, runoff channels). The location and number of samples for standing water bodies is dependent upon the depth, size, and configuration of the impoundment. Criteria for flowing water include accessibility, stream velocity, location of confluencies, water-supply intakes, and stream geomorphology. As with soils, the objective of the sampling program is to characterize the vertical and horizontal distribution of contaminants throughout the medium and to determine the probability of migration. In general, contaminant mixing (distribution throughout the water column) will occur more frequently in flowing than standing water bodies. Stratification of contaminants in stagnant waters must be considered in the formulation of a surface-water sampling plan.

At remedial-action investigation sites, the surface-water drainage pattern identifies potential receptors of waterborne contaminants. Point sources of contamination to surface waters, such as leachate breakouts and effluent stream channels, are sampled to determine concentration changes downgradient to the source. Background and upgradient samples should always be included when sampling surface waters. Field measurements of parameters such as pH, conductivity, temperature, and dissolved oxygen should be noted for each sample. Sample locations, preferably on a scale map, should be carefully documented using permanent fixtures and landmarks.

Two types of samples are collected for surface waters: grab and composite. The type of sample collected depends upon the variability of flow, variability of water quality, the accuracy required, and the availability of funds for conducting the sampling and analytical programs. A *grab sample* is a discrete sample collected over a period of time not exceeding 15 min, while a *composite sample* is a sample formed by mixing a number of discrete samples taken at periodic points in time. Grab samples are used to characterize water quality at a particular time, provide information about minimum or maximum concentrations, meet a requirement of a discharge permit, and corroborate composite samples. Composite samples are used to show the depth integration of a contaminant or its average concentration in the water column.

Sample-collection techniques for water samples are also similar to those for soil sampling. Statistical sampling approaches using standard deviations and confidence-level requirements are used to determine the number of samples and their locations.

Recommended sampling devices for surface-water sampling include the pond sampler, peristaltic pump, Kemmerer depth sampler, and the weighted-bottle sampler. All were previously described (see under *Hazardous-Waste Sources,* on p. 12.17).

Air Monitoring

Air monitoring at hazardous-waste sites is useful as an indicator of potential safety problems and as a means of screening for the presence of possible airborne contaminants. Monitoring is also important as a means of determining the specific identity and concentration of airborne toxic pollutants on site and the extent of their migration off site, for both worker and public-health risk assessments. Explosive vapor clouds, oxygen-deficient atmospheres, and a variety of toxic gases or vapors can occur not only during initial site investigation, but also during remedial-action cleanup. Therefore, continuous monitoring of site conditions, as well as periodic sampling, is recommended. Finally, radiation monitoring is also considered an essential element of air surveillance.

Air-monitoring instruments must have the following characteristics to be useful at remedial-action investigation sites:

- Portability
- Ability to generate reliable and useful results
- Sensitivity and selectivity
- Inherently safe

Necessary attributes of a portable air-monitoring unit are ease of mobility, the ability to withstand the rigors of use, quick assembly, and short checkout and calibration time. The instrument should have a short response time (the interval between an instrument "sensing" a contaminant and generating data) and should be direct-reading. The operating range of the instrument must be sufficient to detect levels significantly below and above action levels. The device must be selective to the constituents being monitored. Finally, the device must be inherently safe. All instruments should indicate the class, division, and group of hazardous atmosphere for which they are approved for use.

Air-monitoring equipment is used to identify three types of atmospheric hazards at remedial-action investigation sites:

- Oxygen deficient
- Explosive
- Toxic

The major types of air-monitoring equipment are described here.

A portable oxygen monitor has three principal components for operation: the air-flow system, the oxygen-sensing device, and the microammeter. The most useful monitor for ambient measurements is the 0 to 25% oxygen-content readout. Many instruments have alarm modes which can be set to activate at a specified oxygen concentration. Normal oxygen concentration required for respiration is 20.9%.

A combustible-gas indicator (CGI) consists of three primary components: the sensor, signal processor, and readout display. The sensor produces a signal which is processed and displayed as the ratio of the combustible gas present to the total required to reach the lower explosion limit (LEL). Combustible-gas detectors are used to determine the potential for combustion or explosion of unknown atmospheres. Along with the oxygen detectors and radiation-survey instrumentation, they are the first monitors used when entering a hazardous area.

Various portable monitoring devices are available for the detecting of most organic vapors: *photoionization detector (PID), flame-ionization detector (FID), and field gas chromatograph (GC).* Each has certain limitations. Although the PID can detect certain inorganic vapors, it cannot detect certain toxic gases and vapors with high ionization potential. The FID does not respond to particulate hydrocarbons such as pesticides, PNAs and PCBs. The GC is capable of obtaining data on concentrations of certain volatile organic compounds in ambient air at waste sites; however, performance may be limited by such factors as interferences, ambient conditions, and operator experience.

Integrated air samples are usually collected during remedial-action investigations. These samples are compared to established standards, such as the threshold limit value (TLV) to determine the ambient air quality on or near a site. Two types of sampling systems are used for the collection of integrated samples; active and passive. Active samplers mechanically move the air through the collection medium, while passive samplers rely on natural forces. *Passive samplers* include diffusion samplers and permeation dosimeters, and they are usually used as personal-exposure monitors.

Active sampling systems usually consist of an electrical pump, a sampler consisting of an appropriate sampling medium, and flexible tubing to connect the sampler to the pump. Sampling media include solid sorbents, low- or high-volume polyurethane-foam samplers, gas absorbers, colorimetric tubes, and sampling bags. Silica gel is the most commonly used solid sorbent. Terax tubes and activated charcoal are also used frequently. Polyurethane-foam samplers are used for volatile and semivolatile organic compounds. Impingers and butted bubble tubes are the most-used devices for capturing gas vapors. Colorimetric detector tubes are specific for individual compounds. Finally, air samples are often collected for detailed GC/MS laboratory analysis as part of site documentation during remedial-action investigations.

Three types of radiation are of major concern at a remedial-action investigation site: alpha particles, beta particles, and gamma rays. Radionuclides have three distinct characteristics that are useful for identification purposes: half-life, type of radiation, and energy. All radiation-survey instruments work on the principle that radiation causes ionization in the detecting media. The ions produced are "counted" electronically and a relationship established between the number of ionizing events and the quantity of radiation present. Four major types of radiation detectors are commonly used: *ion-detection tubes, proportional-detection tubes, Geiger-Mueller detection tubes, and scintillation-detection media.* Federal guidelines have placed the acceptable exposure rate for workers at 5 rem/yr, or 2.5 mrem/h.

12.1.5   DATA MANAGEMENT, ANALYSIS, AND REPORTING

Sampling at remedial-action investigation sites is usually performed in phases, and the sampling program at a site can be a year in duration. This activity is called *monitoring. Monitoring* is the collection, storage, retrieval, and analysis of environmental data. The process begins with the evaluation of existing data and continues through the completion of remedial action. The establishment of standard operating procedures for data management, analysis, and reporting for all phases of monitoring at a remedial-action investigation site is crucial to manage the extensive information generated.

Two types of information need to be managed: technical and project tracking. *Technical data* include field and laboratory documentation of sample collection and analyses. Field documentation includes project and field log books, sample tags, sample data sheets, and chain of custody records. Laboratory documentation includes receipt of sample forms; log books; and laboratory data, calculations, and graphs. *Project-tracking data* include plans and reports generated during the investigation. Plans include sampling, health and safety, quality assurance, and project management. Reports include various assessments: contamination, environmental, public health, and endangerment.

Two procedures are recommended to accomplish effective management of environmental-monitoring data: institution of a document-control system for files and reports, and the use of automated data processing (ADP) for analytical data.

The document-control system should be based on serial numbering of documents. The system may be manual or automated. File structures should be determined at the beginning of the investigation and based on the scope of the project. Many hazardous-waste analytical laboratories utilize ADP not only for data storage, but also data processing, such as with gas chromatograph and mass spectrometry (GC/MS) systems. These systems generally employ a standard electronic data-transmission format.

The goal of these standardized formats is to accommodate any qualitative or quantitative environmental measurement in any medium for any method of analysis.

Hazardous waste is one of the most difficult analyses for an analytical laboratory to perform. There are several reasons for this:

- The medium may be gaseous, liquids, solid, or a combination.
- The constituent may have parameters in the range from parts per billion to percentage.
- There are strict disposal requirements under the Resource Conservation and Recovery Act (RCRA) for high-hazard samples.
- A wide range of analytical equipment is required for complete characterization of the waste.
- There are extensive QA/QC requirements for institutional programs, such as the Contract Laboratory Program (CLP).

This section of Chap. 12 will not attempt to discuss the analytical procedures required for most hazardous wastes. The reader is referred to the EPA's methodology in *Methods for Chemical Analysis of Water and Wastes,* U. S. Environmental Monitoring and Support Laboratory, Cincinnati, Ohio. However, the four basic factors which affect the quality of environmental data are sample collection, sample preservation, analysis, and recording.

Regardless of the nature of the sample, complete stability of every constituent can never be achieved. At best, preservation techniques can only retard the chemical and biological changes that inevitably continue after the sample is removed from the parent source. The changes that take place in a sample are either chemical or biological. Methods of preservation are relatively limited and are intended generally to retard biological action, retard hydrolysis of chemical compounds and complexes, reduce volatility of constituents, and reduce adsorption effects. The sampling plan should identify the analytical procedures that will be employed to assure that the proper sample-preservation techniques are used during sample collection.

The reporting format for data analysis should be specified in the Sample Analysis Request Form. However, certain critical elements should be discussed by the sample-management team and the laboratory long before the sample is brought to the laboratory for analysis. The data-quality requirements should be agreed upon and incorporated into the quality-control protocol. As a minimum, requirements should be specified for detection and/or quantitation limits, turnaround time, and precision and accuracy for all types of measurements. Precision and accuracy can be monitored by using replicate samples, field and method blanks, spikes, and standards.

Data reporting includes documentation, data reduction, and data validation. The analytical laboratory should provide adequate documentation in its report to allow an independent evaluation of its complete analytical procedures. Data-reduction portions of the reports should describe the standard operating procedures employed by the laboratory for calculation procedures for analytical methods, transfer of data to various forms and computer systems, and the generation of final reports. Finally, data-validation techniques, such as intercomparisons, test for suitability of data, data plots, regression analysis, and tests for fitness and outlines, are essential elements of the laboratory's report to sample management.

12.1.6 EVALUATION OF REMEDIAL-ACTION ALTERNATIVES

Remedial actions are designed to control, contain, treat or remove contaminants from uncontrolled hazardous-waste sites. Remedial actions are divided into surface controls, groundwater controls, leachate controls, direct-treatment methods, gas-migration controls, techniques for contaminated water and sewer lines, and methods for contaminated-sediment removal. A flowchart for the selection of remedial actions is shown in Fig. 12.1.8. The steps in this process include

1. Evaluation of the nature and extent of contamination
2. Collection of site-specific data
3. Determination of remedial options
4. Comparison of remedial options to site characteristics and selection of best remedial actions
5. Making preliminary recommendations for site cleanup

Sampling at remedial-action investigation sites contributes important information in both steps 1 and 2. As discussed in Subsec. 12.1.1 on p. 12.3, feasibility studies (step 3) are conducted concurrently with the collection of site-specific data. The collection of site-specific data provides information on the specific me-

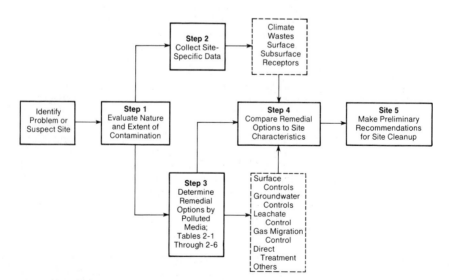

FIG. 12.1.8 Flowchart for selection of remedial actions.

dia, and includes general site characteristics that may inherently affect the choice of remedial measures. Characteristics include quantity and quality of waste material, characteristics of site cover, the climate of the area, subsurface geology, proximity to various receptors, existing land use, and others.

The comparison of site characteristics to remedial options (step 4) requires consideration of the data requirements for alternative remedial options during the sampling-plan development phase. The objective of site characterization is to collect and analyze enough information to determine the necessity, extent, and feasibility of remedial action (step 5). For a description of remedial-action options available, the reader is referred to *Handbook, Remedial Action at Waste Disposal Site,* Municipal Environmental Research Laboratory, Cincinnati, Ohio.

12.1.7 BIBLIOGRAPHY

Almich, Bruce P., William L. Budde, and W. Randall Shobe, "Waste Monitoring," *Environmental Science and Technology,* vol. 20, no. 1, 1986.

American Conference of Governmental Industrial Hygienists, *TLVs Threshold Limit Values for Chemical Substances and Physical Agents in the Work Environment with Intended Changes for 1984–85,* ACGIH, Cincinnatai, Ohio, 1984.

Cralley, Lewis, and Lester Cralley, *Patty's Industrial Hygiene and Toxicology,* volumes I–III, 3d ed., John Wiley and Sons, New York, 1981.

DeVera, Emil R., Bart P. Simmons, Robert D. Stephens, and David L. Storm, *Samplers and Sampling Procedures for Hazardous Waste Streams,* EPA 600/2-80/018, U.S. Environmental Protection Agency, Municipal Environmental Research Lab, Cincinnati, Ohio, 1980.

Ecology and Environment, Inc., *Standard Operating Procedures for Sampling of Hazardous Waste Sites,* SOP-III-2, Buffalo, N.Y., June, 1985.

Environmental Monitoring and Support Laboratory, *Handbook for Analytical Quality Control in Water and Wastewater Laboratories,* EPA 600/4-79/019, U.S. Government Printing Office, Washington, D.C., 1978.

Environmental Monitoring and Support Laboratory, *Handbook for Sampling and Sample Preservation of Water and Wastewater,* EPA 600/4-82/029, U.S. Government Printing Office, Cincinnati, Ohio, 1982.

Environmental Monitoring and Support Laboratory, *Methods for Chemical Analysis of Water and Wastes,* EPA 600/4-79/020, U.S. Government Printing Office, Cincinnati, Ohio, 1983.

Environmental Monitoring Systems Laboratory, *Preparation of Soil Sampling Protocol: Techniques and Strategies,* EPA 600/54-83/020, U.S. Government Printing Office, 1983.

Evans, Roy B., and Glenn E. Schweitzer, "Assessing Hazardous Waste Problems," *Environmental Science and Technology,* vol. 18, no. 11, 1984.

Ford, Patrick J., Paul J. Turina, and Douglas E. Seely, *Characterization of Hazardous Waste Sites—A Methods Manual,* vol. II, *Available Sampling Methods,* EPA 600/4-83/040, U.S. Government Printing Office, Las Vegas, Nev., September 1983.

Hazardous Response Support Division, U.S. Environmental Protection Agency, *Hazard Evaluation and Environmental Assessment,* U.S. Government Printing Office, Cincinnati, Ohio, October 1982.

Hazardous Response Support Division, U.S. Environmental Protection Agency, *Sampling for Hazardous Materials,* U.S. Government Printing Office, Cincinnati, Ohio, November 1983.

Hazardous Waste Engineering Research Laboratory, U.S. Environmental Protection Agency, *Guidance on Remedial Investigations Under CERCLA,* U.S. Government Printing Office, Cincinnati, Ohio, 1985.

Municipal Environmental Research Laboratory, *Handbook for Remedial Action at Waste Disposal Sites*, EPA 625/6-85/006, U.S. Government Printing Office, Cincinnati, Ohio, 1985.

National Enforcement Investigations Center, *NEIC Manual for Groundwater/Subsurface Investigations at Hazardous Waste Sites*, EPA 330/9-81/002, U.S. Office of Enforcement, Denver, Colorado, 1981.

National Institute for Occupational Safety and Health, *Occupational Health Guidelines for Chemical Hazards*, DHHS (NIOSH) Publication No. 81-123, Superintendent of Documents, U.S. Government Printing Office, Washington, D.C., 1981.

New Jersey Department of Environmental Protection, *Field Sampling Procedure Manual*, Environmental Measurements Section, Trenton, N.J., November, 1985.

Roy F. Weston, Inc., *Hazardous Materials Response Operations Course*, Prevention and Emergency Response Division, West Chester, Pa., 1984.

Sax, N. I., *Dangerous Properties of Hazardous Materials*, 6th ed., Van Nostrand Reinhold, New York, N.Y., 1984.

Scalf, Marion R., James F. McNabb, William J. Dunlap, Roger L. Cosby, and John Fryburger, *Manual of Ground Water Quality Sampling Procedures*, EPA 600/2-81/160, Robert S. Kerr Environmental Research Laboratory, U.S. Environmental Protection Agency, Ada, Oklahoma.

U.S. Coast Guard, *A Condensed Guide to Chemical Hazards (CHRIS)*, vols. I and II, U.S. Department of Transportation, Washington, D.C., 1978.

U.S. Environmental Protection Agency, *Interim Guidelines and Specifications for Preparing Quality Assurance Project Plans*, QAMS-005/80, Office of Monitoring Systems and Quality Assurance, Washington, D.C., 1980.

Windholz, M. (ed.), *The Merck Index: An Encyclopedia of Chemicals and Drugs*, Merck and Co., Inc., Rahway, N.J., 1976.

SECTION 12.2
TREATMENT AND CONTAINMENT TECHNOLOGIES

Stephen C. James
Environmental Engineer
U.S. Environmental Protection Agency
Office of Research and Development
Hazardous Waste Engineering Research Laboratory
Cincinnati, Ohio

Paul J. Rogoshewski
Chemical Engineer
Environmental Consulting Services
Jamaica Plain, Massachusetts

The Comprehensive Environmental Response Compensation and Liability Act of 1980 (CERCLA) and the Superfund Amendments and Reauthorization Act of 1986 (SARA) established the legislative means for providing the U.S. Environmental Protection Agency (EPA) with the authority and responsibility for cleanup at uncontrolled hazardous-waste sites. One aspect of this legislation enables the EPA's Office of Research and Development to document existing research and develop emerging remedial-action technologies. This section provides insight into containment and treatment technologies.

12.2.1 CONTAINMENT TECHNOLOGIES

Containment technologies are used to contain waste or contaminated groundwater or to keep uncontaminated water from entering the waste. This is accomplished by three commonly used methods: (1) groundwater pumping, (2) subsurface drains, and (3) low-permeability barriers.[1,2,3]

Groundwater Pumping

Groundwater-pumping techniques involve the active manipulation and management of groundwater in order to contain or remove a plume or to adjust groundwater levels in order to prevent formation of a plume.[1] Types of wells used in management of contaminated groundwater include well points, suction wells, ejector wells, and deep wells. The selection of the appropriate well type depends upon the depth of contamination and the hydrologic and geologic characteristics of the aquifer. Figure 12.2.1 shows the use of extraction wells.

Well systems are very versatile and can be used to contain, remove, divert, or prevent development of plumes under a variety of site conditions.

Pumping is most effective at sites where underlying aquifers have high intergranular hydraulic conductivity. It has been used with some effectiveness at sites with moderate hydraulic conductivities and where pollutant movement is occurring along fractured or jointed bedrock. In fractured bedrock, the fracture patterns must be traced in detail to ensure proper well placement.

Where plume containment or removal is the objective, either extraction wells or a combination of extraction and injection wells can be used. Use of extraction wells alone is best suited to situations where contaminants are miscible and move readily with water; where the hydraulic gradient is steep and hydraulic conductivity high; and where quick removal is not necessary. Extraction wells are frequently used in combination with slurry walls to prevent groundwater from

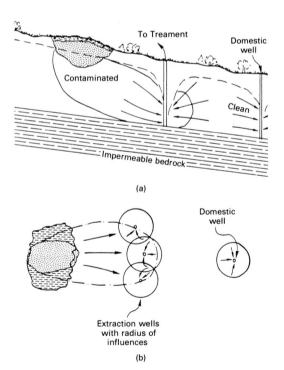

FIG. 12.2.1   Containment using extraction wells. (*a*) Cross-seciontal view. (*b*) Plan view.

overtopping the wall and to minimize contact of the leachate with the wall in order to prevent wall degradation. Slurry walls also reduce the amount of contaminated water that requires removal, so that costs and pumping time are reduced.

A combination of extraction and injection wells is frequently used in containment or removal where the hydraulic gradient is relatively flat and hydraulic conductivities are only moderate. The function of the injection well is to direct contaminants to the extraction wells. This method has been used with some success for plumes which are not miscible with water. One problem with such an arrangement of wells is that dead spots (i.e., areas where water movement is very slow or nonexistent) can occur when these configurations are used. The size of the dead spot is directly related to the amount of overlap between adjacent radii of influence; the greater the overlaps the smaller the dead spots will be. Another problem is that injection wells can suffer from many operational problems, including air locks and the need for frequent maintenance and well rehabilitation.

Extraction or injection wells can also be used to adjust groundwater levels, although this application is not widely used. In this approach, plume development can be controlled at sites where the water table intercepts disposed wastes by lowering the water table with extraction wells. In order for this pumping technique to be effective, infiltration into the waste must be eliminated and liquid wastes must be completely removed. If these conditions are not met, the potential exists for development of a plume of contaminants. The major drawback to using well systems for lowering water tables is the continued costs associated with maintenance of the system.

Groundwater barriers can be created using injection wells to change both the direction of a plume and the speed of plume migration. By creating an area with a higher hydraulic head, the plume can be forced to change direction. This technique may be desirable when short-term diversions are needed or when diversion will provide the plume with sufficient time to naturally degrade so that containment and removal are not required. Each of the well types used in groundwater pumping have their own specific applications and limitations as well.

Well-point systems are effective in almost any hydraulic situation. They are best suited to shallow aquifers where extraction is not needed below approximately 7 m (20 ft). Beyond this depth, suction lifting (the standard pumping technique for wellpoints) is ineffective. Suction wells operate in a similar fashion to well points and are also depth-limited. The only advantage of suction wells over well points is that they have higher capacities. For extraction depths greater than 7 m (20 ft), deep wells and ejector wells are used. Deep-well systems are better suited to homogeneous aquifers with high hydraulic conductivities and where large volumes of water may be pumped. Ejector wells perform better than deep wells in heterogeneous aquifers with low hydraulic conductivities. A problem with ejector systems is that they are inefficient and are sensitive to constituents in the groundwater which may cause chemical precipitates and well clogging.[4]

Subsurface Drains

Subsurface drains include any type of buried conduit used to convey and collect aqueous discharges by gravity flow. Subsurface drains essentially function like an infinite line of extraction wells. They create a continuous zone of influence within which groundwater flows towards the drain. Subsurface drainage is illustrated in Fig. 12.2.2

The major components of a subsurface drainage system are:

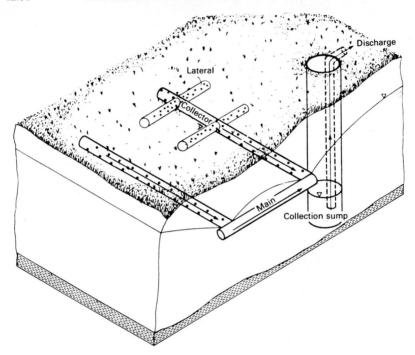

FIG. 12.2.2 Components of a drainage system.

- *Drainpipe or gravel bed*—for conveying flow to a storage tank or wet well. Pipe drains are used most frequently at hazardous-waste sites. Gravel bed or french drains and tile drains are used to a more limited extent.
- *Envelope*—for conveying flow from the aquifer to the drainpipe or bed.
- *Filter*—for preventing fine particles from clogging the system, if necessary.
- *Backfill*—to bring the drain to grade and prevent ponding.
- *Manholes or wet wells*—to collect flow and pump the discharge to a treatment plant.

Since drains essentially function like an infinite line of extraction wells, they can perform many of the same functions as wells. They can be used to contain or remove a plume, or to lower the groundwater table to prevent contact of water with the waste material. The decision to use drains or pumping is generally based on a cost-effectiveness analysis.

For shallow contamination problems, drains can be more cost-effective than pumping, particularly in strata with low or variable hydraulic conductivcity. Under these conditions, it would be difficult to design and it would be cost-prohibitive to operate a pumping system to maintain a continuous hydraulic boundary. Subsurface drains may also be preferred over pumping where ground-water removal is required over a period of several years, because the operation and maintenance costs associated with pumping are substantially higher.

One of the biggest drawbacks to the use of subsurface drains is that they are generally limited to shallow depths. Although it is technically feasible to excavate a trench to almost any depth, the costs of shoring, dewatering, and hard-rock excavation can make drains cost-prohibitive at depths of less than 13 m (40 ft). However, in stable, low-permeability soils where little or no rock excavation is required, drains may be cost-effective to depths of 32 m (100 ft).

The most widespread use of subsurface drains at hazardous-waste sites is to intercept a plume hydraulically down-gradient from its source. Frequently, these *interceptor drains,* as they are commonly called, are used together with a barrier wall. There are two primary reasons for the interceptor drain–barrier wall combination. In the case where a subsurface drain is to be placed just up-gradient to a stream, the drainage system would reverse the flow direction of the stream and cause a prohibitively large volume of clear water to be collected. The addition of a barrier wall would prevent infiltration of clean water from the stream, thereby reducing treatment costs.

In another application, where the primary remedial action involves installation of a down-gradient barrier wall to contain wastes, an interceptor drain can be installed just up-gradient of the wall to prevent overtopping and to minimize contact with wastes which may degrade the wall.

Subsurface drains can also be placed around the circumference of a waste site in order to lower the groundwater table or to contain a plume. A circumferential subsurface drain may be part of a total containment system which consists of a barrier wall and a cap in addition to the subsurface drain.

In addition to depth, other limitations on the use of subsurface drains include the presence of viscous or reactive chemicals which could clog drains. Conditions which favor the formation of iron manganese or calcium carbonate deposits may also limit the use of drains.

For applications at hazardous-waste sites, pipe drains are most frequently used. French or gravel drains can be used where the amount of water to be drained is small and flow velocities are low. If used to handle high volumes or rapid flows, these drains are likely to fail as a result of excessive siltation, particularly in fine-grained soils. Tile drains have not been widely used in hazardous-waste site applications.

Low-Permeability Barriers

The term *subsurface barriers* refers to a variety of methods whereby low-permeability cutoff walls or diversions are installed belowground to contain, capture, or redirect groundwater flow in the vicinity of a site. The most commonly used subsurface barriers are slurry walls, particularly soil-bentonite slurry walls. Less common are cement-bentonite slurry walls, grouted barriers, and sheet piling cutoffs. Grouting may also be used to create horizontal barriers for sealing the bottom of contaminating sites.[2]

Slurry walls are the most common subsurface barriers because they are a relatively effective method of reducing groundwater flow in unconsolidated earth materials.[5] The term *slurry wall* can be applied to a variety of barriers all having one thing in common: they are all constructed in a vertical trench that is excavated under a slurry. This slurry, usually a mixture of bentonite and water, acts essentially like a drilling fluid. It hydraulically shores the trench to prevent collapse, and, at the same time, forms a filter cake on the trench walls to prevent high fluid losses into the surrounding ground. Slurry-wall types are differentiated by the materials used to backfill the slurry trench. Most commonly, an engineered soil mixture is blended with the bentonite slurry and placed in the trench to form a soil-bentonite slurry wall. In some cases, the trench is excavated under a slurry of portland cement, bentonite, and water, and this mixture is left in the trench to harden into a cement-bentonite slurry wall.

Soil-bentonite slurry walls are backfilled with soil materials mixed with a bentonite and water slurry. Of the major types of slurry walls, soil-bentonite walls offer the lowest installation costs, the widest range of chemical compatibilities, and the lowest permeabilities. At the same time, soil-bentonite walls have the highest compressibility (least strength), require a large work area, and, because the slurry and backfill can flow, are applicable only to sites that can be graded to nearly level.

One of the first steps in considering a soil-bentonite slurry wall for a given site is to review the configuration options available and determine which best meets the goals of the remedial action. In the vertical perspective, slurry walls may be "keyed-in" or hanging. Keyed-in slurry walls are constructed in a trench that has been excavated into a low-permeability confining layer such as a clay deposit of competent bedrock. This layer will form the bottom of the contained site, and a good key-in is essential to adequate containment. Hanging slurry walls, however, are not tied to a confining layer but extend down several feet into the water table to act as a barrier to floating contaminants (such as oils and fuels) or migrating gases. The use of hanging slurry walls in waste-site remediation is relatively rare, and most installations are of the keyed-in variety.

One factor that can limit the use of a soil-bentonite wall is the site topography. Because both the excavation slurry and the backfill will flow under stress, the trench line must be within a few degrees of level. In most cases, it is possible to grade the trench line level prior to construction, but this is an added expense.

Cement-bentonite slurry walls share many characteristics with soil-bentonite slurry walls, but are also different in some respects. The principal difference between the two is the backfill, and this produces differences in applications, compatibilities, and costs. To avoid duplication of material already presented, this discussion will highlight the factors that distinguish cement-bentonite walls from soil-bentonite walls.

Cement-bentonite walls are generally excavated using a slurry of portland cement, bentonite, and water. This slurry is left in the trench and allowed to set up to form the completed barrier. For extremely deep installations, a normal bentonite slurry is used for excavation, then replaced by cement-bentonite.

Cement-bentonite walls offer the same configuration options as soil-bentonite walls. They are more versatile than soil-bentonite walls in two ways. First, because the slurry sets up into a semirigid solid, this type of wall can accommodate variations in topography by allowing one section to set while continuing the next section at a higher or lower elevation. Second, because the excavation slurry is commonly the backfill too, this type of wall is better suited to restricted areas where there is no room to mix soil-bentonite backfill. Also, cement-bentonite is stronger than soil-bentonite and so is used where the wall must have less elasticity, for instance, in places adjacent to buildings or roads.

Cement-bentonite slurry walls are limited in their use by their higher costs, somewhat higher permeability, and their narrower range of chemical compatibilities. Cement-bentonite walls average over 30% higher in cost than soil-bentonite walls. The permeability of a cement-bentonite wall is normally around 1×10^{-6} cm/s while a well-designed soil-bentonite wall is capable of achieving 1×10^{-8} cm/s.[5] Cement-bentonite backfills are also more susceptible to chemical attack than most soil-bentonite mixtures. Cement-bentonite is susceptible to attack by sulfates, strong acids and bases (pH $< 4 > 7$), and other highly ionic substances.[6]

Grouting refers to a process whereby one of a variety of fluids is injected into a rock or soil mass where it is set in place to reduce water flow and strengthen the formation. Because of costs, grouted barriers are seldom used for containing groundwater flow in unconsolidated materials around hazardous-waste sites. Slurry walls are less costly and have lower permeability than grouted barriers. Consequently, for waste-site remediation, grouting is best suited for sealing voids in rock. Even in cases where rock voids are transmitting large water volumes, a grout can be formulated to set before it is washed out of the formation. The various types of grouts available are discussed below followed by discussions of the various ways grouts may be employed.

Cement has probably been used longer than any other type of material for grouting applications.[7] Cement grouts utilize hydraulic cement which sets, hardens, and does not disintegrate in water.[8] Because of their large particle size, cement grouts are more suitable for rock than for soil applications. However, the addition of clay or chemical polymers can improve the range of usage. Cement grouts have been used for both soil-consolidation and water-cutoff applications, but their use is primarily restricted to more open soils. Typically, cement grouts cannot be used in fine-grained soils with cracks less than 0.1 mm wide.[7]

One of the greatest potential uses for grouting in hazardous-waste site remediation is for sealing fractures, fissures, solution cavities, or other voids in rock. Nonetheless, rock grouting at waste sites is uncommon and no actual applications were found in the literature.

Rock grouting may be applied to a waste site to control the flow of groundwater entering a site. In theory, grouting could also control leachate flow in rock, yet in many cases contaminants interfere with grout-setting reactions and/or reduce grout durability. In many cases, the waste-grout interaction and compatibility cannot be predicted and extensive testing is required.[6]

Grout curtains are subsurface barriers created in unconsolidated materials by pressure injection. The various methods of forming a grout curtain are described below.

Grout barriers can be many times more costly than slurry walls and are generally incapable of attaining truly low permeabilities in unconsolidated materials. A recent field-test study of two chemical grouts revealed significant problems in forming a continuous grout barrier because of noncoalescence of grout pods in adjacent holes and grout shrinkage. This study concludes that conventional injection grouting is incapable of forming a reliable barrier in medium sands.[9] Therefore, grout barriers are rarely used when groundwater control in unconsolidated materials is desired.

Grout curtains, like other barriers, can be applied to a site in various configurations. Circumferential placement offers the most complete containment but requires that grouting take place in contaminated groundwater down-gradient of the source. As discussed earlier, this could easily cause problems with grout set and durability. As with other techniques, this requires extensive compatibility testing during the feasibility study. Another limitation of curtain grouting is the problem of gaps left in the curtain as a result of nonpenetration of the grout. A few small gaps in an otherwise low-permeability curtain can increase its overall permeability significantly.

In addition to slurry-wall and grouted cutoffs, *sheet piling* can be used to form a groundwater barrier. Sheet piles can be made of wood, precast concrete, or steel. Wood is an ineffective water barrier, however, and concrete is used primarily where great strength is required. Steel is the most effective in terms of groundwater cutoff and cost.[3]

Steel sheet piling can be employed as a groundwater barrier much like the others discussed in this chapter. Because of costs and unpredictable wall integrity, however, it is seldom used except for temporary dewatering for other construction, or as erosion protection where some other barrier, such as a slurry wall, intersects flowing surface water.

One of the largest drawbacks of sheet piling, or any other barrier technology requiring pile driving, is the problem caused by rocky soils. Damage to or deflection of the piles is likely to render any such wall ineffective as a groundwater barrier.

Bottom sealing refers to techniques used to place a horizontal barrier beneath an existing site to act as a floor and prevent variations of grouting or other construction-support techniques. While technically feasible, no application to site remediation has been made to date.

Emplacement of a bottom seal by grouting involves drilling through the site, or directional drilling from the site perimeter, and injecting grout to form a horizontal or curved barrier. One such technique, *jet grouting,* involves drilling a pattern of holes across the site to the intended barrier depth. A special jet nozzle is lowered and a high-pressure stream of air and water erodes the soil. By turning the nozzle through a complete rotation, a flat, circular cavity is formed. The cavity is then grouted, with intersecting grouted masses forming the barrier. The directional drilling method is very similar to curtain grouting except that it is performed in slanted rather than vertical boreholes. Because these techniques are developmental, no detailed analysis of applications, limitations, design, or construction considerations is possible.

Block displacement is an experimental technique for isolating and raising a contaminated block of earth. A perimeter barrier is first constructed by slurry trenching or grouting. Grout is then injected into specially notched holes bored through the site. Continued grout or slurry pumping causes displacement of the block of earth isolated by the perimeter barrier and forms a bottom seal beneath the block. This technique has been field-demonstrated at a nonhazardous site. However, there has been no application to a hazardous-waste site.

12.2.2 TREATMENT TECHNOLOGIES

In Situ Treatment

One alternative to waste excavation and removal and conventional pump-and-treat methods is to treat the wastes in situ. In situ treatment entails the use of chemical or biological agents or physical manipulations which degrade, remove, or immobilize contaminants; methods for delivering solutions to the subsurface; and methods for controlling the spread of contaminants and treatment reagents beyond the treatment zone. In situ treatment processes are generally divided into three categories: biological, chemical, and physical.[10,11] In situ biodegradation is based on the concept of stimulating the naturally occurring organisms to degrade the contaminants of concern. In situ chemical treatment involves the injection of a specific chemical or chemicals into the subsurface in order to degrade, immobilize, or flush out the contaminants. Physical methods involve physical manipulation of the soil using heat, freezing, or other means. In many instances a combination of in situ and aboveground treatment will achieve the most cost-effective treatment at an uncontrolled waste site. Table 12.2.1 highlights these processes.

In situ treatment technologies are not as developed as other currently available technologies for restoring contaminated areas. However, there are some in situ treatment technologies that have demonstrated success in actual site remediations. In addition, most of the methods are based on standard waste-treatment technologies that are conceptually applicable as in situ treatment methods. Applicability of in situ methods must generally be determined on a site-specific basis using laboratory- and pilot-scale testing.

Bioreclamation is a technique for treating zones of contamination by microbial degradation. The basic concept involves altering environmental conditions to enhance microbial catabolism or cometabolism of organic contaminants, resulting in the breakdown and detoxification of those contaminants. The technology has been developed rapidly over recent years, and bioreclamation appears to be one of the most promising of the in situ treatment techniques.

Considerable research conducted over the past several decades has confirmed that microorganisms are capable of breaking down many of those organic compounds considered to be environmental and health hazards at spill sites and uncontrolled hazardous-waste sites. Laboratory, pilot, and field studies have demonstrated that it is feasible to use this capability of microorganisms in situ to reclaim contaminated soils and groundwater.[12,13,14]

The method that has been most developed and is most feasible for in situ treatment is one which relies on aerobic microbial processes. This method involves optimizing environmental conditions by providing an oxygen source and nutrients, which are delivered to the subsurface through an injection well or infiltration system to enhance microbial activity. Indigenous microorganisms can generally be relied upon to degrade a wide range of compounds given optimal

TABLE 12.2.1 Natural Processes That Affect Subsurface Contaminant Transport

Physical processes	Advection
	Hydrodynamic dispersion
	Molecular diffusion
	Density stratification
	Immiscible phase flow
	Fracture media flow
Chemical processes	Oxidation-reduction reactions
	Radionuclide decay
	Ion exchange
	Complexation
	Cosolvation
	Immisicible phase partitioning
	Sorption
Biological processes	Microbial population dynamics
	Substrate utilization
	Biotransformation
	Adaptation
	Cometabolism

Source: Ref. 13

conditions. Specially adapted microorganisms may also be added to the treatment zone.

Anaerobic microorganisms are also capable of degrading certain organic contaminants. Methanogenic consortiums, groups of anaerobes that function under very reducing conditions, are able to degrade halogenated aliphatics (e.g., PCE, TCE), while aerobic organisms cannot. The potential for anaerobic degradation has been demonstrated in numerous laboratory studies and in industrial waste-treatment processes that use anaerobic digestors or anaerobic waste lagoons as part of the treatment process. Using anaerobic degradation as an in situ reclamation approach is theoretically feasible.

The feasibility of bioreclamation as an in situ treatment technique is dictated by waste and site characteristics. More specifically, those factors which determine the applicability of a bioreclamation approach are

- Biodegradability of the organic contaminants
- Environmental factors which affect microbial activity
- Site hydrogeology (specifically hydraulic conductivity)

Generally, organic and inorganic contaminants can be immobilized, mobilized for extraction, or detoxified. Technologies placed in the category *immobilization* include precipitation, chelation, and polymerization. The category encompassing methods for mobilizing contaminants for extraction is termed *soil flushing*. Flushing agents include surfactants, dilute acids and bases, and water. Detoxification techniques include oxidation, reduction, neutralization, and hydrolysis.

These categories do not define the limits of each technology. For example, a treatment method that immobilizes a contaminant may also serve to detoxify it. A flushing solution that mobilizes one contaminant may precipitate, detoxify, or increase the toxicity of another.

The feasibility of a particular in situ treatment approach is dictated by site geology and hydrology, soil characteristics, and waste characteristics. Since the application of many chemical in situ treatment techniques to reclamation of hazardous-waste disposal sites is conceptual or in the developmental stage, there is little hard data available on the specific site characteristics that may limit the applicability of each method.[15]

Most of the treatment approaches discussed in this section involve the delivery of a fluid to the subsurface. Therefore, the same factors that limit the use of injection and extraction wells, drains, or surface gravity application systems for bioreclamation will limit the applicability of most in situ chemical treatment approaches. Minimal permeability requirements must be met if the treatment solution is to be delivered successfully to the contaminated zone. Sandy soils are far more amenable to in situ treatment than clayey soils. Further, the contaminated groundwater must be contained within the treatment zone. Measures must be taken to ensure that treatment reagents do not migrate and themselves become contaminants. Care must be taken during the extraction process not to increase the burden of contaminated water by drawing uncontaminated water into the treatment zone from the aquifer or from hydraulically connected surface waters.

A number of methods are currently being developed which involve physical manipulation of the subsurface in order to immobilize or detoxify waste constituents. These technologies, which include in situ heating, vitrification, and ground freezing, are in the early stages of development.

In situ heating has been proposed as a method to destroy or remove organic contaminants in the subsurface through thermal decomposition, vaporization, and distillation. Methods recommended for in situ heating are steam injection[16] and radio-frequency (RF) heating.[17]

The radio-frequency heating process has been under development since the 1970s. Field experiments have been conducted for the recovery of hydrocarbons. The method involves laying a row of horizontal conductors on the surface of a landfill and exciting them with an RF generator through a matching network. The decontamination is accomplished in a temperature range of 300 to 400°C, assisted with steam, and requires a residence time of about two weeks. A gas- or vapor-recovery system is required on the surface. Excavation, mining, drilling, or boring is not required. Field tests found that leakage radiation levels did not exceed the recommended ANSI Standard C-95. Preliminary design and cost estimates for a mobile, in situ RF decontamination process indicate that the method is 2 to 4 times cheaper than excavation and incineration.[17] This method appears very promising for certain situations involving contamination with organics, although more research is necessary to verify the effectiveness in situ.

Artificial ground freezing involves the installation of freezing loops in the ground and a self-contained refrigeration system that pumps coolant around the freezing loop.[18] Although never used in an actual waste-containment operation, the technology is being used increasingly as a construction method in civil engineering projects. Artificial ground freezing is done not on the waste itself, which may have a freezing point much lower than that of the soil systems, but on the uncontaminated soil surrounding the hazardous waste. It renders the soil practically impermeable but is useful only as a temporary treatment approach because of the thermal maintenance expense.[18]

In situ vitrification is a technology being developed for the stabilization of transuranic contaminated wastes, and is conceivably applicable to other hazardous wastes.[19] Several laboratory-scale and pilot-scale tests have been conducted, and a large-scale testing system is currently being fabricated. The principle of operation is joule heating, which occurs when an electrical current is passed through a molten mass. Contaminated soil is converted into durable glass, and wastes are pyrolyzed or crystallized. Off gases released during the melting process are trapped in an off-gas hood. The depth of the waste is a significant limiting factor in the application of this technology: 1 to 1.5 m of uncontaminated overburden lowers release fractions considerably.[19]

Aqueous-Waste Treatment

Aqueous waste streams resulting from the cleanup of hazardous-waste sites vary widely with respect to volume, level, and type of contaminants and level of solids. The major sources of aqueous wastes include

- Leachate plumes which have been pumped to the surface or collected via subsurface drains
- Contaminated water generated during dredging operations
- Contaminated runoff collected in impoundments or basins
- Contaminated water generated from equipment cleanup
- Aqueous waste generated from sediment or sludge dewatering
- Highly concentrated wastewater streams generated from certain treatment processes applied to aqueous wastes (e.g., backwash from filtration, concentrate from reverse osmosis)

Because these waste streams are so diverse in volume, type, and concentration of contaminants, a wide variety of treatment processes will have application to hazardous-waste site cleanup. Rarely will any one unit treatment process be sufficient for aqueous-waste treatment. Therefore, a combination of unit treatment processes will generally be used for site remediation. The unit treatment processes applicable for use at remediation sites include

- Activated carbon
- Activated sludge
- Filtration
- Precipitation and/or flocculation
- Sedimentation
- Ion exchange
- Reverse osmosis
- Neutralization
- Gravity separation
- Air stripping
- Chemical oxidation
- Chemical reduction
- Extraction technologies

Aqueous-waste treatment at hazardous-waste sites can be accomplished using one of four general approaches:

- On-site treatment using mobile treatment system
- On-site construction and operation of treatment systems
- Pretreatment followed by discharge to a publicly owned treatment works (POTW)
- Hauling of waste to an off-site treatment facility

Mobile treatment systems and systems construction on site have broadest applicability. Wastewaters discharged to POTWs often require extensive pretreatment in order for the facility to meet its National Pollutant Discharge Elimination System (NPDES) permit conditions. Other factors which determine the feasibility of POTW discharge include whether the facility has the hydraulic capacity to handle the waste, whether accepting the waste will result in additional monitoring requirements or process changes, and the potential for opposition in the community. The option of hauling wastes off site for treatment is limited for all but very small wastewater volumes.

The reader is directed to other chapters in this Handbook for more detailed discussions on particular treatment processes.

Solids Treatment

Methods used to separate solids from slurries, and/or to classify contaminated soils or slurries according to grain size are discussed in the following paragraphs. The objective of separating solids from slurries is to attain two distinct waste streams: a liquid waste stream that can be subsequently treated for removal of dissolved and fine suspended contaminants, and a concentrated slurry of solids and minimal liquid that can be dewatered and treated.

Classification of particles according to grain size may be undertaken for one of two reasons. The first reason is that more efficient use can be made of equipment and land area by taking advantage of the differences in settling velocity; high-rate gravity settlers could then be used to remove fine-grained particles.

Second, there is recent evidence to suggest that classification by grain size is important in managing hazardous-waste contaminants to adsorb preferentially onto fine-grained materials such as clay and organic matter. The separation of solids by grain size and level of contamination could prove to be extremely beneficial to the overall management (treatment, transport, and disposal) of contaminated soil material. Whereas relatively noncontaminated soils and sediments may be disposed of in ordinary sanitary landfills or discharged back into the stream, the highly contaminated solids must be disposed in a hazardous-waste landfill, incinerated, or treated to render them nonhazardous.

The most appropriate solids-separation method for a given site depends upon several factors, including the following:

- Volume of contaminated solids.
- Composition of soils or sediments, including gradation, percent clays, and percent total solids.
- Types of dredging or excavation equipment used, which determine the feed rate to solids separation and, in the case of slurries, the percent solids.
- Site location and surroundings: the available land area and ultimate or present land use may limit the type of system that can be utilized.

Solids-separation methods include sieves and screens, hydraulic and spiral classifiers, cyclones, settling basins, and clarifiers.

Solidification and Stabilization

Solidification and *stabilization* are terms which are used to designate a technology employing additives to alter hazardous waste to make it nonhazardous or acceptable under current land disposal requirements;[20] these technologies accomplish one or more of the following:

- Improve waste-handling or other physical characteristics of the waste
- Decrease the surface area across which transfer or loss of contained pollutants can occur
- Limit the solubility or toxicity of hazardous-waste constituents.

Solidification and stabilization methods can be categorized as follows:

- Cement solidification
- Silicate-based processes
- Sorbent materials
- Thermoplastic techniques

- Surface encapsulation
- Organic polymer processes

These types of processes are currently being evaluated for application to site remediation. Possible uses include in-place stabilization and/or solidification of lagoon bottoms and contaminated soils and application to various organic and inorganic wastes which could later be landfilled.[20,21] Solidification processes are discussed in Sec. 7.8 of the Handbook.

Thermal Processes

Thermal destruction is a treatment method which uses high-temperature oxidation under controlled conditions to degrade a substance into products that generally include CO_2, H_2O vapor, SO_2, NO_x, HCl gases, and ash. The hazardous products of the thermal destruction or incineration, such as particulates, SO_2, NO_x, HCl, and products of incomplete combustion require air-pollution control equipment to prevent release of undesirable species into the atmosphere. Thermal destruction methods can be used to destroy organic contaminants in liquid, gaseous, and solid streams.

The most common thermal technologies applicable to hazardous wastes include:

- Liquid injection
- Rotary kiln
- Fluidized bed

These thermal processes are described in Chap. 8 of the Handbook.

12.2.3 EMERGING AND ALTERNATIVE TECHNOLOGIES

Emerging technologies for treatment of hazardous waste include

- Molten-salt systems
- Molten-glass systems
- Wet-air oxidation
- Plasma-arc torch
- Circulating-bed combustor
- High-temperature fluid-wall reactor
- Electric infrared systems
- Supercritical-water oxidation
- Advanced electric reactor
- Vertical-tube reactor

These technologies are discussed in other chapters in this Handbook. The EPA, under the Superfund Innovative Technology Evaluation (SITE) program, plans to conduct demonstrations of these and other technologies as they emerge at Superfund sites. To date, six demonstrations have taken place. Reports on these projects will be available in Fall 1988. The SITE Strategy and Program Plan Report[22] and the SITE Report to Congress[23] further explain the program and current activities.

12.2.4 CONCLUSION

Potential solutions for site remediation will almost always include a variety of technologies. Containment technologies and treatment technologies will invariably be used together. An example is the slurry-wall and groundwater-collection and treatment system at Nashua, New Hampshire.[24] In addition, various technologies will be used to solve problems associated with the treatment of liquids, solids, and sludges at the same site. However, in addition to containment and treatment technologies, other aspects such as materials handling, worker safety, protection from accidental releases, and protection from future financial liability must be incorporated into the complete plan for site remediation.

12.2.5 REFERENCES

1. U.S. Environmental Protection Agency, *Leachate Plume Management*, EPA 540/2-85/004, U.S. EPA, Cincinnati, Ohio, 1984, chap. 5.
2. U.S. Environmental Protection Agency, *Guidance Manual for Minimizing Pollution from Waste Disposal Sites*, EPA 600/2-78/142. U.S. EPA, Cincinnati, Ohio, 1978.
3. U.S. Environmental Protection Agency, *Handbook—Remedial Action at Waste Disposal Sites (Revised)*, EPA 625/6-85/006, U.S. EPA, Cincinnati, Ohio, 1985, chap. 5.
4. J. P. Powers, *Construction Dewatering—A Guide to Theory and Practice*, Wiley, New York, 1981.
5. U.S. Environmental Protection Agency, *Slurry Trench Construction for Pollution Migration Control*, EPA 540/2-84/001, U.S. EPA, Cincinnati, Ohio, 1984.
6. U.S. Environmental Protection Agency, *Compatibility of Grouts with Hazardous Wastes*, EPA 600/2-84/015, U.S. EPA, Cincinnati, Ohio, 1984.
7. R. Bowen, Grouting in Engineering Practice (2d ed.), Wiley, New York, 1981.
8. R. E. Kirk and D. F. Othmen, *Encyclopedia of Chemical Technology*, 3d ed., vol. 5., Wiley, New York, 1979.
9. J. H. May, R. J. Larson, P. G. Malone, and V. A. Boa, Jr., "Evaluation of Chemical Grout Injection Techniques for Hazardous Waste Containment," in *Proceedings of the Eleventh Annual Research Symposium on Land Disposal of Hazardous Wastes*, EPA 600/9-85/013, U.S. EPA, Cincinnati, Ohio, 1985.
10. T. P. Brunsing and R. B. Henderson, "A Laboratory Technique for Assessing the In-Situ Constructibility of a Bottom Barrier for Waste Isolation," in *Proceedings of the Fifth National Conference on Management of Uncontrolled Hazardous Waste Sites*, Hazardous Materials Control Research Institute, Silver Spring, Maryland, 1984, pp. 135–140.
11. U.S. Environmental Protection Agency, *Review of In-Place Treatment Techniques for Contaminated Surface Soils*, EPA 540/2-84/003a, U.S. EPA, Cincinnati, Ohio, 1984.
12. U.S. Environmental Protection Agency, *Systems to Accelerate In-Situ Stabilization of Waste Deposits*, EPA 540/2-86/002, U.S. EPA, Cincinnati, Ohio, 1986.
13. J. F. Keely, M. D. Piwoni, and J. T. Wilson, *Evolving Concepts of Subsurface Contaminant Transport, J. Water Pollut. Control Fed.*, **58**(5):349–357 (1986).

14. E. Heyse, S. C. James, and R. Wetzel, "In Situ Aerobic Biodegradation of Aquifer Contaminants at Kelly Air Force Base," *Env. Prog.*, AIChE, **5**(3):207—211 (1986).

15. R. C. Sims and K. Wagner, "In-Situ Treatment Techniques Applicable to Large Quantities of Hazardous Waste Contaminated Soils," in *Proceedings of Fourth National Conference on Management of Uncontrolled Hazardous Waste Sites*, Hazardous Materials Control Research Institute, Silver Spring, Maryland, 1983, pp. 226–230.

16. D. Hoogendorn, "Review of the Development of Remedial Action Techniques for Soil Contamination in the Netherlands," in *Proceedings of the Hazardous Material Spills Conference*, Government Institutes, Inc., Rockville, Maryland, 1984.

17. H. Dev., J. E. Bridges, and G. C. Sresty, "Decontamination of Hazardous Waste Substances from Spills and Uncontrolled Waste Sites by Radio Frequency In Situ Heating," in *Proceedings of Hazardous Material Spills Conference*, Government Institutes, Inc., Rockville, Maryland, 1984.

18. I. K. Iskander and J. M. Houthoofd, "Effect of Freezing on the Level of Contaminants in Uncontrolled Hazardous Waste Sites," in *Proceedings of the Eleventh Annual Research Symposium on Land Disposal of Hazardous Wastes*, EPA 600/9-85/013, U.S. EPA, Cincinnati, Ohio, 1985, pp. 122–129.

19. V. F. Fitzpatrick, J. L. Buelt, K. H. Oma, and C. L. Timmermen, "In Situ Vitrification—A Potential Remedial Action Technique for Hazardous Wastes," in *Proceedings of the Fifth National Conference on Management of Uncontrolled Hazardous Waste Sites*, Hazardous Materials Control Research Institute, Silver Spring, Maryland, 1984, pp. 191–194.

20. U.S. Environmental Protection Agency, *Guide to Disposal of Chemically Stabilized and Solidified Waste*, EPA SW-872, U.S. EPA Office of Solid Waste and Emergency Response, Washington, D.C., 1982.

21. M. J. Cullinane, and L. W. Jones, *Handbook for Stabilization/Solidification of Hazardous Waste*, EPA 540/2-86/001. U.S. EPA, Cincinnati, Ohio, 1986.

22. U.S. Environmental Protection Agency, Superfund Innovative Technology Evaluation (SITE) Strategy and Program Plan, EPA 540/G-86/001, Office of Research and Development and Office of Solid Waste and Emergency Response, Washington, D.C., 1986.

23. U.S. Environmental Protection Agency, Superfund Innovative Technology Evaluation Program: Progress and Accomplishments—A Report to Congress, EPA 540/5-88/001, Office of Research and Development and Office of Solid Waste and Emergency Response, Washington, D.C., 1988.

24. M. J. Barvenik, W. E. Hadge, and D. T. Goldberg, "Quality Control of Hydraulic Conductivity and Bentonite Content During Soil/Bentonite Cutoff Wall Construction," in *Proceedings of the Eleventh Annual Research Symposium on Land Disposal of Hazardous Wastes*, EPA 600/9-85/013, U.S. EPA, Cincinnati, Ohio, 1985, pp. 66–79.

SECTION 12.3
SITE REMEDIATION

Roger S. Wetzel

Senior Project Manager
Science Applications International Corporation
McLean, Virginia

Site remediation incorporates the use of specific technologies such as capping, slurry trenching, and groundwater treatment to address specific problems identified in the site-investigation process. This section will summarize the currently accepted practices and U.S. Environmental Protection Agency (EPA) guidance on the steps to be taken in developing, selecting, and implementing approaches for site remediation. These steps include remedial investigation (RI), feasibility study (FS), corrective action, and closure. Technologies that may apply to site remediation, such as those discussed in other portions of this manual, must be kept in mind when planning the RI in order to provide valid data to support their evaluation in the FS. These technologies are then combined to form remedial alternatives to completely address site problems. The selected alternative is the one which best addresses site problems based on RI data and factors such as cost and public-health impacts which are discussed later in this section. Design of the selected remedial alternative involves developing detailed specifications and construction drawings necessary for incorporation into bid packages and for implementation by the successful bidder. Implementation involves successfully carrying out the design by installing treatment facilities, earth moving, and/or isolating contamination using containment technologies such as capping or slurry trenches. Figure 12.3.1 shows a slurry trench being installed.

12.3.1 REMEDIAL INVESTIGATION AND FEASIBILITY STUDY

The ultimate goal of the RI and FS is to develop data to support the selection of an approach for site remediation and then to use this data in a structured procedure that results in a well-supported recommended approach.

FIG. 12.3.1 Installation of a slurry trench (*SAIC file photo*)

The EPA has developed guidance on the RI/FS Process[1,2] and for specific types of remedial actions such as those for decontaminating buildings, slurry-trench construction, and leachate-plume management (see Subsec. 12.3.6, *Further Reading*).

Remedial Investigation

Examples of the types of data needs are given in Table 12.3.1.[1] The remedial investigation must establish site characteristics such as media contaminated, the extent of contamination, and the physical boundaries of the contamination. An additional need of the remedial investigation is to provide as much detailed data as possible to aid in selecting specific technologies while dropping others from further consideration. An example is the provision for enough soil borings along the waste-disposal boundary to evaluate soils and relatively impermeable layers at depth for evaluation of slurry-trench applications. Additional technology-specific data can be gained from treatability studies, which are optional for the RI/FS. Focused data collection will keep RI costs reasonable and will provide the detailed data needed to support key site remedies.

For surface impoundments and tanks, definition of the boundaries for the investigation is obvious. However, other problems such as presence of horizontal layers of liquid wastes having very different removal and treatment or disposal requirements can be very serious if not identified in the RI.

One or more key contaminants are normally selected as indicators to determine the movement and extent of contamination. This key contaminant must be selected based on persistence and mobility in the environment and the degree of hazard. An added criterion should be ease of identification. For example, frequent choices are TCE and PCE, which are present at numerous waste sites; are persistent, toxic, and very mobile; and can be easily identified at low concentrations in groundwater. Relatively new technologies, such as soil-gas monitoring,

TABLE 12.3.1 Site and Waste Characteristics

Site characteristics	
Site volume	Depth to bedrock
Site area	Depth to aquicludes
Site configuration	Degree of contamination
Disposal methods	Direction and rate of groundwater flow
Climate	Receptors
Precipitation	Distance to:
Temperature	Drinking-water wells
Evaporation	Surface water
Soil texture and permeability	Ecological areas
Soil moisture	Existing land use
Slope	Depth to groundwater or to plume
Drainage	
Vegetation	

Waste characteristics	
Quantity	Infectiousness
Chemical composition	Solubility
Carcinogenicity	Volatility
Toxicity—chronic and acute	Density
Persistence	Partition coefficient
Biodegradability	Safe levels in the environment
Radioactivity	Compatibility with other chemicals
Ignitability	
Reactivity/corrosiveness	
Treatability	

Source: Ref. 1.

can be used to identify likely areas for locating groundwater-monitoring wells. This reduces guesswork and is likely to reduce the number of monitoring wells needed to produce RI data, thus saving time and resources.

Contamination identified in the RI must eventually be evaluated for receptor exposure. An estimate of contaminant level reaching human or environmental receptors must be made. Then existing standards and guidelines such as drinking-water standards, water-quality criteria, or other criteria accepted as appropriate for the situation may be used to evaluate effects on human receptors who may be exposed to contaminants above appropriate standards or guidelines. This information will be used to evaluate the "no action" alternative in the FS. Qualitative estimates of exposure are sometimes necessary when the level of contamination reaching the receptor cannot be quantified. Reduction in level of exposure by various remedial alternatives will be used as a screening technique in the FS.

RI data must be adequate to estimate costs for remedial alternatives within a +50% and −30% accuracy. The accuracy of costs depends on site-specific and remedial technology-specific factors. For example, slurry-trench construction costs frequently assume an average depth to an underlying clay layer, the minimum length of trench needed to contain contamination, and minimal or no site-preparation costs, while specific RI data can allow much more detailed cost estimating.

Feasibility Study

The FS process is given in Fig. 12.3.2, while an EPA-suggested outline of an FS report is shown in Table 12.3.2.[2]

The purpose of the feasibility study under the Comprehensive Environmental Response Compensation and Liability Act (CERCLA) is to document the problem(s) identified in the RI, determine the range of possible solutions, and select the best solution to waste-site problems. The National Oil and Hazardous Substances Contingency Plan (NCP) requires establishment of response goals (target cleanup levels); identification of alternatives; both preliminary and more detailed assessment of technical feasibility (will it work at this particular site?); determination of public-health and environmental impacts; and cost and institutional analysis for screening of remedial alternatives.[3] The EPA's guidance follows these NCP requirements. Early identification of potentially applicable technolo-

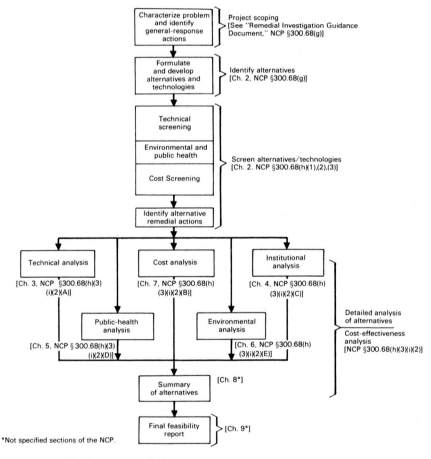

FIG. 12.3.2   Flowchart of the feasibility-study process. NCP sections identified refer to those cited in the *Federal Register*, July 16, 1982. (*Guidance on Feasibility Studies under CERCLA*, U.S. EPA, EPA/540/G-85/003, June 1985)

TABLE 12.3.2   Format for Feasibility-Study Report

Executive Summary

1.0 Introduction

    1.1 Site background information
    1.2 Nature and extent of problems
    1.3 Objectives of remedial action

2.0 Screening of remedial-action technologies

    2.1 Technical criteria
    2.2 Remedial-action alternatives developed
    2.3 Environmental and public-health criteria
    2.4 Other screening criteria
    2.5 Cost criteria

3.0 Remedial-action alternatives

    3.1 Alternative 1 (no action)
    3.2 Alternative 2

      .
      .
      .

    3.N Alternative N

4.0 Analysis of remedial-action alternatives

    4.1 Noncost-criteria analysis

        4.1.1 Technical feasibility
        4.1.2 Environmental evaluation
        4.1.3 Institutional requirements
        4.1.4 Public-health evaluation

    4.2 Cost analysis

5.0 Summary of alternatives

6.0 Recommended remedial action (optional)

7.0 Responsiveness summary (in final version only)

References

Appendices

Source: Ref. 1.

gies is important to the planning of the RI. Establishment of a clear, concise statement of the problem (such as contamination shown to be in excess of established target levels for cleanup) is critical to focusing the FS.

Limitations on cost and time for conducting an FS limit the number of alternatives that should be fully evaluated.

The scope of alternatives that control the source of the migration of contamination include:

- Removal of contamination source and off-site treatment and disposal
- Removal of contamination source and on-site treatment and disposal
- Containment of contamination source instead of removal
- In-place treatment instead of removal and treatment

The "no action" alternative must also be evaluated on the same basis.

For compliance with the Resource Conservation and Recovery Act (RCRA) and current EPA policies, technical and cost issues must both be considered when evaluating removal options. All other considerations being equal, alternatives featuring destruction of wastes are favored by the EPA over redisposal or containment. In fact, an alternative that features destruction of wastes will probably have a higher cost than redisposal or containment, but the higher cost can often be justified by the benefit of increased reliability (effectiveness over time) of the completed remedial action.[4]

Contaminated groundwater poses special remediation problems because of the potential for migration of contaminants over long distances from the source and the potential for exposure of a greater number of people to the contamination. The risk involved and remediation considered are affected by the general decrease in concentration with distance from the source and increase in the volume of contaminated groundwater with distance from the source. The scope of alternatives for contaminated groundwater include:

- Removal and treatment, followed by reinjection, surface discharge, or other use
- Containment or alteration of gradient
- No action with contaminated groundwater, but provision of another source of water to those at risk

The "no action" alternative must also be evaluated on the same basis.

For groundwater treatment, Part 264 Subpart F of the RCRA regulations requires removal and treatment of groundwater until the concentration of contaminants reaches either background levels or maximum permissible concentration limits (MCLs).[5] The MCLs for various contaminants are given in Table 12.3.3.

TABLE 12.3.3 Maximum Concentration of Constituents for Groundwater Protection

Constituent	Maximum concentration, mg/L
Arsenic	0.05
Barium	1.0
Cadmium	0.01
Chromium	0.05
Lead	0.05
Mercury	0.002
Selenium	0.01
Silver	0.05
Endrin[a]	0.0002
Lindane[b]	0.004
Methoxychlor[c]	0.1
Toxaphene[a]	0.005
2,4-D[e]	0.1
2,4,5-TP[f] Silvex[f]	0.01

[a]1,2,3,4,10-hexachloro-1,7-epoxy-1,4,4a,5,6,7,8,9a-octahydro-1,4-*endo, endo*-5,8-dimethanonaphthalene.
[b]1,2,3,4,5,6-hexachlorocyclohexane, gamma isomer.
[c]1,1,1-trichloro-2,2-bis(p-methoxyphenylethane).
[d]$C_{10}H_{10}Cl_6$ (technical chlorinated camphene, 67–69% chlorine).
[e]2,4-dichlorophenoxyacetic acid.
[f]2,4,5-trichlorophenoxypropionic acid.
Source: Ref. 5.

Planned future use for groundwater, site-specific factors, public-health impacts, and costs enter into the decision of whether to treat, contain, or take no action.

Initial screening of alternatives eliminates from further consideration those that are not technically feasible, that inadequately protect public health or the environment, or that have higher costs compared to other alternatives.

Two to six alternatives generally remain for more detailed evaluation after initial screening. Generally, the "no action" alternative and the construction of an on-site RCRA landfill are included.[4]

Evaluation in detail of technical feasibility and environmental and public-health impacts must be made for each alternative, with cost effectiveness also being a key factor.

Technical feasibility considerations include the careful study of any problem that may prevent a remedial alternative from mitigating site problems. Therefore, site characteristics from the RI must be kept in mind as technical feasibility of the alternative is studied. Specific items to be addressed are reliability (operation over time), safety, operation and maintenance, ease with which the alternative can be implemented, and time needed for implementation.

Environmental considerations are whether there are adverse impacts on the environment from the implementation of the alternative. If there are adverse impacts, mitigative measures are planned and costs are estimated.

Public-health considerations include the level of exposure of the population to hazardous substances. Inputs are the population at risk and the route the hazardous substances take to reach the exposed populations. The result of this analysis will be the degree to which each alternative is expected to reduce exposure.

Cost (capital and annual) must be estimated over time so that present-worth costs can be calculated.

Under CERCLA, the public must have input to the RI/FS process. Public acceptance of the alternative selected for implementation is very important.

The EPA makes the final selection of any remedial action under CERCLA, and public comments must be considered before selection of a remedial action is

FIG. 12.3.3 Mixing cement-kiln dust with waste to provide low-cost stabilization (*SAIC file photo*)

final. Following detailed evaluation, the lowest-cost alternative that will effectively mitigate site problems is generally selected. Figure 12.3.3 shows the mixing of cement-kiln dust with waste, a low-cost stabilization alternative where cement-kiln dust is readily available.

12.3.2 SUGGESTIONS FOR IMPROVING THE RI/FS

Criticisms of past RI/FS projects by the EPA have been that they take too much calendar time to complete, supporting data is insufficient or minimally adequate, key alternatives are not evaluated, and evaluation and rationale for the alternatives presented are insufficient.[4] Some common problems that lead to these criticisms can be avoided by adhering to the following criteria:

- Alternatives that bear serious consideration must be identified early in the RI process, allowing the planning and development of adequate data for complete evaluation.
- Quality and scope of data must be adequate to support the screening and detailed evaluation of alternatives.
- The "no action" alternative must be completely evaluated as required by the NCP.
- Cost estimates must be complete, and result in adequate support of the evaluation of relative costs and cost-effectiveness of alternatives.
- Alternatives presented by outside groups such as community representatives must be evaluated with other alternatives.
- On-site RCRA landfill requirements and groundwater-treatment requirements must be included in the evaluation of alternatives.[4]

12.3.3 CORRECTIVE ACTION UNDER RCRA

The corrective-action program applies to "owners and operators of facilities that treat, store, or dispose of hazardous waste in surface impoundments, waste piles, land treatment units or landfills".[5]

The corrective-action program under RCRA is in effect when the RCRA groundwater-monitoring program (for the uppermost aquifer) identifies contaminants significantly above permitted levels. Permitted levels are background levels, MCLs for certain contaminants which are given in Table 12.3.3., or alternative concentration levels proposed by site owners and operators and set by the EPA. The permitted levels generally reflect human-health or environmental effects. Corrective action is taken only when a threat exists to human health or the environment: therefore when permitted levels are significantly exceeded, corrective action is taken.

All RCRA hazardous-waste facility permits issued after November 8, 1984, include a schedule for implementing and completing any necessary corrective action or they include a schedule for gathering information to determine the extent of corrective action required and assurances of financial responsibility for completing the corrective action.[6] The cleanup program goal is to lower the concentration of contaminants to within permit levels or background levels. Cleanup will generally require removal and treatment or treatment in place. The remedial action may be determined as part of a limited investigation of alternatives or the permitting agency could require an action equivalent to an RI/FS. Effectiveness of the corrective action must be proven through additional groundwater monitoring.[5] The EPA believes that air, surface water, and soils also will be covered by corrective action. Additionally, owners and operators must clean up beyond their facility boundary as necessary to mitigate migrating contamination that may threaten human health and the environment, although the EPA has not yet proposed regulations on this issue.[6]

12.3.4 CLOSURE

Waste-site closure consists of final design of the remedial action, implementation, and postclosure monitoring and maintenance. The EPA, U.S. Army Corps of Engineers, states, and/or private industry may take the lead in conducting design and implementation of remedial measures. Figure 12.3.4 illustrates an ongoing drum-removal operation.

Final design involves taking the remedial action selected in the RI/FS process from the conceptual stage to the complete construction drawings and specifications that can be used in bid packages for the selection of a suitable construction contractor. It is not unusual for additional, very specific field data to be required in the final design stage. The need for data at the design stage can be minimized by using foresight in the RI/FS process. Once the final design is complete, bid packages are prepared and issued. Responses are evaluated and a contractor for implementation of the remedial action is selected.

Implementation of the remedial action includes all construction and

FIG. 12.3.4 Drum removal (*SAIC file photo*)

construction-oriented activities needed to successfully mitigate the problems associated with a given site. Independent quality control and on-site inspection are important to ensure that remedial action is implemented as designed and that appropriate construction practices are followed.

Postclosure monitoring and maintenance must be conducted as long as the site is considered a threat. Specific requirements including the length of time and resources to be allocated for monitoring and maintenance are spelled out in agreements between responsible parties and the EPA or states for CERCLA actions and in permits or supplemental agreements with regulating agencies for RCRA corrective action.

12.3.5 REFERENCES

1. U.S. Environmental Protection Agency, *Guidance on Remedial Investigations Under CERCLA,* EPA 540/G-85/002, U.S. EPA, Cincinnati, Ohio, June 1985, pp. 1-1 to 1-6, pp. 2-3 and 2-4.
2. U.S. Environmental Protection Agency, *Guidance on Feasibility Studies Under CERCLA,* EPA 540/G-85/003, U.S. EPA, Cincinnati, Ohio, June 1985, pp. 1-1 to 1-10.
3. *Code of Federal Regulations,* Title 40, Part 300, "National Oil and Hazardous Substances Contingency Plan," Office of the *Federal Register,* Washington, D.C., July 1, 1985, pp. 688 to 693.
4. Bixler, D. Brint, and J. W. Hanson, "Selecting Superfund Remedial Actions," *Proceedings of the 5th National Conference on Management of Uncontrolled Hazardous Waste Sites,* Hazardous Materials Control Research Institute, Silver Spring, Maryland, November 1984, pp. 493 to 497.
5. *Code of Federal Regulations,* Title 40, Part 264 Subpart F, "Groundwater Protection," Office of the *Federal Register,* Washington, D.C., July 1, 1985, pp. 428, 430–431, and 438–439.
6. *Federal Register,* Washington, D.C., vol. 50, no. 135, pp. 28,711–28,717, July 15, 1985.

12.3.6 FURTHER READING

Boutwell, S. H., *Selection of Models for Remedial Action Assessment,* U.S. EPA, Washington, D.C., 1984.
Cohen, R. M., and W. J. Miller III, "Use of Analytical Models for Evaluating Corrective Actions at Hazardous Waste Sites," *Proceedings of the 3rd National Symposium on Aquifer Restoration and Groundwater Monitoring,* National Water Well Association, Dublin, Ohio, May 1983.
Cole, C. R., R. W. Bond, S. M. Brown, and G. W. Dawson, *Demonstration/Application of Groundwater Technology for Evaluation of Remedial Action Alternatives,* U.S. EPA, Washington, D.C., 1983.
Ford, P. J., P. J. Turina, and D. E. Seely, *Characterization of Hazardous Waste Sites—A Methods Manual,* vol. II, *Available Sampling Methods,* EPA 600/4-83-040, NTIS No. PB84-126920, U.S. EPA, Springfield, Virginia, 1983.
Porcella, B., *Protocols for Bioassessments of Hazardous Waste Sites,* EPA 600/23-83-054 and PB-83-241737, U.S. EPA, Springfield, Virginia, 1983.
Repa, E., and C. Kufs, *Leachate Plume Management,* EPA 540/2-85/004, Cincinnati, Ohio, November 1985.

Shuckrow, A. J., A. P. Pajak, and D. J. Towhill, *Management of Hazardous Waste Leachate,* SW-871, U.S. EPA, Cincinnati, Ohio, 1983.

U.S. Environmental Protection Agency, *Characterization of Hazardous Waste Sites—A Methods Manual,* vol. III, *Available Laboratory Analytical Methods,* EPA-600/4-84-038, NTS No. PB84-191048, U.S. EPA, Springfield, Virginia, 1984.

U.S. Environmental Protection Agency, *Compatibility of Grouts with Hazardous Wastes,* EPA 600/2-84-015, U.S. EPA, Cincinnati, Ohio, 1984.

U.S. Environmental Protection Agency, *Design and Development of a Hazardous Waste Reactivity Testing Protocol,* EPA 600/2-84/057, U.S. EPA, Washington, D.C., 1984.

U.S. Environmental Protection Agency, *EPA Guide for Identifying Cleanup Alternatives at Hazardous Waste Sites and Spills: Biological Treatment,* EPA-600/3-83-063, U.S. EPA, Springfield, Virginia, 1983.

U.S. Environmental Protection Agency, *Geophysical Techniques for Sensing Buried Wastes and Waste Migration,* EPA 600/7-84-064, U.S. EPA, Springfield, Virginia, 1984.

U.S. Environmental Protection Agency, *Guide for Decontaminating Buildings, Structures and Equipment at Superfund Sites,* EPA 600/2-85-028, U.S. EPA, Cincinnati, Ohio, 1985.

U.S. Environmental Protection Agency, *Handbook for Evaluating Remedial Action Technology Plans,* EPA 600/2-83-076, U.S. EPA, Cincinnati, Ohio, 1983.

U.S. Environmental Protection Agency, *Handbook: Remedial Action at Waste Disposal Sites (Revised),* EPA/625/6-85/006, U.S. EPA, Cincinnati, Ohio, October 1985.

U.S. Environmental Protection Agency, *Protocol for Soil Sampling; Techniques and Strategy,* EPA-600/54-83-002, U.S. EPA, Springfield, Virginia, 1983.

U.S. Environmental Protection Agency, *Modeling Remedial Actions at Uncontrolled Hazardous Waste Sites,* EPA 540/2-85-001, U.S. EPA, Springfield, Virginia, 1985.

U.S. Environmental Protection Agency, *NEIC Manual for Groundwater/Subsurface Investigations at Hazardous Waste Sites,* EPA 330/9-81-0021, U.S. EPA, Denver, Colorado, 1981.

Remedial Response at Hazardous Waste Sites, Summary Report, EPA 540/2-84-002a, *Case Studies,* EPA 540/2-84-002b, U.S. EPA, Cincinnati, Ohio, 1984.

U.S. Environmental Protection Agency, *Review of In-Place Treatment Techniques for Contaminants in Surface Soils,* vol. 1, *Technical Evaluation,* EPA 540/2-84-003a, and vol. 2, *Background Information for In-Situ Treatment,* EPA 540/2-84-003b, U.S. EPA, Cincinnati, Ohio, 1984.

U.S. Environmental Protection Agency, *Slurry Trench Construction for Pollution Migration Control,* EPA 540/2-84-001, U.S. EPA, Cincinnati, Ohio, 1984.

U.S. Environmental Protection Agency, *Vadose Zone Monitoring for Hazardous Waste Sites,* EPA 600/X-83-064, U.S. EPA, Springfield, Virginia, 1983.

CHAPTER 13

SAMPLING AND ANALYSIS TECHNIQUES FOR HAZARDOUS WASTES

CHAPTER 13
SAMPLING AND ANALYSIS OF HAZARDOUS WASTES

Larry D. Johnson
U.S. Environmental Protection Agency
Environmental Monitoring Systems Laboratory
Research Triangle Park, North Carolina

Ruby H. James
Southern Research Institute
Birmingham, Alabam

Early attempts to cope with environmental pollution problems generally dealt with one or two specific substances. Flue-gas emissions were primarily studied and regulated by the classical "SO_X, NO_X, and rocks" approach. As awareness of potentially harmful effects of more and more materials grew, research studies began to attempt to characterize larger numbers of pollutants from a variety of sources. This required development of sampling and analysis methods with broader scope and the ability to produce information as cost-effectively as possible.[1,2,3] One of the first regulatory programs to deal with increased numbers of pollutants was the U.S. Environmental Protection Agency's (EPA's) Effluent Guideline Program. It was necessary to develop and validate new sampling and analysis procedures to support that program.[4,5,6]

Today, regulators, the regulated community, and researchers alike are still in the process of adjusting their thinking and procedures to hazardous-waste problems involving hundreds of different compounds in solid, liquid, gaseous, or mixed media. Major programs to develop and validate sampling and analysis methods have been started and are still under way. Many of the methods required are still state-of-the-art technology and require considerable expertise on the part of those who apply them. Both sampling programs and the subsequent analysis efforts are often complex, difficult, and expensive. Success depends on careful planning and execution as well as thorough knowledge of the field. This chapter is an overview and guide to the very complicated field of sampling and analysis of hazardous waste and related products. The references have been selected because of their general utility and because they may serve effectively to guide the interested reader to more detailed literature.

13.1 SAMPLING OF HAZARDOUS WASTE AND RELATED PRODUCTS

Sampling of hazardous waste may include a number of very diverse activities. It may be necessary to sample waste contained in tanks or drums, in ponds, in piles, or from various processing or transporting equipment such as conveyor belts. In addition, it may be necessary to sample diluted or transformed waste following leaching, spills, or various forms of treatment technology.

Two of the major tasks related to sampling of wastes in any form are planning of the sampling strategy and selection of the detailed tactics necessary for support of the strategy. The sampling strategy must be consistent with the goals of the overall project, and must include selection of major waste impoundments, containers, or streams which must be characterized. In addition to selection of these critical elements for characterization, a decision is always necessary as to the proper degree of sampling resolution.

For example, one might choose to characterize liquid in a waste-storage tank, flue gas from an incinerator stack, and water in a scrubber-effluent holding pond. It is then necessary to decide whether one needs an average value for each component of interest in each of these wastes, or whether more detail is needed. It might be necessary to know whether the distribution of materials within each of these units is homogeneous, and the spatial variation within the unit.

Once the sampling strategy has been developed, it is possible to begin planning detailed tactics. This planning of tactics includes decisions concerning the number of replicate samples to be taken, whether to combine samples into composites or analyze separately, selection of sampling methods or hardware, selection of sample-packing and shipping methods, and many other seemingly innocuous details. In truth, each of these details may be critical to the success of the program's goals, and must be given careful thought and consideration.

Sampling Hazardous Waste

The few basic references in this field invariably point out that hazardous wastes may be complex, multiphase mixtures with a great variety of physical and chemical properties. As mentioned earlier, the waste itself may be contained in a wide variety of vessels or in ponds or spread throughout sizable areas of soil. Because of all these possibilities, it is impossible to present one standard protocol for the sampling of hazardous waste. Each project will require considerable planning and tailoring of the sampling approach to meet the overall objectives.

One thing that soon becomes evident to the reader in this field is that the strategies and tactics of planning and carrying out the sampling are a great deal more complex than the equipment with which the samples are usually taken. The equipment itself is usually exceptionally simple and inexpensive, especially when compared with the stack-sampling hardware to be discussed later.

The methods of selecting the sampling grid, the number of replicates needed, and the expected results in terms of precision and accuracy, draw heavily on statistics and are quite difficult to discuss in a general overview such as this. Some of the basic procedures often used are simple random sampling, stratified random sampling, systematic random sampling, authoritative sampling, and composite sampling. In a *simple random sampling,* all locations in the batch of waste are identified, and a suitable number of these are randomly selected for sampling. *Stratified random sampling* differs in that the waste is known to be randomly heterogeneous. In this case, the population of locations may be stratified to isolate the source of nonrandom distribution, and each stratum may be sampled by simple random sampling. In *systematic random sampling,* the first unit is sampled randomly, and all others are selected at fixed intervals from the first. These intervals may be spatial or time intervals. *Authoritative sampling* must be based on a thorough knowledge of the waste. Samples are selected by an individual and are tailored to be consistent with the overall sampling strategy as well as the known distribution of chemical impurities. This procedure is very seldom recommended for hazardous waste, and almost never in a regulatory situation.

Reference 7 points out that when little is known about the distribution of chemical pollutants in waste, then simple random sampling is usually the best choice. As one gains more information about the distribution, it may become possible to employ stratified random sampling, systematic random sampling, or even authoritative sampling. For certain applications such as plume definition, geostatistical sampling appears to be a very promising technique.[8] The references on waste sampling[7,8,9,10,11] contain more detailed discussion of the procedures involved.

Table 13.1 provides examples of sampling equipment commonly used in sampling waste in various containers or impoundments.

The illustrations of waste-sampling equipment at the end of this chapter (Figs. 13.1 through 13.5) and the short descriptions of various waste-sampling devices given here are transferred from Ref. 7 and serve to demonstrate the nature and uses of some of the more common hardware employed.

Composite Liquid-Waste Sampler (Coliwasa). The Coliwasa (Fig. 13.1) is a device employed to sample free-flowing liquids and slurries contained in drums, shallow tanks, pits, and similar containers. It is especially useful for sampling wastes that consist of several immiscible liquid phases.

The Coliwasa consists of a glass, plastic, or metal tube equipped with an end closure that can be opened and closed while the tube is submerged in the material to be sampled.

Weighted Bottle. This sampler (Fig. 13.2) consists of a glass or plastic bottle, sinker, stopper, and a line that is used to lower, raise, and open the bottle. The weighted bottle samples liquids and free-flowing slurries. A weighted bottle with line is built to the specifications in ASTM Methods D270 and E300.

Dipper. The dipper (Fig. 13.3) consists of a glass or plastic beaker clamped to the end of a two- or three-piece telescoping aluminum or fiberglass pole that serves as the handle. A dipper samples liquids and free-flowing slurries. Dippers are not available commercially and must be fabricated.

Thief. A thief (Fig. 13.4) consists of two slotted concentric tubes, usually made of stainless steel or brass. The outer tube has a conical pointed tip that permits the sampler to penetrate the material being sampled. The inner tube is rotated to open and close the sampler. A thief is used to sample dry granules or powdered

TABLE 13.1 Examples of Sampling Equipment for Particular Waste Types

					Waste location or container				
Waste type	Drum	Sacks and bags	Open-bed truck	Closed-bed truck	Storage tanks or bins	Waste piles	Ponds, lagoons, & pits	Conveyor belt	Pipe
Free-flowing liquids and slurries	Coliwasa	N/A	N/A	Coliwasa	Weighted bottle	N/A	Dipper	N/A	Dipper
Sludges	Trier	N/A	Trier	Trier	Trier	*	*	*	
Moist powders or granules	Trier	Trier	Trier	Trier	Trier	Trier	Trier	Shovel	Dipper
Dry powders or granules	Thief	Thief	Thief	Thief	*	Thief	Thief	Shovel	Dipper
Sand or packed powders and granules	Auger	Auger	Auger	Auger	Thief	Thief	*	Dipper	Dipper
Large-grained solids	Large trier	Large trier	Large trier	Large trier	Large trier	Large trier	Large trier	Trier	Dipper

*This type of sampling situation can present significant logistical sampling problems, and sampling equipment must be specifically selected or designed based on site and waste conditions. No general statement about appropriate sampling equipment can be made.

Source: Ref. 7.

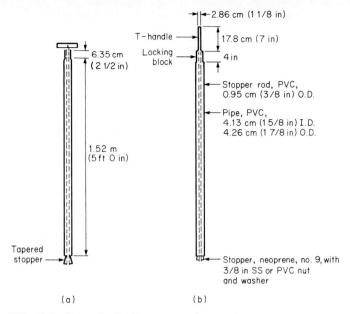

FIG. 13.1 Composite liquid-waste sampler. (*a*) Sampling position. (*b*) Closed position. (Coliwasa).

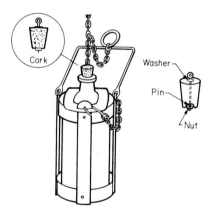

FIG. 13.2 Weighted-bottle sampler.

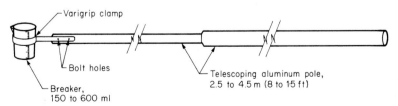

FIG. 13.3 Dipper.

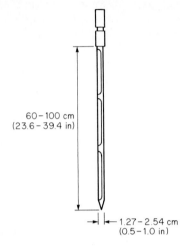

60 – 100 cm
(23.6 – 39.4 in)

1.27 – 2.54 cm
(0.5 – 1.0 in)

FIG. 13.4   Thief sampler.

wastes whose particle diameter is less than one-third the width of the slots. A thief is available at laboratory supply stores.

Trier.   A trier (Fig. 13.5) consists of a tube cut in half lengthwise with a sharpened tip that allows the sampler to cut into sticky solids and to loosen soil. A trier samples moist or sticky solids with a particle diameter less than one-half the diameter of the trier. Triers 61 to 100 cm long and 1.27 to 2.54 cm in diameter are available at laboratory supply stores. A large trier can be fabricated.

Auger.   An auger consists of sharpened spiral blades attached to a hard metal central shaft. An auger samples hard or packed solid wastes or soil. Augers are available at hardware and laboratory supply stores.

Scoops and Shovels.   Scoops and shovels are used to sample granular or powdered material in bins, shallow containers, and conveyor belts. Scoops are available at laboratory supply houses. Flat-nosed shovels are available at hardware stores.

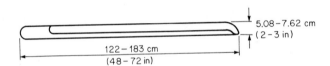

5.08 – 7.62 cm
(2 – 3 in)

122 – 183 cm
(48 – 72 in)

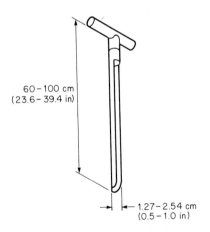

60 – 100 cm
(23.6 – 39.4 in)

1.27 – 2.54 cm
(0.5 – 1.0 in)

FIG. 13.5   Sampling triers.

Bailer.   The bailer is employed for sampling well water. It consists of a container attached to a cable that is lowered into the well to retrieve a sample. Bailers can be of various designs. The simplest is a weighted bottle or basally capped length of pipe that fills from the top as it is lowered into the well. Some bailers have a check valve, located at the base, which allows water to enter from the bottom as it is lowered into the well. When the bailer is lifted, the check valve closes, allowing water in the bailer to be brought to the surface. More sophisticated bailers are available that remain open at both ends while being lowered, but can be sealed at both top and bottom by activating a triggering mechanism from the surface. This allows more reliable sampling at discrete depths within a well. Perhaps the best-known bailer of this latter design is the Kemmerer sampler.

Bailers generally provide an excellent means for collecting samples from monitoring wells. They can be constructed from a wide variety of materials compatible with the parameter of interest. Because they are relatively inexpensive, bailers can be easily dedicated to an individual well to minimize cross contamination during sampling. If not dedicated to a well, they can be easily cleaned to prevent cross contamination. Unfortunately, bailers are frequently not suited for well evacuation because of their small volume.

Suction Pumps.   As the name implies, suction pumps operate by creating a partial vacuum in a sampling tube. This vacuum allows the pressure exerted by the atmosphere on the water in the well to force water up the tube to the surface. Accordingly, these pumps are located at the surface and require only that a transmission tube be lowered into the well. Unfortunately, their use is limited by their reliance on suction to depths of 6 m to 7 m (20 to 25 ft), depending on the pump. In addition, their use may result in out-gassing of dissolved gases or volatile organics and is therefore limited in many sampling applications. In spite of this, suction methods may provide a suitable means for well evacuation because the water remaining in the well is left reasonably undisturbed.

Positive Displacement Pumps.   A variety of positive displacement pumps is available for use in withdrawing water from wells. These methods utilize some pumping mechanism placed in the well that forces water from the bottom of the well to the surface by some means of positive displacement. This minimizes the potential for aerating or stripping volatile organics from the sample during removal from the well. Reference 7 includes further details about pumps as well as descriptions of vacuum extractors, pressure-vacuum lysimeters, and trench lysimeters. The latter three types of devices may be useful for sampling surface water and groundwater.

Reference 11 indicates that brushes and vacuum cleaners are both of utility when sampling dust from hard surfaces.

It is not particularly desirable, and probably not even possible, to list all possible tools or devices that may be used for sampling hazardous waste. As stated earlier, the waste itself may exhibit a great variety of properties and may be contained in a wide variety of vessels or impoundments. Performing the sampling properly may call for ingenuity as well as knowledge.

Sampling Waste-Combustion Products

The increasing importance of combustion as a disposal technique for hazardous waste has resulted in considerable interest in methods for sampling flue-gas emissions. In addition to incineration, waste may be destroyed by cofiring it in industrial boilers or other units such as lime kilns.

Sampling equipment for flue gases is usually more complicated than that utilized for waste itself, or even for soil and groundwater. This is because the flue gas is a multiphase system and is usually at an elevated temperature. Since any of the phases may contain emissions of interest, and many of the combustion products interfere with collection of others, it is necessary to perform filtration, cooling, and various forms of solvent scrubbing or sorption to solid substrates. It is always necessary to measure gas volumes sampled, and this is inherently more difficult and cumbersome than measurement of volumes or weights of liquid or solid samples.

A wide variety of sampling devices has been utilized over the years but only a few will be discussed here. These pieces of equipment are widely used, versatile, commercially available, and their performances are generally reasonably well known for the more common pollutants. Readers interested in more specialized or more exotic sampling approaches can easily locate them in the literature with the aid of the general references given.

In order to ensure that a representative sample is obtained, it is usual practice to move the sampling probe from place to place in the stack according to a prearranged plan. Details as to design of the sampling pattern are given in EPA Method 1.[12] This practice is called *traversing the stack* and compensates for stratification of material in the stack.

Another important concept is that of *isokinetic sampling*. If the velocity of the sample gas drawn into the probe is closely matched to that passing past the probe, the sampling is said to be *isokinetic*. A mismatch in these velocities results in *anisokinetic* sampling. Inertial effects during anisokinetic sampling may result in overcollection or undercollection of particulate matter, depending upon its size and mass. This effect does not occur for gases and is generally of marginal importance for particulate material below 2 mm in diameter.

The Source Assessment Sampling System (SASS) (Fig. 13.6) was developed for environmental assessment programs and is still the train of choice when large amounts of samples are necessary for extended chemical analysis or biotesting. The SASS includes cyclones for particle sizing, a glass- or quartz-fiber filter for fine-particle collection, a sorbent module for collection of semivolatile organics, and impingers for collection of volatile metals. The SASS operates at 110 to 140 L/min (4 to 5 cfm) and is usually operated long enough to collect 30 m^3 of flue gas.

The large size of the SASS makes traversing inconvenient but not impossible unless precluded by physical arrangements at the sampling site. The particle-sizing cyclones require constant gas flow through them for proper operation, and thus limit flow adjustments necessary for true isokinetic sampling. The SASS has been operated, without the cyclones, in the full traverse and isokinetic sampling mode, but the difficulty of this option makes it unattractive. It is usually operated at a single point in the stack under pseudo-isokinetic conditions. Under most circumstances, this mode of operation results in samples that are indistinguishable from those taken under true isokinetic conditions. Since there is always some finite chance that the collection will be less quantitative or representative, samples taken in this manner are somewhat less defensible. For further details on this subject, see Refs. 13 and 14.

Potential corrosion of stainless steel in the sorbent module of the SASS has prompted development of glass sorbent modules which appear to perform adequately. One of these is described in Refs. 7 and 15, which also give detailed in-

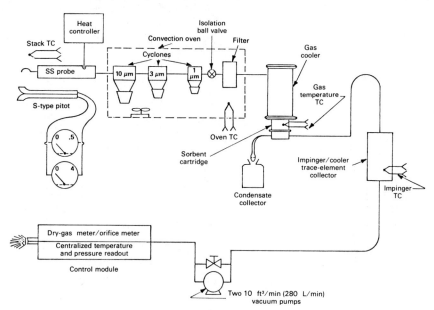

FIG. 13.6 Source Assessment Sampling System (SASS). TC = temperature control.

formation concerning construction and operation of both the SASS and the MM5 trains.

The Modified Method Five train (MM5), or Semi-VOST, is conceptually very similar to the SASS but operates at a lower flow rate, usually 14 to 28 L/min (0.5 to 1 cfm). The MM5, shown in Fig. 13.7, does not include particle-sizing cyclones and is usually constructed of glass rather than stainless steel. The MM5 results from a very simple modification of any of the commercial sampling trains available which conform to the requirements of EPA Method 5. The sorbent module with cooling capability is simply inserted between the filter and the first impinger. The sorbent module must be positioned vertically so the gas and any condensed liquids flow downward through it.

The sorbent of choice for most sampling jobs for both the SASS and MM5 is XAD-2. For further discussion of the reasons behind this choice, as well as sorbent-module placement in the train, see Ref. 14.

Because of its more convenient size, glass construction, and ready availability, the MM5 is usually chosen over the SASS for incinerator sampling unless larger samples are needed for lower detection limits or extensive analysis requirements.

Either the SASS or MM5 provides collection ability for particulate material, acid gases such as HCl, gaseous metal compounds (if appropriate collection liquids are chosen), medium-boiling organics (b.p. 100 to 300°C), and high boiling organics (b.p. greater than 300°C). Organics with boiling points between 100 and 120°C require individual attention during the sampling planning stage and may require decreased sampling times to prevent volumetric breakthrough. Volumetric breakthrough is related to the migration rate of sorbed material through unsaturated sorbent beds. For further discussion of this important concept, see Ref. 14.

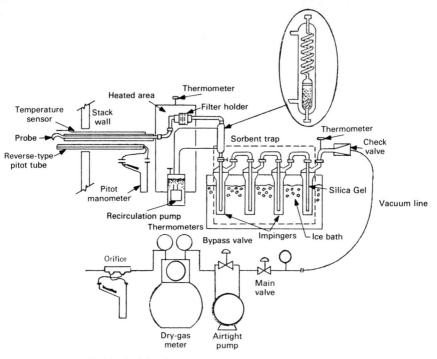

FIG. 13.7   Modified Method 5 train.

Considerable field experience has been gained with the SASS and MM5 trains. In recent years, much of the confidence in these trains' abilities to collect organics rested on knowledge of the behavior of sorbents with respect to collection and recovery. Reference 15 discusses this point in more detail. Recent validation studies for selected compounds show excellent collection and recovery and reinforce the general conclusion that the trains are quite effective.[16,17,18] The MM5 and SASS are not generally quantitative collection trains for organics with b.p. less than 100°C. For these low-boiling compounds, the recommended methods are plastic sampling bags, glass sampling bulbs, or the newly developed volatile-organic sampling train (VOST). The ambient air at incinerator sites may exhibit relatively high levels of volatile organics. This greatly increases the difficulty of obtaining an uncontaminated sample of low-concentration volatile compounds from the stack. All sampling methods for low levels of volatiles are subject to potential contamination and require a great deal of care as well as adequate blanks. All of the above methods have shortcomings, but they are much less severe than the shortcomings and limitations of alternate approaches.[12]

The VOST is shown in Fig. 13.8. This train and the analysis approach applied to the samples resulting from its use were developed in order to address stack concentrations as low as 0.1 ng/L. A sorbent tube containing 1.6 g of Tenax is positioned early in the train in order to remove organics from the gas and liquid stream as soon as possible. A second sorbent tube containing 1 g of Tenax and 1 g of charcoal follows the condensate collector in order to act as a backup in case of breakthrough. The charcoal provides added stopping power for the very low boilers such as vinyl chloride. The train was designed to use six pairs of sorbent

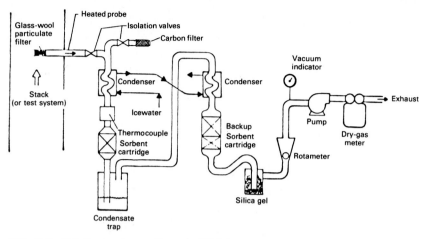

FIG. 13.8 Volatile-organic sampling train (VOST).

tubes sequentially, each operating for 20 min at 1 L/min. The ability to concentrate the organics from all six sets of tubes onto one analytical tube and subsequently heat-desorb into a GC or GC/MS makes it possible to achieve very low levels relative to the stack-gas concentration. For higher stack-gas concentrations, it is possible, and sometimes necessary, to operate the VOST at lower flow rates and longer sampling times. There is no fundamental reason why this piece of equipment should not be useful in a number of operation modes as long as excessive volumes of gas are not pulled through a single tube, since low boilers break through the sorbent after relatively low volumes.

The development and use of the VOST has been discussed in several publications;[7,14,19,20,21] Refs. 7 and 20 are particularly useful to users of the train. Preliminary evidence that the train is effective is given in these references, and an EPA validation project further supports that position.[22,23]

Various types of plastic sampling bags have been used in the past with very mixed results. It is quite possible to get good results with this approach, but it is essential that the sampling and storage characteristics of the specific compound, relative to the specific types of bags to be used, be well-known. For example, organics such as alcohols usually exhibit poor storage characteristics in bags.[24] It is also essential that field blanks be included in the sampling strategy, since all known methods for volatile organics from incinerators are subject to potentially severe contamination in the field and during transit.

When sampling relatively high concentrations of volatiles, glass sampling bulbs with secure seals may be the best choice for certain types of projects. Although somewhat inconvenient and lacking in sample-concentrating ability, the glass bulbs do show better sample-storage characteristics than plastic bags.[24] Because of the lack of integration capability of glass bulbs, they may not be appropriate for regulatory purposes unless taken in great numbers.

In addition to the methods discussed, special procedures may be necessary for certain pollutants. Formaldehyde is an example of a compound that requires special handling.[25] Reference 10 gives guidance in such cases. A number of other less satisfactory general sampling approaches exist, but have not been recommended. For further discussion of these techniques and their shortcomings, see Ref. 13.

Sampling methods have been discussed which are generally applicable to incineration and to processes closely related to incineration, such as cofiring of waste in industrial boilers and burning of contaminated heating oil. Although some of the methods are relatively new and all require a great deal of care and attention, it is possible to produce excellent results through their application.

It should be kept in mind that obtaining a representative sample from the stack, particularly in the case of organic emissions, is a complicated and technically difficult process. The most consistent and well-defined result will be obtained by following detailed, written, validated procedures such as those in Ref. 7. The tendency to modify such procedures or replace them with new ones should generally be resisted.

Sampling Solid and Liquid Effluents from Combustors and Control Devices

The methods and equipment previously discussed for use with waste often will be applicable.[7,10] In addition, it will sometimes be necessary to take liquid samples from taps. In certain instances, sampling will be difficult to accomplish and may require special operation of the equipment. Sampling of bottom ash from combustors is often difficult to carry out and particularly difficult to relate to a given charge of waste. Sometimes it is possible to accomplish this end with pilot units by shutting down operation and collecting the ash from an entire run. Similar operations have been carried out to obtain boiler-tube soot for an entire batch of cofired waste.

Sample Containers, Shipping, and Storage

It is important to stress that every step in the sampling and analysis chain of events is critical and cannot be carried out carelessly without detriment to the final result. Container-selection guidance is given in the references[7,10] and usually recommends glass containers protected against light and excessive heat. Glass is usually recommended because it is relatively inert and easy to clean thoroughly. In certain circumstances, plastic containers may be preferable and are recommended. The containers must also have secure seals to prevent both sample loss and contamination.

Sample-preservation techniques are discussed in the references[7,26,27] and are specific to the various methods to be used for analysis. Organic pollutants are generally not treated by additives for preservation purposes, but are cooled and protected from light.

Shipping must always be done in compliance with U.S. Department of Transportation regulations. In addition, thought must be given to keeping delays before analysis within acceptable limits and to potential for degradation of the samples. For example, it is not acceptable to transport samples from a VOST in the same truck compartment with organic-solvent cans used for spiking waste or for field cleanup operations.

Storage conditions after receipt of the samples from the field are just as important as during shipping, and are subject to many of the same guidelines. Organics are often refrigerated and are always protected from light. Other samples should also be treated with care and not exposed to extreme conditions or high levels of chemical contamination.

Quality Control and Quality Assurance

Although quality-control and quality-assurance (QA/QC) procedures and planning are more often associated with analysis than with sampling, it is still important not to lose sight of these crucial procedures during the sampling phase. These are important aspects of the sampling planning operation and the implementation as well. The use of field blanks, field spiking procedures, and other aspects of QC have a great deal of influence on field sampling operations. The references[7,10] have detailed discussions of this important subject. Audit procedures and samples have been developed for some of the sampling methods[28,29] and their use is highly recommended even when not actually required.

Safety

Safety of the personnel involved is the most important consideration in any sampling activity. The hazardous nature of the waste involved may mean that special clothing or equipment such as respirators is required. Once again, several of the references have discussions of this subject and more detailed information is available.[7,9,10] Efforts should also be made to ensure that safety procedures are not violated during field sampling, since the desire to get on with the job is sometimes contrary to the sampling team's own better judgment. Perhaps the most essential part of the safety program dealing with hazardous-waste-related sampling is a thorough and effective personnel training program. This should include general training in procedures as well as possible risks to be encountered. In addition, a specific, safety-oriented discussion should be carried out with the field crew prior to each sampling trip. This discussion or meeting should deal with specific wastes and hazards that are likely to be encountered at this particular site.

13.2 ANALYSIS OF HAZARDOUS WASTE AND RELATED PRODUCTS

Background

Methods of measuring and defining toxic and hazardous emissions in the environment have been developed as a result of the Toxic Substances Control Act (TSCA) the Resource Conservation and Recovery Act (RCRA), and the efforts of the EPA and industry. The term *hazardous substance* has a very specific meaning according to the Comprehensive Environmental Response Compensation and Liability Act (CERCLA). Section 101(14) of CERCLA defines a hazardous substance as

1. Any substance designated pursuant to Section 311(b) of the Clean Water Act (CWA)
2. Any hazardous waste having characteristics identified under or listed pursuant to Section 3001 of the Solid Waste Disposal Act, otherwise known as the Resource Conservation and Recovery Act (RCRA)
3. Any toxic pollutant listed under Section 307(a) of the CWA
4. Any hazardous air pollutant listed under Section 112 of the Clean Air Act (CAA)

5. Any imminently hazardous chemical substance or mixture with respect to which the Administrator of EPA has taken action pursuant to Section 7 of the Toxic Substances Control Act (TSCA)

6. Any element, compound, mixture, solution, or substance the Administrator determines to be hazardous pursuant to Section 102 of CERCLA.

There are currently 717 hazardous substances (HS) composed of 611 unique chemical compounds and 106 waste streams.[33] This total does not include chemicals or mixtures that exhibit characteristics of ignitability, corrosivity, reactivity, or toxicity according to 40 *CFR* 261.20. The toxic and hazardous category contains many different substances; however, organic compounds are the most complex and difficult to measure. The selection of appropriate analytical methods depends on a number of important considerations. These include the compounds of interest, the type of waste, the source type, instrument selectivity and sensitivity, and the desired level of detection. Consideration of cost, though secondary, as well as the intended use of the data, can be equally important in the selection process. The complex nature of organic compounds dictates the complexity of the analytical methods. Also, organic substances can enter the environment via gaseous, liquid, or solid media. Although each medium may require different sampling techniques (see Subsec. 13.1) and sample-preparation techniques, the analytical methods for most samples contain many steps that are essentially the same.

This section attempts to address the considerations that are most important in the selection of analytical methods for the analysis and management of hazardous waste. Each step of the various processes, from the testing of a waste to the disposal of the waste must be monitored to ensure that each process is environmentally sound. The analytical methods applicable to the analysis of different types of hazardous waste are described or summarized here and in the tables at the end of the chapter. A brief description of the technology and a summary of each method is included with references.

Analytical Methods

Because of the complex nature of organic hazardous waste, no single analytical technique is applicable to all wastes. For most samples, the analyst should employ proven analytical techniques that are selective and sensitive. The most universal methods for trace analysis are gas chromatography (GC), liquid chromatography (LC), mass spectrometry (MS), and a combination of the above methods. More specifically these include gas chromatography and mass spectrometry (GC/MS), liquid chromatography and mass spectrometry (LC/MS).

The minimum detectable quantity (MDQ) for GC depends on the detector used and ranges from 10^{-9} to 10^{-13} g per injection.[34] Samples must be volatile (at least 20 torr at 300°C) but can be gases, liquids, or solids. High-molecular-weight compounds, some ionic compounds or highly polar compounds, and thermally unstable compounds cannot be analyzed directly by GC. In some cases derivatization or pyrolysis techniques can extend the useful range of the GC. Gas chromatographs are widely used. They are (1) not expensive, (2) easy to operate, and (3) give excellent quantitative results. However, GC can confirm the identity of a substance only by retention time. This is a major concern when complex waste samples are analyzed and several components could have the same retention time. The MS in a total ion current (TIC) scanning mode may be used to identify the presence of a compound. If the identity of a compound is known, selected ion monitoring (SIM) can be used for quantitation to extend the MDQ and for confirmation and quantitation. The sensitivity of the MS ranges from 10^{-9} to 10^{-12} g. Samples that can be analyzed by GC can be analyzed by MS. The GC/MS is a universal technique, handling all sample types, and it is specific in that it will usually confirm the structure (identity) of a compound. The GC/MS is preferred for analysis of trace organics but it must be noted that the instrument, though readily available, is moderately expensive and requires an experienced operator. It has been estimated that approximately 20% of all organic compounds can be determined by GC/MS.[35] Fortunately a much higher percentage ($\cong 80\%$) of those compounds currently of environmental concern are amenable to GC/MS analyses.

Liquid chromatography is recommended for those compounds not amenable to gas chromatography. The application of high-performance liquid chromatography (HPLC) to the analysis of waste samples has increased significantly. The high-performance liquid chromatograph can be used for the identification of a wide range of less volatile compounds not amenable to GC. Chemically bonded stationary phases for HPLC have greatly improved the separation of a wide variety of compounds. Samples can be liquid or solid, organic or inorganic, and range in molecular weight from 18 to 6 million.[34] Selectivity is dependent upon sample type and the detector used. The ultraviolet and visible (UV/VIS) detector is considered the most universal and can measure 10^{-9} g of most species. Aliphatic hydrocarbons are a notable exception as the UV/VIS detector is less sensitive to this class of compounds. This lack of sensitivity for hydrocarbons can frequently be used to advantage as most environmental samples contain a higher percentage of such compounds than of the analyte of interest. Other LC detectors are more selective and sensitive. For example, for selected compounds the fluorescence detector can measure 10^{-12} g, and the electrochemical detector can measure 10^{-10} g. Although not yet widely used in waste analyses, the combination of LC and MS allows separation of complex mixtures and extends the selectivity of the MS to a large class of organic compounds not amenable to GC/MS analysis. Table 13.2 summarizes the instrumental techniques widely used for trace analysis.

Reliable analytical measurements of environmental samples are an essential ingredient for making sound decisions involving many facets of society including advances in technology, safeguarding the public health, and improving the quality of the environment. The American Chemical Society's Committee of Environmental Improvement (CEI) charged its subcommittee on Environmental Monitoring and Analysis with the task of developing a set of guidelines to improve the quality of environmental analytical measurements.[36] These guidelines aid in the evaluation of analytical measurements and in the intelligent choice of methods that meet the requirements of a specific measurement. Analytical objectives often require the measurement of parts per million in hazardous-waste samples and parts per billion and even parts per trillion levels in samples of effluents from hazardous-waste combustion. Advances in analytical methodology continue to lower the levels of detection to meet these needs. Many factors are of critical importance at these levels and influence the outcome and reliability of environmental measurements. Good planning is essential to ensure that the results are valid and provide a basis on which a process or regulatory decision can be made. The analyst cannot assume that the person requesting an analysis will also be able to define the objectives of the analysis properly.

A protocol which describes the analytical process in detail should include a

TABLE 13.2   Techniques for Analysis of Trace Organics

Technique	Detector	Estimated sensitivity, g	Selectivity
GC*	Thermal-conductivity detector	10^{-9}	Universal
	Flame-ionization detector	10^{-11}	Universal
	Electron-capture detector	10^{-13}	Selective
	Nitrogen-phosphorus detector (nitrogen mode)	10^{-12}	Selective
	Nitrogen-phosphorus detector (phosphorus mode)	10^{-13}	Selective
	Flame-photometric detector (sulphur mode)	10^{-9}	Selective
	Flame-photometric detector (phosphorus mode)	10^{-11}	Selective
	Hall electrolytic conductivity	10^{-11}	Selective
	Photoionization detector	10^{-11}	Selective
LC†	Ultraviolet/visible	10^{-9}	Universal
	Electrochemical detector	10^{-10}	Selective
	Fluorescence detector	10^{-12}	Selective
MS‡	Total ion current	10^{-9}	Universal
	Selected ion monitoring	10^{-12}	Selective

*GC = gas chromatography.
†LC = liquid chromatography.
‡MS = mass spectrometry.

description of the quality-assurance and quality-control requirements; the sampling plan; and the analytical methods, calculations, and documentation and report requirements. If analytical data is to be used for a screening program or to adjust an operating parameter, then an unvalidated method may be adequate. If on the other hand, regulatory compliance is involved, a validated analytical method is usually required. The CEI recommended that a minimum of three different concentrations of calibration standards must be measured in triplicate. The concentrations of the calibration standards must bracket the concentration of the analyte in the sample. No quantitative data should be reported beyond the range of calibration of the methodology. Internal standards (reference materials) are frequently used for quantitation. Surrogates are frequently used for spiking to determine recovery efficiency during sample preparation. Internal standards are added prior to analyses. Both types of reference materials are chosen to simulate the analyte of interest.

Selection of a proper analytical method is one of the most important factors influencing the reliability of the resulting data. Measurements should be made with tested and documented procedures. Furthermore, each laboratory and analyst must evaluate the methodology using typical samples to demonstrate competence in the use of the measurement procedure.

In order to address the analyses of both hazardous waste and products from the combustion of hazardous waste some assumptions are necessary. Namely, the methods of collection and sample preparation are different but the methods of sample analysis (analytical finish) for the final extracts from the waste samples and combustion sources are similar. Tables 13.3 and 13.4 summarize by method number the methods applicable to the analyses of hazardous waste and hazardous-waste combustion products.[7,10]

Considerations Associated with Hazardous-Waste Analyses

A common expression in the environmental analytical laboratory is "garbage in–garbage out." The integrity of the analytical data is highly dependent on the sample submitted for instrumental analysis. While most instrumental "finishes" yield final data that are of known quality when standard solutions are analyzed, many areas of sample handling prior to analysis are more variable and matrix-dependent and are critical to the accuracy of the resulting data. After a sample has been obtained, the analytical protocol often requires one or more cleanup treatments prior to actual measurement of the analytes. Sample preparation may involve physical operations such as sieving, blending, crushing, drying, and phase separation and/or chemical operations such as dissolution, extraction, digestion, fractionation, derivatization, pH adjustment, and the addition of preservatives, surrogates, standards, or other materials. These physical and chemical treatments not only add complexity to the analytical process but are potential sources of contamination, mechanical loss, bias, and variance. Therefore, sample preparation should be planned carefully and documented in sufficient detail to provide a complete record of the sample history. Further, samples taken specifically to test the quality-assurance system (i.e., quality-assurance samples) should be subjected to these same preparation steps.

The analyst must recognize and be aware of the risks that are associated with each form of pretreatment and take appropriate preventative action for each. This may include reporting, correcting for, or possibly removing interferences from the analytes of interest by modifying the protocol. All changes in the protocol must be documented. In most cases, sample stabilization (either by pH adjustment or by quenching of dissolved chlorine, depending on the compounds present) is not necessary if the samples are to be extracted within 48 h of sampling. However, samples containing phenols and benzidine derivatives require immediate stabilization and samples containing metals and cyanides must be stabilized upon arrival at the laboratory. Both can be stabilized by pH adjustment. Samples are generally stored at 4°C. Purgeable samples should be analyzed within 7 days of collection or as requested in the protocol (7 to 14 days). Extractable organic compounds must be extracted within 7 days and analyzed within 30 days of collection. Metals and cyanides should be analyzed within 30 days.

Ideally, samples of source combustion emissions should be kept at low temperatures and analyzed very rapidly after collection to minimize losses of compounds by vaporization and reaction. Samples should be protected from direct light during and after collection. Most of the literature on source sampling has not provided information on sample preservation. Holding time and preservatives for water pollutants and other analytes have been published in the *Federal Register*. Details of sample-handling containers and preservation and holding times for hazardous wastes are documented in Chap. IV of SW 846.[7]

Many organic samples continue to be reactive after sampling. Therefore, a fresh sample may yield different values for the analyte than when the sample is stored. Because of the complexity of hazardous-waste samples and the limited selectivity of many analytical methodologies, interferences are frequently encountered during analysis. Controls to verify that interferents are not present

TABLE 13.3 Test Methods for Evaluation of Solid Wastes
(Physical and chemical methods from SW 846)

Procedure	SW 846 method no.
Waste-evaluation procedures	
Ignitability	1010:1020
Corrosivity	1110
Electrochemical corrosion method	1120
Extraction procedure toxicity	1310
Multiple extraction procedure	1320
Sample workup techniques	
Acid digestion procedure	
For flame atomic-absorption spectroscopy	3010
For furnace atomic-absorption spectroscopy	3020
Acid digestion of oils, greases, or waxes	3030
Digestion procedure for oils, greases, or waxes	3040
Acid digestion of sludges	3050
Alkaline digestion	3060
Separatory funnel liquid-liquid extraction	3510
Continuous liquid-liquid extraction	3520
Acid-base cleanup extraction	3530
Soxhlet extraction	3540
Sonication extraction	3550
Reverse-phase cartridge extraction	3560
Column cleanup of petroleum waste	3570
Protocol for analysis of sorbent cartridge from volatile-organic sampling train	3720
Sample introduction techniques	
Head space	5020
Purge and trap	5030

Procedure	SW 846 method no.
Multielemental inorganic analytical methods	
Inductively coupled plasma method	6010
Inorganic analytical methods	
Antimony	
AA, direct aspiration	7040
AA, graphite furnace	7041
Arsenic	
AA, furnace	7060
AA, gaseous hydride	7061
Barium	
AA, direct aspiration	7080
AA, furnace	7081
Beryllium	
AA, direct aspiration	7090
AA, furnace	7091
Cadmium	
AA, direct aspiration	7130
AA, furnace	7131
Chromium	
AA, direct aspiration	
AA, furnace	7191
Chromium, hexavalent	
Coprecipitation	7195
Colorimetric	7196
Chelation-extraction	7197
Differential-pulse polarography	7198
Copper	
AA, direct aspiration	7210
AA, furnace	

TABLE 13.3 Test Methods for Evaluation of Solid Wastes (*Continued*)
(*Physical and chemical methods from SW 846*)

Procedure	SW 846 method no.
Inorganic analytical methods (cont)	
Iron	
AA, direct aspiration	7380
AA, furnace	7381
Manganese	
AA, direct aspiration	7460
AA, furnace	7461
Mercury	
In liquid waste (manual cold-vapor technique)	7470
In solid or semisolid waste (manual cold-vapor technique)	7471
Nickel	
AA, direct aspiration	7520
AA, furnace	7551
Osmium	
AA, direct aspiration	7550
AA, furnace	7551
Selenium	
AA, furnace	7740
AA, gaseous hydride	7741
Silver	
AA, direct aspiration	7760
AA, furnace	7761
Sodium	
AA, direct aspiration	7770
Thallium	
AA, direct absorption	7841
AA, furnace	7841

Procedure	SW 846 method no.
Inorganic analytical methods (cont)	
Vanadium	
AA, direct aspiration	7910
AA, furnace	7911
Zinc	
AA, direct aspiration	7950
AA, furnace	7951
Organic analytical methods	
GC methods	
Halogenated volatile organics	8010
Nonhalogenated volatile organics	8015
Aromatic volatile organics	8020
Acrolein, acrylonitrile, acetonitrile	8030
Phenols	8040
Phthalate esters	8060
Organochlorine pesticides and PCBs	8080
Nitroaromatics and cyclic ketones	8090
Polynuclear aromatic hydrocarbons	8120
Organophosphorus pesticides	8140
Chlorinated herbicides	8150
GC/MS methods	
For volatile organics	8240
For semivolatile organics	
Packed-column technique	8250
Capillary column technique	8270
HPLC methods	
Polynuclear aromatic hydrocarbons	8310
Miscellaneous compounds	8320
Thioureas and other compounds	8330

TABLE 13.3 Test Methods for Evaluation of Solid Wastes (*Continued*)

(*Physical and chemical methods from SW 846*)

Procedure	SW 846 method no.	Procedure	SW 846 method no.
Organic analytical methods (Cont)		*Organic analytical methods (cont)*	
Formaldehyde		Soil pH	9045
Basic medium	8410	Specific conductance	9050
Acidic medium	8411	Total organic carbon	9060
Hierarchical analytical protocol for groundwater	8600	Phenolics	
Total aromatics by ultraviolet absorption	8610	Spectrophotometric, manual 4-AAP with distillation	9065
Total nitrogen-phosphorus gas-chromatographable compounds	8620	(colorimetric, automated 4-AAP with distillation)	9066
Derivatization procedure for Appendix VIII compounds	8630	(spectrophotometric, MBTH with distillation)	9067
		Oil and grease, total recoverable (gravimetric, separatory-funnel extraction)	9070
Miscellaneous analytical methods		Extraction method for sludge samples	9071
Total and amenable cyanide	9010	Cation-exchange capacity	
Method for the determination of photodegradable cyanides	9011	ammonium acetate	9080
Total organic halides (TOX)		sodium acetate	9081
By microcoulometric titration	9020	Compatibility test for wastes and membrane liners	9090
By neutron-activation analysis	9022	Saturated hydraulic conductivity, saturated leachate conductivity, and intrinsic permeability methods	9100
Sulfides	9030	Total coliform	
Sulfate		Multiple-tube permeation technique	9131
Colorimetric, automated, chloranilate	9035	Membrane filter technique	9132
Colorimetric, automated, methylthymol blue, AA II	9036	Nitrate 9200	9200
gravimetric	9037	Chloride	
turbidimetric	9038	Colorimetric, automated ferricyanide AA I	9250
pH measurement	9040	Colorimetric, automated ferricyanide AA II	9251
Paper method	9041	Titrimetric, mercuric nitrate	9252
		Grease alpha and grease beta	9310
		Alpha-emitting radium isotopes	9313
		Radium-228	9320

AA = Atomic-adsorption spectroscopy
AAP = 4-Aminoantipyrine
GC/MS = Gas chromatography/mass spectrometry

HPLC = High-performance liquid chromatography
MBTH = 3-Methyl-2-benzothiazolinone hydrazone

Source: Ref. 7.

TABLE 13.4 Sampling and Analysis Methods for Hazardous-Waste Combustion

Procedure	Method no.*
Sample preparation	
Representative aliquots from field samples	
Liquids (aqueous and organic)	P001
Sludges	P002
Solids	P003
Solvent extraction of organic compounds	
Aqueous liquids	P021
Sludges	P022
Organic liquids	P023
Solids	P024
Drying and concentrating of solvent extracts	P031
Digestion procedures for metals	P032
Sample-cleanup procedures	
Florisil	P041
Biobeads SC-3	P042
Silica gel	P043
Alumina	P044
Liquid-liquid extraction	P045
Analysis	
Ignitability	C001
Corrosivity	C002
Reactivity	C003
Extraction procedure toxicity	C004
Moisture, solid and ash content	A001–A002
Elemental composition	A003
TOC and total organic halogen	A004
Viscosity	A005
Heating value of the waste	A006
Survey analysis of organic content	
Total chromatographable organics	A011
Gravimetric value	A012
Volatiles	A013
Infrared	A014
Mass spectrometric	A015
GC/MS	A016
HPLC/IR or HPLC/LRMS	A017
Survey analysis of inorganics	
Analytical methods	
Volatiles	A101
VOST	A102
Extractables	A121
HPLC/UV	A122–A124
HPLC/fluorescence	A125
Aldehydes/ketones	A131
Carboxylic acids	A133
Oximes	A183
Organometallics	A191
Metals	A211–A245
Anions	A252–A254; A256
Gases	A141

*From Ref. 10.

must be used, and appropriate cleanup procedures should be included to eliminate the interferent.

In the analytical process, the recovery of analytes is influenced by such factors as concentration of the analyte, sample matrix, preservation time, and temperature of storage. In certain cases, compounds can be lost during extraction. Hexachlorocyclopentadiene has a strong tendency to be absorbed on glass, and can be lost during liquid-liquid extraction. The sampling medium can also affect the composition of organic emissions. Artifacts have also been reported when Tenax GC is used as the sorbent in source sampling.[37] There is also evidence for the decomposition of polycyclic aromatic hydrocarbons (PAHs) collected on Fiberglas filters commonly used in source-sampling systems. Teflon or Teflon-coated filters, however, appear to be relatively inert.[38] Other problems are often encountered in sample preparation. Sample contamination can occur from improperly cleaned glassware, improper storage and sample handling, impurities in solvents, or carrier-gas contamination; and cross-contamination can occur from analyzing high- and low-level samples in succession. If a high-level sample is run, a blank should be run to check for carry-over. Cross-contamination can also occur when volatile samples are prepared and analyzed in a laboratory or even in the same building where liquid-liquid extractions are performed.[9,39] Any volatile solvent in use in the laboratory is a potential candidate to contaminate volatile samples. Blank samples do not eliminate the problem but do help to identify the problem. Corrective action is frequently required before the analysis of samples can continue.

Emulsions and foaming may occur during sample preparation and analysis. Emulsions can be broken by stirring, filtration, centrifugation, cooling, or simply allowing them to stand for longer periods of time. Foaming can be reduced by antifoam agents or by specially designed purge vessels or in some instances by addition of cleaned glass melting-point capillary tubes to the purge vessel. Matrix effects can cause a wide variability in recoveries with organic compounds. Therefore, to be valid, recoveries of a spike standard must be determined in the same matrix as the sample.[39]

Analyses of Hazardous Waste

The overall strategy for analyzing hazardous waste includes procedures to determine characteristics of the waste and procedures to determine the composition of the waste. Test procedures, or supporting documentation, is required in each of four major areas:

- Waste characteristics
- Proximate analyses
- Survey analyses
- Directed analysis

Details of the procedures and recommendations for conducting these tests are given in *Test Method for Evaluating Solid Waste, Physical/Chemical Methods, (SW-846) Manual* (Ref. 7), and in the second edition of *Sampling and Analysis Methods for Hazardous Waste Combustion* (Ref. 10). An overview of the analytical approach for waste characterization is given in Fig. 13.9.

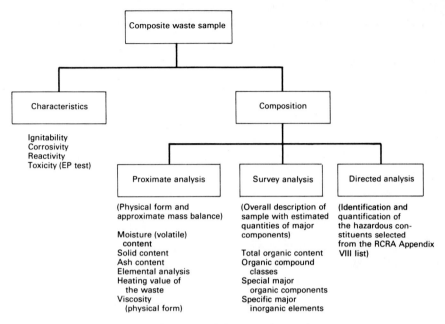

FIG. 13.9   Overview of the analytical approach for waste characterization.

Characteristics

Wastes are classified as hazardous if they exhibit any of the following characteristics:

- Toxicity
- Corrosivity
- Reactivity
- Ignitability

The extraction procedure (EP) toxicity test is currently used to determine if the waste exhibits the characteristic of toxicity. It involves a leaching procedure used to simulate the aqueous leaching of a toxic chemical when the waste is land-disposed. The EP toxicity characteristic has been used to set regulatory levels for eight metals, four pesticides, and two herbicides (National Interim Primary Drinking Water Standards have been established). Any waste that exceeds the EP toxicity thresholds (Table 13.5) is considered hazardous. Details of the EP toxicity procedure are given in Method 1310 in SW 846.[7] A proposed new rule would revise the EP toxicity to include regulatory levels of 38 additional compounds. The new toxicity characteristics leaching procedure (TCLP) expands the leaching to model the behavior of volatile and semivolatile organics. The complete waste evaluation for TCLP will require two extractions, one for volatile and semivolatile compounds and one for metals. Volatiles must be extracted in a zero head-space extractor. Contaminants regulated under the proposed TCLP and regulatory levels are given in Table 13.6.

TABLE 13.5   EPA Toxicity Threshold Levels

Contaminant	Threshold level, mg/L
Herbicides	
2,4-D	10.0
2,4,5-TP	1.0
Metals	
Arsenic	5.0
Barium	100.0
Cadmium	1.0
Chromium(VI)	5.0
Lead	5.0
Mercury	0.2
Selenium	1.0
Silver	5.0
Pesticides	
Endrin	0.02
Lindane	0.4
Methoxychlor	10.0
Toxaphene	0.5

The EPA has proposed the use of the same TCLP for the Land Disposal Restrictions Program under RCRA Subtitle C. The list of constituents considered hazardous for land disposal is given in Table 13.7. Procedures for the analysis of leachates are described here under *Directed Analyses,* and are based on standard methods established for priority pollutants. Methods for the determination of ignitability (Method 1010 and Method 1020), corrosivity (Method 1110), and reactivity are given in SW 846.[7]

Directed Analyses

The directed analyses allow quantitative measurement of designated principal organic hazardous constituents (POHCs) in a variety of samples. The compounds listed in 40 *CFR* Part 261, Appendix VIII, possess a broad spectrum of physical and chemical properties. Because of the high degree of complexity in the field of organic analyses and the parallel complexity of hazardous waste, the only suitable approach to quantitation and confident identification of specific compounds is by either gas or liquid chromatography. Each GC and LC detection method is based on a different operating principle and responds to different compound classes. The MS is a highly sophisticated detector and when coupled with the GC/MS it identifies and quantifies individual compounds by their mass spectra. Analytical methods using these techniques for hazardous waste have evolved from the experience profile of priority-pollutant analysis. Some of the EPA "600" methods shown in Table 13.8 have been adapted to hazardous-waste analysis. Methods 624 and 625 have been extended to the determination of many Appendix VIII compounds.[5,6,7,40] By minimizing the number of analytical procedures and enlisting previous experience with priority-pollutant analyses, the number of compounds that can be analyzed by a single method is maximized. For

TABLE 13.6 Contaminants Proposed for Testing with the TCLP and Proposed Regulatory Levels

Contaminant	Regulatory level, mg/L
Acrylonitrile	5.0
Arsenic	5.0
Barium	100
Benzene	0.07
bis(2-Chloroethyl) ether	0.05
Cadmium	1.0
Carbon disulfide	14.4
Carbon tetrachloride	0.07
Chlordane	0.03
Chlorobenzene	1.4
Chloroform	0.07
Chromium	5.0
o-Cresol	10.0
m-Cresol	10.0
p-Cresol	10.0
2,4-D	1.4
1,2-Dichlorobenzene	4.3
1,4-Dichlorobenzene	10.8
1,2-Dichloroethane	0.40
1,1-Dichloroethylene	0.1
2,4-Dinitrotoluene	0.13
Endrin	0.003
Heptachlor	0.001
Hexachlorobenzene	0.13
Hexachlorobutadiene	0.72
Hexachloroethane	4.3
Isobutanol	36
Lead	5.0
Lindane	0.06
Mercury	0.2
Methoxychlor	1.4
Methylene chloride	8.6
Methyl ethyl ketone	7.2
Nitrobenzene	0.13
Pentachlorophenol	3.6
Phenol	14.4
Pyridine	5.0
Selenium	1.0
Silver	5.0
1,1,1,2-Tetrachloroethane	10.0
1,1,2,2-Tetrachloroethane	1.3
Tetrachloroethylene	0.1
2,3,4,6-Tetrachlorophenol	1.5
Toluene	14.4
Toxaphene	0.07
1,1,1-Trichloroethane	30
1,1,2-Trichloroethane	1.2
Trichloroethylene	0.07
2,4,5-Trichlorophenol	5.8
2,4,6-Trichlorophenol	0.30
2,4,5-TP (Silvex)	0.14
Vinyl chloride	0.05

TABLE 13.7 Constituents Considered Hazardous for Land Disposal

Hazardous constituent	Concentration, mg/L
Acetone	2.0
n-Butyl alcohol	2.0
Carbon disulfide	2.0
Carbon tetrachloride	0.1
Chlorobenzene	2.0
Cresols	2.0
Cyclohexanone	2.0
Ethyl acetate	2.0
Ethyl benzene	2.0
Ethyl ether	2.0
HxCDD (all hexachlorodibenzo-p-dioxins)	0.001
HxCDF (all hexachlorodibenzofurans)	0.001
Isobutanol	2.0
Methanol	2.0
Methylene chloride	1.2
Methyl ethyl ketone	2.0
Methyl isobutyl ketone	2.0
Nitrobenzene	0.09
PeCDD (all pentachlorodibenzo-p-dioxins)	0.001
PeCDF (all pentachlorodibenzofurans)	0.001
Pentachlorophenol	1.0
Pyridine	0.7
TCDD (all tetrachlorodibenzo-p-dioxins)	0.001
TCDF (all tetrachlorodibenzofurans)	0.001
Tetrachloroethylene	0.015
2,3,4,6-Tetrachlorophenol	2.0
Toluene	2.0
1,1,1-Trichloroethane	2.0
1,2,2,-Trichloro-1,2,2-trifluoroethane	2.0
Trichloroethylene	0.1
Trichlorofluoromethane	2.0
2,4,5-Trichlorophenol	8.0
2,4,6-Trichlorophenol	0.04
Xylene	2.0

example the recommended method for volatile organic compounds is similar to EPA Method 624 and the method for extractable (semivolatile) organic compounds is similar to EPA Method 625. Table 13.9 lists toxic organic compounds that are designated as priority pollutants, Superfund (CERCLA) compounds, and selected Appendix IX compounds. This table illustrates that 115 compounds are common to all lists. The GC/MS analysis procedure is the method of choice for organic analysis and has been designated for the determination of many of the constituents from the Appendix VIII list. The GC/MS technique is selective and sensitive and the use of high-resolution capillary columns improves the separation of complex mixtures. The majority of Appendix VIII compounds are amenable to GC/MS analysis. Other compounds may require analysis by HPLC and a few other compounds may require compound-specific or compound class-specific procedures. The GC/MS and LC analysis of organic constituents is briefly summarized below and is discussed later in this subsection (see under

TABLE 13.8 Methods for Chemical Analysis of Water and Wastes

EPA method	Priority pollutants	Method of analysis
601	Purgeable halocarbons	Purge-and-trap (PAT), GC, detection with a Hall (electrolytic) detector
602	Purgeable aromatics	PAT, GC, photoionization detector
603	Acrolein and acrylonitrile	PAT, GC, flame-ionization detector
604	Phenols	Extraction, Kuderna-Danish (K-D) concentration, GC, flame-ionization–electron-capture detection
605	Benzidines	Extraction, concentration, HPLC, electrochemical detection
606	Phthalate esters	Extraction, Florisil or alumina cleanup K-D concentration, GC, flame-ionization–electron-capture detection
607	Nitrosamines	Extraction, Florisil or alumina cleanup K-D concentration, GC, flame-ionization–electron-capture detection
608	Organochlorine pesticides and PCBs	Extraction, Florisil or Alumina cleanup, K-D concentration, GC, detection
609	Nitroaromatics and isophorone	Extraction, K-D concentration, GC, flame-ionization–electron-capture detection
610	Polynuclear aromatic hydrocarbons	Extraction, K-D concentration, GC, flame-ionization detection or HPLC/UV fluorescence detection
611	Haloethers	Extraction, solvent exchange, K-D concentration, GC, electron-capture detection
612	Chlorinated hydrocarbons	Extraction, solvent exchange, K-D concentration, GC, electron-capture detection
613	2,3,7,8-Tetrachlorodibenzo-*p*-dioxin (2,3,7,8-TCDD)	Spiking with labeled 2,3,7,8-TCDD, extraction, solvent exchange, K-D concentration, analysis by GC/MS
624	Volatile organic compounds (purgeables)	Purge-and-trap, analysis by GC/MS
624	Semivolatile organic compounds (base-neutral and acid-extractable compounds)	Extraction, K-D concentration, analysis by GC/MS

TABLE 13.9 Listing of Toxic Organics

Common name	Priority pollutant	Superfund list	Appendix IX
Acenaphthene	X	X	X
Acenaphthylene	X	X	X
Acetone		X	X
Acetonitrile	X		X
Acetophenone			X
2-Acetylaminofluorene			X
Acrolein	X		X
Acrylonitrile	X		X
Aldrin	X	X	X
Allyl alcohol			X
4-Aminobiphenyl			X
Aniline	X	X	X
Anthracene	X		X
Aramite			X
Aroclor 1016	X	X	X
Aroclor 1221	X	X	X
Aroclor 1232	X	X	X
Aroclor 1242	X	X	X
Aroclor 1248	X	X	X
Aroclor 1254	X	X	X
Aroclor 1260	X	X	X
Benzene	X		X
Benzenethiol			X
Benzidine	X		X
Benzo[a]anthracene	X	X	X
Benzo[b]fluoranthene	X	X	X
Benzo[k]fluoranthese	X	X	X
Benzoic acid		X	X
Benzo[a]pyrene	X	X	X
Benzo[a]pyrene	X	X	X
p-Benzoquinone			X
Benzyl alcohol		X	X
alpha-BHC	X	X	X
beta-BHC	X	X	X
delta-BHC	X	X	X
gamma-BHC	X	X	X
Bis(2-chloroethoxy)methane	X	X	X
Bis(2-chloroethyl) ether	X	X	X
Bis(2-chloroisopropyl) ether	X	X	X
Bis(2-ethylhexyl) phthalate	X	X	X
Bromodichloromethane	X	X	X
Bromomethane	X	X	X
4-Bromophenyl phenyl ether	X		X
Butyl benzyl phthalate	X	X	X
2-sec-Butyl-4,6-dinitrophenol			X
Carbon disulfide		X	X
Carbon tetrachloride	X	X	X
Chlordane	X	X	X
p-Chloroaniline		X	X
Chlorobenzene	X	X	X
Chlorobenzilate			X
2-Chloro-1,3-butadiene			X
p-Chloro-m-cresol	X		X
Chlorodibromomethane	X	X	X
Chloroethane	X	X	X
2-Chloroethyl vinyl ether	X	X	X

TABLE 13.9 Listing of Toxic Organics (*Continued*)

Common name	Priority pollutant	Superfund list	Appendix IX	Common name	Priority pollutant	Superfund list	Appendix IX
Chloroform	X	X	X	1,1-Dichloroethane	X	X	X
Chloromethane	X	X	X	1,2-Dichloroethane	X	X	X
2-Chloronaphthalene	X	X	X	1,1-Dichloroethylene	X	X	X
2-Chlorophenol	X	X	X	trans-1,2-Dichloroethylene	X	X	X
4-Chlorophenyl phenyl ether	X	X	X	Dichloromethane	X	X	X
3-Chloropropene			X	2,4-Dichlorophenol			X
3-Chloropropionitrile			X	2,6-Dichlorophenol			X
Chyrsene	X	X	X	2,4-Dichlorophenoxyacetic acid			X
ortho-Cresol		X	X	1,2-Dichloropropane	X	X	X
para-Cresol		X	X	cis-1,3-Dichloropropene	X	X	X
4,4'-DDD	X	X	X	trans-1,3-Dichloropropene		X	X
4,4'-DDE	X	X	X	Dieldrin		X	X
4,4'-DDT	X	X	X	Diethyl phthalate	X	X	X
Dibenz[a,h]anthracene	X	X	A	O,O-Diethyl-0,2-pyrazinyl phosphorothioate		X	X
Dibenzofuran		X		3,3'-Dimethoxybenzidine			X
Dibenzo[a,e]pyrene			X	p-Dimethylaminoazobenzene			X
Dibenzo[a,h]pyrene			X	7,12-Dimethylbenz[a]anthracene			X
Dibenzo[a,i]pyrene			X	3,3'-Dimethylbenzidine			X
1,2-Dibromo-3-chloropropane			X	alpha,alpha-Dimethylphenethylamine			X
1,2-Dibromomethane			X	2,4-Dimethylphenol	X	X	X
Dibromomethane			X	Dimethyl phthalate	X	X	X
Di-n-butyl phthalate	X	X	X	m-Dinitrobenzene			X
m-Dichlorobenzene	X	X	X	4,6-Dinitro-o-cresol	X	X	X
o-Dichlorobenzene	X	X	X	2,4-Dinitrophenol	X	X	X
p-Dichlorobenzene	X	X	X	2,4-Dinitrotoluene	X	X	X
3,3'-Dichlorobenzidine	X	X	X	2,6-Dinitrotoluene	X	X	X
trans-1,4-Dichloro-2-butene				Di-n-octyl phthalate	X	X	X
Dichlorodifluoromethane			X				

TABLE 13.9 Listing of Toxic Organics (*Continued*)

Common name	Priority pollutant	Superfund list	Appendix IX
1,4-Dioxane			X
Diphenylamine			X
1,2-Diphenylhydrazine	X		X
Di-*n*-propylnitrosamine	X	X	X
Disulfoton			X
Endosulfan sulfate	X	X	X
Endosulfan I (alpha)	X	X	X
Endosulfan II (beta)	X	X	X
Endrin	X		X
Endrin aldehyde	X		X
Endrin ketone		X	X
Ethylbenzene	X	X	X
Ethyl cyanide		X	X
Ethylene oxide			X
Ethyl methacrylate			X
Famphur			X
Fluoranthene	X	X	X
Fluorene	X	X	X
Heptachlor	X	X	X
Heptachlor epoxide	X	X	X
Hexachlorobenzene	X	X	X
Hexachlorobutadiene	X	X	X
Hexachlorocyclopentadiene	X	X	X
Hexachlorodibenzo-*p*-dioxins			
Hexachlorodibenzofurans			X
Hexachloroethane	X	X	X
Hexachlorophene			X
Hexachloropropene			X
2-Hexanone		X	X

Common name	Priority pollutant	Superfund list	Appendix IX
Indeno[1,2,3-*cd*]pyrene	X	X	X
Iodomethane			X
Isobutyl alcohol			X
Isodrin			X
Isophorone	X	X	A
Isosafrole			X
Kepone			X
Malononitrile			X
Methacrylonitrile			X
Methapyrilene			X
Methoxychlor		X	X
3-Methylcholanthrene			X
4,4'-Methylenebis(2-chloroaniline)			X
Methyl ethyl ketone		X	X
Methyl methacrylate			X
Methyl methanesulfonate			X
2-Methylnaphthalene		X	X
Methyl parathion		X	X
4-Methyl-2-pentanone		X	X
2-Methylphenol		X	X
4-Methylphenol		X	X
Naphthalene	X	X	
1,4-Naphthoquinone			X
1-Naphthylamine			X
2-Naphthylamine			X
m-Nitroaniline		X	X
o-Nitroaniline		X	X
p-nitroaniline		X	X

TABLE 13.9 Listing of Toxic Organics (*Continued*)

Common name	Priority pollutant	Superfund list	Appendix IX
Nitrobenzene	X	X	X
2-Nitrophenol	X	X	X
4-Nitrophenol	X	X	X
N-Nitrosodi-n-butylamine			X
N-Nitrosodiethylamine			X
N-Nitrosodimethylamine	X	X	X
N-Nitrosodiphenylamine			X
N-Nitrosomethylethylamine			X
N-Nitrosomorpholine			X
N-Nitrosopiperidine			X
N-Nitrosopyrrolidine			X
5-Nitro-o-toluidine			X
Parathion			X
Pentachlorobenzene			X
Pentachlorodibenzo-p-doxins	X		X
Pentachlorodibenzofurans			X
Pentachloroethane			X
Pentachloronitrobenzene			X
Pentachlorophenol	X	X	X
Phenacetin			X
Phenanthrene	X	X	X
Pronamide			X
2-Propyn-1-ol			X
Pyrene	X	X	X
Pyridine			X
Resorcinol			X
Safrole			X
Silvex			X
Styrene		X	X
2,4,5-T			X
1,2,4,5-Tetrachlorobenzene			X
2,3,7,8-Tetrachlorodibenzo-p-dioxin	X		X
Tetrachlorodibenzo-p-dioxins		X	X
Tetrachlorodibenzofurans			X
1,1,1,2-Tetrachloroethane			X
1,1,2,2-Tetrachloroethane	X	X	X
Tetrachloroethylene	X	X	X
2,3,4,6-Tetrachlorophenol			X
Tetraethyldithiopyrophosphate			X
Toluene	X	X	X
Toxaphene	X	X	X
Tribromomethane	X	X	X
1,2,4-Trichlorobenzene	X	X	X
1,1,1-Trichloroethane	X	X	X
1,1,2-Trichloroethane	X	X	X
Trichloroethylene	X	X	X
Trichloromethanethiol			X
Trichloromonofluoromethane			X
2,4,5-Trichlorophenol		X	X
2,4,6-Trichlorophenol	X	X	X
1,2,3-Trichloropropane			X
Tris(2,3-dibromopropyl) phosphate			X
Vinyl acetate		X	X
Vinyl chloride	X	X	X

Documentation, on p. 13.39). Other compound-specific analytical procedures are not discussed in this chapter but all compounds on Appendix VIII must be considered in the listing of hazardous waste.

The analysis method for volatile-organic hazardous constituents is identical to the method specified in EPA Method 624 and Methods 5030 and 8240 given in SW 846. Volatile organic compounds can be determined by these methods after appropriate sample matrix-dependent pretreatment, in a wide variety of samples including water, leachates, wastewater, hazardous waste, and soils. This method uses a purge-and-trap (PAT) system with an inert purge gas to remove the volatiles from the waste and collect them on a sorbent trap. The collected volatiles are then thermally desorbed from the sorbent cartridge to the GC column (SP-1000 on Carbopak B) and analyzed by GC/MS. As with all analytical procedures employing complex matrices, it is essential that the purge efficiencies for each volatile compound of interest be determined.

The analysis of semivolatile extractable compounds is equivalent to EPA Method 625 and SW 846 Methods 8250 and 8270. A capillary SE-54 bonded fused-silica column, recommended in Method 8270, increases the resolution, and the sensitivity of the analytical method. These methods are based on the acid and base extraction of the waste with aliquots of a suitable organic solvent, followed by SW 846 Methods 3500 and 3550 and subsequent concentration of the extracts to a suitable volume. The acid and base extracts are usually combined for analysis. Some organic liquid wastes may be prepared for screening analysis by dilution with solvent and direct injection (Method 3580). Microliter volumes of sample extracts are injected onto the fused-silica capillary column using splitless (or split) injection techniques or on-column injection. The mass spectrometer is usually operated in the electron-impact (EI) mode and the full-scan mode to produce mass spectra for identification. Selected surrogates and internal standards must be added to each sample at the appropriate time as detailed in each procedure.

A generalized HPLC/UV method can be used for some other compounds not amenable to GC/MS analysis and for some compounds such as polynuclear aromatic hydrocarbons amenable to analysis by both procedures. The generalized HPLC methods are based on the use of reversed-phase C_{18} columns. Some compounds are not amenable to the generalized HPLC procedures and require specific HPLC procedures.[7,10] A UV/VIS detector is recommended for use with screening methods because the detector can measure a wide range of wavelengths from 190 to 600 nm. For screening a sample, most procedures recommend that the initial wavelengths be set at 254 nm. For specific compounds, other wavelengths offer both increased sensitivity and selectivity. Various procedural options for the HPLC method are described in the reference.

Analysis of Hazardous-Waste Combustion Products

Effluents from hazardous-waste combustion must be analyzed to determine if principal organic hazardous constituents (POHCs) selected for a trial burn are destroyed in the incineration process and meet the required destruction and removal efficiency of ħ99.99% established for most compounds. Sampling methods for stack effluents were discussed in Subsec. 13.1 under *Sampling Solid and Liquid Effluents from Combustors and Control Devices* (p. 13.1). Other associated streams such as process scrubber water and solid residues (ash) must be analyzed either to determine the presence of POHCs or to determine other residual contaminants where these must be known in order to accomplish safe and regulated disposal of the products. The analysis of incinerator effluents involves the same considerations as the analysis of hazardous waste. As detailed under *Sampling Solid and Liquid Effluents from Combustors and Control Devices* (p. 13.14), the method of sample collection is different, and therefore instead of one composite sample, as is the norm for hazardous waste, there may be several different samples from the effluent of a combustion source as well as other process and residue samples.

Before analysis of combustion effluents, the waste feed must be analyzed. Samples of waste must be taken in appropriate zero head-space bottles for volatile analysis. A second composite sample is required for analysis of semivolatile compounds. The VOA samples are analyzed by the purge-and-trap, GC/MS procedure and the semivolatiles are extracted and analyzed by GC/MS as described earlier.

The waste feed, and the particulates, probe wash, and ash residues from combustion, may be analyzed for metals (Table 13.10) (if metals are present in the hazardous waste). Digestion and analysis procedures are given in SW 846 Methods. The SW 846 Method 3010 is recommended for acid digestion for flame atomic-absorption spectroscopy (AAS); Method 3020 for acid digestion for furnace AAS. The metal-containing constituents and the metals on the Appendix VIII list are analyzed by either atomic-absorption spectroscopy or the techniques of inductively coupled argon-plasma (ICAP) emission spectroscopy.[7] Table 13.11 summarizes recommended SW 846 methods of AAS. The recommended method for ICAP analysis of multielements is SW 846 Method 6010. The sample-digestion techniques depend on the instrumentation employed for the analysis. Mercury is analyzed by a cold-vapor technique. The ash may also be analyzed for selected organic compounds, normally for those POHCs that are designated for analysis in the stack effluent. The ash is prepared for organic analysis by soxhlet extraction with methylene chloride and concentrated in a Kuderna-Danish apparatus and then analyzed by GC/MS or HPLC methods. The ash extract may also be analyzed or screened for other organic compounds by the same general procedures.

Scrubber water and other process waters are analyzed for the same organic compounds designated in the trial burn. These may include both volatile compounds and semivolatile organic compounds.

The volatiles from stack effluents are collected using a VOST system (see under *Sampling Waste-Combustion Products,* on p. 13.9). The volatile POHCs in the stack-gas effluent are collected on several pairs of sorbent tubes. The sorbent tubes are thermally desorbed into the purge vessel of a purge-and-trap instrument and analyzed as described for hazardous waste by EPA Method 624 (SW 846 Method 8240). The purge vessel is necessary to prevent excess water from the sorbent tubes from entering the GC/MS system. The VOST analytical method is directly applicable to compounds with boiling points from 30 to 100°C but with minor modifications can be extended to some compounds with boiling points above and below this range. It should be noted that water-soluble compounds may give poor recovery with this method and extra care should be taken to determine whether the data are acceptable for these compounds or, alternatively, to select a new or modified procedure.[4,5,6,7,10,38]

The semivolatile compounds are collected with a comprehensive sampling train as described under *Sampling Waste-Combustion Products,* on p. 13.9. The samples for analysis consist of probe wash, particulate, sorbent trap, and condensate. Each sample must undergo sample preparation including extraction and concentration (SW 846 Methods 3500 to 3550) before analysis. The extracts from the component parts of the sampling system may be combined for analysis by

TABLE 13.10   Listing of Toxic Inorganics

Common name	Priority pollutant	Appendix IX*	Superfund list
Metals			
Aluminum		X†	X
Antimony	X	X	X
Arsenic	X	X	X
Barium		X	X
Beryllium	X	X	X
Cadmium	X	X	X
Calcium		X†	X
Chromium	X	X	X
Cobalt		X†	X
Copper	X	X	X
Iron		X†	X
Lead	X	X	X
Magnesium		X†	X
Manganese		X†	X
Mercury	X	X	X
Nickel	X	X	X
Osmium		X	
Potassium		X†	X
Selenium	X	X	X
Silver	X	X	X
Sodium		X†	X
Thallium	X	X	X
Tin		X†	X
Vanadium		X	X
Zinc	X	X	X
Miscellaneous			
Cyanide	X	X	X
Fluoride		X	
Phenols	X		
Sulfide		X	

*Derived from Appendix VIII.
†Added from Superfund List.

GC/MS by EPA Method 625 (SW 846 Method 8270) or for specific compounds by HPLC. Individual extracts may be analyzed if the collection efficiency of a particular part of the sampling system is being evaluated. Figure 13.10 is an overview of the analysis scheme for stack-gas samples from the VOST and the MM5 sampling trains.[10]

The EPA does not currently regulate the emissions of most hazardous thermal reaction products, commonly called *products of incomplete combustion* (*PICs*). For the EPA program, compounds are considered PICs if they are regulated compounds (Appendix VIII) that are detected in the stack emissions but not present in the waste feed at concentrations greater than 100 ppm. There are many cases in which a PIC may also be designated as a POHC. The formation of PICs that are monitored as POHC emissions results in anomalously low DREs. Field studies by Trenholm et al. and Castaldine et al. have shown that PICs are emitted from thermal destruction systems for hazardous wastes.[41] Some experimentally

TABLE 13.11 Recommended AAS Analysis Methods

Element	SW 846 method no.	Description
Ag	7760	Direct aspiration
	7761	Graphite furnace
As	7060	Graphite furnace
	7061	Gaseous hydride method
Ba	7080	Direct aspiration
	7081	Graphite furnace
Be	7090	Direct aspiration
	7091	Graphite furnace
Cd	7130	Direct aspiration
	7131	Graphite furnace
Cr	7190	Direct aspiration
	7191	Graphite furnace
	7195	Hexavalent Cr: coprecipitation
	7196	Hexavalent Cr: colorimetric
	7197	Hexavalent Cr: chelation-extraction
	7198	Hexavalent Cr: differential-pulse polarography
Cu	7120	Direct aspiration
	7121	Graphite furnace
Hg	7470	Hg in liquid waste (Manual cold-vapor technique)
Ni	7520	Direct aspiration
	7621	Graphite furnace
Os	7550	Direct aspiration
	7551	Graphite furnace
Pb	7420	Direct aspiration
	7420	Graphite furnace
Sb	7040	Direct aspiration
	7041	Graphite furnace
Tl	7840	Direct aspiration
	7841	Graphite furnace
V	7910	Direct aspiration
	7911	Graphite furnace
Zn	7950	Direct aspiration
	7951	Graphite furnace

observed PICs are shown in Table 13.12. Current data suggest that for thermal destruction facilities where high degrees of DRE have been achieved for POHCs, PIC emissions are also low.[32] It is important in the selection of a POHC for analysis that consideration be given to the potential for the POHC to also be a PIC. For most PICs, the analytical methodology is similar to that described for POHC analysis. Dioxins and furans are notable exceptions and require specific analytical procedures.[10,42,43,44]

To simplify the overall analytical approach, it has been suggested that surrogates be used to evaluate the performance of an incinerator. A compound may

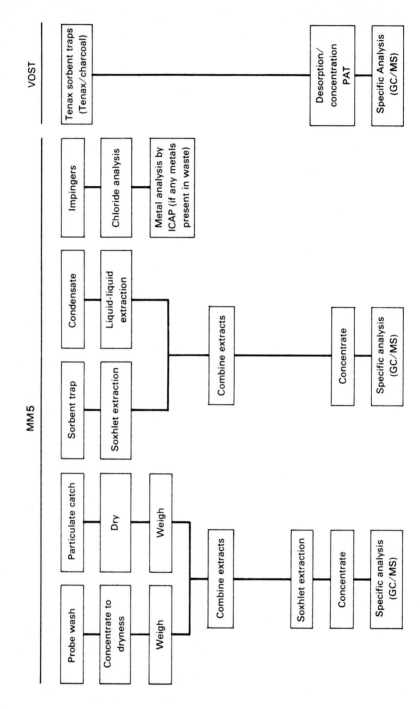

FIG. 13.10 Overview of an analysis scheme for stack-gas samples.

TABLE 13.12 Experimentally Observed Products of Incomplete Combustion (PICs)

Parent (POHC)	Product (PIC/POHC)
Chloroform	Tetrachloroethylene Carbon tetrachloride
Carbon tetrachloride	Tetrachloroethylene Hexachloroethane
Toluene	Benzene
Chlorobenzene	Benzene
Trichlorobenzene	Dichlorobenzene Chlorobenzene
Pentachloroethane	Tetrachloroethylene
Polychlorinated biphenyls	Chlorinated dibenzofurans
Polychlorinated phenols	Chlorinated dibenzodioxins
Kepone	Hexachlorobenzene

act as a surrogate for a specific waste feed, or for any waste feed and any incinerator. An ideal surrogate would be nontoxic and easily monitored and more stable than any other compounds in the waste. Proposed surrogates include total unburned hydrocarbons (TUHC) and carbon monoxide (CO) but destruction of these may not correlate with destruction and removal efficiency for POHCs. A listed compound in the waste or a thermally stable compound added to the waste feed may serve as a surrogate. Selection of a single organic compound (or several specific compounds) for evaluation of DRE would greatly simplify monitoring requirements. Gas-phase thermal stability has been proposed as an alternate to heat of combustion as a ranking scheme for selecting a surrogate compound slightly more stable than any listed compound intended for incineration. Some examples of thermally stable compounds are acetonitrile, hexachlorobenzene, monochlorobenzene, tetrachloroethylene, and trichloroethylene. Halogenated compounds such as CF_4, C_2F_6, C_3F_8, CF_3Cl, $C_2F_3Cl_3$, and SF_6 have also been suggested as possible additive candidates.[11,45,46]

Emissions of CO and TUHC may not correlate with DRE but do tend to act as indicators for any upset conditions. Selecting a thermally stable component of waste feed, or a thermally stable additive, appears to be scientifically defensible and technically feasible. Analytical protocols for the analysis of a few approved surrogates would be more specific and greatly simplify the assessment of incinerator performance.

Documentation

Analytical measurements should be documented to provide easy access to information that will support all claims for the results. Laboratory records should be retained in a permanent file for a length of time set by the government or necessary to fulfill other legal requirements. All data must be recorded in laboratory notebooks (bound notebooks preferred) and cross-referenced to raw data such as chromatograms and mass spectra stored in the raw data file.

Environmental management has a strong documented technical data component. A major problem is the conversion of raw data into information needed for decision making in hazardous-waste management. At present the reporting of results remains a paper-intensive manual system. For example, in 1985 the EPA Contract Laboratory Program (CLP) consisted of approximately 50 laboratories analyzing samples at a rate of 40,000 per year. To meet current requirements for the delivery of data on a single sample requires a 2¼-in thick stack of paper. At this rate, the EPA will have a stack of paper 1½ miles high containing highly technical data to examine this year. These data must be converted into useful information to support environmental management. In 1983, the idea for an improved system for handling and processing data was suggested. Data would be transmitted by electronic or magnetic means and machine-read into appropriate integrated computer data bases.[47] A good start has been made by the EPA in evaluating the feasibility of electronic data transfer, thus substantially improving the efficiency with which the EPA handles data. Data assessment for evaluating hazardous-waste cleanup will be faster and more accurate in the near future.

Note that use of electronic data transfer may be necessary for day to day management of data records from waste-management facilities.

Quality Assurance and Quality Control

Quality assurance (QA) is the highest priority with respect to the analysis of environmental samples. Analytical data from the analysis of soil, air, drinking water, wastewater, sludge, and hazardous waste must be scientifically valid, defensible, and of known and acceptable accuracy and precision. Thus, the QA program must contain procedures that cover program management and personnel responsibilities, facilities and equipment, data generation and processing, data-quality assessment, and corrective actions. Individual QA programs may vary depending upon the nature of the material to be analyzed and the intended use of the final data. However, certain aspects are common to all situations.

Guidance for developing adequate QA program plans may be found in *Guidelines and Specifications for Preparing Quality Assurance Program Plans* (QAMS-004/80).[48] The Quality Assurance Program Plan will stipulate the QA policies, objectives, management structure, responsibilities, and procedures for a total QA program for the organization performing analytical services. Included in the program plan will be information concerning procedures to assure the generation of reliable data; processes for collecting, reducing, validating, and storing data; assessment of data quality (accuracy and precision); corrective actions; and the schedule for implementation of the requirements.

A program plan is usually general in scope. However, a QA project plan gives a more detailed description of how the analytical organization will produce quality data for a specific analysis. Every project that involves environmentally related measurements should have a written and approved Quality Assurance Project Plan. Guidance for QA project plans may be found in *Interim Guidelines and Specifications for Preparing Quality Assurance Project Plans* (QAMS-005/80).[49]

A QA project plan may contain one or more of the following items if they are appropriate for the specific contract.

• Title page with provision for approval signatures
• Table of contents
• Project description
• Project organization and responsibility

- QA objectives for measurement data in terms of precision, accuracy, completeness, representativeness, and comparability
- Sampling procedures
- Calibration procedure and frequency
- Analytical procedures
- Data reduction, validation, and reporting
- Internal quality-control checks and frequency
- Performance and system audits and frequency
- Preventive-maintenance procedures and schedules
- Specific routine procedures to be used to assess data precision, accuracy, and completeness for specific measurement parameters involved
- Corrective action
- Quality-assurance reports to management

QA project plans are usually prepared in a document-control format consisting of information (i.e., section number, revision number, date of revision, and page number) in the upper right-hand corner of each page of the project plan. All 16 items described previously may be considered and addressed. The level of QA will depend on the nature of the project and the end use of the data that are produced. Specific procedures to assess precision and accuracy on a routine basis during the project will be described in each Quality Assurance Project Plan.[7,10]

Conclusion

The disposal of wastes will remain regulated, with greater emphasis on regulation of problem wastes and extremely hazardous wastes as they are identified. There will be fewer landfills and those that do exist will be tightly controlled. There will be more incineration, more material and resource recovery, and more development of methods to destroy and detoxify wastes. Ideally, disposal should be the procedure of last resort. However, proper disposal will be needed even years from now. Many materials have little recycling value, they may be too difficult to degrade, or they contain nonflammable materials difficult to incinerate. Other materials are residues from alternate treatment technologies. Proper burial under RCRA will require attention to creation of properly engineered landfills, and it will require monitoring to make sure that there is no migration into soil or water supplies. Thus, the analysis of hazardous waste will continue to be a vital part of the regulation of hazardous waste. Existing methods are now being validated and extended to a large number of hazardous compounds. New methods are now being studied for application to environmental analysis. Although there will be a lag time before these methods are available for application to the analysis of hazardous waste, the reader would be well advised to consider the merits of new methods as soon as they have been validated and approved for use.

13.3 REFERENCES

1. J. A. Dorsey, L. D. Johnson, and R. G. Merrill, "A Phased Approach for Characterization of Multimedia Discharges from Processes," in *Monitoring Toxic Substances*, D. Schuetzle (ed.), American Chemical Society, Washington, D.C., 1979.

2. F. Briden, J. A. Dorsey, and L. D. Johnson, "A Comprehensive Scheme for Multimedia Environmental Assessment of Emerging Energy Technologies," *Intern. J. Environ. Anal. Chem.,* **9**:189 (1981).

3. D. E. Lentzen, D. E. Wagoner, E. D. Estes, and W. F. Gutknect, *IERL-RTP Procedures Manual: Level I Environmental Assessment (Second Edition),* EPA-600/7-78-201, PB 293795, October 1978.

4. W. A. Telliard, "The Consent Decree Pollutants and their Analysis by GC/MS," *Spectra,* **10**:(4)4 (1986).

5. W. M. Shackelford and J. M. McGuire, "Analysis of Extractable Priority Pollutants in Water by GC/MS," *Spectra,* **10**:(4)17 (1986).

6. *Federal Register,* Friday, October 26, 1984, Part VIII, pp. 43,234–43,544.

7. *Test Methods for Evaluating Solid Waste, Physical/Chemical Methods,* SW-846 Manual, 3d ed., Document No. 955-001-0000001, available from the Superintendent of Documents, U.S. Government Printing Office, Washington, D.C. 20402, January 1987.

8. G. T. Flatman and A. A. Yfantis, "Geostatistical Strategy for Soil Sampling: The Survey and the Census," *Environmental Monitoring and Assessment,* **4**:335 (1984).

9. E. R. deVera, B. P. Simmons, R. D. Stephens, and D. L. Storm, *Samplers and Sampling Procedures for Hazardous Waste Streams,* EPA-600/2-80-018, PB-135353, January 1980.

10. J. C. Harris, D. J. Larsen, C. E. Rechsteiner, and K. E. Thrun, *Sampling and Analysis Methods for Hazardous Waste Combustion,* EPA-600/8-84-002, PB84-155845, February 1984.

11. W. G. DeWees, S. S. Steinsberger, and S. J. Plaisance, *Hazardous Waste Treatment, Storage, and Disposal Facilities: Field Sampling and Analysis Protocol for Collecting and Characterizing Soil Samples from TSDF's,* EPA-450/3-86-014, October 1986.

12. *Code of Federal Regulations,* Title 40, Part 60, Appendix A, 7/1/87 edition.

13. L. D. Johnson and R. G. Merrill, "Stack Sampling for Organic Emissions," *Toxicological and Environmental Chemistry,* **6**:109 (1983).

14. L. D. Johnson, "Detecting Waste Combustion Emissions," *Environmental Science and Technology,* **20**:223 (1986).

15. L. M. Schlickenrieder, J. W. Adams, and K. E. Thrun, *Modified Method 5 Train and Source Assessment Sampling System Operators Manual,* EPA-600/8-85-003, PB85-169878, February 1985.

16. J. Bursey, M. Hartman, J. Homolya, R. McAllister, J. McGaughey, and D. Wagoner, *Laboratory and Field Evaluation of the Semi-VOST Method,* vol. I, EPA-600/4-85-075a, PB86-123551/AS, and vol. II, EPA-600/4-85-075b, PB86-123569/AS, September 1985.

17. J. H. Margeson, J. E. Knoll, and M. R. Midgett, "An Evaluation of the Semi-VOST Method for Determining Emissions from Hazardous Waste Incinerators," submitted to *J. Air Pollution Control Association.*

18. J. Bursey, J. L. Steger, M. Palazzola, D. Benson, J. Homolya, R. McAllister, J. McGaughey, and D. Wagoner, *Laboratory and Field Evaluation of the Semi-VOST Method,* EPA-600/4-86-046, PB87-145934/AS, November 1986.

19. G. A. Jungclaus, P. G. Gorman, G. Vaughn, G. W. Scheil, F. J. Bergman, L. D. Johnson, and D. Friedman, *Development of a Volatile Organic Sampling Train (VOST),* presented at the Ninth Annual Research Symposium on Land Disposal, Incineration, and Treatment of Hazardous Waste, Ft. Mitchell, Kentucky, May 1983; in *Proceedings,* EPA-600/9-84-015, PB84-234525, July 1985.

20. L. D. Johnson, "Development of the Volatile Organic Sampling Train for Use in Determining Incinerator Efficiency," *Hazardous and Industrial Solid Waste Testing: Fourth Symposium,* ASTM STP 886, J. K. Petros, Jr., W. J. Lacy, and R. A. Conway (eds.), American Society for Testing and Materials, Philadelphia, 1986, pp. 335–343.

21. E. M. Hansen, *Protocol for the Collection and Analysis of Volatile POHC's Using VOST*, EPA-600/8-84-007, PB84-170042, March 1984.

22. J. Prohaska, T. J. Logan, R. G. Fuerst, and M. R. Midgett, *Validation of the Volatile Organic Sampling Train (VOST) Protocol*, vol. I, *Laboratory Phase*, vol. II, *Field Validation Phase*, EPA-600/4-861014a, PB-86-145547, EPA-600/4-86014b, PB86-145554, January 1986.

23. R. G. Fuerst, T. J. Logan, M. R. Midgett, and J. Prohaska, "Validation Studies of the Protocol for the Volatile Organic Sampling Train," *J. Air Pollution Control Association*, **37**:(4)388 (1987).

24. K. E. Thrun, J. C. Harris, and K. Beltis, *Gas Sample Storage*, EPA-600/7-79-095, PB298-350, April 1979.

25. K. J. Beltis, A. J. DeMarco, V. A. Grady, and J. C. Harris, "Stack Sampling and Analysis of Formaldehyde," *Proceedings, Ninth Annual Research Symposium on Land Disposal, Incineration, and Treatment of Hazardous Waste*, Ft. Mitchell, Kentucky, May 1983, EPA-600/9-84-015, PB84-234525, July 1984.

26. U.S. Environmental Protection Agency, *Methods for Chemical Analysis of Water and Waste*, EPA-600/4-79-020, March 1979.

27. U.S. Environmental Protection Agency, *Handbook for Sampling and Sample Preservation of Water and Wastewater*, EPA-600/4-82-029, September 1982.

28. R. K. M. Jayanty, J. A. Sokash, W. F. Gutknecht, C. E. Decker, and D. J. von Lehmden, "Quality Assurance for Principal Organic Hazardous Constituents (POHC) Measurements During Hazardous Waste Trial Burn Tests," *J. Air Pollution Control Association*, **35**:(2)143 (1985).

29. R. K. M. Jayanty, S. W. Cooper, C. E. Decker, and D. J. von Lehmden, "Evaluation of Parts-Per-Billion Organic Cylinder Gases for Use as Audits During Hazardous Waste Trial Burn Tests," *J. Air Pollution Control Association*, **35**:(11)1,195 (1985).

30. C. C. Lee, G. L. Huffman, and D. A. Oberacker, "Hazardous/Toxic Waste Incineration," *J. Air Pollution Control Association*, **36**:(8)922 (1986).

31. W. E. Sweet, R. D. Ross, and G. Vander Velde, "Hazardous Waste Incineration: A Progress Report," *J. Air Pollution Control Association*, **35**:(2)139 (1985).

32. E. T. Oppelt, "Hazardous Waste Destruction," *Environmental Science and Technology*, **20**:(4)312 (1986).

33. Comprehensive Environmental Response, Compensation, and Liability Act, (42 USC 9601-9657, Section 101 (42 USC 9601), 1980.

34. H. M. McNair, *Analytical Systems for Trace Organic Analysis*, National Bureau of Standards Special Publication 519, *Proceedings of the 9th Materials Research Symposium*, April 10–13, 1978, Gaithersburg, Maryland (issued April 1979).

35.

36. L. H. Keith, W. Crummett, J. Deegan, Jr., R. A. Libby, J. K. Taylor, and G. Wentler, "Principles of Environmental Analysis," *Analytical Chemistry*, **55**:(14)2,210 (1983).

37. J. H. Johnson, E. D. Erickson, and S. R. Smith, "Artifacts Observed when using Tenax-GC for Gas Sampling," *Analytical Letters*, **19**:(3 & 4)315 (1986).

38. A. F. Weston, "Obtaining Reliable Priority-Pollutant Analyses," *Chemical Engineering*, April 30, 1984.

39. A. F. Weston, "Obtaining Reliable Priority-Pollutant Analyses," *Chemical Engineering*, April 30, 1984.

40. R. H. James, R. E. Adams, J. M. Finkel, H. C. Miller, and L. D. Johnson, "Evaluation of Analytical Methods for the Determination of POHC in Combustion Products," *J. Air Pollution Control Association*, **35**:(9)959 (1985).

41. P. Gorman, W. D. Hathoway, and A. Trenholm, *Practical Guide—Trial Burns for Hazardous Waste Incinerators*, PB 86-190 246/AS, EPA/600/S2-86/050, U.S. EPA, 1986.

42. W. M. Shaub and W. Tsang, "Dioxin Formation in Incinerators," *Environmental Science and Technology,* **17**:(12)721 (1983).

43. *Analytical Procedures To Assay Stack Effluent Samples and Residual Combustion Products for Polychlorinated Dibenzo-p-Dioxins (PCDD) and Polychlorinated Dibenzofurans (PCDF),* prepared by Group C—Environmental Standards Workshop, sponsored by The American Society of Mechanical Engineers, U.S. DOE, and U.S. EPA, September 18, 1984.

44. C. Rappe, "Analysis of Polychlorinated Dioxins and Furans," *Environmental Science and Technology,* **18**:(18)3,78A (1984).

45. B. Dallinger and D. L. Hall, "Surrogate Compounds for Monitoring the Effectiveness of Incineration Systems," *J. Air Pollution Control Association,* **32**:(2)179 (February 1986).

46. B. Dellinger and D. L. Hall, "The Viability of Using Surrogate Compounds for Monitoring the Effectiveness of Incineration Systems," *J. Air Pollution Control Association,* **36**:(2)179 (1986).

47. B. P. Almich, W. L. Budde, and W. R. Shobe, "Waste Monitoring," *Environmental Science and Technology,* **20**:(1)16 (1986).

CHAPTER 14

ENGINEERING CONSIDERATIONS

SECTION 14.1

CONTROL AND DISPOSAL INDICES FOR COMPARING WASTE-MANAGEMENT ALTERNATIVES

Edward Martin
Peer Consultants
Rockville, Maryland

E. T. Oppelt
U.S. Environmental Protection Agency
Hazardous Waste Engineering Research Laboratory
Cincinnati, Ohio

Management and disposal of hazardous wastes are causing a revolution in past sanitary engineering practices. To place chemical, physical, biological, and thermal treatment of hazardous wastes in perspective, one must examine the words that have been used to describe waste-management functions.

Ultimate management of hazardous wastes includes *disposal*—that is, elimination of the waste or the hazard potential of the waste. In classical waste management, *disposal* has more often meant transferring the potential adverse impact from one location to another or from one medium (air, land, water) to another. In the context of hazardous-waste management it is necessary to eliminate the hazard potential to the environment and human health, a stated goal of the Resource Conservation and Recovery Act (RCRA).

In every case where management of hazardous wastes is being contemplated, a comparison of engineering alternatives should be conducted in detail to assess the potential long- and short-term effectiveness of available options. Often such comparisons are made on the basis of cost estimates, which are purported to be based on comparable options. One of the most frequently used is a comparison of costs for incineration versus landfill.

Costs are not examined here. The intent of this discussion is to present a scheme for comparison of management options based on factors other than cost. Once the comparison scheme has been developed for a given case, costs can then be compared on an equitable basis.

14.1.1 APPROACH TO CONTROL AND DISPOSAL INDICES

There are only two objectives possible for hazardous-waste management:

1. *Control* for some period of time
2. *Disposal* or elimination of hazard potential

Each management option may be treated as "black box,"

Where X = Quantity of hazardous component being managed ($X = 1$ unit)
X_1 = Quantity of hazardous component in effluent from the box
X_2 = Quantity of hazardous component in the air emissions
X_3 = Quantity of hazardous component in residue

The black-box approaches for common management options are summarized qualitatively in Table 14.1.1, with a general expectation for X_1, X_2, and X_3 for each approach (based on destruction and/or control efficiencies with $X = 1$).

To develop a useful comparison approach, one must estimate the quantity of expected discharges from the black box and account in some way for the time factor. Also, it is necessary to deal with the relative degree of disposal of hazardous- and toxic-waste components.

In a specific engineering analysis of alternatives for specific wastes at an individual site, estimates could be made for each element necessary in the scheme and applied by using this technique.

For this general analysis, estimates were made for each fractional destruction, removal, and emission rate based on the authors' experience and consultation with others. The actual values of the estimates are not important to the usability of the scheme. Others will have their own estimates, and as we have mentioned, specific waste, site, and management-option conditions should be substituted when available; this should be done to derive meaningful *site*-related analyses. As will be seen, even with wide variations in discharge estimates the relative ranking of control and disposal options will be preserved.

For hazardous-waste management, control is exercised over the widespread exposure of the hazard component of the given waste for some period of time: for example, waste still bottoms are incinerated, or heavy-metal treatment sludges are placed in a landfill for 30 years. Exposure to the environment or exposure in such a way as to have an impact on human health is either through the air or water. These exposure routes are expressed in the black-box approach as X_1 (liquid or other discharge) and X_2 (air emissions). One may therefore define an index to describe the degree of control possible through the application of a black box to a waste in terms of these two parameters of the process and time.

The *control-years index* is defined by

TABLE 14.1.1 The Black-Box Approach

Management option	Effluent X_1	Air emission X_2	Residue X_3
Incinerator	0	Fugitive stack, PIC*	Ash and scrubber water
Secure landfill	Leaks (leachate)	Air emission	Remainder after lifetime
Land treatment	Leachate and runoff	Air emission	Remainder on land
Storage tanks	Leakage	Air emission	Pump-out
Piles	Leaching and runoff	Air emission	Pile residue
SI†	Leaching	Air emission	Residue in impoundment
Containers	Leakage and spillage	——	Container contents
Treatment removal			
(a) Conservative pollutants	Effluent + PIT‡	Air emission	Residue
(b) Nonconservative pollutants	Effluent + PIT‡	Air emission	Residue
Destruction	Effluent + PIT‡	Air emission	Residue not destroyed
Fixation/stabilization	Leachate	Air emission	Fixed solids
Encapsulation	Leachate	——	Fixed solids

*PIC = Product of incomplete combustion.
†SI = Surface Impoundment.
‡PIT = Product of incomplete treatment.
Note: Control relates to management of liquid effluents (or discharges) and air emissions without necessarily destroying components of the wastes. Residues are treated separately. Table 14.1.2 presents the relatively estimated discharge fractions (assuming $X = 1$ unit) for 26 options.

$$CYI = \frac{1}{(X_1 + X_2)(t)} \qquad (1)$$

where t = time in years.
If the time is different for liquid and air emissions, the index becomes

$$CYI = \frac{1}{X_1 t_1 + X_2 t_2} \qquad (2)$$

For purposes of risk analysis the index definition may be expanded to include multiple control efficiencies and exposure times for both liquid and air emissions:

$$CYI = \frac{1}{{}_i X_i t_i + {}_j X_j t_j} \qquad (3)$$

Equation (3) can be used as the basis for a detailed site risk assessment by including all pertinent population or individual exposure routes and compounds.

Values for the CYI for various control technologies are displayed in Table 14.1.2. For the purposes of developing the values for Table 14.1.2, general, non-site-specific estimates were used for X_1, X_2, and X_3. Even with these general estimates, the separation between strategies can be seen. For a given application the time used to calculate the index is the period in years during which the option is being used. In the most general case this period corresponds to the lifetime of the facility [Eq. (1)]. Also presented in the table are the time estimates (in years) used for the comparative analysis.

The results of the CYI calculations ranged over four or more orders of magnitude and are grouped for convenience in the table, into "high," "medium," and "other." High values of CYI reflect a high degree of control of air and other discharges; low values reflect reduced levels of control. It should be emphasized again that the estimates used for this analysis may not be universally applied and are likely to vary significantly for specific cases—i.e., wastes and sites.

True disposal is reflected by the amount of toxic or hazardous components remaining after the option of choice is applied to manage the waste. The disposal index $(1/X_3)$ is also tabulated in Table 14.1.2. As in the case of the CYI, the disposal index may be generalized to account for multiple destruction efficiencies and thereby becomes the basis for site risk assessment.

$$ DI = \frac{1}{\sum_i X_i} $$

A high DI value means that the option exhibits high destruction efficiency or a high degree of true disposal. Some options exhibit a high DI because of high volatilization (i.e., incidental disposal). The same would be true for a leaking landfill. Such options should be evaluated closely to determine the impact on local ambient-air concentrations and groundwater concentrations of toxics.

The best options (independent of cost) are those that exhibit high values of *both* CYI and DI. In this analysis the options that exhibit the highest values of both CYI and DI are

	CYI	DI
Incineration	370	10,000
Destructive treatment	1.6	50

Secure landfill, land treatment for metals removal, long-term storage of nonvolatiles in closed tanks, relatively short-term storage in containers (five years), short-term storage of both volatiles and nonvolatiles in closed tanks, treatment for removal of toxic pollutants, and encapsulation all exhibit medium-range CYIs and low DIs. These values reflect the fact that there are storage options (except for land treatment and treatment by removal, which require further considerations for residue disposal, or in the case of land treatment, assurance of permanent retention in the soil medium). These options can and often are combined for total hazardous-waste management systems. For combined options the analysis may proceed in a similar fashion. The last option in the treatment train is likely to be the controlling one for determination of CYI and DI.

TABLE 14.1.2 Control-Years and Disposal Indices

Management option	Effluent X_1	Air emission X_2	Residues X_3	Time, yr	Control-years index (yr^{-1}) High 10	Med. 1-10	Other 0.1-1	Other 0.01-0.1	Disposal index $1/X_3$ High 10	Other 1-10
Incineration	0	0.00009	0.0001	30	370				10,000	
Secure landfill	0.01	0.01	0.98	50		1				1
Landfill										
Vol.*	0.5	0.05	0.45	30				0.06		2.2
NV and metals	0.2	0	0.8	30			0.17			1
Land treatment										
Vol.	0.05	0.9	0.05	10			0.1		20 *	
NV	0.1	0.05	0.85	10			0.67		1.2	
Metals	0.1	0	0.9	10		1				1.1
Storage (long-term)										
Open-tank (vol.)	0.01	0.98	0.01	20				0.05	100 *	
Closed-tank (vol.)	0.01	0.05	0.94	20			0.83			1
Open-tank (NV)	0.01	0.20	0.79	20			0.24			1.3
Closed-tank (NV)	0.01	0.01	0.98	20		2.5				1
Piles										
Vol.	0.15	0.8	0.05	10			0.1		20 *	
NV	0.15	0.3	0.55	10			0.22			1.8
Surface impoundment										
Vol.	0.05	0.94	0.01	10			0.1		100 *	
NV	0.05	0.25	0.70	10			0.33			1.4
Containers	0.1	0	0.90	5		2				1.1

TABLE 14.1.2 Control-Years and Disposal Indices (*Continued*)

Management option	Effluent X_1	Air emission X_2	Residues X_3	Time, yr	Control-years index (yr⁻¹) High 10	Med. 1–10	Other 0.1–1	0.01–0.1	Disposal index $1/X_3$ High 10	Other 1–10
Storage (short-term)										
Open-tank (vol.)	0	0.05	0.95	20	50					1.1
Closed-tank (vol.)	0	0.001	0.999	20	50					1
Open-tank (NV)	0	0.01	0.99	20		5				1
Closed-tank (NV)	0	0.0001	0.9999	20	500					1
Treatment removal										
(a) Conservative pollutant 1.7‡	$(X-FR)$	0.001	$(FR-0.001)$	30		1.6 †		0.08‡		1.7‡, 1†
(b) Nonconservative pollutant	$(X-FR)$	0.01	$(FR-0.01)$	30		1.1 †		0.08‡		1.7‡, 1†
Destruction	$(X-FR)$	0.001	$[FR-X(1-Y)]$	30		0.6§	0.167¶		50§	4¶
Fixation/stabilization										
Vol.	0.01	0.98	0.01	50				0.02	100*	
NV and metals	0.1	0.03	0.87	50			0.15			1.2
Encapsulation	0.01	0.01	0.98	50		1				1

†At $FR = 0.98$
‡At $FR = 0.60$
§At $FR = Y = 0.98$
¶At $FR = Y = 0.60$

vol. = volatiles.
NV = nonvolatiles.
FR = fraction removed.
Y = fractional quantity destroyed.
*"Disposal" is volatilization.

14.1.2 COST ASSESSMENT

Costs may now be superimposed to arrive at "normalized" control and disposal comparisons. The cost per unit of control-years purchased is given by

$$\text{Unit control cost} = \frac{\text{total control cost}}{\text{CYI}}$$

The cost per unit of disposal purchased is

$$\text{Unit control cost} = \frac{\text{total control cost}}{\text{DI}}$$

Fractional values of either CYI or DI can be interpreted to indicate negative cost benefit when options are considered separately. Unit cost comparisons involving one or more options that exhibit fractional CYI and DI values will favor those options with high index values. A cost advantage accrues to those options with values greater than unity.

14.1.3 CONCLUSION

A method for normalizing performance and costs of hazardous-waste control alternatives has been presented. Incineration and destructive treatment compared favorably to other alternatives in the previous discussion, but there is a very wide disparity in the values of both CYI and DI for the two options. This disparity illustrates a problem that is important to the consideration of data.

Percent removal is the classical treatment parameter and continues as the parameter of choice for monitoring treatment-process efficiency. It is the main reporting parameter for processes reported here. Although 98% removal is considered very high for process treatment efficiency, the 99.99% destruction and removal efficiency for incineration [the current Resource Conservation and Recovery Act (RCRA) regulation] represents three orders of magnitude better control. Even at 99.99%, 1 kg of unburned toxic waste component could be discharged for every 10,000 kg burned. This difference between control effectiveness of the two options is reflected in the index values.

It is important to consider and report in detail process removal and destruction efficiencies in future work. Identification of highly toxic waste components will allow the researcher and designer to limit the range of chemical analyses required to conduct relevant studies and to "track" compounds of most concern. Also, only those treatment processes that are very highly effective on highly toxic compounds—namely, much higher than 99%—will compare favorably to combustion.

SECTION 14.2

COST PERSPECTIVES FOR HAZARDOUS-WASTE MANAGEMENT

Gordon M. Evans

Economist
U.S. Environmental Protection Agency
Hazardous Waste Engineering Research Laboratory
Cincinnati, Ohio

14.2.1   GENERAL COST CONSIDERATIONS

The goal of the hazardous-waste generator is to maximize profits given the constraints imposed by environmental regulation. This challenge presents the need for the timely identification of disposal-cost problems and the implementation of effective corrective actions.

The short-run solution is to turn to the market and purchase the most cost-effective disposal process or service. More comprehensive solutions are possible in the long run, including recycling, waste-reduction efforts, redesigning the production process, or even suspending production altogether. It is the job of management to carefully identify and analyze the cost and benefit of each solution. The steps in making a decision on a disposal strategy are easy to state but involve much work and attention. First and foremost management must develop a set of criteria which reflect the firm's overall goals and objectives. Next, a list of commercially available disposal alternatives must be compiled. Finally, each alternative must be carefully evaluated against the criteria, with this analysis serving as the basis for the final decision.

14.2.2   THE MARKET FOR HAZARDOUS-WASTE DISPOSAL SOLUTIONS

It is best to view the market for hazardous-waste disposal solutions as two distinct submarkets; the on-site waste-disposal market (reflected in the need for the fabrication and installation of disposal equipment) and the off-site waste-disposal market (as seen in the services provided by comprehensive waste-management firms).

The Demand for Hazardous-Waste Disposal Solutions

The market demand for hazardous-waste disposal solutions is a function of the quantity of hazardous wastes generated. This in turn is a function of several factors including the current and future state of government regulations (as these regulations provide the legal definition of a hazardous waste), the underlying demand for goods whose production results in the generation of hazardous wastes, and the long-term trends in waste recycling and reduction efforts (leading to reductions in the total amounts of waste generated).

The adoption of an on-site disposal strategy means the firm must seek the expertise for engineering, fabricating, and installing disposal equipment. The search for a vendor may extend to a national (if not global) market. Firms adopting an off-site strategy require the services of a comprehensive waste disposer. This search is likely to be regional, in large part because of the costs and risks associated with the transport of hazards over great distances coupled with the growing reluctance of states to permit handlers to accept out-of-state waste shipments.

The Supply of Hazardous-Waste Disposal Solutions

The market supply of disposal services is a function of the capacity of the waste-disposal industry. This in turn is a function of the industry's ability to design, fabricate, and install new disposal equipment as well as the ability to site new disposal facilities.

In the short run, this industry's ability to dispose of additional amounts of waste is limited to existing excess capacity. The lead time needed to expand beyond this capacity can range from three to five years. The problems include limited access to capital funds, limits on the availability of technical expertise, the need to establish a market for their solution, public opposition to siting of facilities, and the time needed for permitting. Given a relatively fixed short-run capacity, the costs of obtaining disposal solutions will rise as the demand for these solutions grows.

14.2.3 CHOOSING A DISPOSAL SOLUTION

As was indicated earlier, the market for hazardous-waste disposal solutions is best viewed as two submarkets: markets for off-site and on-site solutions. The firm must adopt a set of criteria which will guide choice between these two options. In preparing the criteria, management should consider the following issues:

- What are the waste stream's physical characteristics? Can the waste be neutralized, concentrated, or otherwise pretreated? How much waste is generated? Are these factors likely to change over time?

- What waste-disposal methods are being used by other generators with similar waste streams? Have their methods been successful? Are they willing to share their experiences?

- What is the corporate position on risk? Have potential liability problems arising from either on-site disposal or off-site transportation been considered? Is insurance available (and affordable)?

- What is the nature of the firm's product market (i.e., competitors, substitutes, price strategies, demand projections)? Can the costs of waste disposal be passed along without significant consequences?

- Is in-house expertise available to handle the technical, managerial, and legal aspects of disposing of hazardous waste? What is the corporate experience with hazardous materials?

- Who is offering disposal solutions? What is their corporate experience and reputation? Have they dealt with waste types similar to those being generated?

- What will be the public's reaction to the adoption of a given disposal strategy? How will the firm deal with this reaction?

Once these questions have been answered, and the firm's criteria established, disposal strategies may then be evaluated.

On-Site Disposal

On-site disposal is accomplished through the design, purchase, and installation of equipment which will reduce, intern, or destroy a given waste stream. Management of on-site disposal activities may be handled by the generator or by an on-site contractor.

Since the generator is familiar with the physical and chemical characteristics of the waste stream it is possible to design a custom treatment process tailored to that waste stream. Waste treatment becomes an integral part of the production process, allowing for efficient and cost-effective disposal. This provides the potential for energy and/or resource recovery (with associated cost benefits). Even a partial on-site disposal or recovery process (e.g., waste reduction) can go a long way toward reducing dependence on an outside waste handler. Keeping wastes on site reduces the possibility of accidents occurring during handling and transportation. Perhaps most important, the firm retains control of both the waste and its disposal costs in the long run.

On-site disposal presents the generator with its own set of problems. Before the firm can even consider the economic feasibility of on-site disposal it must be generating a sufficient quantity of waste to justify the capital expenditure. The on-site disposer must either develop and maintain an in-house staff dedicated to managing the disposal process, or contract out for such service. Some disadvantages lie beyond the control of the firm. Management must be prepared to deal with the public's resistance to the nearby siting of a hazardous-waste facility. And last, given the uncertainty that surrounds the future direction of government regulation, an on-site disposal strategy carries with it the significant risk of committing resources to a technology which could fail to meet future regulatory standards.

Off-Site Disposal

Off-site disposal is accomplished through the purchase of collection, transportation, and treatment services. By employing an off-site disposal service the generator firm, for a fee, has access to the resources of the comprehensive waste manager, relieving itself of much of the administrative burden associated with hazardous-waste generation. There is no longer the need to constantly track changes in technology and regulations. The generator need not commit scarce resources to maintaining an in-house disposal staff. The generator may enjoy the benefits of economies of scale gained through a centralized waste-disposal facility.

There are disadvantages in electing to use an off-site disposer. First, the generator lacks long-run control over costs. Firms seeking the services of a comprehensive waste manager need to recognize that they are purchasing scarce goods which are subject to market forces. As more firms dispose of hazardous wastes, demand for the services of waste handlers will grow, and the prices they charge will rise. Second, once a waste leaves the site, the generator loses control over it. RCRA has established that a generator remains liable for its waste even after the waste has been removed from the site by a third party. The disposer needs to have the corporate experience (and reputation for competence) in managing the waste in question. Even with these assurances, there is the potential for accidental spills to occur during the transport of the waste to the remote site (and the resulting liability issue). Third, the generator should note the growing reluctance of states to allow the shipment of wastes to sites within their boundaries.

14.2.4 ESTIMATING DISPOSAL CHARGES

A distinction needs to be drawn between the concepts of cost and price. *Costs* are those identifiable and accountable expenses incurred in the production of a product or service. *Price* on the other hand is a result of the interaction of supply and demand forces in a market. The difference between price and cost can be simply understood by recognizing that the market *price* reflects the vendor's *costs* plus the vendor's *profit margin*.

Cost

The American Association of Cost Engineers (AACE) has established cost-engineering guidelines which may be followed in estimating construction and operating costs for on-site disposal installations.

Cost-engineering estimates are by nature uncertain. The need for greater certainty means that higher costs will be incurred in producing the estimate. The AACE has established three levels of engineering estimates. The most general is an *order of magnitude* estimate, conducted without the benefit of preliminary design work. This approach provides estimates in the +50 to −30% range. A *budget-level* estimate is available when the design work is 15 to 25% complete and should produce estimates in the +30 to −15% range. A *definitive* estimate is possible when the engineering data is complete. It is the most costly estimate to produce, but is accurate in a +15 to −5% range.

Preparing estimates of capital and operating costs should be a routine task for a cost engineer. Design assumptions will have a major influence on the final cost estimates. These assumptions need to be explicit. These items can include the yearly waste throughput, a facility's degree of utilization (affecting variable costs), scheduled maintenance, equipment replacement, provisions for residue

TABLE 14.2.1 Summary of Relative Prices and Costs for Hazardous-Waste Disposal Options as Reported in Published Sources

Disposal technology	Approx. price range, 1985 $/ton
Incineration	
On-site (ave. costs)	40–475
Off-site: low-BTU liquid organic waste (ave. prices)	250–725
Off-site: high-BTU liquid organic waste (ave. prices)	50–250
On-site: low-BTU liquid organic waste (est. costs)	300–425
On-site: high-BTU liquid organic (est. costs)	100–175
PCBs: concentration less than 500 ppm (ave. prices)	350–450
PCBs: concentration greater than 25,000 ppm (ave. prices)	850–1,350
On-site: fluidized bed (ave. costs)	45–300
Off-site: fluidized bed (ave. price)	375
Dioxin-contaminated soils (est. cost)	600
Landfill	
Drums—local facility (ave. prices)	25–250
Bulk—local facility (ave. prices)	30–100
High-hazard waste—local facility (ave. price)	100–300
West Germany, regional facility, (ave. prices)	20–75
Low-hazard wastes shipped out of state (+300 mi)	100–300
Other	
Landfarming: 35 tons/d (ave. prices)	10–50
Deep-well injection: toxic wastewater (ave. prices)	125–275
Regional facility: Denmark's "Kommunekemi" (fixed price)	100
Warehouse tank farm: 20-year facility life (cost est.)	450–1,000

Note: This table is a compilation of various price quotes taken from published sources. Individual estimates may be found by referring to the sources listed in Subsec. 14.2.5.

handling and disposal, and the need for waste analysis. Insurance costs, often very difficult to project, must be factored in as well. Finally, there is a need to estimate the costs of (and the time to obtain) operating permits.

Price

Obtaining price information is a critical step for the firm that chooses to use an off-site disposer. Prices reflect supply and demand conditions, and therefore there are likely to be regional variations in prices which reflect differences in the availability of disposal services and sites. Availability of alternative disposal technologies will also influence market prices as generators will not be indifferent to competitively priced solutions.

Since final prices in this market are generally the result of negotiations between buyer and seller, the price strategies employed by suppliers can influence the process. In attempts to maintain or gain market shares, a vendor may be prepared to set a price which is less than cost. This may be particularly so when the vendor can gain access to large pools of start-up funds. The new entrant into the market may be trying to establish corporate credibility. The large-quantity waste generator may find that it can negotiate a reduced price with the disposer if there is a steady supply of a specific waste stream, allowing the waste handler to tailor service to that firm. In addition, if that vendor believes that business with a generator has the potential for growth, negotiations may reflect this.

For the firm beginning to investigate the range of market prices, the best source of information, short of obtaining vendor quotes, is through published price information. *The Hazardous Waste Consultant* (published by McCoy and Associates of Lakewood, Colorado) is a bimonthly journal which regularly reviews books, papers, and conference presentations which highlight cost studies. The Further Reading list at the close of this section offers additional sources of price information.

This information is useful in defining the range of acceptable options. Table 14.2.1 offers a summary of the range of published prices. There is no guarantee on the reliability of this information. Since the vendor's motive is to maximize profits, internal cost information becomes proprietary business information with every incentive to keep this information from competitors. True prices will only be obtained through negotiations based on actual wastes.

A second problem associated with published price information is that the assumptions which serve as the basis for quoted prices are often not detailed. The data in Table 14.2.1 offer only the most basic assumptions upon which the quoted prices are based. To have confidence in the quality of published price data, one needs access to the full range of design, engineering, and economic assumptions. Despite these caveats, published cost information can provide useful baseline data.

14.2.5 FURTHER READING

Anonymous, "Burning Hazardous Wastes," *Chemical Week,* **133**:(14)92–95 (1983).

Anonymous, "Commercial Hazardous Waste Management: How The Majors Performed in Fiscal 1984," *The Hazardous Waste Consultant,* **3**:(4)4-1 to 4-15 (1985).

Anonymous, "The Outlook for Commercial Hazardous Waste Management Facilities," *The Hazardous Waste Consultant,* **3**:(2)4-1 to 4-49 (1985).

Anonymous, "Responsible Party Liability: How Far Does it Extend?" *The Hazardous Waste Consultant,* **3**:(3)3-1 to 3-2 (1985).

Anonymous, "Will Hazardous Waste Management Activities Become Uninsurable?" *The Hazardous Waste Consultant,* **3**:(6)4-1 to 4-7 (1985).

Cahill, L., *Assessment of Commercial Hazardous Waste Incineration Market,* EPA/230/02-86/005, U.S. EPA, Washington, D.C., 1986.

DeWolf, G., P. Murin, J. Jarvis, and M. Kelley, *Cost Digest: Cost Summaries of Selected Environmental Control Technologies,* EPA/600/8-84/010, U.S. EPA, Washington, D.C., 1984.

Evans, G. M., "Uncertainties and Incineration Costs: Estimating the Margin of Error," *Incineration and Treatment of Hazardous Wastes: Proceedings of the Eleventh Annual Research Symposium,* EPA/600/9-85/028, U.S. EPA, Cincinnati, Ohio, 1985.

Humphreys, K. K., (ed.), *Project and Cost Engineers' Handbook,* 2d ed., Marcel Dekker, New York, 1984, p. 52.

Jelen, F. C., and J. H. Black (eds.), *Cost Optimization Engineering,* 2d ed., McGraw-Hill, New York, 1983.

Krag, B. L., "Hazardous Wastes and Their Management," *Hazardous Wastes and Hazardous Materials,* **2**:(3)251–308 (1985).

Lehman, J. P., (ed.), *Proceedings of the 1981 NATO/CCMS Symposium on Hazardous Waste Disposal,* Plenum Press, New York, 1983.

Lim, K., R. DeRosier, R. Larkin, and R. McCormick, *Retrofit Cost Relationships for Hazardous Waste Incinerators,* EPA/600/2-84/008, U.S. EPA, Cincinnati, Ohio, 1984.

Lindgren, G. F., *Guide to Managing Industrial Hazardous Waste,* Butterworth Publishers, Boston, 1983.

McCormick, R. J., and R. J. DeRosier, *Capital and Operation and Maintainance Cost Relationships for Hazardous Waste Incineration,* EPA/600/2-84/175, U.S. EPA, Cincinnati, Ohio, 1984.

Neveril, R. B., and W. M. Vatavuk, *Capital and Operating Costs of Selected Air Pollution Control Systems,* EPA/450/5-80/002, U.S. EPA, Research Triangle Park, N.C., 1980.

Rich, G., "Weighing the Factors About Treating Hazardous Wastes," *Pollution Engineering,* **18**:(7)26–29 (1985).

Sitting, M., *Landfill Disposal of Hazardous Wastes and Sludges,* Noyes Data Corp., Park Ridge, New Jersey, 1979, pp. 338–354.

Index